UNSETTLED

A CLIMATE CHANGE DENIER'S HANDBOOK

BY IAN HALL

Copyright © Ian Hall. Phantom Gavel Publications

Ian Hall is a member of the Phantom Gavel team.

ISBN: 9781697991888

All rights reserved, and the authors reserve the right to re-produce this book, or parts thereof, in any way whatsoever.

No part of this book may be reproduced or transmitted in any form or by any means, electronic or mechanical, including photocopying, recording, or by any information storage and retrieval system, without the written permission of the publisher, except where permitted by law.

TABLE OF CONTENTS

Introduction   8
Science and Politics, Science and the Media, Terminology; A Few Words of Warning, The Artificial Construct, Never Fully Trust a Graph, Index of Abbreviations Used, Doing Your OWN Digging.

Section 1; Setting the Stage
The Two Sides, The Argument of 'Settled Science', In Conclusion, The Points to Prove, The USA Temperature Record.

Section 2; The Scientific Method
The Unsettled Science, Science; Cause and Effect, Hyperbole... Science Doesn't Sell, The Hockey Stick Proves It, Climategate... a Storm in a Tea-Cup, Problems Of Climate And Weather, Computer Modelling for Idiots, The Average Global Temperature, The Apparent Global Warming Hiatus, Why Don't Alarmists Face Facts?, The Benefits of Global Warming.

Section 3; Consensus? What Consensus?   96
A Consensus Of 97% Of The World's Scientists, The Real Scientist's Consensus; Parts 1-7.

Section 4; Carbon Dioxide; the Black Pollutant!   123
Found it! Cause, Culprit and Case-Closed!, Old Mother Nature is our Friend!, How Can We Help Cure The Carbon Dioxide Problem?, The Carbon Dioxide Record, CO2... Greenhouse Gas... A Poison to the World, CO2; Amidst the Bleak Midwinter...

Section 5; The Songs of the Oceans...   151
Seas Levels are Rising All Over the World, Satellites and NASA Fraud, Tide Gauges and Stuff, Sea Level Rise... Old School, The Pacific Decadal Oscillation... PDO, The Oceans are Warming at an Alarming Rate, The Oceans are Cooling/Warming?, The Maldives... the Poster Child, Do Lakes Count as Sea Level?, The Great Pacific Ocean Garbage Patch.

Section 6; Poles Apart... and Greenland! 191
Greenland's Melting! 20 feet of Sea Rise!, Antarctica's Melting! We're all going to drown!, Antarctica's Freezing! We're all going to freeze?, Antarctica's Melting! NOT Global Warming?, Arctic Sea Ice Figures, The

Importance of Being Earnest?, The Arctic Sea Ice Predictions, Never Fear, Masie is Here!

Section 7; Global Extreme Weather: Atmosfear! 220
Introduction, Global Warming and Hurricanes, Global Warming and Other Weather, Extreme Weather Sites..., The IPCC's Opinion of Extreme Weather.

Section 8; The Governing Body... The IPCC   236
The IPCC; a Little History, IPCC's Terms of Reference, The IPCC's Process, Ex IPCC Scientists Spill the Beans, The IPCC's Science is SETTLED!, The IPPC's Single Grain of Truth..., We Can't Model What We Don't Understand, Global Greening; the Savior of our Race.

Section 9; The Corruption of NASA   261
Introduction, Homogenization: the Home of the Alarmist, NASA; We Don't Change Data (1-6)..., All NASA Personnel Are Alarmists (1-3).

Section 10; The Sun; it's ALL Cosmic, Man   305
The 600lb Gorilla in the Room   305
Whatever's Happening, Mankind is to Blame!, SunSpots, SunCycles, and SunStuff...?, The Influence of the Universe.

Section 11; Politics Rears Its Ugly Head   320
Political Issue... or Humanitarian Crisis?, Under Pressure to Toe the Party Line, Deniers are Funded by Big Oil, Climate Justice... the New Mantra, Beware the Wolf in the Sheep's Clothing, We Simply HAVE to do Something, Using Chemo-Therapy to Fight a Cold!, The Insurance Policy Principle, Tim Ball's Theory.

Section 12; Climate History... Fixed, or a Fix?   349
Introduction, A Look at Historical Climatology, The Earth's Cooling! ICE AGE 1970!, The World's Hottest Days Were in 1998!, Temperatures today (in the 2000's) are the hottest ever measured.

Section 13; Australia's Role in Global Warming   366
Introduction, The Australian Inconvenient Truth , A True Australian Heroine   , The Great Barrier Reef is Dying, Australia's Deputy PM Speaks Out, Australian Climate Predictions.

Section 14; Why Do Alarmists Love Polar Bears? 383

Introduction, The Polar Bear Population is in Danger, Polar Bears; the Unfortunate Un-Problem.

Section 15; Predictions, Predictions, Predictions…   397
Introduction, We're Really Great at Climate Predictions, U.N. 1989… We Have ELEVEN Years to Stop This…, Earth 2100; an ABC Prediction Special, Future Snow Predictions, The Green New Deal, The Global Warming Challenge.

Section 16; In Conclusion; A Mixed Bag! 420
All The Movies Are on the Alarmist Side!, A Graph of Two Halves, The Subtle Art of the Graph, The Glaciers are All Retreating, The Global Warming Industry, Biomass… the Hoax within the Hoax, Lies to Sell Newspapers?, Wikipedia is NOT Your Friend!.

Section 17; Prominent Deniers of the world   448
More of the World's Deniers, Further Reading; Food for Thought.

INTRODUCTION

"Power tends to corrupt and absolute power corrupts absolutely."
John Dalberg-Acton, (1st Baron Acton)

INTRO 1.1... SCIENCE AND POLITICS
(They're more alike than we think...)

Climate Change is a complex subject... I found that out myself during the writing of this book.
In fact Climate Change is one of the most complex scientific studies ever taken on by mankind's scientific minds.
It is a subject so diverse in subject matter that to investigate it properly you must have a working knowledge of so many scientific fields, it boggles the mind.
In no particular order, you need to be read up on Astrophysics, Astronomy, Geology, Geography, Meteorology, Physics, Chemistry, Geophysics, Ecology, Solar Physics, Botany, Paleontology, Zoology, Mathematics, Hydrology, Archeology, Economics, and there are too many more to mention.

The Science of Politics
"It is as hard and severe a thing to be a true politician as to be truly moral."
Francis Bacon (English philosopher)

Like Francis Bacon, somehow we all accept that politicians have some form of moral illegitimacy inherent in their psyche; politics is a profession in which few are trusted, and fewer still are seen as honest.
The very idea that they want to place themselves in places of power is surely the very fact that they should not be allowed to do so.

We all realize the truth inherent in politicians' motives to earn our vote, yet we joke about their corruption and the lies littered through their campaign promises.
We also know full well that once elected they will go to ridiculous lengths to stay in power; in fact it becomes their main work ethic. One should only look at the career politicians in power at this time, and look at their long lives 'dedicated' to the service to their constituencies.
Yes, once in power, it is the sole purpose of a politician to stay in power, and even advance upwards in their grasp of more. In their political

journey they make deals, alliances, while they conveniently lose their morals, and forget their initial motives... anything to retain power.
There are many examples of such politicians who spend many years in power, then retire with a wealth, a net-worth far beyond their wages while in power.
In politics, money, it seems is a powerful driver of greed, corruption and disregarded idealism.

The Politics of Science
"Politics is more difficult than physics."
Albert Einstein

Throughout history, however, scientists have been viewed as the antithesis of those seeking political power.
Perhaps because of the ethics inherent in the Scientific Method, these 'searchers for Truth' are recognized as being above corruption and immorality.
In fact the Scientific Method itself dictates a scientist's direction and order of work...
Careful observation (1), leads to a formulation of a hypothesis (2). The theory is then proven by experimental duplication (3), and refined (4) to the level where publication (5) is necessary.
Once the theory and workings behind it are released to the world, it is than tested by scientific peers, and verified (6).

In the writing of this book, however, I have found this not to be the case.

It seems the Scientific Method has been corrupted by the very same deity that has so corrupted the world of politics... money.

For most of history, scientists have been thought of as scholars, spectacle-wearing academics working in dingy conditions until their findings made them famous... indeed there are few cases of scientists working for the love of money.

Historically, funding for science was not easy to find.
But that changed immediately after the Second World War.
The Manhattan Project which developed the first atomic bombs, based in Los Alamos, New Mexico, was more than a bomb-making base; it was a scientific body of over 130,000 people who after Hiroshima and Nagasaki suddenly found themselves out of work.

With the end of the war, President Truman needed to keep those scientists under US government employ; the depth and diversity of the scientific personnel was such a security risk, work had to found.
Enter the era of government research contracts.

By 1945, the Manhattan Project had a huge annual budget, and had cost two billion dollars; an incredible sum for the day (about 30 billion today). To keep the Manhattan Project scientists gainfully employed (and out of Russia's hands) they were spread to the huge number of US universities, and settled into a post war mindset; government grants for research became a regular budget item, and both the scientists and the universities cashed in on the new milk cow.

And don't get me wrong, this was predicted by President Eisenhower in his departing address in 1961.

*"Yet in holding scientific discovery in respect, as we should, we must also be alert to the equal and opposite danger that **public policy could itself become the captive of a scientific-technological elite**."*

Sadly today, as Eisenhower predicted, there is now a world full of such scientists, grabbing their share of billions of government dollars, and the new buzz word is climate change.
The USA commits an estimated $300 billion dollars annually to climate research... try multiplying that by 192 countries; it's estimated around one trillion dollars spent worldwide each year.
But such funding is not completely free... it comes with many strings attached.
The governments and private sources providing the billions of dollars are not donating it purely for 'climate research'. To earn a slice of the funding pie, scientists must specifically investigate **'man made'** global warming, now called anthropogenic, hence the new abbreviation... AGW.

The fact that so many studies are looking for man's guilt in climate change dictates who gets the funding and who does not.
Most of the funding is looking for evidence which supports man's involvement. It seems that every other theory of the cause of climate change is ridiculed, unfunded, and cast into the shame of heresy.
The result is, if scientists prove that man is NOT responsible for climate change... the money would dry up... and thousands of scientists would suddenly be out of work.

I know this sounds controversial, but the United Nations International Panel on Climate Change (IPCC) has dictated the specific direction of the research.
From its very inception as the United Nations Framework Convention on Climate Change in 1992, their guideline to research is as follows…

"a change of climate which is attributed directly or indirectly to **human activity** *that alters the composition of the global atmosphere and which is in addition to natural climate variability observed over considerable time periods."*

Alas when the scientists' core purpose is to find evidence of man-made climate change, their endemic purpose is to prove the theory… if they find proof to dispute the theory, their funding will cease.

Now it may not escape your own reasoning that any organization which **dictates** the specific area of research has an agenda… and you'd be right to question it.

All the IPCC needed was a way to get their findings to the people of the world…

INTRO 1.2... SCIENCE AND THE MEDIA
(An alliance forged through common goals...)

"Follow the money."
Deep Throat, (Hal Holbrook), All The President's Men, (1976).

The Initial Concepts of Truth
In a world of conflicting information, and lies upon deceit at every turn, we must accept that unless we see something for ourselves, it's difficult to really *know* of the truth of anything.
The only facts of life are the details which we ourselves live through... everything else is told to us second hand.
While one website convinces us that black is white, the next TV show does the opposite.
One newspaper announces that Antarctica is breaking up, and we are told of greater South Pole ice in the next.
With so many unpoliced opinions, anyone can get 'facts' to support any theory, no matter how ridiculous it is.

There's one rule in this century... fake news rules. And I mean on BOTH sides of a debate.

Let's face it; the media will lie, distort truth and create mountains from the merest molehills just to get the public's attention. It's all about advertising revenue, thus... again we follow the money.
As we've said before, we also know that most politicians tell lies at some point in their career. It's in their DNA.
It's a fact of life. If they don't get elected, they don't stay on the political gravy train and don't get rewarded, again... follow the money.

But before I began the book, I considered scientists to be above that... interested only in the truth, and in proving the facts...
What I did not expect was... in the world of 'science', to come face to face with lies, distortion and fraud on a grand scale.
Scientists shouldn't lie.
Scientists shouldn't commit fraud.
And scientists shouldn't distort the facts to fit their argument.

And yet in the world of Climate Change, it seems they do.

"It is difficult to get a man to understand something, when his salary depends upon his not understanding it."

Upton Sinclair (Pulitzer Prize Winner)

And in a strange way I should have known it from the start.
But I made assumptions that scientists were above the power of the dollar bill, and worked with higher principles... I should have known to follow the money.

One other fact I disregarded as being above a scientists morals... collective reasoning.

"Never underestimate the power of stupid people in large groups."
George Carlin & many others

Group-think is a powerful force; it governs our peer-views, what sports teams we support and what side of the political spectrum we belong in.
It's difficult to buck the trend, trust me.

But there's group-think, and there's mass stupidity.

"On the whole, once we have adopted a belief, we give particular attention to cases that seem to support it, we distort other cases in order to support it, and we ignore or belittle other cases."
Wendell Johnson, (People in Quandaries: The Semantics of Personal Adjustment).

My Version of the Truth
In climate change, uncovering my own understanding of the truth took me two years of research, reading, and investigation. Some of my findings shocked me, some just altered my preconceptions.
What I did not expect to find was... in every facet of the subject, I found corrupt scientists... so many times I told my recent 'finds' to my friends to be met with bewildered faces.
And hand-in-hand with the corrupt scientists (partners in crime) in print, radio and television, sat a corrupt media accepting their every detail, exaggerating the scientists' claims, and churning out nightmare scenarios to sell newspapers or a few minutes of airtime.

With the Alarmists using the world media as a free megaphone, the baffled Deniers cringed in the background.

But the reaction to 'crying wolf' only lasts so long...

Ordinary people are shaking off the scales from their eyes, and digging for their own facts.

This book is the story of my own digging.

Where possible, I show the facts, and the links to the source.

The Science and Media Partnership
The media care little of the content of their pages or airwaves. They care only of the bottom line... readership, audience figures, and profit.
In these days of 24 hour coverage on tens of thousands of television channels, anyone who declares... *"It must be true; I saw it on the news."* is suffering a delusion.

And there lies the problem with trying to change minds, to impart facts, to give a clear account of one side of any argument.
Where do I get good facts? Where do I find honest, un-tampered data?

The media wail with enhanced hyperbole and the scientists obfuscate en mass; I've heard climate change blamed for everything under the sun... from migrating patterns of butterflies to planet-wide droughts.

Some of it may be true; most of it is definitely not.

But one thing remains true to this day... if you're a scientist looking for funding for a project, you'd better link it to climate change, or you ain't getting funded.

With so many thousands of scientists bound to their climate change mantra by a pay-check, it's difficult to get anyone to stand up against the media swell and declare their objections... and yet when many of those scientists retire, somehow they change their minds and rebel against the corporate binds that once held them silent.
Many professors leaving universities, senior engineers leaving NASA, scientists leaving the Intergovernmental Panel on Climate Change (IPCC), have changed their minds and declared themselves to be 'skeptical' of the idea that man is responsible for ALL of the world's climate change.
No longer shackled by corporate rules, they look into other theories, they challenge data, and form alternative positions.

"Worst scientific scandal in history...When people come to know what the truth is, they will feel deceived by science and scientists."

Kiminori Itoh (Ex-IPCC Japanese environmental physical chemist)

There are two sides to every coin, and those believing in man-made (anthropogenic) Global Warming (AGW) are numerous. Their research papers are greedily accepted by the scientific press, peer reviewed by like-minded thinkers, and trumpeted by the fawning media because catastrophe and looming disaster make good headlines.
And a good headline sells newspapers and boosts television viewer numbers.
Called 'Alarmists', these scientists have declared there is a consensus of evidence, they herald catastrophe and doom, and then declare the science 'settled', and the need for further discussion over... hence the title of this book.
With the 'settled' science set in stone, they refuse to debate the skeptic side, refusing to 'bring credence to their 'idiotic' alternative theories'.

And there lies danger.

Refusing to consider other theories, the Alarmists have put all of their eggs in the one basket; carbon dioxide is to blame, and an irresponsible greedy mankind is responsible for its recent rise. The burning of fossil fuels has caused a global catastrophe, and only a 'world' solution can solve the problem.
Having tried and convicted poor defenseless carbon dioxide of the crime, and shaming humans as the cause, they look at just ONE solution... reducing the amount of carbon dioxide in our atmosphere.

The danger in such a simplistic approach is the direction of the solution.

The Alarmists believe that the world must give up fossil fuels entirely.

Lowering emissions is not enough to stem the rise of carbon dioxide, and every country must join in the rush to clean up the atmosphere.

There are, however, two unfortunate facts...
1. The world's highest emitters, China and India, have no intention of cutting emissions just yet, and will not commit to reducing emissions until 2050.
2. It would be economically inept to ask developing countries in Africa, Asia and South America to reduce their emissions; this would only end in further poverty and financial hardship... 70% of Africa's population still

do not have electricity or fresh water... that's ONE BILLION people in the world.

Over the last 40 years, many meetings of the IPCC have faced this problem... and some countries like the USA have actually reduced their carbon emissions, but prominent high-level talks in Kyoto, Copenhagen and Paris have left us with no global workable solution.
An environmental problem has become a morass of conflict, mired in political opinions.

So as you begin reading, remember these three facts...

1. Scientists tell lies.

2. The Media amplifies the lies to a huge audience.

3. Scientists and the Media make money, then they repeat the process.

This is how we have a world full of disinformation, and propaganda, pushed by legitimate scientists, and trumpeted by a media grateful for alarming copy.
In this book, I will show many examples of corruption in the scientific world, the media, and politicians.
In my findings, my belief in the world of science (and scientists) has taken a punch to the stomach. My belief in their lofty positions of purity has taken a few hits.
I hope you will find my information interesting, revealing, and written in plain-speak English

Fortunately there are enough layman-like articles to get me through the morass of scientific terminology, and hopefully I'm enough of a layman to pass my findings on to you.

INTRO 1.3... TERMINOLOGY; A FEW WORDS OF WARNING
(Because in the realms of language, words matter...)

Gases, Greenhouse Gases and Particulates/Aerosols
Gases are usually invisible, usually odorless and exist all around us.
Fundamentally the Earth's atmosphere (a mixture of gases) consists of 78% nitrogen, 21% oxygen, 1% argon, and a bunch of trace gases, of which carbon dioxide (CO_2) is one.
Most of these gases in our atmosphere are harmless, odorless, and non-poisonous. (Kinda speaks for itself, huh?)
(And remember that when you see smoke stacks belching black, brown or grey clouds into the sky... you are NOT seeing carbon dioxide... those smoky clouds are particulates/aerosols)

However, the atmosphere is a complex volume of gases whose above concentrations are mixed with about 1% of greenhouse gases, like water vapor, carbon dioxide, methane, nitrous oxide and ozone.
They all vary in concentration around the world, for instance, water vapor is more prevalent over oceanic areas.
In very simple terms, a greenhouse gas has the ability to transfer reflected sunlight into heat. This heat is what keeps the planet at a temperature in which we can live.
Without greenhouse gases, we'd all freeze to death.

To further complicate the atmosphere, there are microscopically small piece of solid matter, extremely fine dust particles (called particulates or aerosols) suspended in the atmosphere. Pollen, sea spray, carbon, Sulphur, smoke and dust from industrialization, volcanic eruptions and dust storms gather in the atmosphere and also add to the reflectivity of the atmosphere. In general particulates block some of the sun's energy and reduce the temperature of the planet (but mostly by just a degree or so to fractions of a degree, depending on the amount of particulates).

Carbon or Carbon Dioxide?
Carbon in is found in many forms, depending on the molecular construction... in its purest form it is wood, coal, graphite and diamond.
In gaseous form, carbon dioxide, as we've read earlier, is an odorless, colorless gas that makes up approximately 0.04% of the actual atmosphere.
Since that's a very small number, to make calculations easier the figures are usually given in 'parts per million', (PPM).
This makes the make-up of the atmosphere as follows...

(Out of a million parts of atmosphere)
Nitrogen is 780,000ppm
Oxygen is 210,000ppm
Argon is 9,300ppm
Water vapor is 10,000ppm
Carbon dioxide is around 410ppm (and rising slowly)
(The numbers are rounded for ease of remembering)

According to many of the world's climate experts, the increase of carbon dioxide is the main/sole cause of the world's slight warming.
(In my opinion, a rise of a degree or so temperature rise or fall per century is not considered 'catastrophic' by any means)
But here are a few facts about carbon dioxide…
1. Carbon dioxide (CO_2) is a colorless, odorless gas that helps plants grow and photosynthesize.
2. Carbon dioxide is ESSENTIAL for life on Earth; without it, plants would die. And without carbon dioxide in the atmosphere, the Earth would be a colder place.
3. Levels of CO_2 have been higher than 10,000ppm in the past.
4. It is (kinda) accepted that the level of CO_2 in the 1850's before mankind's age of industrialization was around 280ppm.
5. It is accepted fact that if CO_2 levels were to drop below 150ppm, plants could not exist.
6. The world has significantly greened in the last 50 years (NASA source).
7. Crop yields around the world have increased in the last 30-50 years… sometimes by as much at 100%… so food production has increased because of the CO_2 increase.
8. Farmers artificially RAISE the CO_2 levels inside greenhouses to over 1000ppm to get greater crop yields.
9. In coffee shops, because of people's breath, the coffee machines, etc, CO_2 levels can reach 800 or higher.
10. In concerts and auditoriums, CO_2 levels can top 2000ppm.
11. In submarines the CO_2 alarm goes off at 5000ppm.
12. In Apollo missions (where CO_2 scrubbers were not used or limited) the CO_2 levels spiked to 20,000ppm.
CARBON DIOXIDE IS NOT A POISON.

However, considering the above, for political reasons, carbon dioxide has been positioned as a pollutant.
A Scientific American headline said…

"The Worst Climate Pollution Is Carbon Dioxide."

In 2009, the EPA's Lisa Jackson was the first federal official to declare carbon dioxide a pollutant.

"carbon dioxide and five other greenhouse gases pose a threat to human health when concentrated in the atmosphere"

To me, this is propaganda of the most insidious kind.
Carbon dioxide is one of the building blocks of life. Without carbon dioxide, we would die. There is no need for either politicians or the media to spread this kind of pseudo-science.
So when we talk of the so-called down side of carbon dioxide, we have many other factors to take into consideration.

However, the world's media and politicians have not stopped there.
Deeming the population too dim or simple to understand the concepts above, they have changed the narrative.
They now just use the one word… **carbon**… as the demonic guilty party.
Beware of this fact when in discussion.
And please correct those using just 'carbon' in debates and discussions.

Politicians are now talking about the ability to invent carbon sequestration; to make millions of machines to literally filter CO2 from the atmosphere, and bury it in the ground.
I've never heard of such rubbish in my life.

INTRO 1.4... THE ARTIFICIAL CONSTRUCT
(Scientific shenanigans set to deceive...)

Introduction
As I wrote this book, I encountered a few bumps along the way.
The subject matter is expansive and there are scholars out there (on both sides) who are out to confuse, obfuscate, hoodwink and be-dazzle you with their technical terms.
I've broken these down into the following... you can skip this part, but trust me, you'll need it later.

The Art of the Artificial Construct
Yup, there are such things... and you have to be aware of the way they're talked about.
There are facts in life... how many spoons of sugar you like in your coffee, the distance from home to work, the number of kids in a school.
Those are very solid, constant numbers... REAL numbers

But there are other numbers that are not constant; they vary ALL THE TIME... the distance from Earth to the sun, how many beers you have on a Saturday, the amount of people at a Sunday football game.
Those numbers change all the time... they are VARIABLE numbers.

And then there are estimates at numbers, made up to look like real numbers, because the actual number itself is unknown... the amount of stars in our galaxy, the number of grains of sand on Earth, etc.
We can estimate/guess these numbers to give us some idea, but the actual number is unknown.

And there is the artificial construct.
These are numbers which without a basis, a starting point, or just a plain guess... (numbers which mean nothing at all) and there are TWO of them which feature prominently in the climate change debate.

1. Global Average Temperature (GAT)
This is the artificial construct that has the world in turmoil... supposedly it's rising out of all control, and yet it doesn't exist.
It can't exist.
To know the average temperature of a cup of coffee (CAT), we'd have to measure the temperature at the coffee's surface, the coffee at the bottom, the coffee on all sides of the mug, and every single molecule in between.

Only then (by instantaneously measuring every single molecule of coffee) could we provide a figure (CAT) for the average temperature of the coffee... and this figure would only be valid at the instant we took the measurements.
And seconds later the recorded temperature would be meaningless, because the coffee will have cooled slightly since we took the readings.

This is what the Global Average Temperature is.

Satellites travel around the world, taking MILLIONS of temperature readings every day from MILLIONS of places on the Earth.
(Well, the troposphere really, the air from the area where we live)
This gives us the GAT of one single day.
But considering the points measured change every day, and Earth's temperatures change every day, and the seasons change every day, the individual figures are pretty meaningless.

And it does get worse... see Section 2.8.

So remember when someone says the "Average Global Temperature", it's just a constructed number.

HOWEVER, this globally averaged temperature anomaly depends TOTALLY on the earth's climatic systems being in total equilibrium at all times.
And that NEVER happens.

2. The Global Average Sea Level (GASL)
You may have read in the newspapers or seen it on the news... the world's seas are rising.
But don't run for the hills just yet... NASA gives the figure of just 3.4mm per year.
That's an inch and change per decade, about ONE FOOT PER CENTURY.
(And those are NASA figures.)
Al Gore doesn't have to sell his beach property just yet... nor does ex-President Obama.

But again, like the Global Average Temperature, the Mean Sea Level is measured from space... millions of hits onto our continuously moving water surfaces, giving an average level.

But a level in relation to what?

To one place? To an average place? To the Mariana Trench?

Because the actual LAND is also sometimes moving up or down depending on tectonic shift, ice shift, earthquakes or industrialization, it's impossible to give an actual average height.
For instance…
Boston's sea level is rising, but in actual fact most of the 'rise' is actually the sinking of the land due to the large heavy buildings.
In Norway, they have both sea level rise and fall, depending on where you live.

The sea level is a very local figure, and some scientists believe that the actual 'rising' measured by the satellites is not accurate. The satellite radar can measure to an accuracy of just 4 inches.
Yet they can detect just one tenth of an inch rise on a substance that never sits still for a moment?

INTRO 1.5... NEVER FULLY TRUST A GRAPH

Graphs
Graphs are fantastic devices to show statistics in picture form. They're easy to interpret, and often far better than a thousand words on the subject.

However... unless the maker of the graph has intentions of purest gold, any graph maker is showing you a picture that he wants you to believe. A graph maker is trying to put figures into a particular perspective; he has a point to make... so be very, very wary of graphs... graphs are propaganda... and both sides use them well.

Let's take a look at a few of the pitfalls I encountered in my search for the truth...

1. NEVER fully trust a graph unless it has real numbers on BOTH axes. And I mean real figures like dates, temperature, ppm, polar bear numbers etc... be very wary of graphs with vague axes... graphs that have axis labelled as 'average/mean', and variations from the '1979-2010 mean' kind of thing).

2. NEVER fully trust a statistic if it involves some kind of comparison to an average or mean value. Statements like *"The sea level has fallen 3 feet below the 1850-1900 mean."* mean nothing unless accompanied by some solid reference point.

3. NEVER fully trust a graph to show a trend unless you've got the latest and most complete version of it. Scientists cherry-picking a starting and ending point for their graph can show any trend they want it to.

4. NEVER fully trust a graph if its source of data changes in the middle of it. The Hockey-Stick graph is a solid example... 80% proxy data, with the last 20% measured by thermometers.
The graph maker has 'probably' fudged to get the two separate graphs to join nicely.

5. REMEMBER... a few temperature graphs are not actually real figures at all, they are termed 'anomalies', i.e. they are the difference in measured temperatures in comparison to a pre-determined zero line. This is the easiest graph to deceive readers.

Fraudulent Graphs

A graph has to start somewhere, I know this. But some graph-makers choose a starting point which 'helps' confirm their conclusion; taking care while choosing the starting point of a graph can change the trend of the data.

However, if you choose a starting point to ignore the data just before your starting point, and this changes the trend in your favor, that's a fraudulent practice.

Let's look at a couple of examples...

US Temperature Record

The temperature record for the US starts at 1880, right? I mean, that's the farthest back temperature readings go, right?

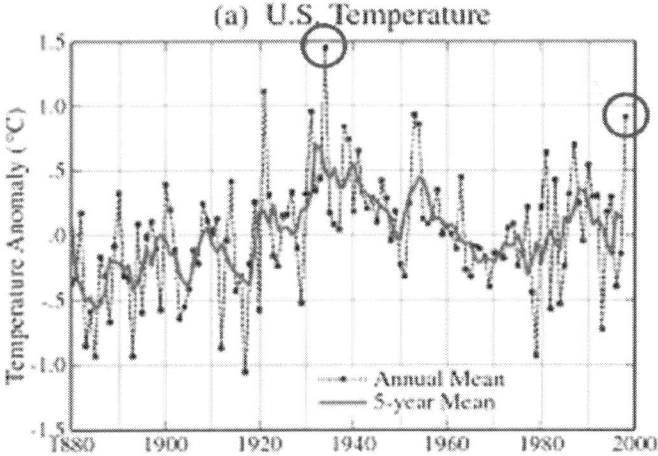

Example above... the NASA graph chooses a starting point when temperature readings started.
Wrong.
What would happen if we had temperature data before 1880?
Well? We do.

Below is a Tony Heller graph which utilizes NASA's pre-1880 data. As you can see, the temperatures around 1880 and before are actually quite high.
Tony has found archived newspapers (from both the US and Australia) which tell of record high temperatures in 1876 and 1877. They tell of a winter in Minnesota with no snow, and summer temperatures above 120 degrees F.

Well, it turns out there was a 'Global Famine' from 1876 to 1878, driven by high temperatures all over the world.
Some newspapers reported that over 50 million people died.

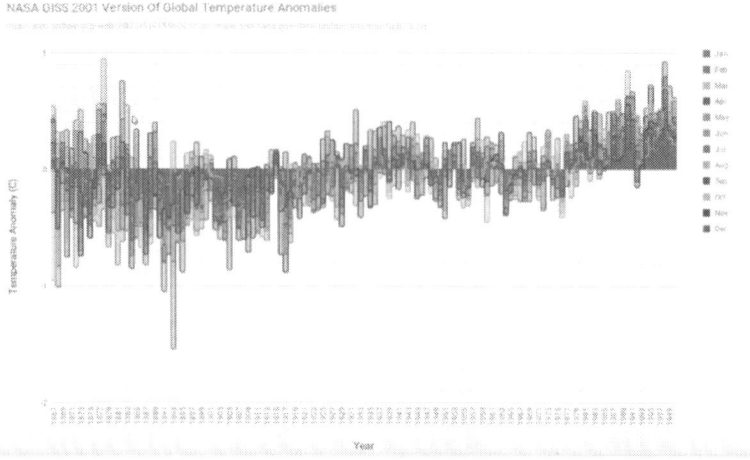

It's extremely convenient of NASA to miss out on these high temperatures. They make the 'catastrophic' warming of modern times pale into the average.
But you never see these graphs on the media.

Tony's video on the subject can be seen here...

https://www.youtube.com/watch?v=cxQjOasRvGE

Arctic Sea Ice
NASA, NOAA, and NSIDC all state that satellite data of Arctic Sea Ice all begin in 1979.
But NOAA has satellite data going back to 1973. Why don't they start their graphs there?

NASA's declaration is here...
https://www.ncdc.noaa.gov/sotc/summary-info/global/201207

At the section labelled...

"Polar ice highlights: July ... Although the melt area is the largest since the beginning of the satellite era in 1979"

However, despite NOAA's above 'admission', the IPCC First Assessment Report (FAR) back in 1990 shows this NASA graph shown below...

As we see, the first data point in 1973 was much lower than the data point in which the current NSIDC graphs are projected from.

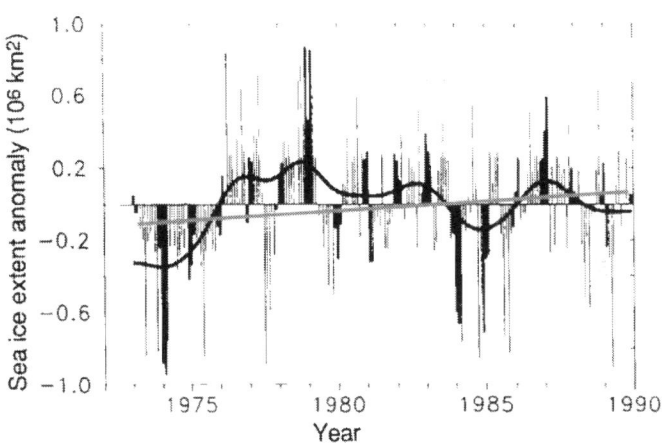

IPCC (1990) *Observed Climate Variation and Change* 7

You see, starting the Arctic Ice Sea graphs in 1979 shows a gradual decline in ice... exactly the point Alarmists are trying to support.
If you start the graph in 1973, a whole cyclical component comes into view.

Upper Atmosphere Temperature
Benjamin David Santer, an American climate researcher, was the Lead Author for chapter 8 of the 1995 Second Assessment Report (SAR). He looked at the graph below, and decided it did not suit his own narrative of global warming.

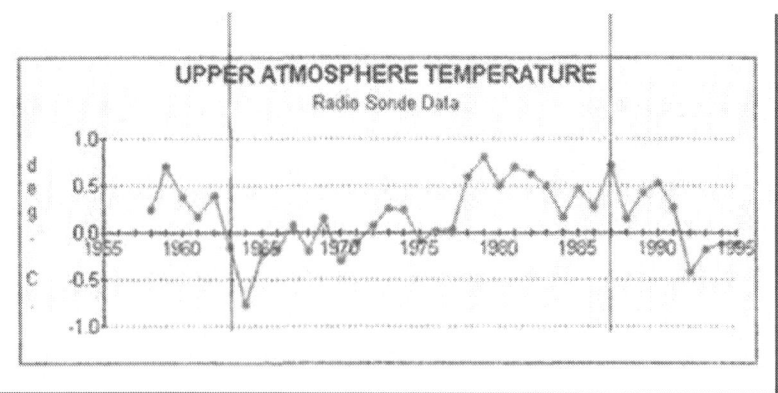

Of the whole graph, he subsequently chose the areas between the lines (added) and published the version of the graph below.

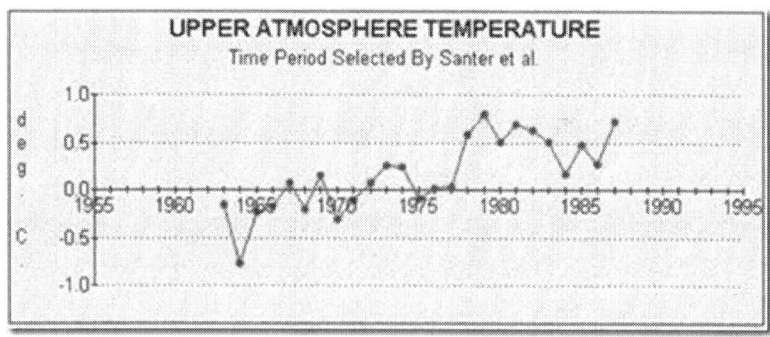

Making a very nice upwardly moving line, supporting his theory.

Here's a couple of other terms to become familiar with.

Data Homogenization
As I dug into the intricacies of terminology concerning climate change, I came across a term totally new to me.
Homogenization.
From a dictionary, the term means...

1. A process by which the fat droplets from milk are emulsified and the cream does not separate.
2. The process of making things uniform or similar.

NASA/NOAA and their myriad of underling departments (and there are many) use this term loosely.

They take raw data from one source, and mix it with that of a nearby data source, and to achieve a 'balanced' result.
Now, to me, that sounded like a reasonable thing to do.
BUT... as NASA/NOAA change the data under the guise of homogenization, they also change the historical recorded data.

Yup, I know this is hard to believe, but scientists all over the world have been found guilty of data tampering, and most of the affected data is in the past, and therefore (to me) written into stone tablets somewhere.
But this doesn't stop the scientists... because homogenization has another use other than giving an average over a few sites.

With so many temperature recording sites to choose from, a scientist can gather data from particular sites in order to achieve the results he was looking for.

It's called data manipulation... but 'they' call it homogenization.

Urban Heat Island Effect
This is a particularly vital part of temperature recording, and has caused many encounters between the two sides in this debate.
In the 1880's when the first nationally used weather stations were being set up in the USA, siting the thermometers didn't take much thought.
The public gave them little notice, and they were originally set up in grassy areas in University playing fields and peoples gardens.
However, the world has changed.
Villages turned into towns, and towns into cities.
Very soon it was realized that the temperature readings from some of the sensors set in urban areas were rising, while the nearby sensors in the countryside were not.
Concrete and asphalt soak in heat and radiate it to the surrounding area. This raises the temperature of sensors nearby during the day.
Concrete and asphalt also hold that warm temperatures longer than grassy areas and dirt. This means that the nighttime temperatures were also affected.

Urged by many 'concerned scientists', and in particular by ex TV meteorologist, Anthony Watts, the Urban heat Island effect has been investigated on a lot of the US's weather stations, and many have been taken from the national grid or moved.
There are now a strict set of guidelines issued by the World Meteorological Organisation (WMO) to deal with such problems.

BUT... even in my home town of Topeka, Kansas, I found problems...
(see Intro 1.5)

The full WMO standards can be found here...

https://www.wmo.int/pages/prog/www/IMOP/SitingClassif/Canada/Siting%20Classification%20System7%20-%20Sep%202012.pdf

There is a small caveat to the story of weather stations.
To keep the stations free from the public's tampering, they are mostly cited in airports, in restricted areas. It makes some sense, as most airports have a weather guy on hand to supervise the station.
However, most weather stations at airports are set near runways, and in the bigger airports, jet engines are passing frequently.
(See Intro 1.6)

Scientists Say
The phrases, 'scientists say' or 'scientists state' are used often in the media. They're useless ambiguous phrases used by lazy journalists to call your attention to 'power', to some arbitrary level which the common man cannot question.
An un-named scientist has more knowledge than poor lowly 'us', and we must bow to their integrity, data, knowledge and suchlike.
But it's a bogus term used when the journalist wants to make a scientific point but either has no proof, or is too lazy to go looking for it.
Never get taken in by this or any other such phraseology.
If a journalist or media presenter has a point to make, let them prove it; make them cite their sources as they spout their politically-filled rhetoric.

I hope you'll note that I cite my references often.

[?]

INTRO 1.6... INDEX OF ABBREVIATIONS USED
(And who doesn't need one of those...)

With so many organisations and technical phrases used in the debate, the need to quickly recognize any of the abbreviations is essential, so here's an alphabetical list of those most commonly used...

AGW (Anthropogenic (man-made) Global Warming)
AWG-KP (Ad Hoc Working Group under the Kyoto Protocol)
BOM (Bureau of Meteorology), Australia
CNES (National Centre for Space Studies), France
CRU (Climatic Research Unit) at the University of East Anglia (UEA), England
CSIRO (Commonwealth Scientific and Industrial Research Organization)
(DMI) Danish Meteorological Institute
(ENSO)El Niño Southern Oscillation
GCI (Global Change Institute, Australia)
GISS (NASA's Goddard Institute for Space Studies)
GWPF (The Global Warming Policy Foundation)
HCCP (Hadley Centre for Climate Prediction, part of the UK Met Office)
IEA (International Energy Agency)
IPCC (Intergovernmental Panel on Climate Change) a body within The United Nations
JMA (Japan Meteorological Agency)
MBH (Hockey stick co-authors Raymond Bradley, Malcolm Hughes, and Michael Mann)
MOHC (Met Office Hadley Centre)
NASA (National Aeronautics and Space Administration)
NCAR (National Center for Atmospheric Research)
NCAS (National Centre for Atmospheric Science), based in UK
NCDC (NOAA's National Climatic Data Center)
NIWA (The National Institute of Water and Atmospheric Research, New Zealand)
NSIDC (National Snow & Ice Data Center)
UEA (University of East Anglia, Norwich, England)
UNIPCC (The United Nations... Intergovernmental Panel on Climate Change)
UHI (Urban Heat Island)
WMO (World Meteorological Organisation)

INTRO 1.7... DOING YOUR OWN DIGGING
(Gumshoe Ian Hall takes to the road...)

In these days of internet saturation, can you do your OWN investigations?
Can one person make a difference?
The answer is... a definitive... YES!

Friendly Investigation
On a balmy morning in July I drove to our town (Topeka, KS) Weather Station, and was met by a very friendly Meteorologist. I asked if I could take photographs of the actual weather station. (I was investigating my local weather station for possible Urban Heat Island (UHI) effects, but I did not mention this.)
This is as close as I was allowed to the weather station... it's there, right in the middle of the shot...

The meteorologist explained, while walking outside towards the nearest point to the distant sensor, that the actual weather station that broadcasts temperatures to the central NOAA hub was not accessible, as it was close to the runway of Billard Airport, and therefore a restricted area.
It turns out that many of the world's weather stations are set at airports, near runways, due to their inaccessibility to the public, and airports usually have a meteorologist at hand.

Then he said something which made me feel extremely vindicated for my trip. This information was given to me, un-asked for, and I was shocked.
"There's sometimes as much as two degrees difference between the Airport sensors and ours..." The meteorologist pointed behind his own building, where 'their' sensors were located on grass, next to a corn field.
"That's a lot," I replied, "Two degrees! Is that the heat island effect?"
He then pointed to the Airport's distant station. *"You can't see it from here, but theirs is set on gravel, there's always a difference between the two."*
The sensors were just a few hundred yards apart, but if ONE temperature gauge was 'out' by up to two degrees... that's ALL the world's warming right there!

Thank goodness for modern digital zoom lenses...

Now, the World Meteorological Organisation (WMO) has a set of strict guidelines for the siting classifications of all weather stations used in national figures.
They have five classes for thermometers, and Class 1 is reserved for the most reliable sites. These are the criteria for Class1:
(a) Flat, horizontal land, surrounded by an open space, slope less than 1/3 (19 degrees)
(b)Ground covered with natural and low vegetation (<10cm) representative of the region
(c) Measurement point situated:

(i) At more than 100m from heat sources or reflective surfaces (buildings, concrete surfaces, car parks, etc)
(ii) At more than 100m from an expanse of water (unless significant of the region)
(iii) Away from all projected shade when the sun is higher than 5 degrees.
A source of heat (or expanse of water) is considered to have an impact if it occupies more than 10% of the surface within a radius of 100m surrounding the screen, makes up an annulus of 10-30m or covers 1% of a 10m radius area.

Well, as I found out, and as the above photo shows, the Topeka weather reporting temperature gauge hit two or three of those points...
1. the gravel surrounding and under the sensor...
2. the concrete runways nearby...
3. the building sited far too close to the sensor...

Now, in the WMO guidelines, the criteria for *"Ground covered with natural and low vegetation"* goes all the way to Class 4 ...
(And in class 4 it states... *"Sensor located at more than 10m from significant heat sources or reflective surfaces")*... this makes the Topeka site a Class 5 site... and therefore practically useless for climate work.

Just remember that there was sometime as much as two degrees difference compared to the sensor near the Weather Bureau building sited on grass... that shows a distinct Urban Heat Island effect for the Topeka sensor site.

The full WMO standards can be found here...
https://www.wmo.int/pages/prog/www/IMOP/SitingClassif/Canada/Siting%20Classification%20System7%20-%20Sep%202012.pdf

Me, Gumshoe detective Ian Hall, had uncovered evidence of the Urban Heat Island Effect in my home town.
I wonder what you could find out?

But back to my visit...

Climate Change
Now, as we walked, the meteorologist and I also talked about man-made global warming...

My friendly scientist told me that in his opinion man had to be influencing the climate, just by us being here, but that there was no way to measure the figure of the actual extent.

I asked if he thought man's influence was large, and he shook his head in a definitive 'no'. *"There are far too many natural influences at play here."*

He mentioned the atmosphere, and the oceans which hold/capture more carbon than we can measure.

In fact he indicated that he didn't place too much hope in the carbon dioxide theory at all... we talked about feedback and clouds and CO_2 saturation... it was a really great visit.

I was talking to a scientist, and basically interviewing him.

On our way back to the car park, he said (owing to not being sure of the source of the warming). *"We'd be far better off dealing with adaptation rather than anything else."*

Hmm... a scientist doubted carbon dioxide as the main/only source of the warming, and he recommended us adapting to any changes that occurred.

Just goes to show... you can do research on your own.

And we all need to do our little part.

Because the science is NOT settled.

Let's get this book started, shall we?

SECTION 1.
SETTING THE STAGE

"Skepticism is the first step towards truth."
(Philosopher Denis Diderot)

1.1... THE TWO SIDES

First, let me introduce the two sides...

The Alarmists
This august group has cornered the market on Global Warming, got into bed with politicians, governments, universities and the media. They hail themselves as the so-called *"consensus of 97% of scientists"* who agree that man-made carbon dioxide is the major cause of all Global Warming, and anyone who disagrees is a kook, skeptic, heretic, denier (a far from subtle reference to 'holocaust denier'), flat-Earther, and a proclaimer of fringe science.

"When the debate is lost, slander becomes the tool of the loser."
Socrates.

The left wing of the Alarmist movement (like Greenpeace) wants every country on Earth to stop burning fossil fuels immediately, or at the very least to decrease emissions to pre-1970 levels.
(And some want them back to 1870 levels!)

The Alarmists (including the IPCC) are fixed in their idea that man-made CO_2 is the culprit, and will hear no argument to the contrary. They have literally nailed CO_2 to their flagpole, and have chosen to ignore every other possible alternative theory.
They're going to toe the party line, no matter what evidence stacks up to the contrary.
Having such a stance, they feel that the only solution to combat Global Warming is to target CO_2 emissions and decrease/eliminate them.
Basically if we take the Alarmist camp's stance to the end of the line... they want to rid the world of fossil fuels, and that includes cars, air travel and all but nuclear ships.
The only problem with their stance is...

1. Emerging nations like India and China have not agreed to reduce their carbon footprint, and are indeed increasing the building of coal-fired power stations to satisfy a growing need for electricity.
2. Most of Africa is still woefully without electricity (and therefore basic human needs), and if we curb their development, we are literally condemning millions to die.

Regardless of its environmental roots, Global Warming is now a politically charged subject, and it's extremely difficult to separate the science from the politics.

The Deniers
Misnamed from the outset as 'skeptics' (because inherently every scientist should be a skeptic), this is a growing group...

1. some who question the causal link between CO2 and climate change, 2. some who question the statistical level of mankind's involvement,
3. and some who question the integrity of the scientists and data flowing from the Alarmist camp.

Deniers work or have worked for NASA, IPCC, and major universities. Some have been Alarmists or neutral, and have been intimidated into the denier camp. Some simply have theories of their own.

Two things are common, however... all deniers are unanimous in thinking...

1. The science behind Global Warming is NOT settled.
2. NO science can be run by a consensus... it's just not the way science should be run.

These are the core tenets which inspired this book.

Deniers question the 'secret' figures and adjustments used by the Alarmists, and want more investigation, more debate, and more measurement before deciding to tax the developed world back into the poverty of the eighteenth century.

Deniers are convinced that data, theories and climate models are being manipulated for political reasons, and do not represent all the possible scientific thinking needed to find exactly what is going on with our present climate.

The Alarmists apparent disregard of every other theory except carbon dioxide anthropogenic global warming is off-putting, non-scientific, and downright dangerous. Before we race into nightmare scenario solutions to a problem we fully don't understand, we need to completely rid the science of the huge level of politically charged rhetoric.

Science is a search for the truth, and no place for politics... no place for party or prejudice.

Science is a search for the absolute truth, openness of ideas and showing your work to the public... no place for secret data, shady tampering or re-writing of history.

Because if science is conducted behind closed doors using secret formulas, then presented as 'settled' to the public, this is chicanery and snake-oil, not science.

Because if historical data is changed for the sake of science, then where do we stop altering data?

Because if historical FACTS, newspaper stories, old diaries, ancient chronicles, paintings, and archeological findings are ignored in order to 'prove' a theory... then the theory was bogus to begin with.

If we can't trust scientists to be above-board, honest and open, then we are indeed left with Eisenhower's nightmare... a 'scientific-technological elite', answerable only to themselves.

1.2... THE ARGUMENT OF 'SETTLED SCIENCE'
(Science is never settled...)

At the root core of the Denier's argument lie eight distinct irritations;

1. **The '97% consensus' figure** has been debunked from as totally misleading to pure propaganda, yet Alarmists still quote it widely today. The simple idea that 97% of all the world's scientists agree with the Alarmist stance is preposterous in the extreme. Tens of thousands of scientists the world over have already nailed their colors to the mast and declared their chosen side... BOTH SIDES!
If you have scientists arguing over basic *data and theories*, the science can hardly be considered settled.

2. **The 'hockey stick graph'** has been debunked as fake, manufactured, twisted, enthusiastically wrong, and out of date. However, the Alarmists have an ally in the media who love their 'catastrophe headlines', and cling to the hockey-stick graph, cling to the same corrupt data, and make more graphs in the same shape.
If you have scientists arguing over *data and graphs*, the science can hardly be considered settled.

3. **The percentage of Anthropological Global Warming** (global temperature rise due to man) CO2 emissions is somewhere between zero and 100%; that's a fact. But many scientists believe some claims are wildly exaggerated, and this can be shown in discrepancies in the new graphs containing the both the 2000's predictions and the measured data since the first climate projections were made. Neither camp is willing to give a specific figure on AGW, and there are as many guesses as there are scientists.
If you have scientists arguing over *CO2 percentages*, the science can hardly be considered settled.

4. **The Alarmists data is a secret,** and they are reluctant to hand over any kind of data for other groups to check and verify. The Deniers choose to look at data directly from the satellites because it's untouched, unchanged by secret formulas. The Alarmists homogenize data, and admit to doing so; they're making adjustments on historical data that they can defend, but they won't tell anyone exactly what they're doing.
In my view, if the research has been paid for by our governments, the data, metadata, and scientists' adjustments to that data should be a matter of public record.

Because every group now has their own super-computer doing their own climate model, the formulae behind the computer models are guarded and kept highly secret. In a way, that's perfectly understandable; they have a lot of cash tied up in those programs, however, this builds wariness among Deniers of the whole gamut of Alarmists data.

If you have scientists arguing over *data*, and getting different graphs from the same data, the science can hardly be considered settled.

5. The Alarmists change the historical data. This sounds distinctly dodgy, but it is a fact, and they admit to doing so. The Australian Bureau of Meteorology has been forced to make public many thousands of such 'adjustments'. Other Alarmist organisations (primarily NASA/NOAA) keep changing the old accepted 1930-1950 temperature figures and 'revise' them down. (To my knowledge, these 'old' temperature 'highs' have NEVER been 'revised' upwards) In other graphs they also smooth out any historical highs, in order to more accentuate modern warming figures... the 'hockey stick' must survive at all costs.

If you have scientists changing accepted *scientific temperature data*, the science can hardly be considered settled.

6. Alarmists put a lot of trust in computer models of climate change. Over 120 computer models are now accepted all over the world... most major universities have one, most major countries who consider themselves as players have one.

However, computer models cannot simulate the chaotic complexity of the climate on the Earth.

We have 120 different nightmare scenarios regarding the rise of the globe's temperature, and EVERY SINGLE ONE is wrong to the physical data, and above what the temperature actually is.

If we have 120 different climate computer models, and they're ALL WRONG, then the science can hardly be considered settled.

7. Alarmists blame Carbon Dioxide as the primary source for the warming; it's easy to blame, it has doubled in concentration in the last 100 years, and the graphs correlate quite well.

Alarmists have also found the source of CO2 for the warming... mankind. Man pumps billions of tons of CO2 into the atmosphere every year.

The Alarmists say the case is closed; walk along, nothing to see.

Thinking in this closed-minded way, they are prepared to ignore any alternative theory put forward by other scientists, and this is dangerous ground to be on. With all our eggs in the one basket, we may make the wrong decisions in the long run.

If you have scientists not agreeing on even the *source* of climate change, the science can hardly be considered settled.

8. Since the high temperatures of 1998, (with a big assist from a larger than average El Niño) the two camps have grown even further apart. The Alarmists quote surface temperatures, and claim higher temperatures every year, the Deniers quote NASA's own troposphere temperatures and say that increasing warming has stopped, and we are now in a period of 'pause/hiatus'.

With their secret datasets the Alarmists try every trick in the book to divert, mislead and obfuscate. The last thing they need is the truth to get out and get in the way of their cash cow. The Deniers doggedly stick to GISS (NASA) figures and plan to wait until the global temperatures drop so much that the pause cannot be ignored.

If you have scientists not agreeing on even the *most basic facts* (temperatures rising/the pause), the science can hardly be considered settled.

9. Science is not consensus. Lastly, the simple fact that the Deniers have been labelled as such is a distortion of the scientific process. Being labelled as such links them to 'Holocaust Deniers', and we all take some level of disdain at the name calling.

Initially, we were Global Warming Skeptics, then as the pause/hiatus hit home, and the Alarmists began to call the 'phenomena' climate change... we became 'deniers'.

Deniers do not 'deny' climate change. The above paragraphs explain their belief system.

Deniers do believe, however, that the debate should be open, not closed.

Any scientific 'fact' should be able to stand on its own, not hide behind a rhetoric of abuse and disdain.

The Deniers consider debate to be useful, and examples of Alarmists refusing to debate with Deniers do their side no credit.

If you have scientists not agreeing on the very basics of *the scientific process,* the science can hardly be considered settled.

1.3... IN CONCLUSION
(My own personal position...)

Both sides agree that the Earth is warming, but disagree on both the cause and the proposed method to cure the planet's ill. They both admit that a certain percentage of the world's CO_2 is man-made (anthropogenic), but differ widely on what that level could be. The IPCC (Inter-governmental Panel on Climate Change) assert that man's CO_2 is more than half to blame.

First of all, I have to state quite clearly I'm not a scientist, but then less than half the personalities that dabble in the morass of Global Warming are. We've got TV pundits, famous singers, Hollywood actors, politicians, English Lords and housewives... all sticking their two cents in, and each believing in the validity, accuracy and truth of their verbiage.

And secondly, I'd like to say that I'm not against science, I think I can focus my thoughts to; "I'm against the distortion and betrayal of the Scientific Method used in the main arguments".

But I am a debater... I like an argument, I love discussion, and I'm an avid collector of facts.
In fact, the only fact that is somewhat undisputed in the whole of the debate is that the Earth's temperature is rising... slightly.
And if we were to ask every climatologist, they would have to admit that no one knows exactly why.
Now, everyone has their own belief or theory, but for sure no one actually KNOWS for certain.

Science has never been about having a consensus of opinion; it's been about theories, backed by scientific experiments, tried and tested. Facts verified in the Scientific Method, peer reviewed, then set aside as the 'theory of the day'.
And as scientific achievement advances, sometimes these theories are disputed, forgotten, disproven and condemned to the rubbish bin of 'could have been'.

But here lies one of the main problems in the science behind Global Warming.
We only have one Earth.
We don't have the luxury of doing experiments on a global scale, and here we find the next biggest problem... Earth is a huge place. Its population is

about 7 billion, a huge number, and yet these people could all fit into New York City, standing up, and they could all have a single room in Texas. How can we possibly bring all the myriad of parameters together to even begin to understand the question we're asking?
And then there's the question of the Sun's effects.

It's a fiery ball in the sky, we orbit at around 93 million miles, and if it weren't for the existence of the Sun, we wouldn't have a climate at all. In fact, every facet of climatology is driven by the energy of this sun, and its reactions with the atmosphere, surface and oceans of our planet.
The sun gave the energy for the original sources of life, and shone masterfully over every diverse part of it.

And yet it's ignored, dismissed by every Alarmist out there.

1.4... The Points to Prove
(CO2 is a boon to plant life...)

Why do I think the risk of Global Warming is being exaggerated by the Alarmist scientists and the media?
My reasons are thus;

1. Doom and Gloom
All the Alarmist's predictions of impending Global Warming are doom-laden. All animals are going to die, humans are going to fry, and when the global ice melts, anyone left will drown in the seas, perhaps one mile higher than today. It's a catastrophic series of possibilities that sells newspapers and glues folks to their 24-hour television news cycle.
But, come on, not every side to the so called 'crisis' can be downbeat. What about the growth of northern hemisphere land area available for cultivation due to the larger areas of Canada and Russia that would come into cultivation in a warmer world?
What about higher food yields because plants grow better in high CO2 levels? The benefits of higher CO2 levels are rarely spoken about. Our planet has 'greened' in the last 30 years, crop yields are up by 35%.
(Although some of this increased production can blamed on better seeds, better growing practices, there have been studies which put the actual global greening by CO2 at between 15-20%)
What about the reduction of water needed for plants to grow in a higher CO2 world?
Not all news needs to be doom and gloom... but doom and gloom make headlines, whether it's in Nature magazine or on CNN.

2. Climate Models
The climate models used and relied upon by the Alarmist scientists have been consistently wrong for almost thirty years. Of the 120+ models now used throughout the world and quoted by the IPCC, over 114 of them are already SO wildly wrong, they are being constantly 'fixed' to suit the actual climate conditions.
Climate models simply cannot cope with the amount of data needed, so modelers ignore clouds, sun cycles and such.
Because of the tampering and the lack of data input, Deniers don't trust climate models; they prefer to rely on measured data.
If the climate models have been 97% wrong in the last 20 years, then why are we using them to predict the global climate in 2100? If they have to adjust them to correlate with present and past levels, than why are we using them in the first place?

If climate models are wrong within a few years, then why do we still look at their projections for 80 years in the future?

The Earth's climate is a complex chaotic system. Even climate modelers admit that there's no computer on Earth big enough to do it properly.

3. Climate Sensitivity

The climate scientists have used a level of climate sensitivity (the level that the climate reacts to internal elements, greenhouse gasses, pollutants, etc) that is obviously too high, but have resolutely refused to change it in their calculations.

In 2019, all of the 120 different climate models developed worldwide are already above the actual temperature readings. With these inaccuracies, why would we consider their extrapolations of the next 100 years?

4. Why Carbon Dioxide?

Climate Alarmists have focused their attention on CO2. Why?

I think the answer is a very basic one; because the most obvious other possible cause, the Sun for instance, cannot be used economically by the world's governments. The Alarmists need something to be able to change; they need to be in control.

This theory is controversial, but trust me, there's evidence to back it up.

5. Follow the Money.

Lastly, the Alarmist climate industry has a vested interest in keeping the alarm/crisis going. If their predictions were "nothing much happening here", their grants would cease, their contracts would soon dry up, and their importance shrink as quick as their ice cream melted. An estimated 300 billion dollars of government (taxpayers) money will be spent on investigating Global Warming in 2019 alone. That's a heck of a lot of money to throw away to a couple of thousand scientists.

Like Upton Sinclair said...

"It is difficult to get a man to understand something, when his salary depends upon his not understanding it."

1.5... THE USA TEMPERATURE RECORD.1.
(Over 1200 weather stations can't be wrong...)

The world is warming, right?
The media tell us so, the scientists tell us so, and the politicians tell us whatever they think will make us vote for them.

It's a game, and everyone knows it.

Out of a standard group of 100 people who read this book, 86 will be American, 9 will be European, and 1 will be Australian.
That's a fact; a statistic.

So let's slide with another set of facts.

The USA has more weather stations than any other country... by far.
The USA has been keeping temperature records since the days of Washington and Lincoln, but continuously since 1880.
The USA is part of the planet, has the third biggest population after China and India, and occupies over three and a half million square miles.

If you wanted to measure temperature, you'd look for a good start here... right?

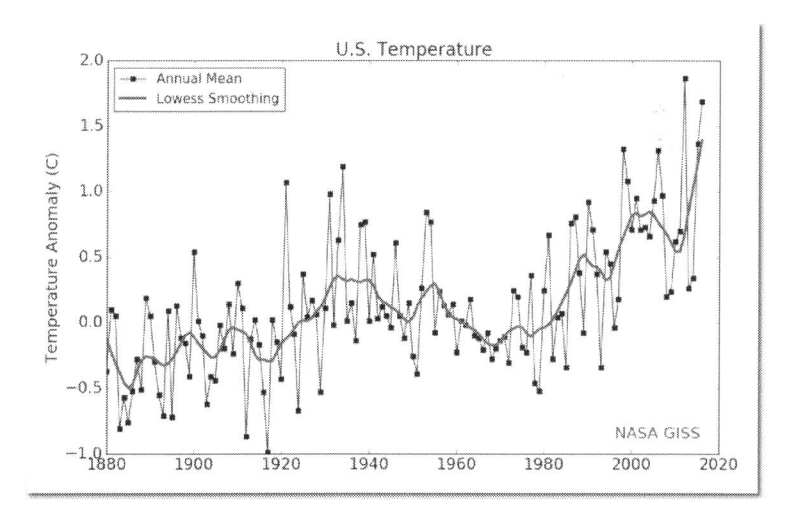

And here we begin. A NASA/GISS graph of average/mean temperature in the USA.

Wow, it's risen 3°C in 140 years... yup it's getting warmer alright... the rest of the world is too... game over... I don't know why I continue writing.

But here's where the differences begin...

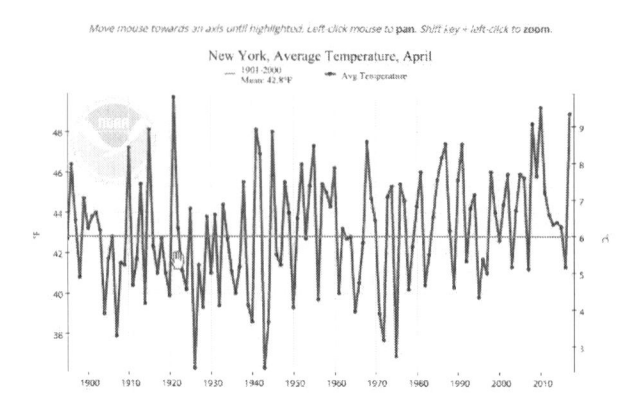

This is the NOAA average temperature for New York for roughly the same time period as the NASA/GISS one for the whole US.
Yup... there's NO UPWARD TREND in New York.

Now, as a student of mathematics, if New York shows NO TREND, and the first graph of the whole USA shows an UPWARD TREND, then to make the average go UP, somewhere else must have a BIGGER AVERAGE TREND.
That's just mathematics. Right?
So let's travel around the USA looking for the HUGE upward trend.
Seattle, (USA's top left hand corner)

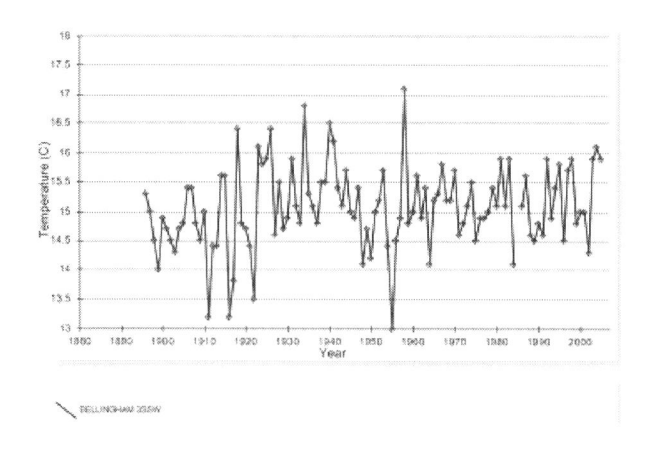

The graph above is the only one I could find for Seattle, there's a slight upward trend... maybe, but it's close to level...
Yup... no upward trend.
My quest continues.
Virginia (USA right hand side, middle)

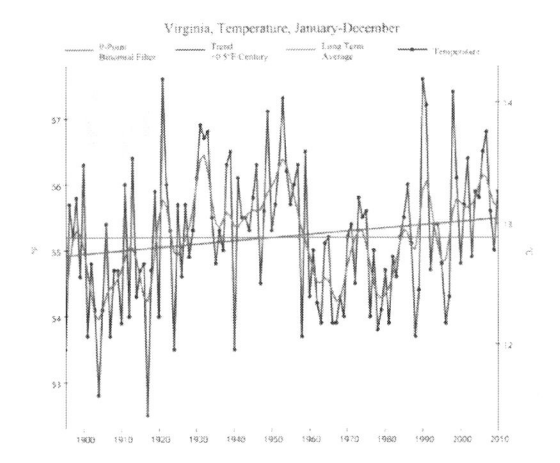

Virginia... now here's a small upward trend, but nowhere near to bring the graphs above to show such a HUGE TREND.

And this is just the tip of the iceberg.
Here's an NDSC graph of Florida (USA's bottom right hand corner)

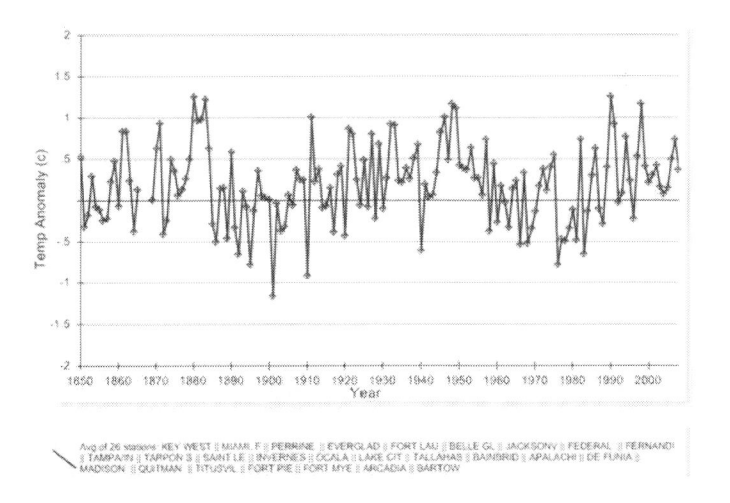

NO trend at all... level.

47

Now I need somewhere near the bottom left of the country...
But these graphs are hand to find.
Since I included the graph of the whole of Florida, I'm including one for the whole of California.
Yes, there's a trend, maybe 1°F in 100 years

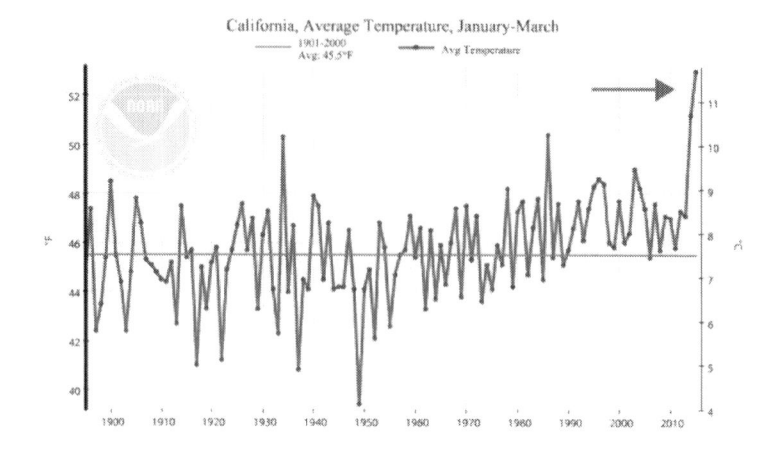

California, Average Temperature, January-March
1901-2000 Avg: 45.5°F

And here it is... the one state that has risen in average temperature from 1950 to 2015 by a STAGGERING 12°F.
This California graph is pure fiction.

Sorry for the mathematicians out there; if I can't find the now HUGE trends upwards, I call the chapter's very first graph pure bull-crap.

If I can't show evidence to justify the first graph's upward trend... I begin to doubt the whole process.

The science is nowhere near settled.

And...

Glacier National Park
This famous National park is situated on the US/Canadian border, in Montana, Alberta and British Colombia.
It's famous for its glaciers... and its weather.
On January 23rd 1916, long before anyone could have blamed man for influencing the weather with his dirty global warming, a world record temperature drop of 100°F (56°C) occurred in only 24 hours.

But if we fast-forward to today, Park Officials have been quietly removing signs, and altering signs and government literature which had been telling visitors that the Park's glaciers were all expected to disappear by either 2020 or 2030.

You see, faced with retreating glaciers, it had been proven that the culprit was man, and he should be castigated publically at every turn of the parks twisting roads.

Until recent years the National Park Service had prominently featured brochures, signs and movies which boldly proclaimed that ALL glaciers in the park were melting away.

However, a recent survey shocked everyone.

The survey showed that many of the glaciers in the park were actually advancing... including the famous glaciers such as the Grinnell Glacier and the Jackson Glacier.
And we're not talking small potatoes either; the Jackson Glacier may have grown as much as 25% or more over the past decade.

A story in the Canadian Free Press said...

"As recently as September 2018 the diorama in the visitors center displayed a sign saying GNP's glaciers were expected to disappear completely be 2020."

And let's face it, with the approach of that fateful year of 2020, the sheer existence of these growing glaciers was proving very embarrassing.

During the winter of 2018 workers replaced the diorama's 'gone by 2020' engraving with a new sign giving a much longer timeframe of the glaciers extinction... contrary to the new reports of the glaciers' increasing, they stuck to Alarmist talking points and insisted that the glaciers will disappear in 'future generations.'

To be honest, it seems that both the New York Times and the National Geographic magazine repeated the 'gone by 2020' story, but no one mentioned the new growth of the glaciers.

It's typical mainstream media cover... pick the pieces that support your own theory, and ignore/bury those that don't.

Are the glaciers at Glacier National park increasing?
If you paid attention to the media, you'd never know it.

Look at the Wikipedia entry... there's no mention of the reversal.

Gianna Kelly, who worked at the Whitehouse under President Obama visited the park in 2008, the first year of his first term. Noting from the park's signs that the glaciers would be gone by 2020, she duly reported back to the President.
But it seems the glacier's doomsayers were a bit too forthright.
From 2010 onwards, the glaciers stopped retreating, and even began to push back.
Despite the newspapers proclaiming the 'eight hottest years ever were in the first decade of the century', the glaciers contradicted the Alarmist evidence, and set new records.
Snowfall in the winter of 2017/18 was at a record high in places.
Next spring, there was a 30-year snow record at Glacier National Park, with crews in April struggling with drifts of 10 to 20 feet.
East Glacier Park Village measured an official 284 inches of snow.

It was one of the coldest winters on record... record wind chills, and 32 days below freezing in February.

But hey, we know the science is settled... whatever.

1.6... The USA Temperature Record.2.
(Facts never hurt anyone, right?...)

US Temperature Records
The USA, having the world's best temperature monitoring system by far, is the standard for most of this book. It's not bias, it's just straight forward common sense.
Europe may have longer records, but not from as diverse stations spread over 3 million square miles.
But in this section, we're dealing with HISTORY.
Here's the graph from the previous section...

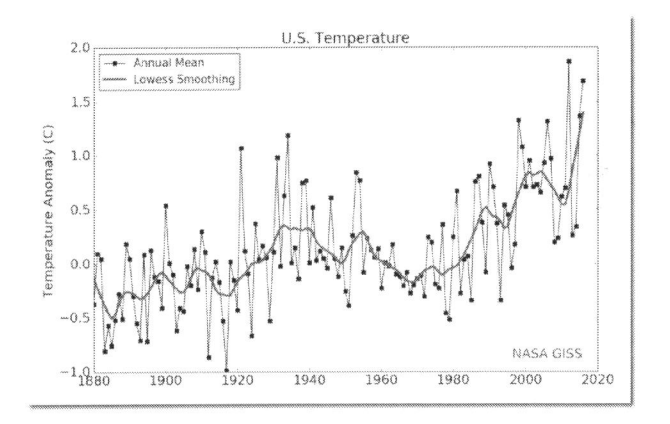

It's NASA/GISS so beyond reproach... right?
The US average temperature has risen over two degrees in 120 years or so.
Let's look at a newspaper headline, shall we?
On July 28th, 2017, the **NEW YORK TIMES** ran with...

"It's Not Your Imagination. Summers Are Getting Hotter."
By Nadja Popovich and Adam Pearce

(https://www.nytimes.com/interactive/2017/07/28/climate/more-frequent-extreme-summer-heat.html)

And they told us unequivocally...

"Extraordinarily hot summers — the kind that were virtually unheard-of in the 1950s — have become commonplace. This year's scorching summer events, like heat waves rolling through southern Europe and temperatures nearing 130 degrees Fahrenheit in Pakistan, are part of this broader trend."

The idea came from work by James Hansen... a HUGE player in the Alarmist world who has been forecasting doom since 1983.
(Oh, and the 'doom' has never arrived)

And let's face it, the article is in the New York Times, so when they state they're talking about *"local summer temperatures"*, they're surely insinuating New York temperatures, right?

Hansen's main point is that in recent years, there have been more days of 'extreme' temperatures...

*"summer temperatures have shifted drastically, the researchers found. Between 2005 and 2015, two-thirds of values were in the hot category, and nearly 15 percent were in a new category: extremely hot.
Practically, that means most summers are now either hot or extremely hot compared with the mid-20th century."*

Basically, he's spreading the Alarmist party line... the Earth has reached the tipping point, and has arrived at runaway Global Emergency.

However, let's stop the hyperventilating, calm down, and get dirty with some real history.

Luckily, in the USA, NOAA and the National Weather Service keep really detailed records... as long as they've not 'revised' them.

New York City
Home of the New York Times. (Home of the 'hottest ever summers')

Let's list New York highs by the month, shall we?

Hottest May... 99°F, on May 19, 1962.
Hottest June... 101°F, on June 29, 1934, and June 27, 1966.
Hottest July... 106°F, on July 9, 1936.
Hottest August... 104°F, on August 7, 1918.

Hottest September... 102°F, on September 2, 1953.
Hottest October... 94°F, on October 5, 1941

Wow... By now are you seeing a trend?

ALL New York's monthly highs were set from 1918 to 1966, and the highest temperature set was in 1936.

So much for 'local' summers getting hotter.

Ah, maybe it was the AMOUNT of hot days?

Yup that's what James Hansen (the great prophet climatologist) said.

Let us (the lowly non-climate experts) look for ourselves, shall we?

From information from NOAA

(https://www.weather.gov/okx/100degreedays)

I got the following.

Amount of days of 100 and above in New York... Listed by decade...
2010's... 5
2000's... 1
1990's... 8
1980's... 2
1970's... 3
1960's... 4
1950's... 12
1940's... 8
1930's... 8
1920's... 2
1910's... 3
1900's... 2

Hmm. The New York Times claim looks shaky.
There's hardly an increase in hot days in New York. In fact, the number of recent hot days is WAY less than the 1930's to 1950's...

And remember Hansen's claim...

"Extraordinarily hot summers — the kind that were virtually unheard-of in the 1950s — have become commonplace."

WAIT! Maybe it was the AMOUNT of days over 90 degrees?

Yeah, there's a webpage for those figures too…
Amount of days of 90 and above in New York… Listed by decade…
2010's… 192
2000's… 123
1990's… 197
1980's… 195
1970's… 163
1960's… 171
1950's… 175
1940's… 201
1930's… 187
1920's… 128
1910's… 115
1900's… 101

So, basically, the headline in the New York Times is a crock of rubbish from start to finish.

Summers have NOT gotten HOTTER… they've got cooler or stayed about the same.

NOW… in my head, that brings in one HUGE question.

If the summer temperatures in New York have been relatively constant… how did NASA/GISS come out with the graph at the beginning of the section?

Where's the huge TWO DEGREE RISE?

That's what this book is all about.

Welcome to the world of Unsettled Science.

Section 2.
The Scientific Method

"Many people think the science of climate change is settled. It isn't."
(Jan Veizer; award-winning isotope geochemist)

2.1... The Unsettled Science
(By definition, science is never settled...)

There is a story is that Einstein was shown a German newspaper that claimed *"One hundred German physicists claim Einstein's theory of relativity is wrong."*
Einstein's reply was supposedly, *"Why 100? If I were wrong, it would only take one."*

And it is this kind of talk that makes the Alarmists' claim of 'settled science' a sham.
In fact, the simple idea that any branch of science is 'settled' (i.e. everything is known, nothing new to discover) would come as a surprise to most scientists, whose very work is to push the limits of science further.
And as the scientists push... science knowledge undoubtedly changes.
And climate science, in these years of its infancy, is no different.

But... to close the door to any dispute of their ideology, the Alarmists say *"the science is settled"* in the same way as policemen divert people from looking at accidents... *"move along... nothing to see here"*.

Many scientists in a myriad of disciplines are adding their own facets of information to the 'climate science' picture every single day.
Sucked in by the estimated one billion dollars spent every DAY in climate investigation, their theories are legion.

But, the Alarmists would encourage you to think that the 'science is settled', and ignore all theories that discourage the idea that man-made carbon dioxide is alone responsible for our current climate change.

Kind of arrogant, huh?

The 'Unsettled' Scientists
In 1995, Fred Singer wrote the 'Leipzig Declaration' on Global Climate Change. It was ratified by a NAMED group of 80 scientists and 25 television meteorologists.
(This book has many examples, and many lists of prominent deniers)

In 1997 as the world's dignitaries headed towards Kyoto to sign the first real climate agreement, a bunch of scientists had another idea.
The first of 31,000 (9000 PhD) scientists signed The Global Warming Petition Project (also known as the Oregon Petition Project).

Teams of retired scientists from NASA have written to both their old employer and to US presidents, and publically denounced Alarmism as nothing more than a sham, decrying their ex-employer (NASA) as a disgrace for its stance on Global Warming.

In the last 20 years since the El Niño enhanced highs of 1998 (as we Deniers say there's been NO significant global warming since then… the hiatus/pause) more skeptics have joined the ranks of those who dispute the claims of the anthropogenic global warming brigade.

Official science bodies from the USA, Poland, Germany, Iceland, Australia, Denmark, and many other prominent countries have been encouraged by their membership to change their Alarmist stance.

You see, when one side (the Alarmists) deny, decry, denigrate and discourage ALL OTHER THEORIES, they're making enemies every single day.

The idea that any kind of science is settled… is pure poppycock.

Yet ALL of the above protests against the Alarmist position are ignored by the Left… they all cling to the "97% of scientists agree" crap.

Anything to stifle debate.

2.2... *Science; Cause and Effect*
(Based on facts, not consensus...)

When we think of science, our mind is sometimes thrown back to schooldays, adding a yellow liquid to a blue one and the resultant mix going suddenly clear, or blue, or blows up.
(Mine goes back to making various kinds of explosives at 17 years old... hey, we all have our own memory of High School Chemistry!)

The Scientific Method usually begins with...
1. Careful observation.
Done either in the laboratory or an examination of a natural process, this should involve a decent amount of skepticism about what is being observed. There should be no rush to judgement, because any early assumptions that we make in the observation can distort later interpretation.
2. Hypothesis Formulation.
From the observation scientists formulate a hypothesis, based on the observation, and embedded in the already 'known' scientific framework.
3. Experimental Duplication.
Where possible, the scientific experiment (or the examination of the natural process under scrutiny) should be duplicated process many times.
4. Refining the Theory.
Once scientists can prove the theory against observation, the theory is processed until the observations, the hypothesis and the test results all link to each other.
5. Publishing the Theory.
Once the scientist has proven the theory, it is then opened to other scientists, allowing them to 'peer review' the findings.

It all sounds good... but see... it all seems to fall down for Global Warming. The Alarmists (including the Intergovernmental Panel on Climate Change (IPCC) have already chosen their 'fall guy' to take the blame for Global Warming/Climate Change... man-made carbon dioxide.
(The IPCC has 'man made' in its core tenet; they were originally set up ONLY to investigate man's effects)
This poor trace gas in our atmosphere has been derided, ostracized, then cast into role of a pollutant, regardless of its basis of all life on planet Earth.

Led by the IPCC's 'guidelines', Alarmists will listen to few other theories with any real attempt at honest review.
If you look at the media versions of the Alarmist scientists' findings, when scientists posit other causes of the supposed warming, the Alarmists' usual reaction is to deride, defend, then attack.

"We have a scientific consensus!" they roar, happy in their little bubble.
"97% of all scientists agree!"

"The Science is Settled", they drone, and brook little sway from their purpose of beating down ANY other theory which comes to light.

Basically, if there are competing theories about any scientific subject, by definition there can be NO CONSENSUS, and the science is NOT SETTLED.

In fact, even the mantra of *"97% of Scientists Agree"* has that fatal flaw.
Remember, there was a time when 99.9% of all ancient 'scientists' believed the Earth was flat and that the Sun went round the Earth!

Cause and Effect
Most scientific theories begin with a subject that changes form in some way or another.
Most of the theory is concerned with the cause of that change... and the effect of that change on the subject or its surroundings once the change has been completed.
So cause and effect are endemic in the scientific method,

Unfortunately, most of the 'unsettled science' on the subject of Global Warming/Climate Change involves the principles of 'cause and effect'.

The Earth has warmed and cooled for all of its existence.
Everyone agrees on that core principle.
The Earth's atmosphere has changed for all of its existence.
Again, we all agree.
The level of carbon dioxide has changed up and down for all of Earth's existence.
Agreed.
There have been historical times of high and low temperatures, and high and low CO2 levels.
However, here is kinda where the two sides split and follow their own roads.

Temperature

Alarmists believe the historical temperature record is not as 'up and down' as historians have researched... they deny the temperature highs of Ancient Egypt, the Roman Empire, and the Medieval Warm Period, no matter what the historical records say.

"Humans have never lived in a climate so hot", they chant.

Alarmists deny the low temperatures of the Dark Ages, the Little Ice Age, and the cold times of the 1970's.

Climate Cycles: Temperatures Trending Down

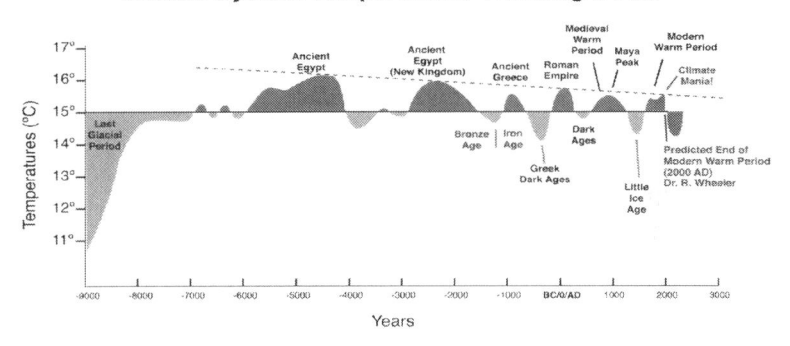

Climate Deniers have looked at all aspects of the historical data... writings, paintings, literature, etc.... and found evidence of historical temperature highs in both the north and southern hemispheres.

They embrace the evidence of grapes being grown near Hadrian's Wall in the first centuries AD. They investigate the Vikings' farming the coasts of Greenland for hundreds of years in the 1100's.

Rate of Temperature Change Higher Now than in Past 1,000 Years

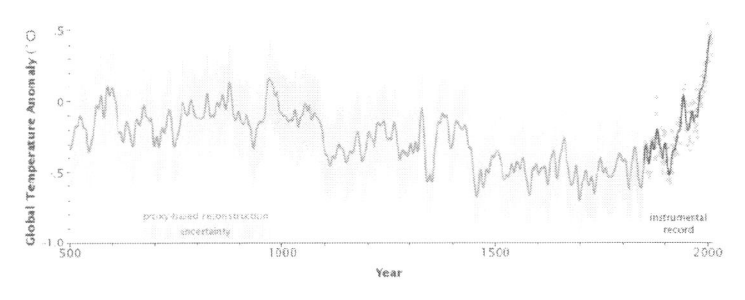

Source: NASA

Alarmist scientists like James Hansen, Michael Mann and the IPCC have settled on the infamous 'Hockey-stick', smoothing out all the historical data in one fell swoop.

Gone are the warm periods in Earth's history, gone the freezing dark ages... replaced by a hockey stick which rises coincidentally when carbon dioxide rises.

You can't wrap lies in a blanket of fog, then pass them off as truth!

Carbon Dioxide

Alarmists believe the increase of Carbon dioxide, (so long riding at 280ppm and now over 400ppm) is completely man-made... anthropogenic in origin.

Alarmists believe the rise of CO_2 during and after the Industrial Revolution, and accelerating after 1950, is the CAUSE of today's temperature rise.

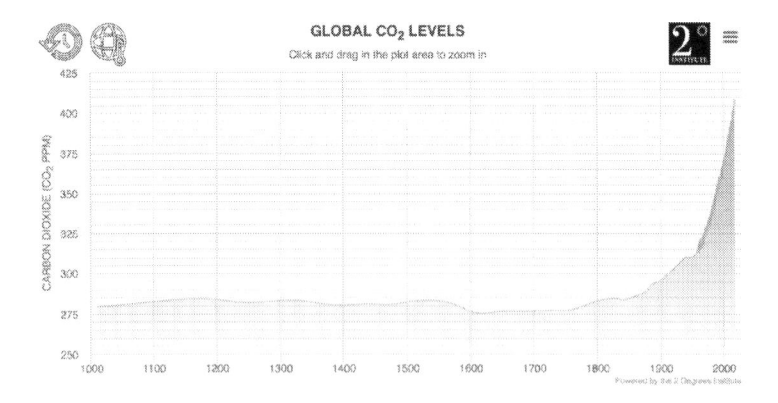

Alarmists believe the rise of modern temperatures is the EFFECT of this CO_2 rise.

Deniers, on the other hand, believe the CO_2 levels in the atmosphere have been caused by temperature fluctuations.

Deniers believe temperature rises first, then is followed by CO_2 rises, sometimes by as much as 1000 years later.

Man-Made Carbon Dioxide

Alarmists believe man-made CO_2 is responsible for much, most or ALL of the recent rise in CO_2... and they DO NOT ACTIVELY look for any other source to blame.

Alarmists believe that unless man's CO2 emissions are curbed drastically, the rise in CO2 will cause unstoppable and runaway Global warming.
The IPCC has stated in their most recent report...

*"Human-emitted greenhouse gases are **extremely likely** (at least 90% chance) responsible for more than half of Earth's temperature increase since 1951."*

Deniers consider much of the recent CO2 rise is natural, but DO NOT DENY the existence of man-made CO2 and its place in the recent rise.
Deniers debate the amount of man-made CO2 in the modern atmosphere to be in the region of 4% to 50%. And even Alarmists can't agree on an exact figure... so much for 'consensus'.
Deniers believe that nothing we can do to curb man's CO2 emissions will bring a significant change in the world's temperatures, if any at all.
Deniers believe that the world cost to attempt this could be better spent elsewhere.
Deniers believe that CO2 as a 'plant food' is beneficial to the Earth as a whole.
Deniers believe that Alarmists have got their mathematics wrong, and a doubling of CO2 will not bring unstoppable Global warming.
Deniers believe there may be other CAUSES of the recent warming, that it may be cyclical, and that we may be heading into a cooling period due to the Sun's recent minimalist cycles.

So much for the science being settled.

2.3... HYPERBOLE... SCIENCE DOESN'T SELL
("Nothing happening here", doesn't sell newspapers...)

Since the arrival of the internet, and all its selling sites (eBay, Craig's List, Facebook Marketplace, etc), newspapers have seen a significant decrease in revenue in the private ad sections.
They need to sell newspapers, and let's face it, controversy sells.

There are many distinct filters on the basic information before it percolates down to the common 'man on the street' from the scientist.
1. The scientist makes a claim, and follows it with his facts, theory, and findings or test results.
2. The scientists decide at this point whether to write a book on the discovery, or write an article.
3. The scientist takes the story to a science journal, who publishes the scientist's words.
(In more recent times, even the journals like Nature and Science are being called 'sensationalist' and 'Alarm Journals' by some scientists.)
4. The scientist's story is picked up by the more mainstream media and brought to newspaper, journal, and magazine.
5. The story hits the headline news, the 24-hour news stream or radio show.

It is well worth noting that there are now at least three separate levels of hyperbole existing in the story's content, message or headline.
This means at least three levels of hype added to the story to get it past the hundreds of other available worthy stories on the news feed.

So, my first caveat is... don't always blame the scientist for scare tactics.

Anyway...
Without doubt, the most common buzzword in the Alarmist's lexicon is 'catastrophic'. Yes, we've had catastrophic warming, we've been promised catastrophic sea level rises, we've been forecast catastrophic weather events, and of course, we've been bombarded with threats of a catastrophic runaway global warming that will wipe all life from our poor defenseless planet before the big game on Sunday.

But these 'buzzwords' are not new. They've been in the English language for hundreds of years, and newspaper companies have sold countless newspapers on the back of them.

Let's face it, 'no news' is boring… it doesn't sell newspapers, and it sure doesn't attract funding.

In fact, the media fan the flames of the 'Alarmists' predictions for their own benefit. What could be better than exciting doom and gloom to fill their long day of empty air time? The television channels need content for their 24 hour TV coverage, and if there's no exciting sports event, no recent horrific terrorist attack, no available coverage of natural disasters… then why not wheel out the nutcase that says Australia will sink by the year 2020!

Even better, get a guy with a PhD, a couple of graphs, a nice television manner, and a university professorship to give him credibility.

Eddie Izzard, a British comedian, once said that England conquered the world by the *'cunning use of flags'*. That was good back in the days of the empire, planting a flag and calling the land 'yours', and today's media are little different. They need two things… pictures of polar bears and graphs… but let's face it; even the graphs have to be exciting…

Enter the Alarmist… and his 'exciting' graph.

A graph so famous, it even has a name.

The Hockey Stick.

2.4... THE HOCKEY STICK PROVES IT
(Conclusive proof of alarming warming...)

Michael Evan Mann is an American scientist with a PhD in geology & geophysics. He is currently director of the Earth System Science Center at Pennsylvania State University, and is seen as one of the main movers and shakers in the 'Alarmist' personnel.

Back in 1998, along with Raymond S. Bradley and Malcolm K. Hughes, he compiled a graph of the mean temperatures of the Northern Hemisphere. Mann & co primarily used tree rings, (proxy data) and then put them all together in a smorgasbord of a graph, shown below.

Primarily in the mix of proxy data was a weighted number of Bristlecone Pine values, which (just incidentally) had a huge growth spurt in the modern era.

The proxy data samples above did all the work for him, until the recent thermometer readings took over around 1880. With the biased proxy data added to the thermometer record, the graph shows temperatures in the 20th Century rocketing skyward, and the resultant graph was figured in the IPCC's Third Assessment Report (2001), and gave the whole 'Global Warming' movement its official *"I told you so"* graph.

The rest, I'm afraid, is history.

The result of man's inhumanity to our own planet was irrefutable, the cause incontrovertible, and the culprit had been caught red-handed.
Mankind itself had caused this catastrophe, and mankind had to do something about it!

The graph I'm referring to?

Yup... it's the infamous 'Hockey Stick'.

It's approaching its 20th birthday.

It's been refuted, disproven, debunked and ridiculed its entire life, yet even in modern debates, Mann and his Alarmist acolytes still trot it out onto the stage like an aging one-trick-pony.

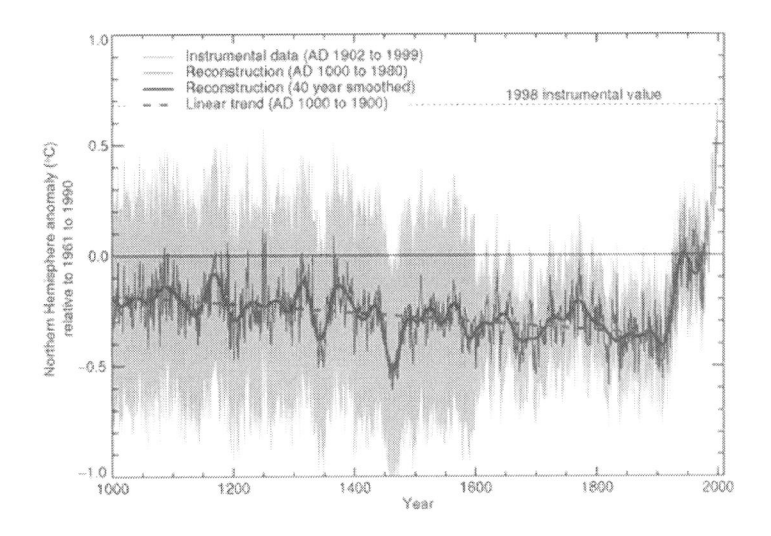

It 'shows' how the average temperature in the Northern Hemisphere dropped a whole quarter of a degree Celsius from 1000 to 1902 (the dotted red line), and then how it rose 1.2°C from 1902 to 1999.

The thick grey bar is the area of uncertainty of the proxy (tree ring) figures, and give a 0.5°C area on either side of the thick black average.

However, nowhere does the area of uncertainty in the historical data come close to the 1998 highs, and the high temperatures of the 1930's and 1950's in the USA (remember this is a graph of the Northern Hemisphere) are not featured at all.

The simplicity of the graphs lines and story are almost too good to be true... and of course, they are.

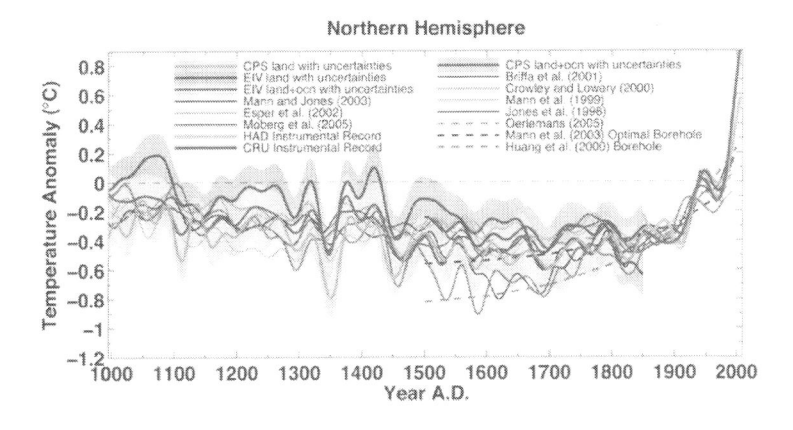

This is the same graph, again by Mann et al, showing a gradual decline in temperature over the millennium, and a hockey stick blade at the end.

However, something's not right, because somewhere along that slowly declining line are meant to be the very identifiable 'Medieval Warm Period' and 'Little Ice Age'.

Difficult to see, huh?

That's because, in the graph-makers urge to get a straight handle for their hockey stick, they selected specific tree ring samples which fitted in with their narrative. I mean, they don't want the inconvenient truth of factual discrepancy to interrupt their theory.

"The great tragedy of science - the slaying of a beautiful hypothesis by an ugly fact." Thomas Huxley

The actual process Mann et al have used is called 'Confirmation Bias'. It is basically the setting out to achieve a result which you have predetermined, and giving bias to all evidence which supports your position.

In the Climategate emails, Phil Jones of the UK's CRU expresses doubt to Michael Mann about the artificially smoothed decline...

"Keith didn't mention in his Science *piece but both of us think that you're on very dodgy ground with this long-term decline in temperatures on the 1000 year timescale."*

Let's take a look at another graph of a similar timescale...

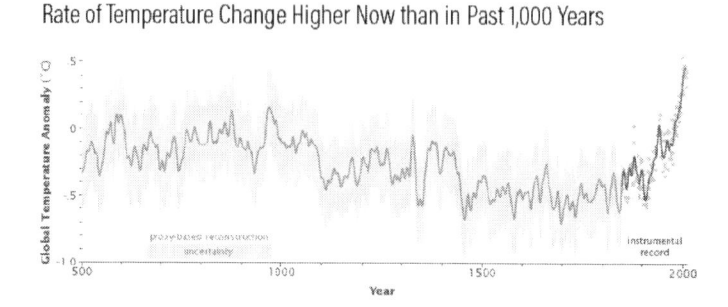

Rate of Temperature Change Higher Now than in Past 1,000 Years

Source: NASA

Above is a 2003 NASA graph, and yes, to a certain extent, even GISS (Goddard Institute for Space Studies) inside NASA has taken on Mann et

al's figures. Now, at least NASA had the decency to actually state on the graph the two different measuring methods, green for proxy data, and blue for the so-called instrumental record. However, like Mann's graph, diminished are both the 'Medieval Warm Period' and the 'Little Ice Age'... and sorry...
Gone also is NASA's scientific credibility.

Considering that it's widely accepted (maybe even a consensus) that temperatures in the Medieval Warm Period were up to two ° above the median (certainly warmer than modern times), and that the Little Ice Age was a degree and a half colder than the median, it's pretty obvious that GISS at NASA has been taken over by 'Alarmist' hacks, and how can we trust anything else they say?
Rather than research their own data, it has become the norm to share datasets from other scientists. Real data crunchers have a mantra about this.

GIGO... Garbage in... garbage out.

"Hold on a minute, Ian!" I hear the astute reader say. *"How can you be so obstinate? Do you have data to show these times of warm and cool?"*
Well, dear reader, it happens I do.

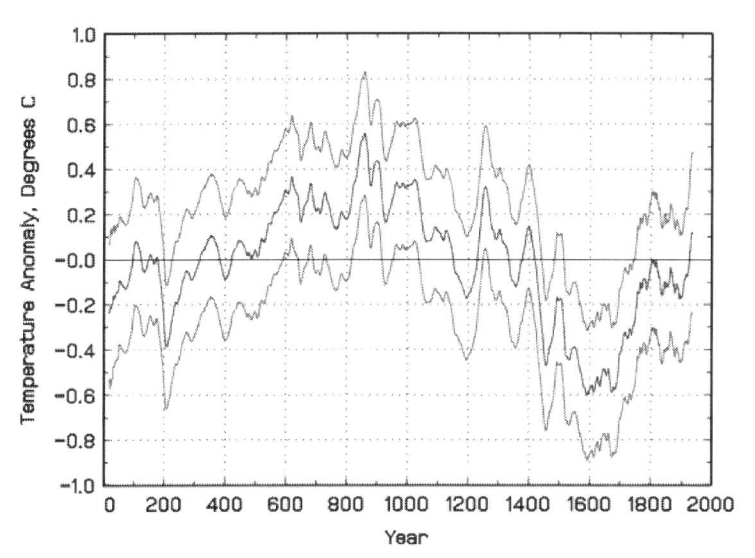

Apart from the anecdotal records of writing, paintings and such, we have an emerging science called dendrochronology.

Tree-ring analysis has come on in leaps and bounds since the 1990's. Using old, dead and naturally preserved trees, there are now interlocking tree-ring databases of many varieties of tree going back tens of thousands of years.

The above 2008 graph, by Craig Loehle and Malcolm McCulloch, shows a one degree rise in the 800's, continuing into the mid 1300's (the temperature rise that allowed Vikings to farm Greenland's coastal areas for two hundred years). Then, in a two degree Celsius dip, it drops abruptly into two hundred years of cooling, the 'Little Ice Age' (the cold period that forced those intrepid Vikings out of Greenland).

The graph above was crafted using multiple tree ring proxies and tree varieties. (with no predetermined bias)

These 'historical' lows and highs are also supported in thousands of historical memoirs spanning the last 2000 years.

And look at the difference in peak value around 870AD... it's a clear 0.4°C hotter than the present values... and no hint of CO2 driving that particular warming. This means other factors had to be driving the earlier heating and cooling.

The graph bears out all the historical evidence of the Medieval Warm Period, the geographical growing records for grape vines, the settlement of Greenland by the Vikings, the great harvests in Europe that led to the age of cathedral building, and thousands more examples, able to fill many books just on that subject alone.

The above graph is what climate aficionados should be in awe about, not the hockey stick.

There are over 400 peer-reviewed papers from scientists and historians all over the world which demonstrate clearly that the Medieval Warm Period did indeed exist, that it was mostly warmer than today's temperatures, and in some cases by up to 0.91°C warmer. The papers show a warming period ALL OVER THE WORLD... the Medieval Warm Period was truly global in nature.

The IPCC (The United Nations Intergovernmental Panel on Climate Change) issues a report, every five/six years. Included in their reports in 1990 and 1995 were graphs showing a distinct Medieval Warm Period and a Little Ice Age. When their next report was published in 2001 (just two years after Mann's publication of the hockey stick) they had changed to Mann's graph.

This is proof positive that the IPCC is political in nature, publishing propaganda rather than researched data, and totally prepared to fudge the facts in order to continue the 'Alarmist' narrative.

Toeing the party line is paramount.

2.5... Climategate... A Storm in a Tea-Cup
(A glance behind the wizard's curtain...)

In November 2009, a hacker broke into some of the email servers of the Climatic Research Unit (CRU) at the University of East Anglia (UEA). Stolen were some fragmentary emails dated from 1996 to 2009, communications between scientists at the CRU and their world-wide associates.
Leaked to the public just a few weeks before the Copenhagen Summit on climate change, the crime was christened 'Climategate' in reference to Nixon's Watergate scandal.

Big names were 'involved'...
Michael 'Hockey-stick' Mann (Penn State University)
Raymond S. 'Ray' Bradley (University of Massachusetts)
Ross McKitrick (University of Guelph)
Phil Jones of the CRU, East Anglia, UK
Kevin Trenberth (University of Illinois)
...and many smaller players too...

The hacking crime was reported by the media, but despite the cries of the alarmists as the flames crept close, the fire soon died down, and many subsequent 'enquiries' have deemed the leaked evidence to be nothing short of a molehill in search of a mountain.
However, these investigations were either carried out at a cursory level, or done by organizations sympathetic to the alarmist movement.
However, in the court of popular opinion, the results stand...
If it was a soccer match, the commentary would run something like this...

The fragmentary state of the hacked emails contained NO EVIDENCE of MAJOR CRIMES committed by anyone. GOAL!

Alarmists United 1, Deniers City 0.

Most of the emails were incomplete, some were unlabeled, and some without context made no sense to the average man in the street... the language used was unexciting, boring, sometimes technical, and most times bland and somewhat ambiguous. No news headlines here. GOAL!

Alarmists United 2, Deniers City 0.

The media reporting the 'Climategate Scandal' had been sympathetic to the alarmist movement for the last ten years... they had a point to prove, they had their own soldiers to defend. With a few exceptions, the coverage was far from investigative and never came close to 'fair and balanced'. GOAL!

Alarmists United 3, Deniers City 0.

However, everyone likes a conspiracy, and these leaked emails did cast doubt upon the supposed unbiased academics crunching the numbers, and the scandal and media coverage did turn a lot of alarmist friends away from their 'climate gods'. GOAL!

Alarmists United 3, Deniers City 1.

When the dust had settled, the scandal wasn't close to the 'win' the deniers had envisaged. The climate organizations, universities and professors all closed ranks, shouted... *"Nothing to see here! Move along!"* and continued on their merry way as if nothing had happened.
The deniers' plaintive cries caused a few of the above to call for investigations to be carried out, but in the face of 'no crime committed', I believe the actual enquiries were either biased, cursory and lackluster (or a combination thereof).
I don't really blame them; in the court of the media, the 'Alarmists' had won, why muddy the water, and further bore/confuse the public.

Wikipedia sums it up remarkably well...
"Eight committees investigated the allegations and published reports, finding no evidence of fraud or scientific misconduct."
However, in talking of the 'verdict' of these investigative committees, Wikipedia states...
"Verdict- Exoneration or withdrawal of all major or serious charges"

Does that mean the professors actually were charged with non-serious changes? A slap on the wrist?
We'll never know.

However there is one thing to be taken away from the fragmentary hacked emails; formerly little-known names were brought to the public eye, and the event did mark a decided point in which the world chose their side.

The 'Alarmists' trotted out their age old (and refuted) graphs, and carried on grabbing money from every hand holding cash.
The 'Deniers', however, seemed to take energy from the scandal, and many of the world's climate doubters can date their moment of 'turning' to that winter of 2009.

In actual fact, if you read the emails (the web address of the full email list and a commentary is given at the end of the chapter), they shed something of a fly-on-the-wall look into the professors mindsets, their thought processes, and their ability to ignore the truth in search of data and graphs to suit their purpose.
And if there was any benefit from the deniers' side, it was the mention of one phrase... 'the pause'.

The Pause
Global temperatures haven't much gone up or down since 1998, and 'everyone' knows it... even Alarmists.
It's been christened as 'The pause', or hiatus.
In an email dated January 5, 2009 (email 1231190304), Phil Jones writes to Tim Johns, Chris Folland, and Doug Smith...

"I hope you're not right about the lack of warming lasting till about 2020. I'd rather hoped to see the earlier Met Office press release with Doug's paper that said something like—"half the years to 2014 would exceed the warmest year currently on record, 1998"!"

Not exactly the email you'd want leaked... huh? Jones continues...

"Still a way to go before 2014.
I seem to be getting an email a week from skeptics saying "where's the warming gone"? I know the warming is on the decades scale, but it would be nice to wear their smug grins away."

Not only do the 'Alarmist' scientists agree about the absence of warming, but the information/data is getting to the 'Denier' side too. And we were starting to push them for answers.
Then, in October 2009, just days before the CRU emails were hacked, Kevin Trenberth, of the University Corporation for Atmospheric Research emailed...

"The fact is that we can't account for the lack of warming at the moment and it is a travesty that we can't."

In following email, he asks the question...

"Where did the heat go?"

Trenberth then admits that three sets of vital data are inconclusive, meaning they CANNOT understand how the climate functions...

"the resulting evaporative cooling means the heat goes into atmosphere and should be radiated to space: so we should be able to track it with sky temperature data. That data is unfortunately wanting, and so too are the cloud data. The ocean data are also lacking..."

The 'Alarmist' crew are admitting to themselves that they do not know where the 'warming' is going.
HINT... maybe it wasn't there to begin with!

His emails are summarized by Tom Wigley, who answers...

*"Kevin,
I didn't mean to offend you. But what you said was "we can't account for the lack of warming at the moment". Now you say "we are nowhere close to knowing where energy is going"."*

So they fudged more data and made the new figures suit the media needs, and they carried on as usual.
But of course, we all know that these guys were not interested in finding the actual CAUSE for their 'lack of warming at the moment', they were more concerned with hiding that 'lack of warming', and finding some other series of 'catastrophic' stories to sell to the eager press.
INSTEAD OF GOING BACK TO FIRST PRINCIPLES, LOOKING AT FRESH DATA, AND STARTING AGAIN.
Once again, toeing the party line is more important than the truth.

But it gets worse...

In an email dated July 5, 2005 (email 1120593115)
Phil Jones writes to climate scientist John Christy:

"This quote is from an Australian at the Bureau of Meteorology Research Centre, Melbourne (not Neville Nicholls). It began from the attached article. What an idiot. The scientific community would come down on me in no

uncertain terms if I said the world had cooled from 1998. OK, it has, but it is only 7 years of data and it isn't statistically significant."

This last statement almost has to have a word-by-word study done on it... Jones is afraid of changing the global warming narrative in case the *'scientific community'* would descend on him?? AND it would 'come down' on this incredibly famous scientist *'in no uncertain terms'*.

The follows...

"'Bottom line: the 'no upward trend' has to continue for a total of 15 years before we get worried.'"

And lo and behold, as I write in the spring of 2019, it has 'continued' for a total of TWENTY-ONE YEARS.
Anyway...

Changing Global Warming to Climate Change
After six years of pause (no warming since the El Niño assisted 1998 high), the Alarmist conspirators were getting a bit of flak from the public saying *"Where's the Global warming gone?"*
Something had to be done... a different label had to be found.
In a 2004 email, from the Minns/Tyndall Centre on the UEA (University of East Anglia) campus...
"In my experience, global warming freezing is already a bit of a public relations problem with the media"
Swedish Alarmist Bo Kjellen replied, stating...
*"I agree with Nick that **climate change** might be a better labelling than **global warming**"*
And in a nanosecond, the goalposts changed, the media had a new slogan, and nobody cared...

Hide the Decline
Back in 1999, the usual suspects (i.e. Michael Mann & co)were trying to get their hockey stick graph into the IPCC's Third (2001) Assessment Report (TAR). (see section 2.4)
But they'd run into some problems... their proxy tree ring data sets were not making the right shape of graph... they showed warm periods in the planet's recent history, times when the planet was just as warm as it is today... and that made for a whole load of awkward questions.
They needed to 'get rid' of those warm periods... smooth out all that earlier warming and cooling.

I mean, if carbon dioxide was the ONLY driver of global warming, then how did the planet warm in Roman times? In the Medieval period? Basically it was like an episode of Scooby Do... these Alarmists would have gotten away with stealing the Medieval Warming Period and the Little Ice Age if it hadn't been for those pesky kids (Deniers) reminding the world about actual recorded history!

On September 22, 1999, (email 0938018124) Keith Briffa writes to Mike Mann, Phil Jones, Tom Karl, and Chris Folland)

"I know there is pressure to present a nice tidy story as regards "apparent unprecedented warming in a thousand years or more in the temperature proxy data" but in reality the situation is not quite so simple... I believe that the recent warmth was probably matched about 1000 years ago."

YUP! He just admitted the temperatures in the Medieval Warming Period were... WARM.
AS WARM AS RECENT TIMES... I. E. TODAY!
Compromises had to be made, but it made for a few awkward discussions...

"This is the problem we all picked up on—everyone in the room at the IPCC was in agreement that this was a problem and a potential distraction/detraction from the reasonably consensus viewpoint we'd like to show..."

More work had to be done...

"Otherwise, the skeptics would have a field day casting doubt on our ability to understand the factors..."

So after fudging/massaging the data that apparently didn't fit their narrative to get the right, 'consensus' viewpoint, the hockey stick, (ignoring the Medieval Warming Period and the Little Ice Age,) got into the report.

Then...
In the leaked email dated November 16, 1999, (email 0942777075), Phil Jones (talking to Ray Bradley, Mike Mann, Malcolm Hughes, Keith Briffa, and Tim Osborn) 'admits' to fudging graphs. It is, without doubt, one of

the key excerpts, and a look behind the wizard's curtain to see what's actually happening in the world of 'climate change'...

"I've just completed Mike's Nature *trick of adding in the real temperatures to each series for the last 20 years (i.e. from 1981 onwards) and from 1961 for Keith's to* **hide the decline***."*

Of course, the left have their excuses, their reasons for all the emails, and dismiss every one.
And the climate scientists concerned were exonerated in nine separate enquiries... blah.
Can someone say, whitewash?

The Final Parting Shot
And as if a screenwriter had written the final screenshot of the movie himself, Michael 'Hockeystick' Mann writes on the final hacked email...

"As we all know, this isn't about truth at all; it's about plausibly deniable accusations."

I have to stop there... the above phrase has sucked all the air out of my small room.

Most of the emails (edited and annotated by John Costella) can be read here...

https://www.lavoisier.com.au/articles/greenhouse-science/climate-change/climategate-emails.pdf

2.6... Problems Of Climate And Weather
(If we can't predict one month away, how can we...)

Michael Fish
If you're older than 45 or so, it's difficult for anyone living in Britain in 1987 to forget the 'Great Storm'.
Well-known BBC Weatherman and meteorologist, Michael Fish, said in a TV forecast a few hours before the storm's landfall on October 15th...

"Earlier on today, apparently, a woman rang the BBC and said she heard there was a hurricane on the way. Well, if you're watching, don't worry, there isn't!"

Later that evening, the worst storm to hit South East England for three centuries caused record damage and killed 19 people.

Yup, even the experts can't get it right all the time.

The live TV 'flub' has been quoted so many times, the term "the Michael Fish effect" is now applied to public forecasts on any topic, which end up being embarrassingly wrong.

On a side-note, Fish later admitted there had been no 'woman caller' to the BBC; it was his own introduction to the subject.

Ultimately, you can count the reasons for such an error in forecasting on one hand; the obvious one is the lack of tracking such weather over the seas, and that comes down to lack of scientific measurement.

National Weather Service
The people of the continental USA are provided with such up-to-date weather forecasting that sometimes local weather events can be predicted down to the minute.
This didn't happen by itself.
There are hundreds of weather stations, constantly calling in their data to the central hub of the National Weather Service (NWS), a division of the National Oceanic and Atmospheric Administration (NOAA), itself a branch of the US Department of Commerce.
There are numerous different automated systems monitoring barometric pressure, temperature, dew point, wind speed and direction, precipitation, lightning, cloud cover, and many more.

A Doppler radar system was completed in 1997, and now the USA boasts the world's best weather monitoring system.

But that's all on land.
How does this service compare to what happens at sea?

Well, to be honest, there's no comparison.
The USA has weather reports from shipping, and from hundreds of buoys along the coast, but in reality, predicting weather over the world's oceans is far more difficult that over land... mainly... you know... because of... THE WATER.

Let's take hurricanes for example.

Hurricane Prediction
Storms hit the coast of the USA every year.
They begin as small cyclones out in the mid-Atlantic and/or Caribbean Sea, and mainly hit the eastern and Gulf shores of the mainland USA.
Because of their destructive capabilities, they are covered by the media almost around the clock, and there are four questions asked repeatedly.

What Hurricane Category will it be at landfall?
Exactly where will it hit?
When will the hurricane hit land?
How high will the storm surge be?

And because of the unpredictability of the water below the storm, and exactly what water it will travel over on route to land, the meteorologists make different guesses every hour.
You see, there are more than 40 different large scale components to a hurricane's size and strength, it's IMPOSSIBLE to predict any of the three repeated questions with any accuracy.
In no particular order, barometric pressure, directional wind speed, angular wind speed, wind shear, storm temperature, storm density, water depth, water temperature, tidal motions... and many, many more.

And here lies the contrast between 'weather', and 'climate'.

If qualified meteorologists with every modern computer available CANNOT predict the size, speed and direction of a storm just 100 miles offshore... how the heck can they predict climate over 100 years?

Single hurricanes have over 40 components... the world's climate has MILLIONS!

If qualified meteorologists with every modern computer available CANNOT predict the WEATHER in a square 100 miles wide... with JUST the variabilities of a SMALL square on Earth's surface... when ALL the eyes of EVERY media in the most technological country and the eyes of EVERY qualified meteorologist in the world are trained at that small square... what chance have climate scientists got, when their field of observation is 317 million square miles??

2.7... COMPUTER MODELLING FOR IDIOTS
(Not sure who the idiots are exactly...)

Since the first 'real' computer (Bletchley Park in 1942) the electronic saviors of humanity have gotten smaller and more powerful by the decade.
In fact computers, processors and data speeds change so quickly that fully 95% of all home computers are less than 10 years old.

Because of the recent availability of the component part, computer modelling is a relatively new science. In the field of climate change, the first modelling was attempted in the 1970's, looking into the short-termed panic about global cooling.
By the 1990's climate modelling was a full-time occupation for many climate 'scientists'.
But of course, we're talking about monitoring the data from 317 million square miles of the Earth's surface, and we're only talking about the data from 3000 weather stations (and that's a VERY generous number).

So... each weather station supposedly monitors 100,000 square mile of the Earth's surface... that's hardly 'global'.

Today there are more than 120 DIFFERENT computer model programs running and being maintained by large corporations, universities and governments... all reporting to the IPCC.
It seems that every country and every university looking to be engaged in the climate debate, 'has' to have one.
And they all use the same data... because they all have the same agenda.
To be all-inclusive to their political base, the IPCC use 90 of the 'best'.

There is an old adage in the world of statistics referring to basic data being fed into a computer... GIGO... *'Garbage In, Garbage Out'*.
And any computer programmer worth his salt knows full well that if you already 'know' the result you're looking for, you can tweak your input data to 'find' the appropriate output. As mentioned before, it's called 'Confirmation Bias'; the art of having a pre-determined outcome, and fudging the data to show which result you actually want.

Now we all know the 'result' the Alarmists want. They want 'real' climate change, and they want it now!
Unfortunately for them, the pause/hiatus in the rising temperature record has them temporarily stumped.

Below is a graph showing all 90 computer modelling programs being run for the IPCC.
Please note, although most have the same trend, all have used slightly different values for the carbon dioxide feedback loop, and so have all arrived at slightly differing results.

The solid/dotted upper line is average of the 90 computer models.
The solid/dotted low line is the CRU's temperature data.
The solid/dotted lowest line is UNALTERED satellite data.

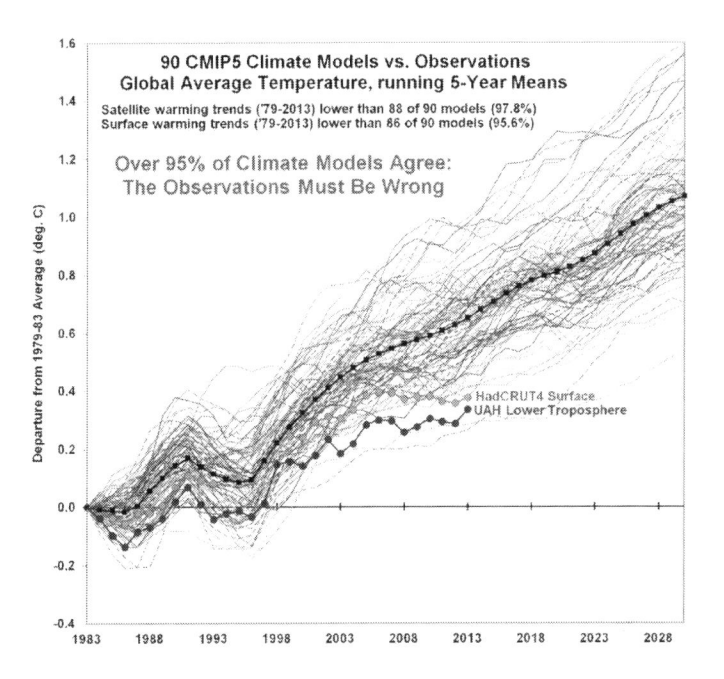

90 CMIP5 Climate Models vs. Observations
Global Average Temperature, running 5-Year Means
Satellite warming trends ('79-2013) lower than 88 of 90 models (97.8%)
Surface warming trends ('79-2013) lower than 86 of 90 models (95.6%)

Over 95% of Climate Models Agree:
The Observations Must Be Wrong

HadCRUT4 Surface
UAH Lower Troposphere

On each new report the IPCC attempt to 'correct' their graphs, but nothing seems to be able to deal with the actual temperature data.
The dotted upper line is an 'average' of all 90 computer models, the dotted green line is the current ground temperatures, the dotted blue line is the satellite feeds from the lower Troposphere.

AND PLEASE NOTE, all of these computer models have been updated...
LOOK AT THE GRAPHS FROM 1983 to 2013...

These records are in the PAST.

If the computer models cannot be adjusted to accurately tell the past temperatures; how can they possibly see into the future?

But this is only one look at the lunacy of computer models... the satellite feeds from the Troposphere are the only ones which are made public without any secret changes by the data's 'owners'.

THAT, dear reader, is a huge problem to me.

The HadCRUT4 surface temperature (Hadley, Climate Research Unit... the ones who had the Climategate leaks in 2009) are sorted, massaged, homogenized, and distorted before the public gets the data... and like a bad schoolboy, they DON'T make their alterations public.

The University of Alabama, in Huntsville, led by state climatologist John Christy and professor Roy Spencer have used the satellite feed of the lower troposphere figures since 1978.
The data is available from their website, and is raw, original and non-adjusted.

Here's what the ACTUAL global warming looks like...

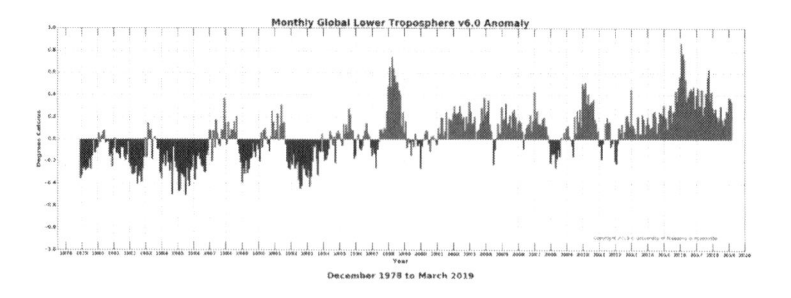

Monthly Global Lower Troposphere v6.0 Anomaly

December 1978 to March 2019

It's hardly the hockey-stick you're used to seeing in the media.
The UAH team also graphed the spaghetti trails of all the IPCC computer models. The thick red line is their average.

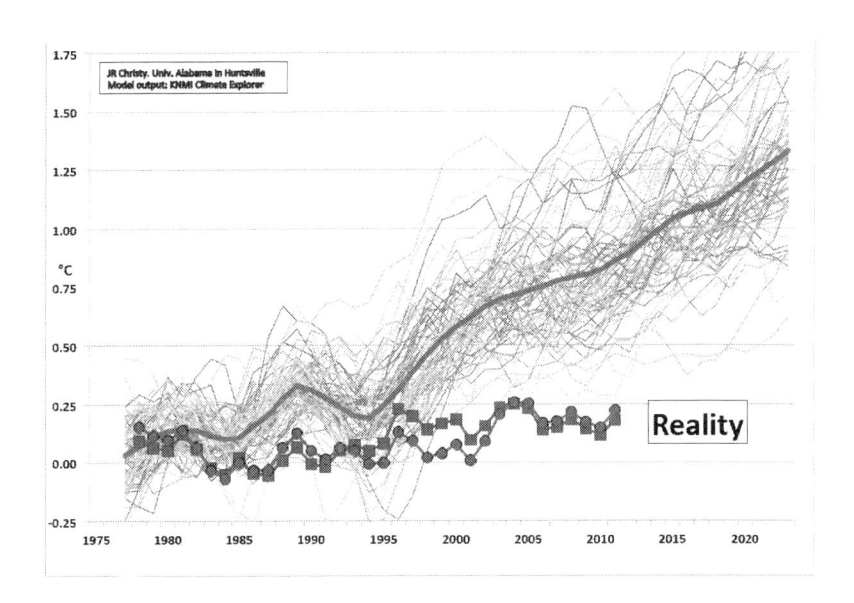

Yes, dear reader... by 2012, ALL of the IPCC climate models had been proven to be WRONG.
Yet the IPCC sticks to its guns, announces a 95% degree of 'certainty' in their figures/data. They predict an average of a 1.35 rise by 2024.
Look at the data from 1998 (the El Niño driven peak) onwards...
THAT'S the pause/hiatus.
THAT's what the IPCC and Alarmists CANNOT explain away.
And THAT's why I wrote this book...

If the IPCC is REALLY trying to predict the global temperature in the year 2100, and is making political statements of the catastrophes that will follow under their predictions, then at least make the predictions REALISTIC.

But they don't want realistic predictions... they're boring, dull, non-catastrophic and very un-newsworthy.

Why would we put the future in the hands of a forecaster with such a bad record?

If the above graph were a prediction of stocks and shares... would you invest in this company?

No.

The climate of our planet is a chaotic system, with millions of variables.
One of the alarming things is… the scientists don't FULLY understand some of the most basic ones.
Let's take a look at some of that variables that scientists don't FULLY understand, and therefore cannot include them fully in their climate models…

1. Changes in seasonal irradiation from the sun.
2. The energy flow/transfer between ocean and atmosphere.
3. The energy flow/transfer between land and air.
4. The natural balance between water, water vapor and ice.
5. The full impact of clouds.
6. The full mechanics of the planet's ice sheets.
7. The full dimension of change of ice sheets, global sea level, and glaciers.
8. The ability to fully factor hurricanes, tornadoes, droughts, and storms.
9. The impact of vegetation on temperature on a local or global perspective.
10. The tectonic movement of plates on the ocean floor.
11. The differential rotation between the Earth's surface and the planet's core.
12. The interaction of the solar system's magnetic field and its gravitational pull.

You see, until we begin to input ALL the data in the world, we will not have an accurate climate model.

I could repeat… "THE SCIENCE IS FAR FROM SETTLED!"

But we all know that this is not science… it is political dogma, driven by adjusted data, with an end-game agenda in mind.

2.8... THE AVERAGE GLOBAL TEMPERATURE
(What does this actually mean?)

The Problem
Talking about the average global temperature is like discussing how many different farts (human carbon dioxide/methane outgassing) happen at a wedding.
Historically it's extremely difficult to maintain details on a non-constant system, complex to work out mathematically, and equally difficult to qualify and keep scientific integrity.

The scope of the problem is the size of the Earth's surface, the diversity of the heat and cold at any one time, the tilt of the earth in relation to the sun, the day/night aspect, and the simple fact that land only makes up 30% of the surface of the planet.

Add to that, every square inch of the Earth's surface changes temperature every single minute.

At the NASA website, there is a page devoted to explaining what is meant by Surface Air Temperature (SAT). The explanation is extensive, so I'll quote briefly.

(https://data.giss.nasa.gov/gistemp/faq/abs_temp.html)

"Q. What exactly do we mean by SAT?
*A. **I doubt that there is a general agreement how to answer this question.** Even at the same location, the temperature near the ground may be very different from the temperature 5 ft above the ground and different again from 10 ft or 50 ft above the ground. Particularly in the presence of vegetation (say in a rain forest), the temperature above the vegetation may be very different from the temperature below the top of the vegetation. A reasonable suggestion might be to use the average temperature of the first 50 ft of air either above ground or above the top of the vegetation. To measure SAT we have to agree on what it is and, as far as I know, no such standard has been suggested or generally adopted. Even if the 50 ft standard were adopted, I cannot imagine that a weather station would build a 50 ft stack of thermometers to be able to find the true SAT at its location."*

So even the folks who keep the actual data agree that they DON'T have a standard.

The Actual Data

There are three ways in which the world's climatologists measure Surface Air Temperature data...

By satellite data, by weather balloon, and by ground station data.

The satellite data has far more data points, sampling millions of different parts of the earth on a daily basis. The surface data is limited by the amount of surface stations, and is added to by estimations of non-station areas and sea temperatures.

There may even be more estimation of non-station areas than actual real data. (we'll never know, remember the data crunchers keep their formulas secret)

There are basically five differing systems being used by the climatologists today, each with their own acronym, and each with their own 'quality' stamp...

1. Quality Class 1... The Satellite Lower Troposphere Data

These are readings taken from space of the lower troposphere (the area of the atmosphere where we live). Many millions of records are taken each day by more than one satellite, and transmitted to the climate scientists on the ground.

There are two sets universally recognized and both derive their data from the same satellite sources... UAH (University of Alabama, Hunstville), and RSS (Remote Sensing Station), a private company in California.

The University of Alabama, Huntsville put all these records online, and they are pristine, not altered, not touched, not massaged, not changed, not adjusted for any reason.

As far as I can determine, the RSS data is adjusted slightly before they release it to the public.

2. Quality Class 2... UK MET Office Surface Data

The HadCRUT temperature data set is a mix of sea surface temperatures from the MET Office Hadley Center, and surface temperatures from the Climatic Research Center, University of East Anglia. The data is now in its THIRD version of revision, and the latest set is called hadCRUT4.

3. Quality Class 3... The NCDC and GISS Surface Estimates

This data is universally known as the most flawed/tampered. With this in mind, remember this is used by NOAA, NASA, and the IPCC. The adjustment formulas are constantly messed around with and the formulas used are kept secret. Even past/historical and supposedly settled data is revised. Recently, after being questioned as to the 20 year hiatus/pause, NOAA massaged the figures again, giving a 36% rise in modern temperatures.

The Graphs
As can be seen in the graph below, all five different data sets show slightly different trends.
The satellite data is immune to the 'administrative changes' of the surface data.

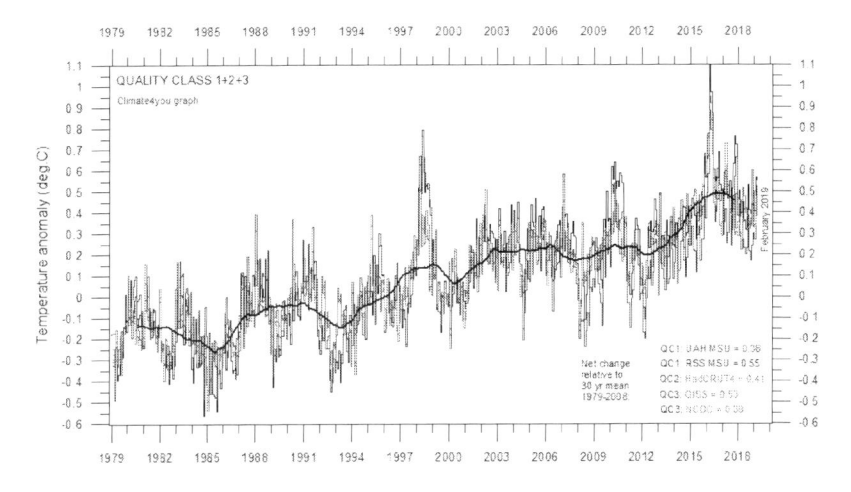

The satellite data sets show less warming than the surface temperature data.

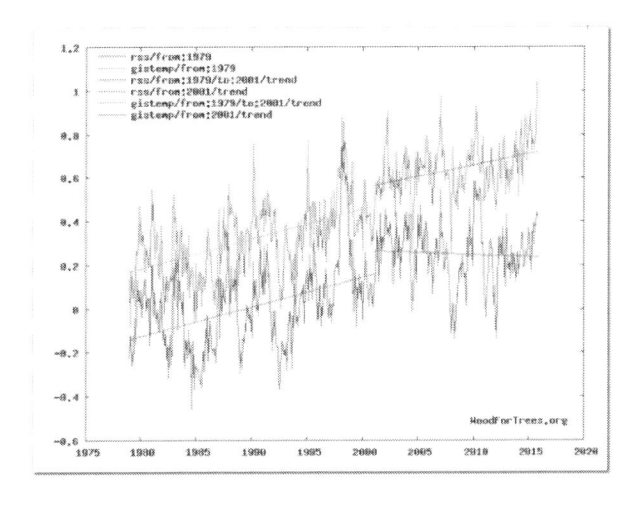

The above graph is of RSS data (LOWER) against GISS data (UPPER).
But, there's something pretty wrong with this graph.

From 1979 to 1997, both records were close, with the GISS (NASA) data showing a clear higher value (from 0.2 to 0.4 degrees). But the trend lines shown are simply wrong.
In 1998 (the huge El Niño assisted spike) the temperatures come together...
Why in that one year?
The records show EXACTLY the same value. How is that possible?

From that 1998 peak, the GISS surface temperature and RSS satellite data show very different trends.
The GISS data shows a slight change from their 1979/2000 trend, but still upward.
The RSS data show a DOWNWARD trend... a DECREASE in global temperature.
What happened to change the trend/data sets?

By 2013, the figures differ by almost 0.6 degrees.

The GISS/NASA (green) graph shows successive peaks after the El Niño high of 1998... FOUR separate peaks higher than 1998. This is not shown in ANY other graph I've seen... 1998 is a high, then there's the next 2016 El Niño high.
In the RSS data (red) graph, none of the successive years reach the 1998 peak.
Let's look at the RSS graph separately...

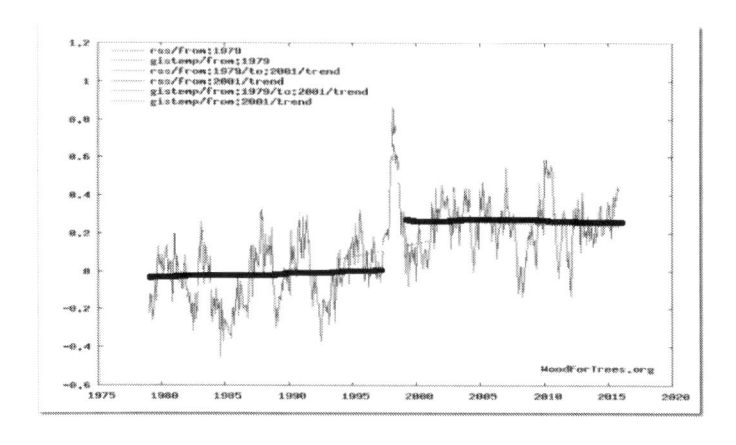

Unlike the original graph which shows warming to 1998 then a cooling....
If we discard the 1998 data as an anomaly, there's no gradual warming...
Look at the graph above... there's a pretty straight 1980 to 1997, then a El Niño peak at 1998, then a slightly higher level/downward trend.

Even if I were to be an independent viewer...
Someone is either telling lies, fudging data, or something happened to the mathematic adjustments of the satellite feeds in 2000.

Either way... there's no possibility that the science is anywhere near settled.

2.9... THE APPARENT GLOBAL WARMING HIATUS
(There's little 'APPARENT' about it...)

Kevin Trenberth (famous for his part in the Climategate emails) told Nature...

"The 1997 to '98 El Niño event was a trigger for the changes in the Pacific, and I think that's very probably the beginning of the hiatus... it will switch back in the other direction"

Wikipedia states...

"It is thought that there have been at least 30 El Niño events since 1900, with the 1982–83, 1997–98 and 2014–16 events among the strongest on record."

So to investigate this 'hiatus', let's take a look at global temperatures in the last 20 years or so.
If we look at the ClimateBet graph based on the University of Alabama in Huntsville troposphere figures (straight RAW data from NASA satellites, no phony 'adjustment' here).

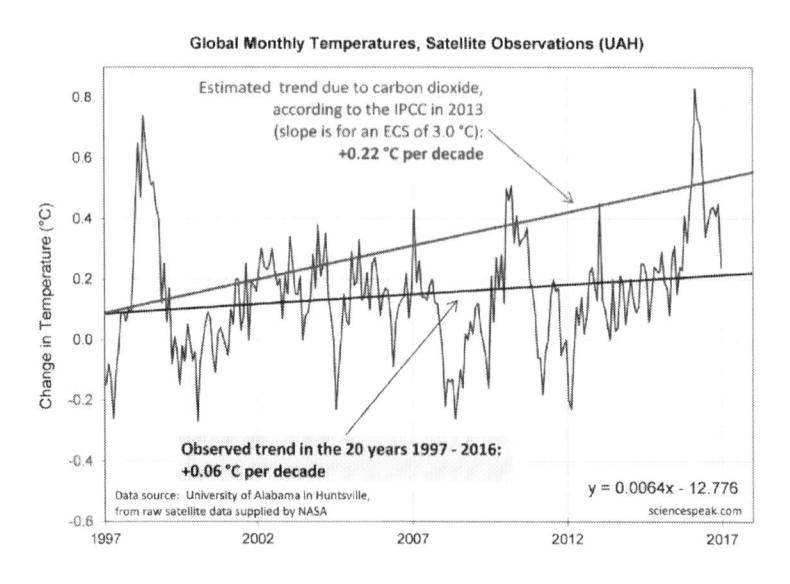

Global Monthly Temperatures, Satellite Observations (UAH)

Estimated trend due to carbon dioxide, according to the IPCC in 2013 (slope is for an ECS of 3.0 °C): +0.22 °C per decade

Observed trend in the 20 years 1997 - 2016: +0.06 °C per decade

Data source: University of Alabama in Huntsville, from raw satellite data supplied by NASA

y = 0.0064x - 12.776
sciencespeak.com

In blue are the monthly global temperatures.
YES, monthly.

We see the elevated peaks in 1998 and 2016, both very strong El Niño event years.
The upper sloping line is the IPCC estimate of global warming due to carbon dioxide to give projected figures of +0.22°C rise per decade, and a roughly 3.0°C rise in a century.
The lower straight line is the actual trend of the plotted graph, giving 0.06°C rise per decade, and a 0.6°C rise in a century.
AND the data for 2018 is in, and they are lower than average, just a fraction under the 2017 figures.
Go take a look... http://www.theclimatebet.com

The latest graph from ClimateBet shows 2018's figures.

Let's take a look elsewhere...
This long term graph from the Japan Meteorological Agency (JMA)

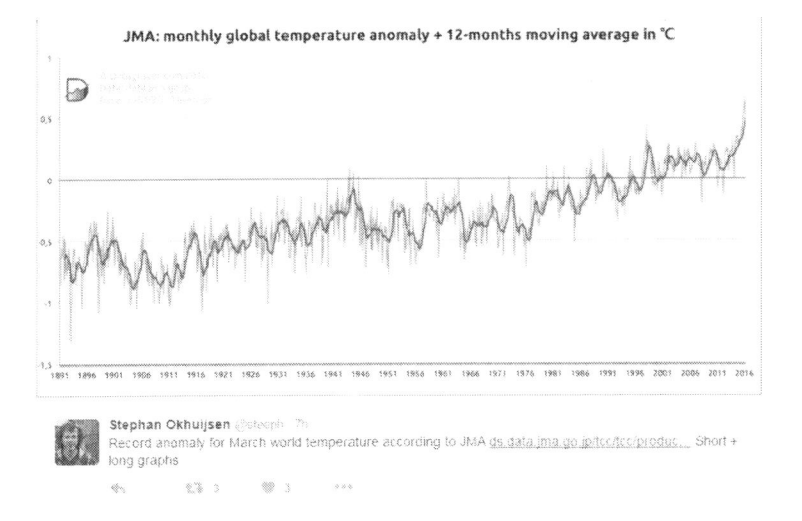

It seems to agree with the one degree per century model, and shows the strong El Niño peaks in 1998 and 2016.
AND YOU KNOW... If I was being REALLY pedantic, I would have pointed out that the ONE DEGREE temperature rise was actually over nearly 130 years!

But let's take a look at another set of figures.

The Hadley Centre for Climate Prediction and Research is the UK's own MET Office climate study centre, and about 200 of the MET office's 1500 staff work at Hadley.

They are well known for both unadjusted data sets, and unfortunately homogenized/adjusted data too, but let's take a look at the Met Office Hadley Centre (MOHC)'s own look at the recent temperature figures...

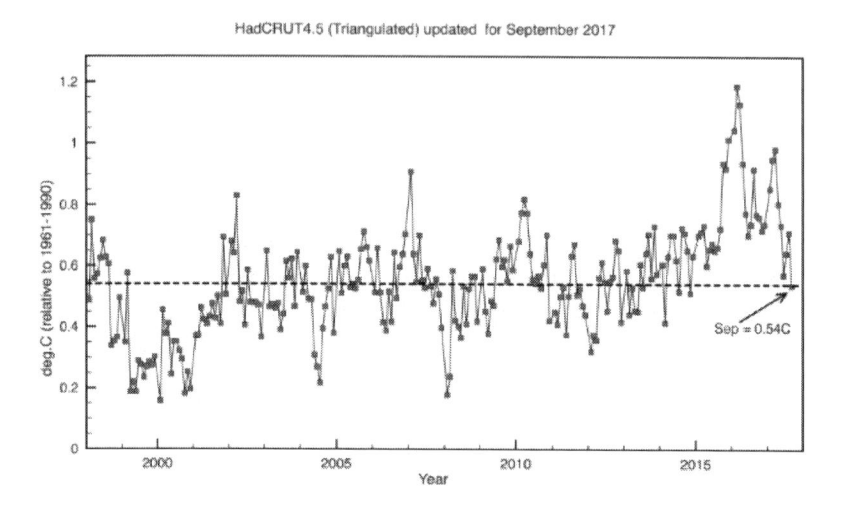

Not showing the El Niño 1998 highs, it lists 2002, 2007, and 2010 as mini-peak years, then the 'meteoric' 2016 El Niño assisted high.
It's not a bad dataset, and I'm okay with the presentation, but it does NOT show any form of increasing temperatures, just a variance.
With 2018 falling below their base line, I'm afraid the MET office confirms the hiatus.

2.10... WHY DON'T ALARMISTS FACE FACTS?
(The art/deceit of running with the ball...)

I get asked one question more than any other...
"If Global Warming is a con, then why do Alarmists keep going, and not face the so-called facts?"

Well, I worked out a football analogy, and it goes like this...

A football team, let's call them Alarmists United, are playing a game, and when they have the ball, the local crowd are cheering, making them feel great, and obviously enjoying both the atmosphere and the money they'll be earning for their stellar football performance.
Now when they're on the defensive, they don't get cheered as much... they don't like that, so they decide to take command of the football and start running around the football field, just keeping possession of the ball. There's no direction, there's no strategy, but they control the ball, run rings around the opposition, and keep the crowd cheering wildly.
Now... there's no need to score... they have possession of the ball, and as long as the crowd's cheering, and the money keeps coming in, they're happy.
Even when the facts change, they don't care; they've got both the crowd's attention and the cash firmly in their pockets.
They're intellectual hedonists, they're shallow, they're in control.

Until it becomes rather obvious that half the crowd have become silent... they're beginning to receive news about the lack of warming since 1998... the number of reputable scientists who have started defecting to the other side, due to the lack of any evidence for their team.
But the players for Alarmists United don't care, they keep running with the ball, aimlessly going in circles; the only criteria in their feeble minds is they keep possession of the football.
As time ticks on, they realize that the only thing keeping the other team from winning is the fact that Alarmists United have the ball... you can't score if you don't have the ball, right?

That's what the Alarmists are doing right now... they're keeping possession of the football, keeping the media hype going, knowing that if they drop the ball... the bubble bursts. Without forward momentum the juggernaut train slows to a halt; that's why the name keeps changing, that's why the stories get regurgitated, fed to the masses on a daily basis.

Let's face facts, shall we?

1. There's been no 'real/significant' global warming since 1998; twenty-one years of 'hiatus', 'pause', whatever you want to call it.
2. The Alarmists' graphs and data are being hailed as phony science; the hockey stick has been debunked, and the 97% consensus has been proven bogus.
3. More opposition scientists are coming out of the woodwork every day; even scientific organizations are polling their members, re-deciding their stance.
4. There's been little sea level rise in the last hundred years, and today's trends (one foot per century) is hardly catastrophic.
5. There's been no acidification of the oceans; the Ph. of the oceans varies by a full Ph. point depending on location, season and time of day.
6. There's been NO drop in food production; in fact there's a huge increase!
7. World leaders are losing patience with the various 'accords' and more and more people are beginning to see the light.

And soon... very soon... the ball is going to be grabbed by the other side.
Deniers City will have the ball, and without the constant repetition from the press, and the adulation of the leftist crowd, the Alarmists are going to get a sound beating.
The bubble can only take so much inflation (hot air) before it bursts.

2.11... THE BENEFITS OF GLOBAL WARMING
(And we've got medical proof...)

Human Health
We're often told by the media headlines that 'Global Warming Kills', and the globe's 'catastrophic' temperature rise will cause the deaths of millions of people...
John Kerry, the USA's last Secretary of State under President Barrack Obama, said of Trump's refusal to sign the Paris Accord in 2017...

"People will die because of the president's decision,"

This is typical Alarmist rhetoric.
In the 1970's (no matter what some Alarmists will try to tell you) there was a scare of Global Cooling... the consensus at that time was of an imminent Ice Age.

Nowadays we have a Global Warming scare on our hands.
But there are scientists who study the Sun (and believe the Sun is the major cause of our Global Warming) who are warning us of an imminent dip in the Sun's output, and telling us that despite the rise in carbon dioxide the Earth may be heading into a cooling phase.
So, faced with the two alternatives... which would we rather have?
A slightly cooler Earth, or a slightly warmer one?

In the case of human health, the answer is very clear.
In the Lancet (the UK premier medical journal) in 2015, a peer-reviewed article made the following claims...
"Warmer temperatures save lives."
The article went on to state that a staggering SEVEN PERCENT of ALL human deaths are caused by non-optimal temperatures.
Approximately 4 million people die EVERY YEAR because of cold or cool temperatures.
And COLD TEMPERATURES kill TWENTY TIMES more people than warm/hot temperatures.
In my schoolboy arithmetic, that makes warm/hot weather killing 200,000 people every year as compared to the 4 million killed by the cold.
Yeah... Global Warming... killing humans slower than cold weather.

Carbon Dioxide and Global Crop Production

In the case of human welfare on the Earth, two primary concerns rise to the fore.

Food and water are the primary factors in determining human health and welfare.

We have enough water. Over 70% of the world's surface is covered in water, and we seemingly fear it rising... so all we need are a few de-salinization plants, and we've conquered the global water shortage.

Food output in a cooler world would decrease as growing seasons get shorter, frost events become more common, the optimum growing period cools.

However, food production in a warmer world (and we're seeing the benefits already!) means the opposite. A warmer climate would lengthen the growing season, decrease frost events, and promote plant growth.

The increase in the planet's carbon dioxide is also a factor, and with increased CO2 levels, plants grow better, faster, and give far bigger harvests.

You see... there ARE benefits of a warmer increased CO2 Earth... but we never hear the media talk about it.

ALL OVER THE WORLD... crop production is THRIVING, breaking all types of records of production PER ACRE.

If only the media would tell the truth instead of the Alarmist lies and propaganda.

(For more info on Global Greening and crop increase, go to sections 8.8 and 8.9)

SECTION 3
CONSENSUS? WHAT CONSENSUS?

"There is no consensus in science. Science isn't a vote. Science is about facts."
(John Coleman, Founder, The Weather Channel)

3.1... A CONSENSUS OF 97% OF THE WORLD'S SCIENTISTS
(Such a Majority Can't Be Wrong... Can it?)

This is probably the single most quoted 'fact' intended to thump us deniers into the sidewalk; it's quoted by scientists, comedians and politicians (or did I repeat myself)... yet at its core lie the very root of the twisted propaganda the Alarmists peddle so well.

"97% of scientists agree that climate change is real"

(Here I'd like to 'quote' a line from Mel Gibson's Braveheart... *"they couldna agree on the color o' shite!"*)

Let's face it, take 100 people of any profession, you'd hardly expect 97% of them to agree to much, would you?

The above phrase, *"97% of scientists agree that climate change is real"* is an object lesson in large scale obfuscation.
Equivocation means, *"using the same term in different and contradictory ways"*.
Basically, the 'Alarmist' camp is trying to say that the majority of scientists are on their side... 97%.
But they phrase is chosen carefully. What idea does the word 'real', actually represent here?

We ALL know that climate change exists... it's been around for as long as there's been a climate.
So the wording of the phrase sets out a banana skin for the 'Denier' camp. How can we disagree? Basically we can't, because we all, technically, agree. Climate change is real.
But what does the statement actually imply?

Well, for the Alarmists it means that if we disagree with it, we are flat-Earthers, holocaust deniers, and kooks of the highest degree.

We all know what we 'Deniers' believe is a complex mixture of knee-jerk reactions to the climate change agenda.
Let's look at the various ° of separation between us. What statement best fits your own beliefs?

1. The whole idea can be explained by natural cycles.
2. Anthropogenic (man-made) climate change exists, but has little or no effect on the climate.
3. Man-made emissions are affecting the climate on a substantial basis, but there's little or nothing we can do to stop it.
4. Man-made emissions are totally responsible for the rising carbon dioxide levels, and we should curb our own CO2 emissions down to pre-industrial levels to mitigate the damage.

Quite a differing field of thought, yet each one holds a specific level of 'Denier' truth for some of us.
But there is a fifth option...

5. Alarmist climate scientists are telling outright lies, altering the figures, hiding the truth, or a combination of all three.

Yeah, I'm in that camp, definitely.

Before we look at history and see where the bogus 97% figure began, let's talk about a small detail that is endemic to these '97%' stories...

The Abstract
Historically the abstract of any scientific paper was a brief summary of the whole paper, but in more recent internet times, the 'abstract' has been used as a list of buzz words, a list of internet search terms to entice more readers to the article. Some of these terms have even little or no relevance in the article, but are used as buzz, key or catch words. It becomes an advertisement for the article instead of a summary.
So... the abstracts of many articles list associated facts NOT necessarily included in the subject being studied.

Naomi Oreskes, 2004
(From an op-ed article in Science 306, *Beyond the Ivory Tower: The Scientific Consensus on Climate Change*)

Naomi is a socialist historian, and her article was NOT peer reviewed, rather was included in Science as an opinion article.

Naomi claimed to have examined 928 abstracts from papers published from 1991 to 2003, and concluded that 75% of these abstracts gave the opinion that anthropogenic climate change was implicit.

However, her article was not peer reviewed, and her review process led to many claims of confirmation bias... being liberal to begin with, she achieved the result she wanted to find.

Harris Poll, 2007
In 2007, Harris Interactive (the survey poll people) randomly selected 489 members of the American Meteorological Society and the American Geophysical Union. The answers of the Harris poll were tabulated by the Statistical Assessment Service (STATS) at George Mason University. They received the following results for their questions...

97% agreed... "Global temperatures increased during the past 100 years?"
84% agreed... "Personally believe human-induced warming is occurring?"
74% agreed... "There is currently available scientific evidence that substantiates its occurrence?"
Only 5% believed... "Human activity does not contribute to greenhouse warming?"
41% thought... "The effects of global warming would be near catastrophic over the next 50–100 years?"
44% thought... "Effects would be moderately dangerous?"
13% thought there was... "Relatively little danger?"
56% agreed that... "Global climate change is a mature science?"
39% agreed that it was... "An emerging science?"

Heck, I agree with most of the above myself.

Doran/Zimmerman, 2009
In 2009, grad student Maggie Kendall wrote a paper citing the 97% statistic. Her professor, Peter Doran (a PhD Professor of Geology and Geophysics at Louisiana State University), added his name to it to give it weight, and it was published in the January 27, 2009 issue of the American Geophysical Union, (a weekly magazine of Earth science).

The article attempted to show that active climate researchers unanimously agreed that human CO_2 emissions had a 'significant' impact on the Earth's climate.

AGU Climate Science Survey Results

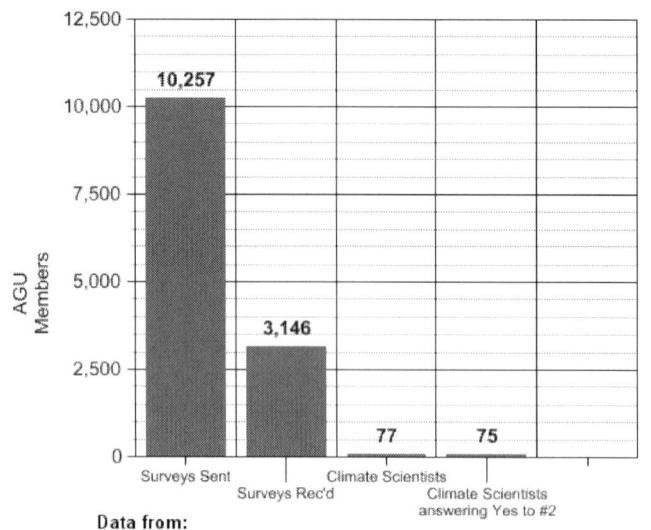

Bar chart showing:
- Surveys Sent: 10,257
- Surveys Rec'd: 3,146
- Climate Scientists: 77
- Climate Scientists answering Yes to #2: 75

Data from:
Doran, P. T. and M. K. Zimmerman (2009), Examining the Scientific Consensus on Climate Change, Eos Trans. AGU, 90(3), 22, doi: 10.1029/2009EO030002.

10,257 surveys were sent out by email to members of the American Geophysical Union, asking two questions...
(It was a two minute survey, with two simple questions)

1. When compared to pre-1800 levels, do you think that mean global temperatures have generally risen, fallen, or remained relatively constant?
2. Do you think human activity is a significant contributing factor in changing mean global temperatures?

They received 3,146 replies, and subsequently threw out 3,069, deciding that the members were not climate scientists. (As if only actual climate scientists have a thought, belief or any skin in the game).
Of those 77 surveys Doran and Zimmerman kept, only 2 had not agreed with the proposition.
Making a percentage of those members who agreed to the questions... 97.40%.
Well I'll be damned...

They threw out THREE THOUSAND replies.

Now, this is where any confirmation bias could fit in. They simply throw out any opinion which does not conform to their expected outcome. Then they count the high percentage of data left over.
Yup, that's transparency...

William R. Love (Anderegg et al.) 2010
Love, a college student, writes a paper, *Expert Credibility in Climate Change*, his professor adds his name, and it is published in *Proceedings of the National Academy of Sciences.*
Love found that 97% of scientists publishing in the climate field were proponents of anthropogenic global warming.
He took the names of the top 50 prolific scientific writers, and found that 98% of them agree with the IPCC.
That's like opening every jar of strawberry jam and finding they contained 97% strawberry jam. It means nothing.
By 2010, everyone knows that skeptic scientists were finding it difficult to get their articles placed.

John Cook, 2013
One of the most-cited surveys (and there were a couple) whose results 'confirmed' the dreaded '97%' figure was conducted by John Cook, a blogger, a climate communications fellow for the Global Change Institute (GCI) in Australia.
In 2013, he and his 'helpers' supposedly read through 11,944 abstracts from science based literature, looking for any mention of 'global climate change' or 'global warming'.

OKAY, WOAH!
I have more than a little problem with this.
1. How long did it take to actually collect 11,944 articles?
2. How long did it take for Cook and his minions to get through them all?
3. How many internet search buzz words were contained in the 'abstract' that were not main subjects of the scientific papers?

Anyway, back to the story...
Of those he and the minions read, he found that 66.4% of them had no mention of Anthropogenic Global Warming (AGW), so these were discarded, not fitting his agenda.

YUP! Yet again, data NOT supporting the theory were thrown out.
If this isn't a case of confirmation bias, I don't know what is.

Of the 33.6% of those articles that did mention a human element, 97.1% agreed that humans were complicit in at least part of the problem.
HOWEVER... that was 97.1% of 33.6% of his initial survey... NOT 97.1% of ALL the papers.
So 32.6% of the 11,944 abstracts studied agreed that humans were complicit in adding to existing global warming.
Not exactly setting the newspapers alight with that figure, were they?

AND... Although in Science & Education in August 2013, David Legates re-examined Cooks initial survey material, and came to the conclusion that only 0.3% had come to the conclusion that *"most warming since 1950 is anthropogenic"*, it no longer mattered.
The media and the 'Alarmists' had already picked up the ball and ran amok with it.

But in real terms, it was hardly a consensus, was it?

In fact, it wasn't even remotely honest.

But just think of the biggest coincidence in Global Warming...

Using a completely different set of climatologists, and a completely different set of questions, Australia's John Cook and his cohorts had arrived at EXACTLY the same percentage as Doran and Zimmermann.

Now, let's get to another couple of real consensuses... or is it consensii?

3.2... THE REAL SCIENTIST'S CONSENSUS; PART 1
(The Leipzig Declaration...)

At the time of writing, Fred Singer is 95 years old, and his brain is still pretty quick.
Born in 1924, Singer is a man of considerable life experience.
He designed mines for the U.S. Navy during World War 2, and in 1948 earned a Ph.D. in physics from Princeton University.
In 1950 he designed one of the first US satellites (the Mouse) and the 'Oriole' rockets of in 1957.
In 1962, he worked for NOAA, then the Department of Interior and the EPA, and is the former Director of the U.S. Weather Satellite Service.
Yup, he's the real deal.

However, in 1990, when he spoke out against Anthropogenic Global Warming (AGW), he was dumped on the 'kook' pile.

In direct opposition to the formation of the IPCC, Fred Singer founded the Science & Environmental Policy Project (SEPP), in 1990.
He wrote the 'Leipzig Declaration' on Global Climate Change, and it was ratified by a NAMED group of **80 scientists and 25 television meteorologists.**
The first version of the declaration was announced in Germany, hence the document's name.
This is the declaration in its complete form.

As independent scientists concerned with atmospheric and climate problems, we -- along with many of our fellow citizens -- are apprehensive about emission targets and timetables adopted at the Climate Conference held in Kyoto, Japan, in December 1997. This gathering of politicians from some 160 signatory nations aims to impose on citizens of the industrialized nations, -- but not on others -- a system of global environmental regulations that include quotas and punitive taxes on energy fuels to force substantial cuts in energy use within 10 years, with further cuts to follow. Stabilizing atmospheric carbon dioxide -- the announced goal of the Climate Treaty -- would require that fuel use be cut by as much as 60 to 80 percent -- worldwide!

Energy is essential for economic growth. In a world in which poverty is the greatest social pollutant, any restriction on energy use that inhibits economic growth should be viewed with caution. We understand the motivation to eliminate what are perceived to be the driving forces behind a potential climate change; but we believe the Kyoto Protocol -- to curtail carbon dioxide emissions from only part of the world community -- is dangerously simplistic, quite ineffective, and economically destructive to jobs and standards-of-living.

More to the point, we consider the scientific basis of the 1992 Global Climate Treaty to be flawed and its goal to be unrealistic. The policies to implement the Treaty are, as of now, based solely on unproven scientific theories, imperfect computer models -- and the unsupported assumption that catastrophic global warming follows from an increase in greenhouse gases, requiring immediate action. We do not agree. We believe that the dire predictions of a future warming have not been validated by the historic climate record, which appears to be dominated by natural fluctuations, showing both warming and cooling. These predictions are based on nothing more than theoretical models and cannot be relied on to construct far-reaching policies.

As the debate unfolds, it has become increasingly clear that –- contrary to the conventional wisdom -- there does not exist today a general scientific consensus about the importance of greenhouse warming from rising levels of carbon dioxide. In fact, most climate specialists now agree that actual observations from both weather satellites and balloon-borne radiosondes show no current warming whatsoever--in direct contradiction to computer model results.

Historically, climate has always been a factor in human affairs -– with warmer periods, such as the medieval "climate optimum," playing an important role in economic expansion and in the welfare of nations that depend primarily on agriculture. Colder periods have caused crop failures, and led to famines, disease, and other documented human misery. We must, therefore, remain sensitive to any and all human activities that could affect future climate.

However, based on all the evidence available to us, we cannot subscribe to the politically inspired world view that envisages climate catastrophes and calls for hasty actions. For this reason, we consider the drastic emission control policies deriving from the Kyoto conference -- lacking credible support from the underlying science -- to be ill-advised and premature. (Leipzig, Germany on Nov. 9-10, 1995)

It refutes the IPCC's so-called 'consensus' of scientists, and was re-issued in 1997, and 2005.

A full list of signatories can be found on the web, here...

https://web.archive.org/web/20060928122018/http://www.sepp.org/policy%20declarations/LDsigs.html

Against the background of the IPCC, Singer set up the **Non**governmental International Panel on Climate Change (NIPCC) after a 2004 United Nations climate conference was held in Milan, Italy.
Singer is the author of over a dozen books on the subject.

So... the fake consensus of consensuses at 97% is looking a bit flimsy.

3.3... THE REAL SCIENTIST'S CONSENSUS; PART 2
(The Global Warming Petition project...)

The Global Warming Petition Project (also known as the Oregon Petition Project) is basically a collection/list of scientists started by Dr Art Robinson (B.S. in Chemistry and PhD in Biochemistry) in response to the false alarm over CO2.
In 1997/98, when the Kyoto Protocol was first being spoken about, Art Robinson was one of the first scientists to see that climate science was being hijacked for political ends. He saw that people were being lied to, and that the 'majority' of scientists did not agree with the idea that man-made CO2 was as important a driver as many would have us believe.

The statement is simple and reads...

"We urge the United States government to reject the global warming agreement that was written in Kyoto, Japan in December, 1997, and any other similar proposals. The proposed limits on greenhouse gases would harm the environment, hinder the advance of science and technology, and damage the health and welfare of mankind.
There is no convincing scientific evidence that human release of carbon dioxide, methane, or other greenhouse gases is causing or will, in the foreseeable future, cause catastrophic heating of the Earth's atmosphere and disruption of the Earth's climate. Moreover, there is substantial scientific evidence that increases in atmospheric carbon dioxide produce many beneficial effects upon the natural plant and animal environments of the Earth."

Contrary to the way the 'Alarmists' 97% 'consensus' was carried out, the Oregon Petition Project list was done using a mailed letter, (emails were not used) and signature/return basis.
By the year 2009, the project had over 31,000 USA signatures, of which over 9000 held PhD's.
While obviously criticized by Alarmists, the petition does serve to give notice that there is NOT a 97% 'consensus' in the climate field or those associated with that field.
The Petition Project webpage lists all 31,000 names and lists their US state. It can be found here...

http://www.petitionproject.org/index.php

3.4... THE REAL SCIENTIST'S CONSENSUS; PART 3
(Members of the American Physical Society (APS)...)

Founded in 1899, The American Physical Society (APS) is the world's second largest organization of physicists. With a membership of over 50,000, it is hardly considered to be anything but a premier organization of scientists in the US.

In 2008, an American Physical Society editor conceded that on the subject of global warming; *"a considerable presence of scientific skeptics exists".*

In 2013, Ivar Giaever, (a Norwegian-American physicist who shared the Nobel Prize in Physics in 1973) gave a presentation in which he included a part of the APS's 2007 statement on Climate Change...

"The evidence is incontrovertible: Global Warming is occurring. If no mitigating action is taken, significant disruptions in the Earth's physical and ecological systems, social systems, security and human health are likely to occur. We must reduce the emissions of greenhouse gases beginning now."

The archived statement can be found here...
https://www.aps.org/policy/statements/07_1.cfm

Giaever resigned from the APS in 2011, and urged his fellow members to do the same.

William 'Will' Happer is an American physicist, specializing in the study of atomic physics, optics and spectroscopy, and professor at Princeton. He is also on the JASON advisory group, (the global climate monitoring satellite).
In September 2018, Happer was appointed senior director of the National Security Council (office for emerging technologies).
He's hardly seen as a second rate scientist.

In July 2009, leading a prominent group of fifty-four APS scientists, Happer tried to get the APS to amend their alarmist statement on man-made global warming. In an open letter in Nature magazine, they urged the APS to reconsider its stance.
Included in the petition was the statement...

"Measured or reconstructed temperature records indicate that 20th - 21st century changes are neither exceptional nor persistent, and the historical and geological records show many periods warmer than today."

Hardly controversial.

The petition was co-authored by the following...
S. Fred Singer, University of Virginia
Hal Lewis, University of California, Santa Barbara
Will Happer, Princeton University
Larry Gould, University of Hartford
Roger Cohen, Durango, Colorado
Robert H. Austin, Princeton University

The petition was rejected by the APS council.

Undaunted, in 2010, and with his petition now containing more than 260 APS scientists, he tried again, but to no avail. Also in 2010, Happer, as chair of the George C. Marshall Institute, testified before Rep. Ed Markey's House Select Committee on Energy Independence and Global Warming. Happer was the sole GOP witness arguing against the global warming consensus.

In 2015, Will Happer testified at a congressional hearing into climate 'dogma' convened by US Senator Ted Cruz, (TX-R) and chair of the Senate Committee on Science.
This is just some of Happer's testimony...

"Along with other witnesses at this hearing, I hope to correct some misconceptions about the trace atmospheric gas, carbon dioxide or CO2. In spite of the drumbeat of propaganda, CO2 is not "carbon pollution." The benefits that more CO2 brings from increased agricultural yields and modest warming far outweigh any harm.

The key issue here is the equilibrium climate sensitivity: how much will the earth's surface eventually warm if the atmospheric concentration of CO2 is doubled? This number has been drifting steadily downward from a youthful Arrhenius's first estimate of about 6 C to the estimate of the International Panel on Climate change (IPCC) of 1.5 C to 4.5 C.

Observations of very small warming over the past 20 years suggest that the sensitivity is unlikely to be larger than 2 C. There are credible estimates that the sensitivity could be as small as 0.5 C.

For many decades the citizens of the USA and of much of the world have been flooded with the message that CO2 is "carbon pollution." We are supposed to trust our government and selfless NGO's for instructions on how to save the planet. Much of the message is false, but its purveyors control key positions in the media, in the government, in scientific societies, in charitable foundations etc. This makes it difficult to get out the truth that climate science is far from "settled." To the extent it is settled, it indicates no cause for alarm or for extreme measures. Indeed, a dispassionate analysis of the science indicates that more CO2 will bring benefits, not harm to the world."

A truly down-to-earth and fact-filled delivery.

For the purpose of full disclosure, a link to the full testimony, including graphs, is given below.

https://www.commerce.senate.gov/public/_cache/files/c8c53b68-253b-4234-a7cb-e4355a6edfa2/FA9830F15064FED0A5B28BA737D9985D.dr.-william-happer-testimony.pdf

Today, 2019, the APS statement on Earth's Changing Climate is...

"On Climate Change:

Earth's changing climate is a critical issue and poses the risk of significant environmental, social and economic disruptions around the globe. While natural sources of climate variability are significant, multiple lines of evidence indicate that human influences have had an increasingly dominant effect on global climate warming observed since the mid-twentieth century. Although the magnitudes of future effects are uncertain, human influences on the climate are growing. The potential consequences of climate change are great and the actions taken over the next few decades will determine human influences on the climate for centuries."

This is a somewhat different statement from the start of the chapter, and shows the change of opinion.

On Climate Science, and Climate Action, their statement has a more conciliatory tone…

"To better inform societal choices, the APS urges sustained research in climate science… the APS further urges physicists to collaborate with colleagues across disciplines in climate research and to contribute to the public dialogue."

3.5... THE REAL SCIENTIST'S CONSENSUS; PART 4
(The German no-longer-silent Majority...)

In August 2009, a coalition of German scientists dissented from the 'consensus' of anthropogenic global warming claims.

"Consensus Takes Another Hit!
More than 60 German Scientists Dissent Over Global Warming Claims! Call Climate Fears 'Pseudo Religion'; Urge Chancellor to 'reconsider' views."

In an Open Letter to German Chancellor Angela Merkel, the 60+ German scientists declared that a

"growing body of evidence shows anthropogenic CO2 plays no measurable role."

The more than 60 signers of the letter (all scientists and university professors and from many disciplines, physicists, meteorology, chemistry, and geology) included several United Nations IPCC scientists.
The scientists declared that global warming has become a pseudo religion, and noted that rising CO2 has had no measurable effect on temperatures.
The German scientists also wrote that the U.N.-led IPCC had lost its scientific credibility, and urged Chancellor Merkel to strongly reconsider her position on global warming. They requested the assembly of an impartial panel (free of ALL ideology) to counter the UN IPCC and review the latest climate science developments.

The complete letter (translated) is given below. It does not hold back punches. I wish I'd written it myself...

Open Letter – Climate Change
Bundeskanzleramt
Frau Bundeskanzlerin Dr. Angela Merkel
Willy-Brandt-Strabe 1
10557 Berlin

Vizerprasident
Dipl. Ing. Michael Limburg
14476 Grob Glienicke
Richard-Wagner-Str. 5a
E-mail: limburg@grafik-system.de

Grob Glienicke 26.07.09

To the attention of the Honorable Madam Angela Merkel, Chancellor of Germany.

When one studies history, one learns that the development of societies is often determined by a zeitgeist, which at times had detrimental or even horrific results for humanity. History tells us time and again that political leaders often have made poor decisions because they followed the advice of advisors who were incompetent or ideologues and failed to recognize it in time. Moreover evolution also shows that natural development took a wide variety of paths with most of them leading to dead ends. No era is immune from repeating the mistakes of the past.

Politicians often launch their careers using a topic that allows them to stand out. Earlier as Minister of the Environment you legitimately did this as well by assigning a high priority to climate change. But in doing so you committed an error that has since led to much damage, something that should have never happened, especially given the fact you are a physicist. You confirmed that climate change is caused by human activity and have made it a primary objective to implement expensive strategies to reduce the so-called greenhouse gas CO_2. You have done so without first having a real discussion to check whether early temperature measurements and a host of other climate related facts even justify it.

A real comprehensive study, whose value would have been absolutely essential, would have shown, even before the IPCC was founded, that humans have had no measurable effect on global warming through CO_2 emissions. Instead the temperature fluctuations have been within normal ranges and are due to natural cycles. Indeed the atmosphere has not warmed since 1998 – more than 10 years, and the global temperature has even dropped significantly since 2003.

Not one of the many extremely expensive climate models predicted this.

According to the IPCC, it was supposed to have gotten steadily warmer, but just the opposite has occurred. More importantly, there's a growing body of evidence showing anthropogenic CO_2 plays no measurable role. Indeed CO_2's capability to absorb radiation is almost exhausted by today's atmospheric concentrations. If CO_2 did indeed have an effect and all fossil fuels were burned, then additional warming over the long term would in fact remain limited to only a few tenths of a degree.

The IPCC had to have been aware of this fact, but completely ignored it during its studies of 160 years of temperature measurements and 150 years of determined CO2 levels. As a result the IPCC has lost its scientific credibility. The main points on this subject are included in the accompanying addendum.

In the meantime, the belief of climate change, and that it is manmade, has become a pseudo-religion. Its proponents, without thought, pillory independent and fact-based analysts and experts, many of whom are the best and brightest of the international scientific community.

Fortunately in the internet it is possible to find numerous scientific works that show in detail there is no anthropogenic CO2 caused climate change. If it was not for the internet, climate realists would hardly be able to make their voices heard.

Rarely do their critical views get published.

The German media has sadly taken a leading position in refusing to publicize views that are critical of anthropogenic global warming. For example, at the second International Climate Realist Conference on Climate in New York last March, approximately 800 leading scientists attended, some of whom are among the world's best climatologists or specialists in related fields. While the US media and only the Wiener Zeitung (Vienna daily) covered the event, here in Germany the press, public television and radio shut it out. It is indeed unfortunate how our media have developed – under earlier dictatorships the media were told what was not worth reporting. But today they know it without getting instructions.

Do you not believe, Madam Chancellor, that science entails more than just confirming a hypothesis, but also involves testing to see if the opposite better explains reality? We strongly urge you to reconsider your position on this subject and to convene an impartial panel for the Potsdam Institute for Climate Impact Research, one that is free of ideology, and where controversial arguments can be openly debated.

We the undersigned would very much like to offer support in this regard.

Respectfully yours,
Prof. Dr.rer.nat. Friedrich-Karl Ewert EIKE
Diplom-Geologe

Universität. - GH - Paderborn, Abt. Höxter (ret.)

Dr. Holger Thuß
22
EIKE President
European Institute for Climate and Energy
http://www.eike-klima-energie.eu/

Signed by
Scientists
Prof. Dr.Ing. Hans-Günter Appel.
Prof. Dr. Dorota Appenzeller Professor of Econometrics and Applied Mathematics, Vice Dean University Poznan, Poland.
Prof. Dr. Wolfgang Bachmann Former Director of the Institute for Vibration Engineering, FH Düsseldorf.
Prof. Dr. Hans Karl Barth Managing Director World Habitat Society GmbH - Environmental Services.
Biologist Ernst Georg Beck.
Dr. Horst Borchert Physicist.
Helgo Bran Former BW parliamentarian Green Party.
Prof. Dr. Gerhard Buse Bio-chemist.
Dr. Ivo Busko German Center for Aviation and Aeronautics e.V.
Dr. Gottfried Class Nuclear Safety, Thermo-hydraulics.
Dr. Urban Cleve Nuclear physicist, thermodynamics energy specialist.
Dr. Rudolf-Adolf Dietrich Energy expert.
Peter Dietze IPCC Expert Reviewer TAR.
Dr. Siegfried Dittrich Physical chemist.
Dr. Theo Eichten Physicist.
Ferroni Ferruccio Zurich President NIPCCSUISSE.
Dr. Albrecht Glatzle Agricultural biologist, Director científico INTTAS, Paraguay.
Dr. Klaus-Jürgen Goldmann Geologist.
Dr. Josef Große-Wördem Physical chemist.
Geologist Heinisch Heinisch.
Dr. Horst Herman Chemist.
Prof. Hans-Jürgen Hinz Former University of Münster Institute for Physical Chemistry.
Andreas Hoemann Geologist.
Geologist Siegfried Holler.
Dr. Heinz Hug Chemiker.
Dr. Bernd Hüttner Theoretical Physicist.
Prof. Werner Kirstein Institute for Geography University Leipzig.

Klaus Knüpffer METEO SERVICE weather research GmbH.
Dr. Werner Köster
Dr. Albert Krause Chemist
Dr. Hans Labohm IPCC AR4 Expert Reviewer Dipl. Business / science Journalist.
Dr. Rainer Link Physicist.
Physicist Alfred Loew.
Prof. Dr. Physicist Horst-Joachim LüdeckeUniversity for Engineering and business of Saarland.
Prof. Dr. Horst Malberg University professor em. Meteorology and Climatology/ Former Director of the Institute for Meteorology of the University of Berlin.
Dr. Wolfgang Monninger Geologist.
Meteorologist Dieter Niketta.
Prof. Dr. Klemens Oekentorp Former director of the Geological Paleolontology Museum of the Westphalia Wilhelms-University Münster.
Dr. Helmut Pöltelt Energy expert.
Klaus-Eckart Puls Meteorologist.
Prof. Dr. Klaas Rathke Polytechnic OWL Dept. Höxter.
Dr. D. Helmut Reihlen Director of the DIN German Institute for Standards and Norms i.R.
Prof. Dr. Oliver Reiser University of Regensburg.
Physicist Wolfgang Riede Physicists ETH.
Mineralogist Sabine Sauerberg Geoscientist.
Prof. Jochen Schnetger Chemist.
Prof. Dr. Sigurd Schulien University instructor.
Dr. Franz Stadtbäumer Geologist.
Dr. Gerhard Stehlik Physical chemist.
Norman Stoer System administrator.
Dr. Lothar Suntheim Chemist.
Heinz Thieme Technical assessor.
Dr. Wolfgang Thüne Mainz Ministry of Environment Meteorologist.
Dr. Dietmar Ufer Energy economist, Institute for Energy Leipzig.
Prof. Dr. Detlef von Hofe Former managing director of the DVS.
Geographist Heiko Wiese Meteorologist.
Dr. Erich Wiesner Euro Geologist.
Dr. Ullrich Wöstmann Geologist.
Prof. Dr. Heinz Zöttl Soil Sciences.
Dr. Zucketto Chemist
Dr. Ludwig Laus Geologist

Here ends the translation of the full German scientist letter.

So as you see, even outside the USA there's hardly a 'consensus'. In fact, it's the opposite; there's actually a large anti-consensus movement!

I salute these brave men and women. It takes a lot of courage to stand up to ridicule and possible repercussions.

From here, that 97% 'consensus' figure is starting to look pretty darn shaky.

3.6... THE REAL SCIENTIST'S CONSENSUS; PART 6.
(The Fallacy of Equivocation...)

By 2008, almost ten years after the El Niño assisted 1998 temperature peak, scientists all over the world were beginning to look at the new pause/hiatus and openly question the theory.
Carbon dioxide was still rising steadily, yet the global temperature rise had basically stopped.

An open letter to President Obama
After his election in November 2008, the soon-to-be President of the USA gave a foreshadowing of his plans while in office.

"Few challenges facing America and the world are more urgent than combating climate change. The science is beyond dispute and the facts are clear."
(President-Elect Barack Obama, November 19, 2008)

In 2009, a group of over 100 PhD scientists gave their answer in an open published letter...

"We, the undersigned scientists, maintain that the case for alarm regarding climate change is grossly overstated. Surface temperature changes over the past century have been episodic and modest and there has been no net global warming for over a decade now. After controlling for population growth and property values, there has been no increase in damages from severe weather-related events.
The computer models forecasting rapid temperature change abjectly fail to explain recent climate behavior.
Mr. President, your characterization of the scientific facts regarding climate change and the degree of certainty informing the scientific debate is simply incorrect."

An open letter to the UN Secretary-General
In December 8, 2009, an open letter to the UN Secretary-General gathered the signatures of 166 international skeptical scientists.
The 166 scientists simply declared "the science is NOT settled."

His Excellency Ban Ki Moon, Secretary-General, United Nations. New York, NY, United States of America, December 8, 2009

Dear Secretary-General,

Climate change science is in a period of negative discovery - the more we learn about this exceptionally complex and rapidly evolving field the more we realize how little we know. Truly, the science is NOT settled.

Therefore, there is no sound reason to impose expensive and restrictive public policy decisions on the peoples of the Earth without first providing convincing evidence that human activities are causing dangerous climate change beyond that resulting from natural causes. Before any precipitate action is taken, we must have solid observational data demonstrating that recent changes in climate differ substantially from changes observed in the past and are well in excess of normal variations caused by solar cycles, ocean currents, changes in the Earth's orbital parameters and other natural phenomena.

We the undersigned, being qualified in climate-related scientific disciplines, challenge the UNFCCC and supporters of the United Nations Climate Change Conference to produce convincing OBSERVATIONAL EVIDENCE for their claims of dangerous human-caused global warming and other changes in climate. Projections of possible future scenarios from unproven computer models of climate are not acceptable substitutes for real world data obtained through unbiased and rigorous scientific investigation.

Specifically, we challenge supporters of the hypothesis of dangerous human-caused climate change to demonstrate that:
1. Variations in global climate in the last hundred years are significantly outside the natural range experienced in previous centuries;
2. Humanity's emissions of carbon dioxide and other 'greenhouse gases' (GHG) are having a dangerous impact on global climate;
3. Computer-based models can meaningfully replicate the impact of all of the natural factors that may significantly influence climate;
4. Sea levels are rising dangerously at a rate that has accelerated with increasing human GHG emissions, thereby threatening small islands and coastal communities;
5. The incidence of malaria is increasing due to recent climate changes;
6. Human society and natural ecosystems cannot adapt to foreseeable climate change as they have done in the past;
7. Worldwide glacier retreat, and sea ice melting in Polar Regions, is unusual and related to increases in human GHG emissions;
8. Polar bears and other Arctic and Antarctic wildlife are unable to adapt to anticipated local climate change effects, independent of the causes of those changes;

9. Hurricanes, other tropical cyclones and associated extreme weather events are increasing in severity and frequency;
10. Data recorded by ground-based stations are a reliable indicator of surface temperature trends.

It is not the responsibility of 'climate realist' scientists to prove that dangerous human-caused climate change is not happening. Rather, it is those who propose that it is, and promote the allocation of massive investments to solve the supposed 'problem', who have the obligation to convincingly demonstrate that recent climate change is not of mostly natural origin and, if we do nothing, catastrophic change will ensue. To date, this they have utterly failed to do.

And that, dear reader, is a pretty good letter!

The Manhattan Declaration
On March 4th (a great date to have a march!) 2008, a group of scientists from over 40 countries gathered at Times Square, Manhattan, New York, to declare their feelings on the whole concept of Global Warming.
The declaration in full...

Manhattan Declaration on Climate Change –
"Global warming" is not a global crisis: We, the scientists and researchers in climate and related fields, economists, policymakers, and business leaders, assembled at Times Square, New York City, participating in the 2008 International Conference on Climate Change, Resolving that scientific questions should be evaluated solely by the scientific method; Affirming that global climate has always changed and always will, independent of the actions of humans, and that carbon dioxide (CO2) is not a pollutant but rather a necessity for all life; Recognizing that the causes and extent of recently-observed climatic change are the subject of intense debates in the climate science community and that oft-repeated assertions of a supposed 'consensus' among climate experts are false; Affirming that attempts by governments to legislate costly regulations on industry and individual citizens to encourage CO2 emission reduction will slow development while having no appreciable impact on the future trajectory of global climate change. Such policies will markedly diminish future prosperity and so reduce the ability of societies to adapt to inevitable climate change, thereby increasing, not decreasing human suffering; Noting that warmer weather is generally less harmful to life on Earth than colder: Hereby declare: That current plans to restrict anthropogenic CO2 emissions are a dangerous misallocation of intellectual capital and resources that should be dedicated

to solving humanity's real and serious problems. That there is no convincing evidence that CO2 emissions from modern industrial activity has in the past, is now, or will in the future cause catastrophic climate change. That attempts by governments to inflict taxes and costly regulations on industry and individual citizens with the aim of reducing emissions of CO2 will pointlessly curtail the prosperity of the West and progress of developing nations without affecting climate. That adaptation as needed is massively more cost-effective than any attempted mitigation, and that a focus on such mitigation will divert the attention and resources of governments away from addressing the real problems of their peoples. That human-caused climate change is not a global crisis. Now, therefore, we recommend – That world leaders reject the views expressed by the United Nations Intergovernmental Panel on Climate Change as well as popular, but misguided works such as —An Inconvenient Truth//. That all taxes, regulations, and other interventions intended to reduce emissions of CO2 be abandoned forthwith. Agreed at New York, 4 March 2008.

The scientists were sponsored by the newly formed International Climate Science Coalition (ICSC). This international group was led by Canadians Tom Harris and Tim Ball, and New Zealanders Terry Dunleavy, and Bryan Leyland. They began with a small group, but by December 2010, their ranks had swollen to over 700 scientists from all over the world.

The Denier ranks were undoubtedly growing.

On the actual declaration, however, because of the large number, only a few signatures were given. I list them here with their particular sphere of expertise in an abbreviated form…

Clive Stevens (Meteorogical Society), Gordon Evans (Meteorology), Oscar Fann (Chief Meteorologist, WTVY-TV), Gary Kubat (Atmospheric Science), Joseph Kunc (Molecular Physics), James Kurtz (Inorganic Chemistry), Rune B. Larsen (Geology, Geochemistry), Asmunn Moene (Meteorology), M. R. Morgan (climate consultant), Travis Littlechilds (Geology), John Llewellyn, (Meteorology), Björn Malmgren (Paleoclimate Science), Emmanuel Malterre (Geophysicist), John Marshall (Dip), Romuald Bartnik (Organic Chemistry), William R. Bennett (Chemistry and Microbiology), Roger Burtner (Geologist), Elizabeth Innes (Chemist), Erwin Lalik (crystallography), Endel Lippmaa (Physics, Chemistry), James Weeg (Geology), Larry D. Agenbroad (Geology), M.I. Bhat, Professor (Tectonics, Geology & Geophysics), Vedat Shehu (Geologist, Engineering Geology, Tectonics), Joseph Silverman (Nuclear Engineering), Jan-Erik

Solheim (Astrophysics), Gösta Walin (Oceanography), Richard Becherer (Optics), Ahmed Boucenna (Physics), Antonio Brambati (Sedimentology), Arthur Chadwick (Molecular Biology), Cornelia Codreanova (Geography), James E Dent (Consultant), Terrence F. Flower (Physics and Astronomy), Lars Franzén (Physical Geography), Katya Georgieva(Physics, Meteorology), Larry Irons (Geology), Leif Kullman (Physical geography), David Manuta (Inorganic/Physical Chemistry), Stanley Penkala (Chemical Engineering), Oleg M. Pokrovsky (mathematics and atmospheric physics), Oleg Raspopov (Geophysics), Jan-Erik Solheim (Astrophysics), Roger Tanner (Analytical Chemistry), Edward M. Tomlinson (Meteorology), Michael G. Vershovsky (Meteorology), Bob Zybach (Environmental Sciences).

This particular group of scientists, however, was destined to go further than some of the others.
The report and list was compiled by Mike Morano (then aide to Senator Jim Inhofe, (OK-R) and later formed the website... www.ClimateDepot.com)
In March 2009, their names were mentioned in a report to the US Senate's Environment and Public Works Committee. The report was entitled...

"U. S. Senate Minority Report: More Than 700 International Scientists Dissent Over Man-Made Global Warming Claims Scientists Continue to Debunk "Consensus" in 2008 & 2009"

The report's introduction began...

"Over 700 dissenting scientists (updates previous 650 report) from around the globe challenged man-made global warming claims made by the United Nations Intergovernmental Panel on Climate Change (IPCC) and former Vice President Al Gore. This new 2009 255-page U.S. Senate Minority Report -- updated from 2007's groundbreaking report of over 400 scientists who voiced skepticism about the so-called global warming "consensus" features the skeptical voices of over 700 prominent international scientists, including many current and former UN IPCC scientists, who have now turned against the UN IPCC. This updated report includes an additional 300 (and growing) scientists and climate researchers since the initial release in December 2007. The over 700 dissenting scientists are more than 13 times the number of UN scientists (52) who authored the media-hyped IPCC 2007 Summary for Policymakers."

Not a bad start.

The whole pdf file, report and list can be found here…

http://104.131.167.144/wp-content/uploads/2012/02/1000_scientists_dissent.pdf?fbclid=IwAR201Yb_nwHK-hQqAI-HyLMIiEbPajAIGKYFV1d1hbxKMToMVVjXiuglOnU

So you see, unlike the media bleats at you, there is no real consensus… there's just opposing points of view… and that's NOT the same thing.

And opposing points of view are okay in the world of unsettled science.

3.7... THE REAL SCIENTIST'S CONSENSUS; PART 7.
(Christians, Japanese, and American meteorologists....)

American Meteorological Society dissent
In 2012, the American Meteorological Society (AMS) conducted a survey within its members.
Asking various questions about global warming, the survey produced quite surprising results.
Remembering that all AMS members are actual meteorologists, and 80% have a Ph.D. or Masters Degree, the results fly in face of the 97% consensus rubbish.
Here's the results...

89% of AMS meteorologists believe global warming is happening.

Of that 89%, only 59% believe humans are the primary cause.

Only 30% of AMS members are 'very worried' about global warming.

Of the 89% who believe global warming is occurring, only 38% say it will be very harmful during the next 100 years.

53% of AMS members thought there was conflict among AMS members on the subject of global warming, 33% believe there is no conflict, and 15% were not sure.

Only 26% of respondents said the conflict among AMS members was 'unproductive'.
(Because 'conflict', or skepticism is GOOD for science, remember?)

Amazingly, only 38% of those who believe global warming is occurring thought it will be 'very harmful' during the next 100 years.

So much for 'settled science' in a society filled to the brim with meteorologists.

Christians signing another declaration
In 2015, a list of 305 Christian scientists added their voices to the growing dissent.
Their website can be found here...

https://cornwallalliance.org/landmark-documents/an-open-letter-on-climate-change-to-the-people-their-local-representatives-the-state-legislatures-and-governors-the-congress-and-the-president-of-the-united-states-of-america/

These 305 initial signers are mostly PhD's in scientific subjects, and since the original signings, the list has grown.

At the time of writing… January 2019, there were over 4000 signatures.

AGAIN, we demonstrate that as the 'pause/hiatus' continues, and carbon dioxide continues to rise, people are falling 'out of love' with the Alarmist point of view.

Japanese dissent
Shigenori Maruyama is a professor at the Tokyo Institute of Technology's Department of Earth and Planetary Sciences. He has authored more than 125 scientific publications, and specializes in the geological evidence of prehistoric climate change.

At the Japan Geoscience Union symposium in 2007, he openly declared animosity to the UN IPCC's Fourth Assessment Report (2007) that detailed that **most** of the observed global temperature increase since the mid-20th century 'is very likely due to the observed increase in anthropogenic greenhouse gas concentrations'.
Basically, the doubt was gone, and man was responsible for MOST of the rising temperatures.
When the question was raised at the Japan Geoscience Union, Maruyama reported that "the result showed 90 per cent of the participants do not believe the IPCC report".

That's one heck of a size of dissent!

On the lack of public support for the 'Deniers' camp, Maruyama said that many scientists were doubtful about man-made climate-change theory, but did not want to risk their funding from the government or bad publicity from the mass media, which he said was leading society in the wrong direction.

So, in Japan, the Deniers are at 90%????

SECTION 4
CARBON DIOXIDE; THE BLACK POLLUTANT!

"Progress in science is driven by skepticism."
Peter R. Leavitt (Meteorologist and President of Weather Information, Inc)

4.1... FOUND IT! CAUSE, CULPRIT AND CASE-CLOSED!
(Well, maybe you're approaching this from the wrong direction...)

Let me ask a few leading questions... and bear with me here, cos I'm actually getting to a point.

Q. What's the biggest source of energy near the Earth?
A. Easy; it's the sun.

Q. How big is the sun?
A. Dunno...

Q. Okay, let's try again. How big is the sun in relationship to earth?
A. A thousand times bigger.

Hmm. Not bad. The sun's radius is 432,000 miles, the Earth's radius is 4000 miles, so in size and volume, it's 100 times bigger than the earth... However, the sun is really dense; it weighs 330,000 times the mass of Earth, and over 1,300,000 Earths would fit inside the Sun.
AND, the Sun contains 99.86% of ALL THE MASS in our solar system.
It's really important. It's really huge!

Sorry, I got diverted... back to the questions...

Q. What's the single biggest part of the earth, apart from the planet itself?
A. More difficult... are we talking about the atmosphere?

Well done astute scientific reader... the atmosphere is huge, beyond reasoning, and beyond the ability of figures to do it justice. It's generally accepted to be about 100km high, and has a mass of around 5.15×10^{18} kg; considering it's a gas, that's pretty heavy, and really big.

Again, back to my quiz.

Q. What covers more of Planet Earth's surface? Land or ocean?
A. Easy again... Ocean, water.

Correct, and remember, the Earth's oceans have 250 times the mass of the atmosphere (and we already know how big that is), and can hold more than 1000 times the heat.

Q. What is the planet's biggest obstruction to light hitting the surface?
A. A bit harder, but I still got it... clouds.

Q. How much of the Earth's surface is covered with clouds at any time?
A. Dunno... it varies.

Actually, the average cloud cover of the Earth's surface is 67%, and since cloud reflects about 30% of all incoming solar radiation, clouds are hugely important.

Okay, we've established that virtually all Earth's energy comes from the sun, the atmosphere is huge, 71% of Earth's surface is water, and very capable of holding CO_2. We know that clouds vary in thickness, magnitude and height.
Now we're making a slight change of approach...

Q. What's the most abundant gas in our atmosphere?
A. Nitrogen, at about 80%.

Q. What's the second most abundant gas in out atmosphere?
A. Oxygen... maybe about 20%.

Okay, you're close, the actual figures are 78.09% Nitrogen, and 20.95% Oxygen, and those are rounded up. Basically after nitrogen and oxygen, it leaves just about 0.97% remaining. Back to the Questions.

Q. What's the third most abundant gas in our atmosphere?
A. Carbon dioxide?

NOPE! (This is where you hear the loud resounding negative buzzer sound from that family TV game.)

Actually, Argon is the next abundant gas, at 0.93%.

Which leaves Carbon dioxide at an extremely small 0.04% of the planet's atmosphere.
That's four parts PER MILLION of CO2.

Now, those figures are correct in the laboratory, in a controlled system, but there is one other 'gas' that is present, but is never measured in these types of conversations because it varies so greatly in concentration from place to place in the atmosphere.
Water vapor actually makes up from 0.001% to 5% of the actual measured atmospherical gases outside of the laboratory. Obviously, it's higher in concentration above water at the hottest parts of the world, and it's lower in concentration above the deserts and the poles.
(Antarctica is actually the driest place in the world. It's so cold, even water vapor finds it difficult to exist in gaseous form)

Let's return to the question and answer game, shall we?

Q. What exactly is a 'greenhouse gas'?
A. Dunno.

Okay. The definition of a 'greenhouse gas' is a gaseous substance that absorbs light energy, and in the absorption, gets slightly warmer.
(Remember Einstein's first law of thermodynamics, also known as the Law of Conservation of Energy, states that energy can neither be created nor destroyed)

Q. What's the most abundant greenhouse gas?
A. carbon dioxide?

NOPE! (Big buzzer noise again)

Water vapor is the most abundant greenhouse gas at about 36% to 72% (depending on geographical location). Carbon dioxide is second from 9% to 26%. Of the other greenhouse gases, only Methane is worth considering; and although its concentrations are much lower than CO2, (4% to 9%) it must be remembered that its radiative effects are 84 times stronger than CO2.

Many 'Alarmists' will try to tell you that carbon dioxide is the 'most important' greenhouse gas; they're simply wrong.
Water vapor, because of its overwhelming concentration is by far the most important greenhouse gas.

(In parts per million, water vapor is 10 to 50,000... carbon dioxide is about 400, and methane a paltry 2)

It is also simply not possible to state that a certain greenhouse gas causes an exact percentage of the greenhouse effect... if anyone tries to, they're estimating, or simply lying.
The reason is simple.
Energy from the sun consists of a variety of wavelengths... the actual white light we see is just a small part of the energy received.
And some greenhouse gases react with particular wavelengths far more than others and some are impervious to some wavelengths.

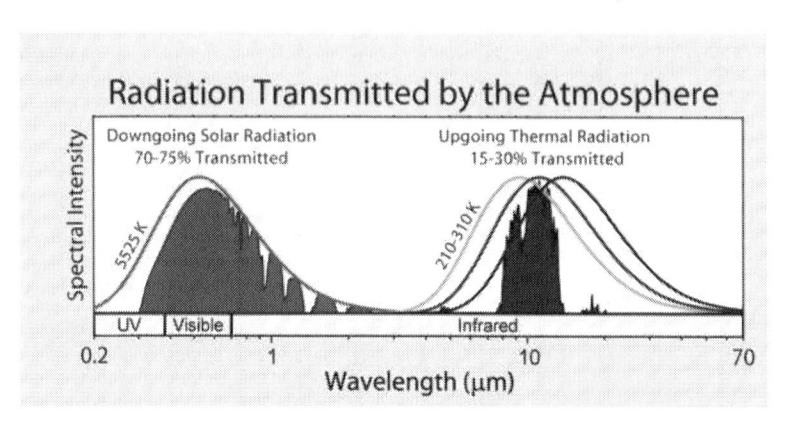

Radiation Transmitted by the Atmosphere

The above graph looks complex, but actually it is not.
It's simply the amount of light energy that our planet receives from the sun that is absorbed by the greenhouse gases, and changed into heat energy.

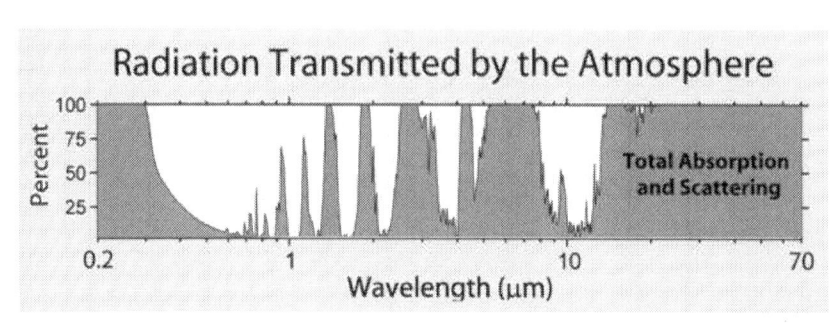

Radiation Transmitted by the Atmosphere

If we look at the information from another angle, the areas in grey are the parts of sunlight's spectrum that are absorbed by greenhouse gases and transmitted as heat contributing to the global warmth of our planet.

Some frequencies are full to the top (saturated)... our greenhouse gases have absorbed all they can, and any excess is allowed to radiate back into space.

If the concentrations of our greenhouse gases were to increase, there would be NO MORE heat transfer at those saturated frequencies than we have at present.

(This, dear reader is one of the most crucial pieces of the CO2 myth, this is why a 'doubling' of CO2 will never produce as much of a temperature increase as you would think... but more of this later.)

As we can see from the areas in white, there are many wavelengths that simply do not react with the greenhouse gases, and are not part of our planet's heat blanket.

If we split these gases into the component forms, we have the graph below.

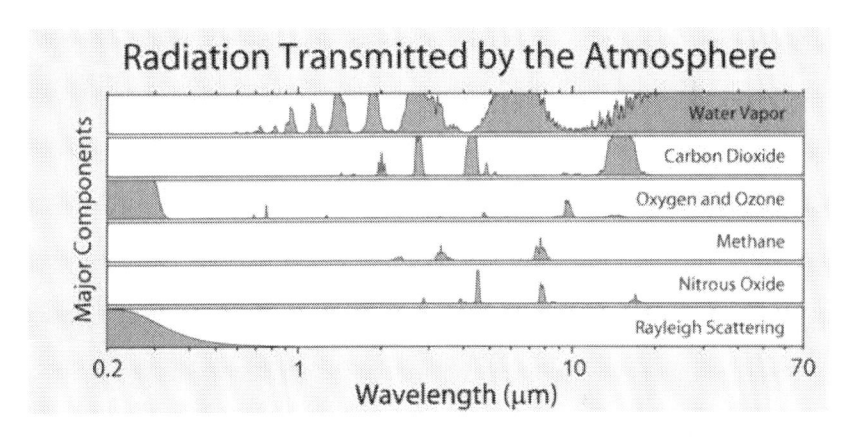

Radiation Transmitted by the Atmosphere

Major Components: Water Vapor, Carbon Dioxide, Oxygen and Ozone, Methane, Nitrous Oxide, Rayleigh Scattering

Wavelength (μm): 0.2, 1, 10, 70

As you can see, most of the heat transfer is taken on by water vapor, and a lot of that is saturated. If the concentration of water vapor were to increase, the heat transfer at those particular saturated frequencies would NOT increase.

Carbon dioxide is the next gas by percentage. As you can see it gives off far less heat than water vapor, and is only active in four main frequency bands. Now, three of those active bands are already saturated.

(Drum roll please....)

So, if the concentration of carbon dioxide were to increase, the saturated frequencies would NOT provide any more heat. The other, non-saturated frequencies would provide more heat, thus warming the planet further. HOWEVER!

Look at the non-saturated frequencies where there is room to grow... they are small, almost inconsequential.

(Louder drum roll please...)

Because of this frequency saturation... if the concentration of carbon dioxide in our atmosphere were to DOUBLE... the heating from the extra CO2 in the atmosphere would NOT SIGNIFICANTLY INCREASE.

This, my dear reader, is the key to understanding that any warming on our planet is NOT related to CO2, and why the CO2 scare is completely fabricated, misguided, and downright fraudulent.
Scientists know this stuff.
Atmospheric scientists, astrophysics scientists, and particle physicists have known it for years... but like conservative actors, who find themselves black-balled from Hollywood, unable to find roles, and their careers are over, their screen-time curtailed because of the loud, active liberal agenda, the scientists have been labelled cooks, deniers, peddlers of fringe science.

Back in the question and answer segment, we find that the Sun is recognized as the earth's largest supplier of energy... almost 100%.

Q. Why then, is more emphasis given to the Sun's part of the 'global warming' research?
A. Because it does not fit into the liberal agenda, who want to take 'control' of the problem and take our money to tax our way to a solution.

Q. Why are the Sun's cycles, orbits, and radiations given more emphasis on Climate Models?
A. We can't blame the Sun because mankind can't control it.

We know that 71% of the Earth's surface is water... and to be honest, we haven't explored 5% of its volume, and yet purport to know all about it. The study of El Niño, the warm phase of the El Niño Southern Oscillation (ENSO), and the cool phase, La Niña, are still in their infancy. Much of the current scientific interest was drawn by the large event in 1982-83. Others like the Pacific Decadal Oscillation and its Atlantic counterpart were only discovered a decade or so ago.
We simply don't know enough about our own oceans and their link to our climate.

Q. Why then, isn't more emphasis given to the Ocean's part of the 'global warming' research?
A. Again... Because it does not fit into the liberal agenda, who want to take 'control' of the problem and take our money to tax our way to a solution.

Q. Why are the Ocean's cycles, CO2 absorptions, 800 years warming and cooling times, given more emphasis on Climate Models?
A. We can't blame the Ocean because mankind can't control it.

We recognized water vapor as the predominant greenhouse gas. It is recognized that it controls up from 76% to 90% of the heat transfer in our atmosphere.

Q. Why then, is more emphasis given to water vapor's part of the 'global warming' research?
A. Yet again... and I repeat myself for emphasis... Because it does not fit into the liberal agenda, who want to take 'control' of the problem and take our money to tax our way to a solution. Mankind cannot control water vapor and never will.

Q. Why is the relative heat transfer information of greenhouse gases not more freely available and understood? Why aren't kids taught this in school... it is science after all.
A. Again, we can't blame water vapor, because mankind can't control it.

So we go to the smallest part of the global warming debate that we CAN control... the lowly 0.04% carbon dioxide.
Mankind, the liberals say with eyes bulging out of their sockets, makes CO2, and if we make it, we can control it.
The liberals demonize CO2; we call it 'pollutant', even though it's ESSENTIAL for life on our planet.
Liberals project images of power station stacks emitting water vapor, and try to make us see pollution. They call CO2 dirty, even though it's invisible and odorless.
Liberals call for a moratorium of CO2 back to pre-industrial levels, even though the planet is greening before our very eyes.
Many Liberals ignore and diminish every natural source of CO2 on the planet, and proclaim...

"Man's CO2 emissions are responsible for ALL global warming."

How arrogant and manipulative 'Alarmists' are.

Just for your information, a recent Institute for Energy Research (IER) article on the Green new Deal said this...

"There are many natural sources of carbon dioxide, which account for about 96 percent of annual contributions to the atmosphere."

So, according to the IER, mankind is responsible for 4% of 0.04% (or 0.0016%) of the world's global warming?

They have focused our attention on carbon dioxide for political gains, and if the world cools, heats up or remains constant for centuries, they will continue to do so.

Saturation Simplified
There are two simplified versions of the above CO2 saturation ideas.
1. Paint an old barn red. From close up, it looks okay, but you can see where the old wood has soaked in the paint. Now walk away a hundred yards. It looks pretty red.
The red color is almost as red as it could be. It's almost saturated. No matter how many times you now re-paint the barn, it won't look much redder.
Carbon dioxide is the same.

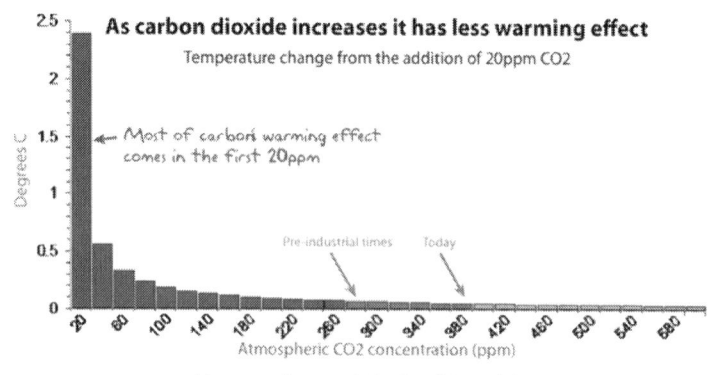

As carbon dioxide increases it has less warming effect
Temperature change from the addition of 20ppm CO2

Most of carbon's warming effect comes in the first 20ppm

Pre-industrial times   Today

Atmospheric CO2 concentration (ppm)

* Assumes a climate sensitivity of 0.15°C/W/m2 following Lindzen and Choi, 2009

Like the above graph shows, most of the Co2's ppm are used up in the first 0-60 ppm. After that, the carbon molecules can't swallow up as much. No matter how much paint you throw at the barn, it's not getting any redder.

2. Think of CO2 as a sponge that soaks in light. The atmosphere CO2 absorbs most of the light it can. However, CO2 only 'soaks up' its favorite wavelengths, and once the first few rays of light have struck the carbon molecule, it can't soak up as much, cos it's already getting full.

Try to hold a real sponge under a water tap. After a while, the water runs past the sponge and down the sink, cos the sponge if full; it's saturated. If you hit the sponge with drops of food coloring, it might go into the sponge, but there's no way that doubling the water flow will make the sponge absorb any more.

4.2... FOUND IT! CAUSE, CULPRIT AND CASE-CLOSED! PART 2!
(Well, maybe a look at the 'actual' temperature record...)

There's another way to look at figures.
For 'some' of us, there's the legendary homogenized Michael Mann Hockey Stick...

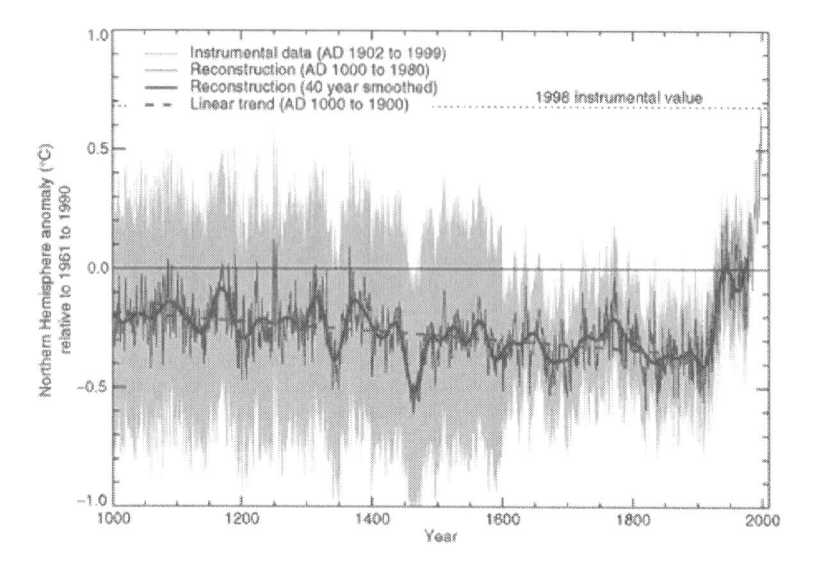

It has no character, no nuance, just a bludgeoning of the history of our planet, hammered into the convenient shape required by its creator.
It shows no Medieval warming, no Little Ice Age, and has been ignored by skeptics/Deniers for as long as it has been worshipped by Alarmists.
It's just rubbish science. (See Section 2.4)

But we turn to the world of science for a moment, and do a little digging.

The Thermometer
The Ancient Greeks dabbled in temperature, but were unable to advance the science.
Hero/Heron of Alexandria in the first century AD was one of the early pioneers of science. As well as inventing the first vending machine (to dispense holy water when a coin was put into a slot), he knew some of the properties of air and knew it could expand and contract, moving along a measured tube.

The great Galileo Galilei in the early 1600's produced the first working thermometer, but it had no scale. He called it a thermoscope.

The first thermometer with a scale (a water device in a glass tube) was produced by Englishman Robert Fludd in 1638.

In 1714 Dutchman Daniel Gabriel Fahrenheit invented the first reliable thermometer. This used mercury, and ten years later, he proposed a temperature scale, which we still use today.

In 1742, Anders Celsius suggested a scale with zero at the boiling point and 100 ° at the freezing point of water.
This is now the world standard, although his scale was immediately reversed.

Enter one of the oldest continuous temperature records in the world.

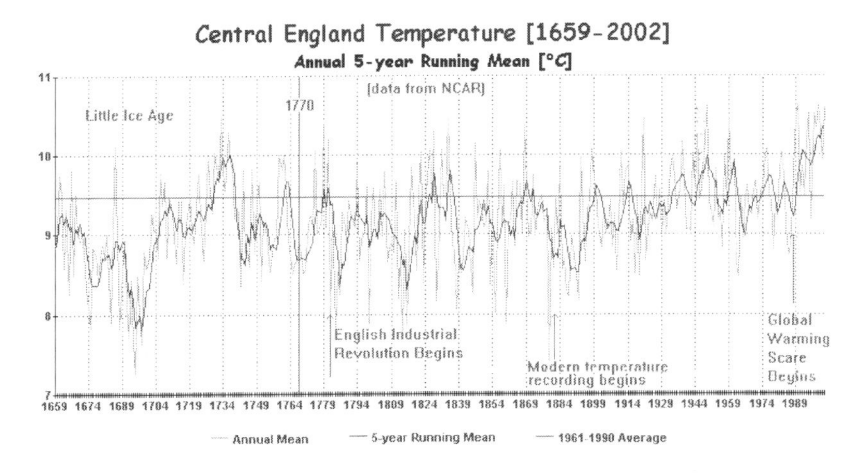

Central England Temperature [1659-2002]
Annual 5-year Running Mean [°C]

[data from NCAR]

Little Ice Age — 1770 — English Industrial Revolution Begins — Modern temperature recording begins — Global Warming Scare Begins

— Annual Mean   — 5-year Running Mean   — 1961-1990 Average

Above is the temperature of Central England from 1659 to 2002.
Please note that the figures come from NCAR (National Center for Atmospheric Research) in Boulder, Colorado.

Relevant dates are conveniently labelled and historically it looks in real good shape.
The Little Ice Age is shown, not rubbed out... there's a couple of troughs around 1812-14 when the Thames River in London froze in winter... there's the warm year after World War 2, in 1946.

But take a look at the yearly average temperature spikes (the peaks of light green); there are high temperatures around 1733, 1779, 1829, 1835, 1869, 1919, and 1946 where the average temperatures are equal or higher than modern 'catastrophic' temperatures.

OOPS!

And where's the English part of today's Global Warming?
In fact if we take the start position of the blue graph (the five year average), and the final 'catastrophic' level, we get a rise of only 1.3°C.
THAT'S JUST 1.3 DEGREES CELSIUS IN 300 YEARS!

We need to take another look at some of the temperature rises on this graph.

1963 to 2002 gives a rise of 1.3°C in just 39 years.
(That nasty CO2 has gone and done it again...)

But wait, let's not be too hasty...

1816 to 1828 shows a rise of 1.3°C in just 12 years!
Extrapolated...
THAT'S A RISE OF 10.8 DEGREES CELSIUS IN A CENTURY!

And there are six or seven peaks in the historical data that if measured would give a similar result.

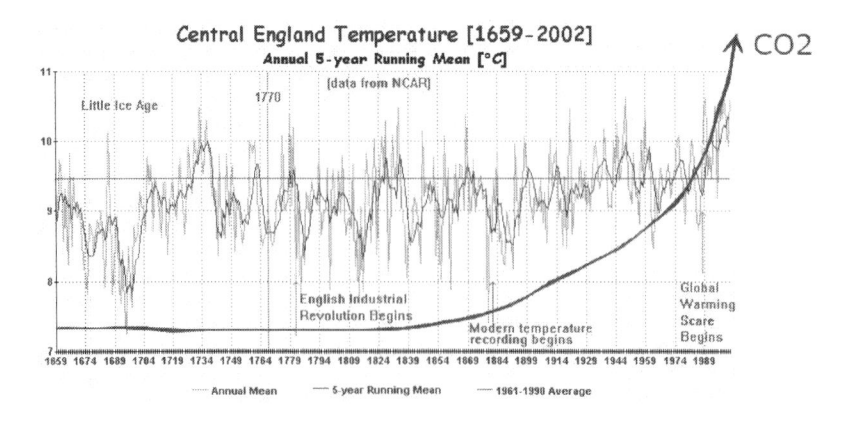

In the above 'alteration' of the original graph, I decided to add (in a nice curved line) the CO2 levels for England. The baseline is around 280ppm, and when it reaches 2002, it is around 400ppm.

Wonderful, you say... "It looks like a Hockey Stick!"

But my 'Hockeystick' CO2 addition is not on Mann's Hockey Stick graph... it's on a historically accurate one.
NUMBER 1... there's NO correlation.
NUMBER 2... IF there is correlation in the modern figures, and we 'admit' that CO2 caused the modern warming...

WHAT CAUSED THE HISTORICAL WARMINGS??????

(And this, dear reader, is EXACTLY the question that Michael Mann and his 'Alarmist' cronies do NOT want you asking.)

And if you think that the NCAR data is unreliable...

CENTRAL ENGLAND TEMPERATURE, 1720-1998
(Data Source: Hadley Centre, UK)

Central England Temperature - Summer
1660 - 2012

And again, in a different Hadley edition.
I leave you to draw your own conclusions...

4.3... OLD MOTHER NATURE IS OUR FRIEND!
(Maybe, but she's a cruel mistress too...)

Man's Gift to Global Warming
Mankind pumps millions of tons of carbon dioxide into the atmosphere every single year.
There you go... it hardly seems a statement to come from a climate denier's mouth, but it's true.
It's a scientific fact.
We emit CO2 when we make electricity in those stinky generating stations.
We emit CO2 when we manufacture virtually everything we use. There are few manufacturing processes that are carbon neutral.
We drive 1.2 BILLION vehicles around the planet, each one venting some form of CO2 in some volume or other. Each used carbon based fuel in their construction.
Our airplanes fly thousands of trips every day, some taking holidaymakers to the sun, some taking climatologists to their yearly conferences.
The cement industry that power our building construction and roads, makes a huge addition to mankind's CO2 emissions.
We, as people, fart multiple times each day, venting methane for all to smell.
We breathe out approximately 1kg (2.3lbs) of carbon dioxide every day just by living... joggers and other athletes breathe out much more. The couch potato is a better friend to the atmosphere than the athlete, trust me.
But remember, DESPITE all of the above, carbon dioxide is still only 0.04% of the atmosphere.

Nature's gift to Global Warming
But whatever mankind does to increase greenhouse gasses, Old Mother Nature has been doing far worse from the beginnings of time.
But let's not dwell on the past; let's just stick to the facts of today's Mother Nature, shall we?
Every leaf from every tree when it falls each autumn, changes from a CO2 guzzler to a CO2 emitter.
Every carbon-based piece of life, from the smallest mite to a raging elephant... when it dies, it produces CO2, and when it decays, it produces methane.

The Earth's oceans are a 'sink' for CO2. When water is warm, it can hold less CO2 and will leak it into the atmosphere. When water is cold, it can hold more CO2, and will absorb it from the atmosphere.
And then there are volcanoes... Nature's natural venting of the liquid core.
If it wasn't for these cracks in our Earth, the planet would explode.
There are about 1500 active volcanoes on land, and perhaps 10 times more under water.

The Startling Conclusion
Regardless of the number of people on the planet, remember they could all stand up in New York City... all 8 billion of them.
So...
On average, it is estimated that EVERY DAY our lovable cuddly Mother Nature puts about TWENTY TIMES more CO2 into the atmosphere than all the human sources put together.
That means today, right now, even if we started with a clean slate, man-made CO2 emissions are responsible for about 4% of the planet's CO2.

Now, remember, that's just 4% of CO2's own small part of the atmosphere (0.04%)... so mankind's contribution to CO2 'pollution'?
A paltry 0.0016% of the atmosphere...

(Oh, and the next time someone says that mankind is responsible for ALL the CO2 and ALL the Global Warming, just say, 'rubbish' and walk away.)

"NO!" shout the 'alarmists'. "Carbon dioxide was at 250ppm in the 1950's, and now it's 410ppm AND RISING, so mankind HAS to have been responsible for the rest!!!!"

But what happened around 1950?

The whole world was recovering from a World War, cities were being rebuilt, forests cleared for all the wood required for the rebuilding, and new areas of the world were being opened up for our blossoming civilization.

And the planet was warming...

Hence the oceans started to give up CO2.

Trust me...

It's not man's fault.
Man can't fix it.
Governments can't tax their way out of it.
And pretty soon we'll be crying about a coming ice age... just to sell newspapers and keep us glued to our television sets.

Just a thought.
Remember when I said we breathe out approximately 1kg (2.3lbs) of carbon dioxide every day?
That's 365kg (840lbs) per year.

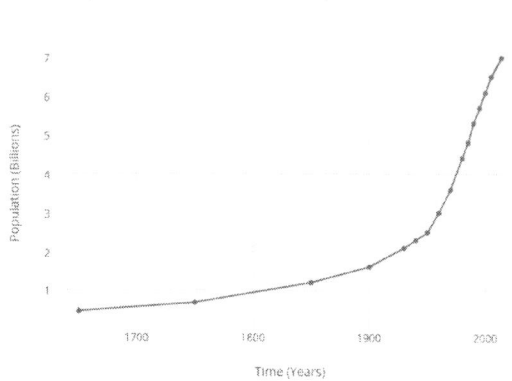

Graph 2: Global Growth of Human Population from 1650 to 2014

The Earth's population has increased from 1 billion people to over 7 billion in just 170 years.
Does the graph above look familiar to you? Kinda hockey-stick-ish?

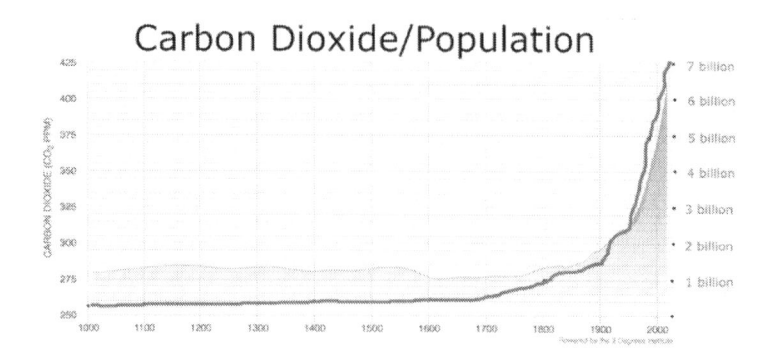

Carbon Dioxide/Population

Above is my own graph.... Interesting huh?
1 person breathes out 365 kg of CO2 per year.

1 billion people breathe out 365,000,000,000 kg of CO2 per year.
That's 40 million tons of CO2 we were breathing out in 1850
That's 280 million tons of CO2 we're breathing out today.
It's just a small hypothesis… but could we (ourselves… the human population) be the cause of Global Warming?
Well, CO2 rise, sorry lost the plot there.

4.4... How Can We Help Cure The Carbon Dioxide Problem?
(Not by taxes, that's for sure...)

The question is simple, but the answer is extremely complex. Carbon dioxide is the food for plants, and therefore the food of us humans too. Even if we wanted to reduce the rise in CO2, it's more difficult than you think.

Governments tax the people of the planet
This is usually called a 'Carbon Tax', and works by our respective governments charging us money for every source of carbon dioxide emitter we personally use... cars, planes, electricity, food, heating, buildings, etc. The government then takes our Carbon Taxes, and does something to mitigate our being bad people and emitting carbon dioxide.
Now, there's a problem here, because we have to trust our governments to actually do something 'green' with this money to reduce atmospheric carbon dioxide... and trust and governments don't go well together.

Carbon offsets
Yeah, another tricky one.
I want to drive a gas guzzling truck, so I pay a company a 'tax' to take my carbon emissions and pay someone else not to drive at all to offset my 'crime'. The recipient of my tax money essentially lives a worse life because I live a better one.
A British economist summed it up perfectly...

"basically, you're paying someone else to die for you"

And then, there's the trust part again... you've got to trust the tax offset company to take your $50 and give it to some tribesman in the Congo, and monitor him NOT buying a truck with it.

I tried this with my brother in law, who can't drink beer because it upsets his tummy.
"You pay for your beer", I said, *"And I'll drink it for you! I'll be your beer offset!"*
Needless to say, he was not amused.

Plant trees
This may sound simplistic, but it actually turns out to be perfectly feasible.

Trees take carbon dioxide out of the atmosphere and change it by photosynthesis into oxygen and water. If we had more trees, we could tackle more CO2... that part just makes sense.
Trees also take water out of the soil, and store it, thus taking some moisture out of the system... less sea-level rise.

Not feasible? Think again.

In the summer of 2017, the Indian government announced a tree planting session.
It would be a record attempting tree planting.
It turns out, more than 1.5 million volunteers turned out to plant 66 MILLION trees in just 12 hours.
That's absolutely incredible.
And if we took that kind of enthusiasm and spread it worldwide, there's no stopping the advantages.
More trees, better than looking at windmills, more wildlife, more parks.

However, there's a downside. Trees are a huge swallower of water. If your trees don't get watered, they'll die. If your trees are in an area of drought, they'll die.

Carbon capture
This is one for the front burner.
Many industrialists are perfecting ways to 'scrub' carbon dioxide from the atmosphere... literally taking it out.
Now the latest studies show the process would cost about $600 per ton of carbon... that's high. Economists have figured on about 4100 a ton being the high end of a fiscally sustainable method.
In a few years, that could happen.
Here's links to a couple of studies about CO2 capture.

https://www.theatlantic.com/science/archive/2018/06/its-possible-to-reverse-climate-change-suggests-major-new-study/562289/

https://interestingengineering.com/new-study-shows-extracting-carbon-dioxide-from-air-is-actually-feasible

4.5... THE CARBON DIOXIDE RECORD
(A short background...)

The graph below is the standard 'accepted' form for carbon dioxide.

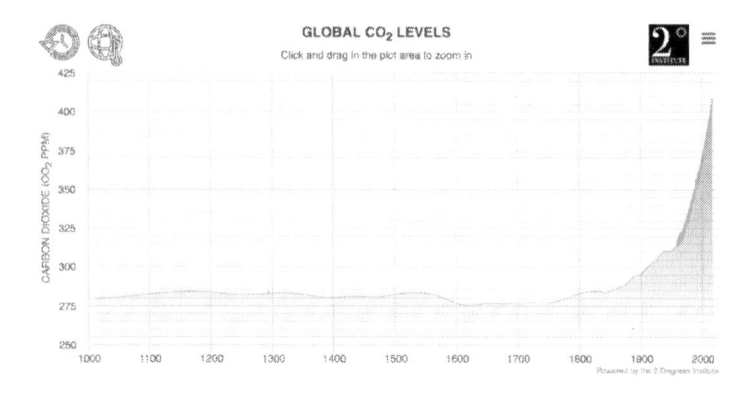

Now, I say 'accepted', because the actual 'modern readings' did not start until the mid-1950's, pioneered by Charles Keeling, an American atmospheric scientist. The infamous 'Keeling Curve' is named after the scientist who began the monitoring program near Mauna Loa in Hawaii, and supervised it until his death in 2005.

I have no problem with the Keeling Curve, or the professor's CO2 measurements, which start at about 315ppm in 1957, and rise in a very formal slope to his final measurements in 2005 of 375ppm.

However, carbon dioxide levels vary from place to place and by wind speed; they are by no means uniform. They can vary as much as 40ppm by global position.
They can vary by as much as 15ppm depending on what time of the day you take your reading.
AND... they vary seasonally throughout the year, by about the same... that's a 20ppm variation just on time, and 60ppm in total.
In a dense forest this variation can be as much as 60ppm from day to night readings.

The 'settled science' tidy hockey-stick graph above of the carbon dioxide record for the past thousand years, is the one regurgitated by most of the media when they want a look at how mankind has totally screwed up the atmosphere of sweet Mother Earth.

Man, that 800 year baseline is amazingly smooth. It's been verified by ice core samples from the Antarctic, and until a few years ago, it was not questioned.

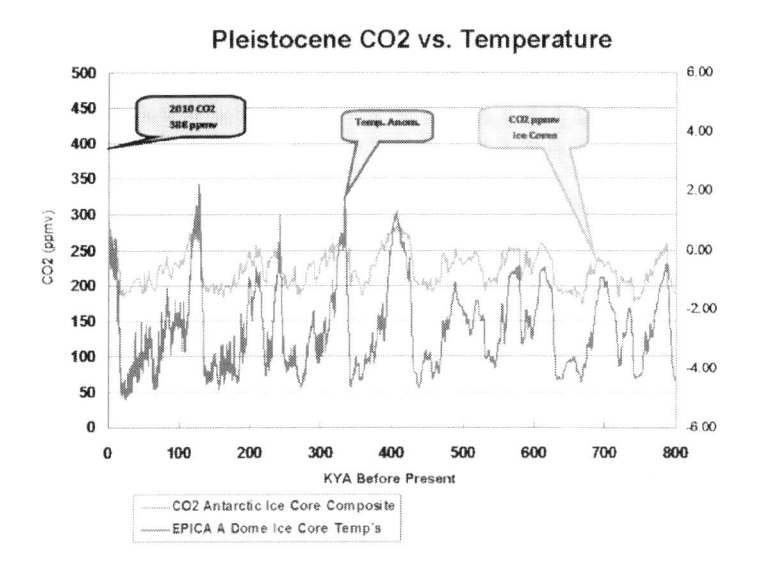

Pleistocene CO2 vs. Temperature

(Graph showing CO2 Antarctic Ice Core Composite and EPICA A Dome Ice Core Temp's over KYA Before Present, with annotations "2010 CO2 386 ppmv", "Temp. Anom.", and "CO2 ppmv Ice Cores")

The above graph is also taken/verified by Antarctic ice core samples, going back 800,000 years.
Take a good look at the variation of CO2 values, rising and falling by over 100 ppm. It's very different from the stable line in the first graph.

We've been conditioned to look at scientists and take everything they say as fact. But we know that science is a perpetually changing field; look at poor Pluto for example; one day it's a planet, now it's not.

Ice cores have been a revelation to reveal the distant past temperatures, past CO2 levels, pollens, plant-life, dust particulates, and more, but the method of determining historical CO2 levels are dependent on one thing... the CO2 being TOTALLY trapped by the ice core.

Because we have no other historical CO2 record to go to.

Or do we?

Carbon Dioxide was first noticed by scientists as far back as 1640. Initially called 'fixed air', in the 1750s by Scottish physician Joseph Black, it was first measured by experiment in 1809.

In the graph below are marked the results of all the recorded results (in experiments) of carbon dioxide from 1809 to 1955. The circled values are those taken by English climatologist Guy Callendar as a baseline... but look at the other test results.

Many other results show just as viable 'baselines' at 350, 400 or even 450ppm.

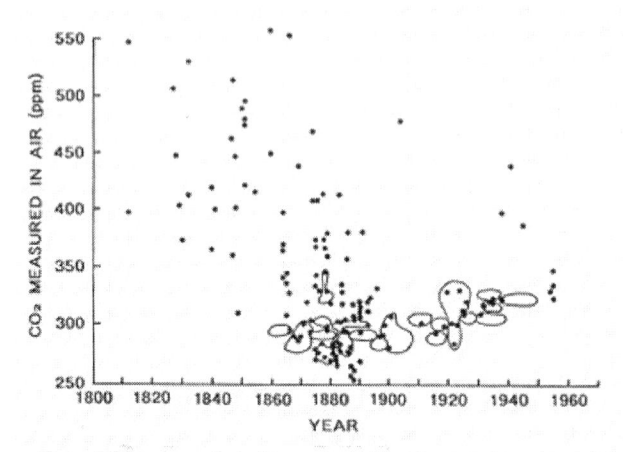

Figure 2. The mean values of atmospheric CO_2 measurements from Europe, North America, and Peru, between 1800 and 1955. The encircled values between 1860 and 1900 were arbitrary selected by Callendar [12] for estimation of 292 ppmv as the average 19th century CO_2 concentration. Slocum [11] demonstrated that without such selection these data average 335 ppmv. Redrawn after [3].

As you see, most of Callendar's circled results (the ones used for the baseline figure) are at the bottom end of the CO2 measurements, but many others are in the +400ppm level. Without his circled values, the average CO2 is 335ppm, way above the now 'accepted' pre-industrial baseline of 270ppm.
Now, remember, Keeling's first readings were around 315, and he extrapolated backwards to Callendar's figures.
If we took 335ppm as a floating baseline, it makes the 'modern carbon curve' a much less ominous thing.

Ice Core Sampling
The science is in a very early stage. Many cores have been dug, and the results give a fantastic look into CO2 values millions of years ago.
But the problem in ice core sampling is the actual pristine condition of the ice core itself.

If any gases have escaped from the core at any stage in the thousands or millions of years they have remained frozen, the results of the core samples would be useless.

Sample contamination is the science's biggest enemy.
The ice core is laid down as a series of snowfalls, which can take as much as 70 years to be compressed into ice... that's a long time for CO2 to leach from the snow and escape... it's a gas held inside a flaky solid for goodness sake!

Law Dome is a scientific expedition by the Australian Antarctic Program to drill the ice and find ice core records.
The records are drilled and examined at the site, sometimes going miles under the ice to free the long-frozen samples.
Like a huge ice version of tree rings, the core record is meticulously cut, brought to the surface, recorded, then sawn into blocks for testing.
The results give us the graph shown at the start of the chapter.
It shows lots of ppm variation throughout the millennia... but little change over the last two thousand years or so.

This is perhaps because the ice core results are an average over a long period of time. The ice cores are good at showing long term CO2 trends, but specific volcanic eruptions don't show easily in the CO2 sampling.
(Particulates like carbon and sulphur, however, do show up, so the ice cores can be dated by these shallow volcanic chemical markers)
Like the tree ring project to establish an unbroken historical record across species, the ice core sampling is essential to the scientific understanding of the world's climate.

However, many scientists have long considered the ice core records of CO2 to be on the low side. They have cited many scientific studies which question the integrity of the readings, as over hundreds of thousands of years the huge pressure of a mile of ice can squeeze the gaseous bobbles out of the ice.
And until recently, there has been little proof.

Plant Stomata
Basically a plant's stomata are microscopic holes in the leaf, stem or needle, pores that enable the plant to breathe. The stomata aids photosynthesis, breathes in CO2, and releases water vapor.

However, in recent years it has been found that frozen or fossilized stomata can be tested for a CO2 signature, revealing records of the CO2 levels at the time the plant lived.
A complete picture record is not available, as these findings are rare, but the results are consistently higher in historical CO2 than the ice core samples.

A growing number of scientists now believe that plant stomata records may be a huge key to opening a new look at the CO2 record.
(Funnily enough, though, at the time of writing, Wikipedia has no record of this scientific breakthrough)

The graph below is a combination of all three different CO2 records.

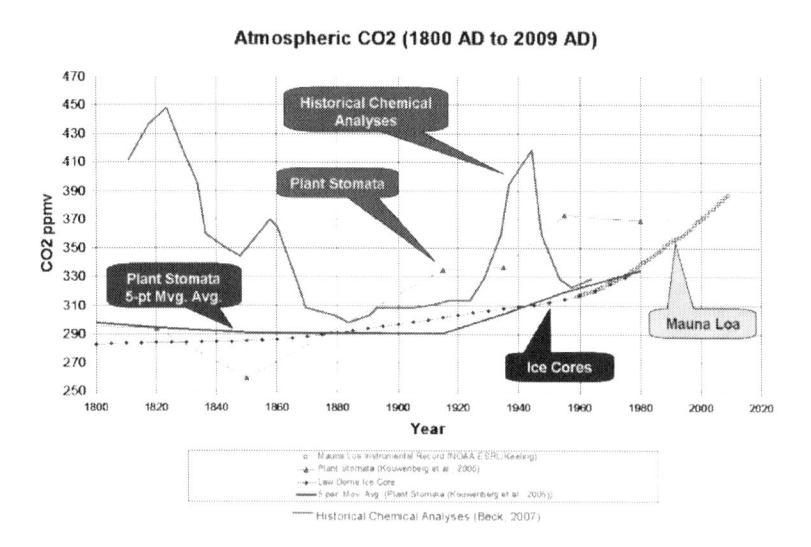

Atmospheric CO2 (1800 AD to 2009 AD)

Ice core data provides a consistent estimate of atmospheric CO2, but recently with satellite observations, it has been noted that CO2 at the poles is sometimes more than 20ppm lower than at the tropics.

As the earlier graph shows, CO2 levels from the Pleistocene through to pre-industrial times were highly variable and definitely not stable as the Antarctic ice cores would have us believe.

Take a look at the 2004 graph below, by Kouwenberg, Ljlungqvist, and Moberg. It shows two things…

Plant stomata readings of CO2 in the last 2000 years are appreciably higher than the ice core samples, AND the CO2 peaks and troughs lag the temperature record by 230 years.

It has long been proven amongst 'Denier Scientists' that temperature leads CO2, and if this is the case, (the carbon cycle lags behind the climate cycle) it proves that CO2 does not drive the climate cycle.

If through further testing, and new finds, this continues to be the case, then the accuracy of the old familiar historical CO2 record must called into question.
If these results hold up, then perhaps the whole man-made CO2 controversy would be split asunder.

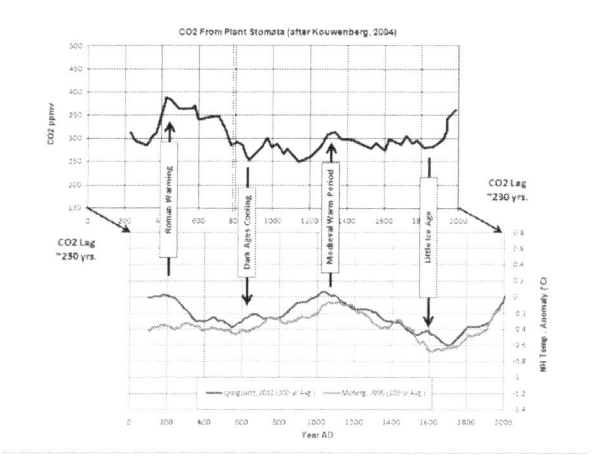

Either way, science is advancing; new finds are being researched, proven, disproven, agreed on or thrown away.

Either way, the science is nowhere near being 'settled'.

4.6... CO2... GREENHOUSE GAS... A POISON TO THE WORLD
(Nope, it's the food of life on Earth...)

For many hundreds of years, carbon dioxide was rightly seen as one of the building blocks of life on Earth. It fed plants, the plants grew, giving us food, and as a by-product produced oxygen for us humans to breathe.
If it hadn't been for CO2, life on Earth wouldn't exist.
In the ages before animal life, CO2 levels were through the roof... perhaps 20,000 parts per million in places, feeding the lush green vegetation that covered the earth from pole to pole.

YES! Even Antarctica was covered in trees... no ice whatsoever.
Then, as the animal populations increased, carbon dioxide declined, and in the last few thousand years, diminished to a dangerously low level.

Why do I say 'low'?

Carbon dioxide is essential for plant growth; without it, plants die.
The lowest level for basic plant growth is around 150 parts per million, 150ppm. Below that, and the plants wither and die, and we humans start to run out of food.
Levels a few thousand years ago dropped to almost that level, (180ppm) but in the last 2-4000 years, they have been steadying out at around 280ppm.

There's no scientist in the world that could tell the 'truth' behind the recent CO2 rise, but most would postulate that man's burning of fossil fuels have something to do with it.
Even I'm in that camp. I just don't know the extent to which we've been responsible. Natural causes can't be ruled out...
For one, the planet's oceans are a huge 'sink' for carbon dioxide. As the oceans heat up, they can't hold as much CO2 as they did when they were colder, so as the world heats up, the oceans give out a far larger amount of CO2.
But where we stand on the CO2/Global Warming situation all depends on what we believe LEADS the process... what DRIVES it.
Around 1850, as the Industrial Revolution spread globally, CO2 levels began to rise from the 'accepted' base 280ppm level. Then, after World War 2, in the blossoming post-war economic boom, it began to spike, rising past 400ppm in 2010.
So far so good huh?

Enter the leftist, commie, greeny, revolutionary, pinko, neo-fascist, determined to put a kybosh (a stop) to all our fun biology text books...

It happened like this...
Increased greenhouse gasses (water vapor, man-made CO2 and methane mainly) cause the Earth to become warmer. (And a good thing for us too, otherwise we'd freeze our asses off! Without greenhouse effect, planet Earth is just too cold to support human life)
Because CO2 is a greenhouse gas (its reaction to reflected sunlight causes heat energy) it has been labelled as a poison, a pollutant, and been cast in the climate movie as the bad guy cowboy wearing the black Stetson.
(I know, water vapor causes 85% of warming due to greenhouse gasses, and poor CO2 only 15%, but we can't label 'water vapor' as a poison... CO2 sounds far more sinister, and therefore MUST be the culprit)
(AND it's man-made, so we inherently have someone to blame)

The 'Alarmists' had a fall-guy, they had a cause of this 'warming' they had found, and were determined to correlate the two together... no matter what some other scientists were saying.

Carbon dioxide was causing Global Warming.
The science was settled.

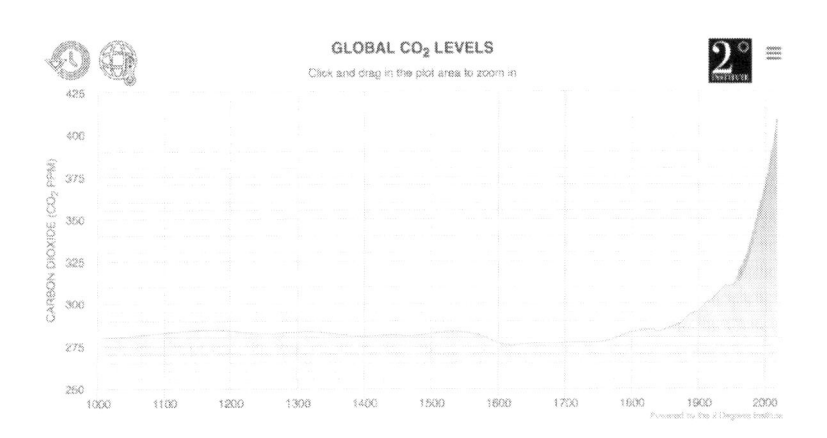

GLOBAL CO₂ LEVELS
Click and drag in the plot area to zoom in

Carbon dioxide was a terrible, terrible thing.

Except for one disturbing fact...

Plants grow better in higher CO2 concentrations.

Farmers have known this for years. In almost every commercial greenhouse they burn propane, and pump the resultant nasty, smelly, black smoke back into the greenhouses.
(Please excuse my sarcastic description of the bad guy CO_2, I know it's both invisible and odorless)
Farmers actually raise their greenhouses to 1000ppm and for some varieties of crops, over 1000ppm to get the best growth, quality of yield, and size of fruit.

"But we can't live in 1000ppm!" the Activists roar, running out of Starbucks, and down the street to the car parks full of Prius's. *"We're all gonna die!"*

Except for another disturbing fact... when humans congregate in one place, (like COFFEE SHOPS!) breathing out our dangerous, smelly, black CO_2 from our diseased lungs (yeah, more dripping syrup-thick sarcasm... I just can't help it)... the CO_2 concentration soon rises.
In a coffee shop, (or schoolroom) it might be between 600 and 1200ppm!
In larger rooms with bigger concentrations of humans, auditoriums, concert halls, night clubs, bars, etc, the level might reach 2000ppm.
And as we dance to the music in our favorite night clubs, get all worked up and excited, we breath out even more CO_2 doing so.
Submariners can cope with 5000ppm, and astronauts over 15000ppm.

And this alarmism of doom and gloom only exists if one thing is true...

What drives the climate?

Because, even there, dear reader, the science is NOT settled.

4.7... CO2; AMIDST THE BLEAK MIDWINTER...
(Just bad news for the bears?)

One other piece of evidence of the IPCC's involvement in blaming humans for the rise in global warming...
THEY NEVER DISCUSS THE PLUS SIDE OF GLOBAL WARMING.

Carbon Dioxide (CO_2) is the lifeblood of the planet.
Most living organisms are carbon based, including us villains in the piece, humans.

CO_2 is a requirement of life; it's one of the building block essentials of the photosynthesis cycle, and its availability in the atmosphere causes plants to breathe it in and emit oxygen for most animals to breath.

CO_2 has been at the root of the fossil fuel industry for millennia.

Ever since the first ape rubbed two sticks together, mankind has derived energy, heat and safety from our carbon fuels.
Until the nuclear age, fossil fuels have powered mankind's rise to prominence.
Wood, then coal and coke gave way to oil and natural gas deposits have driven technology, given us resources, food, houses, and roads. From the lowly donkey to the Lear Jet, CO_2 was at the very core of our exploration, travel, and it will be for many years to come.

Fossil fuels are a cheap source of energy for the developing world, and until we develop some new form of 'wonder-fuel', we'll have to make do with burning hydrocarbons to exist.

The environmentalists, led by the U. N., Greenpeace, the UNFCCC and the IPCC villainize fossil fuels, and plan their downfall.
If it ever happens, we are in for a dark time in humanity's history.

CO_2 has increased from around 300ppm or so during the pre-industrial age to over 400 today.
Of that I'm pretty sure.

One of the main contentions between the Alarmists and Deniers is the effect of the doubling of CO_2, and the subsequent effects on our global temperatures.

But rarely (if ever) do the IPCC set out any details about the benefits of a CO_2 doubling.

Food growers use an increased CO_2 levels in their greenhouses, and many fruits and vegetables have shown 50%+ in both growth and food yield in CO_2 levels of 1000ppm and above.

Many surveys show a rapid greening of the planet, visible since the 1970's and our first real good quality photographs of the planet from space.
(If you don't believe me, just Google 'global greening')
Oh, okay, don't bother moving to Google… we'll do a segment on it later.

Section 5
The Songs of the Oceans...

"To me, consensus seems to be the process of abandoning all beliefs, principles, values and policies."
(Margaret Thatcher, Ex-Prime Minister of the United Kingdom)

5.1... Seas Levels are Rising All Over the World
(Measurement of a variable volume that's in constant motion...)

This is one of the big deals touted by the left...
YES! SEA LEVELS ARE RISING!
But the rate of rise is in serious dispute.

Let's take a look at how mean sea levels were accurately measured through history...
They weren't!

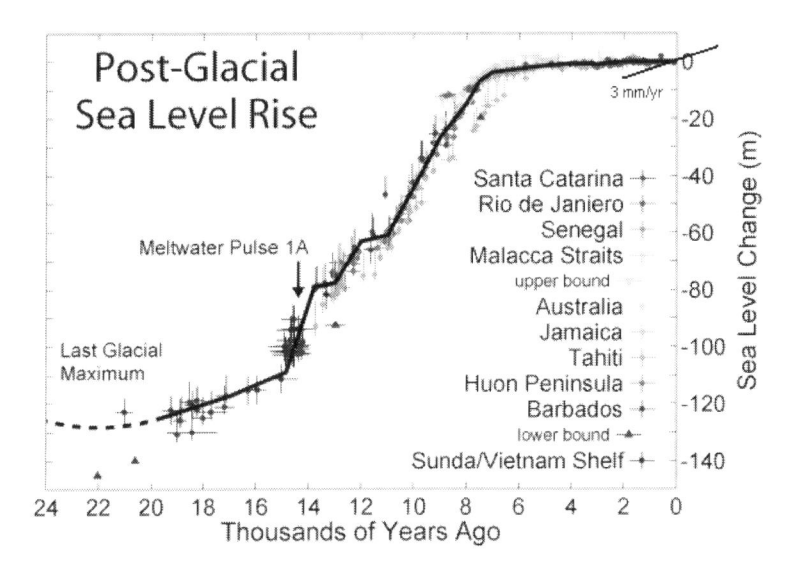

Since the height of the last ice age, around 22,000 years ago, mankind has been putting up with rising seas as the result of melting ice, then somewhere around 8000 years ago, the meltwater lessened significantly. A figure of 3mm per year has been the 'consensus' in the scientific world for many years... that's 30mm per decade... and 300mm per century.

Now while those figures might sound 'alarming, let's look at them for the non-metrics in the world... that's 1/8" per year, an inch and a quarter a decade, and a staggering ONE FOOT PER CENTURY!

From the days of the ancient Phoenicians to the end of the Victorian Era (1900), sea level rise measurement has been purely an arbitrary sport. It has been a collection of average high tides, average tide levels, and a very rough correlation between points on the earth that kept those records.

Enter the world of the sea-level measuring satellite...
Although mankind has used tide gauges at ports for centuries, it wasn't until the launch of weather satellites (primarily the TOPEX/Poseidon in 1992) that any semblance of an 'accurate' base level was established.
The TOPEX/Poseidon was a joint venture by NASA and the French based National Centre for Space Studies (CNES). It was the first targeted atmospheric examination of the world's oceans. A malfunction ceased its use in 2006.
Accuracy?
Well here lies the convoluted world of statistics.
The TOPEX/Poseidon was a revolution in measurement from space. Monitoring at an orbital height of 1330 kilometers, it could measure down to 3 centimeters, 30mm... just over an inch.
Not bad huh?
"But!" (I hear you shout) "The figures below are given to an accuracy of a TENTH of a millimeter!"
Yeah... just how do they do that?

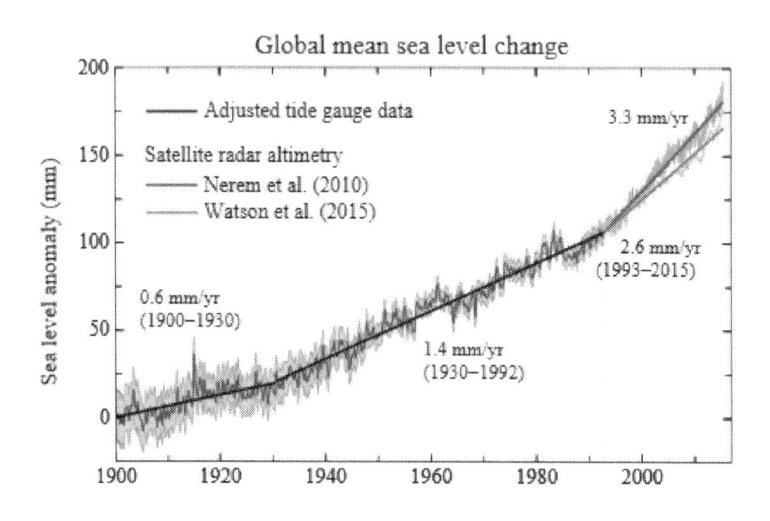

The graph above shows the last hundred or so years... looks daunting yeah? But it's not. The lines to 1995 are the average/mean values taken all over the world in tide figures. The following lines are a statistician's version of the TOPEX Poseidon figures, crunched a few times.
But even those figures are 'okay'.
The most 'catastrophic' rise is just 3.3mm per year, just over a foot per century. (And remember we were using a satellite accurate to just one inch.)
In 2001, a new satellite was launched by NASA and CNES, called Jason1. This was followed by Jason 2 in and Jason 3 in 2017.

However, measuring the average sea level of the oceans to a tenth of a millimeter, even Jason 3 just has an accuracy of 13mm, (0.4 inches).
Below, a brand new NASA report shows quite clearly that by satellite imaging the oceans are rising by an average of 3.4mm per year (That's about 1/8 inch). That's going to be 340mm (ONE FOOT) by the year 2100. That slides well under Al Gore's dire warning of 20 feet by 2100 from his movie, *An Inconvenient Truth*.

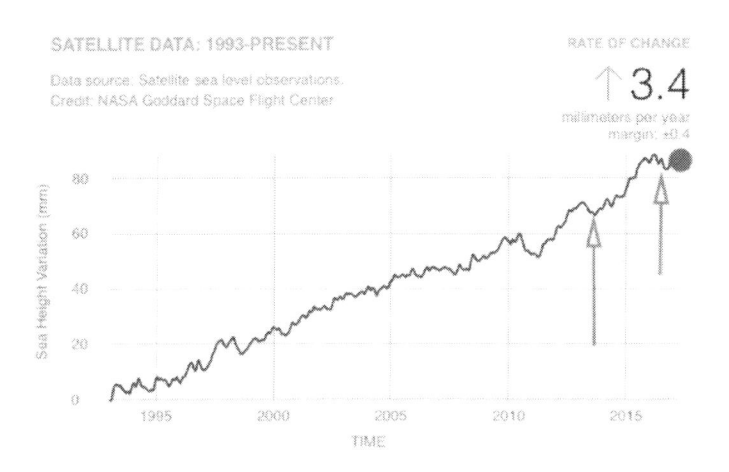

SATELLITE DATA: 1993-PRESENT
Data source: Satellite sea level observations.
Credit: NASA Goddard Space Flight Center

RATE OF CHANGE
↑ 3.4
millimeters per year
margin: ±0.4

Now, I hear alarmists say... *"That's from NASA! It has to be SUPER ACCURATE!"*
But look at the graph. Even if we take it at face value, look at the point they've taken their data from; it's a steep part, over four years. The first point (2013-ish) is around the 68mm level, and the second (2017-ish) around 82. That gives us 14mm in 4 years, and (almost) directly to their 3.4mm per year figure.

But what of we'd taken it from 2015 to 2017? That gives us 2mm in 2 years; that's 1mm per year or 82mm until 2010, or a century's rise of 3 inches.

Let's take a few readings from the graph.
NASA 2012 to 2017...   3.4mm
IAN1 2015 to 2017...   1mm
IAN2 2005 to 2010...   2.8mm
IAN 1993 to 2017...   3.5mm

Try it yourself... basically, with statistics, you can make many figures work and be just as accurate and correct as NASA was.

Now, let's forget the world of statistics, and get VERY accurate and VERY picky. Take a look at another NASA graph, one way more recent, one that shows the actual part of the graph that NASA took for its 3.4mm figure, AND the next year.

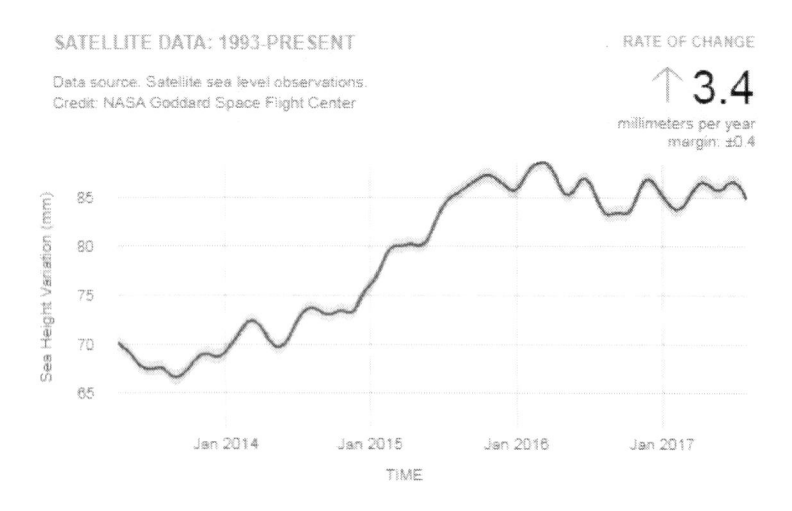

It shows no sea level rise in the last 2+ years. In fact it shows a drop in mean sea level.
Just saying...

Measuring Sea-Levels by Satellite
But let's take a look at the science of NASA measuring the levels of the oceans... FROM SPACE. There are many aspects to consider.

Distance

They're measuring a moving body of water from a distance of 1300km (800-odd miles) with a machine accurate to 13mm, and giving us readings down to 0.1mm. Does that sound right to you? They say that by taking many millions of readings it both averages out AND gets more accurate... convinced?

It's like giving you a yard stick with no graduations, and asking you to cut a three inch piece of wood.

Tidal Motions

Yup, the tides all over the world are governed mainly by the pull of the moon, but the sun affects them too, and for that matter, so do other celestial bodies, Jupiter, Mars, and Venus. Tidal mathematics is a science of its own. There are also different types of tides... yup, like clouds, tides are complex.

Wave Motions

Come on... even on a relatively calm day, try looking at the water at a tide gauge. The water's in constant motion. How do you take a reading, or even go for an average?

Land Motions

Yup, it's a thing. Very few land masses are sitting at a constant height from the earth's core. Most are rising or falling slightly. Now, I do mean 'slightly', but when you're measuring tenths of a millimeter over decades, this land motion does come into effect.

Also, there's the effect of man's building on sea land. There are a few tide gauges in harbors and ports, but if you've built heavily on the surrounding land, the weight of your activity can lower the land levels... enough to show on a tidal scale.

Tectonic Plates

Sometimes known as Continental Drift, this is the motion of the continents in relation to each other. The relative movement of the tectonic plates range from zero to 100 mm each year... you try to tell me that doesn't affect sea levels!

That's one thousand times the smallest measuring decimal point!

Thermal Expansion

As water heats up, it expands, and therefore rises slightly. Now again, this is a very small rise globally, but when measuring to the tenth of a millimeter, this thermal expansion must be taken into account.

Underwater Disruption

On average there are half a million earthquakes on our planet each year. The majority of these happen underwater. These earthquakes cause disruption to the sea bed.

The number of underwater volcanoes is unknown. These spew ash and magma into the seas creating atolls, islands, and countries. If anyone can estimate the change in average sea level due to these eruptions, I'd like to read the scientific paper.

Uneven Water Levels

Most of the world's water surfaces are not level. That might sound stupid, but go look at your fish tank... near the edges a curved meniscus of water slides up the side. It turns out continents 'pull' on nearby bodies of water, making 'humps, hills of water.

The weather also affects the sea surface; low and high pressure distorts the surface, and many weather patterns like storms and hurricanes compress and lift the sea.

Add to that the oceanic currents which swell the water above them, and you've got a pretty uneven surface to begin with.

Satellite Orbit

Yeah, Jason 2's orbit is kept accurate by three other satellites, but even according to NASA data, they are only accurate to a few inches... and they're measuring water height to the tenth of a millimeter??

The Mathematics

Statistics can be made to do anything, we all know that. But let's look at the numbers... thousands (millions?) of readings per day taken from 820 miles away in space... 'adjusting' for tide, wave height, currents, and local weather... the care taken to keep a stable orbit.

Let's be honest, it's not like looking at the side of a fish tank, is it?

As you've read above, the reasons for sea level rise are many. And the measurement is difficult to perform.

When it comes down to it, we do the best we can, be as accurate as we can, and admit no one knows for sure... it's all theory and measurement of a variable volume that's in constant motion.

Records show that world sea levels are rising... I have no problem with that. Scientific estimates however, vary from 1.2mm per year to NASA's 3.4mm per year.

But one thing's for sure... Al Gore's prediction of 20 feet rise in the world's oceans by 2100 sounds a bit on the 'alarmist' side.

5.2... SATELLITES AND NASA FRAUD (1)
(Agreeing to disagree...)

As previously discussed, satellites are supposedly the new truth for detecting sea level rise. There are now at least five satellites that have this capability, and two are dedicated to sea level rise alone.

I don't want to go over old ground, but measuring the level of a constantly moving sea while speeding around the earth at a height of 800 miles is a huge undertaking.
The fact that the instruments are accurate to 13mm (4 inches) and are measuring changes in the moving sea of 0.1mm is difficult for me to understand.
The task of keeping the satellites at a constant orbital height to enable such accuracy is astounding; especially as the height of the stuff they're passing over changes constantly.

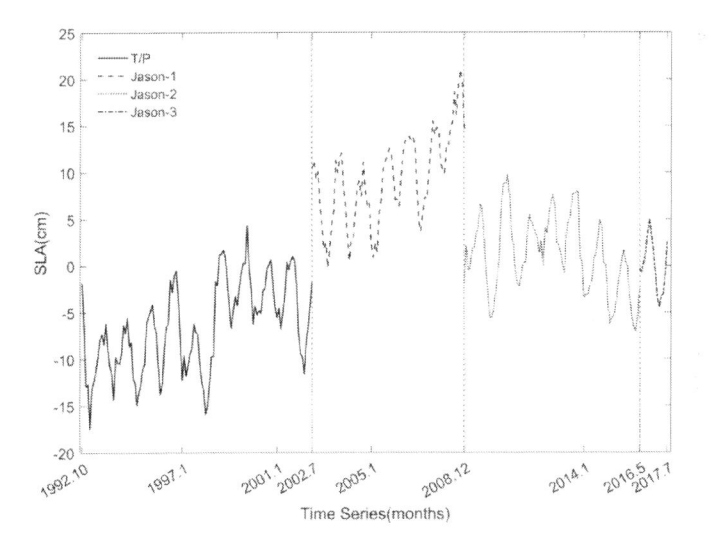

Now, it may not look it on first glance, but the above graph is one of the most terrifying things in the science of satellite telemetry.
The y-axis has a scale, in cm (multiply it by ten for millimeters), and the x-axis goes from 1992 (the launch of Topex/Poseidon) to 2017 (the launch of Jason 3).
"But why do each of the four sections not meet?" you ask.
Good question.

Unbelievably, the satellite experts do not have a reason for this; neither NASA nor the CNES (the French space authority) can account for the 'bias' between their four satellites.
Now wait a minute... the satellites are flying at the same height, guaranteed by computers on board, but their data does not correlate?

There's about 120mm between Topex's data and Jason 1.
There's about 160mm between Jason 1's data and Jason 2.
There's about 25mm between Jason 2's data and Jason 3.

The satellite experts can't get their satellites to measure the same levels of sea?
And the experts expect us to take their data seriously?

And yet, that's not the weirdest thing about satellite sea level measurements.

EVERY SINGLE PIECE OF SEA LEVEL DATA ON EVERY SINGLE GRAPH YOU SEE HAS ALREADY BEEN ADJUSTED BY NASA.

Glacial Isostatic Adjustment
According to NASA/NOAA, the glacial isostatic adjustment...

"... is the ongoing movement of land once burdened by ice-age glaciers."

Basically, when the ice retreated from the last ice age (mainly in the northern hemisphere), the land is still rising and falling from that pressure of the ice 16,000 years ago.
This is called the Glacial Isostatic Adjustment (GIA).

Picture the scene... you're sitting on a comfortable sofa. The cushion you're sitting on is compressed by the weight of your body; its surface has fallen.
The edges of the cushion, however, are possibly higher (depending on the size of the cushion and your bottom).
When you stand up, the cushion falls back into shape... but it's NOT a speedy process. The middle of the cushion rises slowly to its pre-you position, and the edges slowly fall.
This is what's STILL happening to the Earth's land surfaces, 16,000 years after the ice has melted.

So, back to the subject of sea level figures.

And you may wonder why I took time out from the satellite sea level saga to talk about a weird thing called the Glacial Isostatic Adjustment.

But...

The raw satellite data is sent from the satellites to earth.
NASA and other space related entities look at the raw data... and that's not good enough for them...
...then they add the multiplier of the Glacial Isostatic Adjustment.

WAIT A MOMENT!

We think that the satellite data is pure, raw data.
We think that NASA would just tell it as it is.
But no... even in the world of satellite data, between the satellite and our poor eyes who can't deal with raw data, there's a fudge factor.

ENTER GRACE.

GRACE is the Gravity Recovery and Climate Experiment.
It's a joint mission of NASA and the German Aerospace Center involving two identical twin satellites. They've taken detailed measurements of anomalies in the Earth's gravity field, and they suggest an 'adjustment' to the raw satellite data to bring it 'into alignment' with the tide gauge data.

I had to read the last paragraph at least three times, and it still didn't make sense.

GRACE is a multi-billion dollar satellite system, and NASA is adding a fudge factor to satellite data in order that it CONFORM to the same trend as the tide gauge data.

It seems like that NASA and Germany spent a whole lot of money, when they could have just done it the old way and read ALL the tide gauges in the world and took an average.

Doesn't that seem a logical answer?

Now, it's difficult for anyone to get their hands on raw satellite data... and even if I could, I'd need the right computer program to unscramble it...

Here's a screenshot from a lecture by Willie Soon...

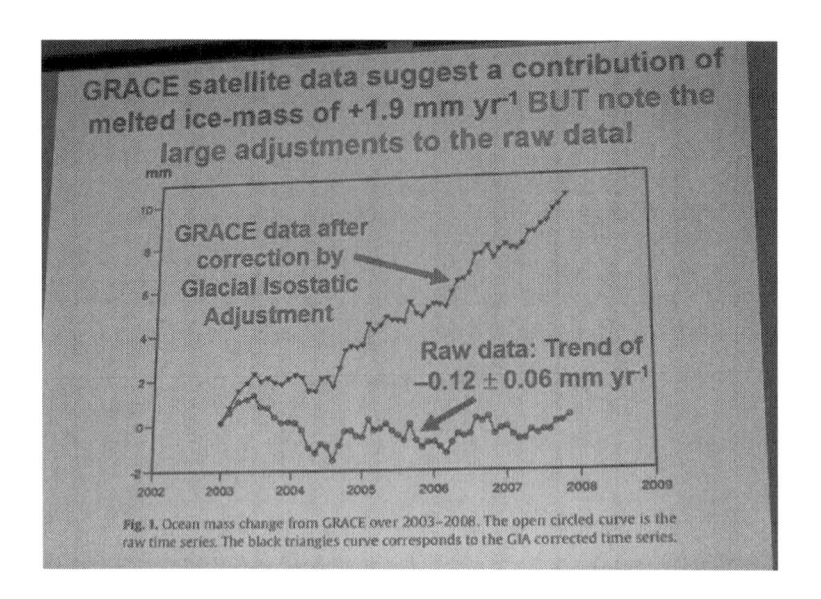

> **GRACE satellite data suggest a contribution of melted ice-mass of +1.9 mm yr⁻¹ BUT note the large adjustments to the raw data!**
>
> *GRACE data after correction by Glacial Isostatic Adjustment*
>
> *Raw data: Trend of −0.12 ± 0.06 mm yr⁻¹*
>
> Fig. 1. Ocean mass change from GRACE over 2003-2008. The open circled curve is the raw time series. The black triangles curve corresponds to the GIA corrected time series.

The graph shows raw data and the data shown to us by NASA etc.

The raw data shows a lowering trend of -0.12mm per year.
The data (after GIA correction) is also shown.

Now, I'm just an ordinary guy... but why would raw data have to be adjusted for continental plate height adjustment?

Ah... now I get it... NASA is just trying to show us that the seal level from global warming (ice melt etc) is GOING DOWN.

But sea level is actually rising from the tectonic plates getting used to the weight of ice not being there!

Thanks NASA!

But there's more... be ready to enter the world of satellite fraud...

NASA Fraud
Below is a graph of data from the Topex/Poseidon satellite from 1992 to 2000.
The graph's y-axis has a center (zero) line and a scale in centimeters.
The graph has a series of points of sea level data, and a trend line.

Now this next part is VERY important.

The trend line starts at tide cycle number 11, (-0.1cm (-1mm)), and ends at tide cycle number 276, (+0.5cm (5mm)).

Now, imagine a drum roll in your head because this is hard to believe...

The trend is 0.6cm (6mm) in eight years.

THAT'S LESS THAN 1mm PER YEAR.

ACTUALLY IT'S 0.75mm PER YEAR sea level rise.

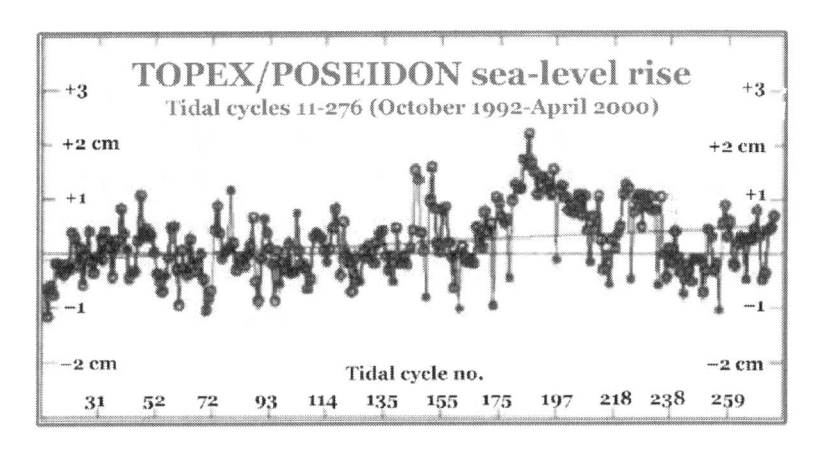

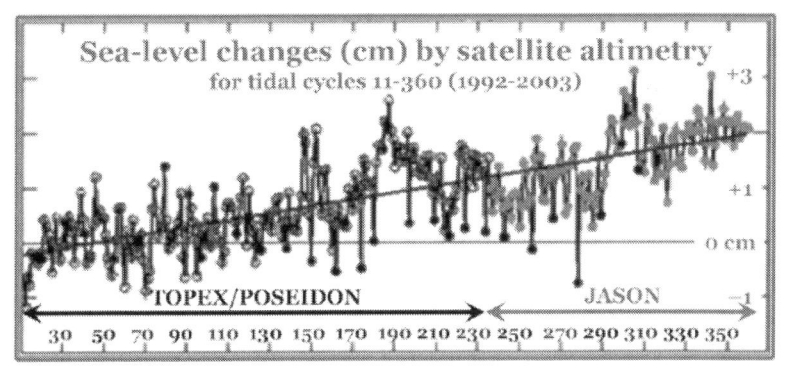

And here's the same figures extrapolated to 2003 with JASON's figures added. Now, just think... this graph was made just THREE YEARS LATER. See how they join seamlessly?
BUT IS IT THE SAME GRAPH????
Look at the figures of tide 11 and tide 276 (-1mm and +5mm from the last graph).
On this new graph, tide 11 is at -2mm, and tide 276 is at +1.5mm.

THEY'VE CHANGED THE TREND FROM 6mm in 8 years to 17mm in 11 years.
They've almost TREBLED the sea level rise rate.

If you study it real carefully, they've kept the start point of the Topex/Poseidon figures the same, but tilted the whole plotline/trend up by about 2-4mm to match the JASON ones.

Look back at the old graph, look at the highest one (it's about tide number 190)... now look forward to the next low one (number 197)... it falls below the zero line.
If you find the corresponding data on the second graph, you'll find the low one after the high point is ABOVE the zero line.
Yup, they've given all the old figures a little shift to make the graphs meet. However, the trend of this new graph is still only 22mm over 11 years, that's still only 2mm per year.

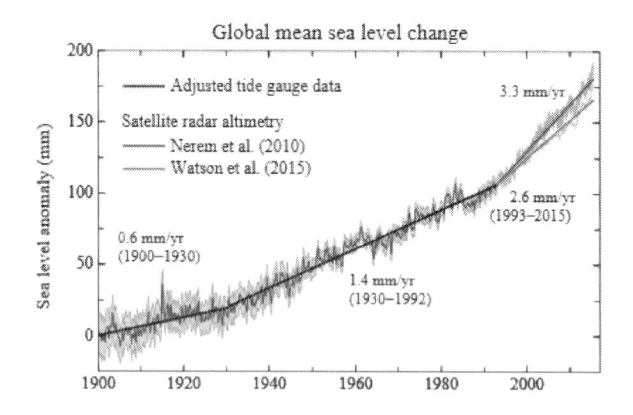

Look... I may be a layman here, but three graphs ago, the sea level rise was 0.75mm per year.
One graph later it had changed to 2.13mm per year.
And yet the graph above still shows a satellite data rise of either 2.6mm or 3.3mm per year?

AND THE ALARMISTS SAY THE SCIENCE IS SETTLED?
I call FRAUD.
And not for the first time.

5.3... SATELLITES AND NASA FRAUD (2)
(Looking at the data...)
In the previous section we looked at possible fraud in the sea level figures... now we'll look at them again, but with emphasis on the sea level rise trends.

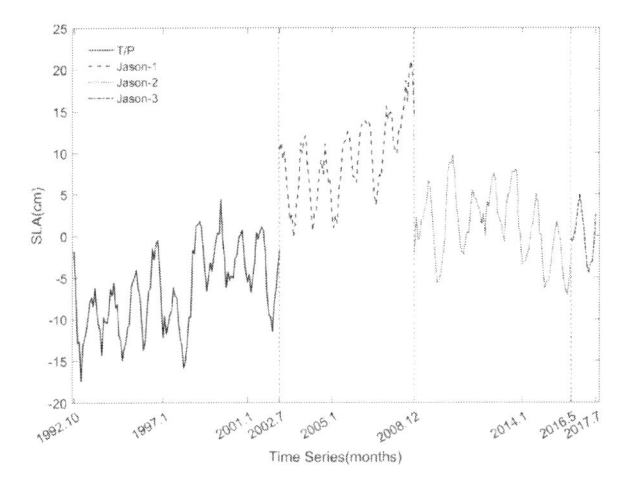

Remember this un-joined graph of satellite data? I joined it, without adjusting ANY of the trend lines.

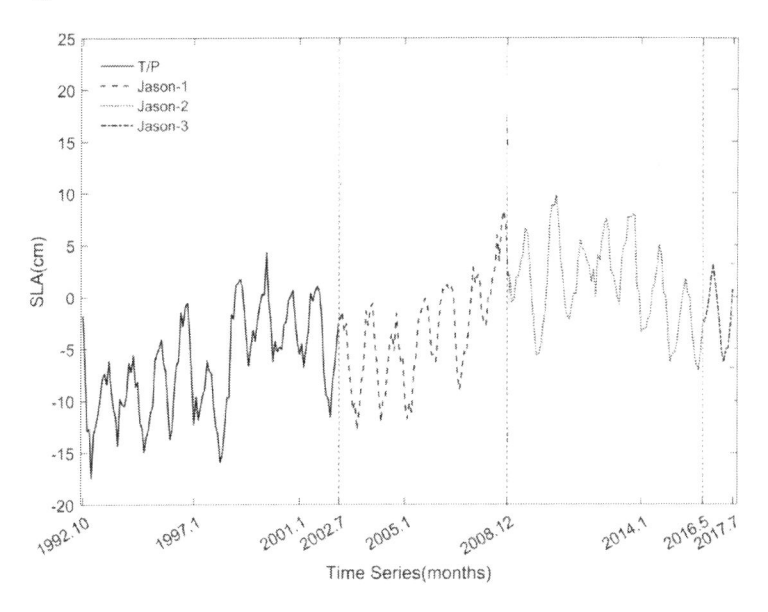

It all matches up now, and I probably did it cheaper than it cost NASA.

Now, look at the start point in 1992... I'd say the trend would be at about -10cm. (and I'm being generous)
The graph ends 25 years later at zero.
That's 100 mm in 25 years.
That's 4mm per year.
And that's NOT the figure NASA quotes on their website. They give an average increase of 3.2-3.4mm per year.

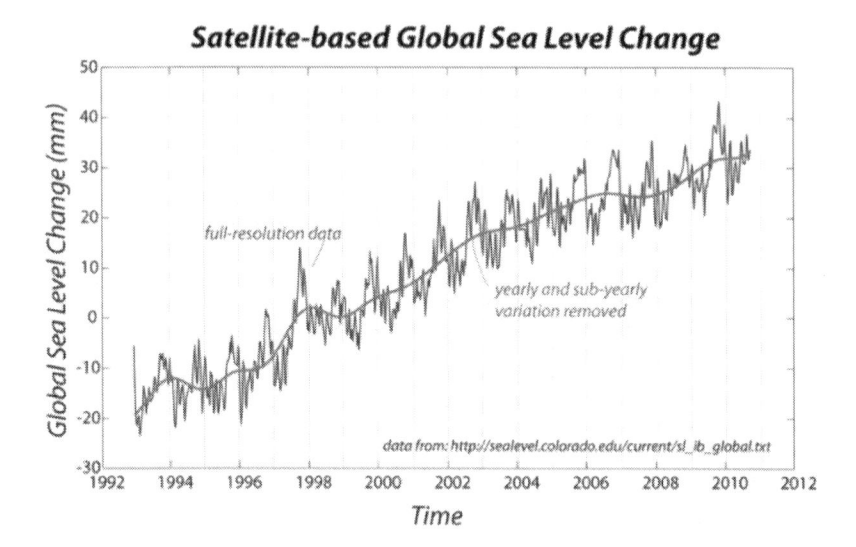

Satellite-based Global Sea Level Change

The above is a graph from a webpage from Penn State, Michael Mann's home university.

https://www.e-education.psu.edu/earth103/node/731

They state quite clearly the trends of the graph...

"The figure above show the details of the satellite-based measure of sea level change, spanning just 18 years... giving us a rate of rise of about 2.8 mm/yr, which is somewhat higher than observed from the tide gauge data over the last 100 years"

They show this graph below, and they explain this graph like this...

"This figure shows the change in annually averaged sea level from 23 tide gauges with long-term records (Douglas, 1997)... the overall trend is fairly easy to see, indicating a rise... about 2 mm/yr over the last 100 years. The

red line shows changes in global sea level as measured by satellites, and you can see that there is very good agreement."

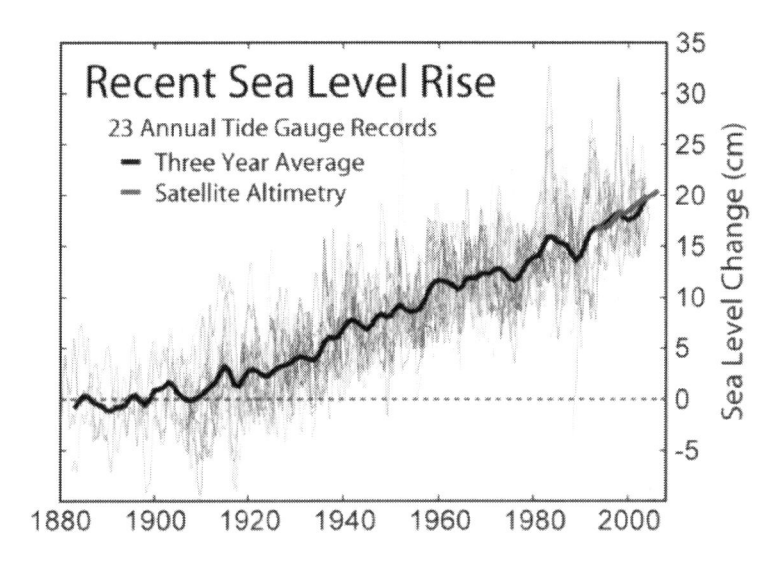

Now, just as an epilogue on this subject, I'll show you this graph again...

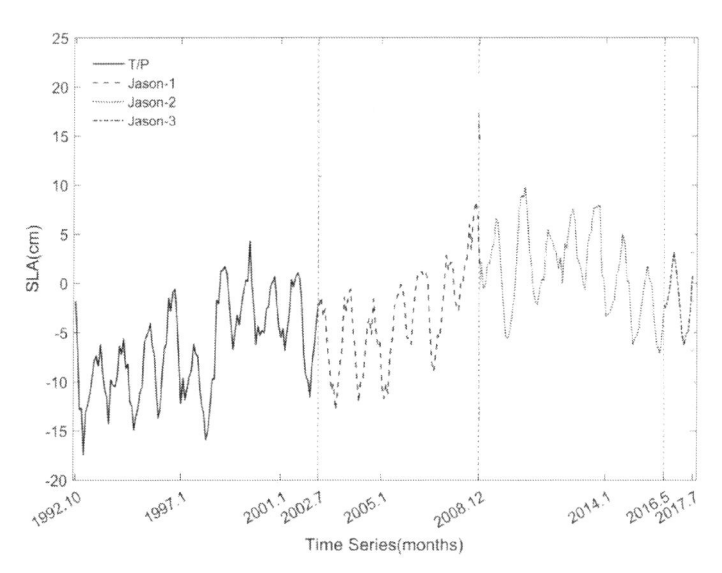

Look carefully at the trend between 2008 and 2017... the trend is DOWNWARD!
Nine years of FALLING sea levels, measured by satellite.

And remember I was generous at giving the starting point at -10cm?

Let's NOT be generous... 1992's figure clearly gives the sea level at -2.0cm.
The end reading is zero.
That's 20mm sea level rise in 25 years... that's less than 1mm per year.
So, to conclude

Okay then... people of the 'settled science'... and the expensive satellites that can't talk to each other...

What satellite figure for sea level rise is correct?

4.0mm, 3.4mm, 2.8mm, 2.0mm or >1mm.

Whatever it is, it sure shows that even in the world of satellite telemetry...

The science AIN'T settled.

5.4... TIDE GAUGES AND STUFF
(Continuing the other side of the story)

Long before the age of satellites, the Permanent Service for Mean Sea Level (PSMSL) was established in 1933 in Liverpool's National Oceanography Centre (NOC), in the UK.
Its only responsibility is the collection, publication, analysis and interpretation of sea level data from the global network of tide gauges.

They can be found here... https://www.psmsl.org/

This is their graph of 120 years of sea level rise in Taiwan. Please notice NOAA's stamp too.

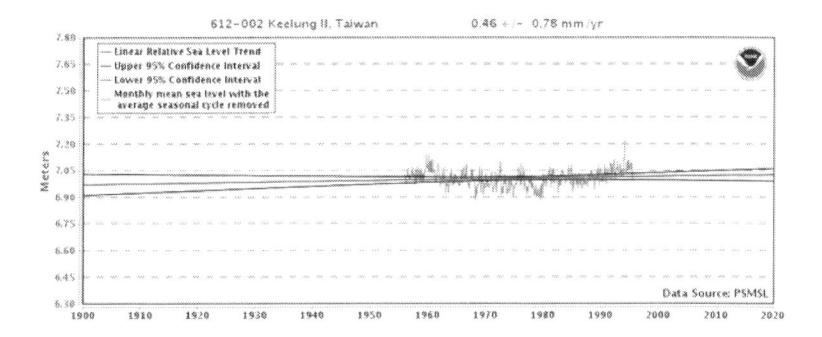

Yup, you noticed it too... there's not much of a trend. In fact the 120 year rise is just 15cm (about 6 inches).
That's virtually nothing.

And certainly NOT the 1 meter promised to us in the coming century.

In fact, a story in the notrickszone in 2018 has this headline...

"Bewildered Scientists... A Global Warming Crisis Fails To Appear: Sea Level Rise Grinds To A Crawl"

A report by Frederiske et al. in 2018 showed...

"The global-mean sea level reconstruction shows a trend of 1.5 ± 0.2 mm yr−1 over 1958–2014"

And that 1.5mm per year is less than half of NASA's figures. It's also close to a 2007 report by Holgate...

"The rate of sea level change was found to be larger in the early part of last century (2.03 ± 0.35 mm/yr 1904–1953), in comparison with the latter part (1.45 ± 0.34 mm/yr 1954–2003)."

This means that if Holgate's figures are correct, the rate of sea level rise is DROPPING!
Jevrejeva et al. in 2008 states the same...

"The fastest sea level rise, estimated from the time variable trend with decadal variability removed, during the past 300 years was observed between 1920– 1950 with maximum of 2.5 mm/yr."

And another four studies show the same.

So why are the satellite figures, the ones we now seem to rely on, so different? But NASA's not alone; he Potsdam Institute for Climate Impact Research cites a rise of one meter by the year 2100.
On the figures alone, this doesn't make sense... unless you have a pre-determined agenda.

Well, one reason could be that measuring sea level from satellites has turned out to be far more complex than first thought. As we showed in the last chapter, the four satellites Topex/Poseidon, Jason 1,2 and 3 seem to have a larger uncertainty than had been expected, and fudge factors have had to be introduced to make the figures join together and make sense.

In no uncertain terms, evidence in these homogenizations is piling up and showing that there is a chance that satellite data has been overstating sea level rise.
For example when measuring sea level rise using tide gauges, the rise has even been far slower.
Swedish sea level expert Axel Mörner's research on tide gauges in 2017 showed an observed sea level rise rate of only 1.5 – 2.0 mm/year.
That's about half the NASA satellite levels.

On the website for The Permanent Service for Mean Sea Level (PSMSL), their 2017 maps show rises of 1-2mm for most of southern Europe, but sea levels around Scandanavia have been FALLING by as much as 4mm.

Whatever the levels... there's one thing for certain... the science is certainly not settled.

https://www.quora.com/How-are-rising-ocean-levels-expected-to-affect-Taiwan

https://notrickszone.com/2018/02/02/bewildered-scientists-a-global-warming-crisis-fails-to-appear-sea-level-rise-grinds-to-a-crawl/

And just for you data hounds, here's a word map of tide gauges and their plotted graphs, courtesy of those wonderful PSMSL guys...

https://www.psmsl.org/data/obtaining/map.html#plotTab

5.5... SEA LEVEL RISE... OLD SCHOOL
(Dipping our toes in the water)

Remember the old NASA story, they spent 20 million dollars developing a pen that wrote upside down and in zero gravity... and the Russian Cosmonauts just used a pencil?
Well, I seem to recall the whole story was a hoax, but it does bring the idea of 'Old School' to mind.
And there's one way to find out what the sea level rise is without paying billions of bucks for expensive, erratic, mistake/error prone satellites.

Tide Gauges
In every world harbour with a tidal element, there is a tide gauge.
Here's the one from New York...

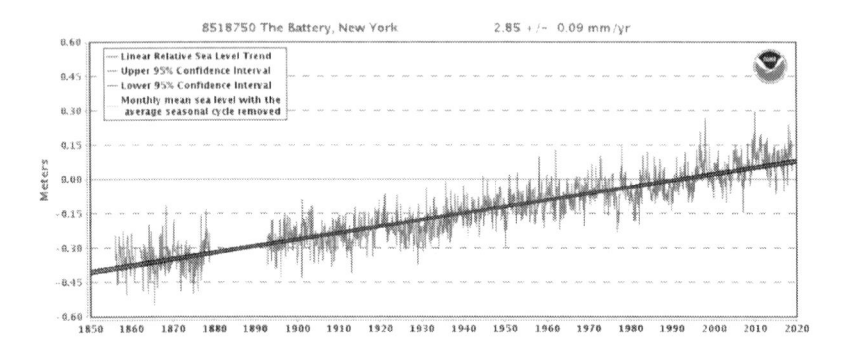

This is a NOAA graph.
From 1855 to the present day, sea level rise in New York has been a fairly constant 2.85mm per year.
That's 11 inches PER CENTURY.

Wait a minute... there's no acceleration after 1950, in fact the graph's more recent figures show a slowing in that rise. If you graphed 2000 to 2019, there would be a hiatus/pause.
Wait a minute... the graph shows the same rate of rise for BEFORE AND AFTER the 1950's industrial explosion that started the CO2 rise.
Wait a minute... this should be PROOF POSITIVE that increasing CO2 has nothing to do with sea level rise.

But I digress. (Although I may revisit this topic)

It's time to look somewhere else for a tide gauge.

This is another NOAA graph.

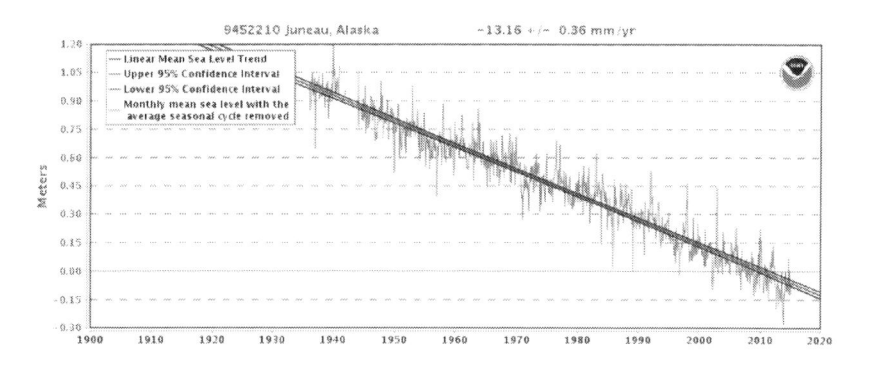

Juneau Alaska's sea level rise... WOAAH!
From 1938 to 2015, there's a DROP in Sea Level of 1020mm.
(I'm reeling)
Now, even I'm not rabid enough to run with this one. Alaska's land level is obviously rising slightly.
Let's try again...

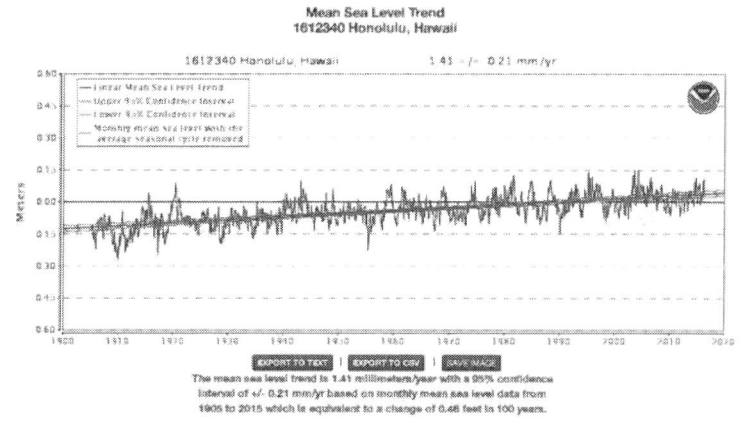

Honolulu, Hawaii's sea level has risen a staggering 200mm in 115 years...
No, I'm exaggerating; that's just a rise of 7 inches per century.
And NO SIGN of any increase, decrease in rate around 1950... NO sign of any deviation with the industrial CO2 cataclysm. They're LINEAR!

Are you seeing a pattern here?
These Hawaii Sea Levels have...
1. A steady, constant rate of rise
2. No correlation with CO2 rise

3. No acceleration at 1993 as indicated by NASA satellites

And you see, I'm just starting.

Because everywhere I look, NOAA has stacks of these tide gauge graphs, and all have the same LINEAR characteristics.

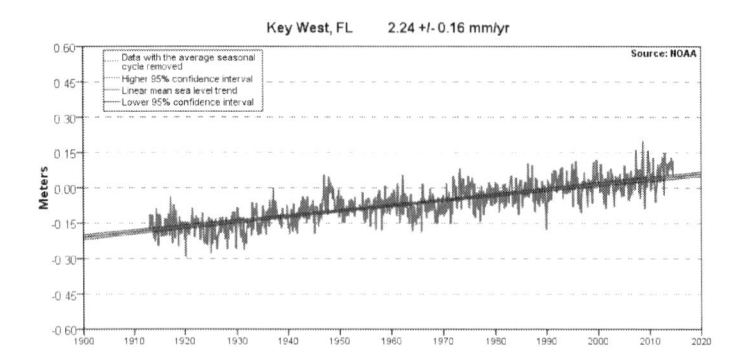

Key West, Florida... a constant and steady 2.24mm rise. (9" per century)

OKAY, let's move away from the USA, because maybe it's just a US anomaly.

NOPE!

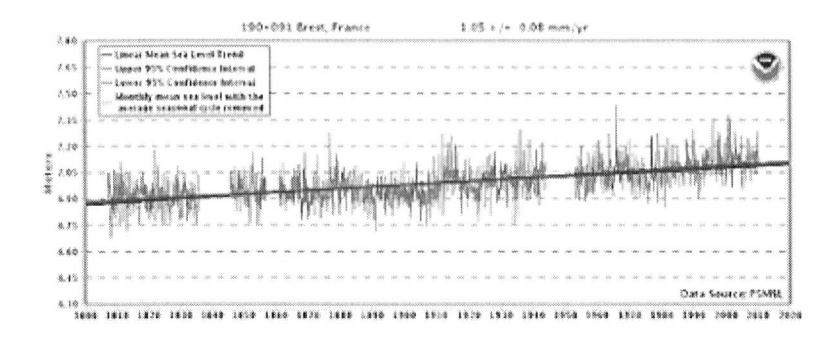

A 200 year tide gauge from Brest, France, shows a 0.4mm rise in the 1800's, 1.2mm rise in the 1900's. (2-5 inches per century)

The next graph is Liverpool, England, the home of the PSMSL people. It shows a 250mm rise in over a hundred years... that's only TEN INCHES per century.

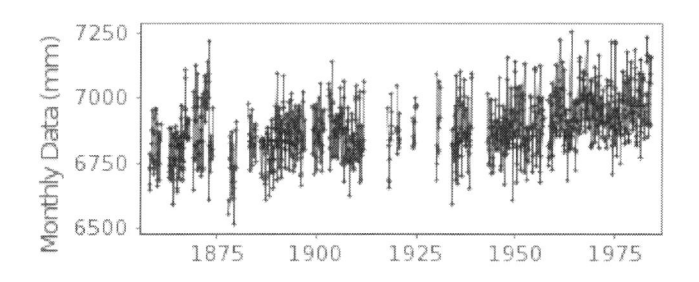

"Okay, that's Europe", I hear some readers cry in panic. "You should go further afield…"

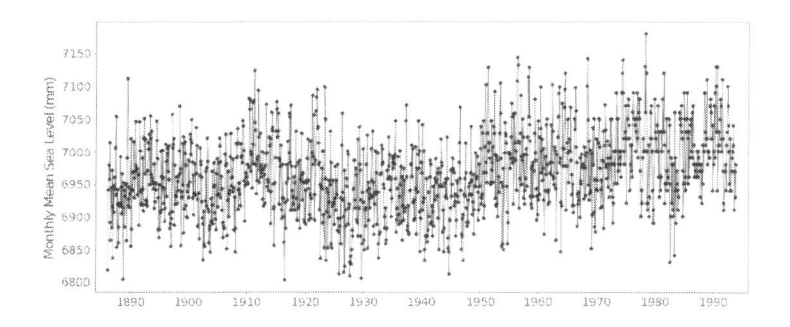

A 100 year tide gauge graph of Sydney, Australia.
It shows a 200mm rise (and I'm being generous) in over a hundred years (8 inches per century)… and still no real rise in the post ww2 industrial age.

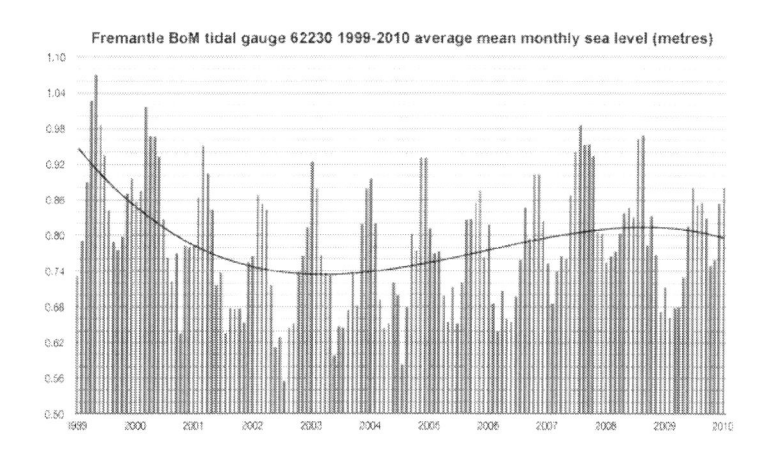

Above is Australia's Bureau of meteorology's tide gauge in Perth. Despite rising for most of the 1900's, it now shows a falling trend.

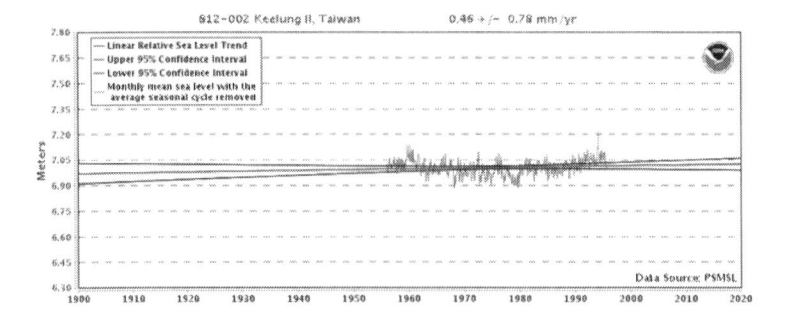

And I reprise the tide gauge from Taiwan from the previous chapter... 0.46mm per year, 46mm per century... less than TWO INCHES per century.

You see... the Alarmists say that we Deniers are the ones who argue semantics, rhetoric, and hyperbole.

But like every leftist who's ever debated... they've turned the argument on its head... Deniers argue the case with FACTS, undistorted, un-homogenized, and undoctored.

Give me a slide-rule against any satellite operated by slick individuals any day.

5.6... THE PACIFIC DECADAL OSCILLATION... PDO
(Yup, it's a kinda new science...)

The world's oceans cover 71% of the planet's surface, and we've sailed most of them since before history began.
But that doesn't mean we know them.
Some scientists reckon that only 5% of the ocean bed has been surveyed adequately, and the sea temperatures have only been measured since the middle of the 20th Century. (This involved dropping buckets into the ocean from moving ships, and sitting a thermometer in the bucket on deck... not a particularly scientific methodology)

We've also mentioned earlier that the Earth's oceans have 250 times the mass of the atmosphere, and can hold more than 1000 times the heat.
The oceans soak in carbon dioxide when they're cool, and vent CO2 into the atmosphere when they warm.
The first sailors noticed many currents at an early date, and used them to aid navigation and passage, but some of the world's ocean's secrets have taken a long time to come to light.
Only in the 20th century did it become known, for instance, that the oceans currents had a time pattern, periods of stronger and weaker flow, times when their paths moved slightly then returned.
Yup, our ocean currents have cycles.

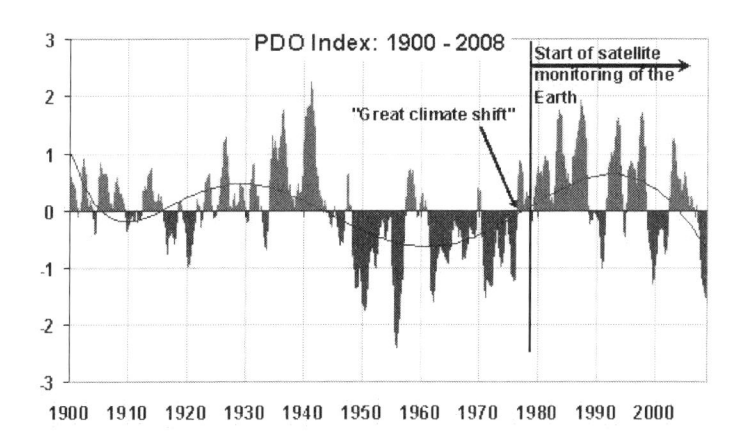

Pacific Decadal Oscillation (PDO)
The Pacific Ocean covers about one third (33%) of the world's surface, and it is only recently that it was discovered to have its own current cycle. This was named in 1997 by marine biologist Steven R. Hare, who

first noticed it while studying salmon. He called it the Pacific Decadal Oscillation, and it has a cycle of approximately 30 years.

Above is one of those 'bastard' graphs, (graphs of two halves) the first two thirds measured by random buckets of ocean water, the last third monitored by satellite. (But at least this time, the two factored graph is not trying to prove a point)

El Niño / La Niña

This Pacific Ocean phenomenon is a cycle of heat/cool transfer from the western Pacific to the eastern Pacific. El Niño is the passage of warm water, and La Niña is the opposite cooling.

Both the Pacific Decadal Oscillation (PDO) and El Niño / La Niña have been extrapolated back in time to form a century old pattern of ocean current activity that can be seen on land weather, effecting the west coast of the whole Americas.

However, El Niño (the cyclical current flow in the Pacific Ocean from Australia to South/Central America) is not a new discovery.

As far back as the 1920's Gilbert walker took notes from Peruvian Fisherman who described a warm current from the west, but only once every four or five years. He called it the Southern Oscillation.

In 1960, Jacob Bjerknes mapped this and found it to be a circular current, coming from the Philippines/Australia, moving to South America, then moving north, up the USA coast.

The fishermen called the warm current, 'El Niño', Spanish for the 'boy child' as the temperate waters usually came around the Christmas period, bringing rain to the dry Peruvian deserts.

It has been more recently discovered that El Niño is the warm phase of what is now known as the El Niño Southern Oscillation (ENSO). After each warm phase, the heat in the upper layer of the Pacific dissipates and the deeper colder water gets to the surface. This is called La Niña ('girl' or 'gift giver' in Spanish), bringing cold, nutrient-rich water for marine life, supporting a larger fish population for the Peruvian Fishermen.

Now, if you look at the biggest El Niño's on record, we see the winter dates of the largest as... 82-83, 97-98, and 2014-16.

Linking these high El Niño's to temperature records, we see a significant rise of temperature on the years of high El Niño, in fact it seems to be a very big player in the Global Temperature Record.

Every one of these 'double years' has resulted in a 'peak'.
Droughts, famines, floods, and even civil wars have been correlated with these ocean currents, and it must be said that investigations into both anomalies are in their infancy.
The huge spike in global temperature in 1998 was an immediate reaction to a high El Niño year, and the recent high in 2016 can be attributed to the same.

Atlantic Multi-decadal Oscillation (AMO)
On the other side of America lies the Atlantic Ocean, and its famous Atlantic Drift current, sometimes referred to as the 'Great Conveyor'.
In 1994, a cyclical effect in the Atlantic Ocean was identified and named the Atlantic Multi-decadal Oscillation (AMO). As yet, not much is known of the effects of the cycles, but again, it has been extrapolated back into the 1800's.

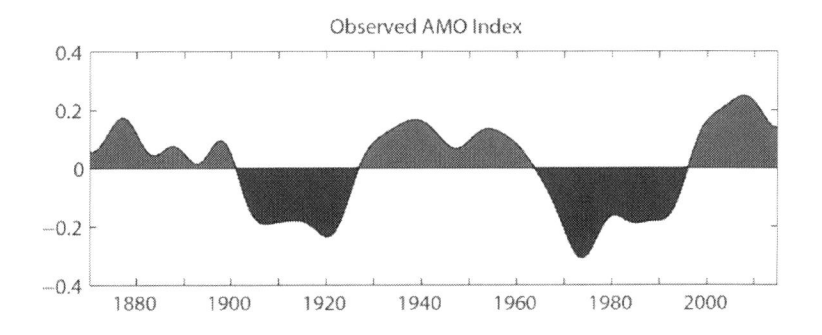

Observed AMO Index

With the Pacific Decadal Oscillation (PDO) on one side of our planet, and the Atlantic Multi-decadal Oscillation (AMO) on the other, we actually have some form of out-of-phase sine wave correlation between the two graphs.
Looking at the AMO graph above, we see that the Pacific Decadal Oscillation (PDO) peaks at the AMO's change-over times, and vice-versa.
Throw El Niño / La Niña in there to mix it up a little, and we've got ourselves a see-saw system with a very clear pattern.
So, to summarize, we've established that the Oceans are a HUGE variable in the climate system, yet many NEW facts are being discovered all the time.

I think as far as the Earth's oceans are concerned, the science is FAR from being settled.

5.7... THE OCEANS ARE WARMING AT AN ALARMING RATE
(Oops... well perhaps they're not...)

Let's go to an article in the scientific publication, Nature.
August, right? Well-known, right? Scientific, right?

In the July 2017 edition of Nature journal, Gavin A, Schmidt, Director of the NASA Goddard Institute for Space Studies, Laure Resplandy of Princeton University, Ralph Keeling of the Scripps Institute of Oceanography and seven other scientific authors published an article citing figures of 'climactic ocean warming', and finding that

"Earth's oceans have absorbed 60 percent more heat per year than previously thought".

This would, of course, imply that the Earth's global warming was 'immeasurably' worse than thought.
The article was widely quoted in the press, and the world shook in its boots.

However, enter scientist and blogger, Nic Lewis.
Unfortunately Lewis knew the mathematics behind the figures and went on to check them.
Not a good day for the scientists concerned.
Lewis found that they had made many mistakes in the basic math formulae. (The full disclosure can be read in the address at the end of the chapter, for the math is way beyond my ken, and downright boring)
The authors of the report admitted their errors (rare, I can tell you), and their 'findings' have been retracted. They now admit that Lewis is correct, and have corrected their results.

A discussion on this is available on Youtube...

https://www.youtube.com/watch?v=wpLVRPJnL4M

https://judithcurry.com/2018/11/06/a-major-problem-with-the-resplandy-et-al-ocean-heat-uptake-paper/

5.8... THE OCEANS ARE COOLING/WARMING?
(How do we know which one?)

The subtitle of this book, *'Unsettled Science'*, speaks for itself.

Science should NEVER be settled. I can think of nowhere else in the scientific community where such a statement would be made and accepted.
Ocean temperatures are one such parameter in which no-one could claim they know EXACTLY what's going on...

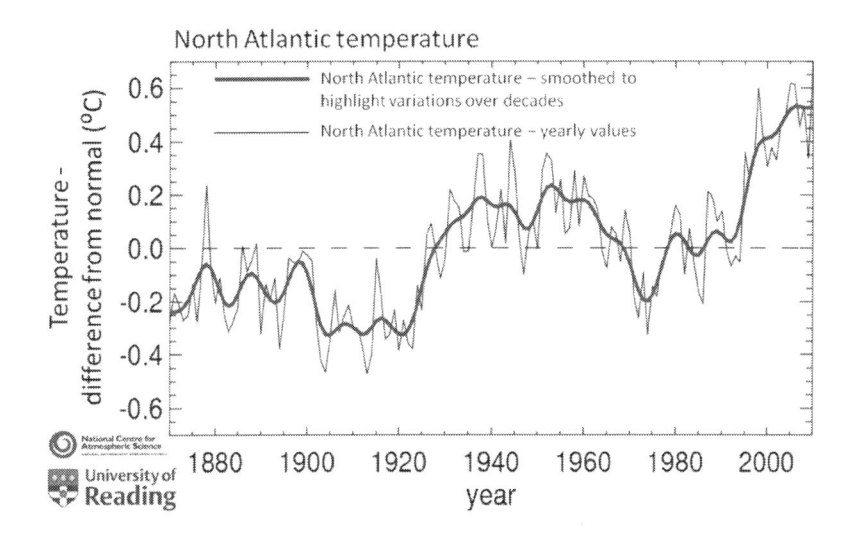

North Atlantic Ocean
Above is a graph of North Atlantic Ocean temperatures, published by the University of Reading in the UK, and endorsed by the National Centre for Atmospheric Science (NCAS), an authority spread around many of the UK's universities.

The graph shows a low in the 1970's (Cod War years), but fail to provide actual temperatures on the y-axis... that's a big no-no. They must have had that data to begin with.
What does the y-axis title, "difference from normal" actually mean? What's 'normal'?

Anyway, the graph starts at -0.2°C from norm, and rises in 140 years to +0.5°C ... in anyone's language that's a 0.7°C warming in a century and a half.

183

That's not a significant rise, in fact it's in line with the rise in surface/air temperatures.

But NCAS isn't the only boat on the ocean.
Scientists Femke de Jong and Laura de Steur of the NIOZ Royal Netherlands Institute for Sea Research have their own graph of the temperatures of the North Atlantic Ocean.

The graph, given below shows pretty much exactly the same time period as the CRU one above, but the values and trends are very different.
For one, they have given exact temperature figures in the y-axis (why wouldn't you do that?) and their graph trend is a cooling one.
Taken from the Irminger Sea, off the coast of Greenland, you'd expect different values than the CRU graph, but you'd expect the same trends.

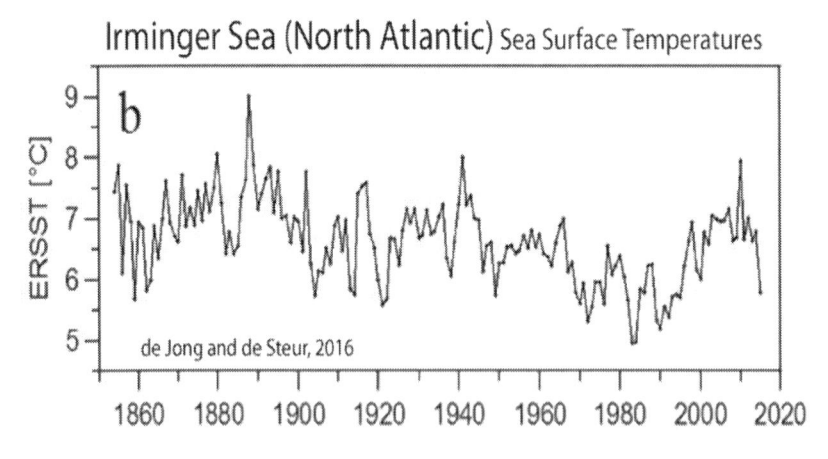

Irminger Sea (North Atlantic) Sea Surface Temperatures

The de Jong and de Steur graph starts at 7.5°C in the 1850's (and 1870 has the same 7.5°C figure), and finishes around 5.75°C in 2016.
The FALL in ocean temperature is 2.75°C in the same 140/150 years.
Those figures are 3.2°C apart.

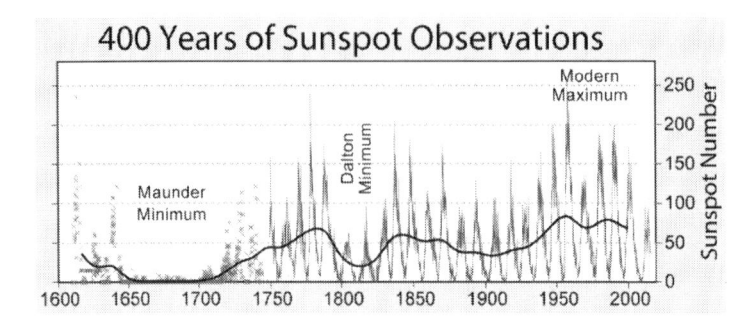

400 Years of Sunspot Observations

And look at the graph above of SEEMINGLY unconnected data.

There's a dip in the sine graph peaks in 1970.
Solar cycle 20 is a low value.
The seas cooled in 1970.

But move along, there's NO correlation between the two.

Oh, but the Dalton Minimum in the early 1800's?
Wasn't that when the Thames River in London froze for the last time??

NO! No correlation here.
Pure coincidence.

DON'T TELL ME THE SCIENCE IS SETTLED!

5.9... THE MALDIVES... THE POSTER CHILD
(The canary in the coal mine...)

The Maldives are a group of islands off the south-western coast of India. The country is comprised of 1192 islands, all at a very low height.
The Maldives is the lowest country on earth, and at an average height of about 6 feet above mean sea level, they're basically the first in line to succumb to sea level rise.
They're the subject of so many of the Alarmist's propaganda, it's a wonder they're still there.
In fact, when you consider the IPCC's projected sea level rises, why haven't they been evacuated already?

But no, the country's capital, a small island surrounded by a sea wall, is home to 100,000 people, a fifth of the population, and shows no signs of evacuating just yet.
In fact, the island's main international airport has just added a new runway to cope with the increase in tourist flights. And it's no small enterprise; they operate frequent medium-haul services to Beijing, Shanghai, and Hong Kong.
(And we know what happens when you build stuff on land near water, huh?)

But from the beginning of the Alarmists' assault in the mid-eighties, the Maldives has been on the lips of every green, liberal Alarmist die-hard... it's almost as if they wanted the islands to be over-run by the rising sea levels.

In 2004, an underwater earthquake off the coast of Sumatra caused a tsunami that spread over the whole Indian Ocean, hitting the Maldives just two hours after the initial shock.
There was no time for emergency procedures, but the atoll island chain was not badly hit.
The wave caught the Maldives at low tide, and had been broken up by deeper water off the coast.
Compared to the 130,000 confirmed deaths in Indonesia, and the 35,000 confirmed in nearby Sri Lanka, the Maldives got off lightly with just 82 confirmed dead.

So much for the supposed fragility of this low-lying island country.

5.10... Do Lakes Count as Sea Level?
(No...)

But just asking the question allows me to write a small chapter on lakes, and the knee-jerk tendencies of the media to blame everything on climate change...

And we KNOW when the media say 'climate change', they're meaning the man-made, anthropogenic version, and pointing the finger at YOU, because YOU'RE guilty.

In 2013, the prodigious National Geographic quotes...

"Climate Change and Variability Drive Low Water Levels on the Great Lakes."

Oh dear... the 'Great' lakes are in danger, and of course, Climate Change is to blame. And we all know the reference is not talking about 'natural' climate change, they're implying the man-made stuff.

And the National Resources Defense Council was not far behind to back up their colleagues...

"Climate change is lowering Great Lakes water levels.
It's no secret that, partially due to climate change, the water levels in the Great Lakes are getting very low."

Now, I don't know what your take on the adjective, 'very', is, but to me, 'very' means a whole lot.

And it gets worse...

On 25th June, 2013, President Barack Obama gave a speech on Climate Change at Georgetown University, Washington DC...

"The 12 warmest years in recorded history have all come in the last 15 years. Last year, temperatures in some areas of the ocean reached record highs, and ice in the Arctic shrank to its smallest size on record -- faster than most models had predicted it would. These are facts."

Yes, they are facts... and he nailed them. He continues...

"The fact that sea level in New York, in New York Harbor, are now a foot higher than a century ago -- that didn't cause Hurricane Sandy, but it certainly contributed to the destruction that left large parts of our mightiest city dark and underwater."

Yup, that foot higher water sure did add to it.

The same day, the second-ranking Senate Democrat, Dick Durbin of Illinois, said this...

"To ignore it is to ignore reality. Lake Michigan, when measured just a few months ago, was at its lowest depth in any measured time in recent history. What we are seeing in global warming is the evaporation of our Great Lakes."

On Minnesota Public Radio, they echoed the theme...

"Scientists at the Oceanic and Atmospheric Administration [are] studying the interplay between low water levels, shrinking ice cover and warm water temperatures, Gronewold said. They have already concluded that climate change is playing a role in determining Great Lakes water levels."

So it's official... climate change is playing a role.

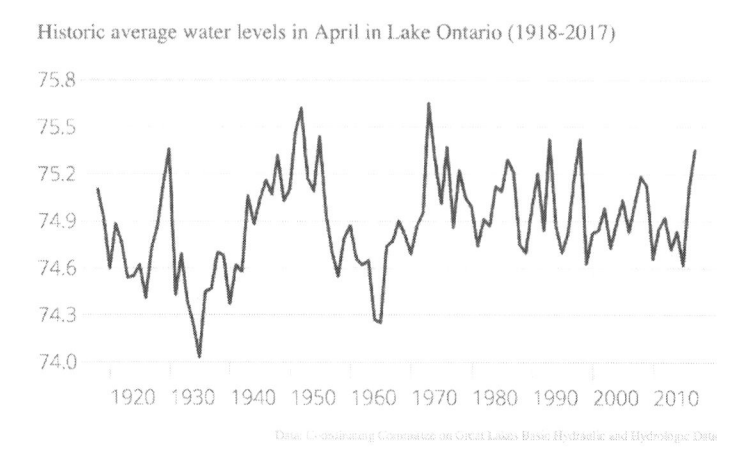

Historic average water levels in April in Lake Ontario (1918-2017)

So, tell me... when did our man-made climate change of Lake Ontario actually begin? From the graph above, you'd be hard-pressed to make a presentable case.

Now National Geographic, National Resources Defense Council, and Senator Durbin were correct… water levels in Lake Michigan were lower in January 2013 than at any time since records began in 1918… they were SIX feet lower than their peak from 1986.

The thing I object to is the constant mantra… "Climate Change is to blame… man-made Climate Change."

Now, let's fast-forward just FOUR YEARS.

It's 2017… let's see what the news was saying.

On 22nd May, 2019, the Wall Street Journal headlined…

"High Water Levels on Great Lakes Flood Towns, Shrink Beaches.
Lakes Erie, Superior are among Great Lakes expected to reach all-time highs this summer."

Okay… we have a wee bit of flooding due to higher precipitation levels.

But the constant mantra of "Climate Change" when we get a small shift in weather patterns is getting me down. Ask any expert, they'll admit that the climate works in cyclical patterns.

Glory be… it got wetter in the 1990s: climate change!
Then it got drier roughly from 2000 to 2013… and the media cry; climate change!
Then it got wetter again and it continues to be that way… yet still more climate change!

Here's a graph of water levels in the Great Lakes… and the base level is the 1918-2018 level… a HUNDRED YEARS of base level!
And a hundred years of cyclical levels due to NATURAL STUFF!

Lake Michigan and Lake Huron

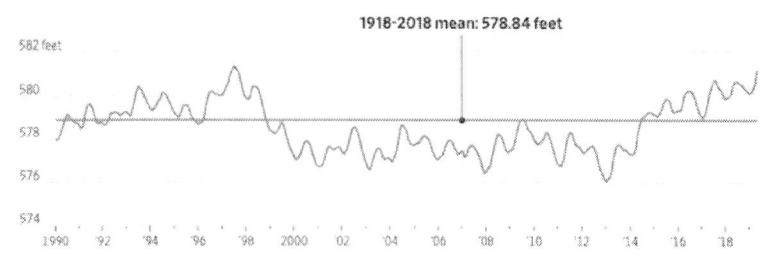

1918-2018 mean: 578.84 feet

It's about time the media started to look at these and writing the truth instead of scare-mongery.

What if the headlines that once said this...

"Climate change is lowering Great Lakes water levels."

"It's no secret that, partially due to climate change, the water levels in the Great Lakes are getting very low."

Said this instead...

"Great Lakes water levels are low, but heck they've rose and fallen for a hundred years... it's a cyclical thing."

"It's no secret that, partially due to the lakes' cycles, the water levels in the Great Lakes are getting very low at this time. Just wait a decade; we'll be complaining about a flood!"

Whatever the levels of the Great Lakes... the science is never settled.

5.11... THE GREAT PACIFIC OCEAN GARBAGE PATCH
(So Thick... You can walk over it...)

The Media Story
We've all heard of the huge patch of floating plastic in the Pacific Ocean, right? Maybe even more than one patch!
It's got many names, but we all know the facts, right?
A layer of plastic bottles, bags, straws, and Walmart sacks... It's so thick you can walk over it.
We've seen men in canoes trying in vain to paddle through it, clogged with plastic bottles.
It's so big... men have invented machines to scoop it all up, make thousands of bucks, cleaning the planet.
Oh, and it's so big, you can see it from space.

But you see... it's ALL a myth.

Yes, and I can hear even the staunchest climate denier shout at me here.

I mean, there it is in all its glory.

IT'S ALL A MEDIA MYTH.

The Myth
It's a media myth to make us feel guilty and donate to eco-fraudsters.
And who better to debunk it but NASA/NOAA themselves. (Links at end of chapter)
From NOAA's own 'Office of Response and Restoration', I quote...

"...some reports about the Great Pacific Garbage Patch would lead you to believe that this marine mass of plastic is bigger than Texas..."

But you see, it's just a media myth, and I repeat myself on purpose.

"For NOAA, a national science agency, separating science from science fiction about the Pacific garbage patch (and other garbage patches) is important when answering questions about what it is, and how we should deal with the problem."

"Garbage patches aren't a solid patch. The name conjures images of a floating landfill in the middle of the ocean, with miles of bobbing plastic bottles and rogue yogurt cups."

The media are lying, and here comes the truth...

"...much of the debris found in these areas are small bits of plastic, or microplastics, smaller than 5mm in size that are suspended throughout the water column.
The debris is more like flecks of pepper floating throughout a bowl of soup, rather than a skim of fat that accumulates or sits on the surface.
Microplastics are nearly ubiquitous today in the marine environment and may come from larger pieces of plastic that have broken down over time, from fleece jackets or plastic microbeads added to face scrubs."

And we're not done.

"Although garbage patches have higher amounts of marine debris, they're not "islands of trash" and you definitely can't walk on them.
The debris in the garbage patches is constantly mixing and moving due to winds and ocean currents.
This means that the debris is not settled in a layer at the surface of the water, but can be found from the surface, throughout the water column, and all the way to the bottom of the ocean."

And now we present the truth... the busted myth coming your way in the form of a LYING MEDIA.
A media so full of agendas, you'd think they were politically driven.

And this is the SAME MEDIA we see pushing climate change in our faces, building a wall of myths so high, we can't see over it.

Now, NOAA fully admits there is an area called the North Pacific Gyre, full of plastic particles... but...
The concentration is about FOUR sand-sized particles in 260 GALLONS of water.
And we've ALL swam in worse than that at the beach!

The Scam
The truth is, the photos that you see of the Pacific Garbage Patch, blamed on the poor old USA, is actually an ecologic scam.
You see, there are many companies out there trying to get you, the public, to feel so ashamed of your country, that you'll send your hard-earned cash for their clean-up projects.

It's ECO-FRAUD in a huge way.

I remember watching a video showing a young man inventing a machine to scoop it all up, making millions in the process; he got gifted millions of dollars to start building his new machine.
There's a cute pair of guys on TV right now, asking for donations because you'll get a plastic wristband made of recycled/captured plastic for your 'donation' to the cause of non-littered beaches.

BUT IT'S ALL A LIE.
Every single scheme aimed at scooping the GARBAGE PATCH is a lie!

Because, as NASA/NOAA have stated... if the whole 'plastic garbage patch' consists of fine particles, NOT all floating on the surface, then MOST of the 'projects' that we unwittingly donate our money to simply wouldn't work.

And why don't the media call the scam artists out?
They're quick to disclose 'facts' about the Red Cross, The Samaritans, and such, citing the director's huge profits.
But they won't debunk this because at the end of it all, every penny donated to the cause is 'worthy' in some way.

Look, I'm not a bad person.
I'm NOT saying there isn't trash in the oceans.
I'm not saying we shouldn't clean it up.
But I am against media fraud.
Oh, and the pictures used by our darlings in the media?
They're from around the area of Manilla Bay, because the Philippines dump their trash DIRECTLY into the Pacific Ocean.
The USA has banned this type of dumping for many decades.

The USA, the world's third most populated country, who manufacture ONE QUARTER of everything made on the planet, contribute just 2% of the garbage in the Pacific Ocean.

In a Forbes article from 2018, they stated...

"China, Indonesia, Philippines, Thailand, and Vietnam are dumping more plastic into oceans than the rest of the world combined."

So, you see, the media's lying to you intentionally.
Just don't believe everything you see/read.

Because as you know by now... the science isn't settled.
Some of it is even lies.

https://response.restoration.noaa.gov/about/media/how-big-great-pacific-garbage-patch-science-vs-myth.html

https://response.restoration.noaa.gov/about/media/debunking-myths-about-garbage-patches.html?fbclid=IwAR3JGUPWgA3KJHomt9sbxutVnXKx7BcT0zDgjVaHb-K_7121thTSXzuvYj8

SECTION 6
POLES APART... AND GREENLAND!

"Nothing is more obstinate than a fashionable consensus."
(Margaret Thatcher Ex-Prime Minister of the United Kingdom)

6.1... GREENLAND'S MELTING! 20 FEET OF SEA RISE!
(Yup, Greenland's melting, but just the edges...)

20 feet of sea level rise
If Greenland melts, there will be 20 feet of sea rise.
It's a horror story bandied about by every alarmist that's ever raised his voice in debate... but... amazingly; this catastrophic story is absolutely true.
I checked it myself.
I took the average height of ice on Greenland (the world's biggest non-continental island) and fed it into the ocean volume... even I got a sea level rise of close to 20 feet.
But...
It's here that you must rethink every geographical 'fact' you ever thought you knew about Greenland.

Greenland facts
Greenland is not mountainous... just the edges are.
AND... it's not really an island... it's a ring of islands.
(And here's where it gets weird...)
Greenland's huge ice sheet that covers 81% of the country has pushed the center of the 'island' down so far... most of the 'island' is actually lower than sea level... the lowest part being almost 1000 feet below sea level.
Greenland is actually a basin, holding together the world's largest ice cube!
The 'cube' holds about 10% of the world's solid water (ice), and even if Global Warming ran out of control and the global temperature rose by FIVE DEGREES, it would take many, many decades to melt the ice mass... possibly even hundreds of years.

Greenland's melting?
If you hold the world's biggest ice cube, and the global temperature rises a bit, you'd expect it to melt... and it has.
But much of Greenland's melting is a seasonal process, the weight of the huge ice mass presses down, forcing water to escape.

Official records show that between 1994 and 2005 (at the peak of global warming) the island actually gained snow/ice covering at a rate of 3 inches per year.
(Incidentally, Greenland has often been used as the 'poster child' of the chant... the world was warmer a thousand years ago', and indeed it seems there's more evidence of this fact. Apart from the Vikings settling there (AND FARMING!) from 1100 to 1300, recent ice core samples from Greenland have revealed DNA of trees, plants, spiders and insects INCLUDING BUTTERFLIES, from a depth of 1.2 MILES under the ice.)

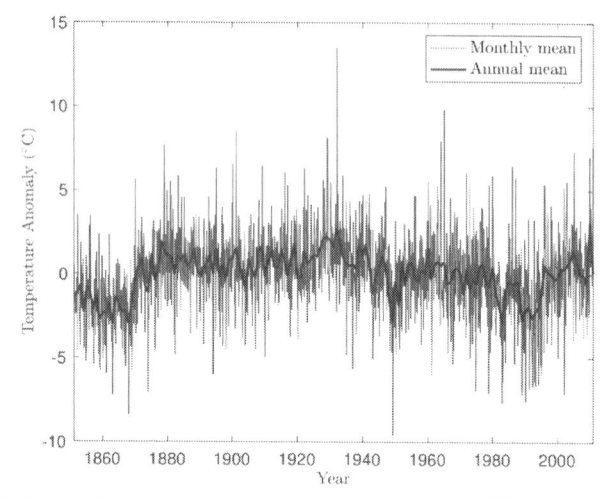

Figure 2. Reanalysis data showing monthly mean surface temperature anomaly (blue curve) over the area 68°N – 80°N, 25°W – 60°W covering a large part of Greenland. The red curve is the annual mean surface temperature anomaly; this has observed variance $\sigma_{T,da} = 1.55$ °C².

The above is a graph of the average surface temperature of Greenland from 1850 to 2010.
The initial temperature is -1°C below the zero line, and the end point is just above zero.
The graph itself gives the temperature rise as 1.55°C.

THAT'S 1.55°C RISE IN 160 YEARS.

That kinda pours cold water over the idea that the Arctic temperature is rising at double the rate of the rest of the world.

Anecdotal evidence
Let me tell you a story...
In 1942, in the middle of the Second World War, six P-38 planes and two B-17 bombers took off from the USA to participate in the European theatre of conflict.

The planes, on a route planned across Newfoundland, Greenland, Iceland and Scotland, ran into weather difficulties after taking off from the west of Greenland. The pilots and crew had to force landings on Greenland's high ice plateau, abandon their planes, but all personnel made it back safely.

50 years later, in 1992, an expedition was sent north onto the ice shelf to try and find the planes. After various different expeditions and using ground penetrating radar, the planes were eventually located, and a shaft was dug to reach them.

It turns out that over the intervening 50 years, 268 feet of ice had accumulated on top of the planes. One P-38 (now named glacier girl, and pictured above) has been rescued and restored and now flies in air shows all over the world.

Let's recap...
268 feet of ice accumulated in just 50 years. That's Greenland's ice shelf GAINING 5 feet of ice every year.

THAT HUGE ICE SHELF IS GETTING THICKER!

There's nothing like boots on the ground measuring to drive a point home.

6.2... ANTARCTICA'S MELTING! WE'RE ALL GOING TO DROWN!
(Oh shut up and have another latté ...)

There must be some sort of knee jerk reaction to every icy part of the world... as far as the Alarmists are concerned, ALL ice and snow is melting... it's ALL catastrophic, and it's All our fault.

Well, Antarctica is the planet's biggest lump of ice.
In fact it is estimated that between 85-90% of our planet's ice lies on our southern polar continent, and of course, because of this, most of the planet's fresh water.
At its edges vast chunks break off every year, and at its core (the actual polar region) ice accumulates at a level that has been going on for 10,000 years.

In 1956, a group of USA Seabees constructed a hut camp at the South Pole, which was christened the Amundsen-Scott Station in honor of the first explorers to reach the bottom of the Earth.
By 1960, just four years later, the wooden huts were already 6 feet deep in snow.

A second camp was built next to the first (the dome on the left of the pic above), but again, much of this new building was soon under snow.

In 2003, an elevated station was built, on extendable stilts.
Yup, the Antarctic is gaining ice mass by about eight inches per year, and the accumulation shows no signs in stopping.

And yet this does not stop the media making ridiculous claims about the serious plight of our southern continent.
In Jan 2019, a story appeared in the Washington Post...

"Ice loss from Antarctica has sextupled since the 1970s, new research finds (An alarming study shows massive East Antarctic ice sheet already is a significant contributor to sea-level rise)"

Not only does the headline herald the fantastic news that Antarctica is doomed, but labels the poor continent as the final spike in sea-level rise, which as all of you 'know' will rise by 20 feet by next Tuesday and overrun most of Florida before tea-time.

The Washington Post story then goes on to make various spurious claims, throws in a few 'catastrophic' terms, (alarming, accelerating, tells of a six fold rise, and yet mentions few facts to prove or substantiate the theory behind the story. It mentions studies done in the 1970's (when such studies had not yet taken place) then throws in some blood-curdling numbers... Antarctica is losing 'gigatons' of ice each year.

"The findings are the latest sign that the world could face catastrophic consequences if climate change continues unabated"

Again with the *'catastrophic'* hyperbole!
The study, by Eric Rignot, an Earth-systems scientist for the University of California at Irvine and NASA, states that Antarctica *"lost 40 billion tons of melting ice to the ocean each year from 1979 to 1989"...* and that is an alarming figure.
But please remember, that despite NASA's homogenized new temperatures of the 1970's, the planet was in a cooling period. The planet has warmed since 1979; this we all know. Yet it does not deter the study from their 'catastrophic conclusion...

"That figure rose to 252 billion tons lost per year beginning in 2009, according to a study published Monday in the Proceedings of the National Academy of Sciences."

Yeah, when the planet is warmer, the ice melts more. Kindergarten kids could tell you that.

But despite the article's menacing GIGATON usage, it turns out to be hardly an atom-splitting scientific advance. The 252 gigatons of ice would not raise the sea-level even ONE MILLIMETER!
(The accurate figure is 0.7mm… that's just THREE HUNDREDTHS OF AN INCH!)

But of course, this does not stop the Washington Post from running off at the mouth…

"In addition to more-frequent droughts, heat waves, severe storms and other extreme weather that could come with a continually warming Earth, scientists already have predicted that seas could rise nearly three feet globally by 2100 if the world does not sharply decrease its carbon output"

It's been scientifically proven that neither droughts, heat waves, nor 'severe' storms have statistically globally increased in the last 100 years.
But why let facts get in the way of a good story.
The Post carries on its alarming message, because he's got talking points to mention… he's got all the 'alarmist' nails to tap firmly with his hammer…

"That kind of sea-level rise would result in the inundation of island communities around the globe, devastating wildlife habitats and threatening drinking-water supplies. Global sea levels have already risen seven to eight inches since 1900."

Yup… the world's sea levels have risen seven to eight inches IN OVER A CENTURY!
That's the only fact in this whole piece!

And I repeat… The 252 gigatons of ice would not raise the sea-level even ONE MILLIMETER!
(Please remember the actual figure is 0.7mm… that's just THREE HUNDREDTHS OF AN INCH!)
Yeah, the rise of the world's oceans by the thickness of a thin piece of card can do all that devastation in seconds.
Man, they really pulled the stops out on this story, didn't they!

And yet this is what the poor unsuspecting, uneducated public has to deal with every day. The only talking points they missed were… *"women, children and the poor will be worst affected".*

An unrelenting onslaught of lies, drivel, and alarmist garbage disguised as fact, filled out with their left-wing mantra... *"stop bad CO2, save the planet".*

Nowhere in this story does the author actually state the sea level rise figures... just how many BILLION TONS of ice plunged into our already rising oceans.
Nowhere does this story mention the cyclical effect that the seasons bring on poor Antarctica, melting in summer, etc.
The story does NOT mention the RISING ice levels on the rest of the continent (and this water vapor has to come from somewhere to end up as snow falling on the South Pole... PROBABLY from those rapidly-rising oceans).

One millimeter of sea level rise is no thicker than this line ----------

It's hardly going to 'devastate' anything or inundate island communities, or threaten drinking-water supplies, is it?

But it isn't just the Washington Post.

6.3... ANTARCTICA'S FREEZING! WE'RE ALL GOING TO FREEZE?
(Let's hear it from the lion's den, shall we...)

I'm going to begin this chapter with some facts that will hopefully give you some idea of how big Antarctica actually is.

Antarctica has an area of 5.4 MILLION square miles, almost one and a half times the size of the USA. (USA has an area of 4 MILLION square miles.)
Antarctica is nearly one and a half times larger than Australia.
In some places in Antarctica the ice is THREE miles high.
The population of Antarctica is 1,106.
The first person to use the name, Antarctica, was Scottish cartographer John George Bartholomew in the 1890's.

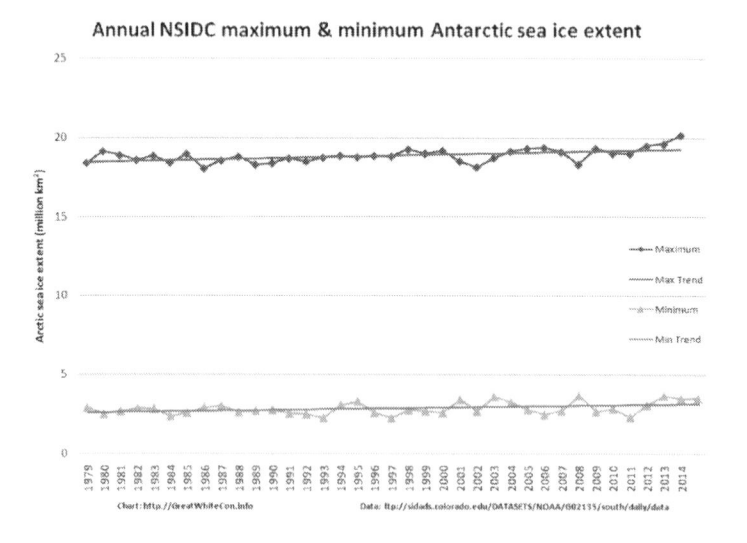

Annual NSIDC maximum & minimum Antarctic sea ice extent

The above graph from the National Snow & Ice Data Center) NSIDC, shows both the max and mins of Antarctic ice.
The maximum is rising.
The minimum is rising.
YUP... from 1979 to 2015 Antarctica is GAINING ICE.

NASA brings good news! (Part 1)
According to a 2015 NASA story, citing a NASA study, featured on the NASA website, Antarctica might be helping the world...

"NASA Study: Mass Gains of Antarctic Ice Sheet Greater than Losses"

Woah! In the last chapter (6.2) the Washington Post in 2019 told the opposite story, citing examples from surveys in 2009. So the Washington Post is right. Right?
Wrong.
The NASA story, written by Jay Zwally, a NASA glaciologist at the Goddard Space Flight Center in Greenbelt, Maryland, was also published in the publication, the Journal of Glaciology, Oct 30th, 2015, and goes into great detail...

"A new NASA study says that an increase in Antarctic snow accumulation that began 10,000 years ago is currently adding enough ice to the continent to outweigh the increased losses from its thinning glaciers."

That's good news... Antarctica is fighting to keep the sea level from rising, and if MORE ice is being added to the continent than is falling into the sea, then surely Antarctica is making sea levels DROP!

"The research challenges the conclusions of other studies, including the Intergovernmental Panel on Climate Change's (IPCC) 2013 report, which says that Antarctica is overall losing land ice."

Holy Crap, Batman!
The hallowed NASA is actually saying that their figures CONTRADICT the 2013 IPCC report.
(Well, I know, it doesn't actually say 'contradict', it says 'challenges', but hey, it's surely the same thing. Right?)

In their 2013 report, the IPCC asserted (with no way of measuring accurately) that Antarctica contributed 0.27mm per year to the sea level rise.

"According to the new analysis of satellite data, the Antarctic ice sheet showed a net gain of 112 billion tons of ice a year from 1992 to 2001. That net gain slowed to 82 billion tons of ice per year between 2003 and 2008."

So, dear reader, this story totally debunks the last chapter, and provides a long-lasting debunk of ALL stories about Antarctica melting... right?
Jay Zwally, NASA glaciologist, and my newest best friend, continues...

"The extra snowfall that began 10,000 years ago has been slowly accumulating on the ice sheet and compacting into solid ice over millennia, thickening the ice in East Antarctica and the interior of West Antarctica by

an average of 0.7 inches (1.7 centimeters) per year. This small thickening, sustained over thousands of years and spread over the vast expanse of these sectors of Antarctica, corresponds to a very large gain of ice – enough to outweigh the losses from fast-flowing glaciers in other parts of the continent and reduce global sea level rise."

Yup, it sure looks like in this case the science is settled. (Man, I loved writing that one.)
But of course, the story cannot just give us good news and leave it at that. It concludes in typical Alarmist style, leaving a slightly bad taste in your mouth.

""The good news is that Antarctica is not currently contributing to sea level rise, but is taking 0.23 millimeters per year away," Zwally said. "But this is also bad news. If the 0.27 millimeters per year of sea level rise attributed to Antarctica in the IPCC report is not really coming from Antarctica, there must be some other contribution to sea level rise that is not accounted for.""

And I'm quite certain they'll find something that mankind has done to make up for the 'missing' amount.

https://www.nasa.gov/feature/goddard/nasa-study-mass-gains-of-antarctic-ice-sheet-greater-than-losses

Washington Post cites same study!
Now... the story should have stopped there...
But on November 5th, 2015, just six days after NASA published the original story, the study was reported in the Washington Post.
It was headlined...

"A controversial NASA study says Antarctica is gaining ice. Here's why you should be skeptical"

WAIT! You take everything alarming NASA says as verbatim, you churn it out as rote, and you keep us in a mood of panic... but when we *have* 'good news' for the planet, you call it controversial, and tell us to be skeptical?
Yup. The study debunks many other previous studies which supposedly proved that Antarctica was losing ice, so let's look at some of the Washington Post's verbiage, shall we?

"the surprising claim..."

Yeah, it's 'surprising to you, because you've published nothing but Alarmist crap for thirty years!

"It contradicts numerous prior scientific claims, including a 2012 study in Science by a small army of polar scientists..."

Oh dear, this has turned into a kind of pissing contest... one poor NASA scientist is suddenly up against the WHOLE ARMY of scientists brought in by the totally neutral Washington Post.

"also contradicts assertions by the leading consensus body of climate science, the U.N.'s Intergovernmental Panel on Climate Change... a number of researchers have been quoted expressing skepticism..."

The Washington Post story mentions that the readings were taken by NASA's ICEsat (Ice, Cloud and land Elevation Satellite), then goes on to state that the readings dispute the figures taken by the GRACE satellite, and spends a few paragraphs praising GRACE, and denigrating ICEsat.

The story, then of course doubles down on the alarmist rhetoric, citing a 'new study' which predicts a HUGE Antarctic ice shelf breaking off, and giving us a TEN FOOT rise in sea level!

Yup... there's no news like fake news.

https://www.washingtonpost.com/news/energy-environment/wp/2015/11/05/a-controversial-nasa-study-says-antarctica-is-gaining-ice-heres-why-you-should-stay-skeptical/?utm_term=.b731515ed461

NASA brings good news! (Part 2)
With the benefit of hindsight, let's bring this story more up to date, shall we?
In the three years that we travel from NASA story in 2015 to the NASA story 2 in 2018, the Washington Post's 'new study' never came to pass, and the sea level didn't rise ten feet.

In December 2018, a new study was featured on the NASA website...
(Now remember, this was a NASA story, citing a NASA study, featured on the NASA website)

Antarctica's contribution to sea level rise was mitigated by snowfall.

Wonderful. I'm glad we worked that out.
And that's surely a good thing.
Well it turns out that one NASA survey has trumped another...

A new NASA-led study has determined that an increase in snowfall accumulation over Antarctica during the 20th century mitigated sea level rise by 0.4 inches. However, Antarctica's additional ice mass gained from snowfall makes up for just about a third of its current ice loss

So in all, what do we really know?
Nothing.
Science is changing every day.
New satellites produce new data, and our view of the world changes as a result.
One thing is sure, just using these two surveys as an example...

The science is not settled.

6.4... ANTARCTICA'S MELTING! NOT GLOBAL WARMING?
(You can't make this stuff up...)

For nearly forty years we've been told that Antarctica's glaciers are melting, and it's caused by Global Warming, and the melt is going to undermine the ice sheet, and soon a chunk of ice the size of Australia is going to slide from the frozen continent and give us a half mile high global sea rise before breakfast...
(Takes breath)
So imagine my incredulity when I noticed this headline in the UK's newspaper, The Independent, 29th June 2018...

"Active volcano discovered beneath Antarctic ice sheet could be contributing to rapidly melting glacier"

What? The culprit isn't global warming? Carbon dioxide's NOT to blame? Let's read some more.

"An active volcano has been unexpectedly discovered beneath a rapidly vanishing glacier in Antarctica by a team of scientists studying the melting ice shelf."

This is literally ground-shaking news. We know so little of the southern ice mass, that it's not entirely surprising that we find new stuff. But, of course, with the headline already enticing us to smash the connection between CO2 and the melting Antarctic glacier, they soon revert back to form...

*"While **rising temperatures** in the polar oceans are still **by far the biggest contributor** to Antarctic **melting**, volcanoes buried deep beneath the ice are a "wild card" that are likely also playing a role.*
*Scientists have warned this hidden source of heat must be taken into consideration when predicting **future sea level rise**."*
(Emphasis added)

Yup... no matter how many studies are done by NASA, showing that Antarctica is NOT MELTING, the collusionist Alarmist media trot out the old Antarctica melting talking points, making sure that this new source of heating is minimized as much as possible.
Because let's face it, these volcanoes didn't appear overnight did they?
And they were around when the ice formed over the top of them, right?

"Dr Robert Bingham, a glaciologist at the University of Edinburgh who has discovered dozens of Antarctic volcanoes, said the melting was probably not a major contributor.
"I don't think people should look at volcanoes underneath west Antarctica as a serious concern for causing instability," he said."

Yup, it takes a dour Scot like me to make us face facts... don't panic, move along, nothing to see here.

But it does bring to mind the idea that Antarctica is thought to have over 100 active volcanoes, and there are many hundreds around the Pacific Ocean, all churning up the sea bed and warming the oceans.
How much study has been done there, I wonder.

I mean, in this world of unsettled science.

https://www.independent.co.uk/environment/volcano-antarctic-ice-melting-pine-island-glacier-sea-level-climate-change-global-warming-a8423131.html

6.5... ARCTIC SEA ICE FIGURES
(The real figures...)

General Ice Sheet Information
The Arctic Ice Sheet (also called the Arctic Ice Pack) is the name given to the 'pancake' of frozen ice that floats around the North Pole on the surface of the Arctic Ocean.
It has a seasonal cycle, decreasing to a minimum area around September, and a maximum area around February/March.
The ice is a thin veneer on the surface of the ocean, and varies between 65 feet thick on the higher older ice ridges to just a few feet thick in others. The average thickness is around 13 feet (4 meters).

The Arctic Ice Pack is important to the Earth's climate as its whiteness reflects a lot of the Sun's light back into space, not allowing the solar radiation to heat the planet. This is called its albedo effect.
If the ice sheet were to disappear altogether, the albedo of the resultant sea water would not reflect nearly as much of the sunlight, and the water would heat up in the process, thus adding to whatever other heating effects we already have.

The Ownership
Five countries have land masses inside the Arctic Circle, and thus 'own' parts of the Ice Sheet; USA, Canada, Russia, Greenland (Denmark), and Norway.
They share responsibilities of conservation, of polar bear monitoring, and data of the ice sheet... and of course, they look forward to the day the ice melts totally, because then they can drill for oil in the Arctic Ocean.

Records of Sea Ice
The earliest records of Arctic Sea ice were measured by the UK's MET Office (now the Hadley Centre for Climate Prediction and Research) and some data goes back to the turn of the 20th century.
However, despite the advent of aircraft, it was not until the first satellites were launched that we got a full view of the ice sheet.
There are two main pieces of data used by climatologists to assess the state of the Arctic Ice Sheet; the winter maximum and the summer minimum.
And here's where we find a few alarming facts.
This is NOT the first time the Arctic Sea Ice has decreased...
In 1922, newspapers reported that the Arctic Ocean was warming, and the seals were finding the water 'too hot'.

Yeah, even back in 1922 there were people trying every trick in the book to sell newspapers.

Here's the graph that tells the story...

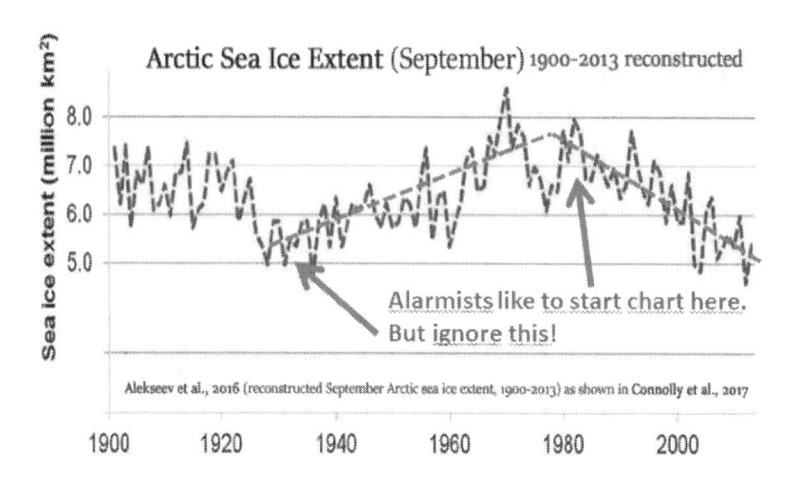

Arctic Sea Ice Extent (September) 1900-2013 reconstructed

Alarmists like to start chart here.
But ignore this!

Alekseev et al., 2016 (reconstructed September Arctic sea ice extent, 1900-2013) as shown in Connolly et al., 2017

The Northwest Passage

The Northwest Passage is a trade route across the north of Alaska and Canada that allow a really short journey from the North Pacific Ocean to the North Atlantic Ocean.

The possibility of Europe having a quicker east-west route to the Orient prompted Columbus's 1492 discovery of America. The continent's discovery was a huge breakthrough in exploration, and many fleets were sent west by the greedy European nations, mostly taking a southern route into the Caribbean Sea, hoping for a quick route to the Orient.

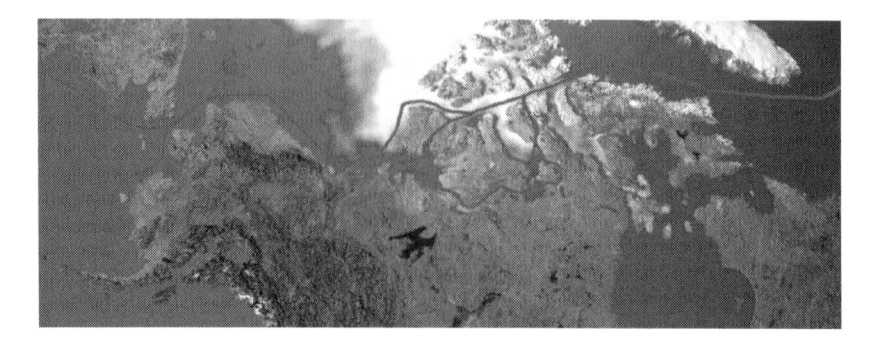

The first ever recorded attempt to discover the possibility of a Northwest Passage was John Cabot. He was sent by Henry VII (the father of Henry

VIII) in 1497. Subsequent English expeditions discovered Canada, Newfoundland, Nova Scotia, Baffin Island and Greenland.

Although the route was discovered by the British (Robert McLure) in 1854, the first Northwest Passage was sailed by Norwegian explorer, Roald Amundsen. It took him three years to make the journey, from 1903 to 1906.
(Amundsen then went on to be the first man to reach the South Pole, in December 1911)

So, although there is a threat of further warming if the Arctic Ice Field melts, it would make the Northwest Passage into a popular commercial sea route.

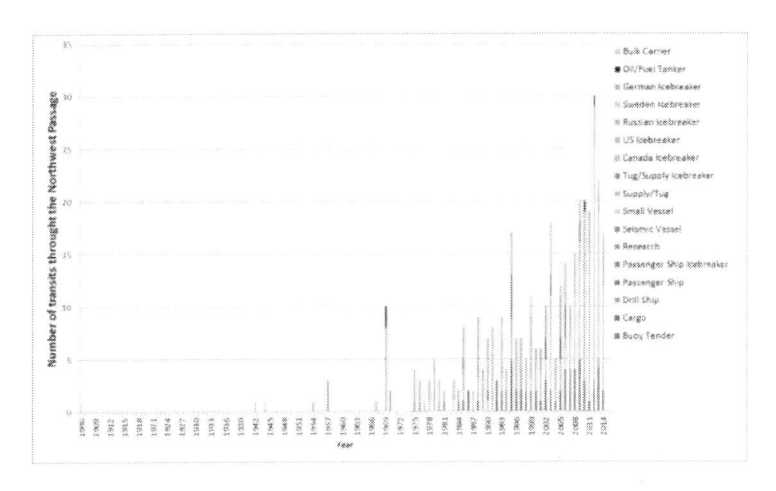

This, however, gives us another way to measure sea ice in the north.
Every year, some ships make the passage, and their journeys are recorded (see previous graph). This number of ships gives us an indication of the level of ice (or the lack of it) at the far Canadian north.

In 2014, when the Arctic ice was at its record low, only 9 ships sailed the northwest passage successfully; 2 from west to east, 7 from east to west. That's hardly a major sea route huh?

Let's take a look at the Arctic Sea Ice, shall we...

We'll begin with some 'facts'...

Sea Ice Measurement.
Since we have five countries measuring these parameters, we'll all be working from the same figures... right?
Very wrong.

(BTW... In 2008, an article in New Scientist predicted the Arctic could be ice-free that summer. (Yup, scientists are good at predictions))

Let's see some summer graphs, shall we?

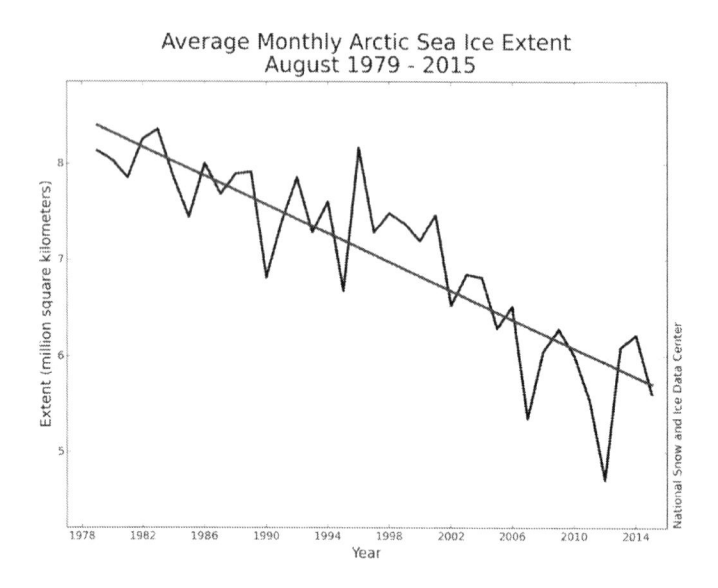

Average Monthly Arctic Sea Ice Extent
August 1979 - 2015

The graph above is the latest area graph I could find of August sea ice, from the National Snow & Ice Data Center (NSIDC), dated 2015. (USA figures)
From 1978 to 2014, the ice sheet (on the average blue line figures) declined from 8.4 million sq km to about 5.8 million sq km. That's 31%... a HUGE drop. Even I'm not going to sugar-coat that.

Arctic Sea Ice Volume
Data Source: Danish Meteorological Institute (DMI)

Aug 23, 2003 - 2018
Edited by @KiryeNet

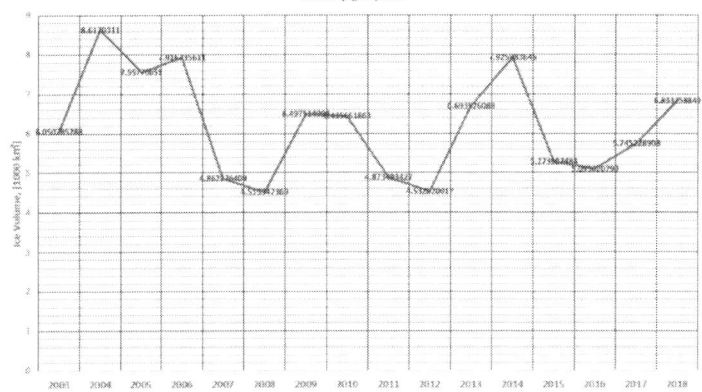

However, here's a graph by the DMI (Danish Meteorological Institute) showing a pretty stable volume from 2003 to 2018. In fact it actually shows a growth in the summer sea ice from 2003 to 2018!

So...my first reaction would be... "Where's the panic?"

But let's look at the full set of Danish results from 1979.
The Danish 1979 figure is 9.15 million sq km.
The Danish 2018 figure is 6.54 million sq km.
The decrease in Arctic Sea Ice by Danish figures is 28%.

That's pretty close to the USA figure of 31%.

And they're NOT from the same data.
It's verified RAW data, like good science should be... those are figures I can put some trust in.

Winter Data
Let's look at the NSIDC graph of WINTER ice, below (from 2014)...

From 1979's 16.2 (blue line average) million sq km, to 2014's 14.8 million sq km, gives us a decrease of 8.6%.
Wait... the winter ice has decreased JUST 8.6% in 36 years?

Can I ask a really stupid question... is there anything else that has decreased by 8.6% in 36 years that has caused so much media furor?

Average Monthly Arctic Sea Ice Extent
March 1979 - 2014

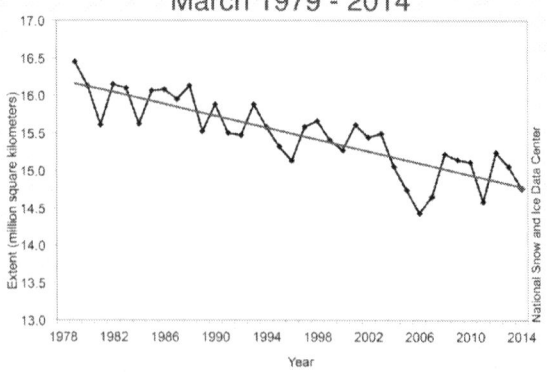

Now I don't have a pre-made graph of the DMI winter (March) ice, but I do have some figures from their awesome website...

1979... 15.64 million sq km. (that's lower than the NSIDC)
2001... 14.76 million sq km. (again, lower)
2018... 14.05 million sq km.
That gives us a decrease of winter ice of 10.2%.

USA winter figures are a 8.6% drop.
Danish winter figures are a 10% drop.

USA summer figures are a 31% drop.
Danish summer figures are a 28% drop

Trust me, those two sets of figures are pretty tight with each other...
Like I said before, verified data from two independent sources is something I can put trust in.

The Danish DMI website tells us...
"The ice extent values are calculated from the latest reprocessed ice concentration data set (OSI-409), from The Ocean and Sea Ice, Satellite Application Facility (OSISAF). Ice extent corresponds to the area where ice concentration is greater than 15%."

Okay, so the Danish dataset comes from a European Satellite funded by 30 European countries. The USA is not included.

"Possible differences between this sea ice extent estimate and other estimates are most likely caused by differences in the applied algorithms and in the definition of ice extent."

So there it is... the algorithms between the Americans and the Danes/Europeans may also be different.
The Europeans have three geosynchronous satellites measuring Arctic Ice, the USA/NOAA have their own monitoring satellites (the POES system).

Can anyone remember during the Monica Lewinski debacle in the USA, President Clinton's statement... "It depends upon what the meaning of the word 'is' is."
Well the difference between the Danish figures and the USA ones may depend on the definition of what constitutes 'ice'.
ALL the data and figures are available on the Danish website here...

http://ocean.dmi.dk/arctic/sie_monthmean.uk.php

It's time to take a closer look at the sea ice figures, because... yup, you got there before I did... the science is NOT settled.

6.6... THE IMPORTANCE OF BEING EARNEST?
(Or where you start your graph...)

Coincidentally, NASA, NOAA, and NSIDC all state that satellite figures all begin in 1979.
But NOAA has satellite data going back to 1973. Why don't they start their graphs there?

NASA's declaration is here...
https://www.ncdc.noaa.gov/sotc/summary-info/global/201207

At the section labelled...

*"Polar ice highlights: July
... Although the melt area is the largest since the beginning of the satellite era in 1979"*

Despite NOAA's above 'admission', the IPCC First Assessment Report (FAR) back in 1990 shows this NASA graph...

IPCC (1990) *Observed Climate Variation and Change* 7

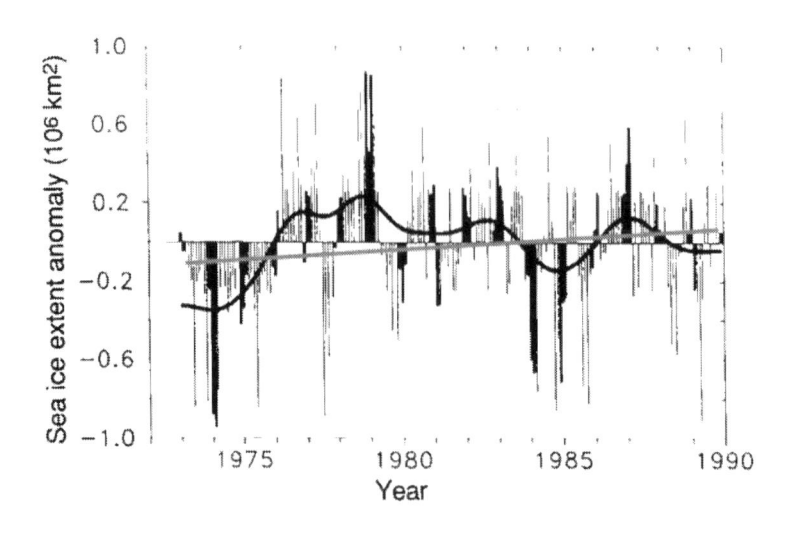

As we see, the first data point in 1973 was much lower than the data point in which the current NSIDC graphs are projected from.

Now if I wanted to be cynical here, I'd point out that if NASA wanted to hype the decline figures, they'd pull that old graph trick, and choose a high starting point to make the decline look worse, like 1979, for instance...

Because, if we take our 'beginning of the satellite era' at 1979, instead of 1973 (when it actually began) it makes any ice loss appear to be far greater in size.
(The difference in starting points is 0.5 million sq kilometers)

QUESTION: Why do the US's weather departments use false statements, and resort to quasi-science?

Remember the NSIDC summer data set way above showed a 31% loss in ice?
This new start point (1973) would change that to just 27%. That's still a substantial loss, but not as bad as current NSIDC figures.
AND that 27% figure is surprisingly close to the Danish figures of 28%.

The new start point in 1973 changes the USA winter figures from 8.6% drop to just 5.7%.

We're looking at just a 5.7% winter Arctic Sea Ice drop in 46 years.
That's literally a drop in the ocean...

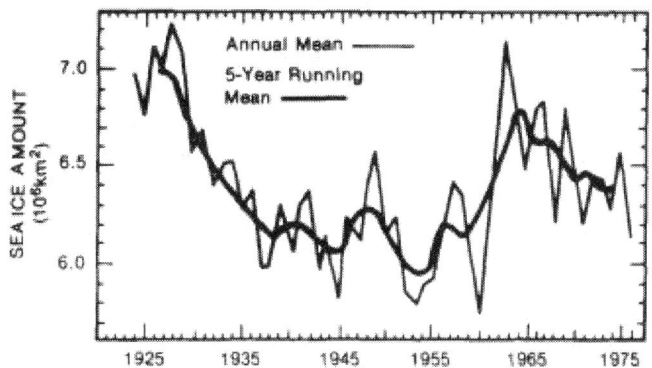

Figure 5.2. Annual mean and 5-year running mean sea ice amount in the Arctic Ocean from 1920–1975 (data from Vinnikov et al. [1980]).

In a December 1985 US Government's Department of Energy report (entitled Projecting The Climactic Effects Of Increasing Carbon Dioxide), the graph above was featured.

It shows the summer area of the Arctic Ice Sheet. The highs in 1927 are around 7.2 million sq km, and the low in 1960 is given as 5.7 million sq km.

"Hold On!" I hear you shout… "Those y-axis figures are about the same as today!"

Yeah, I'd like to make a great hoo-ha about it, but until 1973, we had no real idea of the sea ice extent. When the first satellite measurements began (in 1973), they showed that Vinnikov (et al) had probably underestimated the figure by about 1 million sq km.

Facts and Figures? Or Hyperbole?
The 'facts' of the Arctic Sea Ice are quite astounding.
As we've seen, the winter ice is pretty stable (declining 5.7% in 46 years), expanding to around 13/14 million sq km every single year.
It shrinks to a paltry 6 million sq km or so every summer.

But that means that about 8 million sq km of Arctic Sea Ice surface melts every SIX MONTHS, an area the size of the continental USA… EVERY SIX MONTHS.
Disregarding the change-over periods at Summer and Winter, the Arctic Ice Sheet melts (and freezes) for about 150 days every year… It's always on the move.
Depending on the season that means it melts (and freezes) at a rate of 60,000 sq kilometers every single DAY.
That means an area the size of Georgia (The US State) disappears… (or appears when the Arctic is freezing)… EVERY SINGLE DAY.

Knowing some of the 'real' trends of Arctic Sea Ice, let's look at some of the hyperbole that newspapers have thrown at us…

"… a record disappearance of Arctic summer sea ice in 2007."
and
"Arctic summer sea ice hit striking new lows, with sea ice volume dropping 72 percent from the 1979-2010 mean, and ice extent falling by 45 percent from the 1979-2000 mean."
UK Guardian February 2013.

"The North Pole will be hit by an unprecedented heatwave this Christmas because of man-made climate change, scientists say.
The centre of the Arctic will be 20°C hotter than average, at around 0°C freezing, on Christmas Eve."
UK Independent, 2016.

The one above is so utterly SO BOGUS it boggles the mind... *"20°C hotter than average"* ??? In hindsight the actual temperature was a paltry 1.5°C hotter than average... (it was an extremely high El Niño year), and no need for senseless hyperbole.

In 2013, the UK Independent published this headline...
"The good news: there has been a dramatic increase in Arctic sea ice. The bad news: it's still half the level it was in the 1980s"

Yup, there was an increase; 2012 was the lowest year on record (6.48million sq km's).
Was it *"Half the level of the 1980's"*??
Nope... let's look at the facts... the 2013 level was 8.11, and the level in 1980 was 10.06.

The newspaper then goes on to mix up their own rhetoric...
"The volume of the sea ice floating in the Arctic Ocean increased by about 50 per cent in October compared to the same period of 2012, which was one of the lowest on record, scientists said."

Neither of which figure was remotely factually correct.
It's about time the media got their figures straight... instead of grabbing them out of thin air.

The scientists have proven that they can tell the truth, they just have to do it consistently.

Oh, and a new rumor in...
The UK Guardian newspaper has reportedly told its writers to stop using 'Climate Change' in their climate change stories. From now on they've to use the phrases, 'Climate Emergency' or 'Climate Crisis'.

How about that for telling the news straight.

6.7... THE ARCTIC SEA ICE PREDICTIONS
(They're 'good' at predictions...)

Arctic Ice has always been the poster child of the Alarmist movement.
It's nearby; all in the northern Hemisphere where 'most' people live.
It's between America and Europe.
Its disappearance could open trade routes, and that means people will make money.
It's where the polar bear live, and they're fluffy, cuddly, and make for great newsprint and television.

So predictions of doom/gloom and Arctic apocalypse have been rampant from an early time...

I want to take you back to the halcyon days of Global Warming, before Climategate, and before fervent denialism... 2007.

We'd had the El Nino enhanced 1998 temperature spike, the Alarmists were going bananas over record hot temperatures, and the hiatus/pause hadn't been noticed.

It was time for some predictions of Arctic warming... enter the predictors!

In Feb 2008, Dr. Olav Orheim, head of the Norwegian International Polar Year Secretariat said...
"The Polar Cap may well disappear this summer... If Norway's average temperature this year equals that of 2007, the Ice Cap in the Arctic will all melt away."

A bold prediction indeed, considering he was forecasting just 6 months in the future, but one well worth proposing as the summer sea ice had been reducing for ten years or so.

In June 20th, 2008, the National Geographic took on the claim, and ran with it. David Barber of the University of Manitoba told them...
"We're actually projecting this year that the North Pole may be free of ice for the first time..."

Well, as you can probably tell, it wasn't.

In December of 2007, Mark Sereeze, a senior scientist at the National Snow and Ice data Center (NSIDC) in Boulder, Colorado was quoted by Associated Press reporter, Seth Borenstein in the Star News...
"The Arctic is screaming."
He said that the Arctic sea ice was melting so fast, it could disappear completely by 2040.

Now that's better, put your predictions so far out there, you'll be retired before you're proven wrong...
WAIT!
Nope, in the same article, NASA climate scientist Jay Zwally said...
"At this rate the Arctic ocean could be nearly ice-free at the end of the summer of 2012..."

And like a lemming to the brink, BBC Science reporter, Jonathan Amos, rubber-stamped the end of the ice, but pushed it out one more year. He quoted Professor Wieslaw Maslowski...
"Our projection of 2013 for the removal of ice in summer is not accounting for the last two minima, in 2005 and 2007... So given that fact, you can argue that may be our projection of 2013 is already too conservative."

On June 24th, 2008, 'climate prophet', James Hansen, of GISS at NASA told the world...
"We're toast,"
The followed up with more details...
"The Arctic is the first tipping point and it's occurring exactly the way we said it would."
He said the Arctic would be ice free in the summer in 5 to 10 years.
(However, he also predicted that Lower Manhattan would be underwater, there would be tape across New York's windows to fend off high winds, and there would be no birds.)

In December 14th of 2009, ex Presidential candidate turned Global warming guru, Al Gore, was a little more precise...
"Some of the models suggest that there is a 75 percent chance that the entire north polar ice cap during some of the summer months will be completely ice-free within the next five to seven years,"

Either way you sliced and diced the news; it wasn't looking good for summer ice.

To cap it all, President Obama's science czar John Holdren, also predicted ice-free Arctic winters on a CBC interview ...
"... if you lose the summer sea ice, there are phenomena that could lead you not so very long thereafter to lose the winter sea ice as well."

So you see, in just a couple of years, it was predicted that the Arctic was toast.
And this carried on so long, a whole generation of Children see no way around these predictions.
And as each apocalyptic date comes round, of course, the doomsday year gets projected further out... it's a game.

It's so lodged in the brains of scientists, when they are faced with opposing news, they can't quite take it all in.

When the Jakobshavn glacier in Greenland surprisingly reversed its shrinking, and began to grow again, the news was greeted with shock by research scientists... they quickly dismissed the growth as a temporary 'problem'.

In 2010 in Glacier National Park, some of the glaciers also changed direction, growing steadily.
It took a whole eight years for the scientists and wardens to remove the old 'glaciers will be gone by 2020' signs all over the park.

When good news hits Alarmists, they cannot handle it, as most of their plans include catastrophe, not good news.

And when unsettled science hits their radar, their poor tummies get unsettled too.

6.8... NEVER FEAR, MASIE IS HERE!
(Yup, no sarcasm... I like her too...)

Let's get this right.
NASA is now in charge of all the US Government's climate departments... (Including NOAA)
NOAA is in charge of the US Government's weather departments... (Including the NSIDC)
And NSIDC is in charge of a research initiative called MASIE.

MASIE is an initiative by the NSIDC (and therefore by NASA) and stands for 'Multisensor Analysed Sea Ice Extent'.

MASIE is basically an operational ice data base, supplied from multiple sources to provide the most accurate description of Arctic ice for the ships operating in the region.
MASIE is ultra high resolution. Compared to the previous SII system (still operating) it is based on 4 km sided cells and 40% ice coverage, which means that if an area of 16km square has 40% ice coverage, it is marked as 'having ice cover'.
(The old system, SII, measured 25km sided cells, and a 600km squared grid, so the new system is far more precise.)

Masie is fed by up-to-date information from satellites and ships at sea sailing near or in the ice fields.
I like this two fold system, and so far they haven't trod the party lines when allowing the public to see their data.

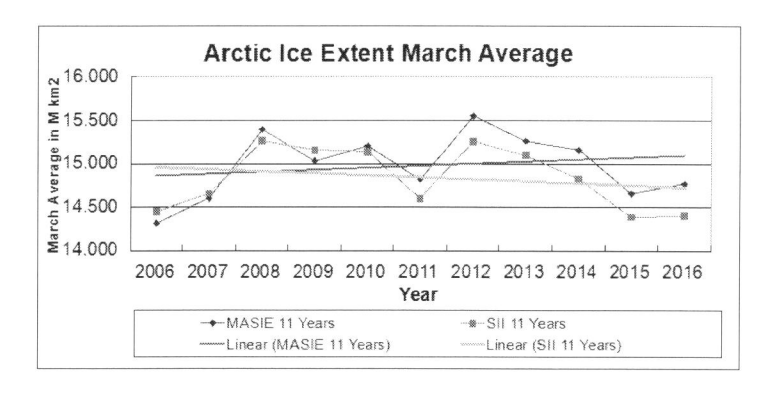

The above is a graph showing the sea ice in the Arctic Sea in March, from 2006 (When Masie was formed) to 2016. It's pretty much in line with the old SII system until more recent years.

Look at the long term trends... SII showed a slight decrease in ice levels, MASIE shows a slight INCREASE in ice levels.

AND more importantly, it maps almost EXACTLY the information given out by the Danish Meteorological Institute...
Danish figures are available here...
http://ocean.dmi.dk/arctic/sie_monthmean.uk.php

Basically Masie seems to be quite unique in the world of climatology; a US Government resource that is trustworthy, accurate and open to the public.
You can go to their website, and there, on the front page is TODAY's extent of Arctic Sea ice extent.
(Today's figure, as I write states, *"June 29th, 2019, 9.8 million square km"*.)
(As I do the final edits, on Sept 13th, the ice coverage is 4.1 million square kilometers)

Now, as a rejoinder, the figures are NOT all computer-driven, there is a human judgement element added by humans when they are given information from shipping, but hey, it maps almost directly with the Danish figures, and that means TWO INDEPENDENT SOURCES agree for once.

Let's be thankful for small mercies.

So here, although the main science is not settled, there is one small part that is making a difference.

Congrats to MAISIE.

(Oh, and if you didn't think NASA was behind MASIE, on the site Kate Heightley, Deputy Manager, and Amanda Leon, Manager, are both listed as being from the NASA Distributed Active Archive Center (DAAC).
Well done guys on a great product)

Section 7
Global Extreme Weather: Atmosfear!

"Whoever is careless with the truth in small matters cannot be trusted with important matters."
Albert Einstein

7.1... Introduction
Man, this is going to be a short chapter, because despite the media rhetoric linking climate change to extreme weather, facts are against the idea.
Yup... sometimes the smell of the coffee is just... coffee.

Oh, and this may be a reasonable time to let this little penny drop.
The IPCC itself... in its last major report in 2013... found that there had been

"no discernible increase in extreme weather events, such as hurricanes, floods and extreme heatwaves".

However, smirking slightly, I digress...

In a 2017 paper entitled, '*Atmosfear: Communicating the Effects of Climate Change on Extreme Weather*', the scientists actually state that constant reference to climate change's mitigation in extreme weather is actually damaging for the theory in point.

"atmosfear misstates and oversimplifies scientific understanding and policy choices"

And

"...we argue that attribution or linking climate change to the frequency or intensity of extreme weather events (what we call atmosfear), although perhaps a scientifically interesting question, is not effective as a means to motivate or legitimize climate policies."

However,
Ever since the first man looked up into the sky questioning the weather, we have been fascinated by the subject.

In today's modern world we have a 24-hour media that has to be filled with news of something, and fantastic television footage of extreme weather is mesmerizing to watch... and it's cheap TV. There's even full channels dedicated to bring weather to your home, every second of the day.

It's easy as we sit in out air-conditioned living rooms watching a bombardment of images to consider that weather is getting worse... I mean, we now see much more of it than our folks or grand-folks did.

And when every rain battered presenter moans about how the extreme weather is global warming induced, or how climate change is responsible, it's difficult for the average man to refute. So the non-facts become ignored, then become ingrained... before too long, we're repeating them.

With the graph of global catastrophes below, it's difficult to make a case for the Denier camp, but there is a slight glimmer of hope... well reason.

Again, the advent of 24-hour news coverage means reporters are actively looking for weather events to cover... it makes for good television.
And look carefully at the graph... every time there's a large El Niño, there's a spike in weather catastrophes; 1998, and 2006.

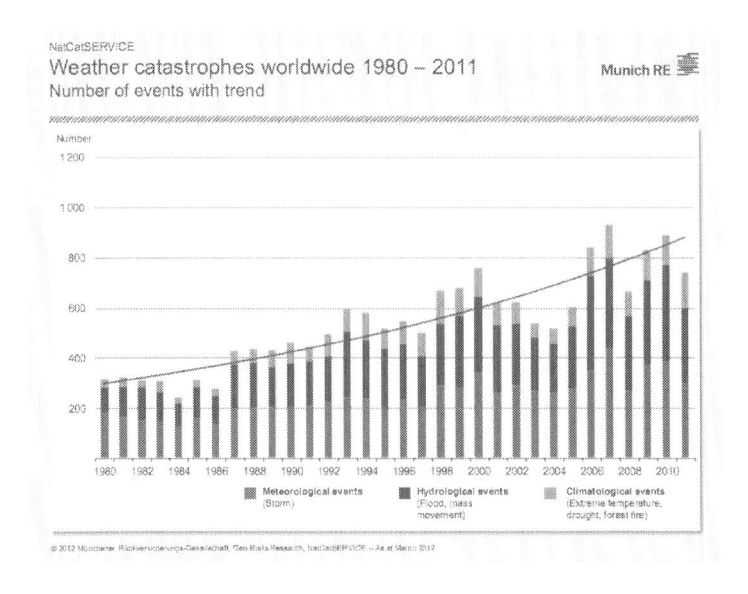

So let's depart from the world of weather, and take a look at a totally unrelated subject... or is it?

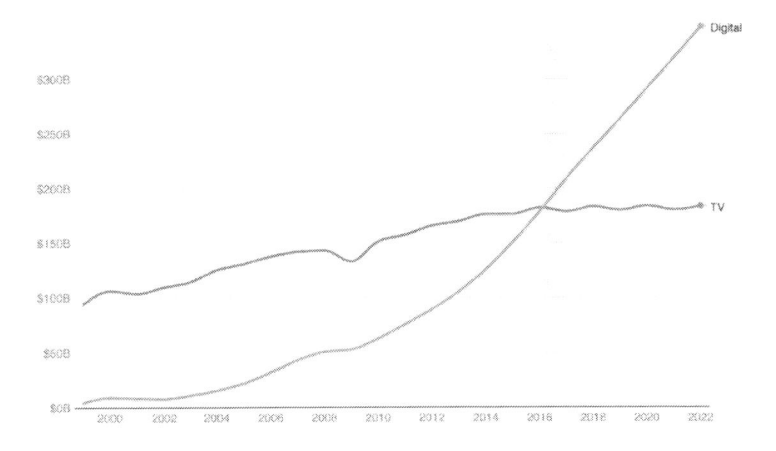

Television stations been growing exponentially since we all had just three or four in the 1970's. However, a silent partner has recently come to the fore… the advent of DIGITAL television… not only available 24 hours per day, but each station accessible from any computer, tablet and phone.
2017 was the first year that advertising revenue on digital accessible systems exceeded terrestrial cable/satellite television. If you linked the above graph with the rise in weather catastrophes, are we really witnessing a rise, or just a rise in the public perception?

Like so many intrepid reporters before me, I'm going to follow a well-trodden mantra… *"follow the money".*

And if you look at the trend line of that graph… I could also overlay the increase in global population on that line. (and CO2 rise)

Hmm… it's almost as if as human population rises, the 'need for television rises…
Weather television is fun, addictive and pretty cheap for the television companies to collect…
More people get into 'weather' on the media, so the money spent on media advertising rises…
The mantra of 'climate change permeates every single clip…
And folks think there's a link.
Hmm… those guys are good.

The Alarmists began with *Global Warming* as their slogan, but when the pause/hiatus hit, they changed their cause to *Climate Change*. When that began to wear out, they diversified a little… *Climate Justice, Climate Anomalies, Climate Weirding,* and the latest in the line… *Extreme Weather.*

But just how much is 'Global Warming' changing the climate for the worse?

Most weather on the planet is caused by air heating up at the equatorial regions and moving north. The difference in temperatures between the equator and the poles dictate the 'extremeness' of the weather.
So... when the Alarmists claim that the poles are warming... how can the weather be getting worse...

It's time to take a close look at some of the rhetoric on the subject, and see what we can dig up.

7.2... GLOBAL WARMING AND HURRICANES
(The pretty spirals that keep on coming...)

Hurricanes, or tropical cyclones, are rapidly rotating storm systems characterized by a low-pressure center. They usually form in the tropical regions over water. They build higher and gain momentum due to the prevailing winds and the heat of the water below them.
When they make landfall, they cause destruction, death and chaos.
They're a particularly dangerous part of Mother Nature that we'd rather not have to deal with.
But are they getting worse or stronger due to climate change?
Well, not according to Wikipedia... and that surprised me.

"While the number of storms in the Atlantic has increased since 1995, there is no obvious global trend..."

Now, trust me, if left-leaning Wiki says *"no obvious global trend"*, the truth is probably, 'there is a downward trend'.
However, Wikipedia alone doesn't stop every Tom Dick or Harry weather reporter from mentioning 'Global Warming' along with their particular facet of extreme weather.
Perhaps they got it from the 2007 IPCC report...

"It is likely that storm intensity will continue to increase through the 21st century"

That implies there was already an increase in the 20th century...
I think it's time for a graph, don't you?

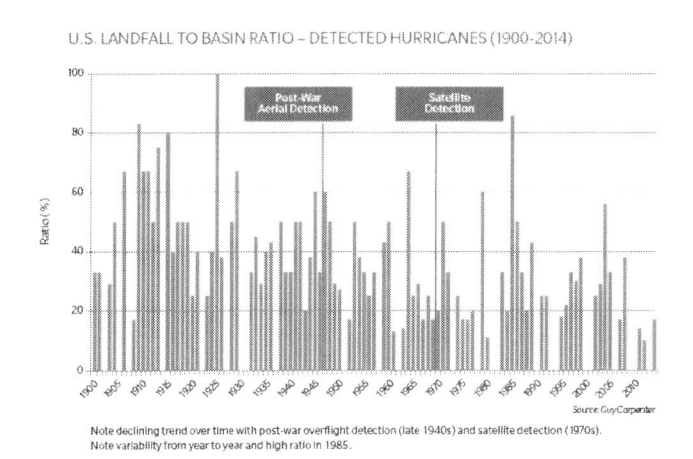

U.S. LANDFALL TO BASIN RATIO – DETECTED HURRICANES (1900-2014)

Source: GuyCarpenter

Note declining trend over time with post-war overflight detection (late 1940s) and satellite detection (1970s).
Note variability from year to year and high ratio in 1985.

The 2013 graph shown above is detected US hurricanes from 1900 to 2015...
Does it look like storms are increasing as the IPCC states?

Our detection methods have improved over the last 120 years, and we now have numerous weather satellites reporting 24-7.
But that implies that before the new-fangled detection techniques there must have been some hurricanes that went unreported, so the historical figure may be even higher.

If you look at recent times, even with the help of satellites, from 1990 onwards (a time of HUGE global warming, CATESTROPHIC climate change, and RECORD HIGH carbon dioxide) there's actually a significant decline.
Let's take a look at NOAA.

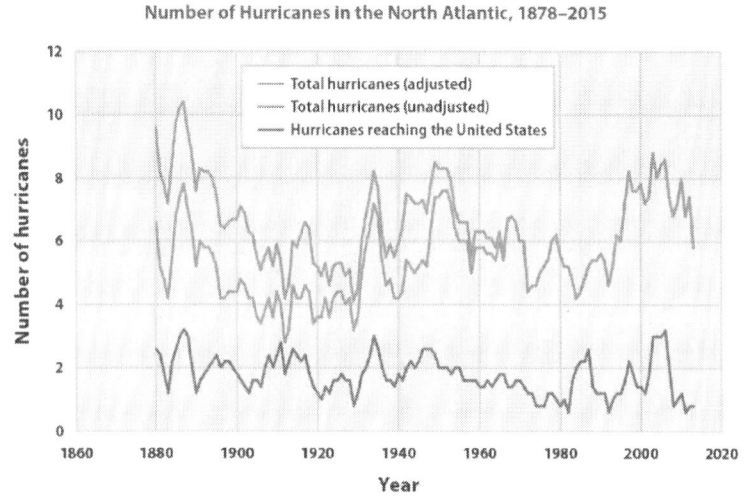

Number of Hurricanes in the North Atlantic, 1878–2015

Data sources:
- NOAA (National Oceanic and Atmospheric Administration). 2016. The Atlantic Hurricane Database Re-analysis Project. www.aoml.noaa.gov/hrd/hurdat/comparison_table.html.
- Vecchi, G.A., and T.R. Knutson. 2011. Estimating annual numbers of Atlantic hurricanes missing from the HURDAT database (1878–1965) using ship track density. J. Climate 24(6):1736–1746. www.gfdl.noaa.gov/bibliography/related_files/gav_2010JCLI3810.pdf.

For more information, visit U.S. EPA's "Climate Change Indicators in the United States" at www.epa.gov/climate-indicators.

I can see little evidence of an increase in trend. There's no huge upturn, there's no hockey-stick here.

But weathermen, TV anchors, and Alarmists still push the connection at every conceivable moment.

Roger Pielke Jr, a professor of environmental studies at the University of Colorado in Boulder, published a book, *The Rightful Place of Science: Disasters and Climate Change.*
Pielke gave evidence regarding extreme weather and climate change to the Senate Environment and Public Works Committee, using various graphs taken from IPCC and NOAA data.
In his opening statement, Pielke said...

"It is misleading, and just plain incorrect, to claim that disasters associated with hurricanes, tornadoes, floods or droughts have increased on climate timescales either in the United States or globally."

To further compound his fate, he went on...

"Hurricanes have not increased in the US in frequency, intensity or normalized damage since at least 1900... Floods have not increased in the US in frequency or intensity since at least 1950... Tornadoes have not increased in frequency, intensity or normalized damage since 1950... Drought has for the most part, become shorter, less frequent, and cover a smaller portion of the U. S. over the last century."

Oops...

And I have another couple of dozen graphs that would bore you, because they're all the same.
North Atlantic hurricanes and hurricanes hitting USA have declined.

Maybe the trend that everyone is quoting is global?

But look below... we have a graph which contains details of both northern hemisphere storms and global ones...

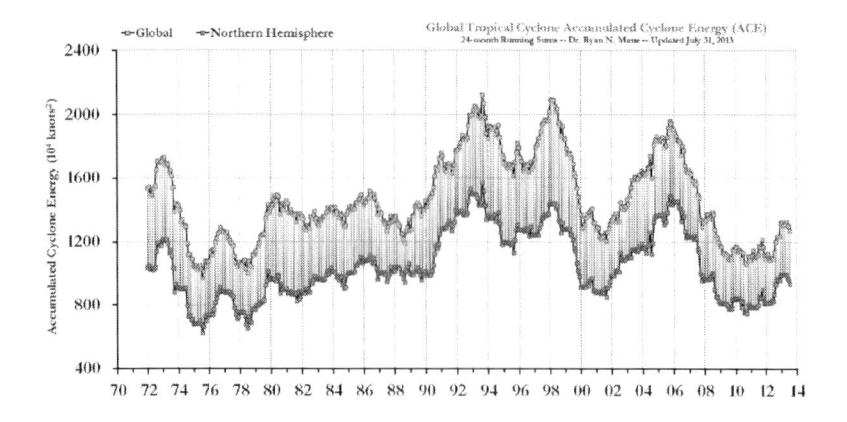

And if you can see a trend there, you're a better person than me.

With satellite detection, we won't miss a single storm. And with 24-7 multi-channel newsfeeds, the media certainly won't miss bringing one into your homes.

But as far as I'm concerned, the case of storms/hurricanes being exacerbated or increased in frequency is certainly…

NOT PROVEN!

7.3... GLOBAL WARMING AND OTHER WEATHER
(Is global weirding actually a thing?)

Droughts
It doesn't matter where you go... there are now droughts everywhere! Even when NASA has written columns on Global greening, those cracked desert floor pics keep on resurfacing.

Yup. The media tell us about it ALL the time.
Droughts...
But again, is it true? Or is it just in your face by the media 24-7?

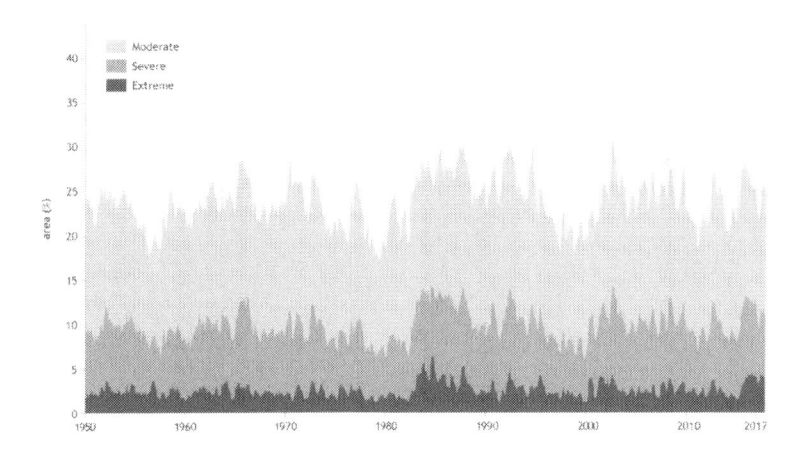

The graph above is taken from NOAA, (climate.gov) and shows world droughts since 1950.
Now, I'm not a master statistician, but it looks pretty level to me.
This NOAA graph shows that droughts are NOT linked to global warming or CO2 levels.

https://www.climate.gov/news-features/featured-images/2017-state-climate-global-drought

Tornadoes
Twisters are on the increase! Storm chasers show us fantastic pictures every tornado season.
(And trust me, I live in Topeka, Kansas, and we just had a run of them last night.)

However, just because you see the TV screen plastered in tornado footage, doesn't mean there's an increase.

The NOAA graph below shows EF3 and above, since 1950.
Amazingly the average is going down despite them being linked to climate change, CO2 and 'global warming'.
Look at the graph… despite the new radar systems all over the USA, tornadoes are DECREASING!

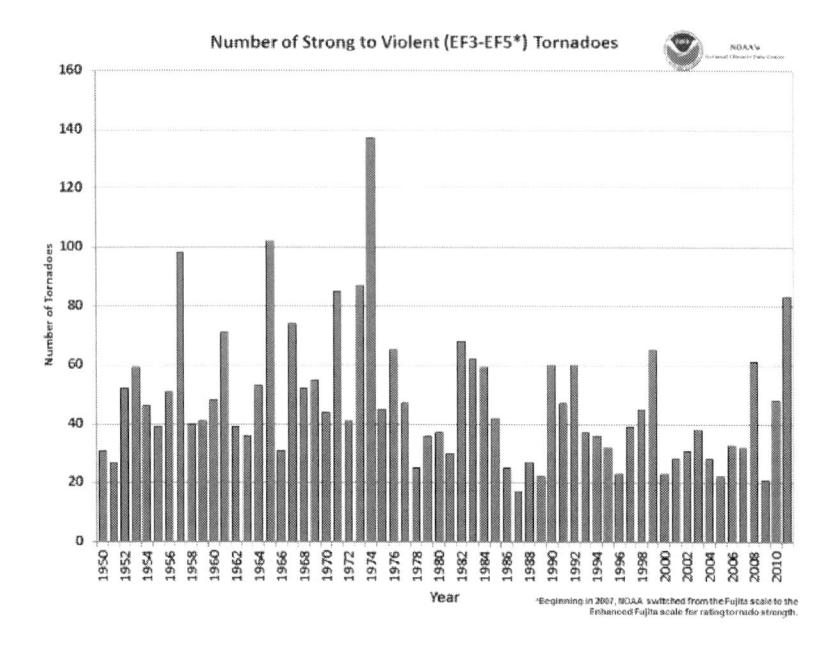

There's nothing like getting the alarmists to go to their own data for refutement.
Revenge… best served cold.

Freezing Weather; the Polar Vortex
As I'm writing this book (Jan 2019), there's a huge Polar Vortex coming south from the Polar Regions, and giving us a few dozen days of chilly weather.
"Unprecedented", one TV station trumpets.

I was driving in my car a couple of days before the vortex hit us, and the news at the top of the hour on my own radio station in Topeka said the imminent cold was…

"life threatening"

Oh dear, I thought, then drove on to McDonalds... and I have to admit, it got cold for a couple of weeks.
But...
We had no burst pipes (we've had those in the past).
We had no frozen fuel lines on our vehicles (we've had those too).
When the hyperbole started, everyone was saying the weather was 'out of control'... on the coldest day, I almost believed them.
Almost.

So, cold vortex cold waves are increasing huh?
I could hardly wait to investigate.
Unfortunately, there was one small problem for us 'deniers' using this information to refute the frequency of cold vortex snaps to any alarmists...

The article was featured on https://wattsupwiththat.com , (yes, a denier website), it was written by Roy Spencer, a solid scientist (but GW denier), and used data from John Christy (a solid GW denier) from the University of Alabama Huntsville (UAH).

But I'll show you John Christy's graph, and give you a link to the full story anyway.

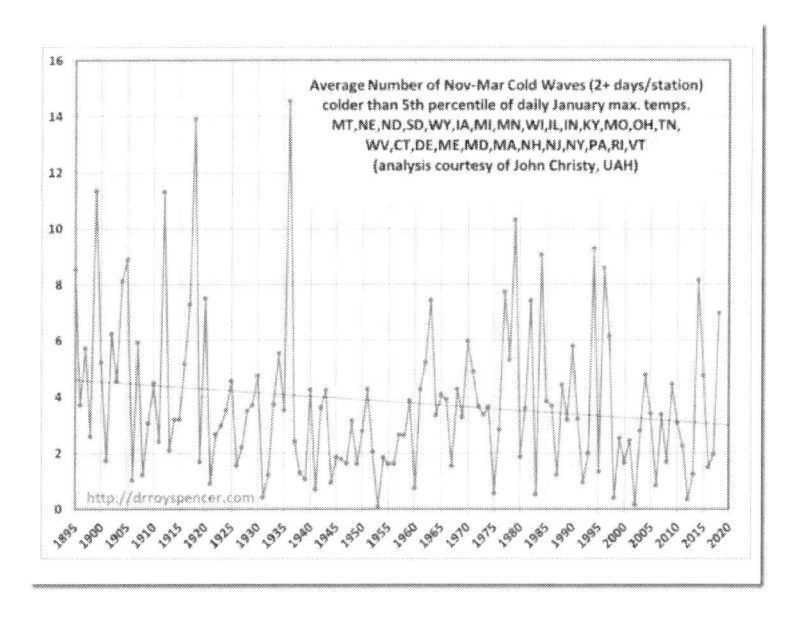

Average Number of Nov-Mar Cold Waves (2+ days/station) colder than 5th percentile of daily January max. temps.
MT,NE,ND,SD,WY,IA,MI,MN,WI,IL,IN,KY,MO,OH,TN, WV,CT,DE,ME,MD,MA,NH,NJ,NY,PA,RI,VT
(analysis courtesy of John Christy, UAH)

Going back to the 1800's, John has provided a great illustration of cold snaps in the US lasting 2 days or more.

I'm now going off to wrap up, cos the forecast looks bad again…
Here's the whole article…

https://wattsupwiththat.com/2019/02/01/if-the-polar-vortex-is-due-to-global-warming-why-are-u-s-cold-waves-decreasing/?fbclid=IwAR3FN5GZ0c_kOUMF-zddqvE5FToSMvZSy7dGAYTJFkWV1OcbZubNsmWeP-g

Hot Days
Below is a graph of the percentage of US days over 90°F.
It's getting warmer… right?

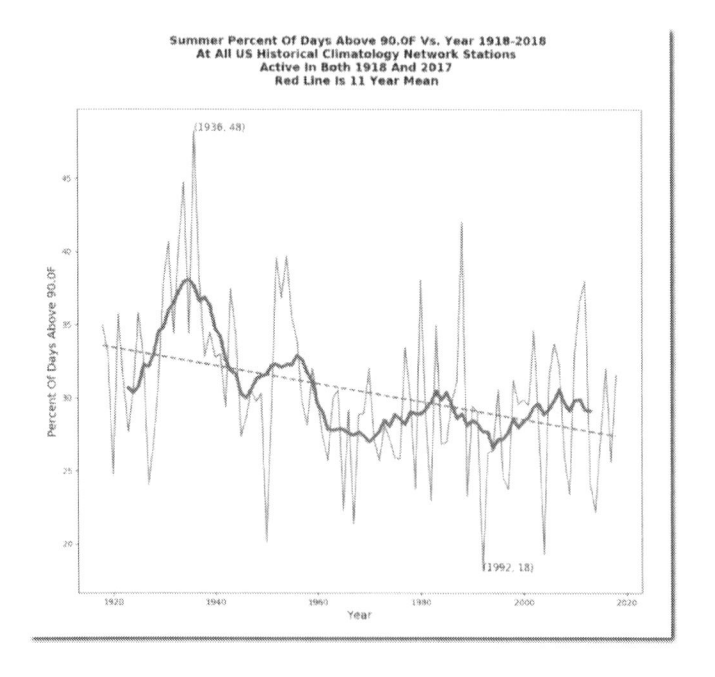

Well, actually… no.
It looks like the number of days over 90 ° has actually fallen.

Who'd have thunk it?
And you've never seen that graph on the media, EVER!

So The USA, with all its THOUSANDS of weather stations, it being the MOST COVERED land mass on the PLANET, is showing less hot days now that 100 years ago.

So the rest of the world is warming, and the USA's going to catch up exponentially, right?

Yeah, I know, sugar-dripping sarcasm is the lowest form of wit.

And now I should say something about science not being settled... but I'll resist the temptation.

7.4... EXTREME WEATHER SITES...
(The champions of extreme weather...)

Amongst others, there is a website called Carbonbrief.org that specializes in plotting extreme weather events on a world map, and showing just how many of these events have been influenced by climate change.

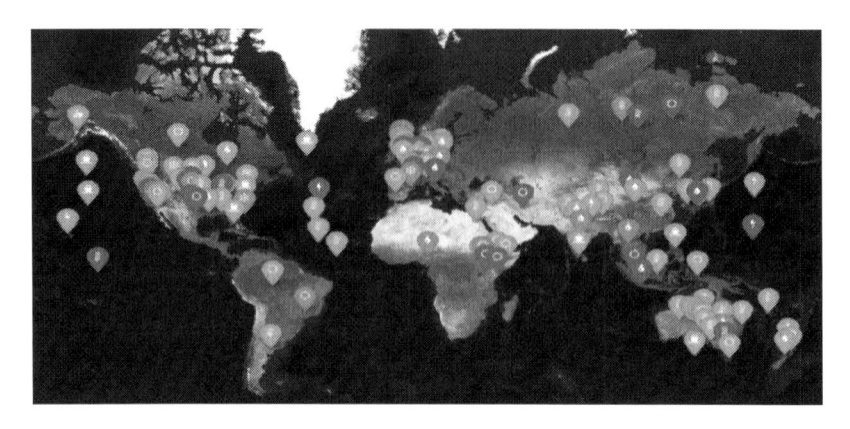

This is a simplified version of the map which supposedly shows information from over 140 peer-reviewed articles. Each article has occurred in the last 20 years, and the causes of the extreme weather are investigated.

One little fact hit me immediately; most of these 'extreme weather events' on the map happened in areas of high population. Look at USA compared with Canada and Mexico. Look at Europe compared to Russia. Others happen in areas of high weather station coverage; USA, Europe, look at the concentration in Australia.
Yeah... there's more to this map than just extreme weather.

On the full interactive map (web address at end of chapter), you can toggle to a single year.
2011, for example, has just 13 weather incidents... not a great deal of extreme weather that year, you see.
In one example (flooding in the UK), they quote a 2011 article...

"In nine out of ten cases our model results indicate that twentieth-century anthropogenic greenhouse gas emissions increased the risk of floods occurring in England and Wales in autumn 2000 by more than 20%, and in two out of three cases by more than 90%"

Wow, they can do that with climate models? Most IPCC climate models are wrong as soon as they're published.

AND... their featured 2011 case study is from 2011, but sites a flood in 2000???
Let's look at their 2011 weather extremes a bit closer, shall we?
Hurricane Katrina & Ivan are mentioned.... That's one event.
9 out of the 13 events mentioned are 'hot temperatures'... hardly extreme weather events.
And there are three flooding matters.
Blah.
Hardly the stuff of intrepid weather reporters, huh?
However, their small sound-bites on each weather occurrence show us a whole lot about scientific thought.

On the carbonbrief website, rainfall in New Zealand in 2012... is shown as an EXTREME weather event.
A scientific paper by Dean, S. et al., (2013) from the American Meteorological Society states...

"This analysis indicates that the total moisture available for precipitation in the Golden Bay/Nelson extreme rainfall event of December 2011 was 1%–5% higher as a result of the emission of anthropogenic greenhouse gases."

Looking at the full transcript, Dean considered the deluge to be a 500 year event... a rainstorm that happened every 500 years or so.
But New Zealand has just been colonized for 200 years; how could he make that assertion?
And what's more, how could he possibly estimate a figure of 1% - 5% of human global warming effects?
Yup... they'd forecasted the extreme rain event.
And they concluded that the rain was influence by 1% - 5% by human-caused greenhouse gasses... even though there was never a human report about the last such storm.
Climate models proved it.
How flipping quaint of them.

https://www.carbonbrief.org/mapped-how-climate-change-affects-extreme-weather-around-the-world

7.5... THE IPCC'S OPINION OF EXTREME WEATHER
(The IPCC has an opinion...)

You know, there's nothing more satisfying when debating than beating our opponent with their own facts.
The IPCC (Intergovernmental Panel on Climate Change) are the go-to people for the so called 'facts'.
Every few years they release a report about the world's condition. These 'Assessment Reports' are numbered, and the latest one (AR5) was completed in 2014.
AR5 is a huge document, of over 1000 pages. Let's take a look at Chapter 2, (Entitled; "Observations: Atmosphere and Surface") shall we?
(And trust me, I'm just citing the summaries, because the text is so dry and boring, you'd stop reading!)

"Flooding: In summary, there continues to be a lack of evidence and thus low confidence regarding the sign of trend in the magnitude and/or frequency of floods on a global scale."
(AR5, page 214)

Wow, that surprised me... most media blame AGW for flooding events.

"Droughts: In summary, the current assessment concludes that there is not enough evidence at present to suggest more than low confidence in a global-scale observed trend in drought or dryness (lack of rainfall) since the middle of the 20th century, owing to lack of direct observations, geographical inconsistencies in the trends, and dependencies of inferred trends on the index choice."
(AR5, page 215)

And now they summarize 'normal rainfall'...

"In summary, confidence in precipitation change averaged over global land areas is low for the years prior to 1950 and medium afterwards because of insufficient data, particularly in the earlier part of the record. Available globally incomplete records show mixed and non-significant long-term trends in reported global mean changes. Further, when virtually all the land area is filled in using a reconstruction method, the resulting time series shows less change in land-based precipitation since 1900."
(AR5, page 202)

And now regarding HEAVY rainfall (watch the 'careful' wording here)...

"In summary, further analyses continue to support the AR4 and SREX conclusions that it is likely that since 1951 there have been statistically significant increases in the number of heavy precipitation events (e.g., above the 95th percentile) in more regions than there have been statistically significant decreases,"
(AR5, page 213)

Wait... with all the statistics of the world at their disposal, with 'thousands' of scientists working on the project, they CAN'T state an outcome categorically... just 'likely'?

The IPCC also doesn't link tornadoes to climate change...

"There is low confidence in observed trends in small-scale phenomena such as tornadoes and hail because of data inhomogeneities and inadequacies in monitoring systems."

Here's the deal, dear reader, if the Alarmist's own organization won't commit, why should we?

No matter what they say, the science sure isn't settled, is it?

Section 8
The Governing Body... The IPCC

"We're not scientifically there yet. Despite what you may have heard in the media, there is nothing like a consensus of scientific opinion that this is a problem. Because there is natural variability in the weather, you cannot statistically know for another 150 years."
Tom Tripp (lead author of the UN IPCC)

8.1... The IPCC; A Little History
(Some stories, some facts...)

The Background
From what you hear/read in the media, the United Nations Intergovernmental Panel on Climate Change (UN-IPCC) is an august body (a panel) of scientists who, through their research, diligence, and dedication to humanity, keep us informed about our climate, and how its changes are going to affect us.
This, however, is simply not true.
Wikipedia's definition is as follows...

"The IPCC provides an internationally accepted authority on climate change, producing reports which have the agreement of leading climate scientists and the consensus of participating governments."

Yup, it looks good, but as 'Deniers', we have to choke on most of the wordage...
For instance, the "internationally accepted authority" use data we don't agree with, and their scientists list conclusions which a whole host of independent scientists round the world challenge on a daily basis. The "agreement of leading climate scientists" is a point of contention, and the word "consensus" is, of course, present.

In actual fact the IPCC is a panel of politicians under the control of the United Nations (UN), one panel member from each UN country, usually about 190 delegates in all. Very few actual members of the Panel, if any, are scientifically literate, and all are politicians or diplomats responsible to their own governments and parties, not humanity in general.
The panel is advised by a group of specially chosen scientists, but nowhere are 'denier' scientists allowed to participate.
(The main reason you will read below)

For the IPCC, "the science is settled", and they will not allow anyone to break the party ranks.

The IPCC present a report every few years or so, called "Assessment Reports", and these are usually presented at delegate conferences and greeted with huge fanfare by the fawning media.
Each year the IPCC select their group of scientists, (called 'Working Groups') all of whom are willing to write their 'science' under the IPCC list of instructions (terms of reference).

However, many former IPCC scientists have expressed their views to the public. Tales of coercion, of 'being told what to write', and some views having been edited out completely are common.
Here's a few dissenting voices…

"Worst scientific scandal in the history…When people come to know what the truth is, they will feel deceived by science and scientists."
Kiminori Itoh (Ex-IPCC Japanese environmental physical chemist)

"We're not scientifically there yet. Despite what you may have heard in the media, there is nothing like a consensus of scientific opinion that this is a problem. Because there is natural variability in the weather, you cannot statistically know for another 150 years."
Tom Tripp (lead author of the UN IPCC)

"There was a global warming in medieval times, during the years between 800 and 1300. And that made Greenland, now covered with ice, christened with a name (by the Vikings) that refers to land green: 'Greenland'."
Rosa Compagnucci (Argentinian El Niño expert and former IPCC author)

The Role of the IPCC
Most members of the world's concerned population believe the IPCC's job is to collate information, and inform the public.
But in that they are absolutely wrong.

In actual fact, the aim of the IPCC has little to do with climate change. And that statement might alarm you.
The IPCC's aim is to build a case for **man-made** climate change; Anthropogenic Global warming (AGW) and to mitigate scientific solution to reduce the problem.

And this sounds like conspiracy, kooky, Denier bullcrap, but it's actually not.

In the world of corporate industry, any commissioned report usually has guidelines to give the report-makers an idea of the parameters of the scope of enquiry.

The scope of the enquiries a report is called the 'terms of reference', and this limiting can be used in two ways; to constrict or focus an enquiry/report.

UNBELIEVABLY, the IPCC have restricted their 'terms of reference' to include **ONLY** man-made global warming.

At their core, they fully intend to ignore all natural causes.

This simple fact alone sounds incredible, but it's true.

The 'focus and research' of the United Nations Framework Convention on Climate Change (their guideline) is as follows…

*"a change of climate which is attributed directly or indirectly to **human activity** that alters the composition of the global atmosphere and which is in addition to natural climate variability observed over considerable time periods."*

In 1990, in the first Assessment Report, the first sentence of the introduction is…

*"Many previous reports have addressed the question of climate change which might arise as a result of **man's activities**."*

So, as you see, as a core of the IPCC's doctrine, it has no interest in the 'real' causes of climate change, whatever they may be, just the human ones.

AND THAT IS WHERE THE IPCC'S PROBLEM REALLY LIES.

From its inception, the IPCC has only been looking for possible 'man-made activities' that have been influencing climate change, and forcibly rejecting hypotheses and theories that look for causes away from mankind.

That's dangerous thinking, and may (if the culprit is NOT carbon dioxide) lead to us moving in the wrong direction to mitigate the problem.

Read on, dear avid reader… read on.

8.2... *IPCC's Terms of Reference*
(Setting up the straw man...)

Around half of the US population considers Government outreach to be a bad thing.
Around half of the US population considers Government involvement in their lives to be a good thing.
And never the twain shall meet.
(Well, except the one thing both groups hold some form of agreement on... *"you can never really trust what a politician actually says".*)

If there's a subject of controversy, sometimes a "Commission of Enquiry" is the best way forward. Choose a few neutral investigators, let them loose and see what they dig up.

But of course, when you need something investigated, you have to tell them what's to be investigated. You give details of the subject, and most times you give limits to the direction, scope, or depth of the enquiry.
(This is common practice... let's face it you don't want the members of the investigation team going off in a 'silly' direction, do we?)

These guidelines are called 'terms of reference', and are the limits to the investigation.

In 2017, when the Special Counsel investigation (led by Robert Mueller) was set up to investigate Russia's possible involvement of the 2016 US Presidential campaign, the 'terms of reference' were not firm and tight... a point made by Republicans at the time.
The investigation included collusion between President Trump and the Russian government and... *"any matters that arose or may arise directly from the investigation."*
That's pretty loose.
Many Republicans moaned at the oversight, and a few sidelined conspirators were indicted on side charges, but in the end, Trump was exonerated.

When Earl Warren was given the task of investigation into the Kennedy assassination, he was asked why he didn't investigate the role played by Jack Ruby (The man who shot and killed Lee Harvey Oswald).
He simply replied... *"It wasn't in my terms of reference"*

What many people don't know (and this is the case with the framework given by the IPCC to its team of scientists) is that if you set up the 'terms of reference' tight enough, you can forcefully guide the investigation in the direction you want... effectively the terms of reference dictate the result of the enquiry.

For instance... If the IPCC sets out the terms of the investigation...

*"scientific, technical and socio-economic information relevant to understanding the scientific basis of risk of **human-induced** climate change, its potential impacts and options for adaptation and mitigation."*

They are basically telling their chosen scientists to list all human-caused global warming information, and nothing else.
They'd be as well saying... *"Just look at **human** warming effects... nothing more."*

If we use a tennis analogy...

The IPCC have gotten their own 'terms of reference' into their paperwork without many people knowing the truth...
POINT TO THE IPCC.

The IPCC scientists' list only human effects on climate change... that's what they've been told to do by the IPCC's 'terms of reference'...
GAME TO THE IPCC.

The media list only 'human effects on climate change' stories, because that's all they've been given, and let's face it, it allows them to use all those wonderful 'catastrophic' headlines...
SET TO THE IPCC.

The average man in the street sees only human effects being talked about in the media... he perhaps sees in the periphery some guys talking about water vapor, sunspots, clouds, cosmic rays... but surely they're kooks, because the IPCC would have told us if these side-lined scientists were telling the truth... And then he calls everyone else who does not toe the popular line... a Denier.
In tennis we'd call that...
GAME, SET AND MATCH TO THE IPCC.

Staying with the tennis analogy, it's like the subject of the enquiry is the tennis ball.
The 'terms of reference' are *"You are allowed to investigate the tennis ball, its interior, and its solid ball shape. You are not allowed to investigate outside the tennis ball."*

You'd get some weird report about a cold ball going back and forth with no explanation as to why (as the game takes place). Then the ball heats up a little. Then the ball stops moving (as the game ends).
The report is factually correct, but you have no idea of the actual game, because the details outside the ball were never asked for.

8.3... THE IPCC'S PROCESS
(Ultimate power corrupts...)

The purpose of the IPCC is to provide reports for the larger organization; the United Nations Framework Convention on Climate Change (UNFCCC), which writes the main international treaty on climate change.

Since its inception in 1988, the IPCC has published five comprehensive reports to the UNFCCC, (called Assessment Reports).
These reports are a review of the latest climate science, (limited by the IPCC's terms of reference).
These are known as...

1990... First Assessment Report (FAR)
1995... Second Assessment Report (SAR)
2001... Third Assessment Report (TAR)
2007... Fourth Assessment Report (AR4)
2014... Fifth Assessment Report (AR5)
And the IPCC is currently preparing the Sixth Assessment Report (AR6), which is due to be published in 2022. (Yes, each volume takes three or four years to produce)

The process of each assessment report is as follows.

Three Working Groups of scientists are chosen from the 'world of scientists', each group to study different facets of **man-made** climate change.
Working group I is responsible for... "The Physical Science Basis."
Working Group II is responsible for... "Impacts, Adaptation and Vulnerability."
Working Group III is responsible for... "Mitigation of Climate Change."

The work is done in Working Group Order; the science is done first, then the second group adds their ideas on the science's impacts, then the third group adds their ideas on what can be done.

This Working Group report is never published... because there's lots of work still to be done.
The working Groups are sent home, their job done.

There is now a long editing process.

Editing 1
A different group of scientists (still operating under the UNFCCC's original terms of reference) review, edit, and change the first scientists' work. These scientists are not required to keep anything that the Working Groups have written. They may have very different opinions from the Working Groups.
(**Fraud 101**; This is the first time that the 'scientists' data can be corrupted or changed)

Editing 2
This is what's called the government/expert review stage. Both the new group of scientists (still operating under the UNFCCC's terms of reference) and high-ranking U. N. reviewers re-write the Working Group report, once again.
(**Fraud 102**; the second time that the Working Group scientists work can be adapted/corrupted/changed)
This new document is now called… "Summaries for Policymakers", and these 'summaries' can be in excess of 800 pages long.
But it's still not in its finished form.

Editing 3
With the scientists now completely out of the process, there is another government review. These reviewers have no restrictions in their editing, and can make sweeping changes in any of all of the conclusions and wording agreed upon so far.
(**Fraud 103**; this is the third time the report can be adapted, corrupted, or changed, under the excuse of 'polishing')
When the final document is deemed 'ready' for the public to see, it has been changed considerably from the bare work of the Working Groups, the actual scientists concerned.
(Many IPCC scientists have complained to the IPCC that their words have been changed, skewed or omitted)
The three volumes are then completed by a synthesis report, that re-writes the working groups' contributions and combines them together along with any 'special data' that has been produced/discovered during the time the Working Groups have been compiling their reports.
(Remember, this process takes 3 or 4 years)

Now, this is not the way to conduct science, and the constant review process opens the door to many charges of corruption both from the public, interested parties, and from some Working Group scientists too.

Case in Point... Benjamin Santer
Benjamin David Santer is an American climate researcher at Lawrence Livermore National Laboratory. He has also worked at the UK's CRU (University of East Anglia's Climatic Research Unit), the source of the leaked emails in the Climategate scandal of 2009.

Chosen by the IPCC chairman as the Lead Author for chapter 8 of the 1995 Second Assessment Report (SAR), Santer started work at Livermore in August 1994. The first draft was presented in Sweden in October 1994. As Lead Author, Santer had complete control of the final content of his chapter, and had chosen a particular graph of upper atmosphere temperature to illustrate a point.

This is the original graph...

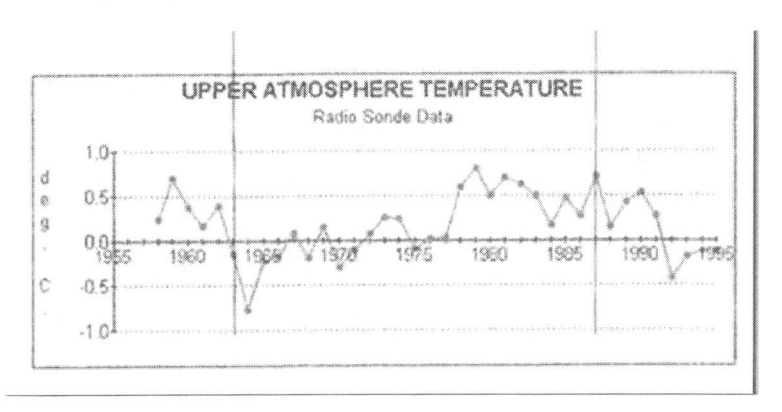

It shows a relatively straight line graph, with no exact temperature on the y-axis. A pretty mediocre level graph with little or no upward or downward trend.

The graph below is what was published in the final SAR report.

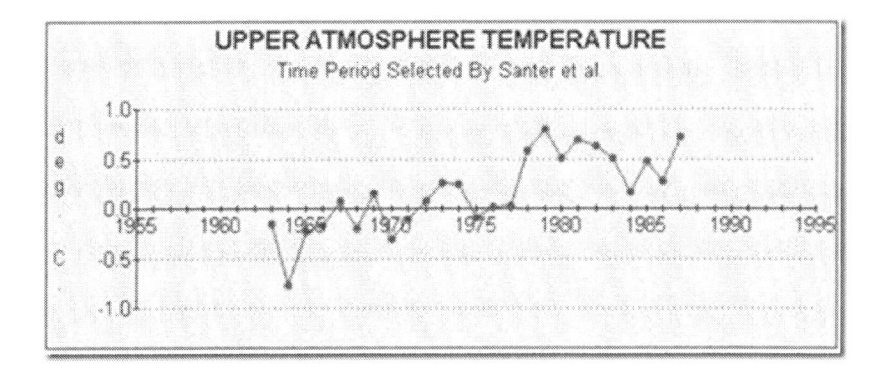

The start of the first graph, and its end have been chopped to show a definite rising average trend.
Look at the subtitle... "Time Period Selected By Santer et al"

I'm sorry, I don't care what color of cowboy hat you wear... what the IPCC did with the graph was pure manipulation of data, fraud and many other terms I could use.

Part of Santer's group's original text was...

"None of the studies cited above has shown clear evidence that we can attribute the observed (climate) changes to the specific cause of increases in greenhouse gases."

"While some of the pattern-base discussed here have claimed detection of a significant climate change, **no study to date has positively attributed all or part of climate change to man-made causes.**"

"Any claims of positive detection and attribution of significant climate change are likely to remain controversial until uncertainties in the total natural variability of the climate system are reduced."

Basically, he's saying we don't know, it's not clear that it's human caused, and we're not sure.
However, what appeared in the final draft was...

"There is evidence of an emerging pattern of climate response to forcing by greenhouse gases and sulphate aerosols... from the geographical, seasonal and vertical patterns of temperature change... **These results point to a human influence on global climate.**"

"The body of statistical evidence in chapter 8, when examined in the context of our physical understanding of the climate system, **now points to a discernable human influence on the global climate.**"

Yup, Santer's words have changed from "We don't know", to "It's man made..." thus keeping the report within the terms of reference given to him by the IPCC and the UNFCCC.

And the media ran with the story... and the rest is history.

"The IPCC 'policy summaries', written by a small group of their political operatives, frequently contradict the work of the scientists that prepare the scientific assessments. Even worse, some of the wording in the science portions has been changed by policy makers after the scientists have approved the conclusions."
Peter Friedman (mechanical engineer at University of Massachusetts, Dartmouth.)

8.4... Ex IPCC Scientists Spill the Beans
(You can tell the truth once you've left...)

Many scientists chosen by the IPCC to fill their Working Groups have 'come out' after the fact and told the 'truth' about the IPCC's malpractices...

"We're not scientifically there yet. Despite what you may have heard in the media, there is nothing like a consensus of scientific opinion that this is a problem. Because there is natural variability in the weather, you cannot statistically know for another 150 years."
Tom Tripp (lead author of the UN IPCC)

Professor John Christy was a Lead Author for the IPCC. His resume reads...
1973; B.A. in Mathematics from California State University.
1984/87; M.S. and Ph.D. in Atmospheric Sciences from the University of Illinois in 1984.
1991; Awarded NASA Exceptional Scientific Achievement Medal (with Roy Spencer)
2000; Appointed Alabama's state climatologist.
2001; Lead Author of a section of the IPCC.
He is currently in the University of Alabama, Hunstville, and is in charge of collating and disseminating the only unhomogenized data set (from satellite feeds), commonly called the UAH record.
He has testified before both US and Canadian Senate Committees that global warming has paused, and is NOT related to CO2 build-up.

"Worst scientific scandal in the history...When people come to know what the truth is, they will feel deceived by science and scientists."
Kiminori Itoh (Ex-IPCC Japanese environmental physical chemist)

"There was a global warming in medieval times, during the years between 800 and 1300. And that made Greenland, now covered with ice, christened with a name (by the Vikings) that refers to land green: 'Greenland'."
Rosa Compagnucci (Argentinian El Niño expert and former IPCC author.)

"Gore prompted me to start delving into the science again and I quickly found myself solidly in the skeptic camp... Climate models can at best be useful for explaining climate changes after the fact."
Hajo Smit (Dutch meteorologist and former member of the Dutch UN IPCC committee.)

"Most of the extremist views about climate change have little or no scientific basis. The rational basis for extremist views about global warming may be a desire to push for political action on global warming."
Gerd-Rainer Weber (German scientist of Meteorology/Atmospheric Sciences and listed a reviewer of IPCC reports.)

And these are not the only ones.
A chapter at the end of the book gives a long list of scientists who do not believe the IPCC's talking points

Scientists like German Physicist and meteorologist Klaus Eckart Puls…
"Ten years ago I simply parroted what the IPCC told us. One day I started checking the facts and data – first I started with a sense of doubt but then I became outraged when I discovered that much of what the IPCC and the media were telling us was sheer nonsense and was not even supported by any scientific facts and measurements. To this day I still feel shame that as a scientist I made presentations of their science without first checking it. The CO2-climate hysteria in Germany is propagated by people who are in it for lots of money, attention and power."

Yes, even in Germany, the science is not settled.

8.5... THE IPCC'S SCIENCE IS SETTLED!
(Oh no it's not...)

The glib dismissive phrase *"the science is settled"* slips like warm coal over hot butter from the mouths of all Alarmists. When those words are spat at us, it declares the conversation is over.
If we continue to argue, we're kooks, deniers, fringe scientists, and more. To us climate Deniers, the words are annoying, jarring, and downright rude.
And yet... are they true?

Why don't we take those words, and stuff them right back in the Alarmists oral cavity!

Example 1, IPCC Settled Science
The icon of Alarmist, the IPCC, should be a little more careful with their own rhetoric.
Because...

In 1990... First Assessment Report (FAR)
The IPCC did not state at any page any level of global warming attributed to mankind. No anthropogenic warming.
(Remember, this was eight years before the El Niño enriched warming spike of 1998)

In 1995... Second Assessment Report (SAR)
The IPCC stated...
*"The balance of evidence suggests a **discernable** human influence on climate."*
(Still before the warming spike of 1998)

In 2001... Third Assessment Report (TAR)
The IPCC changed its mind...
*"Human-emitted greenhouse gases are **likely** (67-90%) responsible for more than half of Earth's temperature increase since 1951."*
(Two years after the 1998 El Niño enriched warming spike of 1998... there had been three years of high temperatures)

2007... Fourth Assessment Report (AR4)
Well, the IPCC altered its position once again...
*"Human-emitted greenhouse gases are **very likely** (at least 90% chance) responsible for more than half of Earth's temperature increase since 1951."*

2014... Fifth Assessment Report (AR5)
And in the latest IPCC report, it goes even further...
*"Human-emitted greenhouse gases are **extremely likely** (at least 95% chance) responsible for more than half of Earth's temperature increase since 1951."*

Now come on... in the course of 24 years, and through 5 IPCC reports, the IPCC change their mind not once, but FIVE TIMES!

Well, if the IPCC can change their minds on such a critical subject... how the heck can the IPCC call the science SETTLED?

It's almost as if it were a TV series, and the writers keep hyping the drama a little bit each show, just to hold our attention.

Example 2, IPCC Settled Science
In the first IPCC Assessment Report in 1990, the following graph was introduced...

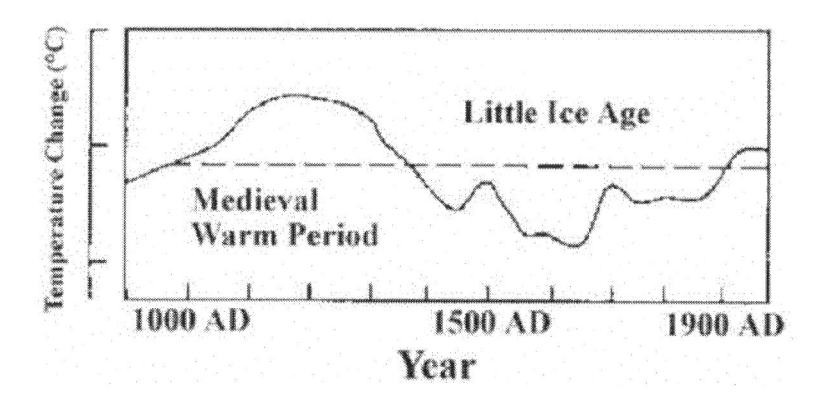

It shows the 'lows' of the Little Ice Age.
It shows the 'highs' of the Medieval Warm Period.
It shows the Medieval Warm Period being significantly warmer than the present day.

In 1990, few climate scientists disputed these historically-researched periods in the world's temperature record. They are backed up by proxy data, by anecdotal evidence and by written records.

Then, in 2001, in the IPCC's Third Assessment Report (TAR), Michael Mann et al introduced their much-disputed hockey-stick graph.
Created from a very obvious choice of proxy data, it showed no Little Ice Age, and no Medieval Warm Period.

It was a blatant change of 'accepted' history, and a watershed of belief and trust in the IPCC.

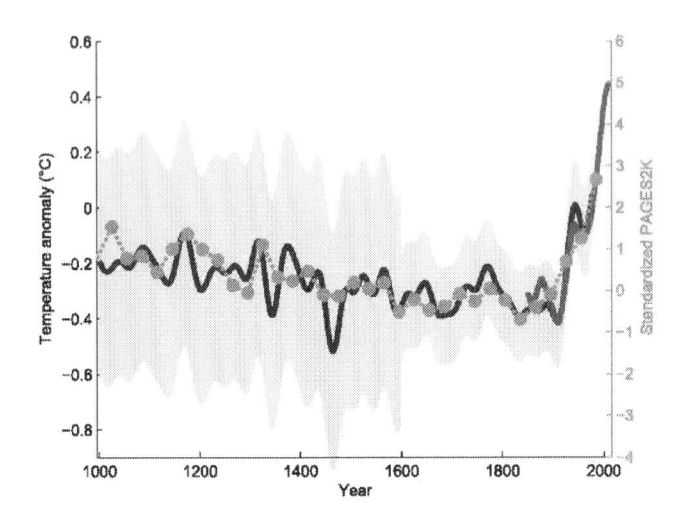

And as I mentioned before...

If the IPCC itself can change its mind on the amount of anthropogenic global warming...

If the IPCC itself can change its mind on the graph and data of the historical climate record...

Then, how they heck can the IPCC say the science is SETTLED?

8.6... THE IPPC'S SINGLE GRAIN OF TRUTH...
(The gorgeous slip-up...)

This little beauty of a sentence popped up in the IPCC's Third Assessment Report in 2001...

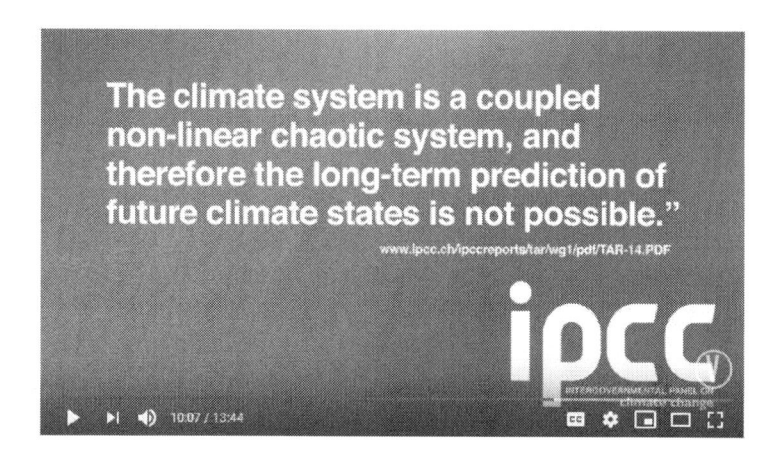

> The climate system is a coupled non-linear chaotic system, and therefore the long-term prediction of future climate states is not possible."
>
> www.ipcc.ch/ipccreports/tar/wg1/pdf/TAR-14.PDF

Now, before you go all... "THAT'S NOT THE WHOLE QUOTE!" on me, here's the 'whole' quote...

"The climate system is a coupled non-linear chaotic system, and therefore the long-term prediction of future climate states is not possible. Rather the focus must be upon the prediction of the probability distribution of the system's future possible states by the generation of ensembles of model solutions."

Does this sound contradictory to you?
In fact, does it sound like the IPCC are saying climate predictions are impossible?

Let's examine it more closely.

"The climate system is a coupled non-linear chaotic system"

This part is absolutely true, and that very fact makes it impossible to examine inside a box (a climate model software program)

"and therefore the long-term prediction of future climate states is not possible."

Yup... that's what the IPCC have actually said. Now, that's a fact. Because of the complexity of the climate system, many parameters are cut out of the climate models solely because they cannot be duplicated in a computer program.

Let's now read on and see if the IPCC get themselves out of the hole they dug.

"Rather the focus must be upon the prediction"

That's educated guesswork, right?

"of the probability distribution"

And that's statistics speak for 'looking at the average'.

"of the system's future possible states by the generation of ensembles of model solutions."

Hmm.

So... what the IPCC are saying is this...

Climate is too complex to predict... but if you play with enough climate models (the BEST of which have been proven woefully wrong in the past) and take an average of them all, you'll get an average prediction of the parameters we allow ourselves to input at the start.

Not too scientific, is it?

Let's take a look at the complexity of the climate, shall we?

8.7... WE CAN'T MODEL WHAT WE DON'T UNDERSTAND
(So your Climate Models are crap then...)

Despite what the Alarmists will tell you, mankind's grasp of the workings of the climate of our planet are nowhere near complete. Earth's climate is a highly complex subject, and is commonly called a 'chaotic' formula; as many of the variables are so complex, even modern computers don't have the computing power to process them all.
So, to radicalize the chaotic nature of climate factors, in practically every computer model used by our frantic Alarmist friends, they have left out four HUGE components. The reason why?

Well, it could be said that they leave out these four factors because they don't help the cause of CO2 as the main climate driver, but the reason is less conspiratorial than that.
The Sun, Oceans, Clouds and Cosmic Rays affect the Earth's climate in so many ways; NO ONE knows exactly how much and in what way these four factors affect climate.
The science is not settled in any of these scientific disciplines.
Let's take a look at each of them in turn.

The Sun
The biggest influence in our planet's climate is the sun. No one really disputes that; without our friendly ball of fire in the sky, we wouldn't exist; nothing would.

The sun influences our planet's climate system in many ways.
We get heat and light energy from the sun; but we also get every part of the electromagnetic spectrum thrown at us too... gamma rays, x-rays, high and low frequency waves.
There are professors who could tie you in knots with the sun's emissions' complexity.
Events on the sun's surface influence our climate too. Solar flares (coronal eruptions) rise millions of miles into space, causing such an energy dispersal they interrupt radio communications on Earth. Because of the sporadic pattern of these flares, they are not taken into consideration in climate models.

Sunspot activity has been known since the first telescope, and these flares and sunspots alter our climate; high sunspot activity means temperatures go up, yet sunspot activity does not figure in climate models because of their chaotic and erratic nature.

The Sun also has an eleven year sine wave cycle (primarily driven by sunspot activity) which affects its emission levels, and ejections. High sunspots mean the cycle is at its greatest, and low sunspot activity usually means the end of a cycle.

This is not a new area of science. It was first measured in 1755, and since then we have had 24 solar cycles. In 2019, we're currently at a solar minimum, at the beginning of Solar Cycle 25.

Scientists have also measured other oscillations; there's the 87 year Gleissberg cycle, the 210 year Suess cycle, the Hallstatt cycle and many more.

So the Sun's radiation ebbs and flows… are these fluctuations fed into the climate models?

Of course not.

Why not? Well most Alarmist climatologists would argue that even under very different sun cycles, the amount of energy transmitted to the Earth does not vary by much.

And then there's the Solar Wind, a stream of highly charged particles given off by the Sun's corona. Again, this data is not featured in the climate models because it's a highly specialized area of science.

The Oceans

Our oceans account for approximately 71% of the surface of our planet, and 97% of the world's water. It goes without saying that this huge percentage of water would influence the land it surrounds. Climate models have included ocean temperatures since their outset, yet only in the last few decades have extensive temperature measurements been available from satellite in low earth orbits.

The Earth's oceans can absorb 1000 time the carbon dioxide as the atmosphere, and this is commonly held in just the first few hundred feet of water. To get the warming down to the deep currents takes hundreds of years. It's currently understood that the oceans are presently acting on atmospheric warming or cooling many hundreds of years in the past.

To say that man understands the earth's oceans fully would be a fallacy; scientists are still making discoveries about current flow and the oceans ability to dissolve and give out carbon dioxide.

The Clouds
On average, clouds cover about 60% of the Earth's surface at any one time, and stand as one of the highest reflectors of sunlight reaching our atmosphere.
Clouds reflect light back into space, cooling the planet.
Predicting their exact location, density, size, height, and thickness is however, not possible.
Because of this complete unpredictability, they are unable to be used well in computer climate models.
So climate modelers either leave them out or include yet another fudge factor.

The Cosmic Rays
It may seem bizarre, but cosmic rays actually play a part in our climate.
Cosmic rays are high-energy particles raining down from exploded stars, far out in our galaxy. Basically they knock electrons out of air molecules. This process produces ions; positive and negative molecules in the atmosphere.
These ions attract aerosols (dust, small chemical particulates and water vapor), and gradually grow in size. These larger particles reflect sunlight, thereby cooling the planet ever so slightly.
Cosmic rays are not a constant force, however, when the sun is less magnetic, they increase, when the sun is active, they are reflected away from our solar system.

There's more about cosmic rays in section 10.4.

8.8... GLOBAL GREENING; THE SAVIOR OF OUR RACE
(You can have your rice cake and eat it...)

Carbon Dioxide is the food that all plants on Earth require to live.
Scientists know if the CO2 levels were to drop below 150ppm, plants would die... and thus, so would we.
NOW... don't worry, like we've said so many times, CO2 levels were about 300 for most of the Industrial Revolution, and are now around 420ppm. The Earth's plants are doing fine.

Well, actually, they're NOT doing fine... they're FLOURISHING in this CO2 'polluted' world.

In the Geophysical Research Letters, dated 2013, an article entitled...

"Impact of CO2 fertilization on maximum foliage cover across the globe's warm, arid environments"

Tells us that...

"Satellite observations reveal a greening of the globe over recent decades..."

(https://agupubs.onlinelibrary.wiley.com/doi/10.1002/grl.50563)

It seems that the rise of CO2 (just up to 2010) has encouraged plants to grow in even dry, arid environments.

Here's what NASA have to say about it...

"From a quarter to half of Earth's vegetated lands has shown significant greening over the last 35 years largely due to rising levels of atmospheric carbon dioxide..."

(https://www.nasa.gov/feature/goddard/2016/carbon-dioxide-fertilization-greening-earth)

Now, that's SIGNIFICANT (using NASA's word here).
Both NASA and NOAA fully agree that the Earth is GREENING significantly.
And NASA says it's LARGELY DUE to CO2 rise.
But don't think that this greening is a drop in the ocean... oh no. They go on...

"The greening represents an increase in leaves on plants and trees equivalent in area to two times the continental United States."

TWO NEW USA's producing GREEN stuff!
Now... when was the last time that NASA headline appeared.... ANYWHERE!

"Results showed that carbon dioxide fertilization explains 70 percent of the greening effect... "The second most important driver is nitrogen, at 9 percent. So we see what an outsized role CO2 plays in this process.""

Man, that article (on NASA's own site), has inspired me so much!

There's an old Frank Sinatra song entitled; *"And then I go and spoil it all, by saying something stupid, Like: "I love you"".*
And that's what happens here, in the NASA article.
After building us up so much, making us buoyant with glee, almost giddy with hope, they hit us with the typical Alarmist clap-trap. (British for bull-shit).
The article ends...

"While rising carbon dioxide concentrations in the air can be beneficial for plants, it is also the chief culprit of climate change. The gas, which traps heat in Earth's atmosphere, has been increasing since the industrial age due to the burning of oil, gas, coal and wood for energy and is continuing to reach concentrations not seen in at least 500,000 years. The impacts of climate change include global warming, rising sea levels, melting glaciers and sea ice as well as more severe weather events."

Yup, even though NASA admit in other articles that the sea isn't rising that much (a foot a century), world sea ice is relatively stable, and even using NASA and NOAA figures NO ONE has linked Global Warming to any severe weather 'event'.
Yeah, go NASA for taking the wind out of our sails.

However, taking some hope from the GREENING OF THE PLANTS side of the NASA article, I actually noticed one thing... the article ALWAYS refers to the greening as vegetation, leaves, plants, leaf area, leaf cover, plant growth and trees.

And I had already instinctively made the jump to another couple of words. Alternative words left out INTENTIONALLY because NASA doesn't want you to link this GREENING to another TYPE of 'vegetation'.

CROPS.
FOOD.

Yeah, not ONCE did the NASA article mention that if the green stuff is growing better, that means foodstuffs are growing better too.

Let me do NASA's homework for them, shall I?

Has crop production gone up in the last 70 years? (Since the CO2 level has been rising?)

Well, would you know it... a graph showing crop yield PER ACRE from 1959 to 2009.

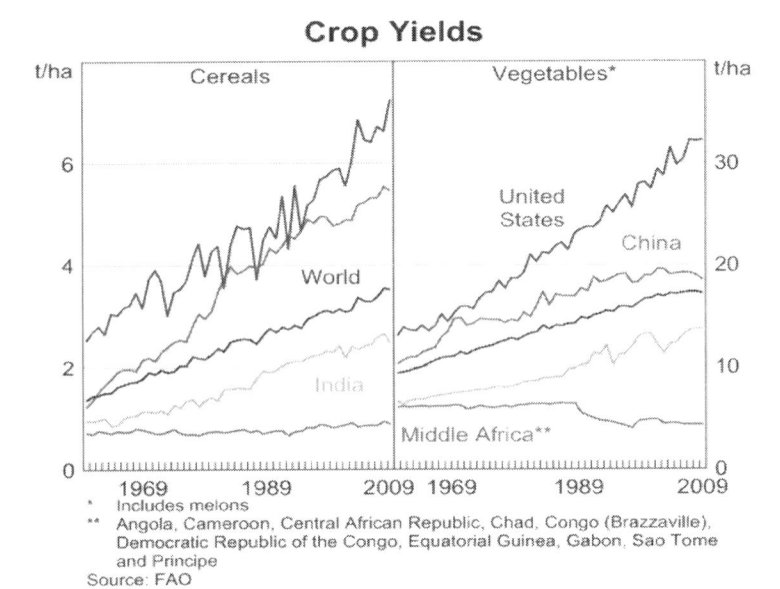

Crop Yields

*Includes melons
** Angola, Cameroon, Central African Republic, Chad, Congo (Brazzaville), Democratic Republic of the Congo, Equatorial Guinea, Gabon, Sao Tome and Principe
Source: FAO

Looks like ALL OVER THE WORLD, even up to the year 2009, crops are up. But the figures tell more than a graph. Let's take a more recent look...
From 1970 to 2010... World vegetable production is up 300%. World fruit production is up 250%. World corn production is up 250%. World rice production is up a staggering 400%.

AND A SIGNIFICANT AMOUNT OF THAT INCREASE IS THE EFFECTS OF HIGHER CO2 LEVELS.

Okay, I know, through science, better crops raised, seeds are more resilient, farming methods are better.
BUT... CO2 levels have risen, and CO2 is playing a part.
These figures are also due to longer growing seasons, less frost events; both BENEFITS of a global warming.
WOW! That's a BENEFIT of a rising CO2 level; you never hear about that in the media, do you? Let's look some more...

In the growing season of 2017-2018, the United Nations Food and Agriculture Organization reported GLOBAL RECORD YIELDS for corn, wheat and rice (the big three).
Five of the six biggest harvests of the 'big three' have been in the LAST SIX YEARS!
When did you read those figures in the media?
(Or does the media tell us that 'Global Warming' is destroying global crop output?)

Let's take a look at crop output in the five most populated countries, shall we? (figures from TheGlobalEconomy.com, 2016)
(And these figures are PER ACRE... we're NOT talking about cultivating more/extra land!)

China...
... set a new record for crop production of the big three in 2016.
... the top four production years were all in the last four years.
... crop production is now 27% than in 2000, and 100% higher than 1980.
Per Acre...!

India...
... set a new record for crop production of the big three in 2016.
... the best six years production has been set in the last six years.
... crop production is now 30% than in 2000, and more than 100% higher than 1980.
Per Acre...!

U.S.A...
... set a new record for crop production of the cereals in 2016.
... the best four years production has been set in the last four years.

... crop production is now 39% than in 2000, and more than 100% higher than 1980.
Per Acre...!

Indonesia...
... set a new record for crop production of the big three in 2016.
... the best five years production has been set in the last five years.
... crop production is now 34% than in 2000, and more than 100% higher than 1980.
Per Acre...!

Brazil...
... set a new record for crop production of the big three in 2015.
... the best five years production has been set in the last five years.
... crop production is now 92% than in 2000, and more than 300% higher than 1980.
Per Acre...!

Those are staggering figures... and a SIGNIFICANT part of those rises are DIRECTLY attributable to Global Warming/CO2 increase.

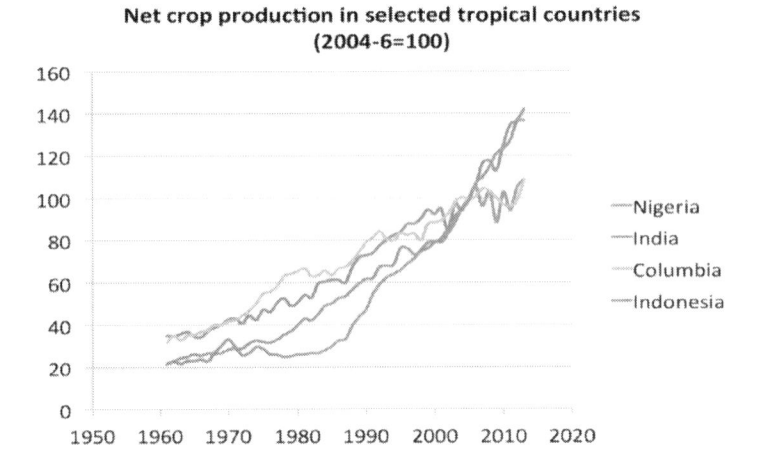

Net crop production in selected tropical countries (2004-6=100)

— Nigeria
— India
— Columbia
— Indonesia

And when did you hear good news of CO2 in the media... anywhere?

SECTION 9
THE CORRUPTION OF NASA

"Torture the data, and it will confess to anything."
(British economist Ronald Coase)

9.1... INTRODUCTION
It's difficult for me to write this chapter, because as a boy and young man, I had nothing but admiration for the men of NASA.
However, their role in the spread of science has changed.
They have thrown their hat firmly into the Alarmist camp, and although they remain close-lipped during their employment years, many ex-employees and retirees are not happy with NASA's direction.

NASA in association with GISS and the GHCN have changed so many data sets, their gall is astounding.
Changing a graph's trend is an easy process, but when your graph is showing historical temperature sets, firm in stone, set on paper, printed many times, how blatant can you be.

How are the historical data figures wrong? They were measured... the numbers were taken down, noted, and tabled.

How are SO MANY of the historical data figures wrong?
There are thousands of temperature records for thousands of stations, simply thrown away, and replaced by estimates, guesses... basically lies.

How are so many of the historical data figures wrong... TOO HIGH?
Well, the answer to that one's easy...
IF WE LOWER THE HISTORICAL DATA, AND KEEP THE MODERN DATA THE SAME (like tilting a see-saw), THE RESULTANT 'RISE' IN TEMPERATURE WILL BE GREATER.

And if they hype the modern figures a little too... the resultant rise in temperature will have us all sweating our cotton socks off... even if it's all just a bunch of made-up 'facts'.

Changing an annual temperature
Historical temperature stations recorded the maximum temperature in each day, and the minimum temperature.
That's over 2500 numbers each day reaching the Weather Center.

For the calculation of the average temperature, these numbers are averaged to get the average temperature of the day.
This happens each day of the year.
That's almost a million numbers being crunched to give an average temperature for a single year.

Now, in the next chapter, we see how NASA adjust historical temperatures.
There's one thing to remember when they change a year's average temperature by half a degree…
They're making ONE MILLION numbers change to get that average temperature.

But let's not get bogged down in hyperbole… let's have a look at some facts…

9.2... HOMOGENIZATION: THE HOME OF THE ALARMIST
(The fiddly-widdly, timey-wimey bits...)

Historical temperature records are available in some parts of the world going back to the first thermometer in 1714.
Historical temperature records are thought by many (including me until I wrote this book) to be raw temperature records as measured by the thermometer's owners, written down, the figures in stone.
I pictured thousands of folks getting to their thermometers at the same time every day, their friends ignoring the geek that measures the weather.
However, as the need for 'accurate figures' is needed, some of these records are being called into question.
(Not by me; I like the idea that old history CANNOT be re-written!)

Anyway...
Raw temperature records are subject to a multitude of possible readings which may have occurred as a result of;
1. Movement or repositioning of measurement site
2. Change of thermometer
3. Change to the physical environment of the site (tree growth or removal, urban encroachment, etc)
4. Change to the time of day measurements were made

I understand this.
Enter the Global Historical Climatology Network (GHCN), established to be the collection center of all data things weather-wise. The GHCN collects data from all over the world, collates the data, and makes it available to everyone. They supply records to GISS, NASA and NOAA.
Unfortunately, in their spare time, they also change historical data.

Their computers troll records continuously, and when they come across a temperature record that's 'out of their monitoring parameters', they dash in and make it conform.
And there lies the problem.
The vast majority of 'adjustments' made to historical data figures are in the downwards direction.
You just have to look at the temperature maps of the US, Australia, and in particular, Iceland, to see the trend.

Let's see what's going on, shall we?

9.3... NASA; WE DON'T CHANGE DATA (1)... WE ADJUST IT...
(No, you beat it to death...)

In the USA, Thomas Jefferson started the ball rolling. While in Philadelphia for the adoption of the declaration of independence, he purchased both a thermometer and barometer from a local merchant. (The temperature on that 'first' 4th of July, he noted, was a balmy 74°F) Jefferson wasn't the only weather geek, George Washington also took regular weather and temperature observations; the last entry in his diary was made the day before he died.

When the Smithsonian Institution supplied weather instruments to telegraph companies in 1848, things got a little serious. With a short break for the civil war, then a transfer from the military to civilian control, the USA had the first weather reporting service, the observations being submitted by telegraph to the Smithsonian, where the data was processed and weather maps are created.

Today there are over 1200 weather recording stations in the contiguous 48 states of the USA. They send data every day, and these figures are crunched into graphs of any kind of combination of data you can imagine.

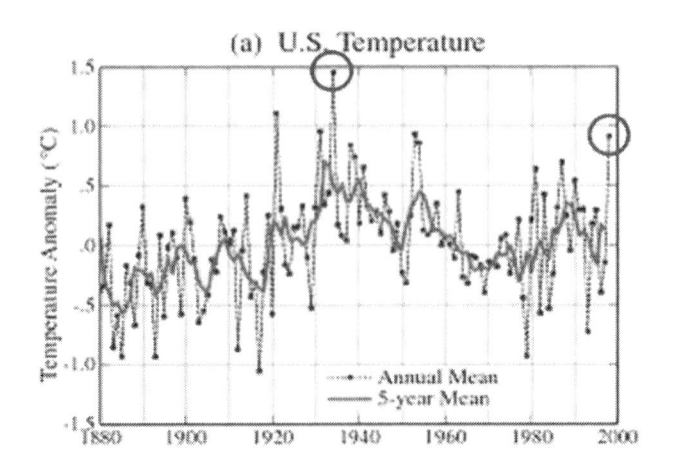

The James Hansen graph above is NASA's, taken from an IPCC document (the IPCC Third Assessment Report (TAR) from 2001), and features the 5 year average from 1880 to 1999.
It's also used in the GISS document, entitled...

Analysis of Surface Temperature Change

J. Hansen, R. Ruedy, J. Glascoe, and M. Sato
NASA Goddard Institute for Space Studies, New York

Here's a link to the doc
(Figure 6, shown at the end of the pdf file)
https://pubs.giss.nasa.gov/docs/1999/1999_Hansen_ha03200f.pdf

There are many historical facts and temperature trends that can be recognized from the graph's yearly average figures (the black dots).

1. Low and medium temperatures in the late 1800's, a gradual warming until 1920, nothing much out of the ordinary.
2. An extra cold year in 1917, historically verifiable.
3. A record hot year in 1934, part of the cause of the dustbowl that displaced tens of thousands from America's Midwest. (The record 1934 figure is 0.4°C higher than the next highest, in 1921).
4. Historical Highs in the early 1950's which sent hundreds of New Yorkers to the hospital; the record was 300 in one day. (The 1950's temperatures are roughly the same as 1998)
5. The dipping low temps of the mid to late 1970's, causing scientists the world over to believe we were heading into another ice age.
6. The 5 year mean figure from 1934 is the highest red point on the graph, 1940 second highest, then 1936, then 1954.
7. From the very first sign of a red line in 1880 (at zero) to the figure at 1997 (0.2), we have a 0.2 degree rise in temperature over 117 years.
8. There is significant cooling between the 1930's and 1979... almost a whole degree Celsius. (This is not consistent with a warming because of carbon dioxide which rose all the way through this graph).
9. There is NO constant upswing from the 1979 figure, in fact it varies up and down.
10. The El Niño-assisted 1998 figure is clearly shown as the FIFTH highest in the 20th century. It is surpassed by 1934, 1921, 1931, and 1953. (remember, 24 of the 50 US states STILL show their record high temperture was in the 1930's)
11. The high 1934 figure is a staggering 0.6C° above the 1998 figure.
12. From 1885 to 1931, there's an average warming of 1.2°C. From 1931 to 1978 there's a cooling of 1.0°C.
13. Basically it's a kind of hump in the middle graph.

It's a wonderful graph.

Now let's look at an updated version (2007) of the same graph, supposedly using the same basic data set, because why would data in the past be changed? Again, the graph is a NASA graph, AGAIN prepared by James Hansen, the same guy from the first graph.

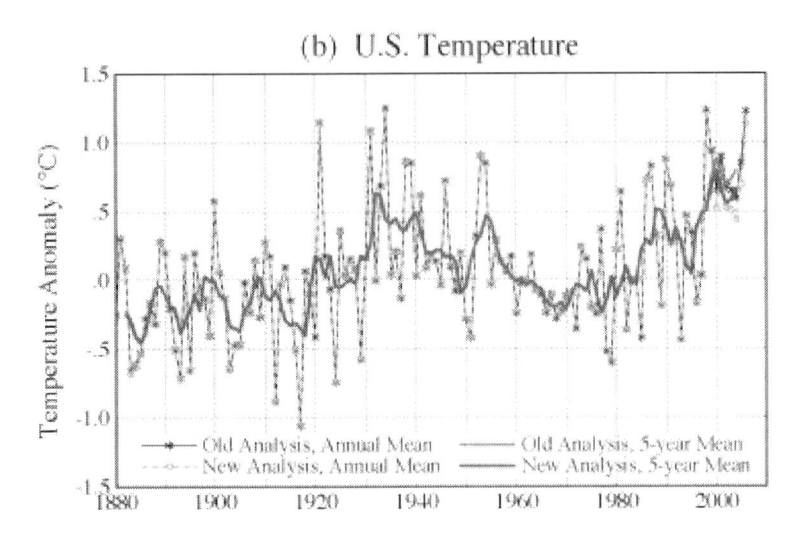

(b) U.S. Temperature

Let's see if the initial checklist stands up.

1. Low and medium temperatures in the late 1800's, a gradual warming until 1920, nothing much out of the ordinary.
CHECK!
2. An extra cold year in 1917, historically verifiable.
CHECK!
3. A record hot year in 1934, part of the cause of the dustbowl that displaced tens of thousands from America's Midwest. (The 1934 figure is 0.4°C higher than 1921).
NO... IT'S CHANGED... THE 1934 FIGURE IS JUST 0.1°C HIGHER THAN THE REST NOW. (In essence, NASA cooled the dustbowl by A QUARTER OF A DEGREE!!!)
4. Historical Highs in the early 1950's which sent hundreds of New Yorkers to the hospital; the record was 300 in one day. (The 1950's temperatures are roughly the same as 1998)
NO, THE 1998 FIGURE IS NOW HIGHER THAN THE 1950'S BY 0.4°C. (NASA achieved their Alarmist agenda... because of Hansen's adjustments... "the warming is unprecedented!")
5. The dipping low temps of the mid to late 1970's, causing scientists the world over to believe we were heading into another ice age.

KINDA. THE 1979 LOW HAS BEEN EQUALLED BY ONE IN 1969! (NASA dumbed down the ice age hype!)
6. The 5 year mean figure from 1934 is the highest red point on the graph, 1940 second highest, then 1936, then 1954.
NO, THE 1934 RED LINE HAS DROPPED, AND THE LATE 1990'S RED LINE HAS BEEN RAISED.
7. From the very first sign of a red line in 1880 (at zero) to the figure at 1997 (0.2), we have a 0.2°C rise in temperature over 117 years.
NO, THE FIRST RED LINE APPEARS AT -0.25°C, AND THE RED FIGURE FOR 1997 IS +0.5°C. THAT'S A 0.75°C CHANGE!
8. There is significant cooling between the 1930's and 1979... almost a whole degree Celsius. (This is not consistent with a warming because of carbon dioxide which rose all the way through this graph).
CHECK!
9. There is NO constant upswing from the 1979 figure, in fact it varies up and down.
CHECK!
10. The El Niño-assisted 1998 figure is clearly shown as the FIFTH highest in the 20th century. It is surpassed by 1934, 1921, 1931, and 1953. (24 of the 50 US states STILL show their record high temperture was in the 1930's)
NO. THE 1998 FIGURE NOW TIES THE 1934 FIGURE. AND SURPASSES THE 1921, 1931, AND 1953 FIGURES!
11. The high 1934 figure is a staggering 0.6°C above the 1998 figure.
NO. THE 1998 FIGURE NOW TIES THE 1934 FIGURE. AN 'ADJUSTMENT' OF 0.6°C!
12. From 1885 to 1931, there's an average warming of 1.2°C. From 1931 to 1978 there's a cooling of 1.0°C.
NO, THE RISE BETWEEN 1885 AND 1931 IS NOW JUST 1.1°C, AND THE COOLING BETWEEN 1931 AND 1978 IS 0.8°C.
13. Basically it's a kind of hump in the middle graph. **NO, THE NEW GRAPH IS A KINDA RISING UP TO THE RIGHT, KINDA GRAPH.**

Basically Hansen and NASA has made a mockery of the US temperature record and US history.

Below is one graph superimposed on the other.
(Sorry it's so blurred)

Why are so many temperature readings changed?
I see five temperature highs at least before 1950 that have been 'revised up', but not the 1934 figure, that's came down.

(a)(b) U.S. Temperature

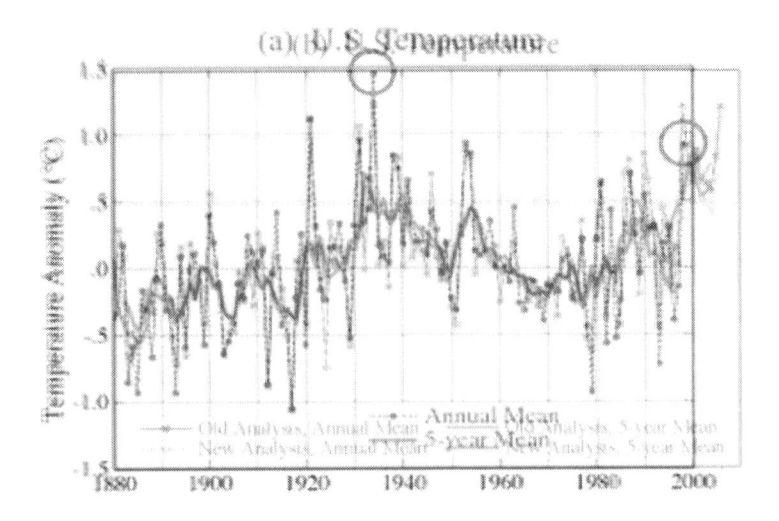

The really low in 1979 has been 'revised up' by 0.3°C... is that to make the ice age scare less credible?
And there are now THREE high late 1990's figures.

Now, we could start asking questions right away. But one spring IMMEDIATELY to mind...
In the second graph, there is a key line (green) that states... "Old Analysis, 5 year Mean".
Then WHY isn't the old 5 year mean drawn in green ALL the way through the graph to show ALL the changes they've made??

In actual fact they've gone out of their way to mischaracterize the key in another way. The black ("Old Analysis, Annual Mean") has been shown EXACTLY THE SAME as the brown line ("New Analysis, Annual Mean").
(And we've just determined that it is definitely NOT the case)
That's just dishonest, misleading and downright wrong/false.

It leaves me asking one other question. Why are there any changes at all?

The answers could be twofold.
1. If we dumb down the earlier temperature highs, it stops embarrassing questions like "1934's hotter than 1998... no global warming".
2. It smooths out the hump, making the graph more hockey-stick in appearance.
3. Third, and most importantly, it allows the Alarmists to postulate...

"The warming in the late 20th century is unprecedented."

But the story is not finished.
NOAA came out with a 'new' graph of historic temperatures in 2012.

Yes, it is yet another 'Contiguous US Average' graph prepared for us by NOAA, but this one is considerably different from the NOAA graph at the beginning of the chapter.

Note that the graph's title suggests no average for the year... This new graph is based on average **July** temperatures.

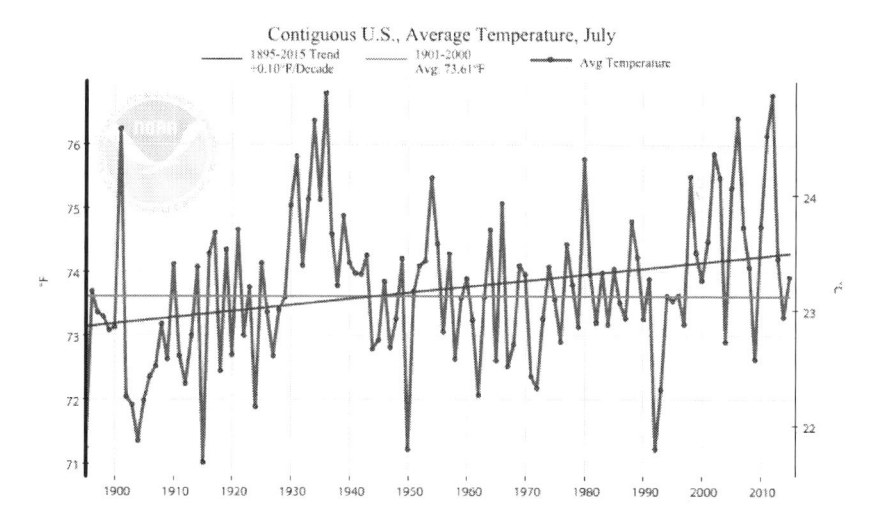

The 1934 high is still there, and the 1950 one. There's a little slip cooler in the 1970's (but no sign of any ice age), and even the 1998 'record high' (now reduced significantly) has been topped four times in the 2000's.

And WHAT is that huge spike in 1900 all about?

That spike doesn't appear in ANY of the previous graphs!
Both the 1999 graph and the 2012 one show that the spike in 1934 is a full degree warmer than 1900.
But, yup, they've changed history again.

And yet, look at the century long temperature increase.
On average (blue line) the US has only risen exactly ONE DEGREE in a CENTURY! (120 years actually)

Hardly catastrophic, is it?

So... historical temperature graphs can actually look different, if they're based on different times of the calendar year.

But also, it seems, they can draw two DIFFERENT graphs of the SAME month in a matter of days.

The Daily Caller wrote a column denouncing NOAA's practice of changing America's history... Citing a Christopher Booker article in the UK's Telegraph, they caught NOAA showing two different graphs

Well, you'd think that NASA would have done with US July temperature graphs in 2012, but it seems they wanted to do another one.

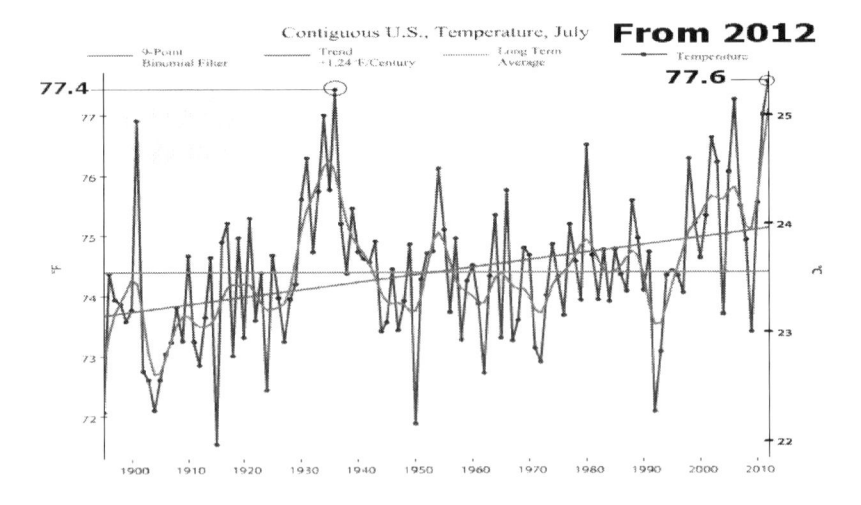

Look carefully... the temperature record is given for the month of July.

The peak in 1936 is given as 77.4°.
The peak in 2012 is listed as 77.6°.

Okay... move along, nothing to see here... NOAA has spoken.

BUT... the Daily Caller states...

"According to NOAA's National Climatic Data Center in 2012, the "average temperature for the contiguous U.S. during July was 77.6°F, 3.3°F above the 20th century average, marking the warmest July and all-time warmest

month on record for the nation in a period of record that dates back to 1895."

"The previous warmest July for the nation was July 1936, when the average U.S. temperature was 77.4°F," NOAA said in 2012."

The daily Caller goes on to say that a meteorologist and climate blogger, called Anthony Watts checked the NOAA data a few days later to find the following graph…

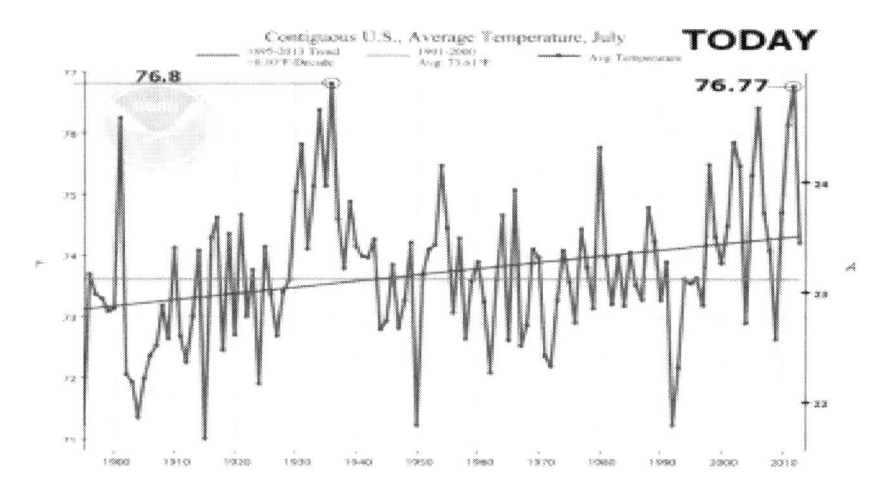

YUP! Watts had found that NOAA had second thoughts, and quietly reinstated July 1936 as the US's hottest month on record. Watts is quoted as saying…

"The past, present, and future all seems to be 'adjustable' in NOAA's world… This is not acceptable. It is not being honest with the public. It is not scientific. It violates the Data Quality Act."

NOAA… people not to be trusted.

BUT… let's look at ALL THREE 2012 graphs again, shall we.
In the first 2012 graph, the 1934 average is shown at about 77°F.
In the second, it's actually stated as being 77.4°F.
In the third, it's stated as being 76.8°F.

Three separate graphs in the same year, by the same organization, showing DIFFERENT historical temperatures.

AND the modern temperatures don't stand up to scrutiny either.
Graph 1 shows the 2012 high as just below 77°F, then 77.6°F, then 76.77°F.
Which graph do you trust?

Digging on my own...
There's nothing like finding your own data, and the internet is crammed full of it.
The following figures are from the government's own websites.

Here are the record top temperatures by US state, graphed against the decade they occurred. You can look these figures up yourself... I did. The graph below is my own work. You'll find only a few states have recent 'highs'... most of the records were set in the 1930's or 1950's.

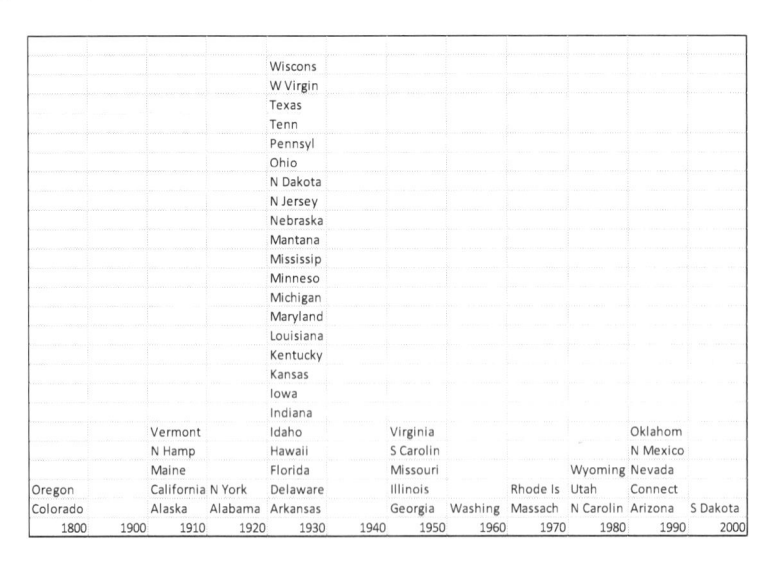

	1800	1900	1910	1920	1930	1940	1950	1960	1970	1980	1990	2000
					Wiscons							
					W Virgin							
					Texas							
					Tenn							
					Pennsyl							
					Ohio							
					N Dakota							
					N Jersey							
					Nebraska							
					Mantana							
					Mississip							
					Minneso							
					Michigan							
					Maryland							
					Louisiana							
					Kentucky							
					Kansas							
					Iowa							
					Indiana							
		Vermont		Idaho		Virginia			Oklahoma		
		N Hamp		Hawaii		S Carolin			N Mexico		
		Maine		Florida		Missouri		Wyoming	Nevada		
	Oregon	California	N York	Delaware		Illinois		Rhode Is	Utah	Connect		
	Colorado	Alaska	Alabama	Arkansas		Georgia	Washing	Massach	N Carolin	Arizona	S Dakota

The actual figures are given below for your perusal.

State	Temp	Date	Station
Ala.	112	Sept. 5, 1925	Centerville
Alaska	100	June 27, 1915	Ft. Yukon
Ariz.	128	June 29, 1994	Lake Havasu
Ark.	120	Aug. 10, 1936	Ozark
Calif.	134	July 10, 1913	Death Valley
Colo.	118	July 11, 1888	Bennett
Conn.	106	July 15, 1995	Danbury
Del.	110	July 21, 1930	Millsboro

Fla.	109	June 29, 1931	Monticello
Ga.	112	July 24, 1952	Louisville
Hawaii	100	April 27, 1931	Pahala
Idaho	118	July 28, 1934	Orofino
Ill.	117	July 14, 1954	E. St Louis
Ind.	116	July 14, 1936	Collegeville
Iowa	118	July 20, 1934	Keokuk
Kansas	121	July 24, 1936	Alton
Ky.	114	July 28, 1930	Greensburg
La.	114	Aug. 10, 1936	Plain Dealing
Maine	105	July 10, 1911	N. Bridgton
Md.	109	July 10, 1936	Cumberland and Frederick
Mass.	107	Aug. 2, 1975	New Bedford and Chester
Mich.	112	July 13, 1936	Mio
Minn.	114	July 6, 1936	Moorhead
Miss.	115	July 29, 1930	Holly Springs
Mo	118	July 14, 1954	Warsaw and Union
Mont.	117	July 5, 1937	Medicine Lake
Neb.	118	July 24, 1936	Minden
Nev.	125	June 29, 1994	Laughlin
N.H.	106	July 4, 1911	Nashua
N.J.	110	July 10, 1936	Runyon
N.M.	122	June 27, 1994	Lakewood
N.Y.	108	July 22, 1926	Troy
N.C.	110	Aug. 21, 1983	Fayetteville
N.D.	121	July 6, 1936	Steele
Ohio	113	July 21, 1934	Gallipolis
Okla.	120	June 27, 1994	Tipton
Ore.	119	Aug. 10, 1898	Pendleton
Pa.	111	July 10, 1936	Phoenixville
R.I.	104	Aug. 2, 1975	Providence
S.C.	111	June 28, 1954	Camden
S.D.	120	July 15, 2006	Kelly Ranch/Usta
Tenn.	113	Aug. 9, 1930	Perryville
Texas	120	Aug. 12, 1936	Seymour
Utah	117	July 5, 1985	Saint George
Vt.	105	July 4, 1911	Vernon
Va.	110	July 15, 1954	Balcony Falls
Wash.	118	Aug. 5, 1961	Ice Harbor Dam
W. Va.	112	July 10, 1936	Martinsburg
Wis.	114	July 13, 1936	Wisconsin Dells
Wyo.	116	Aug. 8, 1983	Basin

The best way to come to your own conclusions is to do some of the work on your own...

Maybe here, in these untouched historical figures, the science is actually settled.

9.4... *NASA; WE DON'T CHANGE DATA (2)... WE 'ADJUST' IT.*
(No... you corrupt it)

The National Oceanic and Atmospheric Administration (NOAA) is the US agency that collects, collates, analyses and publishes temperature data.
They are considered to be one of the most prestigious weather-related agencies in the world.
However they also take raw temperature data from their many US weather stations... and before we, the public, see the data... they change it.
Now, yes, they call the changes, 'adjustments', and they have numerous homogenizations applied to the raw data before the final 'product' is released.
But being blunt, they change the data.

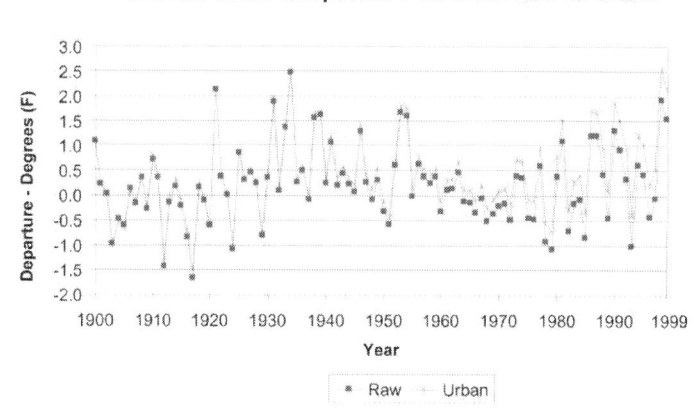

USHCN Mean Temperature 2.5 X 3.5 Raw vs Urban

This is the NOAA graph of raw and adjusted data.
Raw data is shown as a dark square, and unadjusted data is shown as smaller spots.
From left to right, the early 'adjustments' have little to no effect on the raw data, but as we read forward in time, the changes become more pronounced. By 1999, the adjusted data is about 0.7°F higher than the raw data.
To quickly compare to the US graphs (James Hansen) from the previous chapter (9.3) we note...
The 1998 high is shown as the same as the 1934 record.
There are highs in the mid 1950's.
There is a low in the late 1970's (Cold period/Cod War).
Wonderful.

But the data adjustment makes any temperature rise in the late 20th Century 0.7°F higher than raw data would show.

The NOAA adjustments are AMPLIFYING any rising temperature trend.

Some would say their adjustments actually ARE the warming.

For the roots behind the changes, we go to another NOAA graph.

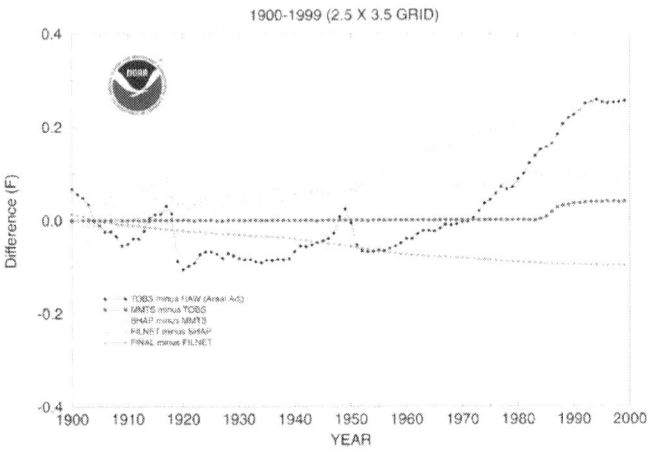

This graph shows the five ways NOAA adjust their raw figures before making them public.
Here are the five adjustments, and my take on them.

1. The top line (the one that produces the most upward bias) is an adjustment for time of day... Time of day Bias (TOBS)

Basically NOAA is saying that many old readings were reported wrong or the time of reporting meant the maximums were actually those of the day before (depending on when the thermometer was reset). NOAA are saying that old folks were lax, and they're looking back in time and correcting their faults.

But, dear reader, look at the bias here... it's a classic 'trick'; the older measurements are lowered, the recent data is raised, giving the whole graph a rising trend.
However, why do we need to make an adjustment to the figures for 'time of day'?

Aren't these readings daily maximums? And we don't care when the maximum happened, we just care about the value of the daily maximum, so we can combine daily figures to give us an annual figure. Right?
I mean, in a million data figures, who cares if a couple are kinda wrong... it'll all average itself out... that's why we look at average figures!

But of course, with all modern adjustment in an upward direction, it suits the warming agenda.

2. The next line (the very straight one) is for a change in the type of maximum/minimum thermometers used... a 0.04°F rise after 1990.

To my method of thinking, the need for this adjustment would indicate that NOAA considered the old type max/min thermometers to be 'off' (reading too high) by 0.04°F, and the new thermometers (now measuring the 'correct' temperature) had to be adjusted upwards by 0.04°F to bring them in line with the old, less accurate temperatures.

Tell me, when you change the type of thermometers in your weather stations, why would you adjust the new more accurate thermometers to read the same 'off', 'inaccurate' temperature figures???

SURELY the whole point in replacing these old thermometers was to make the raw data more accurate? Yes?
Then WHY would NOAA add an adjustment to the new more accurate figures at all???
In actual fact, you're making the new more accurate data, less accurate.
(I just don't get that one)

3. The next line (a gradual upward trend) has been supposedly included for 'changes in station siting'.

Hmm... so NOAA changes the site of a weather station, for whatever reason. And in EVERY CASE they raise the adjustment bias of this new station to read warmer than its predecessor? (This indicates the old station has been reading too low)

Look at the line. There is no time of downward trend at all... not once in a hundred years.

Every time NOAA have changed the site of a weather station, they have added an upward bias to compensate for the old station measuring too low.

So since the trend has been slow and gradual over 100 years, why not just keep the new, more accurate reading, and accept the new reading.

Each small individual adjustment (when a station site was changed) was probably unnoticeable in the grand scheme, so why not just read the new temperature? Don't adjust for the miniscule bias. Why not allow the temperature record stand for itself and swallow the change?

Nope... they adjust EVERY change of site with a rising bias.

By the end of the century this bias has reached 0.2°F.

That's 0.2°F ADDED to every modern NOAA reading of raw data!

4. The next line (which rises from zero at 1900 to about 0.06°F in 2000) is to fill in missing data from individual station records.

Now some stations have missing days; I accept that. But we're talking about a year-long average here... ONE max and min temperature figure for each day of the year. Surely if a station supplied only 354 records one year, it should just be divided by 354 to get that particular station's yearly average. Why do we need to estimate a day's readings?

THEN, when you add all the other stations, and average them too (to get the average figure for the USA), it'd all work out fine.

Nope! NOAA would rather raise the raw data by ANOTHER seemingly useless bias... and lookie!... the line trend goes constantly upward!

Yet another win for the climate Alarmists... adding another 0.06°F to every single modern piece of raw data.

5. The next line (the only line with a downward trend) is supposedly included to compensate for Urban Heat Island (UHI) effects.

NOAA realized that urban areas gave a raising bias (roughly 0.1°F to temperature readings, and allowed a correction for such readings.

That's good.

But surely as NOAA changed the weather stations from the yellow line, to more appropriate sites, this should decrease in value as the century rolls on.
NEWS JUST IN... NOAA have actually decided to remove this correction in the most recent adjustments to the historical temperature record, as it is unnecessary.
By this time I'm reminded of Tom Cruise's verbal slap at Jack Nicholson's rant in the movie,
"I want the truth!"
"You can't handle the truth!" Nicholson hits back.

Surely AGW is the most 'adjusted' hypothesis in history.

Now, with so many adjustments to consider, let's look at NOAA's graph of ALL adjustments added together, shall we?

NOAA call the graph; "Difference Between Raw and Final USHCN Data Sets".
Basically they're saying this is the final data (the figures released to the public) minus the raw data... leaving only the corrections left behind.

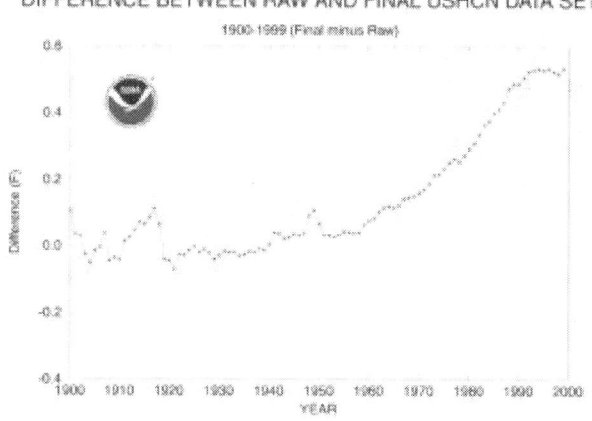

DIFFERENCE BETWEEN RAW AND FINAL USHCN DATA SETS
1900-1999 (Final minus Raw)

Now let's look at the graph of added biases.
It starts roughly at zero, and rises approximately 0.5°F.
Hmm. This graph of ADDED bias looks amazingly like another graph...

In James Hansen's graph below, the rise in temperature from 1990 to 2000 is 0.75°C (1.35°F) .In fact... it all begs one question...

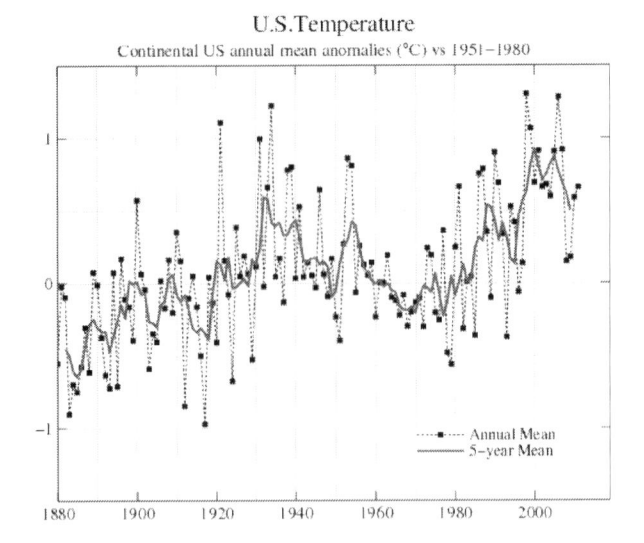

U.S. Temperature
Continental US annual mean anomalies (°C) vs 1951-1980

WHAT HAPPENS if we take the added biases AWAY from the James Hansen graph above?
Supposedly, that would give us the REAL RAW DATA temperature for the USA.

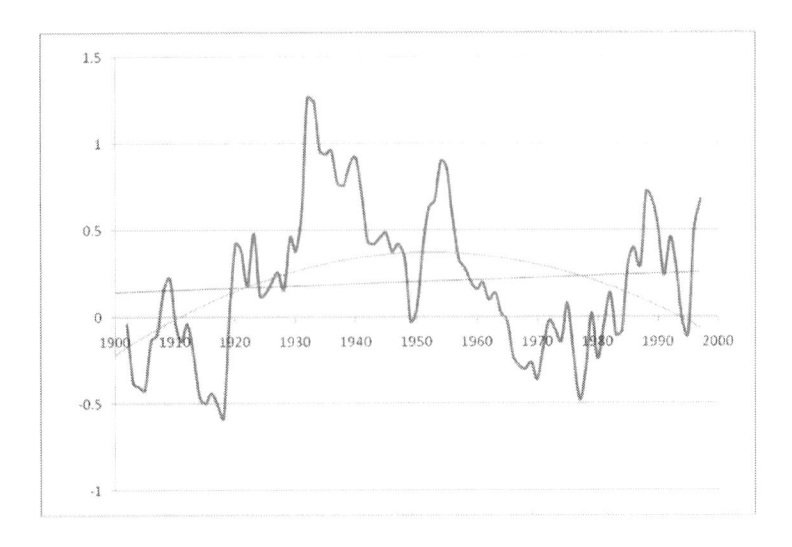

There it is in its majestic glory.

High temperatures in 1930's.

Kinda high temperatures in the mid 1950's.

Major cooling from 1934 to 1979.

Lowest dip in 1979.

It all fits with anecdotal evidence and individual state's temperature highs, and nowhere is there ANY modern warming CLOSE to the 1934 highs.

In my mind, the science seems to be settled... huh?

https://jennifermarohasy.com/2009/06/how-the-us-temperature-record-is-adjusted/

9.5... NASA; WE DON'T CHANGE DATA (3)... WE ADJUST...
(Yeah... try telling that to the Icelandic people...)

Iceland is a proud nation... proud of their heritage, their history, their place in the world... they take it seriously. But when some foreign nation tries to hijack their historical temperature records, they soon get mightily pissed!
Let's start at the beginning...

The Global Historical Climatology Network (GHCN) is a database of temperature records managed by the National Climatic Data Center (NCDC) at Arizona State University.
They are meant to be a collator of records, a keeper... a source beyond reproach.
They supply records to GISS, NASA and NOAA.

So, when an American company (GHCN) look back at the old records, and start changing American temperatures, that's a bit naughty.
Historical records are meant to be IN STONE, unchanging. If they're faulty, let someone point it out... but they're not to be radically changed... especially for political or ideological purposes!
So GHCN changed American historical temperature records.
Like I said, that's naughty.
But when they go to another country and do the same... that's downright cheek!
Let's start at the beginning... again.

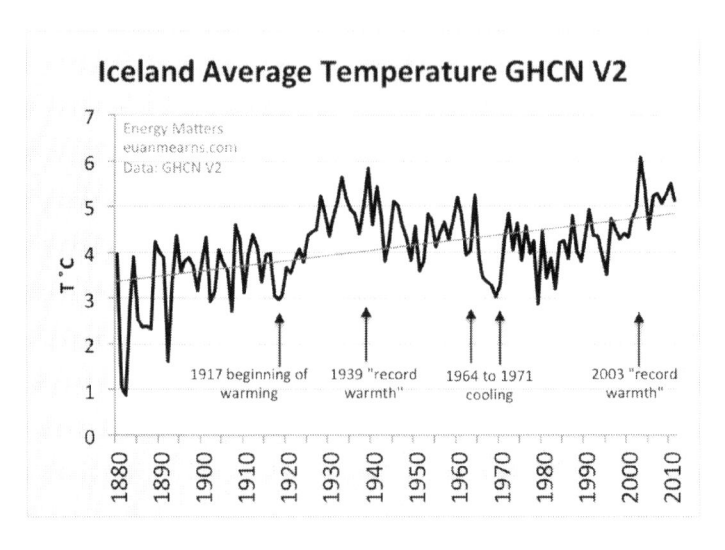

Above is the GHCN Graph of Icelandic temperatures... VERSION 2.
This is an amalgam of SEVEN stations all over Iceland.

Looks like at the end of the little ice age, Iceland (like the rest of the planet), warmed up a bit.
Take a look at the graph...
In fact, from the low in 1880, it warmed all the way to 1939, where it peaked; a record year.
Then it cooled until 1979 (the peak of the cooling, 'Cod War' years with the UK), then it warmed again, to a 2003 record.
Nothing much to see... very much the same as many other places in the world, the USA, UK, etc.
And also 'kinda' follows the historical records in Iceland.
Icelandic warming from 1880 to 2010 is about 1.4°C; about 1 degree per century.
But look carefully at the first forty years after 1880... there's a lot of warming there NOT caused by carbon dioxide.
Let's zoom in a bit, shall we...

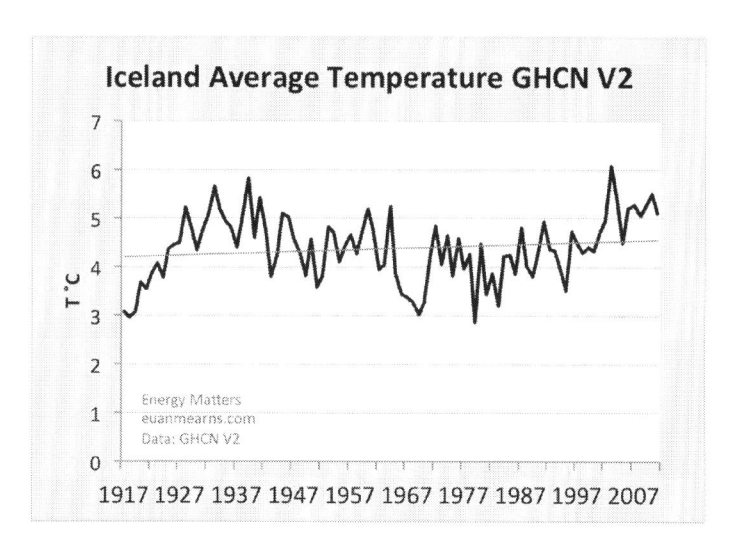

Iceland Average Temperature GHCN V2

This is the same graph, but with a 1917 start.
It shows almost 3°C in the first 20 years, then 2°C cooling in the next 50 years.

For obvious reasons GHCN were not happy with the record. It did not fit the CO2 warming curve narrative, it did not toe the party line, it did not fit the new 'adjusted' data... something had to be done.

So, in typical GHCN fashion they dug into the data, fudged it, homogenized it, tampered with it, and adjusted it... to read 'correctly'.

AND.... (big music build up here)... what do you know...

After their unbiased, neutral tampering, the data showed exactly what they wanted it to show.

Let's take a look at GHCN Version 3, and the V2 together, shall we?

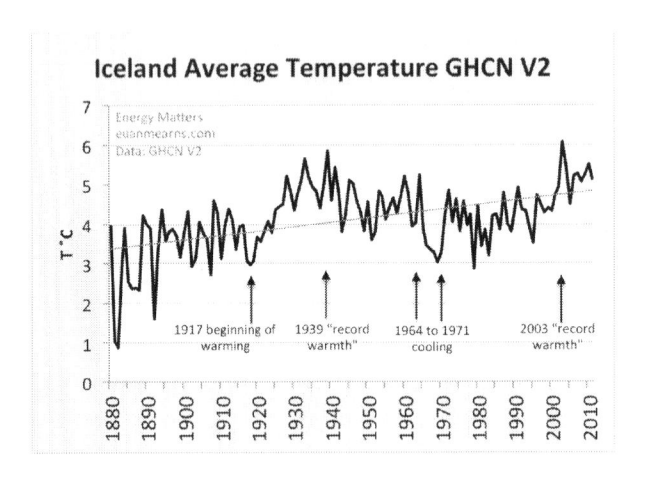

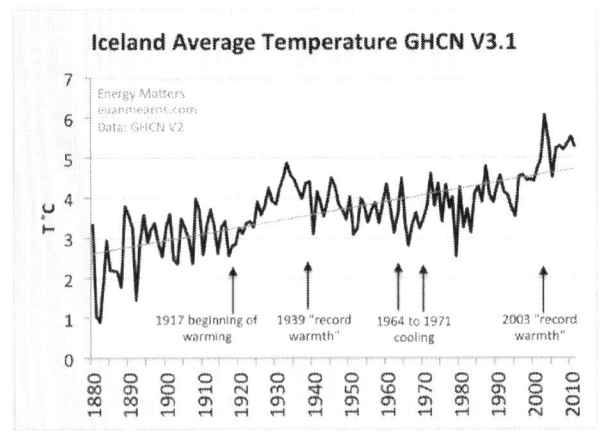

Iceland Average Temperature GHCN V2

Iceland Average Temperature GHCN V3.1

They 'kinda' look the same, but when you examine the differences, a few points jump out.
1. The warming curve over 130 years has gone from 1.4°C to 2.2°C.
2. The 1939 'record', and basically all the values from 1917 to1971 have been reduced. There's NO 3°C rise from 1917 to 1937. There's NO embarrassing 2°C cooling from 1939 to 1985.

3. The 1892 low and 2003 highs have been kept the same.

Yup... I can notice the new trend... a more pronounced rise.
You can see the new trend inserted.

AND... so could the Icelandic climatologists. They refused to accept the American data, and they still do.

9.6... *NASA; WE DON'T CHANGE DATA (4)... WE 'REVISE' IT*
(Yeah, but you ALWAYS 'revise' it for your own purposes!)

To begin this rather complex, yet vital, part of the Denier's arsenal, we have to first look at the two ways temperature has been 'measured' in recent history.

Measurement by Thermometers
We see the data between the 1850's and 1900 on most graphs of 'global temperature', but there's one very important fact that is not well known. Up to 1880, there were very few actual reliably kept weather stations in the world. The map below shows the details of 1880.

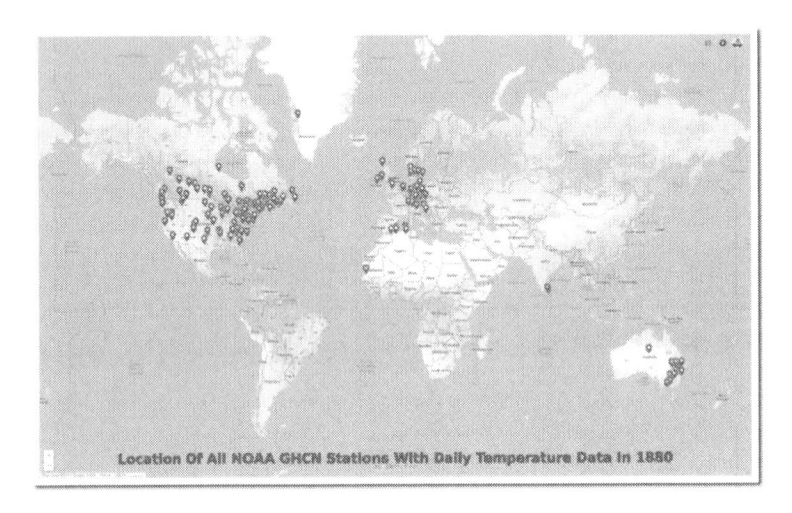

Location Of All NOAA GHCN Stations With Daily Temperature Data In 1880

As you can see, any data before 1880 did not actually show mean 'global' temperature. It showed an average of the USA/Canada, mixed with a few figures from Europe, Australia, and a dribble from India, North Africa, and one from Greenland.

"Yeah, but it got better damn quick, didn't it!" I hear the alarmists say, postulating the coming of the global age of technology.
Well, it did... and it didn't.

The map below shows the weather stations twenty years later in 1900. Yes, lots more stations, yet not really more 'global' coverage; a handful in North Africa, and none in South America or Antarctica.

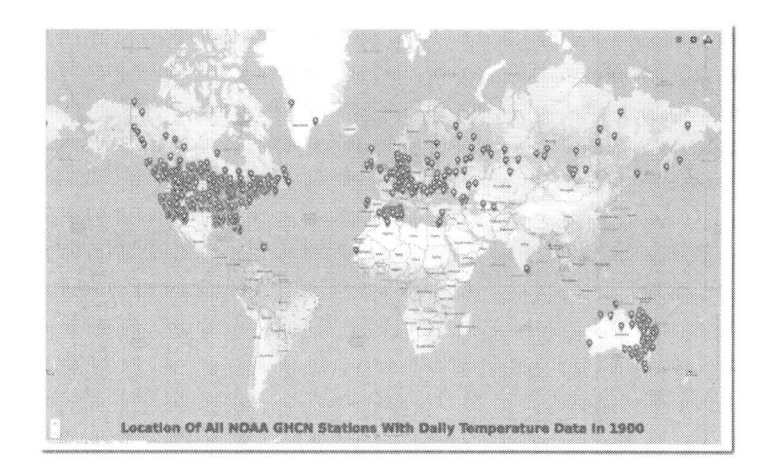

Location Of All NOAA GHCN Stations With Daily Temperature Data In 1900

There are 1000 stations in the USA, some in Europe, Russia, Canada, Australia, but HUGE areas of the globe are not covered. Basically, in 1900, what scientists called 'mean global temperature', was still actually the mean USA temperature.

But detailed records were kept, right? And graphs plotted, right?

Yes, they were, and land temperature stations have been increased globally, but even today, the full globe is not covered.
South America has only a few in comparison to the thousands in the USA, Africa is nowhere near covered, and large parts of the Middle East and Asia show no signs of catching up soon.

Even as near as forty years ago, (1980!) climate scientists couldn't even come close to claim they knew anything about southern hemisphere temperatures. There was simply so little data.
The map below shows the average land surface temperature globally for Sept 2018. This is the most modern map I've been able to find.
There's a lot of data, in nice easily understandable pink and blue squares. Red means 'hotter than average', pink means 'kinda hotter than average', pale blue, 'kinda cooler than average', and darker blue, 'cooler than average'.
Straightforward so far.

The areas in grey (as the map states in the bottom right hand corner, state 'missing data'.
SO... even in 2018, look at the areas of the globe that are NOT covered by land measuring stations.

Look at the huge swathes of Canada, South America, the Middle East and Asia not fully covered. Then look at Africa and Antarctica. We still don't have a full land surface temperature picture, do we?

Look at poor Greenland, the much mentioned 'holder' of a mile thick ice shelf... with little to no coverage.

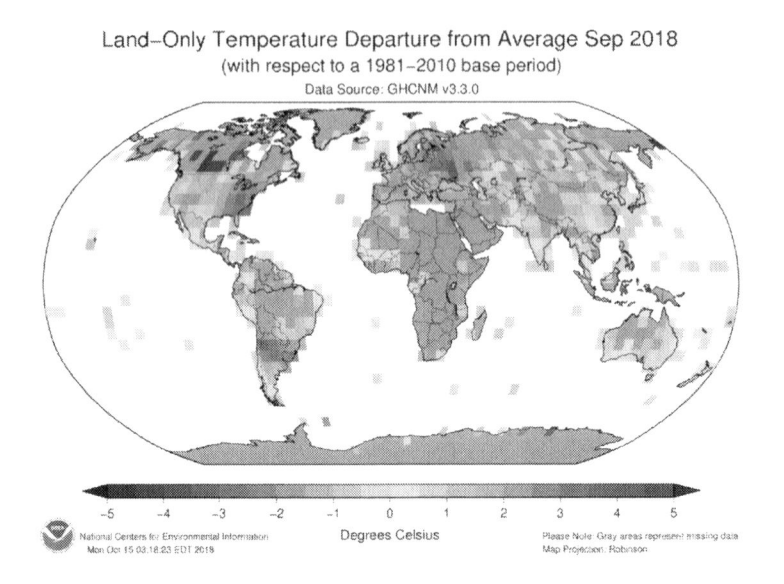

Land–Only Temperature Departure from Average Sep 2018
(with respect to a 1981–2010 base period)
Data Source: GHCNM v3.3.0

There's more grey (uncovered area) than squares (where they have thermometers)!

So much for all the graphs of 'mean global temperatures, right?

"So..." you ask, "How do NASA and NOAA calculate their 'global' average temperatures?"

Good question!

Basically, they take all the readings they have, then for the areas they have no data, they extrapolate, estimate, guesstimate, guess and make it up.

Literally.

And that's just land temperatures.

Sea temperatures between 1880 and 1950 are not available, inaccurate, incorrectly taken, erratically taken or unsubstantiated.

Basically there's no real record of sea temperatures at all.

(Because we all know that for a temperature 'record' to be set, we must have regular readings taken from the same place at regular times).

Since 1950, these readings have been more carefully taken, and now with the new satellite measurement, they are now fully global. As long as clouds and weather don't get in the way, we can 'see' any temperature, anywhere, any time, day of night.

It's a fantastic resource, if it's used correctly, but any extrapolation back in time beyond 1950 is pure guesswork.

Measurement by Observation
By observation, we usually refer to the written word, historical evidence, and the media. (And by media, I'm talking the age of newspapers)
But I want to look further back than 1880…
Back past the newspaper age… as far back as humanly possible.

England has many written records, and they go back to the time of King Henry the Eighth.
Maps were kept to govern local boundaries.
Weather records were kept… whether it rained early in spring, light rain, drizzle, late snow etc… these details were important.
Records were kept of temperature long before thermometers… whether the spring was exceptionally warm, hot, very hot, chilly, etc. These details were kept to calculate the local and national taxes on crop production.

For small landowners, dukes and kings, keeping records was an important job.

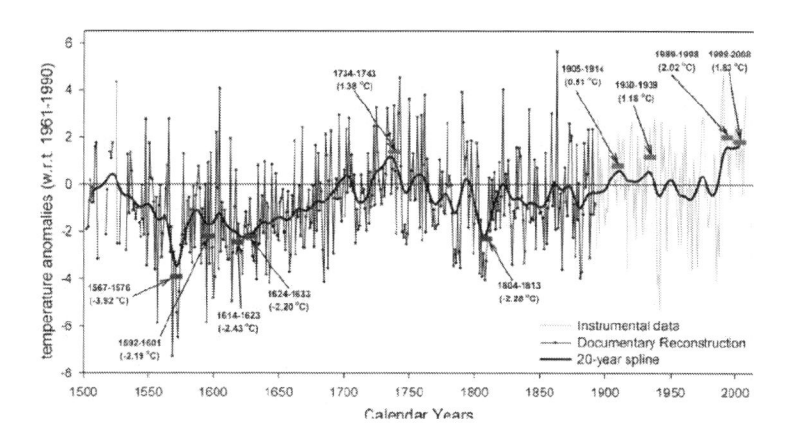

The graph above is a painstaking work that links written records of crop production, warm and cold winters and summers from parish records, from farm and manor records, and from public (government) records.

Looking up at the graph we see the end of the renaissance warm period, and the dip in temperatures when the Thames River in London froze annually for winter fairs (last frozen in 1814).

Not only are these evidences anecdotal in writings of the day, paintings and historical fiction, they are backed up by tree ring evidence and ice cores.

From 1740 to 2000, the average temperature rose half a degree if that.

Yes, it's climate change.

It could even be called English Warming.

But it's hardly catastrophic, is it?

9.7... NASA; WE DON'T CHANGE DATA (5)... WE 'IGNORE' IT
(More data fraud at NASA??...)

NASA's not getting too good a press at the moment... at least not in this book.
But you see, when you publish contradictory work on so many subjects, you leave yourself open for study by folks like me.

In the world of climate change subjects, none has the coastal population quite so on their toes than sea level rise.
NASA, through their own satellites and their link to the European ones, have taken on this work too, so let's take a look at their work on sea levels.

There was a NASA story on their website (web address at end of chapter) dated Feb 2018 which tells more gloom and doom.

"New study finds sea level rise accelerating"

Oh crap, not Florida disappearing again!

"Global sea level rise is accelerating incrementally over time rather than increasing at a steady rate, as previously thought, according to a new study based on 25 years of NASA and European satellite data."

'Accelerating incrementally', huh? That means the rate of rise is increasing on an increasing level.
And 25 years of NASA data eh? Well that can't be good for your beach condo... (But of course, it didn't stop President Obama from just splashing out $15 million on his beach front property at Martha's Vinyard!) I'll let NASA continue...

"If the rate of ocean rise continues to change at this pace, sea level will rise 26 inches (65 centimeters) by 2100—enough to cause significant problems for coastal cities."

TWENTY-SIX INCHES?!?!

Seriously, that's a lot. Even I know that.
So I looked through the article.
NASA blames the thermal expansion of the oceans as they heat up, they blame changes in ice sheets, and ice and glacier melt.

Now I could regale you with facts and figures from my Denier friends, but I'm going to use NASA's own data to pour cold water on the 'new data'.

Let's take a look at some NASA graphs shall we, they're accurate, and include state of the art satellite data available to everyone on the planet.

They're featured on Google, you can collect them at your leisure.

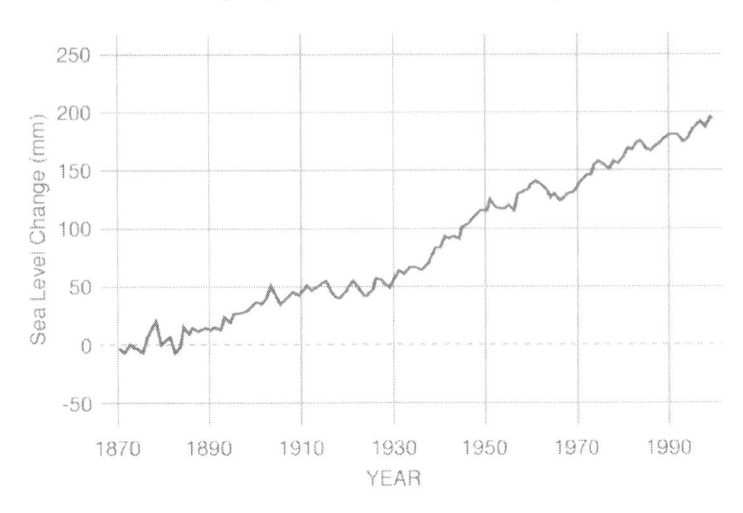

It looks bad... there's 200mm of rise in the last 120 years...
That's 1.66mm per year, 166 per century... around SIX INCHES per century!
But that's okay... the NASA story mentioned new data in the last 25 years, so let's get a more modern NASA graph, shall we?

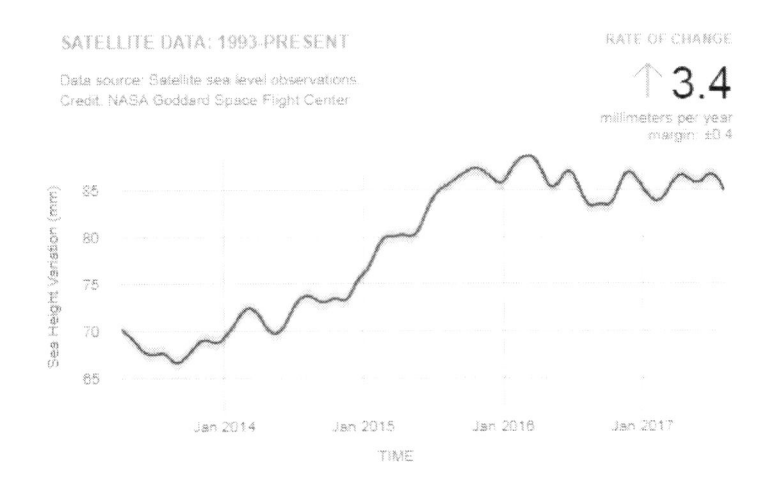

Oh... just 3.4mm per year, that's 340mm per CENTURY!, just about ONE FOOT.
So, it is increasing somewhat...
But look carefully at the last TWO YEARS of data... there's a 'pause'... a 'hiatus'...
How are they EVER going to explain that one away?

Basically I think NASA should look at its own data from time to time before publishing 'Alarmist' dogma.
And NASA is supposedly a non-political organization.

The NASA story can be found here...

https://climate.nasa.gov/climate_resources/163/video-new-study-finds-sea-level-rise-accelerating/

and here...

https://climate.nasa.gov/news/2680/new-study-finds-sea-level-rise-accelerating/

9.8... NASA; WE DON'T CHANGE DATA (6)... WE 'MOLD' IT
(More data fraud at NASA??...)

James Edward Hansen is a stalwart for the 'Alarmist' movement.
He began work at the Goddard Institute for Space Studies in 1967, and from 1981 to 2013, he was the director of the NASA GISS in New York City.
In 1982, along with his colleagues Lebedeff and Gornitz, he plotted a graph of average global sea levels. The graph, including an article, was featured in Science magazine... of course it was... Hansen was the director of GISS at NASA!

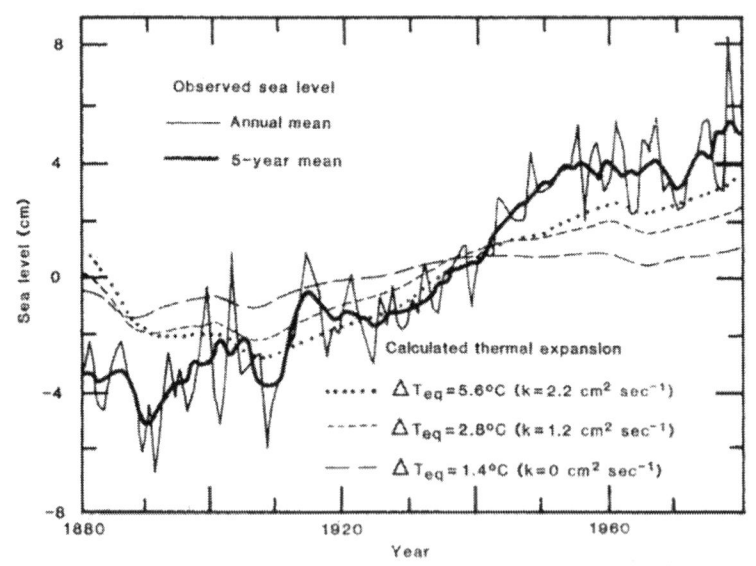

Fig. 2. Global mean sea level trend from tide-gauge data and comparison to the thermal expansion of the upper ocean obtained from the model of Hansen *et al.* (*13*) (see their equation 9 for the heat flux into the ocean). The radiative forcing used was CO_2 + volcanoes + sun [figure 5 in (*13*)], but a similar result would be obtained for other forcings that fit the observed global temperature trend; ΔT_{eq} is the equilibrium sensitivity of the model for doubled CO_2, and k is the diffusion coefficient beneath the mixed layer.

Gornitz et al. 1982
Gornitz, V., S. Lebedeff, and J. Hansen, 1982: Global sea level trend in the past century. *Science*, 215. 1611-1614

As can be seen, there was a steady sea level rise until the 1950's where it levelled off considerably. From 1880 to 1980 (a century), they graphed an 8 cm rise... 80mm, (that's about THREE INCHES per century).
AND, most of the sea level rise happened before the pause/hiatus of the 1950's.

And it should have stayed there... sitting in pride of place, a well-researched graph by a trio of up and coming climatologists.

But along came satellites, (TOPEX/Poseidon was launched in 1992) and a new source of data for NASA to deal with. Below is the new satellite data graph, all spliced together with the old data. You hardly notice they'd changed the data/shape/slope of the new graph at all.

BUT... joining the two graphs seamlessly together... (The older graph has been compressed for the differing scales of the two Y-axis)

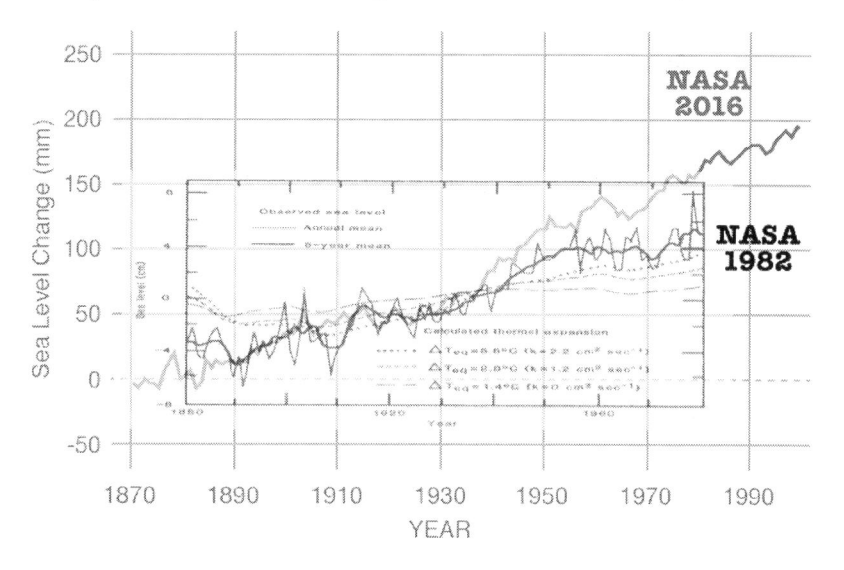

Superimposed, the two graphs are aligned, and show the scale of the deceit.

The pause at the 1950's has been eradicated, and a flowing constantly upward trend is shown. They've done the 'old trick' of decreasing the old levels, and inflating the more modern ones, making the trend steeper.

They have 'faked' the new data to show a higher sea level rise.

How quaint.

Now some of you would ask the question.

"How can they possibly change data that they themselves investigated?"

Yup. Good question.

But you see, on the subject of sea level rise, there's NEVER a solid scale figure on the Y-axis.

Because there's NO accepted value of average sea level.

So SURELY they should have taken the NEW satellite readings, and attached them to the end of their painstakingly researched pre-satellite data... like I did below.

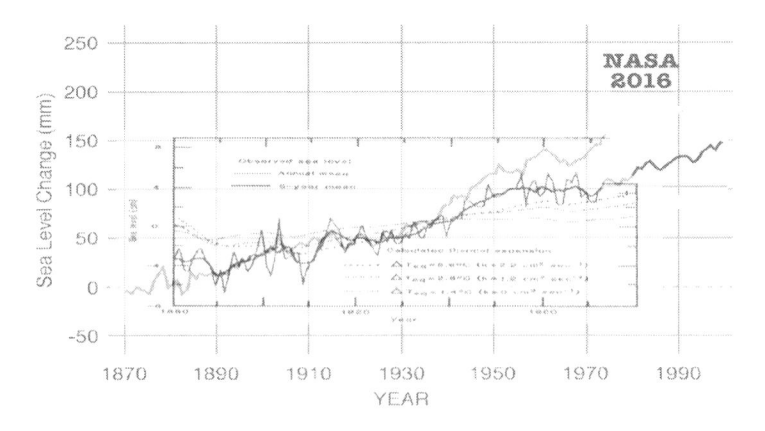

The first satellite figure on sea level rise was recorded in 1973.

So... NASA's 2016 values of sea level rise before 1973 are either pure conjecture or solid, researched data.

The ONLY proper thing to have done to join the two graphs together would have been my own graph above, showing...

A 150mm sea level rise in 130 years... That's 115mm in 100 years.

Or just about FOUR AND A HALF INCHES in a CENTURY.

9.9... ALL NASA PERSONNEL ARE ALARMISTS (1)
(Taking the corporate line bigtime...)

Michelle Thaller is an American astronomer, a research scientist, and the assistant director for Science Communication at NASA's Goddard Space Flight Center.
On a YouTube video dated November, 2018, she states quite categorically NASA's position on the 'closed shop' of current NASA scientists.
Here's her speech, word for word...

*"So your question is how widespread is it within NASA that scientists are convinced that human activity is responsible for climate change? And this is something that is important to say very, very clearly. I have known and worked with hundreds of earth scientists at many different locations in NASA, **all of them, all of them** believe that human activity is responsible for the current climate change that we see going so fast it's almost unprecedented. I want you to think about that.*

*One thing that I take really seriously and I'm very proud of is that **NASA is not a political organization**. We are scientists that work for the American people. We're funded by taxpayer money. And what we do is we make measurements. We have many, many different satellites that are orbiting the earth right now they're looking at things like ice on the oceans and at the poles, they're looking for things like vegetation growth and the change of that, ocean level, is the ocean level rising? Yeah it turns out that it is. So we have many scientists all over the planet studying all of the different ramifications of climate change. **We understand the causes. There actually is no scientific controversy about that.** Humans are releasing greenhouse gases into the atmosphere and this is warming our planet.*

*Now what scientists are researching currently, and **they don't all agree** about, is what are the most important components of driving climate change. Is it carbon dioxide? Could it be something else like methane? When methane gets released that's an even more powerful greenhouse gas. **We don't agree** on how quickly things like the ocean level will rise. **People have different estimates** for how quickly that will happen. So there **still is scientific controversy** about what the most important aspects of climate change are, and how quickly it will go in the future, but there is no scientific disagreement within NASA that humans are causing climate change.*

*Now I started this off by saying that one of the things I'm very proud of is that NASA is not political. And what that means for me is that I cannot advocate for any specific solution to climate change. That's not my job. That's up to policymakers. People might suggest things like having more solar energy or cutting carbon emissions or things like that, but at NASA we really understand that's not us, that's up to the American people, our leaders and leaders around the world. What we do is provide the facts to everybody on the planet. All of our data is actually free to any government, any person, any scientist all over the world that wants to use it. So **we all know what's causing climate change**, we can't tell you what to do about it but we can say, it's time to do something about it."*
(My own emphasis added)

Now that's some speech... NASA has worked it all out.

Greenhouse gases are responsible for climate change.
But I'd like to crave your indulgence for a moment... I'd like to go through her speech in a little bit more detail... if you don't mind.

Take 2... Paragraph 1...
"So your question is how widespread is it within NASA that scientists are convinced that human activity is responsible for climate change? And this is something that is important to say very, very clearly. I have known and worked with hundreds of earth scientists at many different locations in NASA, **all of them, all of them believe that human activity is responsible for the current climate change** *that we see going so fast it's almost unprecedented. I want you to think about that."*

Okay, Michelle, I'm not a scientist, but I am a believer in human nature; either you're lying, or some of the scientists you questioned were lying.
You can't get ten kids to agree on the color of the sky.
(Look back at Section 3, and see some real NASA Deniers)
Maybe there are NASA scientists that are too afraid to come out into the open, because as the following chapter shows... as soon as scientists leave NASA, they sure become climate skeptics super damn quickly.

RECAP... NASA, all of them, ALL of them believe that human activity is responsible for the current climate change...

Anyway... back to her very proud speech...

Take 2... Paragraph 2...

*"One thing that I take really seriously and I'm very proud of is that NASA is not a political organization. We are scientists that work for the American people. We're funded by taxpayer's money. And what we do is **we make measurements.** We have many, many different satellites that are orbiting the earth right now they're looking at things like ice on the oceans and at the poles, they're looking for things like vegetation growth and the change of that, ocean level, is the ocean level rising? Yeah it turns out that it is. So we have many scientists all over the planet studying all of the different ramifications of climate change. **We understand the causes.** There actually is **no scientific controversy** about that. Humans are releasing greenhouse gases into the atmosphere and this is warming our planet."*

Okay, Michelle, I get it; you're proud of your NASA. That's wonderful. But you're also full of talking points... NASA not a political organization... you make measurements...the ocean's rising, you betcha... you understand the causes... *"Humans are releasing greenhouse gases into the atmosphere and this is warming our planet."*

But, hold on there, Michelle... for the past thirty years you've been blaming carbon dioxide... we've looked at CO2 graphs for thirty years... now you've changed your tune to 'greenhouse gases'?
Okay... water vapor is the most abundant greenhouse gas, let's see where this goes...

RECAP... NASA understands the causes. There is NO scientific controversy about that. They have it all worked out... its greenhouse gases... consensus... settled science...

Take 2... Paragraph 3...
*"Now what scientists are researching currently, and **they don't all agree** about, is what are the **most important components** of driving climate change. **Is it carbon dioxide**? Could it be something else like **methane**? When methane gets released that's an **even more powerful** greenhouse gas. We don't agree on how quickly things like the ocean level will rise. People have different estimates for how quickly that will happen. So there **still is scientific controversy** about what the most important aspects of climate change are and how quickly it will go in the future, but there is **no scientific disagreement** within NASA that humans are causing climate change."*

Okay Michelle… my head's about to explode. For thirty years you've linked carbon dioxide with climate change… for thirty years you belittled methane because it exists at far smaller concentrations than CO2.
Now you're copying the 'Denier' scientists who have said for thirty years that methane was a far more powerful greenhouse gas than carbon dioxide.

But wait… water vapor is the most abundant greenhouse gas, existing at sometimes up to 90% compared to the other combined greenhouse gases.
Now you're trying to tell me NASA scientists… *"don't all agree about, is what are the most important components of driving climate change."*

WAIT! You don't know the KEY COMPONENT???

There's not many left…
Depending on geographical location…
Water vapor averages between 36% to 72%.
Carbon dioxide is 9% to 26%.
Methane is only 4 to 9%.
Ozone is a paltry 3 to 7%.

You said you had worked it out! The science was settled! Now you are saying you DON'T know the culprit by name????
But you KNOW it's a greenhouse gas, you said that already. (You're not looking at the sun, for instance)
"Could it be something else like methane?"
And still you ignore the greenhouse gas elephant in the room… WATER VAPOR!

Ah, of course, you can't blame water vapor because mankind CANNOT be held responsible for water vapor. (And you say that NASA is non-political!)
This is starting to unweave like a cheap flying carpet.

RECAP… NASA knows the cause of global warming; it's greenhouse gases, but they're NOT sure what ones, maybe carbon dioxide, maybe methane, but they're sure not even mentioning WATER VAPOR!

Take 2… Paragraph 4…
"Now I started this off by saying that one of the things I'm very proud of is that NASA is not political. And what that means for me is that I cannot

advocate for any specific solution to climate change. That's not my job. That's up to policymakers. People might suggest things like having more solar energy or cutting carbon emissions or things like that, but at NASA we really understand that's not us, that's up to the American people, our leaders and leaders around the world. What we do is provide the facts to everybody on the planet. All of our data is actually free to any government, any person, any scientist all over the world that wants to use it. So we all know what's causing climate change, we can't tell you what to do about it but we can say it's time to do something about it.*

Okay, Michelle, NASA doesn't have solutions, but you do *"suggest things like having more solar energy or cutting carbon emissions or things like that"* (And you say that NASA is non-political!)
And you finish with... *"but we can say it's time to do something about it."* Wonderful finish.

I'm off to get me some free NASA data.

Here's the video on YouTube...

https://www.youtube.com/watch?v=kHp0ph37QRk

9.10... ALL NASA PERSONNEL ARE ALARMISTS (2)
(Nope, there are 'mavericks' everywhere...)

Thinking of Michelle Thaller's 100% air-tight NASA consensus in the last chapter...
In 2012, 49 former NASA scientists and astronauts put their names to a letter of protest to the NASA Administrator, Charles Bolden, Jr, at NASA Headquarters. These disgruntled ex-employees were not all low level personnel; the 49 included seven Apollo astronauts and two former directors of NASA's Johnson Space Center in Houston.
They declared...

"We, the undersigned, respectfully request that NASA and the Goddard Institute for Space Studies (GISS) refrain from including unproven remarks in public releases and websites. We believe the claims by NASA and GISS, that man-made carbon dioxide is having a catastrophic impact on global climate change are not substantiated..."

This was a direct swipe at director Jim Hansen and climatologist Gavin Schmidt, the primary names at the Goddard Institute for Space Studies, who have been extremely vocal as Alarmists. (Even though NASA claims to be a non-political organization)
The letter continues...

"The unbridled advocacy of CO2 being the major cause of climate change is unbecoming of NASA's history of making an objective assessment of all available scientific data prior to making decisions or public statements."

These guys are pulling no punches...
But it doesn't stop there... considering the title of this book, the group of 49 ex-NASA scientists state...

"We believe the claims by NASA and GISS [NASA Goddard Institute for Space Studies], that man-made carbon dioxide is having a catastrophic impact on global climate change are not substantiated, especially when considering thousands of years of empirical data... With hundreds of well-known climate scientists and tens of thousands of other scientists publicly declaring their disbelief in the catastrophic forecasts, coming particularly from the GISS leadership, it is clear that the science is NOT settled."

The names of the signees, their position in NASA, and their years of service is worth listing... (Please note JSC is NASA's Johnson Space Center, in Houston, Texas)

Jack Barneburg, Jack - JSC, Space Shuttle Structures, Engineering Directorate, 34 years
Larry Bell - JSC, Mgr. Crew Systems Div., Engineering Directorate, 32 years
Dr. Donald Bogard - JSC, Principal Investigator, Science Directorate, 41 years
Jerry C. Bostick - JSC, Principal Investigator, Science Directorate, 23 years
Dr. Phillip K. Chapman - JSC, Scientist - astronaut, 5 years
Michael F. Collins, JSC, Chief, Flight Design and Dynamics Division, MOD, 41 years
Dr. Kenneth Cox - JSC, Chief Flight Dynamics Div., Engr. Directorate, 40 years
Walter Cunningham - JSC, Astronaut, Apollo 7, 8 years
Dr. Donald M. Curry - JSC, Mgr. Shuttle Leading Edge, Thermal Protection Sys., Engr. Dir., 44 years
Leroy Day - Hdq. Deputy Director, Space Shuttle Program, 19 years
Dr. Henry P. Decell, Jr. - JSC, Chief, Theory & Analysis Office, 5 years
Charles F. Deiterich - JSC, Mgr., Flight Operations Integration, MOD, 30 years
Dr. Harold Doiron - JSC, Chairman, Shuttle Pogo Prevention Panel, 16 years
Charles Duke - JSC, Astronaut, Apollo 16, 10 years
Anita Gale
Grace Germany - JSC, Program Analyst, 35 years
Ed Gibson - JSC, Astronaut Skylab 4, 14 years
Richard Gordon - JSC, Astronaut, Gemini Xi, Apollo 12, 9 years
Gerald C. Griffin - JSC, Apollo Flight Director, and Director of Johnson Space Center, 22 years
Thomas M. Grubbs - JSC, Chief, Aircraft Maintenance and Engineering Branch, 31 years
Thomas J. Harmon
David W. Heath - JSC, Reentry Specialist, MOD, 30 years
Miguel A. Hernandez, Jr. - JSC, Flight crew training and operations, 3 years
James R. Roundtree - JSC Branch Chief, 26 years
Enoch Jones - JSC, Mgr. SE&I, Shuttle Program Office, 26 years
Dr. Joseph Kerwin - JSC, Astronaut, Skylab 2, Director of Space and Life Sciences, 22 years

Jack Knight - JSC, Chief, Advanced Operations and Development Division, MOD, 40 years
Dr. Christopher C. Kraft - JSC, Apollo Flight Director and Director of Johnson Space Center, 24 years
Paul C. Kramer - JSC, Asst for Planning Aeroscience and Flight Mechanics Div., Egr. Dir., 34 years
Alex (Skip) Larsen – JSC, Chairman of the Shuttle Payload Safety Review Panel, 15 years
Dr. Lubert Leger - JSC, Asst. Chief Materials, Engr. Directorate, 30 years
Dr. Humbolt C. Mandell - JSC, Mgr. Shuttle Program Control and Advance Programs, 40 years
Donald K. McCutchen - JSC, Project Engineer - Space Shuttle and ISS Program Offices, 33 years
Thomas L. (Tom) Moser - Hdq. Dep. Assoc. Admin. & Director, Space Station Program, 28 years
Dr. George Mueller - Hdq., Assoc. Adm., Office of Space Flight, 6 years
Tom Ohesorge, Langley, and JSC, 49 years
James Peacock - JSC, Apollo and Shuttle Program Office, 21 years
Richard McFarland - JSC, Mgr. Motion Simulators, 28 years
Joseph E. Rogers - JSC, Chief, Structures and Dynamics Branch, Engr. Directorate, 40 years
Bernard J. Rosenbaum - JSC, Chief Engineer, Propulsion and Power Division, Engr. Dir., 48 years
Dr. Harrison (Jack) Schmitt - JSC, Astronaut Apollo 17, 10 years
Gerard C. Shows - JSC, Asst. Manager, Quality Assurance, 30 years
Kenneth Suit - JSC, Ass't Mgr., Syst Integration, Space Shuttle, 37 years
Robert F. Thompson - JSC, Program Manager, Space Shuttle, 44 years/s/
Frank Van Renesselaer - Hdq., Mgr. Shuttle Solid Rocket Boosters, 15 years
Dr. James Visentine - JSC Materials, Engineering Directorate, 30 years
Manfred (Dutch) von Ehrenfried - JSC, Flight Controller; Mercury, Gemini & Apollo, MOD, 10 years
George Weisskopf - JSC, Avionics Systems, Engineering Dir., 40 years
Al Worden - JSC, Astronaut, Apollo 15, 9 years
Thomas (Tom) Wysmuller - JSC, Meteorologist, 5 years

The full article, including the letter, can be found here…

https://www.businessinsider.com/nasa-scientists-dispute-climate-change-2012-4

9.11... ALL NASA PERSONNEL ARE ALARMISTS (3)
(Scientists see sense when they leave NASA...)

There seems to be a possible graph building in my mind.
According to Michelle Thaller, just two chapters ago, scientists working at NASA are 100% behind mankind's greenhouse gases being behind ALL of the global warming... *"human activity is responsible for the current climate change".*

But as soon as they leave the strict regimen of their employers, NASA scientists get some nerve, stick their middle finger in the air, and say enough is enough!

Because the 2012 letter to NASA featured in the previous chapter was only the beginning.

In 2016, a group of 25 ex NASA scientists got together and found they ALSO did not agree with the idea of Anthropogenic Global Warming.
And these guys are not fringe kook scientists.

Harold Doiron PhD, joined NASA in 1963. He developed the Apollo Lunar Module landing dynamics simulation software that put men on the moon in 1969.
Jim Peacock PhD, is the website's 'manager', and Jim is a retired NASA aerospace engineer on the Apollo Project, Sky Lab, and the Space Shuttle.

Proud of their days in NASA, and dismayed of the new direction their ex company has taken, they created a website... go look... it's fun.

https://www.therightclimatestuff.com/

These old guys don't hang around. The first video on release is entitled...

"Climate Alarmists Are a Threat to Our National Security"

Man, I love these guys already.

In a presentation to the Heartland Institute they explain that as the Alarmist movement pushes for fewer and fewer fossil fuel emissions, this will lead to the USA being less energy independent, thus threatening our national security.
To be honest, I agree.

(I cheered out loud when President Trump opened ANWR.)

This was their publication…

National Security Depends on Energy Security
TRCS

- **President Trump's America First Energy Plan** and other presentations in this conference make the case that:
    - Our **national security** depends on our **energy security**
    - **US energy dominance** thru exploiting our **vast energy reserves and technology** is a key to global stability and security
- **Anthropogenic Global Warming (AGW) Climate Alarm** unsupported by **physical data** is a **threat to our national security** by:
    - Preventing rational political decisions regarding development of US energy resources
    - Causing our military to plan for unrealistic climate change

And I print it simply because I couldn't say it better or more succinctly than they already have.

In addition to their constantly-updated website, they have written a report, which they have sent to President Trump.
The letter and report, in full, can be found here…

http://nebula.wsimg.com/1ca304a328496c0c011ac02790fc56ed?AccessKeyId=4E2A86EA65583CBC15DE&disposition=0&alloworigin=1

So, climate Alarmists, not all NASA is on your side. In fact it seems that climate Deniers have to retire before coming out of the closet… it's a terrible thing when you have to toe the party line and ignore the truth just to keep a paycheck coming in.

What happened to all that settled science?

Section 10
The Sun; it's ALL Cosmic, Man

"The most difficult subjects can be explained to the most slow-witted man if he has not formed any idea of them already; but the simplest thing cannot be made clear to the most intelligent man if he is firmly persuaded that he already knows, without a shadow of a doubt, what is laid before him."
Leo Tolstoy

10.1... The 600lb Gorilla in the Room
(It's the Sun, dumbass!)

About 93 million miles away from you right now, travels the Sun, the star to which we all owe our lives.
Although our solar system seems to sit in a calm leisurely spiral, it is actually moving at about 750,000 mph through the cosmos. But, of course, just like the mom and child in the speeding jet airplane, they don't feel the motion; they just see the immediate outside moving gently. The earth far below moves by at a snail's pace.
The Sun supplies 99% of the energy soaked in by our planet. Its energies are the sole battery from which we draw our resources, yet this simple fact is ignored by most climate scientists.
(Well, it's not ignored... it's more bypassed because its effects are too complex to figure into the Alarmists' computer models)

The Sun's energy comes in 11 year cycles, first measured in the 1700's.

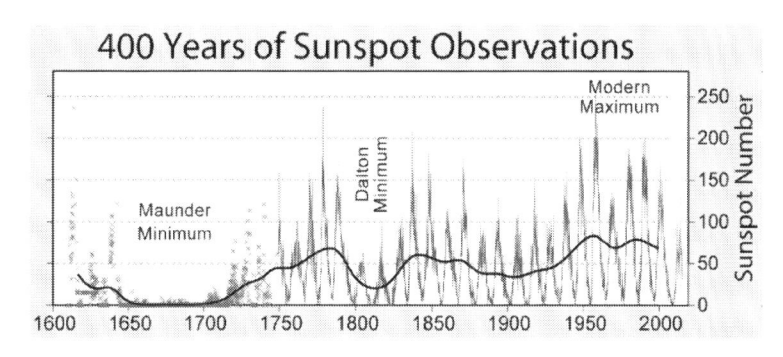

Each cycle is very clearly defined, regularly ends at a time of sunspot minimum, and peaks at a sunspot maximum.

Since its first recording in the 1600's, we are now entering the beginning of Solar Cycle 25.

Looking at the graph above, we cannot miss the seeming correlation between sunspot maximums, and global temperatures.

The Maunder Minimum correlates with the Little Ice Age. The Dalton Minimum lies at the center of the ice freeze on the Thames in London in 1814, and the Modern Maximum lies over the high temperatures after the Second World War, in the 1950's.

Solar Cycle 24 (the most recent, and the far right of the graph) is the lowest in energy output since the late 1800's, and is one of the reasons why some solar scientists think we are heading for some form of cooling period.

Sunspot activity has been gradually decreasing in the last 30 years or so.

Solar Cycle 24 is actually the lowest output since the last cycle of the Dalton Minimum.

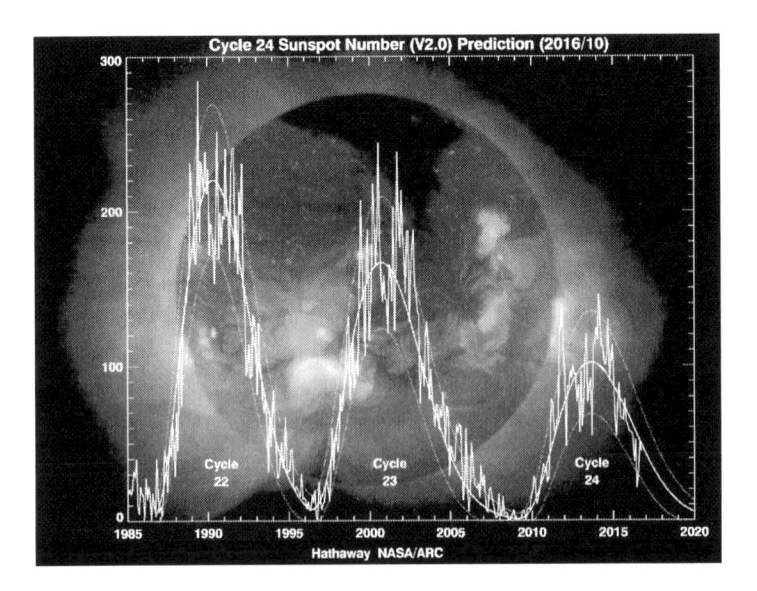

Looking in more detail at the last three solar cycles, 22, 23 and 24, we see the gradual decline more clearly.

Only after reaching 2030 will we clearly see if the low level of cycle 24 is part of a trend.

However... don't take it from me... why not listen to a plethora of newspapers, scientific journals and newsmen who have been running around telling us this for the last 20 years.

Yeah... crickets.

The main reason we haven't heard more is the Alarmists' fascination with carbon dioxide.

The weird thing is... the above data are NOT reports just by the Denier camp.

These are studies and data from ALARMIST SOURCES!

NASA

NASA has reported of the Sun's reduction in output since 2008. At a briefing at NASA headquarters, solar physicists announced that the solar wind is losing power.

Dave McComas of the Southwest Research Institute in San Antonio, Texas said...

"The average pressure of the solar wind has dropped more than 20% since the mid-1990s... This is the weakest it's been since we began monitoring solar wind almost 50 years ago."

NASA solar physicist David Hathaway said...

"if the sun remains in its sleepy state, we could see a replay of the Dalton Minimum..."

If this is indeed the case, we are in for a chilling next few years.

During the Dalton Minimum (from 1790 – 1830), global temperatures plummeted, giving us 1816's *"Year Without a Summer"*. The abnormal cold weather destroyed crops in northern Europe and the northeastern United States/Canada.

Hathaway continues,

"The slowdown we see now means that Solar Cycle 25, peaking around the year 2022, could be one of the weakest in centuries,"

(Well, speaking in 2008, he got the year a bit wrong, sun cycle 24 stuttered a bit... the new peak of sun cycle 25 should be around 2025/26)

However...

If we look at the long term graph again, we see a distinct 100 year sinusoidal aspect going back to the Maunder Minimum. This is known as the Gleissberg Cycle.

Dips in solar output in the 1820's and 1920's produced cooling periods on Earth.

As we approach the 2020's, the idea of a new era of cooling cannot be ruled out.
And despite the CO2 lobby demanding that emissions be cut, the solar experts continue to dole out predictions of cooling…

Sami K. Solanki (Max Planck Institute for Solar System Research) has postulated for years that a repeat of the Dalton Minimum is very likely.

In a study published in Nature.com in 2015, Mike Lockwood (Professor of space environment physics, University of Reading, UK) showed that the Sun is in a rapid decline, not seen in 10,000 years.

Dr. Willie Soon (Astrophysicist at the Harvard-Smithsonian Center for Astrophysics is quoted…

"Right now, we are in the deepest solar minimum of the entire Space Age… less energy means a cooler planet."

Who knows?

Except that Russian scientists have been predicting global cooling due to the Sun's activity for a few years now.
But who listens to them, huh?

When looking for the cause of global warming, it's difficult not to look to the huge white/yellow ball in the sky, our Sun.

And many scientists have said so…
Even Alarmist James Hansen, (NASA/GISS) said in August 1999…

"… changes of solar irradiance (the brightness of the sun) are difficult to dismiss as a mechanism of climate change, because there are observed correlations of solar variability and climate change.
The upshot is that we will be able to understand climate change well only with the help of global climate models that are able to incorporate all of these mechanisms on an equal footing…"

And of course, Alarmists have NOT been putting the solar cycles in their climate models for years… seemingly solar irradiance is too complex to model.

But it's difficult to close this section without drawing attention to a NASA page entitled 'Long Range Solar Forecast'

On this page (https://science.nasa.gov/science-news/science-at-nasa/2006/10may_longrange/) **dated 2005**, there are a few choice quotes by Hathaway...

But one section bothered me.

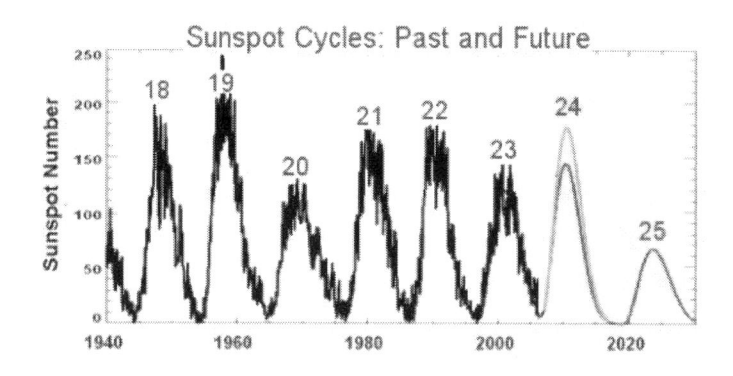

Sunspot Cycles: Past and Future

The above 2005 graph, entitled *'David Hathaway's predictions for the next two solar cycles and, in pink, Mausumi Dikpati's prediction for cycle 24'* shows NASA's predictions for the level of Sun cycle 24.

Looking back, knowing the actual figures of sun cycle 24, his predictions are WAY out.
With hindsight, sun cycle 24 turned out to be MUCH smaller than even NASA had predicted.
If sun cycle 25 is as low as the diagram, it falls below the value set by sun cycle 20, and that brought the cooling in the 1970's that set off the ice age panics.
And Alarmists tell us that the science is settled.

At the blog site, notrickszone, there is a clicky list to over 400 scientific papers that look at the solar forcing to our climate. Here's the link...

https://notrickszone.com/2017/06/12/20-more-new-papers-link-solar-forcing-to-climate-change-now-80-sun-climate-papers-for-2017/?fbclid=IwAR33IsVct6CNZssbEaOLlZpxu5yiZPmKw1Jt4fcoJwzXptdHC8lc3mbyZrQ

10.2... WHATEVER'S HAPPENING, MANKIND IS TO BLAME!
(Well, actually, perhaps we're not...)

Professor Carl-Otto Weiss, advisor to the European Institute for Climate and Energy and former President of the National Metrology Institute of Germany, Braunschweig, holds a particularly unique view of climate change.
He thinks...
Mankind is NOT responsible in any major way at all.

Yup, you did read that last sentence correctly.
In a paper with his associates, Lüdecke and Hempelmann, they compiled a series of temperature records from...

1. Six temperature measuring stations in Central Europe going back 300 years.
2. Tree rings going back 2500 years.
3. Ice cores from Antarctica going back 2500 years

Using the above they combined all the graphs and looked for cyclical spectra to show through. They are convinced that every temperature anomaly in the climate history is cyclical, many natural cycles coming together in a pattern.

The players...
Carl-Otto Weiss... physicist and advisor to the European Institute for Climate and Energy, former President of the German Meteorological Institute, in Braunschweig.

Horst-Joachim Lüdecke... nuclear physicist and emeritus professor at the University of Applied Sciences of the Saarland.
(Hempelmann is a German astronomer)

In a 2017 paper the German physicists doubled down on their discoveries, saying...

"CO2 plays only minor role for global climate"

The published study (in The Open Atmospheric Science Journal, web address given at the end of the chapter) the German scientists used a large number of temperature worldwide proxies to construct a global temperature mean over the last 2000 years (data from five full proxy series were admitted into their calculations).

The results were extremely interesting, and showed far more correlation to the temperature record than hockey sticks or mis-aligned NASA data.

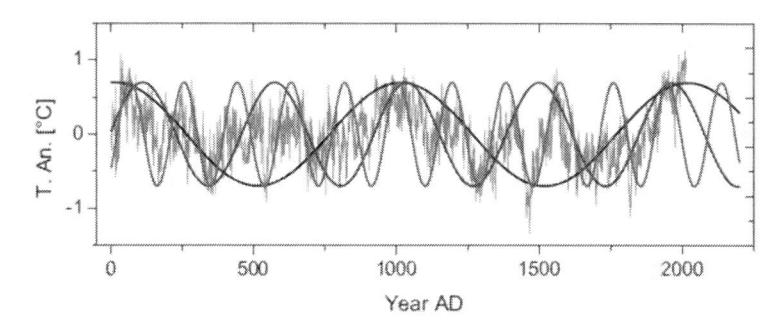

The above graph shows three solar cycles, a 1000 year cycle, a 460 year cycle, and a shorter 190 year cycle, superimposed on top of the 'accepted' global temperatures (Britain's Hadley CRU temperature records) shown in grey.
(Please note that the grey part of the graph shows quite clearly The Roman Optimum (~0 AD), the Medieval Optimum (~1000 AD), the Little Ice Age (~1500 AD), and the Present Optimum.)

If you just look at the blue 460 year cycle, it correlates quite well with the grey temperature record. Just from that relationship alone, I think this study has some merit.

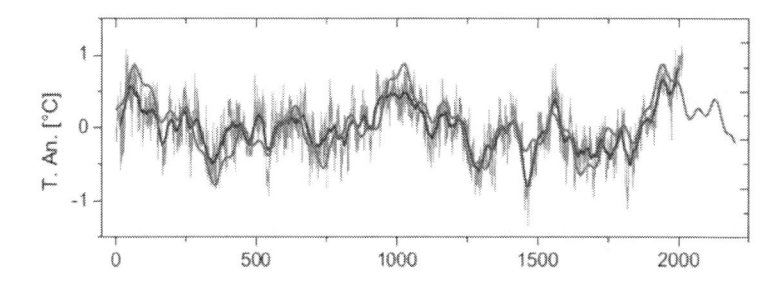

However, when you add the three solar cycles together, we get the red line above... and when you add it to the grey temperature record, shown clearer by the 30-year average blue line, the correlation is a staggering 84%.

That, dear reader, is far more than they expected, and is a figure you should read VERY carefully.

In statistical terms, it's a Pearson Correlation of 0.84.

Moreover, as a result of the combined three natural solar cycles the red solar graph also shows the warming from 1850 to 2000.

The bad news is that it also indicates a significant cooling of global temperatures into the 21st century, perhaps as far as 2070.

More study is probably needed, but since the sun's cycles are pretty-well documented/researched, this should not take long in gaining traction.

If ever there was a canary in a coal-mine situation... this is surely it.
I repeat the scientists' conclusion...

"Thus one can conclude that CO2 plays only a minor role (if any) for the global climate."

The full report can be found here...

https://benthamopen.com/FULLTEXT/TOASCJ-11-44

10.3... SunSpots, SunCycles, and SunStuff...?
(Does this mean what I think it does?)

By now the graph below will be familiar; just another showing of recent decreasing sun spot activity.

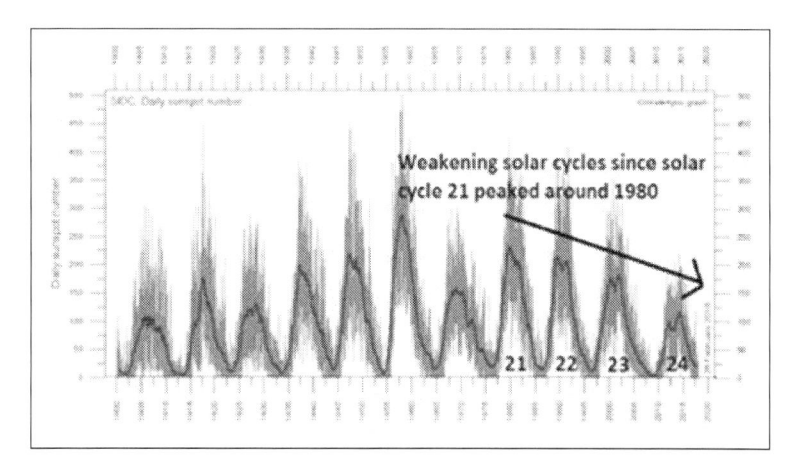

The graph shows the last 11 sunspot cycles, and since suncycle 21 they've got progressively weaker.
And if you've been reading closely, we all know what happens when sunspot activity goes really low... we get minimums. (Look at the solar minimum in the 1970's... Earth cooling, Cod War etc...)
And minimums mean cooling.
Or even Colding. (Did I just invent a new scientific term?)

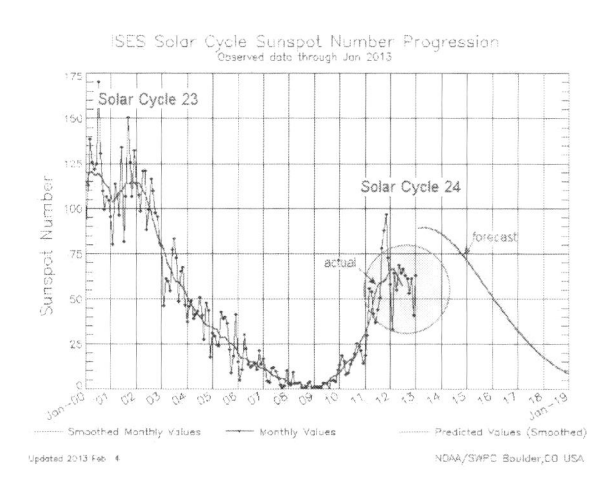

Let's take a closer look at the most recent suncycles...

In the graph above, we see the weakened solar cycle 23 ending in 2009. It didn't peak as forecast, in fact it had a stuttered peak.
So the scientists re-forecast/adjusted the expected suncycle 24 sine wave and waited with baited breath...

The readjusted forecast for suncycle 24 is shown in red, but again the reality failed to live up to the science.
Solar cycle 24 was even worse.
This prompted a trumpeting from the battlements of a 'cooling period'.
(Although in truth, this lessening was expected)

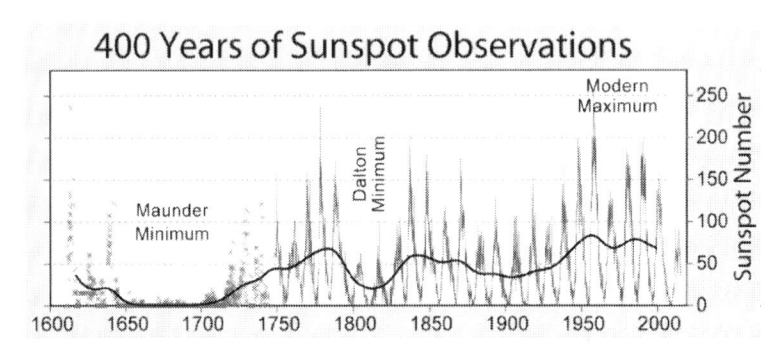

Scientists had already postulated a roughly 100 year cycle. Schwabe and Gleissberg can duck it out as to its proper name, but as we see above, there's minimums at 1700, 1810, 1910, and we're dipping again.

Cycle 25 is going to be of HUGE interest, trust me.

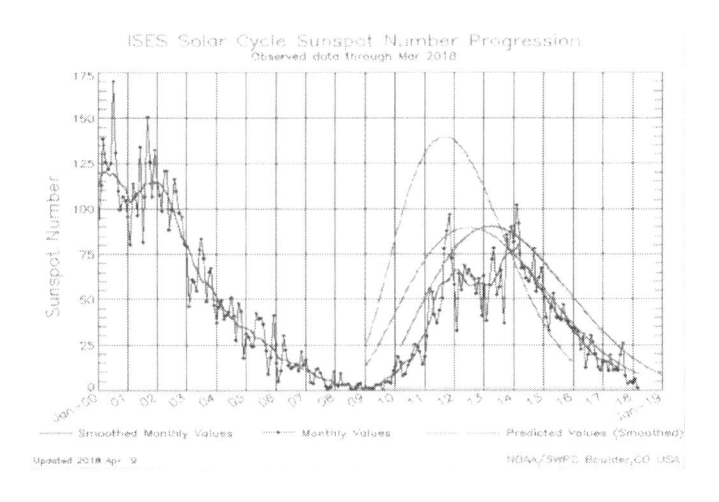

This is the graph showing completion of suncycle 24.
The highest green curve was NOAA's first estimate (before cycle 23 had peaked), and the red curve was their later estimate, and the actuality didn't even meet that prediction.
Solar cycle 24 disappointed in its erratic peaking, and its general amplitude.
Solar Cycle 25 begins in the year that this book was published.

Only the future will hold the answers, but trust me, in the world of solar cycles...

... the science is FAR from being settled.

https://arizonadailyindependent.com/2015/07/09/solar-cycle-changes-indicate-a-cooling-period-ahead-2/

And there's another theory...

The Fall of Empires
The Chinese kept details of temperature and the condition of the sun. Modern Chinese climatologists are now increasingly fearing a new solar minimum, and they have a good reason for doing so...

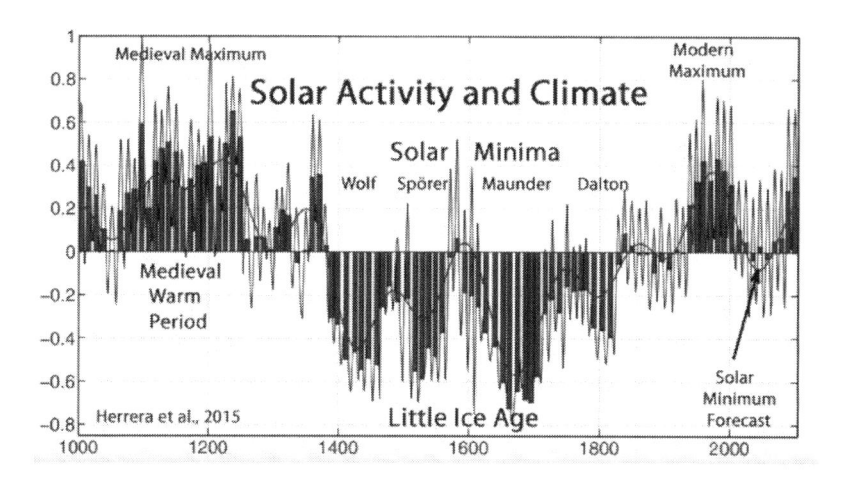

Above is an extension back in time of solar minimums.

It's speculative, yes, because the telescope wasn't invented until the early 1600's, but if you compare this graph with one taken from the Greenland Ice Cores, there is a certain similarity.

Climate: Data from Greenland Ice Cores (GISP2)

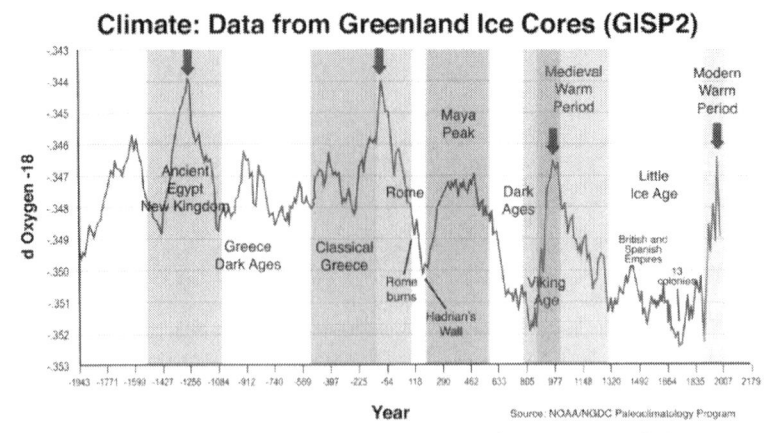

As we can see from these temperature peaks, many of the periods of higher temperatures can be linked to times of empire building, times of plenty.

And, of course, when these times end with a dip in temperature, the converse becomes obvious... when times get cold, hardship steps in, and empires crumble.

This is exactly what the Chinese have noted about their own history...

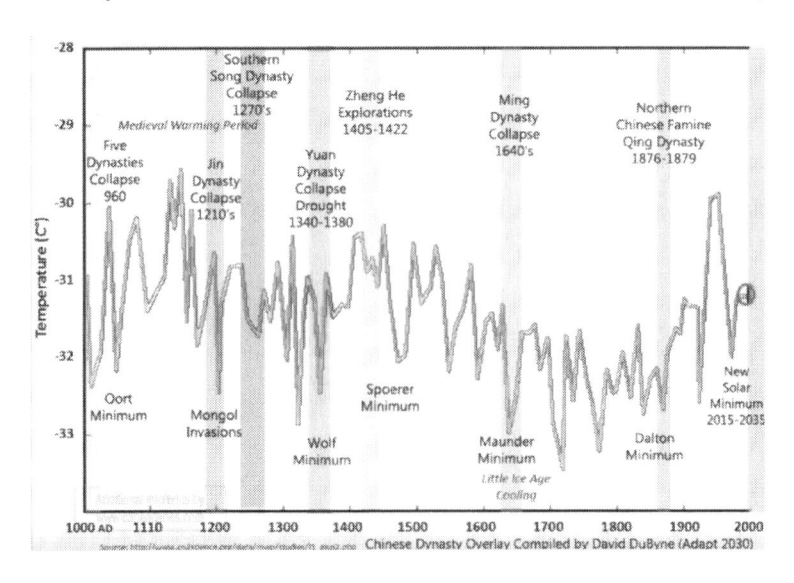

Italian philosopher, George Santayana, once said…

"Those who cannot remember the past are condemned to repeat it."

And looking at the graph, and linking the demise of every Chinese dynasty with a drop in temperatures, the Chinese are running scared…

Very scared.

And that should scare us too.

10.4... THE INFLUENCE OF THE UNIVERSE
(Cosmic Rays, and clouds...)

Yeah, you did read the title correctly.
It might sound weird to talk of cosmic rays when dealing with the Earth's climate, but there is an important correlation that seems to be ignored by most scientists of today.
Cosmic rays are single atom particles (stripped of their electron shells) given off by the sun, supernovae and other stars. They hit the Earth's atmosphere and must be considered as an energy source.

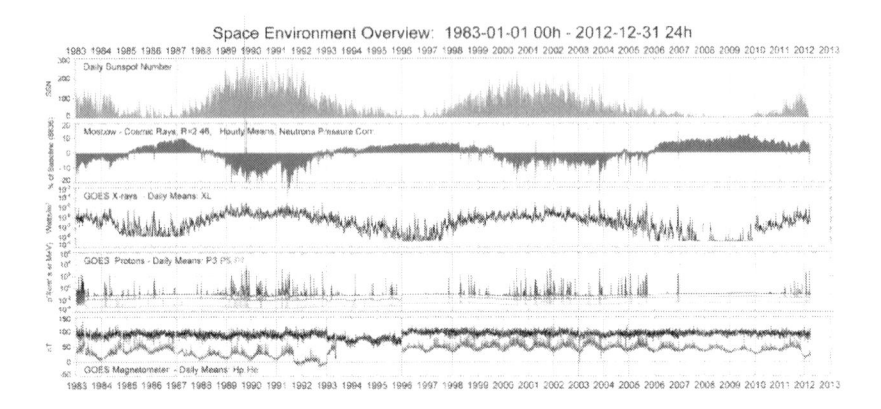

The top two traces on the graph above show the relationship between sunspot activity and cosmic rays.
Seemingly, when sunspot activity/solar irradiance is high, cosmic rays are forced away from earth by the sun's increased activity.
However, when the solar activity is low, cosmic rays increase.

You may ask... "How are cosmic rays connected to climate change?"

Well, Danish physicist Henrik Svensmark (professor in the Division of Solar System Physics at the Danish National Space Institute (DTU Space) in Copenhagen) has proposed that cosmic rays affect the rate of cloud formation and hence they would be an indirect cause of global warming.
Svensmark is one of several scientists who play down the effect of CO2 warming, and argue that the energy given out by the Sun is a more plausible culprit.

His theory posits that energy from the Sun deflects cosmic rays and during times of increased solar radiance, there is a lower cloud cover and density.

As the last Sun Cycle was weak, this theory would mean that a lower solar output would mean more cloud production by cosmic rays, thus increasing the earth's albedo, and reflecting more of the Sun's radiation back into space, cooling the planet.

Svensmark wrote a book, The Chilling Stars, with Nigel Calder, and although the theory still gets little consideration, there's one thing for sure...
If such theories exist... the science is far from settled.

Yup, he's controversial... but so was Galileo.

And yet from the first days that man looked up, we've looked at clouds. And wondered what they were.
Then, as science grew in stature, we learned more about them.
In the 20th Century we saw them from both sides, and from inside, and we thought we pretty well had clouds worked out.

Until Global Warming...

Clouds, today, hold a key place in the science of global warming, and most scientists would admit, that as far as global warming is concerned... (in the words of...Joni Mitchell), we *"really don't know clouds at all."*

At any one time, clouds cover as much as 60% of the world's skies, and their albedo is crucial to keeping the earth cool.
However, since no one really can model clouds to any great extent, their influence in climate models is underplayed.

Clouds, huh?

Cosmic, man.

SECTION 11
POLITICS REARS ITS UGLY HEAD

"If the science was 'settled', why are there more than 120 different computer models?"
(Sharp old scientist on an NIPCC panel...)

11.1... POLITICAL ISSUE... OR HUMANITARIAN CRISIS?
(Oh, it's Political in EVERY Way!)

Although global warming/climate change is an issue to be fought over by scientists, policy decisions will be made by politicians.
So, as the results of scientists' findings will be handed over to politicians for their own private debate, the integrity of the science is paramount to the issue.

One thing is sure... the solution, delivered by politicians, and their recommendations to their respective voters, will be a political one.
There's just no getting away from that point.

So as the scientists argue their particular theories, the politicians have decided to throw their hat in the ring early.
Hence, the whole debate is now heavily mired in politics.

The Alarmists want more government outreach, more government departments, and more government control... maybe even a World Governance on the subject. They demand more taxes from the people to support their 'solutions', more power to the aforementioned government. Unfortunately that seems to be their only solution.
If EVERYONE reduces CO2 output, the world will cool down.

The Deniers are more casual with their argument, content to see what the science finally digs up before coming to any knee-jerk reaction.
The Deniers know the Alarmist solutions will not work, will be ineffectual, and extremely costly.
Even if the ENTIRE WORLD stopped using fossil fuels THIS INSTANT...
... the CO2 already in the atmosphere would NOT disappear for many, many years. You want scientific proof? Look at ANY graph of CO2 build-up... (Denier or Alarmist) carbon dioxide builds quickly, peaks, then recedes slowly. CO2 in the atmosphere can stay there for HUNDREDS of years.

According to James Taylor (senior fellow for environment and energy policy at The Heartland Institute)...

"the solutions are pushing a Leftist agenda that is not supported by sound science."

On 28 January 2009, in the run-up to the Copenhagen Climate Council (COC), in December 2009, the European Commission released a position paper entitled...

"Towards a comprehensive climate agreement in Copenhagen."

The paper addressed three key challenges...
1. *"targets and actions"*
2. financing of *"low-carbon development and adaptation"*
3. building an *"effective global carbon market"*
Now if that's not political, I don't know what is.

In October 2018, a story in the Telegraph (UK) hit my GWBS (Global Warming Bull Shit) radar.
In a story written by hard-hitting journalist Christopher Booker, he tells of the latest report by the IPCC.

Seemingly, to avert *"complete climate catastrophe, with wars, famine and disease spreading across the globe"*, we have to spend 2.4 TRILLION dollars EVERY YEAR until 2035, to build a solar and wind infrastructure.
Added to this huge cost, each country in the world has to reduce their CO2 emissions to *"Net Zero"*. Basically we have to shut down the carbon-fueled industrial process that built our modern civilization.
Oh, and BTW, the IEA (International Energy Agency) has stated that we currently rely on carbon based fuels for 81% of our energy.
That's a heck of a catch up for the next 16 years.

To show how much of the IPCC's report is pie-in-the-sky, it *"allows for very limited amounts"* of coal and gas still to be used; but only on condition that we find ways to capture their *"carbon emissions"* to *"bury them under the ground"*.

They also want to develop the ability to suck vast quantities of CO2 out of the atmosphere.

Unfortunately, both of these technologies, (which the IPCC does admit) have not yet been sufficiently developed to a low cost solution.

Politics gone mad
To make matters worse, one of the report's writers cites *"more extreme weather, rising sea levels and diminishing Arctic sea ice".*

However, the IPCC itself, in its last major report in 2013, found that there had been *"no discernible increase in extreme weather events, such as hurricanes, floods and extreme heatwaves".*

And sea levels have risen about 8 inches per hundred years since the 1600's.

Amazingly, following the media catastrophes, the report shows a graph of a rapidly reducing Arctic sea ice area (distinctly untrue even by NASA's own figures). Again, the IPCC's own climate models show a small reduction in Arctic sea ice, nothing more.

Bluster, panic, alarm, confusion, and pure political hogwash.

I wonder what will settle first, the politics or the science?

11.2... UNDER PRESSURE TO TOE THE PARTY LINE
(Don't change ships midstream... you'll regret it...)

Introduction
Once you get settled in a good-paying job, it takes guts to give it up for your principles.
Sometimes sacrificing your principle means swallowing some level of corporation ideas.
There are many scientists who, after admitting some form of agreement with the climate skeptic position have been driven from their jobs.
Here are a few...

David Bellamy, an English Botanist and Eco-warrior said in 2004 that the idea of man-made global warming was 'poppycock'. This was a huge stand for the TV botanist to take, but perhaps he hadn't thought it all through... one thing he didn't expect... was his ties to the BBC running dry.
He afterwards complained that several of his television ideas were rejected by the BBC because of his anti-global warming stance.

The BBC, of course, denied the claim... however from 1970 to 2004 he appeared on radio and television 149 times.
After his 'outburst' in 2004, he appeared just 9 times.

Lennart Bengtsson, a Swedish Meteorologist, was department head of research at the European Centre for Medium-Range Weather Forecasts from 1975 to 1981, then director until 1990. He then was the director of the Max Planck Institute for Meteorology in Hamburg.
(Not a bad job resume)
Bengtsson submitted papers to the IPCC fourth and fifth assessments, but was rejected.
In a fit of pique, in April 2014, he became a board member of the Global Warming Policy Foundation (GWPF), a climate skeptic organization. He resigned less than a month later, on May 14th, under a withering fire from Alarmist sources.
The Daily Mail on 15 May, reported from his press release, Bengtsson said...
"I have been put under such an enormous group pressure from all over the world... It is a situation that reminds me about the time of McCarthy."
Later Bengtsson admitted...
"I had not expecting such an enormous world-wide pressure put at me from a community that I have been close to all my active life."

He is now a senior research scientist at the Environmental Systems Science Centre at the University of Reading, England.

Timothy Francis 'Tim' Ball was a professor in the Department of Geography at the University of Winnipeg from 1971 to 1996. Hockey-stick inventor, Michael E. Mann, who recently lost a lawsuit with Ball, called him; "perhaps the most prominent climate change denier in Canada".
Since retiring in 1996, Ball has been a tireless speaker, and specializes in historical climatology. In the middle of a European University tour, the tour was cancelled when the Universities found out what he was teaching, and left stranded in Europe.
He has had many threats on his life... *"I had to retire"* he jokes on tour, *"I had so much scar tissue on my back, there wasn't room for any more knives".*

Judith Curry is an American climatologist with a Ph.D. in geophysical sciences from the University of Chicago in 1982.
Specializing in the climate's effects on the oceans, Curry is a former professor of Atmospheric and Oceanic Sciences at the University of Colorado-Boulder, and through her academic career, she has written 2 books, and published over 130 scientific peer-reviewed papers.
She's no lightweight.
Throughout her career she has, however, always encouraged the two sides in the climate change debate to come closer together, and exchange both views and data. She sees the 'closed doors' of the Alarmist camp to be to the detriment of their cause, and their holding their data as 'top secret' to be wrong.
Under pressure to 'toe the line', Curry left her professor's position in 2017, and has been a firm climate denier ever since.
She has testified to the US senate and congress on the topic.

Peter Ridd is a marine geophysicist, once working at James Cook University in Australia. He is an expert on the Great Barrier Reef (GBR), and the inventor of a machine to measure sediment particulates on the coral. *"I probably know more than all other marine experts on coral particulates"* he said in a speech to a liberal think tank.
He was asked by the Institute of Public Affairs (IPA) to write a chapter on the Great Barrier Reef in one of their publications, and was scathing on the propaganda on the subject. He explained that most of the data on the subject was wrong, and sat back to get some feedback on his words.
He was summoned to the Dean's office in JCU, and sacked for misconduct.

The JKU never said that the content of his climate data was the reason for his dismissal, but cited many university misconduct laws he had broken. He countersued, and in 2019 he won his case against wrongful dismissal. However, he maintains to this day that $500 million Australian dollars annually are spent on 'finding out what's WRONG with the Great barrier Reef'... and he maintains that if the money was spent on finding out what's RIGHT with the coral, the world would all be reading the good news.

11.3... DENIERS ARE FUNDED BY BIG OIL
(Nope, but Alarmists are...)

There is a strong rumor that Deniers are funded by large corporations and the conglomerate of 'Big Oil', but it's something that's neither proven nor true.
In actual fact, like so many stories, the opposite is actually the case.

Here's one example you might like, followed by a blue clicky link to another 25+.

The Climate Research Unit (CRU) in the UK is a major player, having ties with both the Met Office (Britain's Government Weather Office) and the Alarmist movement. They were hacked in 2009, and tens of thousands of emails were released. (See Section 2.5 on Climategate)
The Professors at the CRU, located inside the University of East Anglia have openly accepted contracts from BP for reports on climate related topics.
In their corporate webpage, the CRU list the following financial backers...

"This list is not fully exhaustive, but we would like to acknowledge the support of the following funders (in alphabetical order):

"British Council, **British Petroleum**, Broom's Barn Sugar Beet Research Centre, **Central Electricity Generating Board**, Centre for Environment, Fisheries and Aquaculture Science (CEFAS), Climate and Development Knowledge Network (CDKN), Commercial Union, Commission of European Communities (CEC, often referred to now as EU), Council for the Central Laboratory of the Research Councils (CCLRC), Department of Energy, Department of the Environment (DoE, 1970-1997), Department of the Environment, Transport and the Regions (DETR, 1997-2001), Department of the Environment, Food and Rural Affairs (DEFRA, 2001-present), Department of Energy and Climatic Change (DECC), Department of Health, Department of Trade and Industry (DTI), Earth and Life Sciences Alliance, **Eastern Electricity**, **Engineering and Physical Sciences Research Council (EPSRC)**, Environment Agency, Forestry Commission, Greater London Authority, **Greenpeace International**, International Institute of Environmental Development (IIED), **Irish Electricity Supply Board**, KFA Germany, Joint Information Systems Committee (JISC), Leverhulme Trust, Ministry of Agriculture, Fisheries and Food (MAFF), National Assembly for Wales, **National Power,** National Rivers Authority, Natural Environmental Research Council (NERC), Norwich Union, **Nuclear Installations**

Inspectorate, *Overseas Development Administration (ODA), Reinsurance Underwriters and Syndicates, Royal Society, Scientific Consultants, Science and Engineering Research Council (SERC), Scottish and Northern Ireland Forum for Environmental Research,* **Shell***, SQW Consulting, Stockholm Environment Agency,* **Sultanate of Oman***, Tate and Lyle, Tyndall Centre, UK Met. Office, UK Nirex Ltd., UK Water Industry Research (UKWIR), United Nations Environment Plan (UNEP),* **United States Department of Energy***, United States Environmental Protection Agency, Wolfson Foundation and the World Wildlife Fund for Nature (WWF)."*
(Emphasis added)

And this is just one name on the 'they get money from big oil' list.
Here's a link to a list of Climate Alarmist societies and foundations who receive money from rich investors. You won't find many Deniers on the books, trust me.

http://notrickszone.com/2015/02/09/long-list-of-warmist-organizations-scientists-haul-in-huge-money-from-big-oil-and-heavy-industry/

Disclaimer
To be honest, it's difficult to find an educational establishment where Alarmist scientists work that doesn't get some funding from the big backers, big oil, big pharma etc.
The 'donations' are money that drive research and development, and many Universities need such donations and programs to keep themselves afloat.
I'm NOT against this.
It's not all a bad thing… but just remember, all you Alarmists out there… when it comes to taking money from BIG OIL…

…those in greenhouses shouldn't throw stones.

Not until the science is settled anyway.

11.4... CLIMATE JUSTICE... THE NEW MANTRA
(The REAL cause rears its ugly head...)

The cause of the debate always had a title... it just morphed a few times as the circumstances changed.
First it was Global Cooling, the ice age scare of the 1970's.
Then it was Global Warming as the planet 'heated up' with the hockey stick graph and CO2 emissions.
Then, as no real warming was shown after 1998's El Niño assisted peak, the cause became Global Climate Change.
But, you know, the Alarmists were never really happy with that one because it sparked the rejoinder "but the climate's ALWAYS changing".
They tied to link the cause to dangerous and catastrophic weather events, but no snappy slogan from that period ever really caught on.
Then, in recent years, the real monster has finally poked its ugly head around the wizard's curtain...
Climate Justice.

Those who use the term climate justice are trying to involve global warming in ethical and political issues, rather than see it as a separate environmental or scientific issue.
The term was first used in 2000, and recently, several climate justice groups have taken major fossil fuel companies to court, blaming them with global warming damage or sea level rise.

At https://www.peoplesdemands.org/, the Climate Justice elements stated their demands to the world's leaders.

"SIGN THE PEOPLE'S DEMANDS FOR CLIMATE JUSTICE
To: Government representatives to the 24th Session of the Conference of Parties (COP24) of the United Nations Framework Convention on Climate Change,

We urge you to stand with people across the world — not Big Polluters — and immediately take steps to address the climate crisis.

Climate change is the crisis of our time. This December at COP24, you will lay out the rules to implement the Paris Agreement, policies that will affect the lives of billions of people.

The urgency of the climate crisis requires a just response centered on human rights, equity, and justice. We demand you:

Keep fossil fuels in the ground.

Reject false solutions that are displacing real, people-first solutions to the climate crisis.

Advance real solutions that are just, feasible, and essential.

Honor climate finance obligations to developing countries.

End corporate interference in and capture of the climate talks.

Ensure developed countries honor their "Fair Shares" for largely fueling this crisis.

Even the demands are loosely worded and vague in the extreme.
Let's take a look at the above with a bit of introspection from me. Feel free to add your own…

Keep fossil fuels in the ground.
And by doing so bring the world to a standstill, stopping all foreign aid to poor countries; ceasing all production of anything, making billions starve and die of thirst… the consequences of this one demand is the death-knell of civilization as we know it.
But then we KNOW that is the agenda of the far left greenie pinko fascist loonies anyway.

Reject false solutions…
Hmm, that, dear reader is a purely subjective term…
Who decides what solutions are real or false? The unelected? The elected? Who implements the 'real' solutions, and who pays for them?
If the world economy has crashed due to the sudden stopping of fossil fuels, does money exist anymore?

…that are displacing real, people-first solutions to the climate crisis. (Okay, maybe you're smarter than me, but what the heck is an example of a 'people first solution' to the climate crisis? This gobbledygook baffles me)

Advance real…
Again we have the use of this word, 'real'… what do these folks think of our solutions so far? Imaginations?

... solutions that are just...
Justice! Money from the rich to the poor!
And 'just' to who? The top 20 countries? The poor Africans in the Congo?

...feasible.
Now here's one I'd like to see explained... give us ONE FEASIBLE solution to Global Warming, and surely the world would jump at it!

Honor climate finance obligations to developing countries.
This is simply another way to move money from the rich to the poor... are you starting to see a pattern to this?

End corporate interference in and capture of the climate talks.
When the heck have 'greedy corporations' interfered in climate talks?
If ANYONE has interfered in climate talks it's been the United Nations with their collection of political panels... the UNEP, the UNFCCC, and the IPCC.
Those political panels have had the full cooperation of the world's politicians and media since Kyoto, and accomplished absolutely NOTHING!
So COP24 decides to 'eliminate' the corporate side, and just let the politicians talk about it.
When it comes to implementation of the politicians plans (obviously involving corporate organizations), do we ignore the corporations altogether?
Maybe we should throw away the pandering greedy politicians.
Maybe we should throw away the Media mantra of continual catastrophe.
Maybe we should wash away ALL the false climate computer models, and get back to facts and data.
But that's me ranting again... sorry.

Ensure developed countries honor their "Fair Shares" for largely fueling this crisis.
And yup, that's them doubling down on the money from the rich to the poor again.

Websites dealing with this new mantra have sprung up like fresh grass in Spring.
For instance, EarthRights International begins their website like this...

"Climate Change and Climate Justice
The Problem

Burning fossil fuels, such as coal and gas, account for most of the world's global warming pollution. Some say we have harmed the planet sufficiently that there is no going back.

We are already seeing some of the effects—droughts, floods and storms. Climate change is here and now, and they are only going to get worse. Poor communities, especially in developing nations, suffer the most."

Yup, it really is a hodgepodge of talking points, lies, vague innuendo, and twisted rhetoric.

I'll analyze each facet, shall I?

Climate Change and Climate Justice
(Yup, good start, let's link the old name to the new one. And what a word... 'Justice', because what kind of person would deny justice to anyone... you'd have to be a bigoted white male nazi conservative tree-hating nazi! Sorry, that last one got away from me.)
The Problem
(Let's not begin with our corporate statement, let's get straight to attacking poor defenseless, odorless, life-giving carbon dioxide, shall we?)
Burning fossil fuels, such as coal and gas, (but notice, no wood here, no biomass, no biofuels... they're 'clean energy', and not 'really' fossil fuels anyway because they have the prefix 'bio') *account for most* (don't your scientists have a figure? You're just going to go with vague? Really?) *of the world's global warming pollution.* (Oh boy, where to start... those ugly stinky fossil fuels have supplied more than 99% of the world's energy since man rubbed two sticks together. Even now, fossil fuels 'account' for over 90% of the world's energy; they supply the backbone that allow us this rich and lavish lifestyle. Fossil fuels bring aid and comfort, medicines and supplies to all parts of the globe) *Some* (again, piling it on deep with the lack of figures here... really, why don't they take a poll or something?) *say we* (and here they admit their own culpability which allows them to pile more scorn) *have harmed the planet sufficiently that there is no going back.* (Going back to when? Back to the 1900's when kids worked down the coal mines? The 1800's when 50% of kids didn't make it to age 10? The 1700's and the age of world piracy? When was this golden age they want to go back to?)

*We are already seeing some of the effects—(*Oh dear, wait for it, here come the lies) *droughts, floods and storms.* (Yup, even NASA data doesn't prove the correlation of higher global temperatures/CO2 'pollution' to any of these global weather events. Droughts, floods and storms have not increased in the last 100 years, and some have actually diminished slightly) *Climate change is here and now,* (Yup, and it always was) *and they* (the droughts, floods and storms) *are only going to get worse.* (Get worse? Where is the evidence of this? Even NASA data contradicts this shallow throwaway line) *Poor communities, especially in developing nations, suffer the most.* (And here, dear reader, the Climate Justice brigade goes for the tug-at-the-heart, mentioning the poor. What they don't say is that if we instantly STOP all development in Fossil Fuels, the poor developing nations will suffer most! But why let facts get in the way of a good summary, huh?)

Climate Justice
Climate Justice is the new cry. It is the poor people of the world getting back what they deserved. We, the rich nations, have had control of the world's financial and economic reins for far too long. It's time for the poor people to rule for a while.

It's time for the white man to be put on trial... judged... sentenced... and for 'justice' to be done.
And there's not a thing we can do about it, except drop our shoulders, and slip into the realms of obscurity.

But here lies one huge problem.
The Climate Justice Brigade is missing one crucial point (well, they're missing a whole lot of them, but just one comes to mind real quick)...
If we decide to take the reins from the top nations in the world, and turn the new global economy... there's a whole lot of work to be done BEFORE we let these poor, developing nations take control of the world's wealth, energy, weapons, and food.
We have to get them all to declare peace first.

I don't think we want Africa's problems to be the new modern global standard, do we?
There's hardly an African nation that doesn't have some form of political/rebel war going on. Even the more 'civilized' (and I use that word carefully) like Egypt and South Africa have real problems.
Give them more global power and more money and I hate to think what they'd do.

Climate Justice is the radical side of the green, environmental, ecological movement bursting into the public view. They've always been there.
Now they want power.
And the way they're going to get it is to hi-jack the climate change debate, and insert their own political agenda in its place.

11.5... BEWARE THE WOLF IN THE SHEEP'S CLOTHING
(The Deniers denier...)

As 'Deniers', we face enough of an uphill struggle to be heard and not dismissed already.
But when there are official organisations disseminating lies yet calling themselves 'mainstream' and neutral, it makes out task even more difficult.
A February 2019 article in Forbes is one such case. Entitled...

"11 Things Climate Change 'Dismissive' People Say On Social Media"

I was immediately on the defensive.

It seems that the neutral, independent, unbiased Yale Climate Communication group and George Mason University conduct a yearly survey on us Deniers and test our knowledge, resolve, and intent. The last survey, in March 2018, found our 'dismissive' numbers at 11%.
Their opening paragraph in their 'independent' report got my spidey-sense on full-tilt.

"It is clear that climate is changing, and there is a human component on top of the naturally varying system. Most climate scientists understand this, and most logical people do too. The 4th National Climate Assessment report is a good place to find affirmation for these statements."

This neutral, independent, unbiased story begins by insinuating that we Deniers are not logical, then cite the very core of the Alarmist IPCC as their norm?
I read on as they list our main reasons for our denial, then antagonistically shred them... yeah, they were that independent.

"While their numbers are small, they are often very loud, persistent, aggressive and vitriolic in social media."

Hmm... I can't say I'm aggressive, and certainly not vitriolic... I'm calm and assured... confident in my beliefs.
I'm certainly not as aggressive as George Monibot, from the UK's Guardian left-leaning newspaper, who once said...

"Every time someone dies as a result of floods in Bangladesh, an airline executive should be dragged out of his office and drowned."

Wonderfully aggressive, huh?
Anyway, back to the Forbes article...
They cite specific 'talking points' in which we 'dismissives' defend our denier reasoning...

"Ice ages always seem to come up and some statement about natural cycles. It is honestly stunning this happens since most climate scientist are very aware of the various ways that climate changes naturally."

As we know, although 'most' scientists may be aware of Ice ages and natural cycles... but some scientists go out of their way to blatantly ignore these, and launch into the IPCC dogma.

Okay, we have seen the undoctored graphs of the cold spell in the 1970's, and we're old enough to remember the 'Cod War'; the UK against Iceland. Then the article dismisses a decade of fear and growing concern and says this...

"That magazine article from the 1970s. Apparently there was an article in Newsweek in 1975 that ran a story about a "cooling world.""

Yup... I was stupid enough to be duped by one Time magazine cover... (I never even knew there was such a publication; I was in my teens with other things on my mind.)

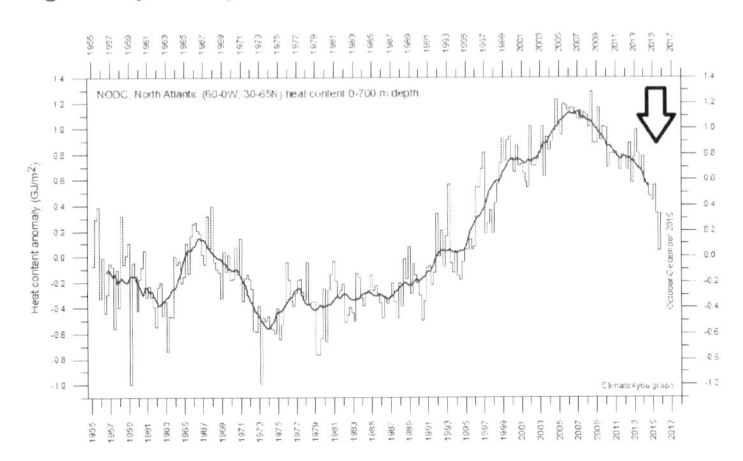

This is Alarmist dogma of the highest degree. Look at the un-homogenized North Atlantic Ocean temperatures of the 1970's (in the graph above). Look at the ice figures around Iceland, and reports of their

harbor's freezing over for the first time in hundreds of years. Then look at the Alarmist web-sites which belittle and dismiss the whole 1970's decade as a 'one magazine article' event.
The rest of the Forbes article slides into the morass of critique, dropping their supposed neutral façade, and belittling our beliefs into the kook area.

Yup, we may be kooks, but we have more ways to defend our position than you think, Yale and George Mason...
And here was me thinking that prestigious universities were places of learning... not houses of indoctrination.

As usual, the article in full can be found here...

https://www.forbes.com/sites/marshallshepherd/2019/02/16/11-things-climate-change-dismissive-people-say-on-social-media/#789bdd6b15e4

11.6... WE SIMPLY HAVE TO DO SOMETHING
(Anything's better than nothing... Right?)

There is a natural human knee-jerk reaction to climate change...

"We simply have to something... right?"

And while I'm experiencing the same feelings, I always take the debate to the next statement...

"What exactly can we do?"

And there lies the rub... as Shakespeare once said, probably.

And there's a huge difference between climate and the environment, and many get the two causes mixed up.
On climate change, most people are focused on CO2, and that's where everyone expects the main emphasis to be.

Because, let's face it, cutting pollution (the CO2 part) is a good thing. Right?

There are huge costs involved in reducing man's emissions. And when we don't have a clue as to how much reduction will mitigate how much temperature rise, the idea of CO2 reduction seems totally silly.

Australia is a case in point.

Many years ago their government signed the Kyoto protocol, and decided to do 'their bit' in reducing the world's CO2.
They invested highly in 'renewables', windmills and solar.
They stopped all their coal-fired power stations even though they had 200+ years of coal left in the ground.
Ten they sat back and watched as their electricity prices went through the roof, three times, four times what they had been paying. Despite the governments promises and their continued investment in renewables, the prices have not came down.
Australians are faced with power outages, power cuts, high prices, high unemployment, for what?

Australian scientists have determined that Australia contributes 1.3% of the world's anthropogenic CO2.

Further cuts are needed to reach their goals, but they hope that they can reduce CO2 emissions by 50% by 2050.

It has been calculated that such a HUGE commitment (at considerable personal sacrifice) will reduce the average global temperature by 0.001 degree Celsius.
We don't have even thermometers in the weather stations that can measure that small an amount.

Does that really make sense to you?

At the time of writing, Australia's population pay the world's highest prices for electricity.
Wait...
They have 200+ years of coal left in the ground...
They ban all coal fired power stations...
They export coal to China...
And Australians pay the world's highest prices for electricity...

Yup, that's climate logic for you... pretty soon they'll be selling their coal to China, and buying electricity back from them...

11.7... Using Chemo-Therapy to Fight a Cold!
(Can't we just adapt to it?)

There is little doubt that 'Climate Change' will never end.
Whatever happens, we'll be left to deal with whatever the climate is doing at any time in the future.
So, in this year of writing (2019) we are facing one of three solutions to deal with our 'crises'.

1. We can simply do nothing...
2. We can fight our onerous carbon dioxide emissions, and tax our nation states back to the 19th century...
3. We can adapt to our situation, whatever it is at any given time...

It seems that mankind is never going to have any real interest in the first option. We're just not built to do nothing in the face of danger; that's what all this global warming hullaballoo is about... we feel compelled to act.

Option two is a track filled with dangers.
If we cut our emissions to make a dent in CO2 (assuming that CO2 is the culprit), our lives will be taken back to some pre-industrial age, China India and a fragmented Africa will take over the world as the new superpowers, and all I see is chaos.
As the title states, it's a huge hammer of a solution that will take a level of mankind-wide trust and political statesmanship that I haven't witnessed so far in the world's history.
Like using chemotherapy to treat a cold, the supposed 'cure' may be far worse than the actual cold.

And that leaves option three as the most effective way out of our current situation of inactivity.
And it's by far the cheapest, the most cost effective, the easiest, and the one least likely to bring our world into political disorder.
We adapt.
Whatever warming, cooling, or stasis occurs, the climate change won't happen overnight, so we have time to adapt to it, and overcome any problems it might make to the planet.
We have a hundred years to adapt...
From solutions as diverse as better farming techniques to a rocket to Alpha Centauri, mankind will push through any problems, and science and the human condition will help.

Humans have undergone ice ages and warming periods before, we can do it again, and it does give us a get-out-of-jail-free card if the proverbial shit doesn't hit the fan.

Consider these FACTS...

We DON'T know for certain what is causing the climate to change. We have theories abounding, but absolutely NO ONE knows for sure. Yeah, some scientists have went for the minutiae, theories of carbon dioxide driving the change, some have gone for the obvious, our Sun. But we don't actually know... for certain.

We DON'T know for certain if the long-time trend is up, down, or a flat-line. Plenty of theories here, but again the jury has to be out on this; even the 'Alarmists' are hunkering down to see what happens after the 'hiatus'/pause has passed.

We DON'T know for sure what the eventual temperature rise will be. There is some form of consensus to the actual rise in temperature in the 21st Century... it's seeming to be in the 1-2°C range... and mankind can cope with that.

We DON'T know exactly why carbon dioxide is increasing. Sure there are many who say mankind is COMPLETELY to blame... but the evidence is not forthcoming. Some scientists say mankind is responsible for most of the increase, some say half, some say 4%, some say none at all. And they ALL have theories and facts to back them up.

We DO know that an increased CO2 level is good for plants, so crop yields will increase, the Earth will 'green' further than we have at present. Perhaps we are heading for 'Optimum', not 'Catastrophe'?

Considering all the stuff that we DON'T know for certain, solution number two seems a little bit overkill, hence the title of the chapter.

11.8... The Insurance Policy Principle
(Only a lunatic hates insurance...)

I remember my dad's words as I bought my first car... *"Buy as good an insurance policy as you can... you never know what's round the next corner."*
Good advice in 1985, and good advice now.
Depending on where you live, anyone who drives a car without insurance is either... a lunatic, a criminal, a ne'er-do-well, or a member of the arrogant youth. Car insurance saves you a catastrophic bill if you crash into someone, it can fix your own car, or even buy a brand new one for you.
Having been in a couple of car wrecks, I know the feeling as I sat in the ambulance... *"at least I was insured".*
However, as you choose your car, insurance can be a buyer's consideration. Car insurance isn't cheap, and if you want all the bells and whistles, it's a fair outlay every year.
I know of no-one who says on Dec 31st, as they suck on their New Years whisky... *"dang, threw away a thousand bucks again this year... didn't have a single crash".*

But in the insurance business there is a term called the 'net gain'.
If the cost of insuring a particular business, project or building is prohibitive, then why insure at all?

Carbon Tax is the way most Alarmists consider to be the best way forward. With such a system, everyone (personal and corporate) pays a lot of money to our respective Governments to pay for research into climate change, cuts in carbon dioxide emissions, and as gifts to the poor, still-developing countries so they can continue to pollute as they catch up to our lofty first-world status.
Estimates of just exactly how much mankind could cut man-made CO2 have varied from 10ppm to 100ppm... and that's with a really hefty carbon tax, beating us back into some pre-industrial age.
In most cases it means no more petrol/gasoline cars, trucks, trains, airplanes, ships... no more air conditioning, no more anything driven by fossil fuels.

And for that mammoth reduction in our livestyles... unfortunately, the largest figure in I could find was a reduction of... 0.1°C.

Yup, the industrial nations give up their world-dominating status, pass it to the Chinese, Indians and the continent of Africa, ALL for a tenth of an degree in overall world cooling.

And I know that as I sat in my cold hut on Dec 31st, lifting up my whisky to wish my country a Happy New Year, I'd be regretting that decision.

I've watched too many documentaries about African warlords and Somali pirates to even consider handing them more power.

Insurance is NOT the way out of this situation.

11.9... TIM BALL'S THEORY.
(A much maligned scientist...)

Canadian, geographic scientist, and historical climate denier, Tim Ball has a theory, and his take on the darker political entrenchment of the IPCC is an eye opener.
If what he says is true, the politics run deeper than I ever thought.
I have attempted to cover the theory here... the bold print highlights are a pathway of crumbs from a 'thinker' in the early 1800's to the takeover of the planet; a bloodless coup resulting in a World Government.

English scholar, **Thomas Robert Malthus** (1766 – 1834) wrote a book, *An Essay on the Principle of Population*, in which he espoused that there were too many people in the world at that time.
When he ran population on a graph, and then drew in a line of resources, he found they crossed, making the perfect population.
Charles Darwin took Malthus' book on his voyage of discovery of evolution aboard the Beagle, where he researched and wrote his infamous *Origins of Species.*
According to Malthus, governments helping the poor did not help. They merely encouraged the poor and unworthy to breed, making yet more people that the world couldn't support.
Unknown to Malthus, he had discovered **Eugenics**.

In 1968, Dr. **Paul R. Ehrlich** took Malthus' ideas, and wrote his own version, called *The Population Bomb*.
Ehrlich said...

"*I would take even money that England will not exist in the year 2000.*"

"*The battle to feed humanity is over. In the 1970's the world will undergo famines... hundreds of millions of people (including Americans) are going to starve to death.*"

"*Before 1985 mankind will enter a genuine age of scarcity... in which the accessible supply of many key minerals will be facing depletion.*"

Economist **Julian Simon** challenged Ehrlich on his 'key mineral' bet, and even allowed Ehrlich to choose the ten commodities that he predicted would become scarce in the next ten years.
Ehrlich chose scientist **John Holdren** (later to become President Obama's science tsar) to pick the ten minerals for the bet.

Ehrlich lost the wager, and was eventually forced to pay up.
In 1977, Paul Ehrlich, his wife Anne Ehrlich, and **John Holdren** co-authored the textbook Ecoscience: Population, Resources, Environment. Tim Ball chooses just one quotation from the book...

"Indeed, it has been concluded that compulsory population-control laws, even laws requiring compulsory contraception, could be sustained under the current Constitution if the population crisis became sufficiently severe to endanger the society."

Holdren and Ehrlich had taken Malthus' eugenics, and ran off with their principles to save the planet.
But of course, there were a few barriers in their way...

The Club of Rome was formed (in Rome) in April 1968 by Italian industrialist, Aurelio Peccei, and Scottish scientist, **Alexander King**. The membership included world leaders and elite environmentalists.
Published as a report to the Club of Rome, the book, *The Limits to Growth*, posited that because of the depletion of resources and overpopulation, economic growth could not continue indefinitely.
The book sold 30 million copies; the best-selling environmental book ever.
In 1991 **Alexander King** then wrote another book, this time with Club of Rome member Bertrand Schneider; *The First Global Revolution*.
Only King, Schneider and their kind could see that mankind was too wrapped up in their parochial problems to unite. Their slogan, *"the common enemy of humanity is man"* bulldozed its way into the minds of environmentalists everywhere.
The Club of Rome was going to save the planet. In 1994 they found the way to transcend national boundaries.

"In searching for a common enemy against whom we can unite, we came up with the idea that pollution, the threat of global warming, water shortages, famine and the like, would fit the bill."

Ball notes, *"They needed something that was going to impact the whole Earth."* Old grudges would be thrown away; the idea that individual countries could cope with such a global threat was ridiculous. Each country had to be controlled, led, ruled, and guided by the one idea.
There was only one organization that would fit the bill as a vehicle for the next phase of their plan.... **The United Nations.**

Insert mournful music interlude here…

Maurice Strong was a Canadian oil and mineral businessman, Club of Rome member, and a diplomat who served as Under-Secretary-General of the United Nations. He is also credited with an early encouraging of Al Gore's climate work. He is quoted as the Chairman of the UN Rio Earth Summit in 1992 as saying…

"Isn't the only hope for the planet that the industrial nations collapse? Isn't it our responsibility to bring this about?"

Strong knew exactly how to bring that threat to bear. The industrial nations need fossil fuels to keep their industrial base working. Taking away the top nations' fossil fuels is the easiest way to bring them down, and he was in a position in the U. N. to begin that work.

Now many of the Deniers have wondered hard and long as to 'why' the environmentalists chose carbon dioxide to blame for global warming. It's not a gas of much proportion in the atmosphere… but here's where the CO2 Alarmists win… CO2 as a culprit can be placed firmly at the feet of the industrial nations of the world.
Man can be seen as the miscreant of the whole problem.

In the same year as the Rio summit, 1992, Strong was made Chairman of Ontario Hydro in Canada. He stripped the state of its coal fired and nuclear power stations, and installed wind turbines. Ball states that Ontarians will be paying a surcharge for 40 years to pay off the damage Strong did.
Elaine Dewar, on interviewing Strong for her book, *Cloak of Green*, asked Strong why he didn't choose to go into politics. He answered that he could get far more power in the United Nations… and there, his power was without accountability.
In her book, Dewar said…

"Strong was using the U.N. as a platform to sell a global environmental crisis and the Global Governance Agenda."

(Incidentally, Strong was at the core of many controversies, perhaps the best noted being the U.N.'s Oil-for-Food Program. Federal investigators showed that in 1997, while working for Kofi Annan, Strong had endorsed a Jordanian bank check for $988,885, made out to "Mr. M. Strong." He died in China, exiled from the USA by a warrant for his arrest.)

Anyway, to summarize Strong's influence in the United Nations...

In 1972, Maurice Strong founded the **United Nations Environment Program (UNEP).**

In 1992 at the Earth Summit in Rio de Janeiro, under Strong's chairmanship, The **United Nations Framework Convention on Climate Change (UNFCCC)** was officially formed.

In the Rio Summit Maurice Strong also launched **Agenda 21** (now Agenda 2030) which Wikipedia describes as *"a non-binding action plan of the United Nations with regard to sustainable development."*

It's basically Eugenics brought up to date.

(Strong also formed an annual conference on global warming called The Conference of the Parties (COP) to discuss only global warming and its consequences. The COP has met every year since 1995. COP3 is possibly the most famous conference, held in December 1997 in Kyoto, Japan. After exhaustive negotiations, the conference adopted the Kyoto Protocol, which outlined the limits of greenhouse gas emissions.)

Many thought that the new United Nation' climate organizations (UNEP, UNFCCC, COP's) core purpose was to work out the causes of Global warming, and eradicate the problem.

However, the UNFCCC's 'official' objective was to...

*"stabilize greenhouse gas concentrations in the atmosphere at a level that would prevent dangerous **anthropogenic** interference with the climate system".*

FROM THE VERY BEGINNING, the UNFCCC's agenda was only linked to MAN-MADE warming... with no regard to natural causes at all.

The IPCC (Intergovernmental Panel on Climate Change) was set up in 1988 by the World Meteorological Organization (WMO) and Maurice Strong's United Nations Environment Program (UNEP).

The IPCC produces reports for the United Nations Framework Convention on Climate Change (UNFCCC).

Basically it's a huge conglomeration of U. N. diplomats getting money for shuffling paper and going to exotic locations to meet.

Again, many people think that the idea behind the IPCC is to determine what's causing the planet's climate change.
Nothing could be further from the truth.

The lead focus of the IPCC is no different from its parent bodies.
The 'focus and research' of the IPCC is...

"*a change of climate which is attributed directly or indirectly to* **human activity** *that alters the composition of the global atmosphere and which is in addition to natural climate variability observed over considerable time periods.*"

And the guidelines for the IPCC's reports are...

"*scientific, technical and socio-economic information relevant to understanding the scientific basis of risk of* **human-induced** *climate change, its potential impacts and options for adaptation and mitigation.*"

In 1990, in the first Assessment Report, the first sentence of the introduction is...

"*Many previous reports have addressed the question of climate change which might arise as a result of* **man's activities**.*"*

So you see...
Man-made Anthropomorphic Global Warming (AGW) was in the bricks at the very beginning of these organisations.

The IPCC is not a climate investigation organization.
The IPCC are not a group of independent scientists.
The IPCC are a group of politicians who tell scientists to investigate human caused global warming within their own IPCC framework and nothing else.
The IPCC are a group of politicians who publish 'proof' of human global warming and nothing else.
The IPCC state that carbon dioxide is the sole cause of modern global warming.
And the IPCC state the science is settled, and belittle anyone else who has an alternative theory.

In essence, the IPCC are not interested in actually PROVING the theory of carbon dioxide induces climate change... they are only interested in SUPPORTING it to the detriment of all other theories.
(And that might be the most important statements in this book)

But of course, Tim Ball's theory goes on...

Democrat **Tim Wirth** was a Colorado Senator and first Under Secretary of State for Democracy and Global Affairs from 1994 to 1997 under Bill Clinton. He was chairman of the United States Delegation at the 1994 Cairo Conference on Population and Development, and was the lead U.S. negotiator for the Kyoto Climate Conference. He was also President of the then newly created United Nations Foundation.
He said in 1993...

"We've got to ride the global warming issue. Even if the theory of global warming is wrong, we will be doing the right thing..."

Now all he had to do was find a lead scientist to 'admit' that human fossil fuel emissions were causing global warming.

Enter **James Hansen**, then working at GISS (NASA Goddard Institute for Space Studies). Hansen was invited to testify before the United States Senate Committee on Energy and Natural Resources on June 23, 1988, Washington's hottest day.
They sneaked inside, opened the committee room windows, and turned off the air conditioner... the perfect conditions for a talk on global warming.

And the Global Governance Agenda had its own manifesto in Agenda 21, it also had its own scientist, bought and paid for.

"It is difficult to get a man to understand something, when his salary depends upon his not understanding it."
Upton Sinclair

Agenda 21
There is no doubt that Agenda 21 is the brainchild of Maurice Strong, bolstered by his Eugenics-loving cronies.
In short, Agenda 21 is an ecologist-centric framework to organize various green programs in the Industrial Nations of the world.

Through environmental policies, its aims are to hamstring further development in those Industrial Nations, and bring about economic change.

Agenda 21, first shown at the 1992 Rio Summit, was agreed by 178 world nations. It is a 350 page document, and shows guidelines in many areas including...

Section I
Social and Economic Dimensions... poverty, food production and diet, health, population control, and decision making.

Section II
Conservation and Management of Resources for Development... atmospheric protection, opposing deforestation, protecting fragile environments, biodiversity, pollution control, management of biotechnology, and radioactive wastes.

Section III
Strengthening the Role of Major Groups... the roles of children, youth, women, the organization of NGOs, local authorities, business and industry, and their workers, strengthening the role of indigenous peoples, communities, and farmers.

Section IV
Means of Implementation... includes science, technology transfer, education, international institutions and financial mechanisms.

Each facet of Agenda 21 looks innocent enough, but every part of its purpose is the deepening of the U. N.'s involvement in every part of an individual country's existence.

It is an official U. N. document to be used in environmental litigation, and already has been used in hundreds of court cases here in the USA.

Thanks to George H. W. Bush, who signed the document at the Rio Summit, the United States is a signatory country to Agenda 21.

In 1993, Bill Clinton signed an Executive Order, and created the President's Council on Sustainable Development (PCSD) to formulate recommendations for the implementation of Agenda 21.

In 1996, the PCSD submitted its report, "Sustainable America: A New Consensus".

See... they even use the same terms... 'consensus'.

In 2012 Glenn Beck co-wrote a dystopian novel entitled *Agenda 21,* which has some basis on the concepts in the U. N. plan

In a protest to activist control, in *BLUE (Blue Beats Green),* movie-maker Jeffrey D. King sets out to air his personal views on the environmental movement, global warming and life.
In his own state, he finds communities, counties hit by a ban on logging to save an owl population.
He cites Agenda 21 as the driver of the ban on logging which collapsed an entire logging industry.
For the Denier, it's well worth searching out on YouTube.

https://www.youtube.com/watch?v=2WW8O_JF77c

But get there quick before it gets removed from YouTube once again.

SECTION 12
CLIMATE HISTORY... FIXED, OR A FIX?

"Lies, damn lies, and statistics."
Mark Twain

12.1... INTRODUCTION

The Archeology of Climate History is available to us in a variety of ways...

1. In early times, farmers had to record harvests in order to pay the landowners their taxes or tithe. Landowners had to record their income in order to pay national or royal taxes.
If they didn't keep good records, they could lose money, or even their land.
2. Clerics and scholars kept diaries, including notes on weather and population.
3. Painters and artists recorded scenes on canvas of climatic conditions.
4. The advent of the printing press gave way to the birth of the newspapers and almanacs. Many of these tell stories of floods, storms, cold spells and heatwaves.
5. In more recent times, television, video and the internet serve as our source for so much information... and it's free... mostly.

With so many claims of official government bodies tampering with historical data, it's good to have these 'older' sources to fall back on for verification.

They can change the data... but they can't change history...

Or can they?

12.2... A LOOK AT HISTORICAL CLIMATOLOGY
(The dark seedy side of the coin...)

Rate of Temperature Change Higher Now than in Past 1,000 Years

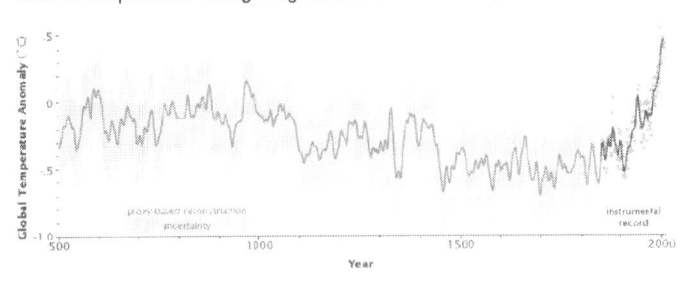

Source: NASA

Above is Michael Mann's 1998 'Hokey-schtick' graph... showing a slight, 0.5°C, cooling from 500AD to the 1800's.
(Please also let the record state that Mann's WHOLE graph shows just 0.5°C rise in 1500 years)

When the graph was first devised, Mann and his cohorts thought the subject dead and buried... they had found the optimum graph of world historic temperature... the science was settled.
(In section 2.4, the graph is disproved)

Compare this to another graph...

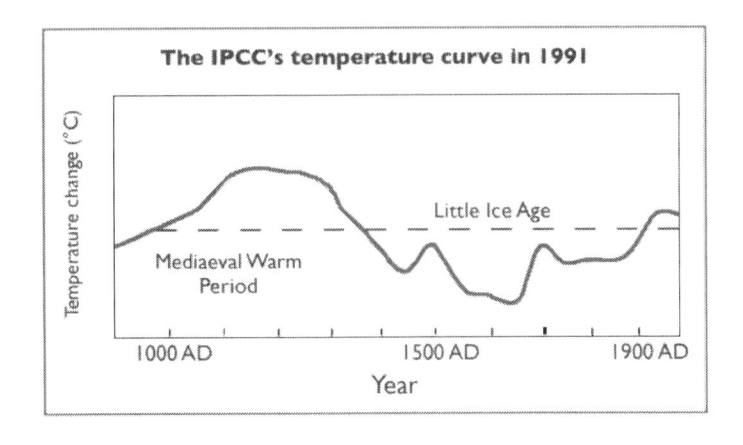

The IPCC's temperature curve in 1991

This was used by the IPCC in 1991, just a short decade before publishing Mann's 'Hokey-stuck'.

Something is wrong between the two graphs.

The 1991 graph shows Medieval times warmer than today, and Mann's graph shows 2016 warmer than any time in history.
They both can't be right.

The Medieval Warming Period (MWP)
Mankind thrives in times of good climate, and finds the Ice Ages particularly difficult. The historical record of the Medieval Warming Period has been taught in schools and universities all over the world for hundreds of years.
It's a fact.
It's been called the Medieval Climate Optimum, or Medieval Climatic Anomaly, but we all know what it actually was… a warm period that lasted 400 years, roughly between 900AD and 1300AD.
There's bookloads of historical evidence to prove it, of which I mention only a few bullet points…
1. In Europe, grain crops flourished, chronicled by landowners. Alpine tree lines rose, and many new cities formed and grew. In 400 years, it is estimate that the population of Europe more than doubled (again, chronicled in hundreds of historical documents).
2. The Vikings began a colony in Greenland, establishing over 600 farms, growing grain, bringing livestock. The population at its zenith is estimated at least 6000, possibly up to 10,000. The new Greenlanders traded for over 300 years. The last evidence is in 1378, when the ice was too thick to sail through.
3. Grape growing spread up European mountains to new altitudes, giving better yields and allowing different varieties to prosper (this is also documented).
4. The 1200's and 1300's are also the times of the great Cathedral building in Europe. Look at most of the huge churches, they were built in a time of food prosperity… due to favorable growing conditions there was a surplus of food to supply workforces of thousands of masons, carpenters and such. When the good weather ceased, so did the cathedral building.

Many of these historical facts can be compared to present day climatic conditions…

1. Vineyards grew to an altitude of 780m implying temperatures of 1.0–1.4°C warmer than the present day (Oliver, 1973).
2. Wheat and oats were grown near Trondheim, Norway, implying temperatures 1.0°C warmer than the present day (Fagan 2000).
3. Sediment samples show the sea surface temperatures in the Sargasso Sea were approximately 1°C warmer than today's (Keigwin, 1996).

But was it linked ONLY to the Northern Hemisphere?

I mean, can a 400 year temperature variation be held to just one half of a weather-filled globe? I think not.

Well, Michael Mann relied on one particular tree sample for his 'hockey schtook', so we'll take a similar approach.
Oxygen isotope studies from Greenland, Ireland, Germany, Switzerland, Tibet, China, New Zealand, and many other places in both hemispheres indicate a warm period around 900-1300AD.
Tree rings from ALL OVER THE WORLD confirm the presence of a global Medieval Warm Period.

There are SEVERAL THOUSAND published papers that confirm, prove and support the idea of a Medieval Warming Period... most of them postulate that the temperatures were WARMER than today.

Let's look at some graphs, shall we?

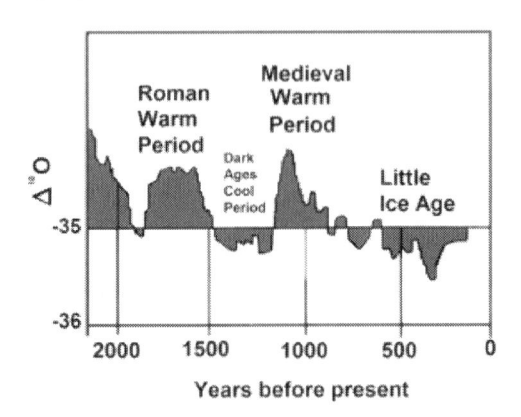

This is a graph compiled from oxygen isotopes from the GISP2 Greenland ice core (Grootes, P.M., Stuiver, M., 1997).

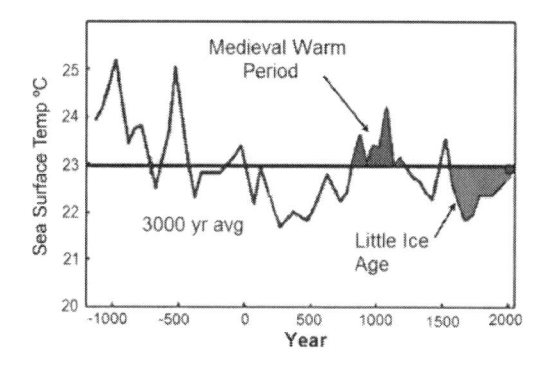

The above shows surface temperatures of the Sargasso Sea (Keigwin, 1996).

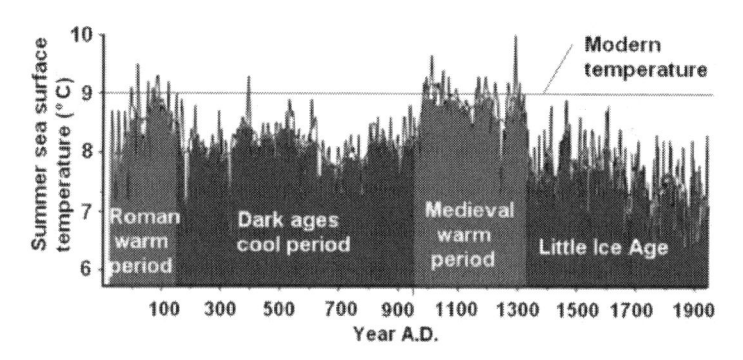

This graph shows summer sea surface temps near Iceland (Sicre et al., 2008).

And look back at the last three graphs... ALL THREE show warming periods further back in history... and coolings.
Let's walk back shall we...
Little Ice Age 1300-1900AD
Medieval Warming Period... 900-1300AD
Dark Ages... 200-900AD
The Roman Warming Period 200BC – 100AD
The Minoan Warming Period 2500 years and 3000 years ago.

Historical Cooling
So far we've dealt with warm periods, but these were punctuated by dark cold times when the cold temperatures affected the world's population. The Medieval Warming Period ended with a thunderclap.

From a position of record harvests all over Europe, the cold came like a blanket. A sudden period of low sunspot numbers called the Wolf Minimum came between about 1300 and 1320 AD.

Three years of torrential rain beginning in 1315 led to the Great Famine of 1315–1317. Crops failed, straw rotted in the fields, livestock and people died.

From around 1600, a growing fur trade from North America became established in Europe. The cold temperatures made the furs valuable and became a huge part of the New World's allure.

In 1683, the River Thames in London froze to three feet thick prompting an ice fair on the river with wooden buildings and stalls. This became a popular winter attraction for many years and is mentioned widely in historical documents, drawings and paintings.

Climatologist Tim Ball says of the Little Ice Age, *"You don't understand the fur trade unless you understand how cold it was in the 17th Century in Europe."*

In the early days of 1709, Europe awoke to a continent frozen solid; Italy, Norway, Germany, France and Russia shared the same fate. In the worst winter in 500 years, tens of thousands of people died in the three month freeze.

Almost all the European rivers froze solid. Venice, Rome, Naples, Paris and Valencia in Spain, all froze solid. Canal networks and ports froze, and heavy snow blocked every road. Marseille on the Mediterranean coast, froze over… the English scholar William Derham wrote: *"I believe the Frost was greater … than any other within the Memory of Man."*

Charles Dickens was born in 1812, just two years before the Thames froze over for the last time. He chronicles that it snowed every winter in London for the first nine years of his life, and it stayed on the ground for three months… long winters indeed.

Many of his books are based in a cold, snow-swept London.

Michael Mann's attempt to 'smooth' out the Little Ice Age is indeed a crime against recorded history.

12.3... THE EARTH'S COOLING! ICE AGE 1970!
(The 'first' consensus...)

The 1970's Ice Age
Back in the 1970's there was a decade long scare, in which scientists predicted an oncoming ice age.
Now, Alarmists would have you believe that it was just a few kook scientists, a rogue bunch... but Alarmists are wrong... they're trying to down-play the whole scare, because it foreshadows the current scare.
Let's take a look at some facts, because some of us lived through it.

Those of us in advancing years (getting to 60 and beyond) can remember the Ice Age warnings from the 1970's. The Earth had cooled a degree Fahrenheit in less than a decade, and both scientists and media were in a cooling knee-jerk frenzy.
Look at the graph of US temperatures (thus the northern hemisphere) below. Look at the high of 1934, then the 20 year decline, a little kick up at 1954, then a fall in temperatures until 1978.
In essence the temperature had cooled for almost 40 years straight.
From the 1934 peak high to the 1979 maximum low, the temperature had dropped 2.5°C in 45 years.
No wonder the scientists were predicting gloom!

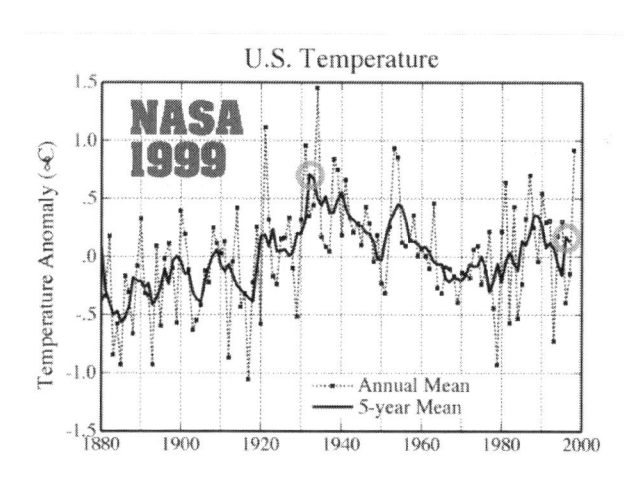

Now we aren't talking about a few loony-fringe idiots shouting about a flat Earth or the universe going round the sun or suchlike. These were prominent scientists of the day, and there sure was a consensus of agreement.

So one little clue… when you're looking at 'graphs' of temperature of the last 50 years or more, look for a degree cooling between 1969 and 1979, and 2.5°C cooling between 1934 and 1979.

Anyway, despite what NASA graphs you've seen of the northern hemisphere temperatures of the last century, in the 1970's there was a 'consensus' that we were headed to an Ice Age…
Global temperatures were dipping low, there was a world energy crisis caused by OPEC oil companies strangling oil distribution, and a miners strike in the UK. There was a fishing 'war' between Iceland and the UK, 'The Cod War', because Iceland's ports were frozen, and they had to use Russian fish tankers to 'land' their catch.
It was a gloomy time.
Let's look at some of the quotes of the day… and there are many hundreds.
And please note… THIS GOES ON FOR ALMOST A DECADE!

New York Times, Jan30, 1961…
"… specialists from several continents seem to have reached unanimous agreement on only one point: it is getting colder"

Global Ecology, 1971…
"continued rapid cooling of the Earth"

Windsor Star, Sep 9, 1972…
"Prof. Hubert Lamb, director of climate research at the University of East Anglia… there may be a few fluctuations from time to time, but these will be offset by the general downward trend"

Science Digest, Feb 1973…
"the world's climatologists are agreed… prepare for the next ice age"

Canberra Times, May 16, 1974…
"global cooling has been recorded by the world network of weather stations"

Christian Science Monitor, Aug 27, 1974…
"the North Atlantic is cooling down about as fast as an ocean can cool… growing seasons in England and Scandinavia are getting shorter"

Canberra Times, Nov 22, 1974…

"A new ice age could grip the world within the lifetime of present generations"

Newsweek, April 28, 1975...
"meteorologists are almost unanimous... the trend will reduce productivity for the rest of the century."

Canberra Times, May 31, 1975...
"Some scientists believe a new Ice Age is on the way. Others insist it has already begun"

Milwaukee Sentinel, July 1975...
"the world's climatologists are agreed... prepare for the next ice age"

International Wildlife, July 1975...
"a new ice age must stand alongside nuclear war as a likely source of wholesale death and misery"

New York Times, Aug14, 1975...
"Many signs pointing to the possibility that Earth may be heading for another Ice Age..."

Science, march 1, 1975...
"the approach of a full-blown 10,000 year ice age"

Science Magazine, Dec 10, 1976...
"toward extensive Northern Hemisphere glaciation"

New York Times, Jan 5, 1978...
"International team of Scientists Finds No End in Sight to 30-year Cooling Trend"

Now, many 'Alarmists' will say it was NOT a consensus, and a bunch of weirdo science nuts had hi-jacked the media... but I can remember on British television, the documentaries of gloom and doom, and lots of snow.
One Alarmist site even states that just 6 stories talked about global cooling in the 1970's, and many more predicted global warming.
This is total hogwash, and an attempt to re-write temperature history.

Don't believe the 'Alarmist' nut jobs... go ask someone who lived thru it all.

Two wonderful facts came rolling out of those newspaper files... one of them is quoted below...

A big player for the Alarmist side is the University of East Anglia's Climate Research Unit (CRU) which was at the center of the Climategate email scandal in 2009.

Let's look at that newspaper report in detail.

This chappie, Hubert Lamb, is indeed a big player; he's ex British MET (Meteorological Office), and the organizer of two international conferences in 1973 and 1975.

He died in 1997, just one year before the 'hottest' year.

However, looking at the article from 1972, this is NOT the evidence of a kook scientist... this is Hubert Lamb, a full-fledged Climate Alarmist in the 80's and 90's... look carefully at the rhetoric he's spouting.

> The Windsor Star - Sep 9, 1972   Browse this newspaper »   Browse all newspapers »
>
> ## There's a new Ice Age coming!
>
> NORWICH, England (AP) — A new Ice Age is creeping over the Northern Hemisphere, and the rest of this century will grow colder and colder, a British expert on climate says.
>
> Prof. Hubert Lamb, director of climate research at the University of East Anglia, had a few comforting thoughts in an interview Sunday:
>
> "The full impact of the new Ice Age will not be upon us for another 10,000 years and even then it will not be as severe as the last great glacial period.
>
> "We are past the best of the inter-glacial period which happened between 7,000 and 3,000 years ago," he continued.
>
> "Ever since then we have been on a downhill float regarding temperature. There may be a few upward fluctuations from time to time but these are more than offset by the general downward trend."
>
> Lamb said temperatures have been slowly dipping for the last 20 years.
>
> "We are on a definite downhill course for the next two centuries," he said. "The last 20 years of this century will be progressively colder. After that the climate may warm up again but only for a short period of decades."
>
> Lamb said climate changes come in cycles determined by astronomical and physical factors. He said one main cause is the amount of radiation received from the sun.

The other beauty I uncovered was a short by James P. Sterba, a journalist for the New York Times. He quotes 'Climatologists' as believing *"that the* earth's climate has moved into a cooling cycle". However, in the yonder days of the 1970's the sides had reversed. If you read the whole of the article, you see skeptics to the Global Cooling mantra saying of the 'coolers' that *"their stuff is right out of fantasy land".*

And the quote, *"how can climatologists predict climate years ahead?"* is a classic.
How things change, huh?
But no matter how they try and tell you that the 70's cooling scare was fringe scientists, we can go to the video archives of the magical YouTube.

In The Coming Ice Age (TV 1978) with Leonard Nimoy, the cooling is discussed, and it's not fringe.

I also recall a UK Documentary with the dialogue by Magnus Magnusson in a similar vein.

Climatologists Forecast Stormy Economic Future

By JAMES P. STERBA
Special to The New York Times

They believe that the earth's climate has moved into a cooling cycle, which means highly erratic weather for decades to come. And that, they say, has profound implications — most of them bad — for world food production, economic stability and social order. With the world's population now so high, the results of even minor year-to-year shifts in climate could be catastropihc, they say.

Skeptical Scientists

Some scientists think all that is nonsense, mainly because climatologists can offer no scientific proof to back up their theories. If meteorologists, using sophisticated computers, can forecast weather only a day or two in advance, they ask, how can climatologists project climate years ahead?

"It's interesting," said one skeptical scientist. "But some of their stuff is right out of fantasy land."

And if there had been nothing else to say, the chapter would have stopped there...

But I was a thirteen year old kid in 1972, and I remember the television stories quite clearly.

The Cod War
The Cod War between Iceland and the United Kingdom was a big thing for us UK schoolkids in the 1970's. Colder sea temperatures had driven the schools of cod southwards, and Icelandic trawlers were encroaching on British waters to fish.
The Royal Navy got involved after a series of near misses between the respective trawlers, and the Icelandic Navy responded in kind.
The TV reports were quite dramatic for a child of 13.
Part of the cause of the Cod War in the 70's was the freezing of Iceland's harbors... their trawlers simply couldn't get into the ports with their catch, so they unloaded the cod onto Russian fish transporters.
The graph below shows the dip in the North Atlantic Ocean temperatures for the period.

But hey, it's pretty easy to make up new graphs to go along with your theories, Michael Mann's made a good living doing so.
It's way harder to try to erase history.
The Cod War... go look it up.

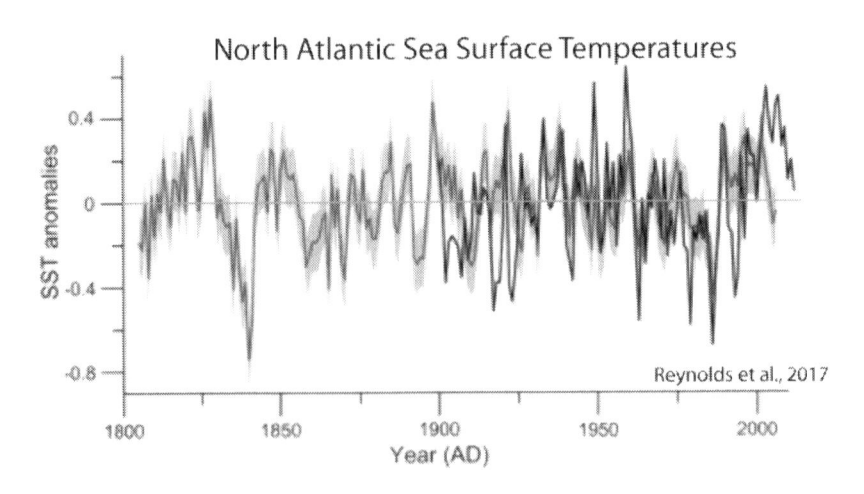

North Atlantic Sea Surface Temperatures
Reynolds et al., 2017

You see, in a time where the Alarmist side is trying to re-invent the past, you have to deny, denigrate and deride ALL stories to the contrary.
But sometimes history is just too big to be changed.

You see, even back then, the science was never settled.

12.4... THE WORLD'S HOTTEST DAYS WERE IN 1998!
(Eh, actually, NO! they weren't...)

Amazingly, despite the GISS graph (below) of world mean temperatures, 40% of all countries had their hottest day, before the 'peak' of 2000's Global Warming.
A staggering 14 countries (11%) had their hottest day during or before the 1940's.
Yet when we look at 'revised' temperature graphs of the last 100 years, every 'revision' has trended the old figures down, and the new figures up.

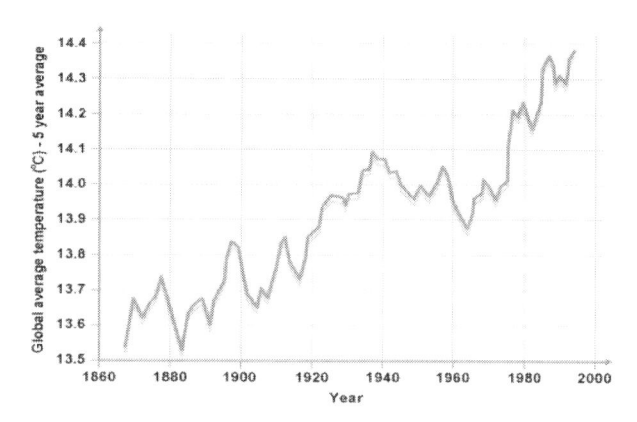

Let's check some stories...
Christopher Booker (From the Telegraph (UK), 2014 and 2015) reported when 2014 was announced as the 'hottest year on record', he did some digging. Paul Homewood, creator of the blog, 'Notalotofpeopleknowthat', investigated three remote weather stations in Paraguay, where NASA's Goddard Institute for Space Studies (GISS) had reported a 1.5°C rise in mean temperature from 1950 and 2015, that's double the world's measured rise. When he checked the actual data, he found was the absolute contrary; the original data still held at the stations showed a DROP in temperature between those dates by a full degree!
Someone was cooking the books!

Let's retrace a bit and start at some accepted basics;
The IPCC rely on five official data records. Three of these are surface based temperature readings, which are compiled by GISS (NASA), the US National Oceanic and Atmospheric Administration (NOAA), and by the University of East Anglia's Climatic Research Unit (CRU). (The CRU work hand in hand with the Hadley Centre for Climate Prediction, which is part of the UK Met Office). The last two records are derived from satellite

measurements, then compiled by Remote Sensing Systems (RSS) in California and the University of Alabama, Huntsville (UAH).

Here's where the data takes two distinctly different paths...
The surface based data shows the very familiar hockey stick graph, show recent years as the 'hottest since records began', citing 2014 as the hottest year ever. (The story is based in 2015)
The satellite RSS feeds, however show the hottest year as 1997, and rank 2014 as only the sixth warmest year since that maximum.
Obviously Deniers and Alarmists take their pick, and argue black is white, while playing with two totally different datasets.

Arctic Cooling...
And that wasn't Homewood's only tussle with RSS feeds.
After looking into South American weather stations, like McIntyre at the start of the chapter, Homewood turned his attention to the Arctic records in 2015. He found that in 'nearly every case' in the original records some adjustment had been made, some by up to 1 degree Celsius.
When Iceland's Traust Jonsson (from the Iceland MET Office) read Homewood's findings, he found that somehow the Goddard Institute for Space Studies (GISS) and the National Climate Data Center (NCDC)'s new 'improved and adjusted' version had managed to smooth over Iceland's "sea ice years". This was a particularly cold period, around 1970, when extreme cooling almost devastated Iceland's economy, and precipitated the 'Cod War' between Iceland and the United Kingdom from 1972 to 1973. Iceland's harbors froze up for the FIRST TIME... hardly an event to forget.

More on the Arctic...
In 2007, mathematician and statistician Steve McIntyre challenged a 1987 paper by scientist James Hansen who worked at (then later ran) NASA's GISS from 1967 to 2013. The original graph showed Arctic temperatures as being highest in 1940, but after subsequent 'adjustments, the early temperatures were toned downwards, and the more recent temperatures adjusted upwards.
Unfortunately it's a rather familiar story.

Down Under...
In New Zealand, Anthony Watts, blogger at the science blog, Whatsupwiththat (WUWT), was taken aback by the NIWA (The National Institute of Water and Atmospheric Research, New Zealand) figures for

the mean New Zealand temperature data. The graph's dotted black line, from 1909 to 2009, showed a 0.9°C warming over 100 years.

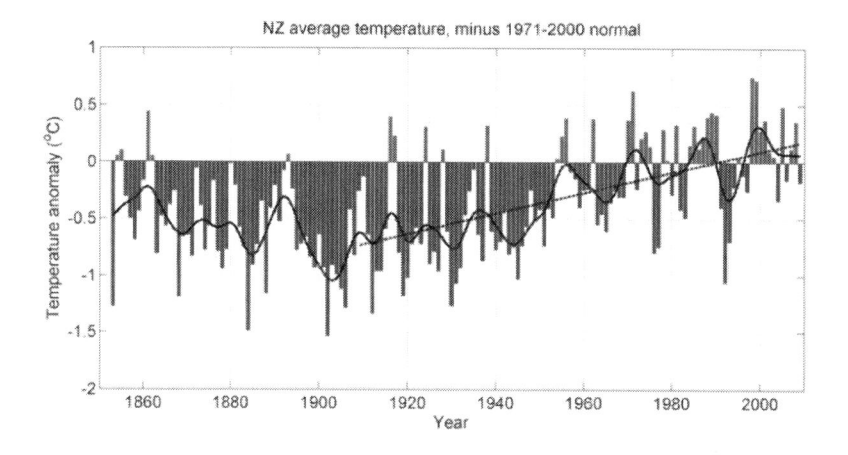

However, the original data is easily downloaded, and it shows that the NIWA used the same techniques to 'adjust' figures as NOAA's National Climatic Data Center (NCDC). The figures above the 'zero' line were enhanced, and the figures below the line were lowered. This was done to highlight the rise in the 100 year span.

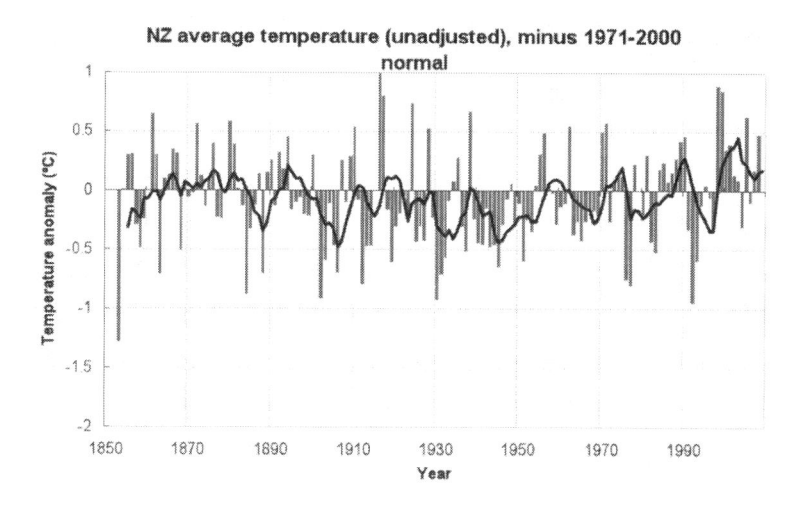

If the original figures are plotted, above, it shows a far more horizontal line with a warming of just 0.06°C in a hundred years.
Of course, Alarmists argue the 'original', 'adjusted' figures rather than the 'raw data'.

12.5... TEMPERATURES TODAY (IN THE 2000'S) ARE THE HOTTEST EVER MEASURED

(Well, hold your horses, that's just NOT true!)

1. When the Earth was being formed, before the crust solidified, trust me, it was way hotter than today! (Okay, I know that's not a legitimate argument, but it sounds funny, so I included it)

2. In the ages of the dinosaurs it was way hotter than today, and the animals still survived.
(Still too vague, okay we'll continue)

3. In the 1100's, Vikings from Iceland discovered Greenland, and called it 'Greenland'. (Because it was green; go figure) They farmed the coastland for nearly three hundred years, but eventually had to leave as the Medieval Warm Period was coming to a close.
It is accepted reluctantly by ALL Alarmists and that ice core samples tell us it was WARMER in the year 1100 than today (the 2000's).
So why are we alarmed?

4. Would it surprise you that out of the FIFTEEN North American countries (listed below), only FOUR had their highest temperature in the 2000's. The USA had its record high in 1913!
Hardly the records you'd expect to find when the Alarmists are crying 'doom' today.
WHY?
To me the solution's easy.
The highest temperature is set in many textbooks; difficult to ignore, difficult to wipe away.
The individual temperature records for each hour, day, week etc, are not so easily available to the public, so therefore far more easy to change, to fudge, to meddle with.
The record highs and lows, however, are public record. Here's the records for North America...

Anguilla, 34.2 °C (93.6 °F), The Valley, 12 September, 2015
Antigua and Barbuda, 34.9 °C (94.8 °F), St. John's, 12 August, 1995
British Virgin Islands, 35.0 °C (95.0 °F), Airport, 22 July, 2016
Canada, 45.0 °C (113.0 °F), Saskatchewan, 5 July, 1937
Cayman Islands, 34.9 °C (94.8 °F), Airport 21 August, 2016
Cuba, 38.8 °C (101.8 °F), Jucarito 17 April, 1999
Dominica, 35.5 °C (95.9 °F), Airport, 3 October, 2015

Dominican Republic, 43.0 °C (109.4 °F), Mao, 31 August, 1954
Greenland, 30.1 °C (86.2 °F), Ivittuut, 23 June, 1915
Panama, 40.0 °C (104.0 °F), San Francisco, 20 March,1998
Mexico, 52.0 °C (125.6 °F), Mexicali, 28 July, 1995
Puerto Rico, 40 °C (104 °F), Mona Island, 2 July, 1996
United States, 56.7 °C (134.1 °F), California, 10 July, 1913
Virgin Islands, 37 °C (99 °F), Saint Thomas, 31 July, 1988

There are some cynics out there that will postulate that in the more 'rustic' Latin American countries they introduced more thermometers in recent times of prosperity... THAT'S why their highest temperatures are more recent. I wouldn't dare say such a thing.

Now, I will agree, the world's getting warmer... out of the 125 world countries who have publically divulged their hottest ever measured temperature, only 51 have NOT been in the 2000's, but it's still a statistic that 40% of all countries had their hottest day, before the 'peak' of Global Warming.
A staggering 14 countries (11%) had their hottest day during or before the 1940's.
"Oh, Global Warming isn't JUST about heat, it's about extreme low temperatures too..." I hear the Alarmists say, trying to recapture lost ground.
NOPE! Out of the 94 countries who have given out their coldest temperatures, 80 are before 2000.
(That's 85% had their coldest day before the Global Warming highs)

Section 13
Australia's Role in Global Warming

"Current atmospheric temperatures and CO2 content are no higher than they have been at various times during the past million years."
Paul Berenson (M.I.T physicist and Scientific Advisor to NATO's Supreme Allied Commander Europe (SACEUR))

13.1... Introduction

It has often been said that Australians are the most down-to-earth people in the world.
If they see poop, they call it 'shite'.
If they see lies, they're quick to point it out.
They're like Texans... solid, dependable, and bullish. And I say all the above with love in my heart for those sturdy ancestors of pioneers.

When the idea of a second front in World War 1 was dreamt up, they chose the ANZAC (Australian and New Zealand Army Corps) to be the spearhead, and the name of Gallipoli in Turkey will remain one of the dirtiest stains in that war's history.

When the SAS (Special Air Service) was formed in the desert in World War 2, the British Army chose from Australians, New Zealanders and South Africans to be the core of the new special unit.

In the war against climate change, the Australians have also led from the front. Not only do they have an extensive amount of weather stations (some established as early as 1885), but they also lay claim to a huge chunk of Antarctica, and because of their proximity, are a major force in the southern continent.

The national weather department in Australia is the Bureau of Meteorology (BOM) established in 1906, and it monitors and supplies all Australian weather data.

So let's take a look at some Australian climate change information.

13.2... THE AUSTRALIAN INCONVENIENT TRUTH
(Temperature tampering down under...)

Australia is the largest island/continent on the planet. It has an area of 7.6 million sq km, with a population of just 25 million people.
It is almost exactly double the size of the USA, with the population of less than the island of Madagascar.
Apart from a few settlements around the coast, it's a big deserted place.
The Australian's weather department is called the Bureau of Meteorology (BOM), and its weather stations have been watching over the country's weather since 1906.
However, there are many weather stations that have continuous monitoring since the 1880's.
The BOM have been issuing a raw data set dating back to that time. It's a matter of record, written into stone, supposedly unchangeable.

But guess what? The Australian BOM have been charged with 'altering' the historical record so many times, that they've actually admitted to doing so, with details reported in the newspapers.

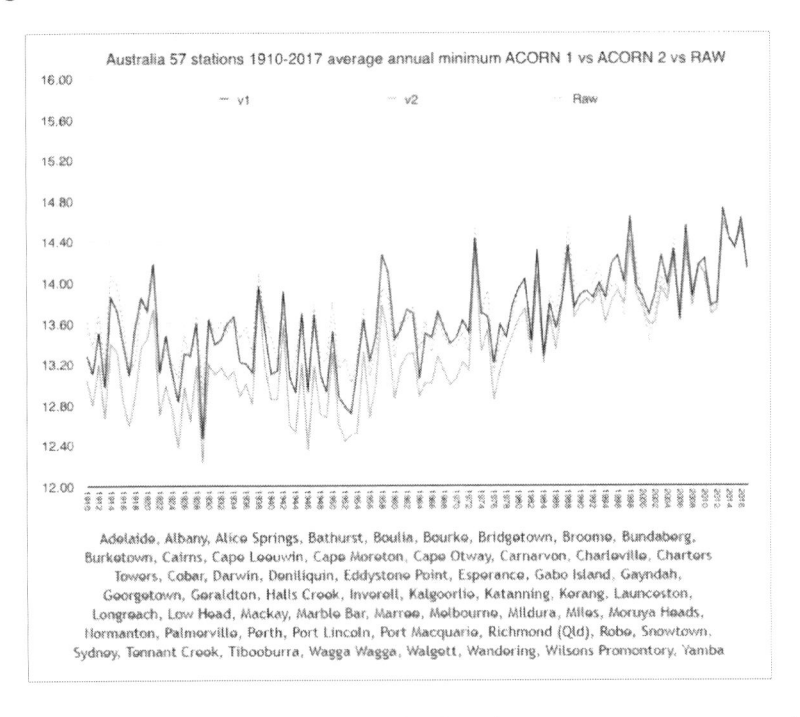

In 2012, the BOM issued a new record of surface temperatures going back to 1910. The data is called ACORN 1 (Australian Climate Observations

Reference Network – Surface Air Temperature), and was greeted with some disdain from the vocal climate 'Deniers' down under.
This re-written (and homogenized) dataset reduced the national 'area averaged' maximum temperatures from the historical records; the older temperature figures were made lower. These adjusted figures made the national global warming figure to look considerably greater and new figures for Australia's warming began to hit the headlines.

Just four years after the introduction of ACORN 1, the BOM issued a new set of data, supposedly 'correcting' the 'wrong' data from ACORN 1.
With the typical creativity of scientists, they called it ACORN 2... and amidst the ripple of distrust, the Deniers claimed that the BOM had doubled down on their homogenization, and lowered the historical temperatures again.

Again in characteristic Australian fashion, these adjustments were challenged by TV stations and newspapers, forcing the BOM to justify their adjustments.
The usual explanations were given out... some stations had been moved, some now disused, some been destroyed by wildlife (cockatoos destroying thermometers), there's even an account of an observer giving 'hot' data to the BOM in order to get sent home from work because the temperature was too high. Australians!
Those challenging the figures don't doubt that there were numerous justifiable reasons for some of the new adjustments in ACORN 1. However, there are enough 'unjustified' adjustments to force the BOM to make a list of statements (given below).
The new combined figures in ACORN 1 and ACORN 2 have risen Australia's global warming by 23%.

In other words... BOM 'adjustments' to the written temperature record in Australia have ALTERED the figures.
That's a lot of heat.
And this isn't just Denier claims...
According to the bureau report, covered in the TV and newspapers, they made 660 adjustments in the ACORN 1 data, and 966 adjustments in ACORN 2.
Now, that's NOT a total of 1626 readings out of 57 stations in a period of 36,000 days.
This is 1626 days with ALL their data changed.
That's the equivalent of FOUR AND A HALF YEARS of data.

For your perusal, here's a list of the first ACORN changes, available thru the BOM website...

http://www.bom.gov.au/climate/change/acorn-sat/documents/ACORN-SAT-Station-adjustment-summary.pdf

13.3... A True Australian Heroine
(The 'bitch' who just won't go away...)

I love this lady!
Jennifer Marohasy is an Australian PhD biologist, columnist and blogger, and has had many peer-reviewed articles published.
She is an avid climate denier, as can be referenced by 70% of her page on Wikipedia being on her *"Public position on global warming"*, and lambasting her by quotes by Gavin GISS Schmitt and Michael Hokey-Stick Mann.
Yup they brought out the big guns there...

In September 2014, an article in Australia's Newcastle Herald by local lawyer Anthony Cox told of Jennifer Marohasy's research into the BOM's homogenized figures.
(Newcastle is just a hundred miles up the Australian coast from Sydney.)

"Recently a group of researchers including scientist and author Dr. Jennifer Marohasy have questioned the accuracy of the record, because the final temperatures from the record differ from the raw data."

He specifically mentions the town of Bourke, just inland from Newcastle. (My bold)

*"Overall, Bourke **raw data** shows a **cooling** maximum temperature trend from 1880 of 1.7° a century. After adjustment, **(ACORN 1)** the temperature record at Bourke shows a **slight warming** temperature trend. This change of maximum temperature trend has a great effect on the whole of Australia's temperature record. The bureau has offered no reason for the adjustments."*

And this is not an isolated incident. He also mentions a suburb of Newcastle, a town called Williamtown, near Newcastle Airport.

"...the temperature site at Williamtown also shows a marked difference between raw minimum temperature data and the temperature record after adjustment by the bureau."

https://www.theherald.com.au/story/2558481/opinion-adjusted-temperatures-need-explaining/?cs=308

On Jennifer's own website, she went into more detail on Newcastle and other Australian temperature records.

She also highlighted a high temperature period in Newcastle in 1878, showing highs well above todays supposed 'near dangerous' levels.

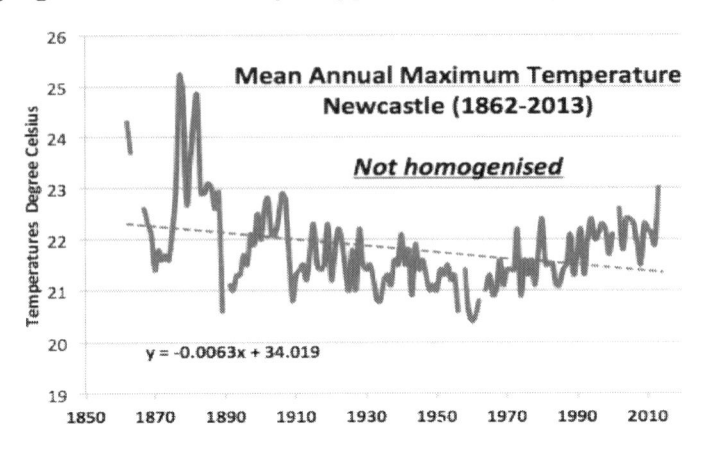

Mean Annual Maximum Temperature Newcastle (1862-2013)
Not homogenised
y = -0.0063x + 34.019

As can be seen by the original data, Newcastle has actually cooled in 150 years.

But the historical highs of the 1870's are never mentioned, despite numerous references of high temperatures and deep drought available in Australian newspapers.

In fact, the new record used is from RAAF (Royal Australian Air Force) at Williamtown (Newcastle Airport).

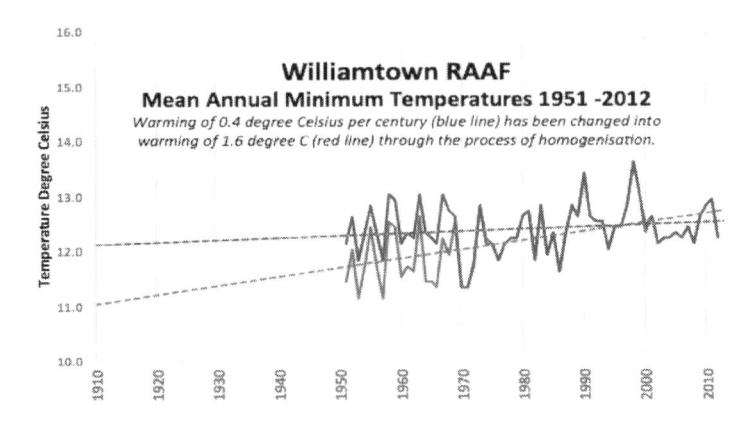

Williamtown RAAF
Mean Annual Minimum Temperatures 1951 -2012
Warming of 0.4 degree Celsius per century (blue line) has been changed into warming of 1.6 degree C (red line) through the process of homogenisation.

And these results have been homogenized too...

Cooling the data between 1951 and 1970 may look like unobtrusive tinkering by the BOM, but it changes the extrapolated local warming (in blue) from +0.4°C per century to a staggering +1.6°C (in red).

Add a few of those figures to the Australian temperature record, and it will make an implementation of carbon taxes far more possible.
Yup... it's political.

The records for Rutherglen were also adjusted.
This adjustment made a slight cooling trend into a warming one.

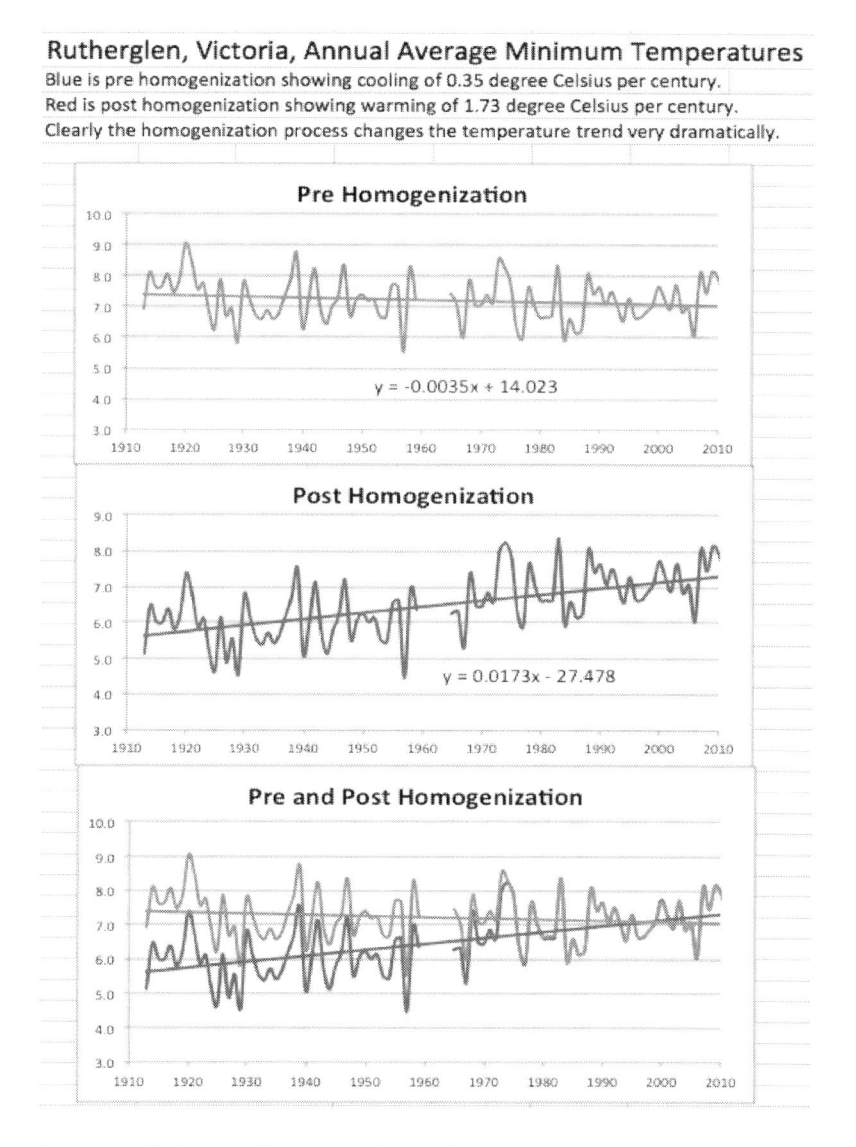

Rutherglen, Victoria, Annual Average Minimum Temperatures
Blue is pre homogenization showing cooling of 0.35 degree Celsius per century.
Red is post homogenization showing warming of 1.73 degree Celsius per century.
Clearly the homogenization process changes the temperature trend very dramatically.

Pre Homogenization
$y = -0.0035x + 14.023$

Post Homogenization
$y = 0.0173x - 27.478$

Pre and Post Homogenization

Marohasy and her colleagues have been meticulous with their own record keeping, even challenging the BOM on their choice of stations and their algorithms used.

But of course, they are called Deniers and dismissed.

Here's the link to Jennifer's own website.

http://jennifermarohasy.com/2014/09/bureau-caught-in-own-tangled-web-of-homogenisation/

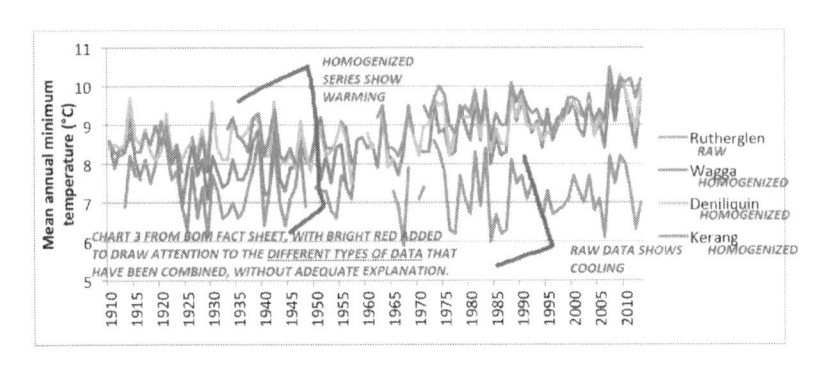

And if you think that's the end of the 'bitch who wouldn't go away', then think again.

In 2017 she took to see a couple of stations personally.

On 2nd July 2017, (Australia's winter), Marohasy literally camped outside the Goldburn Weather Station, noting that the station's own thermometer showing a minimum level (-10.4°C) on what was predicted to be the coldest day.
To her astonishment, the weather station sent the figure of -10.0°C.
(Any record cooling would be listed as a new record low, and would have cooled the winter figures.)

Marohasy actually watched the figures on the screen at Threadbow weather station, and it dropped below 10.4°C. The figure sent to ACORN 1 was recorded as a blank.

The newsman who interviewed Jennifer at Sky News in Australia wrote to the BOM and asked for an investigation.
The reply from Environment Minister Greg Hunt, said…

"he would limit the investigation in case the Bureau's credibility was undermined by it."

As the Aussies say... "What a lot of croc!"

https://www.youtube.com/watch?v=mmg2NTzqL74

But if you think that was the end of it all... Jennifer's like a terrier.
A real fire-cracker.
In Feb 2019, she cites yet more examples.
In one temperature change, a reading from 1933, it was changed from 44.8°C (raw) to 51.2°C (ACORN 1) to 49.5°C (ACORN 2)
Number one... why do they bother to change historical data?
Number two... this is a change of over SIX degrees Celsius.

https://jennifermarohasy.com/2019/02/met-bureau-rewrites-history-again-at-albany/

Go for it, Jennifer. Without persistent and keen-eyed folks like you, the BOM and their ilk would get away with murder!

On a side note, another climate data-miner (Ian George) has looked into a month of BOM homogenization in Bourke, 1939.

"These are the changes to daily temps for Bourke, Jan 1939 between ACORN 2, ACORN 1 and AWAP (raw temps).

Date	A/N2	A/N1	Raw
1/01/1939	38.1	38.4	38.9
2/01/1939	39.3	39.1	40.0
3/01/1939	41.3	41.9	42.2
4/01/1939	37.6	37.9	38.1
5/01/1939	38.1	38.4	38.9
6/01/1939	40.9	41.5	41.7
7/01/1939	40.9	41.5	41.7
8/01/1939	42.7	43.0	43.4
9/01/1939	45.4	45.7	46.1
10/01/1939	47.6	47.9	48.3
11/01/1939	46.5	46.8	47.2
12/01/1939	45.5	45.8	46.2
13/01/1939	45.0	45.3	45.7
14/01/1939	45.4	45.7	46.1
15/01/1939	46.5	46.8	47.2
16/01/1939	46.0	46.3	46.7
17/01/1939	39.3	39.1	40.0

18/01/1939	39.3	39.1	40.1
19/01/1939	39.3	39.1	40.0
20/01/1939	41.0	41.7	41.9
21/01/1939	41.7	42.1	42.5
22/01/1939	43.5	43.8	44.2
23/01/1939	36.3	36.5	36.7
24/01/1939	39.4	39.2	40.3
25/01/1939	36.2	36.5	36.6
26/01/1939	30.0	29.5	29.4
27/01/1939	29.9	29.4	29.3
28/01/1939	29.2	28.9	28.8
29/01/1939	31.3	30.5	30.6
30/01/1939	35.3	35.4	35.6
31/01/1939	38.0	38.3	38.6

The original AWAP maximum mean for this period was 40.4°C.
After ACORN 1 it was down to 40.03°C.
After ACORN 2 it is now 39.9°C.
That's a drop of 0.41°C in two homogenisations.

Note how lower temps have been increased but higher have been decreased (is this a form of covering up so the overall adjustments don't look as bad?). Can't wait for ACORN 3."

Basically what he's alluding to is a systematic 'dumbing down' of old data to make the new graph more angled upwards…

It further enhances the theme of my book…

IF THE BOM CONTINUE TO CHANGE THE HISTORICAL TEMPERATURE DATA, THEN EVEN THE BOM THINK THE SCIENCE ISN'T SETTLED!

Keep up the good work down under.

http://joannenova.com.au/2015/08/the-bom-homogenizing-the-heck-out-of-australian-temperature-records/

13.4... THE GREAT BARRIER REEF IS DYING
(Well, only if you listen to certain people)

The reef is dying... right?
Coral bleaching has decimated the reef... right?
Sediment is smothering the reef... right?
The reef has decreased in size by 60%... right?
Warming seas is killing the coral... right?

Amazingly all of the above claims have been made many times by the world's media.
They have ALSO all been made by reef scientists.

But incredibly, (just like the Pacific Ocean Garbage patch THAT DOES NOT EXIST (NASA words, not mine)) all the above statements are not only false... they're INCREDIBLY false.

The reef is dying... right?
According to at least one expert, the reef is alive and well, and living in the sea on the western edge of Australia. The ecosystem is fine, new corals are growing, and the fish population is doing quite nicely... no decrease in 20 years.

Coral bleaching has decimated the reef... right?
According to at least one expert, coral bleaching is a natural phenomenon, and has been going on for hundreds of thousands of years. Like a wildfire that decimates flora and fauna, it looks terrible over the short term, but it also clears the coral to allow new growth to begin, and the coral recovers rapidly.

Sediment is smothering the reef... right?
According to at least one expert, (the guy helped invent the actual machine that measures coral sediment, so he should know, right?) he reckons he's done more measuring of the sediment on the reef than all other experts put together. He says there's no problem with sediment on the reef.

The reef has decreased in size by 60%
According to at least one expert, that figure is absolute rubbish. Some parts of the reef have shrunk, but only by 5-10%. Between 2011 and 2016 the amount of coral in the southern part of the reef increased by over 250%. Yes, I said increased by two and a half times.

Coral growth has increased by 10% since the 1940's.

Warming seas is killing the coral... right?
According to at least one expert, the same corals that grow on the Great Barrier Reef also grow near Thailand, where the seas are at least two degrees higher, AND the corals there grow 50% faster there due to the 'better' conditions.

Okay, by now you're starting to wonder who my 'expert' is...

Professor Peter Ridd is an expert on Australia's Great Barrier Reef.
He worked for James Cook University in Townsville, Australia, for 30 years, and is considered one of the reefs most passionate advocates.

In 2017, he was asked to write a chapter in an IPA publication (Australia's Institute for Public Affairs) on climate change. Peter's chapter would, of course, be on the Great Barrier Reef.

Now, as you've probably gleaned, Peter is not in the mainstream of scientists who publish daily doom on Australia's most beautiful asset. He has his own views, borne from personal reef experience, and 30 years academic work.

In the book he said something to the effect that... *"In the light of the replication crisis, which shows a roughly 50% failure rate, if organisations use solely peer-review as their only checking mechanism, then their results could not be trusted."*

(The Replication Crisis is a new scientific study into the 'fact' that up to 50% of new scientific theories CANNOT be repeated, and are therefore unable to stand in modern science).

Peter Ridd was charged with MANY cases of serious misconduct by the James Cook University, dismissed from his post, and for two years he fought his case.
To make a long story short, in 2019 he was vindicated, and every charge was thrown out... and he was awarded $1.2 million in damages.
Since the case started, he has become outspoken of many of the organisations who spread the alarmist stories of the reef's condition.

Citing evidence that over $500 million is spent every year on new research into the Great Barrier Reef, Ridd said...

"There's some absolute rubbish being spoken about the reef and people's livelihoods are being put in jeopardy. If nobody will stand up, then this is just going to go on and on and on. It has to be stopped."

The livelihoods Ridd quoted were restrictions on farming, fishing, mining and other onshore industries, who are restricted on what they dump in the sea, because of stories of particulates supposedly damaging the reef.

Ridd concludes that only one piece of particulate in 100 comes from those industries, and is quickly swept out to sea.

In Australia the science is never settled.

13.5... Australia's Deputy PM Speaks Out

I've written before about NASA and government employees keeping their skeptic thoughts to themselves until they retire; their jobs depend on their silence.
Barnaby Joyce, twice deputy prime minister of Australia, hit the backbenches last year, and since he no longer had a party line to toe, he took to social media.

The next two pages is what he wrote...

"The very idea that we can stop climate change is barking mad. Climate change is inevitable, as geology has always shown."

These are the views of New Zealand lecturer of geology, David Shelley. A person vastly more competent than me and the flotilla of others telling the kids the world is going to end from global warming.

The central theme of David Shelley's analysis is that sea levels are rising and have been for thousands of years and will fall during the next ice age which is expected about now, give or take a thousand years.
When the ice age does arrive temperatures will drop around ten degrees. A warmer planet will be a disconsolate chronicle and many, maybe most, will die from starvation as is the usual experience of man or beast in previous ice ages.

The weather is going to brutally win the population problem and the parliament of Australia has no power against it. One may suggest that warmer weather is the better problem of the two.

One of the few graces of being on the backbench is you can be honest with what your views really are. I believe this is one of the greatest policy phantoms, the misguided and quite ludicrous proposition that Australia can have any effect on the climate. If we could we should be the first to make it rain and, more importantly, stop the recurrence of an ice age anytime in the coming millennium.

Politics takes politics to the absurd. We have to absolutely affirm that our domestic settings can deal with a proposition which is stated quite clearly by the Intergovernmental Panel on Climate Change that:

"In climate research and modelling, we should recognize that we are dealing with a coupled nonlinear chaotic system, and therefore that the long-term prediction of future climate states is not possible."

You don't get the feeling when you listen to the political propaganda or the supporting lobbyists that there is any doubt about their capacity to "fix the climate problem" I do get the feeling that you will be tried for heresy if you dare question the zeitgeist so you basically have to lie about your honest assessment of what the hell we are doing to our economy, standard of living, our basic rights and the real future of our children.

Today, more than in the past, the political debate is set within a predetermined paradigm. Participants cannot ague outside these preset boundaries. Maybe it is over cynical but I believe the promotion of the primacy of the state over the individual is very well served by the apparent necessity of climate policy.

Private property rights are removed, by the implementation of vegetation laws, because of "climate action". The state will limit your access to electricity because of "climate action". You will drive an electric car because of "climate action". You will divest the nation of its largest export because of "climate action". Rather than state there is no prospect whatsoever that any action of ours, and most likely of anyone else, will have any effect whatsoever on the trajectory climate is on.

We have instead the congenial narrative that we are all trying to make the world get cooler, but one path or the other path is the better alternative of cooling policies. We will do this by shutting down all our power stations, replacing them with windmills and rejiggering our nation away from our largest exports of mining and agricultural resources to carbon neutral tourism and the knowledge economy. Australia will be the catalyst to a global epiphany and the totalitarian Chinese regime will follow our lead because of our righteousness followed by India and the United States.

No, I don't think that will happen. I hate to say it but I doubt the majority of people on the planet, give a toss about the Paris Agreement. I would be amazed if one percent of the planet could competently explain it.

I will make one prediction; after this is published it will be promptly followed by the remnants of the traditional media in furious pursuit of my heresy. Questions will be asked by the fourth estate and high octane derision will issue forth from the climate change actionistas.

No doubt I will be accused of not knowing what I am talking about, and when it comes to predicting the weather more than a fortnight or so out, that is true. But of those who ask the questions, will any of them truly understand what on earth are they are talking about?

(Ends)

Wow, that's quite the diatribe.
And yup, he was hit by a barrage of Alarmists with nothing but revenge on their pens.

Just goes to show... when skeptics retire, they open fire.

13.6... AUSTRALIAN CLIMATE PREDICTIONS
(The End of Snow...)

Supposedly, due to graphs by the Australian Bureau of Meteorology (BOM) and Commonwealth Scientific and Industrial Research Organization (CSIRO), Australia has been warming for the last 100 years. So much so that predictions of never seeing snow again have plastered the journals... much to the disdain of Australia's winter sports industry. Here's a few...

"A very rare and exciting event... Children just aren't going to know what snow is." Dr David Viner, Senior scientist, climatic research unit (CRU)

"Good bye winter. Never again snow?" Spiegel (2000)

"Milder winter temperatures will decrease heavy snowstorms" IPCC (2001)

"End of Snow?" NYTimes (2014)

Winter sports in Australia happen in the mountains of New South Wales, Victoria, and Tasmania, from June to October.

And CSIRO in 2013 said this...

"By 2020, the average annual duration of snow-cover decreases by between five and 48 days; maximum snow depths are reduced and tend to occur earlier in the year; and the total area covered in snow shrinks by 10-40%"

HOWEVER, rather than the winter sports industry dying due to climate change, for the THREE YEARS, the season has been extended.
This season (2019) is no different, with RECORD snow falling.
In fact, in many areas, it's the deepest snowfall in over 20 years.
As Australians get their biggest snowfall for decades, I leave the subject with one last quote from 2012...

"Enjoy snow now ... by 2020, it'll be gone" The Australian

I love settled science.

SECTION 14
WHY DO ALARMISTS LOVE POLAR BEARS?
(What the heck is it with Alarmists and Polar Bears?)

"Global Warming has a much better correlation with changes in solar activity than CO2 levels".
Patrick Moore (GreenPeace ex-president)

14.1... INTRODUCTION
The polar bear has been the posterchild for the green movement for some time. From their low numbers in the 1970's (mostly due to an open hunting season) they have figured in many posters, campaigns, and movies to promote the environmental, green agenda.

Well, one could list their cuddliness, the soft fur, and some would blame Coca Cola for their continued sponsorship.

Whatever the reason, they are now firmly ensconced as the Alarmists poster child for global warming.
(Who can forget the poor polar bear standing on the lone piece of iceberg in Al Gore's triumphant movie, *An Inconvenient Truth*?)

However, just to dismiss the some of the hype about polar bears, let me hit you with some facts...
1. Don't worry about polar bears drowning... polar bears can swim more than 350km, and have been known to swim from the ice sheet to Iceland.
2. Don't worry about them not finding food... they eat fish, birds, seals, and any whales unfortunate to get caught.
3. Polar bears can also swim underwater for more than three minutes, sometimes chasing tired seals up onto the ice to escape.

Let's deal with the myth first... polar bear numbers.

14.2... THE POLAR BEAR POPULATION IS IN DANGER
(Nope, they're actually increasing)

Polar Bear Population Myth
This one's easy.
In the years up to the 1970's, hunting was the main destroyer of polar bear populations. It is estimated that in 1977, there may have been as few as 7000 left in the wild.
(A report from the Soviet Ministry of Agriculture's S.M. Uspensky, who surveyed nesting sites on a portion of Russian turf and extrapolated an Arctic-wide population of 5,000 to 8,000 in 1965.)

Living mainly above the Arctic Circle, the bears live in nineteen separate groups, with about 60% living in Northern Canada and the Canadian Arctic.
At the time of writing (2019), the estimates of polar bear numbers are between 22,000 and 30,000 worldwide, a huge increase since the nineteen-seventies.

At the meeting of the IUCN's (International Union for Conservation of Nature) Polar Bear Specialist Group in Copenhagen, 2009, scientists reported that of the 19 known subpopulations of polar bears, eight were declining, three were stable, one was increasing, and there was insufficient data on the other seven. This was an increase from five declining, five stable, and two increasing in 2005.

The latest Wikipedia figures (2014) are that of the 19 recognized polar bear populations, three are declining, six are stable, one is increasing, and nine have insufficient data to make a decision.

Now, the U.S. Geological Survey does predict that "two-thirds of polar bears will disappear by 2050", but they base that figure on the assumption of catastrophic Global Warming, and the complete decline of the Arctic ice field.
And we know how good climate scientists are at predictions.

To conclude on a uplifting note, a January 2018 story in Polar Bear Science blog suggests all the stories about polar bear populations decreasing with the supposed shrinking of the Arctic ice field are a little premature.

They quote a flurry of scientists (giving reference notes) saying the Arctic sea ice has been relatively constant since 2007, and that the polar bear population is not just keeping constant, it's actually growing.

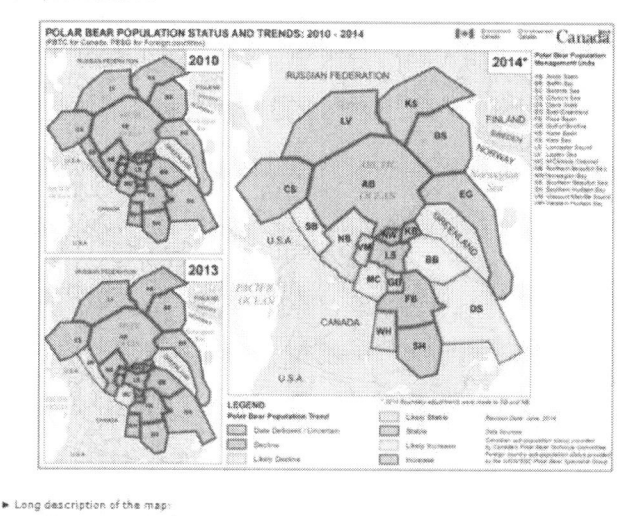

Map 4: Series of Circumpolar Polar Bear Subpopulation and Status Trend Maps 2010, 2013 & 2014

▶ Long description of the map

From six populations in decline in 2010, everything seems fine by 2014. This is mainly because despite the ice sheet shrinking by 38% since 1979, the polar bear population has increased by 16% from 2005 to 2015. Not good numbers if you're prophesying extinction.

You see, it turns out that warmer summers/low ice actually helps polar bears hunt, because the seal population increases.

One small addition… and I quote from their sleeve notes…

"For example, USFWS biologist Eric Regehr stated late last year that the Chukchi Sea subpopulation, that has lived with very low summer sea ice since 2007 (Regehr et al. 2016; Stein et al. 2017), "appears to be productive and healthy.""

There's positive proof that polar bears can live with low summer ice levels, and still be… *"productive and healthy".*

https://polarbearscience.com/2018/01/21/polar-bear-numbers-not-declining-despite-media-headlines-suggesting-otherwise/

But the media just can't help throwing out the dying polar bears to 'prove' global warming... in some form of date order, we find the following...

Polar Bears Dependent on Emissions Cut
In 2012... As you're about to see, the UK newspaper, The Guardian, is hot for polar bears...

"Polar bear could be saved if emissions are cut, says new study"

Cutting the whole WORLD's emissions would cut just 0.01°C from any global warming, and they know that... it's a made up story to throw their Alarmist talking points at us, and they do... in spades.
Despite the scientific fact that CO2 in the atmosphere take a long time to diminish.

Feeding the Polar Bears
Yup, it's from the polar bear newspaper, the UK Guardian, from February 2013 this time. As polar bear populations thrive and increase, this newspaper trots out the old doom and gloom scenario...

"Polar bears 'may need to be fed by humans to survive'"

Interesting headline... but what would they feed them with? Clubbed Seals? No we stopped that. Kansas antelope? No, they're cute. The second headline raises the ante with more hyperbole...

"Drastic measures are required to save the beleaguered animal from extinction, say scientists"

Yeah, cos nothing says 'extinction' more than a well monitored growing population. It's there that the typical Alarmist talking points fly... are you ready?

"an extended ice-free season... as Arctic sea ice disappears... two-thirds of the polar bears in the world could disappear... as Arctic sea ice, continues to disappear in spring, summer, and fall... a record disappearance of Arctic summer sea ice in 2007..."

And for all you seal-lovers, here's the kicker...

"Helicopters could be used to deliver the seals, but the logistics and expense of such a plan would be daunting. Thousands of seals would have to be killed by wildlife officials every summer to meet the needs of hungry bears, who each consume up to five seals a week."

Like Meatloaf said in the 1970's STOP RIGHT THERE...!!!
The idea that each polar bear eats FIVE seals a week is preposterous! A single seal keeps one bear satiated for more than a week! Where the heck does The Guardian get their facts???

"not dealing with greenhouse gases, because without action on that front, there's little that could be done in the longer term to save the species,... if greenhouse gas emissions are not curbed dramatically"

I can think of a few politicians I'd like to volunteer to be polar bear food... maybe ALL politicians... and lawyers, oh and environmentalists... I mean, they're up there anyway.
Heck, maybe spend some of the cash the governments waste on these constant polar bear surveys.

Polar Bears V's Dolphins
The UK Guardian again, this time from June 2015.

"Polar bears eat dolphins as Arctic warms"

Wow, that's a difficult one for conservationists to get their heads round... polar bears eating dolphins.
But of course, it's blamed on global warming...
Because before global warming, a polar bear would never have considered eating a dolphin... the cuddly white bears only did it because they're starving... right?

"Norwegian scientists have seen polar bears eating dolphins in the Arctic for the first time ever and blame global warming for the bears expanding their diet."

Wow... I mean, in the history of the world, a polar bear has never eaten a dolphin... ever.

To be honest, this story's mostly okay, but still drops the global warming blame-game a few hundred times.

Pizzly or grolar bear?
In the UK's Guardian in May 2016, Oliver Millman tells the story of a new grizzly-polar bear hybrid, and like most of his reporting, he can't help but throw in the concluding climate change rejoinder.
The headline reads...

"Pizzly or grolar bear: grizzly-polar hybrid is a new result of climate change"

The climate's changing is blamed for the grizzlies walking north, and the polar bears coming south. Every talking point is dropped into the article...

"Climate change... causing the curious emergence of a new type of bear in the Arctic."

"...the Arctic has warmed at twice the rate of the global average."

"Polar bears are spending more time on land as Arctic ice diminishes, causing them to lose body weight and decline in numbers"

"As temperatures continue to increase – the world has just broken monthly heat records for 12 months in a row"

He just can't help himself, can he?
Incredibly, the same newspaper tells that in the late 1980s, a hybrid narwhal-beluga whale was seen off the west coast of Greenland... so hybridization is hardly a climate change only fact.

Polar Bears Are Close To Extinction
In the UK's Guardian newspaper (story by New York's Oliver Millman again, who seems to like polar bear stories), the bastion of all things Alarmist, a new story hit the headlines on 1st February, 2018.
(If it had been two months later on April Fools' Day, it would have made more sense)
Now, please remember that despite the evidence to the contrary, that polar bear population has expanded from 7000 in the 1970's to between 22,000 and 30,000 today, they cite a survey over a three-year period by the US Geological Survey and University of California, Santa Cruz first published in the magazine, Science.

The Guardian leads with the headline...

"Polar bears could become extinct faster than was feared, study says"

Hold on... can I examine that headline... *"become extinct faster than was feared..."* So, they're saying that they fear that the polar bear is definitely going to become extinct. AND... faster than they initially expected.
That's a heady headline for a protected species that's increased in population 350% in the last 40 years!
Man, they double down...

"The animals facing an increasing struggle to find enough food to survive as climate change steadily transforms their environment"

Oh boy.

Then as the Alarmists throw their newspapers away, running for their safe spaces, I read on...

"A study of nine polar bears over a three-year period ..."

WHAT?!
You followed only NINE polar bears? Then predict polar bear extinction??
The story continues...

"the animals require at least one adult, or three juvenile, ringed seals every 10 days to sustain them"

Okay. Polar bears eat seals. That makes sense. And they require regular meals.
HOLD ON... I'd read earlier (when they were talking about helicoptering seals in for the bears to eat) that polar bears eat *"up to five seals a week"*.
But why should I let facts come before a good story.
I read on in expectation of the story becoming interesting.

"Five of the nine bears were unable to achieve this during the research, resulting in plummeting body weight..."

HOLD ON a cotton-picking minute!
You monitor NINE bears for THREE YEARS, and you notice FIVE bears not catching enough food to keep their body mass up... yet they DON'T DIE (because you would have told us about that).

And you think this is a story worth telling?
How much did this survey cost us poor taxpayers?

It gets worse...
They show a map with polar bear population centers, marking them for increasing or diminishing numbers.

BY THEIR OWN GRAPH... they admit the polar bear population is at 26,000... and out of the 19 or so polar bear groups, they admit that they've found... ONLY TWO GROUPS ARE IN DECLINE!!!

And the survey is postulating that polar bears are going to be extinct faster than predicted!

And if we look at the article carefully, we see that they've picked a declining population to study, the one in the Southern Beaufort Sea, off the coast of Alaska...

Maybe there aren't as many seals there?
Maybe the GPS monitoring collars with cameras (that were fitted to the nine bears) put them off their hunting?
Perhaps you shooting them with hypodermic needles to weigh them from time to time put their bodies out of whack?

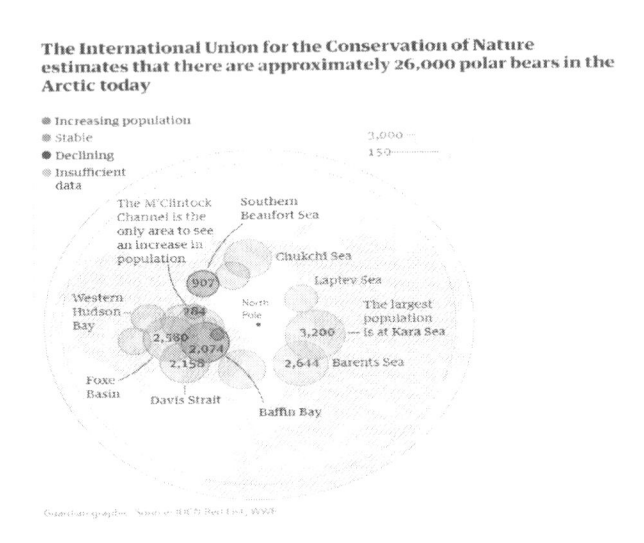

The International Union for the Conservation of Nature estimates that there are approximately 26,000 polar bears in the Arctic today

- Increasing population
- Stable
- Declining
- Insufficient data

The M'Clintock Channel is the only area to see an increase in population — 907

Southern Beaufort Sea
Chukchi Sea
Laptev Sea
The largest population is at Kara Sea — 3,200
Western Hudson Bay — 784
North Pole
2,560
2,074
2,158
2,644 Barents Sea
Foxe Basin
Davis Strait
Baffin Bay
3,000
150

(Incidentally, the only two groups shown to be in decline are in areas where the polar bear is still hunted by the native Nunavut population.

This area accounts for 80% of the polar bear killings, many by game hunters under Nunavut license.)

Anyway...
Honestly, I cringe at the hypocrisy of the storyline.
They continue...

"The Arctic is warming twice as rapidly as the global average..."

NASA has graphs of Arctic Temperatures from 1880... before we had weather stations there... this 'double the global average' is pure lies.

"if we don't change the trajectory of sea ice decline, polar bears will ultimately disappear..."

Aka... mankind is to blame... we didn't turn off the CO2 tap in the bathroom.

"the Trump administration has reversed measures that tackle climate change..."

Aka... now President Trump is to blame... he's a nasty, dirty man, and an enemy of polar bears.

"when measured at its September minimum, Arctic sea ice has declined by around 13% per decade since 1979..."

Wait... Arctic sea ice has decreased by just 13% in 40 years? I thought Gore predicted no Arctic sea ice by 1015?

Hyperbole... Hysteria... Hype...
They all sell newspapers, whether the facts are true or not.

As usual, the story can be found here...

https://www.theguardian.com/us-news/2018/feb/01/polar-bears-climate-change

Polar Bears Invade Russian Town

However, my book would not be complete without the polar bear story from Feb 2019...

Basically, a group of 52 polar bears were left behind on a Russian island of Novaya Zemlya in the Arctic Ocean when the ice retreated. Since December 2018, they have been taking easy pickings from the island's only town, Belushya Guba.

The main-stream media have gone wild with their claims that 'climate change' is to blame.

"Polar bears are affected by global warming with melting Arctic ice forcing them to spend more time on land."
Daily Mail

"The loss of Arctic sea ice due to climate change is the primary threat to polar bears and contributes to nutritional stress..."
ABC News

However, Ilya Mordvintsev, who is a lead researcher at the Severtsev Institute of Ecology and Evolution, sees the news story in a different light. He explains the bear's annual migration routes...

"...they come ashore in the southern part of the archipelago, where the ice is changing. They migrate through Novaya Zemlya heading north, where the ice is solid..."

That seems logical enough... bears migrate... that's what bears do... but why did they stay on the island when the ice did its annual retreat? Mordvintsev explains...

"It is migration from the south to the north. They are staying in that location because there is some alternative food. They could have gone past but for the food. But as there are bins with edible waste, they stop to flock."

So basically, some lazy bears stopped to eat easy-to-get food, and decided to stay to get some more.

Have you heard about Yogi Bear and Booboo?
"Hey Yogi, we gotta go north for the winter."
"I know Booboo, but there's an unguarded picnic basket over there..."

14.3... Polar Bears; the Unfortunate Un-Problem
(Susan Crockford against the established world...)

Dr. Susan Crockford is a PhD Canadian zoologist, and is currently an adjunct professor in Anthropology at the University of Victoria, Canada.
She is basically an expert on polar bears, and has studied them for many years.

Time magazine in 2000 warned the world about the struggles of polar bears. The same magazine again in 2006 advised us to *"be worried... be very worried"*, about the furry creatures' perilous future.
Despite the polar bear population increase, their plight has been hammered into our poor children since Al Gore's 2006 movie.
In 2007, CBS ran a story... *"Scientists: Most polar bears dead by 2050".*

WHAT??

Yup... a bunch of USGS scientists had predicted the downfall of summer Arctic sea ice, and with that 'loss of habitat', a rapid decline of the polar bear population due to a decrease in food availability... seals.
Doubling down on the research, in 2008, Nature released a story... *"Polar Bear numbers set to fall".*

Yup, it seems the scientists had reached a consensus, and in December 2008 the lowly polar bear was included on the 'Endangered Species List'.
Despite numerous other surveys listing rising populations.
But hey, why let facts get in the way of shocking those kids one more time.

The scientists couldn't get enough of the gloom and doom, however, predicting summer sea ice figures of a fall of 42% since 1979 levels, and thus... the end of the bears... TWO THIRDS OF OUR POLAR BEARS WERE GOING TO DIE.

16,000 BEARS! Starving... drowning... weak... defenseless.

The 'Tipping Point' of summer Arctic sea ice was set at 5 million square km... and in 2010, the impossible happened...
The summer sea ice fell below that level.
(Now, okay, Danish figures actually state that the 'record' low did not arrive until 2012, but hey, it's just a game... right?)

All would have been fine for those scheming Alarmist scientists... but like Scooby Doo and his gang to disrupt the scientists' false aura of impending doom... enter zoologist and polar bear researcher Dr. Susan Crockford.

In 2015, Susan put the two facts together...
The ice tipping point for 2050, the record low 5 million square kilometer level of summer sea ice calculated for the 'doom' of TWO THIRDS of our polar bears... had arrived... THIRTY-FIVE YEARS EARLY!

She almost rushed into her personal jet and flew to the North Pole to witness the carnage... almost.

But like any sensible scientists, she set to work to discover exactly what was going on in these doom-laden summers... she wanted to know what was actually happening to her precious polar bears.

Summer sea ice vs. polar bear numbers since 2005

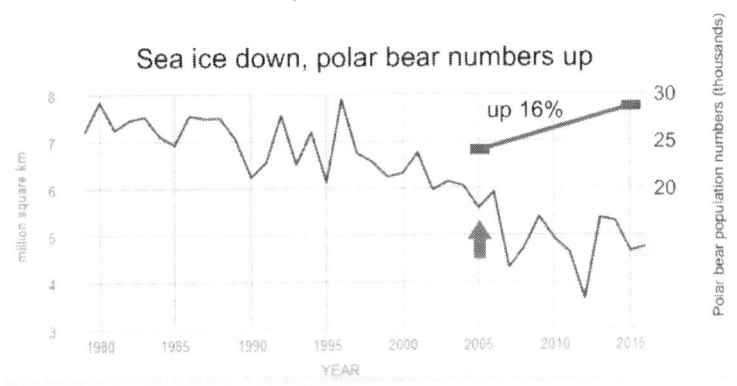

Sea ice down, polar bear numbers up

up 16%

This is Susan's own graph.
To her astonishment, despite expecting to have to wade through waist-high dead polar bears, she found the exact opposite.

In the 'record' smaller summer sea ice area, polar bear numbers were actually rising.

In time, her research showed more... the number of polar bear triplets was increasing, despite the decline in Arctic summer sea ice.

You see, what the Alarmist scientists hadn't realized was that the lower volume of summer sea ice actually INCREASED the seal population.

Because... Ringed and Bearded seals BENEFIT from more open water in summer and a longer summer season.

And polar bears eat seals.
With the facts flying in the face of their impending doom, the polar bear numbers were increasing at about 16%.

NOT ONE POLAR BEAR POPULATION HAD BEEN IN SERIOUS DECLINE OR WIPED OUT.

The Alarmist 'models' and predictions had been TOTALLY WRONG.
So Susan carried on...
She found that...
1. The population in Hudson Bay is stable.
2. The population in the Davis Straits has increased in the last 10 years.
3. The population in Baffin Bay had increased by 36% over previous estimates.
4. The Kara Sea population, previously estimated by the USGS at 2000, was recently 'revised' by Russian scientists to over 3000.
5. The Barents Sea population, north of Norway had increased by 42% from 2004 to 2015.

Wonderful, huh?
But then, in 2017, Susan made the fatal mistake of actually publishing her findings.

"Testing the hypothesis that routine sea ice coverage of 3-5 mkm2 results in a greater than 30% decline in population size of polar bears."

Let's just say that the Alarmist world went berserk.

The same year, Susan was lambasted by a paper denouncing her work.

Internet Blogs, Polar Bears, and Climate-Change Denial by Proxy

JEFFREY A. HARVEY, DAPHNE VAN DEN BERG, JACINTHA ELLERS, REMKO KAMPEN, THOMAS W. CROWTHER, PETER ROESSINGH, BART VERHEGGEN, RASCHA J. M. NUIJTEN, ERIC POST, STEPHAN LEWANDOWSKY, IAN STIRLING, MEENA BALGOPAL, STEVEN C. AMSTRUP AND MICHAEL E. MANN

And just look at the list of co-authors...

Yup... even Michael (hokey-schtick) Mann... who obviously knows a heck of a lot about polar bears.

In stories to follow in the world's press, Susan was labeled a denier, a conspiracy theorist... she was accused of using dubious science, of ignoring science... the list is endless.
Polar Bears International even issued a press release, denouncing the study, the data, and her credentials and integrity.

But there was a good side to the attention.
Her blog-hits went through the roof.
She got some funding.

Susan wrote her findings in a book...

The Polar Bear Catastrophe That Never Happened

And then went on to theorize that the polar bear numbers have ALWAYS been 'low-balled' by the USGS... and perhaps for political reasons.
Using the most modern figures from her research, she posits that there are perhaps as many as 39,000 polar bears on the planet (the estimate is actually 26,000-58,000).

Here's the link to her book...

https://www.amazon.com/Polar-Bear-Catastrophe-Never-Happened/dp/0993119085/ref=sr_1_1?crid=2FL9R31ZGQ67S&keywords=susan+crockford&qid=1556625583&s=gateway&sprefix=susan+cr%2Caps%2C201&sr=8-1

SECTION 15
PREDICTIONS, PREDICTIONS, PREDICTIONS...

"Those who call themselves Green planet advocates should be arguing for a CO2-fertilized atmosphere, not a CO2-starved atmosphere... Diversity increases when the planet was warm AND had high CO2 atmospheric content..."
Burt Rutan, (named "100 most influential people in the world, 2004" by Time Magazine)

15.1... INTRODUCTION
Mankind has been making predictions since he learned to walk upright.
We've predicted the end of our civilization at least 100 times in my lifetime, and the world of climate change is no different.
If I were to list every person who predicted no summer Arctic Sea Ice, I'd be writing for days.
And yet they still trot out the old pony every year.

Why do we like predictions?
Well, it may be the same reason we like science fiction... it's like looking forward and wondering... it's a world of the future... predictions.
Because let's face it, when they don't come true, the author just throws out another extended time period, and people nod, wondering about the next fifty years.
But very few of us actually throw out the doomsayers themselves, do we? I mean, *"he's a climate expert, right?"* We nod, trust his scientific credentials... *"he just got the timing wrong"*... *"the ice will melt eventually... it's inevitable."*

Al Gore gave the summer Arctic Sea Ice until 2015.
James Hansen (climate prophet) gave New York twenty years... in 1988.
Paul R. Ehrlich gave England until 2000.
In June 2019 Prince Charles gave the planet 18 months.
In Spring 2019, Alexandria Ocasio Cortez gave mankind 12 years.
And the list, of course, goes on...

And yet these people above, and their thousands of compatriots STILL are treated with respect by many of us.

Reader. I give you a challenge...
Look up ONE PREDICTION about climate change.

Research it yourself, find out if it came true.
Then email me.
I'll write a new book... trust me.

Do you know why we don't throw these doomsayers out?
Why we never chastise these purveyors of doom-laden predictions that NEVER materialize?

Because we supposedly don't care?

I'm not so sure.
I think part of our reluctance to call these fallen predictors out is a ingrained respect for their supposed scientific credentials.

Because we've become an indoctrinated species.

Let's take a short look at our seats of learning; our Universities.
1. There are more university graduates walking around the planet right now than ever before. EVER.
2. There are more university degree choices available now than ever before. EVER.
3. There are more variety of science degrees available now than at any stage in history. EVER.
4. To summarize... there are more young 'smart' people walking around today than at any point in history. EVER.

And yet when Alexandria Ocasio Cortez stands up with her 'Green New Deal'...
a plan to bankrupt the USA... spend 95 trillion dollars... tax the people back to the middle ages... take away cars, trains, airplanes, ships... and herd us all into cities for easier food distribution...
(ALL this for one TENTH of a degree Celsius reduction in global temperatures in 100 years)
We stand and cheer.

ALL of the Democrat contenders in the upcoming Presidential Election in 2020 stood by the Green New Deal in the first debates... then they started to talk their way back from the precipice.

But still people cheered.
So why are our young people so dumb?
Why are our young people so easy to hoodwink?

"It's easier to fool people than to convince them that they have been fooled."
(oft attributed to Mark Twain)

WHY, when the predictions of the so-called 'green science' have been proven wrong SO MANY TIMES... do our younger population STILL buy into their mantra.

In one word... INDOCTRINATION.

15.2... WE'RE REALLY GREAT AT CLIMATE PREDICTIONS
(Actually, Alarmists, you're NOT. You're really crappy...)

You'd think that after almost zero predictions correct, the Alarmists would stop making them... but they can't. They're simply so confident of their own superiority, (swallowing their own cool-aid helps), that they can't help sticking their necks out.

James Hansen
Global Warming and NASA scientist, Hansen is a poster child for the climate change movement.
On my 29th birthday, June 23rd, 1988, Hansen testified in Washington to the US Senate. In a hot meeting room he told them that Global Warming *"wasn't coming... it had already arrived"*.
(What he didn't admit until later was that he had chosen the hottest day to testify, opened the windows, and turned off the air conditioner to the room; the whole picture of him sweating under testimony was a contrived sham.)

Anyway, he predicted that the US would be considerably hotter as the next thirty years went on, blaming fossil fuels as the major culprit. He forecast record level days over 100°F and 90°F, runaway heating that would only stop if we halted our being shackled to fossil fuels for power and transport.
However, from that record summer of 1988, the number of days over 100°F and 90°F days have **decreased** over the USA.
Not a good start for his predictions... and don't forget that he did say *"The oceans will boil..."*

From that record hot summer, his climate models also predicted heatwaves for the 90's and into the 2000's for the Midwest and Southeast of the USA.
Again, unfortunately for Hansen, the rainfall has actually increased up to 2017 in most parts of those areas, again, Hansen fell short.

He stated that *"heat would cause inland waters to evaporate... lowering the level... of the Great Lakes"*. NOPE! NOAH figures show the current Great Lakes water levels at median or high levels.

Let's run forward twenty years to 2008. Hansen predicted that the Arctic Ocean would be free of summer ice in *"five to ten years"*. Oh dear, although the ice sheet declined in 2012, it bounced back again, and at the time of writing, still covers the top of the globe and it's far bigger and thicker than it was when Hansen made the call in 2008. Hansen is hardly the "Climate Prophet" that Representative Ed Markey (D-Mass) proclaimed.

So, poor James Hansen hasn't done so well. And yet he doubles down on every prediction he's ever made. It seems that despite being wrong so many times, he can defend himself.

However, keeps being invited to talk... follow the money.
He doubles down on everything he's ever said.
And he keeps wearing that hat indoors!

More Predictions
On Sept 19th, 1989, the St. Louis Post-Dispatch newspaper reported...
"New York will probably be like Florida 15 years from now."

In 1990, Michael Oppenheimer, then a member of The Environmental Defense Fund, said...
"By 1996, the Platte River of Nebraska would be dry, while a continent-wide black blizzard of prairie topsoil will stop traffic on interstates, strip paint from houses and shut down computers... The Mexican police will round up illegal American migrants surging into Mexico seeking work as field hands."

Not to be out-done, on October 15, 1990, Carl Sagan, that smooth talking scientist, made a more vague prediction...
"The planet could face an ecological and agricultural catastrophe by the next decade if global warming trends continue."

In the same year, 1990, actress and climate scientist Meryl Streep (I know, she's NOT a scientist, but it didn't stop her), said...
"By the year 2000 – that's less than ten years away — earth's climate will be warmer than it's been in over 100,000 years. If we don't do something, there'll be enormous calamities in a very short time."

In July 26, 1999, The Birmingham Post wrote...
"Scientists are warning that some of the Himalayan glaciers could vanish within ten years because of global warming. A build-up of greenhouse gases is blamed for the meltdown, which could lead to drought and flooding in the region affecting millions of people."

On April 1, 2000, Der Spiegel said...

"Good bye winter. Never again snow?"

On cable TV on March 29, 2001, CNN warned...
"In ten years' time, most of the low-lying atolls surrounding Tuvalu's nine islands in the South Pacific Ocean will be submerged under water as global warming rises sea levels."

(Now, I went to Wikipedia (Sept 2019) and checked that last one... no mention of disappearing islands on the Tuvalu page. They do state...
"At its highest, Tuvalu is only 4.6 metres (15 ft) above sea level. Tuvaluan leaders have been concerned about the effects of rising sea levels. It is estimated that a sea level rise of 20–40 centimetres (8–16 inches) in the next 100 years could make Tuvalu uninhabitable."
But wait, the Maldives are no more than 6 ft above sea level, so Tuvalu will get a bit of warning from them surely?)

15.3... U.N. 1989... WE HAVE ELEVEN YEARS TO STOP THIS...
(More warnings and great predictions...)

Peter James Spielmann is a long time reporter for the Associated Press, and an Editor of their North America desk.

Back in June 1989, he published a story for the AP, and made some terrifying predictions of the condition of the planet.

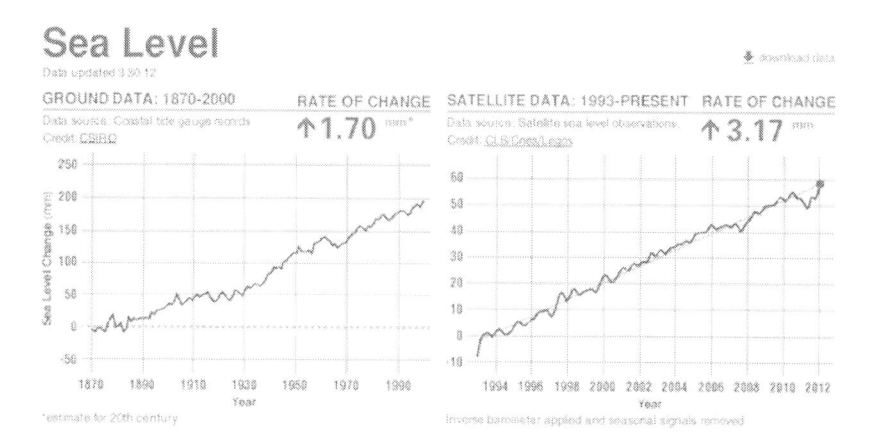

Quoting senior U. N. official, Noel Brown, a director of the U.N. Environment Program, (UNEP),

"By the year 2000",

He begins… (that's only ELEVEN years!)

*"entire nations **could** be wiped off the face of the Earth by rising sea levels"*

(As far as I checked… NO United Nations country has ceased to exist in the last 30 years by drowning! See graph above.)

*"ocean levels **will** rise by up to three feet, enough to cover the Maldives and other flat island nations"*

(30 years later, according to NASA the seas have risen just 100mm, that's JUST OVER THREE INCHES! See graph above.)
(If we take ground figures, the rise is HALF OF THAT!)
(AND, on a side note, the Maldives are still advertising on the tourist front and Saudi princes are buying Maldives real estate)

*"one-sixth of Bangladesh **could** be flooded,"*

(Yes, it 'could'... but it didn't. I also checked for any outbreaks of World Wars Three, Four or Five, Alien Invasion, and World Peace, because they **'could'** have happened too. Nope, I came up empty, sorry)

*"A fifth of Egypt's arable land in the Nile Delta **would** be flooded, cutting off its food supply"*

(I searched for ANY such news story in the last 30 YEARS, but again came up with nothing. As of writing this book, Egypt still has its food supply.)

"UNEP estimates it would *cost the United States at least $100 billion to protect its east coast alone."*

(Wow... it turns out that estimates are another thing the U. N. are crappy at!)

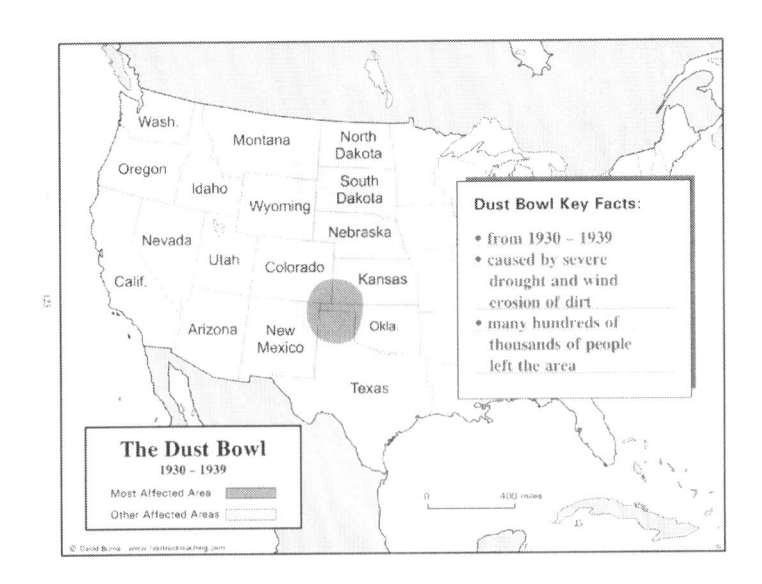

The Dust Bowl
1930 – 1939
Most Affected Area
Other Affected Areas

Dust Bowl Key Facts:
- from 1930 – 1939
- caused by severe drought and wind erosion of dirt
- many hundreds of thousands of people left the area

*"Shifting climate patterns **would** bring back 1930s Dust Bowl conditions to Canadian and U.S. wheatlands"*

(Well, I actually **live** in Kansas, USA, literally the heart of the dustbowl... move along, nothing to see here.)

"even the most conservative scientists "already tell us there's nothing we can do now to stop a ... change" of about 3°C."

In actual fact NONE of the 1989 U. N.'s predictions from this article came true.

NOT ONE.

If poor Peter James Spielmann had been a weatherman, he'd have been out of a job.
Here's the link to this now 'infamous' AP piece.

https://www.apnews.com/bd45c372caf118ec99964ea547880cd0

Now, after an 'old' 11 year prediction, let's take a look at a modern one.

15.4... EARTH 2100; AN ABC PREDICTION SPECIAL
(Yup, it was special alright...)

In 2012, ABC made a documentary looking forward to the year 2100, a whole 88 years in the future.

They looked through the eyes of a child born on June 2nd, 2009, just months after the inauguration of President Barack Obama.
It was seen as a time of uncertainty and poor humanity was painted as foolish...

"We've acted as though we were independent of the environment. We burned fossil fuels. We've overused our renewable resources."

(Can I just ask one question here? When did we OVERUSE our renewable resources? We haven't even used up our non-renewable ones)

Anyway...

Van Jones, (President Obama's Special Advisor for Green Jobs) was quoted...

"People are complaining about the economic crisis we have right now. You haven't seen nothing yet. You know, if we continue down this suicidal pathway that we're on, where we basically turn living stuff into dead stuff and call it economic growth, this will look like the good old days."

The world was seen as a disappearing entity, we were running out of water, oil, and land.
Hard times indeed.

It's easy to look at their predictions, and poke fun... but hey, we've made it through to 2019 without our world doing a single thing against global warming... so let's have a little laugh.

In the movie...
In 2014... there was a dragonfly epidemic.
In 2015... there was an oil shortage... people moved from the suburbs to the cities because cars were becoming rare, expensive.

Heidi Cullen, a climatologist at Climate Central said...

"By 2015, add another 20 million people to the USA population, and then just play that out... see what that does to consumption patterns. The number of people we've got to feed. There's just basically this slow creeping tension for natural resources."

Yeah, I can remember my 'slow creeping tension' in Walmart in 2015.

The documentary shows lines of cars waiting for gas in a fictional future 2015, the lines going round the block... the gas was priced at over five bucks a gallon!
China faces its worst rice harvest in a century.
(Despite China's food production breaking new records)

By 2030, they envisaged that two thirds of the world's population will be under water stress. Water rationing was the norm in the USA. Rallies broke out to protest the increasing price of water.

By 2040, the world's population approached 9 billion.
Millions of Latin Americans mass on the USA's southern border.

Yup... sounds like 2019.

15.5... FUTURE SNOW PREDICTIONS
(We're good at snow... yeah, whatever...)

At a time of huge snowfalls in Australia for the last three winters, let's look at some snow predictions, shall we?

2000
"Winters with strong frosts and lots of snow like we had 20 years ago will no longer exist at our latitudes." Professor Mojib Latif

2000
"Good bye winter. Never again snow?" Spiegel

2004
"Snow has become so rare that when it does fall – often just for a few hours – everything grinds to a halt." Mark Lynas (British author, journalist and environmental activist)

2005
Planning for a snowless future: *"Our study is already showing that that there will be a much worse situation in 20 years."* Christopher Krull, (Black Forest Tourism Association) in Spiegel

2005
"Winter is no longer the great grey longing of my childhood. The freezes this country suffered in 1982 and 1963 are – unless the Gulf Stream stops – unlikely to recur. Our summers will be long and warm. Across most of the upper northern hemisphere, climate change, so far, has been kind to us..." George Monbiot (British environmental and political activism writer)

2006
"Because temperatures in the Alps are rising quickly, there will be more precipitation in many places. But because it will rain more often than it snows, this will be bad news for tourists. For many ski lifts this means the end of business." Daniela Jacob of Max Planck Institute for Meterology, Hamburg

2007
"It Seems the Winters of Our Youth are Unlikely to Return" BBC "One Planet Special"...

2007

"Snowlines are going up in altitude all over the world. The idea that we will get less snow is absolutely in line with what we expect from global warming." Sir John Houghton, (former head of the IPCC and former head of the UK Met Office)

2007
"First the snow disappears, and then winter." Die Zeit

2008
"I don't believe we will see the kind of snow conditions we have experienced in past decades," Christoph Marty, (Swiss Federal Institute for Snow and Avalanche Research)

2012
"Enjoy snow now . . . by 2020, it'll be gone" The Australian

2014
"Nothing besides a national policy shift on how we create and consume energy will keep our mountains white in the winter... This is no longer a scientific debate. It is scientific fact." New York Times

2017
"Climate change puts Australia's ski industry on a downhill slope... Australia's ski resorts face the prospect of a long downhill run as a warming climate reduces snow depth, cover and duration. The industry's ability to create artificial snow will also be challenged, scientists say."
Peter Hannam (global warming catastrophist)

15.6... The Green New Deal
(Democrats gone insane...)

In the US 2018 mid-term elections, the Democrats elected a few outright 'leftists' to their ranks.
One, Alexandria Ocasio-Cortez (D-NY), is causing a stir in her first few months.
On Feb 25th, 2019, on Instagram she said...

"Our planet is going to be a disaster if we don't turn this ship around, and so it's basically, like, there's scientific consensus that the lives of children are going to be very difficult. And it does lead, I think, young people to have a legitimate, you know, should... is it okay to still have children?"

But the *'left'* is strong with this one...
In February 2019, some members of the US Democrat Party (led by Congresswoman Alexandria Ocasio-Cortez and Senator Ed Markey (D-MASS)) have introduced a resolution called the Green New Deal.
(This is a term purposefully reminiscent of Franklin Delano Roosevelt's original New Deal in the 1930s)

This Green New Deal is, by far, the most far-reaching piece of legislation ever attempted, and according to the Sunrise Movement, it will *"transform our economy and society at the scale needed to stop the climate crisis."*

The Green New Deal (GND) begins with an endorsement of the November 2018 IPCC Fourth National Climate Assessment, asserting that...
(And please pardon my sickly-sweet sugar-dripping sarcasm as I give my take on the individual points...)

(1) human activity is the dominant cause of observed climate change over the past century;

A stance which most Deniers will stand up and protest...

(2) a changing climate is causing sea levels to rise and an increase in wildfires, severe storms, droughts, and other extreme weather events that threaten human life, healthy communities, and critical infrastructure;

Yup, there's nothing like hitting the unproven 'catastrophe' button early on. There's proof all over the meteorological world against the rise of

wildfires, droughts, and other extreme weather no matter what the cause... humbug.
Let's not even linger here, because it gets better...

(3) global warming at or above 2°C beyond pre-industrialized levels will cause—

Now please note, the following examples are not given as possibilities, they are stated as FACTS. *"Global warming AT OR ABOVE 2°C beyond pre-industrialized levels WILL CAUSE"*—

(A) mass migration from the regions most affected by climate change;

(B) more than $500,000,000,000 in lost annual economic output in the United States by the year 2100;

(C) wildfires that, by 2050, will annually burn at least twice as much forest area in the western United States than was typically burned by wildfires in the years preceding 2019;

(D) a loss of more than 99 percent of all coral reefs on Earth;

(E) more than 350,000,000 more people to be exposed globally to deadly heat stress by 2050; and

(F) a risk of damage to $1,000,000,000,000 of public infrastructure and coastal real estate in the United States; and

"Holy Warming, Batman!"
The writers must have watched every scary/end of the world B-movie made in the last ten years!
(Although they did leave out earthquakes/volcanoes and Godzilla)

The GND future scenario sounds terrible. What are we to do to avoid such cataclysmic times?

(4) global temperatures must be kept below 1.5°C above pre-industrialized levels to avoid the most severe impacts of a changing climate, which will require—

Okay... you've solved the problem... (I don't know why we didn't think of this sooner)... we just have to keep global temperatures below a rise of 1.5°C.
WHEW! Problem solved, democrats!
No... seriously...
We've just been delivered with a huge punch to the solar plexus, and we need a new leader to tell us what to do to avert such a catastrophic conclusion to our planet.

Tell us, great Alexandria Ocasio-Cortez, what do we (the people) have to do to avert this cataclysmic crisis?

(A) global reductions in greenhouse gas emissions from human sources of 40 to 60 percent from 2010 levels by 2030; and

BUT THAT'S ONLY 12 YEARS!

(B) net-zero global emissions by 2050;

WAIT! Did you say ZERO?

Yes, they did.

Did you say GLOBAL?

Yes, again, they did.

The Democrats want the USA to begin the process to reduce carbon emissions to ZERO in 32 years.
AND they want the rest of the world to follow...

By 2050, they want no fossil fuels anywhere ...
No fossil fueled powered cars or trucks.
No fossil fueled powered trains.
No fossil fueled airplanes.
No fossil fueled ships to carry food, produce, manufactured goods.
No fossil fueled power stations.
No nuclear power stations.

But within 32 years **(while decreasing CO2 emissions to ZERO, AND halting all production/manufacturing)** they want to...

Increase green sources of electrical power production.
Build and install a country-wide network of wind turbines and solar stations.
Build a new high speed rail network to every town.
Build new high speed trains (did anyone see *Atlas Shrugged*?).
Build new electric cars for the new elite (because the normal working man can't afford them).
Build new electric trucks for the new transport structure.
Build new electric tractors for the farmers.
Run a higher current electricity network to every house to have charging ports for cars.
Revamp every building, factory and house to new green standards.
And much more…

With everything being mandated as being electric powered…
How are we (the normal people) going to pay for our new electric cars?
How are we (the normal people) going to pay for our green improved homes?
How are we (the normal people) going to pay for the more expensive food that we have to eat?
How are we (the normal people) going to pay for the millions of new green power installations?

Bad news… how is our manufacturing base and the trucking industry going to keep prices down, while they scrap their fleet of trucks to buy electric ones?
Worse still, how are American businesses going to conduct their business without air travel?
Worse still than that… how is America going to exist with no air travel of goods and people to and from outside the USA?
We can't buy from the rest of the world?
But I digress from one lunacy to the next…

WHERE ARE THE RESOURCES THAT ARE NEEDED FOR ALL THIS NEW PRODUCTION?
WHO'S GOING TO PAY FOR ALL THIS PRODUCTION?

And we're going to do ALL THIS PRODUCTION by 2050, while reducing our carbon emissions to ZERO while we do it.

It's actually a chicken and egg scenario…

What comes first… the electric trucks needed to transport the resources to begin the building of the huge new rail network?
Or the gas powered trucks building the huge new rail network to deliver the resources to build the electric trucks?

Would the rest of the world follow us in this madness?
No.
In the meantime… the world stands by watching our country do this new deal… they wait to see how it turns out, this new experiment.

And the air that we Americans breathe will get more and more CO_2 because the rest of the world are waiting to see if the new deal works, still increasing their dirty fossil fuel output.

You see, America has a population of 330 million out of a world population of 7.7 billion.
That makes the USA just 4.3% of the world's population.
And the democrats want to gamble that their new deal will spark a worldwide 'revolution', and that all other countries will join.

Here's one kicker… you can't 'measure' emissions, you can only estimate.
The USA's estimated CO_2 emissions are between 1.6 and 5 billion tons.
We emit perhaps 15% of the world's CO_2, and our emissions are FALLING… we're now below our 1970 level.
(And this was done without the GND!)

(And please, don't start telling me that the USA emit a disproportionate level of CO_2 for our population… because remember that we PRODUCE over 20% of the world's stuff!)

But the democrats want more…
And the stakes are high.
By mandating this GND, the democrats want to play Russian roulette with the American people.
But what the democrats don't realize…

THE GND IS A LOSE-LOSE SITUATION.

IF THE GND **DOESN'T** WORK, then America becomes a third world country in 32 years. We're powerless, stripped of money, and have loans due to every country in the world to pay for the resources we've used to try to combat CO_2.

We have no Army, Navy or Air Force (all CO2 users), so we can't fight. We're a sitting duck for any country with guns.

IF THE GND **DOES** WORK, because of everyone else's continued emissions, CO2 will increase past the 2.0°C barrier, and (if Alarmists are correct) we'll still get a global catastrophe, and America will have done all that work for NOTHING!

WHY?
Because as the world stands still watching our efforts, THE REST OF THE WORLD will still be LIVING LIFE AS USUAL.
By putting ourselves out as the poster child of a new world order, we have created our own destruction.
By 2050 we may have the best electrical grid in the world, but the first ten countries to jump in their planes and parachute in (does anyone remember *Red Dawn?)* will reap the benefits of it.

15.7... THE GLOBAL WARMING CHALLENGE
(A professor of marketing beats Al Gore...)

Professor J. Scott Armstrong is an award winning PhD from the University of Pennsylvania, Philadelphia. But he's not a climatologist...
He is a professor of marketing, and has written many articles and books on using marketing techniques to predict the outcome of elections, sales strategies, etc.

In 2007, he turned his non-climatology brain to Global Warming.
Through his investigations, he perceived that there was no global warming any more, and predicted that future findings would prove his theory.

To illustrate (and market, and publicize) his claims, he set up theclimatebet.com, and issued a challenge to Al Gore, fresh from his million selling movie, An Inconvenient Truth (2006).
I give you the bet straight from Armstrong's website...

"Scott Armstrong of the Wharton School challenges Al Gore $20,000 that he will be able to make more accurate forecasts of annual mean temperatures than those that can be produced by climate models. Scott Armstrong's forecasts will be based on the naive (no-change) model; that is, the forecasts would be the same as the most recent year prior to the forecasts. The money will be placed in a Charitable Trust to be established at a brokerage house. The charity designated by the winner will receive the total value in the fund when the official award is made at the annual International Symposium on Forecasting in 2018"

His original challenge was not accepted by the ex-vice president.

The challenge was simplified, and again sent to Al Gore, and the deadline for acceptance was extended to March 26th, 2008.
When March 26th passed, Armstrong sent another letter to Al Gore, asking the question...

"When and under what conditions would you be willing to engage in a scientific test of your forecasts?"

Well, as you can probably tell, Al Gore didn't rise to the fly, and the challenge might have been forgotten in the mists of time...

But ClimateBet.com had already begun to plot a graph, starting at the month of the first challenge...

Now here, just for the data junkies, 'alarmists' who are reading, and the still skeptical amongst us, I must go back a ways and give a little back story.

When Armstrong first thought of the challenge, he realized that he would need an unquestionable transparent source of data for his own graph... it couldn't be tainted with IPCC tampering... it had to be both scientifically sound, above reproach, the data had to be open to public inspection.
It couldn't be homogenized, massaged, altered, and full of unknown bias like the NASA/NOAA figures.
AND it had to be neutral...
There was serious money on the line, not to mention a reputation, and there was no room for doubt.

Enter the University of Alabama in Huntsville (UAH).
Since 1979, the university have been taking satellite readings from the atmosphere (where most of the greenhouse warming occurs), and the dataset is available to the public to read.
Each month, the UAH provide the latest average (mean) global temperature from the satellite figures.

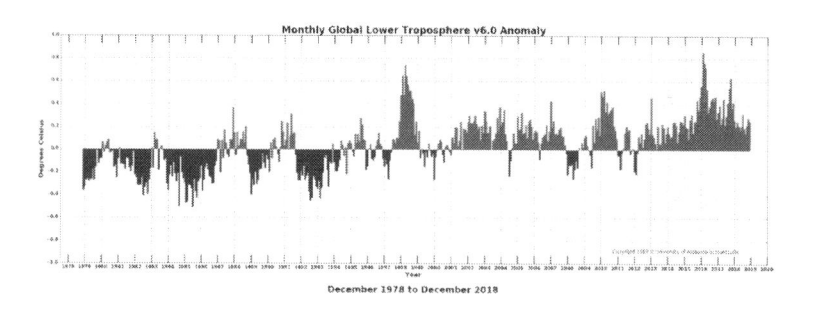

Monthly Global Lower Troposphere v6.0 Anomaly
December 1978 to December 2018

As can be easily seen, the above graph runs from the 'ice age approaching' figures of the late 1970's to today.

For the ClimateBet.com, a decision was taken to use these monthly figures, as they satisfied the main parameters for good scientific data, and would not be able to be criticized from any quarter...
1. The figures are fully disclosed, publically available and untainted.
2. The figures are global and comprehensive; satellite coverage gives a global average of millions of readings.

3. The atmospheric data avoids local 'heat island' effects which so many ground weather stations are criticized for.
4. The atmospheric data avoids inconsistent measurements, as many of the weather stations still use human readings (mercury thermometers).

ClimateBet did not use the Hadley (HADCRUT) data series (which NOAA also uses), which fail on all 4 of the above criteria.

I think the choice is a good one...
The Hadley (HADCRUT) data series (University of East Anglia & the UK Met Office) take manually recorded temperatures from various selected locations around the world and then adjust them.
Here lies the problem in the Hadley series.
1. Nowhere in any public document are the original data available.
2. Ocean temperatures with buckets thrown overboard are still incorporated in the Hadley series.
3. Neither the selection of locations used, the number of selections used, NOR any details of the full adjustment process are fully disclosed.

This is bad science... way to go professor of marketing, making his figures undisputable...
Because if the figures COULD be disputed, you're dang sure Al Gore's lawyers would have had a law suit going well before now!!!
So... how did the figures do?

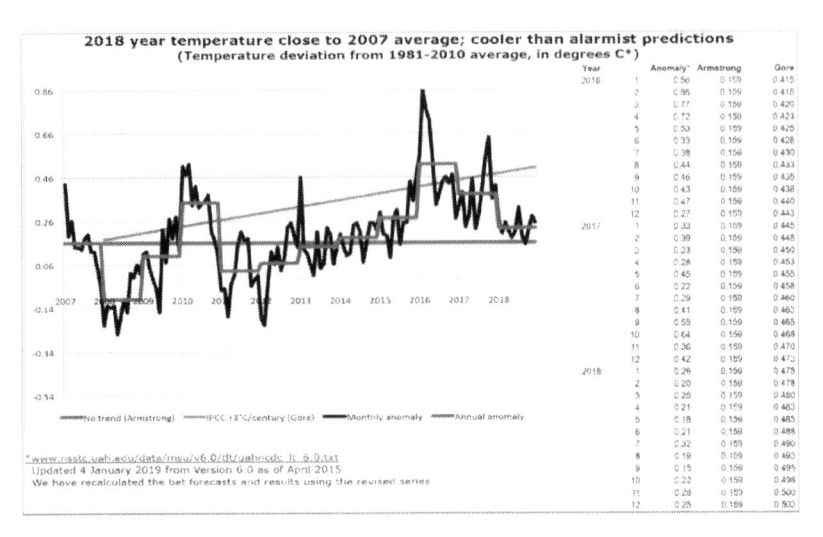

The green line is a 'no rise' baseline from 2007 figures at the start of the bet. The red line is the '3°C per century' minimum quoted by Gore and the

IPCC, the dark blue line is the annual average, and the black line are the monthly values.

Below is a link to the ClimateBet's website. Along the right are all the archives of previous articles and monthly data from UAH.

http://www.theclimatebet.com/

Below is a link to Professor Armstrong's original treatise on why he thinks the IPCC have gone wrong. Dry, but crammed with wonderful data.

http://www.kestencgreen.com/G&A-Skyfall.pdf

SECTION 16
IN CONCLUSION; A MIXED BAG!

"If the definition for a documentary is 'presenting facts objectively without editorializing or inserting fictional matter, as in a book or film,' then I am confident that Al Gore's movie was not a documentary."
J.R. Kirtek, (award-winning Meteorologist of Michigan's WJRT Channel 12)

16.1... ALL THE MOVIES ARE ON THE ALARMIST SIDE!
(Nope, we deniers have our filmmakers too...)

(1990) The Greenhouse Conspiracy
The UK's Channel 4 broadcast this documentary on 12 August 1990, as part of the Equinox series. The film criticized the theory of global warming and made the assertion that scientists critical of global warming theory had been denied funding.
As far as I know, it is the earliest documentary which suggests a conspiracy supporting the 'false claims' of global warming.
Patrick Michaels said, *"It may not quite add up to a conspiracy, but certainly a coalition of interests has promoted the greenhouse theory: scientists have needed funds, the media a story, and governments a worthy cause".*
https://www.youtube.com/watch?v=lvpwAwvDxUU

(2007) The Great Global Warming Swindle
Channel 4 in the UK broadcast the 75min film on the 8th March, 2007.
Made by British television producer Martin Durkin, the film interviews scientists, economists, politicians, writers, all who dispute or doubt the so-called Alarmist 'consensus'.
Supposedly the working title was, "Apocalypse My Arse", but the final title is an homage to the 1980 film, *The Great Rock 'n' Roll Swindle* about the punk band the Sex Pistols.
Of course, it has been highly criticized and lambasted, but it did gain a popular following, and is still available in full on YouTube.
https://www.youtube.com/watch?v=pIRICfZOvpY

(2008) The Cloud Mystery
Directed by Danish director Lars Oxfeldt Mortensen, the film takes an alternative look at the issue of Global Warming. Instead of trotting down

the well-used CO2 path, it explores the published theory by Danish scientist Henrik Svensmark who has investigated how galactic cosmic rays and solar activity affect the amount and type of cloud cover. Clouds, as I have written already are not wholly understood in relation to climate change.
http://www.thecloudmystery.com/

(2009) Not Evil Just Wrong
This documentary film by Ann McElhinney and Phelim McAleer challenges the data and direction of Al Gore's *'An Inconvenient Truth'*. The documentary uses talking heads of the era, who state that evidence of global warming is wholly inconclusive, and challenges the growing legislation would be more harmful than the actual climate change.
It was screened at the International Documentary Film Festival Amsterdam in 2009.

(2010) Cool It
Based on his book, this is a documentary by Danish statistician and political scientist Bjørn Lomborg. He wrote *'Cool It: The Skeptical Environmentalist's Guide to Global Warming'* in 1998, and released the feature-length documentary in the USA.
It has been described as 'urgent and intelligent', and debates both climate change and possible cost-effective solutions to the problem.
https://www.youtube.com/watch?v=w6UMniZqC7k

(2011) spOILed
Directed by Mark Mathis, 'spOILed' is a documentary about energy facts and myths. It tackles alternative energy, wind and solar, and discusses the energy future of mankind.

(2012) An Inconsistent Truth
Directed by Shayne Edwards, the film is the vehicle of radio talk-show host and climate denier, Phil Valentine.
It is a direct slam of Al Gore's movie, and features prominent deniers…
Newt Gingrich, Jim DeMint, James Inhofe, Fred Singer, John Christy, and Roy Spencer.
https://aninconsistenttruth.com/home

(2014) Blue
(Originally called Blue Beats Green)
Written, produced and directed by J.D. King this movie is a look at a so-called environmental movement, gone wrong.

(2015) Climate Hustle

Produced by "CFACT Presents" it is hosted by Climate Depot publisher and prominent denier Marc Morano. Marc spent many years working in the congress, and was an aide to Senator James Inhofe.

The film's press release stated...

"prominent scientists from around the world who used to warn about the dangers of man-made global warming but have reexamined the evidence and have now grown more skeptical..."

"Climate Hustle" premiered on December 7, 2015 in Paris, France, coinciding with the COP21 United Nations summit on climate change.
http://www.climatehustle.com.

(2017) The Uncertainty Has Settled

This is a documentary about climate, energy and agriculture, produced and directed by Dutch filmmaker Marijn Poels. In the film, Poels visits regular climate scientists and skeptical scientists like Piers Corbyn and Freeman Dyson, and does not force the narrative, letting the viewer come to his own conclusions.

The film premiered on February 9, 2017 during the Berlin Independent Film Festival, and was awarded the Best Documentary award.
https://www.youtube.com/watch?v=GuoxLggqI_g

These are all worth a viewing... the viewpoints are quite dynamic.

16.2... A Graph of Two Halves
(The art of graphical lying...)

A graph is a pictorial distribution of facts that illustrate data in an easy-to-digest form.
It has axes which have definitions, and by looking along each axis we can make determinations.
For instance... here are two graphs depicting the data of a car's depreciation in value over time...

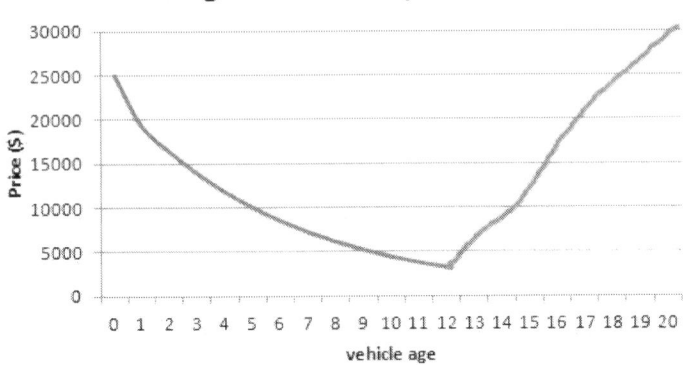

Average Vehicle Depreciation

Average Vehicle Depreciation

We look up the y-axis, finding the cost of the car at $25,000, then looking along the blue line, we scan from left to right (essentially following the x-axis) to see the probable value at whatever age the car is.
Now, the tags on the axes never change, and we would never expect them to do so.

It's here that I draw your attention to the second graph. Something is out of whack... reaching the car's 12th birthday, the value suddenly starts to increase again.
Even the most cursory glance would tell you something was wrong.
However, if I explain...

Graph 1 is a car's depreciation over time.
Graph 2 is a car's depreciation until it gets to 12 years old, when the owner starts to dump Irish mob cash in the trunk (boot).

"But your x-axis label doesn't tell me that!!!" you scream at me.
And you would have every right.
Welcome to the world of the 'Blatant Lying to Your Face', climate graph.

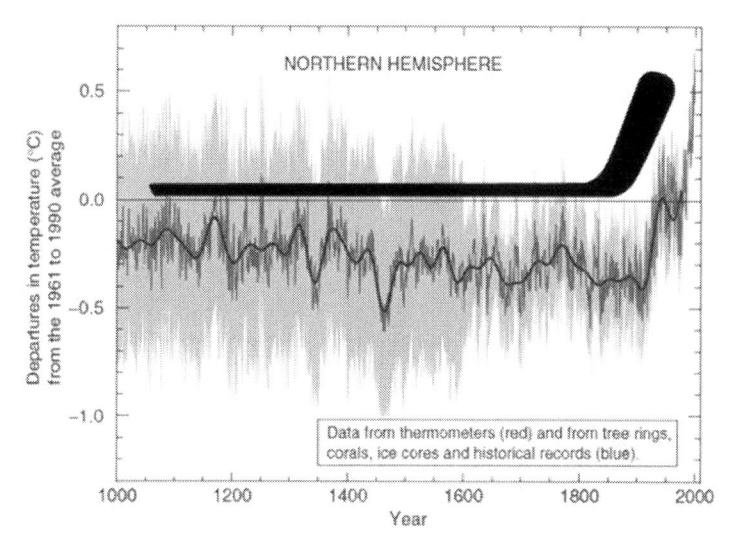

It's our friend, the Michael Mann Hockey Stick graph, randomly taken from the web for the reader's delight (and to illustrate my point).
It allegedly shows the temperature of the Northern hemisphere from the year 1000 to 2010.

What the Alarmists want to show is that the graph above is proof undeniable that 'Alarmists' are right, and I'm wrong. How can I dispute such care and attention? How can I question the motives of a mathematician of such stature? How can I cast nasturtiums at a graph so meticulously plotted?

It looks good (apart from the fact that it misses the Medieval Warming period totally, and I'm pretty sure the Medieval Period happened...

right?). Anyway, the graph has nicely labelled axes, and any 'Alarmist' would be proud to issue it to you as proof positive (or is it spoof positive) that he was right... fait accompli... slam dunk.
Not so fast sunshine...
Look carefully... the small insert states;

"Data from thermometers (red) and from tree rings, corals, ice cores and historical records (blue)."

So you've got data from coral for 1000 years.
You've examined a millennia of tree rings.
You've dug a mile of ice cores.
And, the graph shows, up until the 1880's the temperature is sinking like the titanic with a small hole in the side.
THEN!
SUDDENLY thermometers are invented... (yeah, in 1700, not 1880)
And you insert the thermometer data into the graph, and the world starts warming, and the blade of the hockey stick arrives!
It's like MAGIC... we didn't see your hands move, but you sure changed the name on the x-axis, didn't you?
Just when you wanted the graph to change, you added another parameter to the data, and abracadabra, the graph goes where you want it to go.

Oh, and by the way, Mister Michael Mann, Greenland is also in the Northern Hemisphere, and it was warmer in 1100-1200 (The Medieval Warming Period) than it is today... this is proven by period records, archaeological evidence, and the fact that up to 5000 Vikings FARMED there for 200+ years!
So, mister 'Alarmist', you could have saved yourself a few years 'research' into your suspicious tree rings, and just built it from scratch in a few minutes on Photoshop.

16.3... THE SUBTLE ART OF THE GRAPH
(Glass half empty, half full...)

1. Perspectives and Agenda
How can we manage to cope with this Alarmist phrase...

"16 of the 17 warmest years on record occurring since 2001, and 2016 being the warmest year on recorded history"

With this Denier one...

"There's been no warning trend since 1998... there's a pause/hiatus"

It's actually quite easy... look at the graph below...

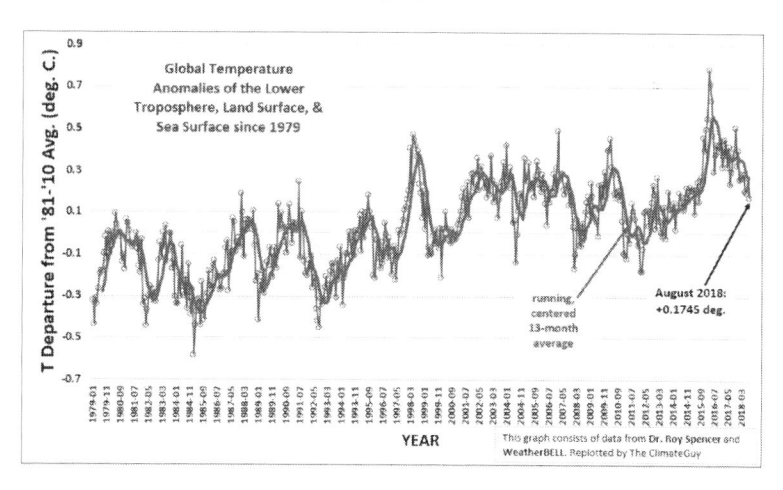

It's a graph of combined warming of atmosphere, surface and sea temperatures.
This is a graph I 'think' I can believe.
And yet, it does show...
ALARMIST CLAIM...
"16 of the 17 warmest years on record occurring since 2001, and 2016 being the warmest year on recorded history"

AND it does back up the second claim...
DENIER CLAIM...
"There's been no warning trend since 1998... there's a pause/hiatus"

How can one graph support two very diverse views?

Simplified, it's very much a glass half empty/half full scenario.
Each person looks at such a graph, and sees something different... they write their rhetoric based on their end-game agenda, and yet BOTH ARE TRUE.

Let's look at the graph again, and make some more 'TRUE' statements of your own... Here's my ones...

"Temperatures did not vary from 1979 to 1997. Twenty Years!"

"The El Niño assisted figure for 1998 is the highest/quickest climb in temperature in recent memory."

"An average temperature drawn between 1998 and 2018 shows a pause, maybe even a drop in global temperatures"

All statements are 'true', all perfectly valid, a nice selection for both sides.

2. The Starting and Ending Points
Look at this one...

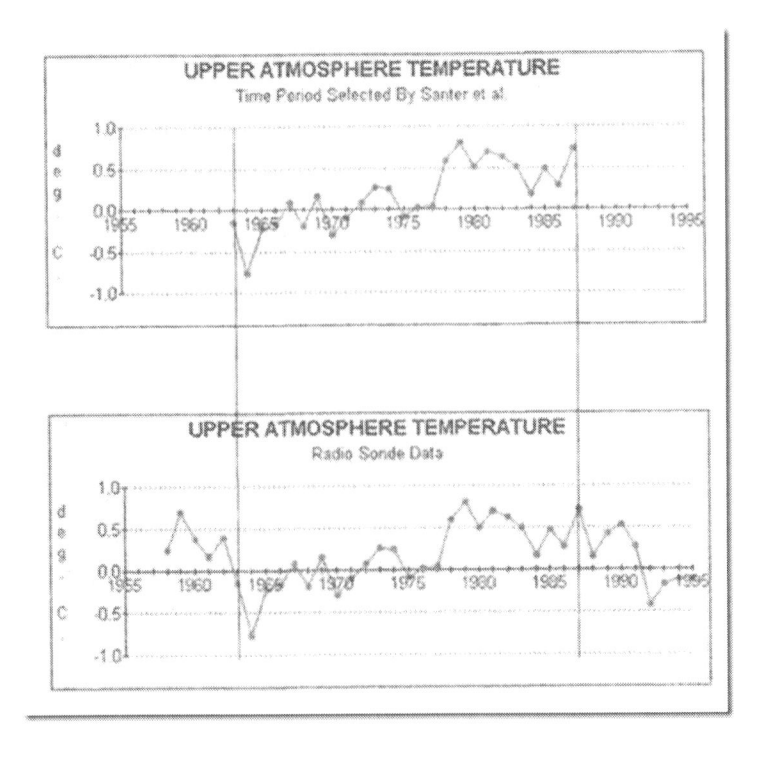

Where we choose to start our graph and where we end it can change the trend of a graph completely.

The lower graph was the one agreed by the authors of chapter 8 of the IPCC's 1995 Second Assessment Report (SAR).
It shows little of any trend, and if there 'was' a trend, it would be slightly downward.

The upper graph is the one chosen by IPCC Lead Author Benjamin Santer for the final draft of the report. The precise starting and end points change the trend of the graph completely.
It's a fraud, a lie, misleading, whatever you want to call it.
It's NOT scientific.

3. Joining Two Different Data Sets
Data on a graph should conform to the same specific criteria for the whole graph.

For example, a graph showing car sales in China should show just cars in China... it should not halfway through show Japanese Imports for a section, then end with just British car sales in China.
The graph below shows variations of the infamous hokey-stick...

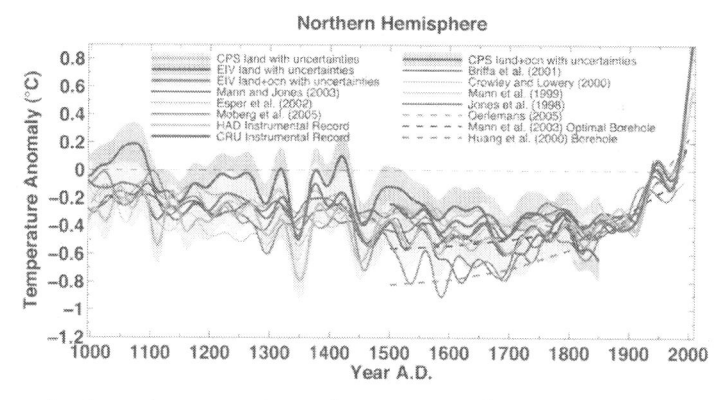

These graphs have been made with numerous batches of proxy data, (the wide green and blue areas being the margin of error in the proxy data) then superimposed is the actual temperature record measured by thermometer.

Now, since the thermometer wasn't invented in the 1300's the data for the graph is tainted.

It doesn't make sense.
And apart from showing a lot of data lines 'agreeing' with each other, it's meaningless.

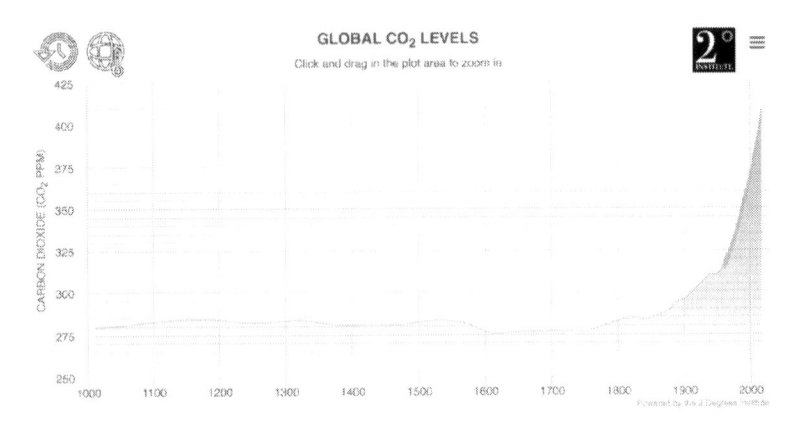

[Graph: GLOBAL CO₂ LEVELS, Y-axis CARBON DIOXIDE (CO₂ PPM) ranging 250-425, X-axis years 1000-2000]

This is another seemingly innocuous example.
But the data before 1955 has been reconstructed by a 'carefully chosen' set of proxy data (probably from tree rings or ice cores).
After 1955, the graph uses the chemically-tested data set from Hawaii. Nowhere on the graph does it inform the reader of the fact that 2 sets were used, then magically joined together.
(And to be honest, I would challenge that 2 very different data sets would join so well together in 1955)

(AND I also challenge the rather smooth period from 1000 to 1800AD. What happened to forest fires, earthquakes, volcanoes??)

4. The Importance of Qualifying the Y-Axis

In most graphs in the climate game, the X-Axis (the horizontal one along the bottom) is usually marked in TIME. This allows the reader to trace the course of the graph's data over a certain historical time range.
The Y-Axis is usually ascribed to whatever the graph is trying to tell us... temperature over time... ice thickness over time... etc etc.
Look at the graph above.
Although this graph actually breaks the previous rule of combining two different data sets, its axis are precisely laid out. At any time between 1000AD and 2012, we can look up the Y-axis and fine the CO2 value for that year.
This kind of graph is showing ACTUAL MEASURED DATA.

You can use these figures for calculations, and you can compare two different graphs together to look for correlation, to look out for errors between graphs, and to superimpose more than one graph onto of the other.

However, there is another type of graph which does NOT qualify any value to the data on the Y-Axis.

Take a look at the graph below.

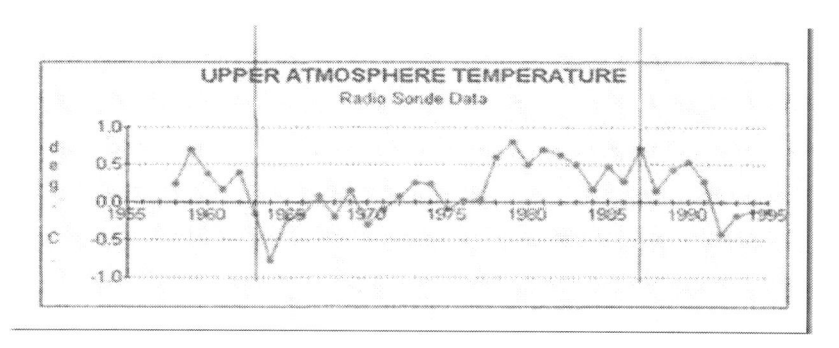

This is a graph of temperature to time.

For some reason the values on the X-Axis are placed in the middle of the graph, but they're usable... they clearly show 1955 to 1995.

But look at the values on the Y-Axis.

They go from -1.0 to +1.0.

The zero level is highlighted (across the middle) but it gives no actual value to the actual red temperature line.

Actually, the red temperature line has been chosen by the graph maker... it's called an anomaly. (And it gives us no indication what parameters were chosen to put the zero level where it is.)

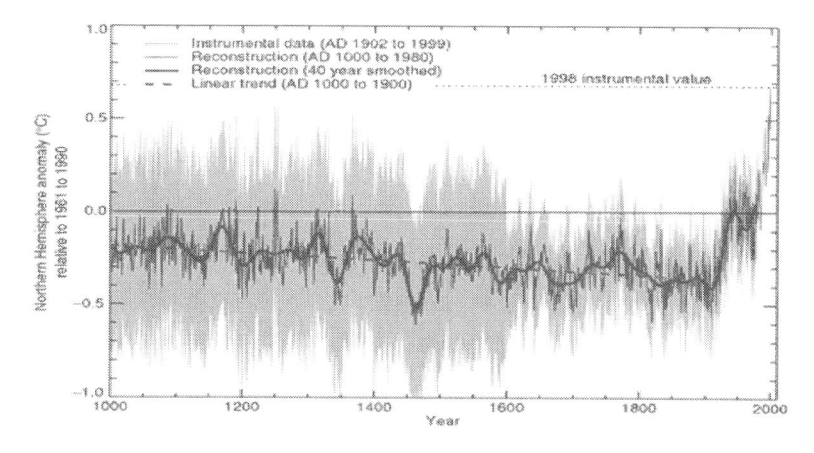

Above is Michael Mann's hokey-schtick.

It shows supposed time along the X-Axis... wonderful.
It shows temperature up the Y-Axis from -1.0 to +1.0...
However it does tell at the left hand side that the zero line is positioned along the 1961-1990 average.
That at least is being accurate and honest.

The technical idea for such a graph is to show an anomaly from a certain pre-determined baseline.
The problem is, the graph maker chooses (in this case) the zero base line.
This, dear reader, is a type of graph to show a TREND.
And we all know that a graph showing a TREND is trying to tell us an opinion.
Because we KNOW that not all trends are honest, legitimate, and truthfully delivered.
On the graph above, why does Mann not show us the ACTUAL temperatures along the Y-Axis?
The 'Distinguished Professor' MUST have had these temperature values in Celsius to draw the graph in the first place!

What is Michael Mann trying to hide? (If anything)

Yup... graphs are funny things, and a good graph-maker can make you believe anything.

16.4... THE GLACIERS ARE ALL RETREATING
(Come on; this is old news!)

Introduction
Dear reader, I'd like to conduct an experiment.
I'd like to ask you two questions, ask you to close your eyes, and guess the answers.
(Answers on the next page)

DON'T LOOK!

1. How many glaciers there are in the world?
2. How many glaciers (as a percentage) are actually monitored?

Eyes open yet?
Got your answers?
Look to the next page...

Okay... let's set the record straight with a few facts.

There are more than 198,000 glaciers on Earth.

That fact alone is quite staggering, but it gets better.

Only about 10% of these glaciers are monitored regularly, and many have never been explored.
Glaciers occupy about 10% of the land on Earth.
If All the glaciers melted, it would raise the average sea level by 4 meters (about 13ft), so this is a substantial topic.

Now, most climate studies would say that "Glaciers are a very short-term trend. They react quickly to climate change".
And they're correct... a glacier is sometimes only a few hundred feet deep... and that amount of ice is easy to melt in warmer times...
But you have to remember that the planet is still coming out of an ice age, and had a 'Little Ice Age' just 300 years or so ago.
Many of the landmasses covered in the ice age are still rising, the pressure of ice released.

But one fact you have to remember... the temperature of the air around the glacier does NOT dictate or limit the flow of the ice.
The ice flows because of massive pressure further up the mountain... it's the new falls of precipitation high up the glacier that propels it downwards.

All Glaciers are Melting
Nope.
Not even close.
(Even though Wikipedia states... *"Currently, nearly all glaciers have a negative mass balance and are retreating"*)
I really don't know why I bother...
The fact is... many glaciers that have been retreating are now static, or have started advancing again.
Let's look at a few examples.

Glacier National Park
This famous National park is situated on the US/Canadian border, in Montana, Alberta and British Colombia.

As I stated in Section 1.5, the National Park Service have had to remove all the signs in the park... they used to state that ALL glaciers were retreating, and would be gone by 2030.
But, lo and behold... the most recent survey shocked everyone.
The survey showed that many of the glaciers in the park were actually advancing... including the famous glaciers such as the Grinnell Glacier and the Jackson Glacier.
And we're not talking small potatoes either; the Jackson Glacier may have grown as much as 25% or more over the past decade.
Snowfall in the winter of 2017/18 was at a record high in places.

Greenland Glaciers
A National Geographic article from March 2019 warns...

"A Greenland glacier is growing. That doesn't mean melting is over."

Wow, just how negative can you be, NatGeo?

"A pulse of cooler water at its edge let part of the glacier gain some mass. But overall, the melting across Greenland continues apace."

Hold on... I thought that a glacier advances when pressure from above (snow falling) forces it down.
(Don't we understand glaciers at all?)
But, as the quotes above indicate, NatGeo are insistent on giving us good news, then hitting us with bad.

"This reverses the glacier's 20-year trend of thinning and retreating. But because of what else is happening on the ice sheet, and the overall climate outlook, that's not necessarily a good thing for global sea level."

Again, the negative outlook. Can we not be HAPPY that the glacier has stopped retreating? It's almost as if NatGeo WANT to give us bad news, and can't help themselves.

"The thinking was once glaciers start retreating, nothing's stopping them," explains Josh Willis, an oceanographer at NASA's Jet Propulsion Laboratory and OMG's lead scientist. *"We've found that that's not true."*

And how do I paraphrase that one?... "the science is NOT settled?"

Oh, and on that subject, can we go back to 1967...

Maynard M Miller was a professor of Geology at Idaho University, and founder of the Juneau Icefield Research Program. Foundation for Glacier and Environmental Research.

He wrote a 1967 article in National Geographic, entitled *'Science find new clues to our climate in Alaska's mighty rivers of ice'* and said this...

"Records point to a remarkable regularity of the climatic cycles over the last two centuries. The records show that Alaska's ice masses advance and retreat in direct correlation with cyclic changes in sunspot numbers. Sunspots coincide with strong electrical discharges from the sun."

Man, I wish I could bottle that one and make Alarmists drink it.

The Petermann Glacier, Greenland's second largest glacier, has been growing about 10 feet a day since 2013. You won't hear that on the news.

16.5... THE GLOBAL WARMING INDUSTRY
(Follow the money...)

In the 1960's, there were perhaps two or three hundred climatologists.
(Well, in those days, they were probably known as physicists, geologists, atmospheric scientists, geologists and such, folks who dabbled in climate as a side to their main thesis... no one considered a career as a climatologist... it was always a second, lesser area of work.)
The USA's total annual budget for climate investigation was less than $75 million.
Climatology was the poor boy of the scientific community.

The sudden cooling in the northern hemisphere from 1954 to 1979, a staggering 2°C COOLING in 25 years changed the career choices of a generation of young scientists.
They had an ice age to look forward to.

Unfortunately, by the time this new generation had studied and graduated from their universities, ready to investigate the new oncoming ice age, the crisis was over... temperatures were rising again.

So... by the early to mid 1980's, there was a generation with PhD's just waiting to study the climate.

Before the IPCC was formed in 1988, those students were joining university, looking for a degree in climatology, and looking for jobs within the industry.

In 1990, the US Government budget for climate science had increased to $170 million, over double what it had been thirty years before.

From the late chills of 1979, the temperatures rose, but peaked at 1988, before falling again.
James Hansen testified before the Senate, and boasted that the 'seas will boil'.
The PhD climatology graduates looked for jobs, served as new professors in the new climate study field, and some found friends in the media.
By the 1990's, a new type of journalist had arrived, fresh from university; the environmental journalist. And this new breed had a new whole 24 hour news cycle to fill.
With reports from polar bears to equatorial droughts, the hyperbole was ratcheted up to fever pitch.

Then the El Niño of 1998 arrived, shot the global temperature past every value since 1954, and the whole world began to read about Global Warming.
We were going to drown, our coastal cities were going to drown, polar bears, seals, and penguins were at risk.

The US Government's budget went through the roof too, breaking TWO BILLION dollars in 2000.

Trust me, if you were to graph carbon dioxide rise with US government's climate budget, it would be a close correlation.

Then true to the rise of the budget, came the logo for the climate world to believe in.
In the IPCC's Third Assessment Report (2001), the Hockey Stick graph by Mann, Bradley & Hughes 1998 (MBH98), was published.
The science was settled...
The dice were cast...
Debate stifled...

And the queue of young scientists struggling to make their name (and some grant money) poured into and out of universities worldwide to add their piece of brilliance to the global warming industry.
Every newspaper had its own Climate Department, advisor, or reporter. Every world government had their own climate bureau, every local council, county and city had their own climate/global warming office.

In just a decade or two, global warming had risen from a germ of an idea to a multi-billion global industry.

When Al Gore released his movie, *An Inconvenient Truth,* in 2006, the annual US government climate budget had doubled to FOUR BILLION dollars.
Thousands of scientists filled hundreds of science journals with every angle of global warming news possible.

By 2012 the US Government was throwing over $10 billion at the problem, that's $10,000,000,000.

But there was a problem...

Somewhere in the mix (probably driven by the IPCC's core statement and such), the threat had been 'found' as carbon dioxide, and mankind had been fingered with the crime.

Anthropogenic Climate Change was here to stay, and governments didn't need anyone to investigate the 'source' of the warming any more... emphasis was shifted (and the grant money too) to SUPPORTING the theory, rather than PROVING it.

As I write this book in 2019, the estimated world spend on climate change research has risen to a staggering ONE TRILLION DOLLARS.

And people wonder why the scientists do it?

16.6... BIOMASS... THE HOAX WITHIN THE HOAX
(It's hard not to ask WTF?)

There's a new source of 'renewable' fuel that's surfaced in the last 20 years.
Biomass.
It's the environmentalist's dream.

Or is it?

What is Biomass?
Well, here's where businessmen and environmentalists begin to disagree. If we were to believe the corporate hype, biomass is tree matter, organic wood, farm waste, industrial waste, and some animal organic waste.
The initial idea seemed sound.
Woods are filled with rotting trees, broken stumps, decaying shrubbery and vegetation. It would be good forest management to clear this mess up, and thin the trees down, it would mean fewer large forest fires.
The initial idea was sound... until the men hit the forest, and the money-men started to look at the bottom line.

Logging logistics
Anyone clearing a forest for biomass production is paid in net tons harvested; logistics are a pain. Getting men and large machinery into the middle of a forest with whilst leaving the trees standing is difficult and costly. The demand for cheap biomass demands shortcuts; why waste time thinning out trees when they can just clear them and harvest the standing ones?
And that's exactly what's happened in a whole lot of cases.
There's little regulation to enforce any re-planting, and certainly no oversight.
And the stumps that would limit new tree growth? They were meant to be part of the main mass taken from the forest.
But stumps are the most resilient part of a tree. They've dug down into the earth, and have kept the tree standing for years; stumps are expensive and difficult to remove, and unlike straight lumber, they're bulky and costly to transport. So in many cases they're left behind.
CANCELLING THE REAL POINT.... RIGHT?

Other biomass sources
Like I said earlier... the idea is good. Getting rid of waste and making electricity is a good thing, right? It's a 'renewable' resource, right?

But we have to remember the collection costs and transport costs for all these different kinds of biomass. Anything has to be packed, trucked, and delivered to the pellet factories. Transportation is NOT a carbon neutral part of the plan. Remember, biomass includes...
1. Twisted trees unsuitable as lumber
2. Broken trees and tree limbs
3. Forest floor shrubbery
4. Industrial/commercial organic waste
5. Farm waste, animal manure

Countless factories around the world grind up the biomass and make small pellets. These pellets are used as a power station fuel.
And these pellet factories are not entirely 'green' either; there are cases of extremely fine dust being emitted into the atmosphere, and locals having health problems because of it.
The biomass pellets can be used in household fires, but mostly they are burned in power stations converted for the process.
The corporate promise is that *"for every tree we harvest, we'll plant another"*... thus qualifying it as a 'renewable' resource.

But we all know what corporate promises are worth when it comes down to economics... nothing.
Environmental blogs are full of cases where land has been purchased, ALL trees cleared, and the land left to die.

This part is indeed a cause worth fighting for.

The WTF moment
Biomass is called a 'renewable' resource.
But to qualify as a 'renewable' fuel, the source of the fuel, (the trees), have to be re-newed. Without re-planting, biomass just becomes another way to cut down trees and burn them.
It becomes just another fossil fuel used to make electricity.

Because it's called 'biomass', the true carbon implications are not fully researched.
There's research that proves that biomass is actually putting MORE carbon dioxide into the atmosphere than the coal and natural gas power stations it was initially designed to replace.

Then there's the logistics... think about this for an example of a carbon footprint.

1. Trees and forest matter are harvested in North Carolina, USA. Not removing all brush, shrubbery, tree stumps, etc. This leaves a partly cleared area, in most cases, new saplings are planted.
(PLEASE NOTE; THIS INITIAL PART is the ONLY part of the biomass process that is in any way environmentally friendly. From now on, the 'biomass' is just ANOTHER FOSSIL FUEL.)
2. The harvested matter is transported by truck to a pellet making factory. This is sometimes more than 200-300 miles from the initial harvesting site.
3. The pellets are made, spilling fine dust into the surrounding atmosphere.
4. The pellets are then transported by railcar or truck to a harbour and loaded into ships.
5. The pellets are then taken by ship/tanker to their destination country… and that might be several thousand miles away. A lot of European countries, including the UK, and Italy both use biomass from Africa and the USA.
6. In ideal conditions, the pellets are unloaded direct to the power station… but in real terms, the pellets are loaded into railcars or trucks and taken to the power station.
7. The pellets are burnt, and carbon dioxide is released. This is hopefully caught in some filter process, just like any other fossil fuel… you know, the bad fossil fuels that the world's environmentalists are trying to phase out.

BIOMASS… it's an environmental scam.

The whole idea is a HOAX!

16.7... LIES TO SELL NEWSPAPERS?
(Dang tootin'!...)

There's little doubt that the Alarmists are beginning to quake in their boots as the pause/hiatus continues into 2019.
But the warming rhetoric continues too.
Let's look at a few stories from the years just gone by...

A Very BBC Summer
A few months ago, the BBC Website posted...

"2018 was the joint hottest summer on record for the UK as a whole, and the hottest ever for England, the Met Office has announced."

This is a serious announcement, causing British millennials to rush to their coffee shops (their 'safe' place) and mull over their over-priced latte.

Well, I almost hung up my denier boots, stuck my skeptic pen in my new hemp shirt pocket, and packed in my whole book project.
Almost...

However, if the only talking point taken from this headline is *"Hottest Summer on Record"*, you're not reading deep or far enough. The quote said *'joint hottest'*. Now that means there are more summers that were recorded as at the same temperature. Let's look at the BBC's next paragraph...

"It (The MET Office) said highs for summer 2018 were tied with those of 1976, 2003 and 2006 for being the highest since records began in 1910."

So after the headline, the waters get clearer.
"TERRIBLE!" I hear you cry. "Three of the hottest summers were in the last 15 years! We're heading off the cliff! We've hit the TIPPING POINT!"

That would make a great headline for the same news.
But like most statistics, they can be manipulated.
Consider another take on the same news...
Because if 2018 tied with 1976, we could have run with a very different headline...

"Global Warming Not Changed in 42 Years!"

Yeah, I can hear all you fellow deniers laughing, but for sure the editor of the BBC site knows his page count would have suffered if he'd used my headline instead of his own.
"But wait!" The 'Alarmists' roar, crashing their empty coffee cups on the Starbucks tables… "The Hockey Stick! Temperatures are rising! The CONSENSUS says so!"

Just calm your ardor, young ardent minds… look at the headline again. This is a British survey, published by a British website, and it ONLY refers to a SUMMER, not a mean/average record of the whole year.
And that's where statistics and local figures can be taken out of context.

USA's lies, damn lies, and statistics
According to NOAA, quoted in USA Today, *"The nationwide average of 73.53°F made 2018 the warmest summer since 2012 and tied for the fourth-warmest on record."*
Hmm… again, it 'tied' for fourth place. Surely if the temperature was rising like the Hockey Stick shows, it would be the hottest, or tie for the hottest? The article then goes on to say, *"The hottest summer in U.S. history remains 1936, at the height of the Dust Bowl."*

So… *"No Warming for 83 Years"*, could also have been the headline if the media wasn't already in the tank.

Yes, I know… my headline wouldn't have sold as many newspapers.

"Lies, damn lies, and statistics." Mark Twain was certainly a visionary.

Time (online) featured a 29th November WHO report that *"2018 Is On Track To Be the Fourth Hottest Year On Record, Scientists Say"*. The article says… *"The figures are based on five independently maintained global temperature data sets."*

Yeah, and we know what datasets they're using, don't we? The term *'independently maintained'* has never been used so well.

And can I stop you right there?
'Maintained'?
I just mentioned global data sets. Figures in stone, never to be touched again… yeah.

Try putting *adjusted/tampered* with in place of maintained. You'd be closer to the truth.

However, doubling down on hyperbole, the article goes on… *"Greenhouse gas concentrations are once again at record levels and if the current trend continues we may see temperature increases 3-5°C by the end of the century."*
It's difficult to stand against this type of verbal barrage, when all we have is our own graphs, and our own figures.
Perhaps the roundabout will turn our way soon.

Canberra's Heatwave

The media hitting the data and twisting the truth is not new, however.
Let's take a look at Australia's capital, Canberra.

In 2013, the mainstream media in both the TV weather news and major newspapers, gave the Canberra locals a little taste of Alarmist spin.
The Australian weather board, the Bureau of Meteorology (BOM) claimed that July 2013 was the warmest ever in Canberra, and it even gave a figure and a graph.

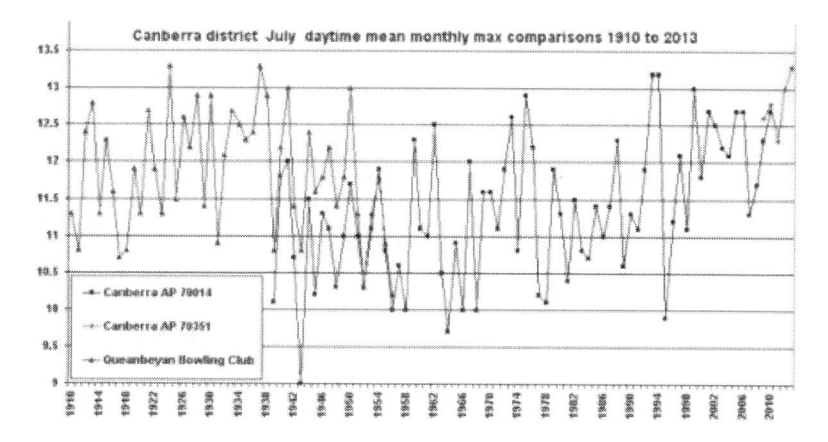

Announcing a 'Record High' with a mean monthly max of 13.3°C, they claimed the end of the world was nigh, and Aussies should take the next transport plane for Antarctica.
However, the Canberrans need look no further than the BoM's own graph.
Look at the highs in 1924 and 1937. Don't they look suspiciously like the same figure?

On investigating the figures from Queanbeyan, just 6 miles from the Canberra City Center, which the BoM had based their graph on, we find that the figures are indeed a tie... 13.3°C. Not only that, 1993 and 1992 are just a tenth of a degree colder, coming in at 13.2°C.
This is the BoM's own data.
So the claim of the headline is dubious to downright misleading.

Yet again, bad/dubious reporting gets in the way of a decent story.
Queanbeyan is a suburb itself, and probably has the same Urban Heat Island problems as the station at Canberra Airport.
So the public get fed mushrooms and kept in the dark.

The Australians are beginning to wake up to the corruption of their weather figures, and we all know when Aussies get riled up... anything can happen.
So given the poor quality of data from Canberra's weather stations – the best you could claim about July is that it was warm and probably similar to a few other July's in the past 100 years. As I have said a few times, everything the BoM says needs examining with a fine toothed comb.

16.8... WIKIPEDIA IS NOT YOUR FRIEND!
(Nope... they HATE deniers...)

Wikipedia is an inexhaustible resource, and one which I use in every one of my books to some degree.
It's helpful, and full of those blue click links for quick navigation between and within topics.
But some things you have to remember...

1. Wikipedia is added to and updated by ANYONE who can log into their 'edit' process.
2. These 'contributors' can write anything they feel like.
3. Yes, there is an auditor/editing process, but it may take time to appear, and may never do so for some pages.
4. Academia (the University system) is generally considered to be 'left' in its political inclination, and this may explain the particular 'left-ish' slant of a fair majority of the postings.
5. The media is 'left; in its political inclination, so again, this may account for some of Wikipedia's 'left' bias.

In a section entitled, *List of scientists who disagree with the scientific consensus on global warming*, there is a statement near the top of the page...

"The scientific consensus is that the global average surface temperature has risen over the last century"

And 'scientific consensus' has a blue clicky.
Yup... this statement actually implies that if you're a 'Denier', you're avoiding the most basic of climate change facts.

On a group level, take a look at some of the 'consensuses' on both sides.
The 97% Alarmist consensuses go virtually unchallenged.
BUT... go to the Denier consensuses, the Leipzig Declaration, The Global Warming Petition Project... and these consensuses get vilified remorselessly.

On a more personal level try looking up Gavin Schmidt, James Hansen or Michael Mann. These Alarmists are treated like Gods.
Do the same with deniers Will Happer, Willie Soon or Fred Singer... that's a different tune.

Yup, because of the way Wikipedia is built up (by notations from the public) we don't have many friends there.

Facebook, Twitter and Google
These organizations are accepted to lean to the left, so it comes as no surprise that they censor the words of the right wing.
There have been so many issues with these social media giants, it's a book in itself... but just remember, when you journey into that world, you are not amongst friends.
There is a reason that when you 'google' climate change stuff, one of the top hits is always 'climateskeptic', a VERY left-leaning Alarmist site.

Section 17
Prominent Deniers of the World

"Science is a beautiful gift to humanity; we should not distort it."
A. P. J. Abdul Kalam (aerospace scientist and 11th President of India)

17.1... More of the World's Deniers
(Scientists, Media, and famous people)

In popular culture, there has been a derisive dismissing of the numbers and the scientific credentials of the world's misnamed 'climate deniers'. We're labelled as the '3% of scientists'; the kooks in the woodpile.
I hope that in this book I've proven that the denier ranks are far bigger than three percent, and that their beliefs are both creditable and scientifically worthy.
So to conclude, here's a 'wee' list I assembled.
The list includes Nobel Prize winners, astronauts, ex-NASA and ex-NOAA personnel, IPCC panelists, IPCC scientists, professors, scientists, weathermen, meteorologists, politicians, and media figures.

To various degrees they either dispute the carbon dioxide 'consensus' or the way the IPCC has gone about its business.
Some dispute the integrity and motives of the Alarmist scientists, some question the very data which they use, and some have a few theories of their own.
Many actively discredit the accuracy of IPCC climate projections, and even if the warming was proven to be real, some simply just don't believe that mankind can change the climate to an extent large enough to solve the problem.

Added to the lists and names already mentioned in this book, this list adds a few more to the roll who consider the science nowhere near settled...

Listed in Alphabetical order...

Habibullo Abdussamatov is a Russian astrophysicist from Uzbek. He is a supervisor of the Russian section of the International Space Station and the head of Space research laboratory at the Pulkovo Observatory in Saint

Petersburg. He believes that global warming is primarily caused by natural processes.

John Stuart Agnew is an MEP for the East of England for the UK Independence Party (UKIP). *"I just wonder when the bubble is going to burst on this whole ludicrous (man-made global warming) saga."*

Alexandre Aguiar is a Meteorologist for the MetSul Weather Center of Ulbra TV in Porto Alegre, Brazil. *"Dubious Connections between Global Warming and Extreme Weather Events."*

Jarl R. Ahlbeck is a Finnish chemical engineer, and ex Greenpeace. *"So far, real measurements give no ground for concern about a catastrophic future warming."*

Arun Ahluwalia is a Geologist from Panjab University, Chandigarh, India. He is active in petroleum exploration. *"The IPCC has actually become a closed circuit; it doesn't listen to others."*

Don Aitkin is a former foundation Chairman of the Australian Research Council. *"Is the warming unprecedented? Probably not. There is abundant historical and proxy evidence for both hotter and cooler periods in human history."*

Syun Akasofu, is a Japanese Geophysicist, and the founding director of the International Arctic Research Center (IARC) of the University of Alaska Fairbanks. *"Even if we spend lots of money on suppressing CO2 release, it wouldn't do any good, because it's a natural change."*

Ralph B. Alexander is a former Associate Professor of Physics at Wayne State University and the author of the book, Global Warming False Alarm. *"I am offended that science is being perverted in the name of global warming."*

William JR Alexander is a retired Professor Emeritus, Dept. of Civil and Biosystems Engineering, University of Pretoria, South Africa. *"No scientifically believable evidence to support the alarmist claims"*

Claude Allègre is a French politician and scientist. Although he wrote in 1987, *"By burning fossil fuels, man increased the concentration of carbon dioxide in the atmosphere..."* he now is quoted as saying *"the cause of this climate change is unknown".*

Charles R. Anderson is a former Department of Navy research physicist who has published more than 25 scientific papers. *"My evaluation is that man's effect upon the global climate is still small compared to the natural forces at work and that they are incapable of causing anything on the scale of a catastrophe."*

David Archibald is a scientist operating in the multiple fields of cancer research, climate science, and oil exploration. *"There are no deleterious consequences of higher atmospheric carbon dioxide levels. Higher atmospheric carbon dioxide levels are wholly beneficial... Anthropogenic Global Warming is so miniscule that the effect cannot be measured from year to year, and even from generation to generation."*

José Ramón Arévalo is professor of Ecology at the University of La Laguna, Spain. *"Climate warming is more an ideology, that I have read is call "Climatism"... so, as an ideology is perfect to me, the problem is when administrators become members of this sect, and then they have to spend millions in demonstrating their ideology."*

J. Scott Armstrong is an award winning PhD from the University of Pennsylvania, Philadelphia. He is a professor of marketing, and used his marketing techniques to challenge Al Gore to a bet on the climate. (See previous chapter entitled; 'The Global Warming Challenge')

Bob Ashworth is a chemical engineer who holds 16 U.S. patents on fuels and emission control techniques and has written 55 technical papers on fuel technologies. *"The lesson to the world here is, when it comes to science, never blindly accept an explanation from a politician or scientists who have turned political for their own private gain. Taxing carbon will have absolutely no beneficial effect on our climate, will hurt the economies of the world, and will be harmful to the production of food because less carbon dioxide means reduced plant growth."*

J.K. "Jim" August is an engineer and physicist formerly of the U.S. Navy nuclear power program. *"Gore's An Inconvenient Truth is not scientifically based... The book denies the legitimacy of science for review. The irony is, of course, the treatise that Mr. Gore uses to make his points, which could only have any value based on some scientific certainty basis, is not based on science nor the scientific method -- nor can scientists even use science to review it, or follow its logic."*

Robert H. Austin is a prize-winning Princeton University Physicist and has published 170 scientific papers. *"It is tragic that some 147 perhaps well-meaning but politically motivated scientists who should know better have whipped up a global frenzy about a phenomena which is statistically questionable at best, and I deeply regret that scientific societies such as the National Academy of Sciences and the American Physical Society have not played leadership roles in giving voice to questions about the depth of our knowledge of climate change and man's ability to change climate."*

Dennis T. Avery is the director of the Center for Global Food Issues at the Hudson Institute, editing Global Food Quarterly. He believes that global warming is a natural cycle and therefore unstoppable, and wrote the book, Unstoppable Global Warming, with Fred Singer.

Ritesh Arya is an Indian geologist who specializes in hydrogeology and groundwater resources in the Himalayas. *"There is a hype of global warming created by western mass media and there is a need to redefine the whole concept."*

Sallie Baliunas is a retired astrophysicist from the Harvard-Smithsonian Center for Astrophysics. She said with Willie Soon… *"But is it possible that the particular temperature increase observed in the last 100 years is the result of carbon dioxide produced by human activities? The scientific evidence clearly indicates that this is not the case… measurements of atmospheric temperatures made by instruments lofted in satellites and balloons show that no warming has occurred in the atmosphere in the last 50 years."*

Timothy (Tim) Ball, (M.A. from the University of Manitoba in 1971 and PhD in climatology from Queen Mary University of London) is an English climatologist, and was sued by Alarmist Michael Mann for defamation… Ball won his case. Ball totally ejects the scientific consensus on climate change… *"CO2 is not a greenhouse gas that raises global temperature."*

Robert Balling, (M.A. and PhD in Geography) is a professor of Climatology at Arizona State University. He describes global warming as *"a vastly overrated threat whose proposed solutions are worse than the problem."*

Joe Bastardi has a degree in meteorology and is a weatherman, and TV presenter. In his book, his views are sharply at odds with the scientific 'consensus' on global warming. *"CO2 cannot cause global warming. I'll tell* you why. It doesn't mix well with the atmosphere, for one. For two, its specific gravity is 1 1/2 times that of the rest of the atmosphere. It heats and cools much quicker. Its radiative processes are much different. So it cannot - it literally cannot cause global warming."

Michael Beenstock is a professor at the Hebrew University and honorary fellow with Institute for Economic Affairs. *"Because the greenhouse effect is temporary rather than permanent, predictions of significant global warming in the 21st century by IPCC are not supported by the data."*

Larry Bell of the University of Houston is the author of the book, Climate Hysteria. *"Many questions remain to be answered regarding the real significance of anthropogenic carbon dioxide as a climate forcing factor and related rising sea level consequences projected by the [UN] IPCC."*

David Bellamy OBE, (PhD in Botany) is an English TV presenter, filmmaker and environmental activist. In 2004, he wrote an article in the Daily Mail in which he described the theory of man-made global warming as *"poppycock".* (That's polite English for 'crap'.)

Lennart Bengtsson, meteorologist, Reading University. Bengtsson has been on both sides, joining the Global Warming Policy Foundation (GWPF), a climate skeptic organization, becoming a board member, then resigning after just one month.

Greg Benson is a geological modeler of 30 years' experience. *"Geologists and climatologists are certain that the Earth has gone through periods both warmer and colder than what we call 'normal' today."*

Paul Berenson is an M.I.T physicist and Scientific Advisor to NATO's Supreme Allied Commander Europe (SACEUR). *"Current atmospheric temperatures and CO2 content are no higher than they have been at various times during the past million years."*

Donna Bethell is a lawyer, and board member of Fred Singer's climate contrarian Science and Environmental Policy Project.

John Blethen is a physicist who runs the global warming skeptic website Heliogenic.blogspot.com. *"The Sun, not a harmless essential trace gas (CO2), drives climate change... Someone should tell these people the globe has been cooling."*

Karl Bohnak is a TV meteorologist in Michigan. *"Water vapor accounts for about 95 percent of earth's natural 'greenhouse' effect. Carbon dioxide gets all the attention because that is what is released in the burning of fossil fuels. Yet it accounts for less than 4 percent of the total greenhouse effect. For the anthropogenic global warming argument to work, water vapor must increase along with CO2. CO2's contribution - natural and manmade - is just not enough to raise global temperatures as much as climate models predict."*

Peter Bonk is a chemist and a member of the American Chemical Society. *"When one realizes all the factors at play that define and shape climate, the idea of hanging all the changes (and saying that they are all bad) on CO2, an important but minor atmospheric component, seem beyond all reason."*

Christopher Booker was a writer on the UK's Sunday Telegraph. *"the West's obsession with catastrophic global warming, at a time when the rest of the world, led by China and India, is taking not the slightest notice, as it continues to build thousands more coal-fired power stations."*

Donald J. Boudreaux is an economist and the Chairman of the Department of Economics at George Mason University. *"I am a global warming skeptic - not of the science of climate change (for I have no expertise to judge it), but a skeptic of combating climate change with increased government power."*

R. W. Bradnock is former Head of Department of Geography at the School of Oriental and African Studies (SOAS). *"In my own narrow area of research, I know that many of the claims about the impact of 'global warming' in Bangladesh, for example, are completely unfounded. There is no evidence that flooding has increased at all in recent years."*

Pal Brekke is a Norwegian solar physicist. *"Anyone who claims that the debate is over and the conclusions are firm has a fundamentally unscientific approach to one of the most momentous issues of our time."*

William M. Briggs is a climate statistician. *"After reading [UN IPCC chairman] Pachauri's asinine comment [comparing skeptics to] Flat Earthers, it's hard to remain quiet."*

Jim Buckee is an astrophysicist from Oxford University, and has lectured on climate change at the University of Aberdeen. *"I think (climate* skepticism) is the dominant view in professional science circles. I know lots of people in universities and so on and quite often they have to retire before they can say what they want because it's so frowned upon."

Nigel Calder is an independent author and television scriptwriter, and was the script consultant for "The Cloud Mystery" (2008). *"It's likely that CO2 has some warming effect, but real proof of that hypothesis is tricky. You have to confirm by observation exactly how the CO2 changes the situation at different altitudes in the atmosphere and in different regions of the world."*

Mark L. Campbell is a professor of chemistry at the U.S. Naval Academy in Annapolis, MD. *"I don't even grant that there is a consensus among scientists; I"s just that the press only promotes the global warming alarmists and ignores or minimizes those of us who are skeptical. To many of us, there is no convincing evidence that carbon dioxide produced by humans has any influence on the Earth's climate."*

Patrick Carroll is a retired Environment Canada meteorologist. *"The IPCC theory of anthropogenic warming is a hoax that is rapidly falling into disfavour among atmospheric scientists."*

Bob Carter was an English paleontologist, and head of the School of Earth Sciences at James Cook University in Australia. *"Climate change shows all the hallmarks of orchestrated propaganda"*

Phil Chapman is an Australian geophysicist and former NASA astronaut. *"All those urging action to curb global warming need to take off the blinkers and give some thought to what we should do if we are facing global cooling instead."*

Randy Cerveny is a geography professor at Arizona State University. *"I don't think [global warming] is going to be catastrophic... Hopefully, our grandkids are going to have a lot better weather information than we did."*

Alan Cheetham is an engineer with 30 years' experience of data analysis, modeling and statistics and runs the website "Global Warming Science". *"My knowledge and training enables me to scientifically evaluate the data and the scientific studies. When I began to look into the science behind the global warming issue, I started to realize that the scientific debate is not over (the political debate may be over, but it shouldn't be) - because the science doesn't match the fearful scenarios portrayed by the media."*

George Chilingar is an American-Armenian Professor of Civil and Petroleum Engineering at the University of Southern California (USC). *"recent global warming of Earth's atmosphere is not due to an increase in anthropogenic carbon dioxide emission but rather to long-term global factors."*

Antonis Christofides (and Nikos Mamassis) of the National Technical University of Athens, Greece. *"We maintain there is no reason whatsoever to worry about man-made climate change, because there is no evidence whatsoever that such a thing is happening."*

John Christy is the state climatologist at the University of Alabama in Huntsville (UAH), and makes public the raw data from satellites. *"I see neither the developing catastrophe nor the smoking gun proving that human activity is to blame for most of the warming we see."*

Petr Chylek (PhD Physics) is a researcher for Space and Remote Sensing Sciences at Los Alamos National Laboratory. *"A common view on the current climate change (global warming) is that it is a result of fossil fuel burning and the following increase in atmospheric concentration of carbon dioxide. In reality there are several factors that produce the current climate change."*

Charles Clough is an atmospheric scientist and Chief of the Atmospheric Effects Team with the Department of the Army. *"Government officeholders at federal and state levels assume that current global warming is chiefly, if not entirely, due to mankind's growing carbon dioxide emissions, but they have not examined the science enough."*

Tom Coburn (former US senator) claimed that sea level rise had been no more than 5 mm in 25 years, and asserted there was now global cooling. In 2013 he said *"I am a global warming denier. I don't deny that."*

Roger W. Cohen retired from ExxonMobil as manager of strategic planning. *"I was well convinced, as were most technically trained people, that the IPCC's case for Anthropogenic Global Warming (AGW) is very tight. However, upon taking the time to get into the details of the science, I was appalled at how flimsy the case really is."*

Tim Coleman is an atmospheric scientist. *"It has now gotten to the point that almost any major weather event is blamed by someone on global* climate change... No one mentions the possible advantages of AGW, if it is occurring."

Rosa Compagnucci is an Argentinian El Niño expert and former IPCC author. *"There was a global warming in medieval times, during the years between 800 and 1300. And that made Greenland, now covered with ice, christened with a name (by the Vikings) that refers to land green: 'Greenland'."*

Piers Corbyn is an English weather forecaster and astrophysicist, who owns WeatherAction. He describes Global warming as... *"utter nonsense"*, and a *"false narrative"*. He stated... *"The whole thing is a manufactured protest and the government wants to hear these things because they have some sort of agenda of increasing taxation."*

Vincent Courtillot is a French professor of geophysics at the University Paris-Diderot. He believes that solar cycles control the climate by influence on cloud formation (the cosmic ray theory of Svensmark et al).

Uberto Crescenti is a geology professor of the University G. d'Annunzio in Italy and is past president of the Society of Italian Geologists. *"I am very glad to sign the U.S. Senate's report of scientists against the theory of man-made global warming... I think that climatic changes have natural causes according to geological data."*

F. James Cripwell is a physicist and former scientist with the UK's Cavendish Laboratory in Cambridge. *"Whatever is causing warming, it is not an increase in levels of carbon dioxide."*

Susan Crockford is a Canadian Zoologist, and adjunct professor in Anthropology at the University of Victoria. She is a specialist in polar bears, and is featured in section 14.3.

Matthew Cronin is a biologist at the University of Alaska, Fairbanks. *"We don't know what the future ice conditions will be, as there is apparently considerable uncertainty in the sea ice models regarding the timing and extent of sea ice loss."*

Javier Cuadros is an Earth Scientist of the UK Natural History Museum and has published more than 30 scientific papers. *"Curiously, it is a feature of man-made global warming that every fact confirms it: rising temperatures or decreasing temperatures, drought or torrential rain,* *tornadoes and hurricanes or changes in the habits of migratory birds. No matter what the weather, some model of global warming offers a watertight explanation... For a scientist like me, this sounds fishy."*

Claude Culross is an organic chemist from Baton Rouge, and signed the ICSC Manhattan Declaration. *"There is no proof that man-made carbon dioxide causes additional warming, or that carbon-dioxide reduction would reduce warming. Problem mitigation and conservation are the right approach."*

Walter Cunningham is an ex NASA astronaut from Apollo 7. *"NASA should be at the forefront in the collection of scientific evidence and debunking the current hysteria over human-caused, or Anthropogenic Global Warming (AGW). Unfortunately, it is becoming just another agency caught up in the politics of global warming, or worse, politicized science."*

Judith A. Curry is an American climatologist, former Alarmist, and former chair of the School of Earth and Atmospheric Sciences at the Georgia Institute of Technology. On Alarmism, she admits... *"I came to the realization that I had fallen into the trap of 'group-think'. I had accepted the consensus based on second order evidence."*

Dave Dahl is a chief TV (ABC) meteorologist. *"Many peer-reviewed scientific papers are now looking at the real possibility that the sun may play the main role in climate variation here on earth."*

Peter Dailey is director of Atmospheric Science at Boston based AIR Worldwide. *"There is now a near consensus that global air temperatures are increasing, however, there is no consensus on how this has affected the temperature of the world's oceans, and in particular in the Atlantic Ocean, or how much of the recent warming trend is attributable to man's activities."*

Robert DeFayette is a chemist and nuclear engineer formerly with NASA and the Nuclear Regulatory Commission (NRC). *"I also strongly object to the IPCC and its use of so-called 'experts'... Civilization may one day cease to exist but it won't be from global warming caused by CO2."*

Chris de Freitas was a New Zealand climatologist at the University of Auckland, New Zealand, and author of over 200 related articles.

Joseph D'Aleo is a past Chairman American Meteorological Society's Committee on Weather Analysis and Forecasting, former Professor of Meteorology, Lyndon State College, and founder of the ICACAP blog. *"Sunspot cycles and their effects on oceans correlate with climate changes. Studying these and other factors suggests that a cold, not warm, climate may be in our future."*

Fred W. Decker is a professor of meteorology at Oregon State University. *"There is no convincing scientific evidence that human release of carbon dioxide, methane, or other greenhouse gases is causing or will, in the foreseeable future, cause catastrophic heating of the Earth's atmosphere and disruption of the Earth's climate."*

David Deming is a geologist and geophysicist, and an associate professor at the University of Oklahoma. *"When compared to the period of time over which human civilization rose, present day temperatures are colder than average. Even if mean global temperature were to rise another degree, it would still be colder than it has been for much of the last 10,000 years."*

Roger Dewhurst is a retired geologist. *"All the warm periods prior to the current one have been warmer (in most cases substantially warmer) than the modern warming that we are having hysterics about. There have been five significant Little Ice Ages scattered between these warm periods. If you can explain how man-made carbon dioxide was the temperature driver for these events please go ahead and tell me."*

Jack Dini is a materials engineer at Lawrence Livermore National Labs. *"Thirty years ago we were supposedly headed into a cooling cycle akin to the Little Ice Age. Now it's an unprecedented heating cycle. If you ask me, that's an awfully quick time for a flip-flop on the weather."*

William DiPuccio is a retired weather forecaster for the US Navy and former meteorologist for the National Weather Service. *"We should be cautious about placing our faith in climate models that vastly oversimplify the actual climate system. Supporting evidence for the IPCC's projections does not warrant the high level (90%-95%) of confidence exhibited by its authors."*

Hal Doiron is a retired NASA physicist. *"You don't make critical decisions based on 'garbage in, garbage out.' Yet our government has been doing that with respect to climate alarm, because too many academics in universities* *are writing papers, drawing conclusions from models that don't agree with physical data."*

Delgado Domingos is a Portuguese environmental scientist, and founder of the Numerical Weather Forecast group. *"Creating an ideology pegged to carbon dioxide is a dangerous nonsense... The present alarm on climate change is an instrument of social control, a pretext for major businesses and political battle. It became an ideology, which is concerning."*

Art V. Douglas is an atmospheric scientist and former Chair of the Atmospheric Sciences Department at Creighton University in Omaha, Nebraska. *"Whatever the weather, it's not being caused by global warming. If anything, the climate may be starting into a cooling period."*

Diane Douglas is a climatologist and paleoclimate researcher who has authored or edited over 200 technical reports. *"The recent 'panic' to control GHG (greenhouse gas) emissions and billions of dollars being dedicated for the task has me deeply concerned that US and other countries are spending precious global funds to stop global warming, when it is primarily being driven by natural forcing mechanisms."*

David Douglass is a professor of physics in experimental condensed matter physics at the University of Rochester. *"Have the climate models been successful in predicting anything? They, of course, predict substantial global warming. This is not surprising given the expressed belief of some of the model builders in the global warming hypothesis and the many parameters in the model that need to be introduced."*

Paul Drallos is a working physicist at Sandia National Labs in Albuquerque and formed Plasma Dynamics Corporation. *"We must conclude that there is no logical basis on which to believe that theory of human caused global warming... global warming caused by human activity? There is good correlation between temperature and solar variations. There is historical evidence of natural climate cycles. There is historical evidence that changes in temperature cause changes in atmospheric CO2. There is scientific basis for each of these observations."*

Nicholas Drapela is a chemist at Oregon State University Chemistry Department. *"My dear colleague Professor Hansen, I believe, has finally gone off the deep end. When you have dedicated the bulk of your career to a cause, and it turns out the cause has been proven false, most people cannot bring themselves to admit the truth."*

John Droz, Jr. is a physicist and environmental activist. *"There are other theories that have been put forth by very qualified scientists, and these alternative explanations are supported by significant scientific data. These other theories also have their weak points, but the fact is that they do explain some facets of the Climate Change situation better than the Global Warming Theory does."*

Geoffrey G. Duffy is a New Zealander professor of Chemical and Materials Engineering. *"Even doubling or tripling the amount of carbon dioxide will virtually have little impact, as water vapour and water condensed on particles as clouds dominate the worldwide scene and always will."*

Jonathan DuHamel is a registered geologist of Arizona with a masters degree. *"I am a geologist familiar with the scientific literature on climate change, but I have yet to see any proof or compelling evidence supporting the assertion that human carbon-dioxide emissions have produced measurable temperature change."*

Freeman Dyson is an award winning English/American theoretical physicist and mathematician. *"The (climate) models solve the equations of fluid dynamics, and they do a very good job of describing the fluid motions of the atmosphere and the oceans. They do a very poor job of describing the clouds, the dust, the chemistry and the biology of fields and farms and forests. They do not begin to describe the real world we live in."*

Don Easterbrook is a professor emeritus of Geology at Western Washington University. *"The total increase in global warming for the century should be ~0.3°C, rather than the catastrophic warming of 3-6°C (4-11°F) predicted by the IPCC."*

Ferdinand Engelbeen is a chemist and process engineer. *"To make a climate model, where a lot of parameters and reactions are not even known to any accuracy, for me seems a little bit overblown. And to speak of any predictive power of such models, which are hardly validated, is as scientific as looking into a crystal ball."*

Dave Epstein is a professor of meteorology at Framingham State. *"Man-made warming fears - one of the biggest myths of modern time. Long-term climate data indicate that world climate varies naturally, and those cycles* are the collective result of scores of interrelated variables, playing out either in consort or not."

Willis Eschenbach is a climate researcher who has published in Energy and Environment, and in Nature. *"I am definitely a critic of the IPCC, they are doing their job abysmally poorly. Rather than advance the cause of climate science, they impede it through their reliance on bad statistics, bad economics, and bad data."*

Robert Essenhigh is a professor of mechanical engineering. *"At 6 billion tons, humans are then responsible for a comparatively small amount - less than 5 percent - of atmospheric carbon dioxide... And if nature is the source of the rest of the carbon dioxide, then it is difficult to see that man-made carbon dioxide can be driving the rising temperatures. In fact, I don't believe it does."*

John R Etherington is an ecologist and author of the book, The Wind Farm Scam. *"We are making some of the most expensive global decisions ever, on the basis of what atmospheric physicist James Peden has described as 'computerized tinker toys with which one can construct any outcome he chooses'."*

Mike Fairbourne is a 40 year veteran TV meteorologist. *"There has been some warming of global temperatures in recent years ... there is still a pretty big question mark... Do we need to be wise stewards (of the Earth)? Absolutely. Do we have to pin everything that happens on global warming? No, we need to have cooler heads."*

Michael F. Farona is a chemist and biochemist at the University of Akron, Ohio. *"Data, numbers, graphs, trends, etc., are generally missing in supposedly scientific reports on global warming. These articles are usually long on opinions and short on hard data. Phrases such as 'scientists agree that' ... 'scientists doubt that' ... do not belong in a scientific article. There are more data in Michael Crichton's novel State of Fear than in all the global warming articles combined that I have read."*

Viv Forbes is an oil scientist/geologist and the chairman of Australian based "The Carbon Sense Coalition". *"There is no evidence that carbon dioxide in the atmosphere is driving surface temperature, and there is plenty of evidence to show that current levels of temperature and carbon dioxide are neither extreme nor of concern."*

Louis H. Fowler is an environmental scientist and physicist who worked with the Texas Air Control Board. *"I have a masters in physics and over 33 years' working in the fields of environmental science, monitoring air quality and overseeing modeling analyses from permit models to hazardous release impact models. Were I to submit such fraudulent types of reports to my clients or to the federal or state regulatory agencies I would be kicked out of my company. Too bad folks like (NASA's) James Hansen can continue lying at taxpayers' expense in efforts that will line his and Al Gore's pockets. Follow the money!"*

Patrick Frank is a chemist who has authored more than 50 peer-reviewed articles. *"There is no falsifiable scientific basis whatever to assert this warming is caused by human-produced greenhouse gasses because current physical theory is too grossly inadequate to establish any cause at all."*

Peter Friedman is a mechanical engineer at University of Massachusetts, Dartmouth. *"The IPCC 'policy summaries', written by a small group of their political operatives, frequently contradict the work of the scientists that prepare the scientific assessments. Even worse, some of the wording in the science portions has been changed by policy makers after the scientists have approved the conclusions."*

Lloyd C. Furer served as a meteorologist for the U.S. Air Force and has authored more than 35 publications. *"I have been told that there is a list of about 650 scientists who have doubts the human-produced CO2 has little or nothing do with the current global warming. You can certainly list me as a skeptic."*

Cyril Galvin is a coastal engineer wrote an analysis for the American Geophysical Union (AGU). *"The prevailing assumption in the press, in legislative considerations, and in climate science is that global warming causes seal level rise, which causes beach erosion… Nowhere on the sandy ocean shores of the world is there a beach whose erosion has been documented to be caused by sea level rise."*

Jym Ganahl is a TV meteorologist and the youngest person to be granted the American Metrological (AMS) Seal of Approval. *"Just wait 5 or 10 years, and it will be very obvious. They'll have egg on their faces… It is remarkable how many people are being led like sheep in the wrong direction… Sunspots - and not carbon emissions - are to blame for the slow warming of the globe. It has nothing to do with us. When there are* *sunspots, like freckles on the sun - dark spots - these are like turning on a furnace and the earth warms. When there are no sunspots, it is like the furnace is in standby and the earth cools."*

David Gee is a geologist and was chairman of the science committee of the 2008 International Geological Congress. *"For how many years must the planet cool before we begin to understand that the planet is not warming? For how many years must cooling go on?"*

Ivar Giaever is a Norwegian-American physicist who shared the Nobel Prize in Physics in 1973. *"Is climate change pseudoscience? If I'm going to answer the question, the answer is: absolutely."*

William C. Gilbert is a published research chemist. *"I am ashamed of what climate science has become today. The science community is relying on an inadequate model to blame CO2 and innocent citizens for global warming in order to generate funding and to gain attention. If this is what science has become today, I, as a scientist, am ashamed."*

Jeffrey A. Glassman is a physicist and former Division Chief Scientist for the Hughes Aircraft Company. *"The consensus of climate mistakenly attributes solar wind warming to man-made carbon dioxide... CO2 concentration is a response to the proxy temperature in the Vostok ice core data, not a cause."*

Stanley Goldenberg is an Atmospheric Scientist of the Hurricane Research Division of NOAA. *"It is a blatant lie put forth in the media that makes it seem there is only a fringe of scientists who don't buy into anthropogenic global warming."*

Harry A. Gordon is a retired meteorologist formerly of the National Weather Service. *"A personal examination of a 100-year period of weather in Kansas City showed a continuous series of short-term warming and cooling periods. Studies from China covered more than a thousand years and confirmed this. No cycles have been discovered that would help in forecasting climate changes."*

Thomas L. Gould is an award-winning chemical engineer with the Society of Petroleum Engineers. *"Even if you accept the alarmist view that the seas will rise and this is a 'Planetary Emergency', why do we think that we can solve this problem with climate control, costing $10's of Trillions? We need to change the debate, and not let the alarmists set the agenda."*

William K. Graham is a professional engineer. *"For a theory to be scientific, it must be testable and falsifiable. The theory of global warming is being tested and data proves it is coming up short."*

Thomas B. Gray is a meteorologist and the former head of the Space Services branch at NOAA. *"I am sure that the concept of a 'Global Temperature' is nonsense... The claims of those convinced that AGW (anthropogenic global warming) is real and dangerous are not supported by reliable data."*

Kenneth P. Green is an environmental scientist of the American Enterprise Institute. *"While I believe that Earth has experienced a mild, non-enhanced greenhouse warming which will continue in the foreseeable future, I think the chaotic nature of the climate system makes projections of the future climate no better than science fiction."*

Guido Guidi is an Italian Air Force meteorologist who managed weather stations. *"If the temperature does not increase again, I see it getting hard for those who support the theory of man-made global warming."*

Kenneth A. Haapala is a quantitative economist and past president of the Philosophical Society of Washington. *"The quantitative models used by the IPCC to make predictions are unreliable and biased in greatly overestimating the influence of carbon dioxide emissions on warming."*

James A. Haigh is an environmental engineer of 36 years' experience. *"My greatest concern is that national policy decisions should not be made in a media-induced 'State of Fear'. Concentrating on CO2 as the sole source of immediate and dire global climate change diverts the focus from the true concern, environmental compliance."*

William Happer is a physicist and adviser to Presidents Bush and Trump. *"I am convinced that the current alarm over carbon dioxide is mistaken... Fears about man-made global warming are unwarranted and are not based on good science."*

Cliff Harris is a Canadian climatologist of the Long Range Weather service. *"In the past 10 years, especially the past couple of years, the Earth's climate has begun to cool, even though CO2 emissions have soared on a worldwide scale. How many years of declining temperatures will it take to finally break up Al Gore's 'global warming consensus'?"*

Dirck T. Hartmann is an aerospace engineer/physicist who worked on the NASA Apollo Space Program. *"Our mainstream media uses every opportunity to hype the hoax of man-made global warming by repeated reporting of data and events that appear to support it, and ignoring those that contradict it... Hopefully man made global warming will come to be recognized for the hoax it truly is."*

Jon Hartzler is a retired science professor from St. Cloud State University in Minnesota. *"The Chinese laugh at the Kyoto Protocol and the 'civilized' world trying to fix 'global warming'. Our puny little effort (but very costly) when China refuses and puts their economy first, makes us seem insignificant."*

Klaus P. Heiss is an ex-NASA space engineer, and ex-US Atomic Energy Commission. *"The entire atmospheric carbon dioxide, of which man-made CO2 is only a fraction of, is not to blame for global warming."*

Marc Hendrickx is an Australian professional geologist, and author of the book, A Climate Change Story For Little Skeptics. *"We're not scared anymore Mr. Gore!... The contention that recent rises in global temperature as measured by satellites are due solely to increased concentrations of CO2 from anthropogenic sources is misplaced."*

Victor Manuel Velasco Herrera is a Mexican geophysicist. *"The models and forecasts of the UN IPCC are incorrect because they only are based on mathematical models and presented results at scenarios that do not include, for example, solar activity."*

Louis A.G. Hissink is an Australian field geologist, and editor of Australian Institute of Geoscientists Newsletter. *"The assumption that humanity, from its burning of hydrocarbons, is raising the surface temperature of the earth by affecting its greenhouse effect, is not supported by theory nor the physical evidence."*

Wayne Hocking is a Canadian physics professor who heads the Atmospheric Dynamics Group. *"Maybe in 10 years' time, it'll all start to freeze over, we just don't know."*

Doug L. Hoffman is a mathematician, computer programmer, and engineer, worked on environmental models and taught at the University of Central Arkansas. *"More often than not, it is outside observers* uncovering the errors, casting doubt on the trustworthiness of all science. This is exactly what happens when you base your arguments on 'consensus science' and not scientific fact."

Dennis Hollars is an astrophysicist from New Mexico State University. "What I'd do with the IPCC report is to put it in the trash can because that's all it's worth... carbon dioxide... rather than being the purveyor of doom it is currently viewed as today, it is needed in order for plants to grow."

Gary M. Hoover is a physicist with operational experience in atmospheric energy absorption, nuclear reactor operations and exploration geophysics. "*Temperature proxy measurements going back hundreds of thousands of years through many ice ages and warm periods indicate atmospheric carbon dioxide concentration to be a result of temperature change and not a cause.*"

Jerome Hudson is a physicist who studies aperture synthesis and optics. "*I am not convinced that CO2 is any threat to the environment, nor am I convinced of any catastrophic warming. You can classify me as a skeptic.*"

William Hunt is a research scientist and has worked for NOAA. "*Scientists and activists alike have jumped on the (global warming) bandwagon. It's become a fad, a trend, a wave of enthusiasm and the scientists are going along with the fad to get research grants and the media limelight.*"

A. Neil Hutton is a former Assistant Chief Geologist for the Western Canadian Basin. "*On the whole, the media have done a remarkably poor job in reporting on global warming. Typically, the reports have been a simple regurgitation of the spin produced by the Intergovernmental Panel on Climate Change (IPCC).*"

Craig D. Idso is a science adviser to the Science and Public Policy Institute. He claims that rising CO2 levels will have mainly positive environmental effects.

Kiminori Itoh is an award-winning Japanese environmental physical chemist, and ex IPCC. Said that warming fears are the "*worst scientific scandal in the history...When people come to know what the truth is, they will feel deceived by science and scientists.*"

Terry Jackson is a physicist and teacher of 30 years' experience. *"There is ample scientific proof that CO2 pollution levels follow temperature rises and not the other way round... The earth has been much warmer in the past than it is today. CO2 concentrations have been much higher from 1885 to 1961 than they are today."*

Steven M. Japar is an atmospheric chemist who worked on the (IPCC) Second (1995) and Third (2001) Assessment Reports. *"Are there other possibilities to explain the (unverified) temperature increase of the last 40 years? Yes! Current warming is consistent with the 300 year trend... Changes in solar activity could explain much of it."*

Hans Jelbring is a Swedish Climatologist of the Paleogeophysics & Geodynamics Unit at Stockholm University. *"The dysfunctional nature of the climate sciences is nothing short of a scandal. Science is too important for our society to be misused in the way it has been done within the Climate Science Community. The global warming establishment has actively suppressed research results presented by researchers that do not comply with the dogma of the IPCC."*

Claes Johnson, Swedish professor of Mathematics. *"AGW alarmism is based on an idea of back radiation or re-radiation from an atmosphere with greenhouse gases, but the physics of this phenomenon remains unclear."*

Daniel P. Johnson is a chemist and engineer. *"I still must admit to a strong relationship between atmospheric CO2 and global warming since 1970 but feel, based on the above, that it is only one potential contributor to global warming and that the changing geomagnetic field is another major player in what has occurred during this same period."*

Norm Kalmanovitch is a Canadian professional geophysicist. *"The temperature record shows that the global temperature has been increasing naturally at a rate of about 0.5°C/century since the Little Ice Age. The forcing parameter is based on the full measured 0.6°C/century without subtracting the natural warming of 0.5°C/century giving a forcing parameter that is 6 times larger than can be attributed to the measured increase in CO2."*

Andrei Kapitsa is a Russian geographer and Antarctic ice core researcher. *"The Kyoto theorists have put the cart before the horse. It is* global warming that triggers higher levels of carbon dioxide in the atmosphere, not the other way round."

Al Kaprielian is a veteran meteorologist, forecasting for 25 years. *"We don't have enough data right now. We'll have to wait and see what future weather brings."*

Geoffrey Kearsley is a geographer and director of Wilderness Research Foundation. *"Science is rarely determined or finalized; science evolves and the huge complexity of climate science will certainly continue to evolve in the light of new facts, new experiences and new understandings."*

Richard Alan Keen is a meteorologist and weather observer. *"Earth has cooled since 1998 in defiance of the predictions by the UN-IPCC..."*

J.R. Kirtek, award-winning Meteorologist of Michigan's WJRT Channel 12. *"If the definition for a documentary is 'presenting facts objectively without editorializing or inserting fictional matter, as in a book or film,' then I am confident that Al Gore's movie was not a documentary."*

Arnold Kling is an economist formerly of the Federal Reserve Board and Freddie Mac. *"I am worried about climate change. In one respect, I may be more worried than other people. I am worried because I have very little confidence that we know what is causing it."*

Paul C. (Chip) Knappenberger is a senior researcher and admin for the skeptical climate website www.WorldClimateReport.com. He states that the website's aim is to... *"Point out the weaknesses and outright fallacies in the science that is being touted as 'proof' of disastrous warming."*

Christopher J. Kobus, Associate Professor at Oakland University, specializes in alternative energy. *"In essence, the jig is up. The whole thing is a fraud. And even the fraudsters that fudged data are admitting to temperature history that they used to say didn't happen... Perhaps what has doomed the Climategate fraudsters the most was their brazenness in fudging the data."*

Tom Kondis is a chemist and a consultant with practical experience in absorption and emission spectroscopy. *"To support their argument, advocates of manmade global warming have intermingled elements of greenhouse activity and infrared absorption to promote the image that* *carbon dioxide traps heat near earth's surface like molecular greenhouses insulating our atmosphere. Their imagery, however, is seriously flawed."*

Takeda Kunihiko is vice-chancellor of the Institute of Science and Technology Research at Chubu University in Japan. *"CO2 emissions make absolutely no difference one way or another.... Every scientist knows this, but it doesn't pay to say so... Global warming, as a political vehicle, keeps Europeans in the driver's seat and developing nations walking barefoot."*

Kanya Kusano is a Japanese physicist and program director of the Japan Agency for Marine-Earth Science and Technology. *"I believe the anthropogenic (man-made) effect for climate change is still only one of the hypotheses to explain the variability of climate... It could take 10 to 20 years more research to prove or disprove the theory of anthropogenic climate change."*

Miroslav Kutilek is a professor of Soil Science and Soil Physics at Czech Technical University in Prague. *"The Earth climate change is influenced by eight factors: (1) Milanković cycles, (2) Solar activity, (3) Continental drift, (4) Greenhouse gases, (5) Thermohaline circulation, (6) Aerosols, volcanoes and asteroids, (7) Vegetation cover, (8) Earth magnetic field."*

Pierre R Latour is an ex-NASA chemical process control and has published more than 70 publications. *"Water vapor and clouds are a powerful link between weather and climate, but historic data for modeling is absent. Water vapor is the dominant greenhouse gas, the inconvenient truth often neglected by AGW alarmists."*

Robert B. Laughlin, physicist, won the Nobel Prize for physics in 1998. *"Please remain calm: The Earth will heal itself. Climate is beyond our power to control...Earth doesn't care about governments or their legislation. You can't find much actual global warming in present-day weather observations. Climate change is a matter of geologic time, something that the earth routinely does on its own without asking anyone's permission or explaining itself."*

Peter R. Leavitt is a meteorologist and President of Weather Information, Inc. *"Skepticism in regard to AGW (Anthropogenic Global Warming) does not mean that the opposite is true, only that there is insufficient hard evidence to conclude that AGW is a significant factor in climate if it is a factor at all. Progress in science is driven by skepticism."*

Kevin Lemanowicz is a meteorologist from Fox 25 TV in Massachusetts. *"I continue to say that we have obviously warmed, but we should not be setting policy based on an uncertain climate future."*

Anatoly Levitin is the head of geomagnetic variations laboratory at the Institute of Terrestrial Magnetism in Russia. *"The energy mankind generates is so small compared to that overall energy budget that it simply cannot affect the climate... The planet's climate is doing its own thing, but we cannot pinpoint significant trends in changes to it because it dates back millions of years while the study of it began only recently. We are children of the Sun; we simply lack data to draw the proper conclusions."*

Geraldo Luís Lino is a Brazilian geologist who authored the 2009 book —The Global Warming Fraud. *"Hundreds of billion dollars have been wasted with the attempt of imposing an Anthropogenic Global Warming (AGW) theory that is not supported by physical world evidences... AGW has been forcefully imposed by means of a barrage of scare stories and indoctrination that begins in the elementary school textbooks."*

Al Lipson is a meteorologist and former lead forecaster at the Weather Channel and Accuweather. *"I don't doubt that climate changes. In that, there is no dispute. However, I must join the ranks of many scientists who dispute that global warming is taking place at such a rate that it will have apocalyptic consequences the alarmist theorize."*

Philip Lloyd is a South African nuclear physicist and chemical engineer. *"The quantity of CO2 we produce is insignificant in terms of the natural circulation between air, water and soil... I am doing a detailed assessment of the UN IPCC reports and the Summaries for Policy Makers, identifying the way in which the Summaries have distorted the science."*

Gerhard Lobert is a German physicist and recipient of The Needle of Honor of German Aeronautics. *"As the glaciological and tree ring evidence shows, climate change is a natural phenomenon that has occurred many times in the past, both with the magnitude as well as with the time rate of temperature change that have occurred in the recent decades."*

Keith Lockitch is a researching physicist in science and environmental issues for the Ayn Rand Institute. *"Despite the constant assertion that global-warming science is 'settled', it is far from certain that we face any sort of catastrophic global emergency."*

Craig Loehle is a climate researcher formerly of the Department of Energy Laboratories. *"The 2000-year (temperature) trend is not flat, so a warming period is not unprecedented."*

John Lott Jr. is a senior research scientist at the University of Maryland. *"Are global temperatures rising? Surely, they were rising from the late 1970s to 1998, but 'there has been no net global warming since 1998.' Indeed, the more recent numbers show that there is now evidence of significant cooling."*

Jon Loufman is a TV meteorologist in Cleveland, Ohio. *"Climate records also show that long before industrialization, the Vikings had settled in Greenland because it was warm enough . . . I think the jury is still out on this."*

William "Bill" Lyons is a retired USAF Meteorologist at Global Weather Central at Strategic Air Command (SAC) Headquarters. *"I decry the creation of the fear mentality which many seem to use to prevent rational study and encourage irrational actions. I am pleased to be considered a 'denier' in this cause if this puts me in the class with those who defied prevailing 'scientific consensus' that the earth was flat and that the earth was the not the center of the universe."*

Allan M.R. MacRae is a Canadian professional engineer with a degree from Queen's and the University of Alberta. *"The IPCC's position that increased CO2 is the primary cause of global warming is not supported by the temperature data."*

Pavel Makarevich is a biologist of the Biological Institute of the Russian Academy of Sciences. *"There are clear cycles during which both temperature and salinity rise and fall. These cycles are related to solar activity... In my opinion and that of our institute, the problems connected to the current stage of warming are being exaggerated. What we are dealing with is not a global warming of the atmosphere or of the oceans."*

Nikos Mamassis (and Antonis Christofides) of the National Technical University of Athens, Greece. *"We maintain there is no reason whatsoever to worry about man-made climate change, because there is no evidence whatsoever that such a thing is happening."*

Francis T. Manns is a Canadian geologist. *"CO2 dissolves in cold water and bubbles out of warm water. That's why CO2 trails natural warming."*

Oliver K. Manuel is a nuclear chemist who has authored more than 100 scientific papers and published research in peer reviewed literature. *"Compared to solar magnetic fields, however, the carbon dioxide production has as much influence on climate as a flea has on the weight of an elephant."*

Luigi Mariani is a professor of the Dept. of Crop Science at the University of Milan, who has authored or co-authored more than 50 peer-reviewed studies. *"The sheer number of contradictions in Gore's charts vs. words makes that all too evident. The graph showing CO2 concentrations over the last 650,000 years alongside temperatures clearly indicates increased CO2 is an effect of warming and not a cause."*

James A. Marusek is a retired U.S. Navy physicist/engineer. *"The anthropological global warming (AGW) hypothesis would have us believe that global temperatures are rising as a result of increased carbon dioxide levels in Earth's atmosphere and that humans are the primary cause of this increase. An opposing hypothesis - natural global warming (NGW) - believes the rise in recently observed atmospheric carbon dioxide levels is driven by natural global warming and by volcanic activity and that humans have little effect in altering Earth's climate."*

Shigenori Maruyama is a Japanese geologist and professor at the Tokyo Institute of Technology's Department of Earth and Planetary Sciences. *"Our nation must pay huge amounts of money to buy carbon discharge rights... This is not reasonable, but meaningless if global cooling will come soon -- scientists will lose trust."*

William F. McClenney is a professional geologist. He states he has done... *"the math and realized that you just can't get to global warming with CO2."*

Mike McConnell is a hydrologist and geologist in the U.S. Forest Service. *"Climate change is a climate system that we have no real control over... Our understanding on the complexities of our climate system, the Earth itself and even the sun are still quite limited."*

Tom McElmurry is a USAF meteorologist and a former tornado forecaster in Kansas City. *"The money we are about to spend on drastically* *reducing carbon dioxide will line the pockets of the environmentalists who have such expertise readily available at the right price. And some politicians are standing in line to fill their pockets with kick back money..."*

Ian McQueen is a chemical engineer a research analyst at Eight Capital, Research Division. *"Carbon dioxide is not the bogeyman - there are other causes that are much more likely to be causing climate change, to the extent that it has changed."*

Jean Meeus is a Belgian meteorologist/astronomer and has authored numerous books. *"My own impression, however, is that the number of those 'remaining' skeptics is increasing! Al Gore has exaggerated, and now comes the reaction."*

Seymour Merrin is a geologist and a research scientist. *"The basic theory states that, as human-produced CO2 increases, Earth temperatures increase. Simple prediction, simple to test. The predominance of 'hottest' years should be in the last 20 years, but that is not true. In fact, the last 10 years have been relatively flat — with 2007 and 2008 having declining temperatures. No correlation at all — actually, a disproval of that particular theory."*

Grant Miles was a member of UK Atomic Energy Authority Chemical Separation Plant Committee. *"There is no credible evidence of the current exceptional global warming trumpeted by the IPCC and there can be no such thing as an average world temperature... The IPCC is no longer behaving as an investigative scientific organization or pretending to be one. It is now showing its true colors in its use of fantasy and propaganda to advance its environmental socialist agenda."*

Timothy R. Minnich is an atmospheric scientist of more than 30 years' experience in design and management. *"Historically, the claim of consensus has been the first refuge of scoundrels; it is a way to avoid debate by claiming that the matter is already settled... Let's be clear: the work of science has nothing whatever to do with consensus... The greatest scientists in history are great precisely because they broke with the consensus."*

Ferenc Miskolczi is a Hungarian scientist and ex-NASA atmospheric physicist. *"Unfortunately, my working relationship with my NASA supervisors eroded to a level that I am not able to tolerate. My idea of the freedom of science cannot coexist with the recent NASA practice of handling new climate change related scientific results."*

Lord Christopher Monckton is recognized as one of the Denier greats. He has toured and talked restlessly for many years and is the inventor of the mathematical puzzle, Eternity. *"The sun has more to do with this than meets the eye."*

Andrew Montford is an English chemist and author of, *The Hockey Stick Illusion*. *"I believe that CO2, other things being equal, will make the planet warmer. The six million dollar question is how much warmer."*

Gregory W. Moore is an aerospace and mechanical engineer who has authored/co-authored more than 75 publications, and the 2001 Version of the NASA Space Science Technology Plan. *"There is definitely no sufficient data which would create a basis for making any kind of conclusions either way... There is no sufficient data in existence anywhere which would allow for making any kind of definite statements on the effect of natural mechanisms on the global warming phenomenon."*

Patrick Moore is an ex GreenPeace president. He is a popular speaker, and believes that global warming... *"has a much better correlation with changes in solar activity than CO2 levels".*

Marc Morano is a Denier with a terrier-like approach. He runs the website www.ClimateDepot.com, made the movie, *Climate Hustle*, and was one of the first people to break the Climategate email scandal.

Nils-Axel Morner is Swedish, and former head professor of paleogeophysics/geodynamics at Stockholm University. He is an open critic of the IPCC, and with 45 years' experience monitoring sea levels maintains they are not rising, if not falling.

Richard Mourdock is a licensed professional geologist and former field geologist. *"I'm scared to death about each of the three (U.S. presidential) candidates and their positions on global climate change."*

R. John Muench is an associate professor of chemistry of Heartland Community College in Illinois. *"Global warming alarmism is more religion than science. The believers have their messiah, Al Gore, who is not a scientist and has refused all monetary offers to debate the science."*

Vincent U. Muirhead was professor emeritus of aerospace engineering at the University of Kansas, teaching 'gas dynamics' for 28 years (he also developed a laboratory model of a tornado). *"The new green left (environmentalist) propaganda reminds me of the old red left (communist) propaganda. The dirty word is now carbon rather than capitalism. The game is simply to intrude and control everything. How much will the carbon tax be for each of us to breathe?"*

Mary Mumper is a chemist at Frostburg State University in Maryland. *"I am an environmentalist,"* but *"I must disagree with Mr. Gore".*

Michael J. Myers is an analytical chemist who specializes in spectroscopy and atmospheric sensing. *"I am troubled by the lack of common sense regarding carbon dioxide emissions. Our greatest greenhouse gas is water. Atmospheric spectroscopy reveals why water has a 95 percent and CO2 a 3.6 percent contribution to the 'greenhouse effect'."*

Nasif Nahle is a Mexican biologist, whose research focuses on solar influences and biology. *"We could fail if we think that the change of temperature was caused by the CO2 when the reality is that the Sun was what heated up the soil."*

Muriel Newman is a mathematician and member of the Northland Conservation Board. *"Around the world, as controversy over climate change continues to grow, it remains very clear that contrary to what the politicians tell us, not only is there is no consensus of scientific thought on this matter, but the science is certainly not settled."*

Jim Nibeck is an environmental chemist who worked in the biomedical research industry. *"My personal skepticism regarding anthropogenic global warming is based on the premise that man-made CO2 is a significant portion of environmental CO2 and that man-made CO2 is the prime cause of global warming."*

John Nicol was Chairman of the Australia Climate Science Coalition and a Lecturer of Physics at James Cook University. *"The claims so often made that there is a consensus among climate scientists that global warming is the result of increased man- made emissions of CO2, has no basis in fact... There is no evidence, neither empirical nor theoretical, that carbon dioxide emissions from industrial and other human activities can have any effect on global climate."*

Perry Ong is an ecologist/evolutionary biologist and the director of the Institute of Biology at the University of the Philippines. *"Climate change* has become a convenient excuse when there are other (environmental) issues that need to be addressed."

David Packham is a former research scientist with Australia's CSIRO. *"I find that I am uncomfortable with the quality of the science being applied to the global warming question."*

Dan Pangburn is a licensed engineer with masters in Mechanical Engineering. *"Most of earth's history carbon dioxide level has been several times higher than the present… The conclusion from all this is that carbon dioxide change does NOT cause significant climate change. Actions to control the amount of non-condensing greenhouse gases that are added to the atmosphere are based on the mistaken assumption that global warming was caused by human activity."*

Tony Pann is a TV meteorologist in Washington, DC. *"I believe the earth naturally goes through warming and cooling cycles. It could very well be that it's colder in five years. We just don't know. Anyone remember the Time magazine cover stories back in the 1970s about us going into a mini ice age?"*

Theodore G. Pavlopoulos is an ex-US Navy physicist/chemist. *"CO2 in air has been branded as the culprit for causing the greenhouse effect, causing global warming. However, regularly omitted is another important greenhouse gas also present in air and in much higher concentration. It is water vapor. In the air, it absorbs infrared radiation (heat) more strongly than CO2."*

James A. Peden is an atmospheric physicist formerly of the Space Research and Coordination Center in Pittsburgh. *"Many [scientists] are now searching for a way to back out quietly (from promoting warming fears), without having their professional careers ruined."*

Robert A. Perkins is an associate professor of environmental engineering at the University of Alaska Fairbanks. *"All the 'science' that you read about global warming is based on models, not observed facts."*

Charles Perry is a research hydrologist of the U.S. Geological Survey. *"Projections show the current warm period may be ending and that the earth's climate may cool to conditions similar to the Little Ice Age…"*

Doug Pettibone is a biologist/neuropharmacologist who has authored 120 scientific publications and holds ten patents. *"There is currently no satisfactory answer to the central question: 'What is the actual proof that humans are causing catastrophic global warming?' All of the climate computer models in the world do not provide the proof. It boils down to a matter of faith that the 30-year positive correlation between man-made CO2 and global temperature provides the proof. But correlations are not proof of cause and effect."*

Victor Pochat was a teacher of water resources planning at Universidad del Litoral in Argentina. *"It is not clear that increases of a few ° in average temperature of the planet is directly related to human activity but could be due to cyclical effects... Scientists that deserve credit for their background say global warming is a climatic variability associated to cycles of warming and cooling of the Earth."*

Boylan Point is a meteorologist/hurricane expert and retired U.S. Navy Flight meteorologist. *"A lot of folks have opinions in which they have nothing to back them up with. Mr. Gore I think may well fit into that category."*

Klaus Eckart Puls is a German physicist, meteorologist and was once an avid Alarmist. *"I became outraged when I discovered that much of what the IPCC and the media were telling us was sheer nonsense and was not even supported by any scientific facts and measurements. To this day I still feel shame that as a scientist I made presentations of their science without first checking it."*

Ed Rademacher is a licensed Professional Engineer. *"To date, global warming alarmists have not come close to providing any valid scientific data that proves humans are the sole source of changes in so-called global average temperatures. Quite simply, correlation between the carbon dioxide levels and the global average temperatures does not prove a causal relationship."*

Art Raiche is a former chief research scientist with Australia's CSIRO. *"The suppression of scientific evidence that contradicts the causal link between human-generated CO2 and climate has been of great concern to ethical scientists both here in Australia and around the world."*

Hilton Ratcliffe is a South African astrophysicist member of the Astronomical Society of Southern Africa (ASSA). *"The whole idea of* anthropogenic global warming is completely unfounded. There appears to have been money gained by Michael Mann, Al Gore and UN IPCC's Rajendra Pachauri as a consequence of this deception, so it's fraud."

John Reid, Atmospheric Physicist, worked with Australia's CSIRO's (Commonwealth Scientific and Industrial Research Organization). *"Global warming is the central tenet of this new belief system in much the same way that the Resurrection is the central tenet of Christianity. Al Gore has taken a role corresponding to that of St Paul in proselytizing the new faith...My skepticism about AGW arises from the fact that as a physicist who has worked in closely related areas, I know how poor the underlying science is. In effect the scientific method has been abandoned in this field."*

John Reinhard is a biologist/biochemist and member of the American Chemical Society who has published 76 papers. *"First, global warming is a theory. Nothing more. It lacks objective data that would normally be needed to advance a theory."*

George Reisman is an economist and Emeritus Professor at Pepperdine University. *"Global warming is not a threat. But environmentalism's response to it is."*

Colin Robinson is an economist and founder of the Department of Economics at the University of Surrey, UK. He has authored 25 books and 160 journal papers with a focus on energy policy. *"One does not have to be a 'climate change denier' to see that a degree of skepticism about the present consensus might be in order. In that sense, I think that the skeptics are right."*

Mark Rose is a surface chemist, Head of Environmental Quality at Qatar Petroleum and has developed the largest Purified Wet Acid Plant in the world. *"This human-centered view of changes in the earth's environment is challenged by many scientists, who argue that climate is, and always will be, controlled by the sun. Man's puny efforts at changing the climate is a mere sideshow compared with the unimaginably vast changes in the sun's solar output."*

Robert Rose is a professor of materials science/engineering at MIT with approximately 50 years of experience. *"Cooler heads (are) needed in global warming debate."*

Caleb Stewart Rossiter is a long-time teacher of statistical modeling. *"The models cited by the IPCC are still primarily statistical curve - fitting exercises rather than mathematical reenactments of the effects of the laws of physics on the interaction of molecules in the hurly-burly of a global system."*

Kenneth Rundt is a chemist and bio-molecule researcher and formerly a teacher at Abo Akademi University in Finland. *"My personal belief is that natural forcings have more importance than anthropogenic forcings such as the CO2 level... It can also be reliably inferred from palaeoclimatological data that no uncontrolled, runaway greenhouse effect has occurred in the last half billion years when atmospheric CO2 concentration peaked at almost 20 times today's value."*

Tom Russell, Chief Meteorologist of CBS affiliate in Harrisburg PA, is certified by both the American Meteorological Society and National Weather Association. *"Despite what you've heard, carbon dioxide is not a pollutant. It's kind of important to our survival. Research now shows that CO2 does not drive temperature but rather temperature drives CO2."*

Burt Rutan, named "100 most influential people in the world, 2004" by Time Magazine. *"Those who call themselves Green planet advocates should be arguing for a CO2-fertilized atmosphere, not a CO2-starved atmosphere... Diversity increases when the planet was warm AND had high CO2 atmospheric content...Al Gore's personal behavior supports a green planet - his enormous energy use with his 4 homes and his bizjet, does indeed help make the planet greener. Kudos, Al for doing your part to save the planet."*

Anthony J. Sadar is a consulting meteorologist and co-author of Environmental Risk Communication: Principles and Practices for Industry. *"Everyone has been conditioned to believe that an extremely complex climate system is largely controlled by a single simple gas - carbon dioxide - even though the biggest single climate regulator on Earth is most likely water."*

Nicola Scafetta is a physicist and active advocate for the Sun as the crucial factor in determining the cause of climate change.

W. M. Schaffer is a professor of Ecology and Evolutionary Biology at the University of Arizona. *"I am mistrustful of 'all but the kitchen sink' models that, by virtue of their complexity, cannot be analyzed mathematically.*

When we place our trust in such models, what too often results is the replacement of a poorly understood physical (chemical, biological) system by a model that is similarly opaque."

Jack Schmitt is an ex-NASA Apollo 17 Astronaut, Geologist and has walked on the moon. *"The 'global warming scare' is being used as a political tool to increase government control over American lives, incomes and decision making. It has no place in the Society's activities."*

Jerome J. Schmitt is an engineer who holds five patents and has authored 30 technical publications. *"The UN's 2001 Climate Change report distorted the historical record by eliminating the Medieval Warm Period in the famous 'Hockey Stick Curve' which, by many accounts, unreasonably accentuated temperature rise in the 20th century."*

Robert L. Scotto is an atmospheric scientist of 30 years air quality consulting experience. *"Proponents of AGW analyses of recent surface temperature records which are suspect at best, as they clearly contradict much more reliable satellite data... the Earth has been cooling since 1998."*

Frederick Seitz was a renowned physicist and former president of the National Academy of Sciences. *"The IPCC is pre-programmed to produce reports to support the hypotheses of anthropogenic warming and the control of greenhouse gases, as envisioned in the Global Climate Treaty."*

Marcel Severijnen is the former head of Environmental Monitoring Department of the Province of Limburg in the Netherlands. *"'Debate closed' is a deadly pitfall, unworthy to integer researchers. Any result of research, be it measurements or modeling should be open to confirmation or denial from other researchers. That is the only way to come closer to the real world."*

David Shelley is a former geologist from the University of Canterbury in New Zealand. *"The very idea that we can stop climate change is barking mad. Climate change is inevitable, as geology has always shown."*

Allan Shepard is former Chief Geologist for Amoco International. *"I conclude that there has been both slight warming and cooling over the past hundred years or more largely driven by the sun and its interaction with clouds, winds, ocean currents etc. CO2's contribution to any warming is negligible."*

Topper Shutt is an ex-CNN meteorologist. *"Some of the effects of global warming have been greatly exaggerated (when the ice cubes in your drink melt does you glass overflow?) and our money may be better spent exploring other avenues in addition to CO2 reduction."*

Allen Simmons is an ex-NASA engineer, and wrote computer systems software for the world's first weather satellites. *"During our research, we discovered a thing missing - hard science and how little of it made it into debate. Further surprise came from the lack of knowledge among the general public and, believe it or not, among some scientists actually involved in climatology. We discovered global warming is a topic much discussed but little understood. We called it 'Global Confusion'."*

Dr. Joanne Simpson is an Atmospheric Scientist, ex NASA. *"Since I am no longer affiliated with any organization nor receiving any funding, I can speak quite frankly.... As a scientist I remain skeptical... The main basis of the claim that man's release of greenhouse gases is the cause of the warming is based almost entirely upon climate models. We all know the frailty of models concerning the air-surface system."*

Siegfried (Fred) Singer trained as an atmospheric physicist and has written many books, papers and articles on global warming. *"One cannot simply argue that just because CO2 is a greenhouse gas it causes warming."*

Hajo Smit is a Dutch meteorologist and a former member of the Dutch UN IPCC committee. *"Gore prompted me to start delving into the science again and I quickly found myself solidly in the skeptic camp... Climate models can at best be useful for explaining climate changes after the fact."*

G LeBlanc Smith is a retired research scientist with Australia's CSIRO. *"I have yet to see credible proof of carbon dioxide driving climate change, yet alone man-made CO2 driving it. The atmospheric hot-spot is missing and the ice core data refute this. When will we collectively awake from this deceptive delusion?"*

Mike Smith is a consulting meteorologist and the CEO of WeatherData Services of Wichita Kansas. *"I don't know what 2009's or 2029's weather might bring, nor does anyone else. The sciences of meteorology and climatology still have a lot of learning to do."*

Robert Smith is a professor of chemistry at University of Nebraska at Omaha. *"The amount of carbon dioxide in comparison to the amount of* water that is affecting global warming is minimal, Smith says. Ice ages are eminent and the next one will happen in the next 2,000 years."

Ron Smith is a former chemistry professor, and the Director of International Relations and Security Studies at the University of Waikato in New Zealand. *"The consequence of suppressing the deviant view may not be simply that we remain in ignorance. It may be that we embark on policies that are likely to be very damaging to us and only marginally advantageous (if at all) to the wider global community."*

Glenn Speck is a chemist with 35 years of laboratory experience in the government and private sectors, testing air, water, fuel, and soil for pollutants and other chemicals. *"Although much of the liberal press and the liberal politicians endorse man-made global warming as a complete, irrefutable reality, there are a substantial number of us scientists that strongly disagree. There is little disagreement that some warming has likely occurred, but many of us think that most, if not all of that change is due to natural planetary processes."*

Roy Spencer is one of the Denier 'greats'. In 1991 he was awarded the NASA Exceptional Scientific Achievement Medal (with John Christy) for his work on the UAH satellite data feed. *"Recent global warming is one of many natural cycles of warming and cooling in geologic history."*

Jim Sprott is a consulting chemist/forensic scientist from Auckland, NZ. *"The projections of the IPCC are simplistic, superficial, and now proven wrong. The whole issue requires a fresh start, based on the mass of irrefutable data which has been assembled."*

Leighton Steward is an ex-NASA award-winning geologist. *"Gore won't debate this subject... but maybe he has come to understand it... maybe that's why he has cancelled the last six months of his tour."*

Peter Stilbs is a Swedish physical chemist, chair of the climate seminar Department of Physical Chemistry at the Royal Institute of Technology (KTH) in Stockholm, and author of more than 165 scientific publications. *"There is no strong evidence to prove significant human influence on climate on a global basis. The global cooling trend from 1940 to 1970 is inconsistent with models based on anthropogenic carbon dioxide emissions."*

David Stockwell is an ecological modeler who has published research articles on climate change. *"Two claims made in the IPCC Chapter 3 Section 3.4 p40 of WG1 are obviously false... Use of dubious evidence and false claims to support a theory indicates the degree of confirmation bias operating in global warming."*

Alex Storrs is an associate professor at the Department of Physics, Astronomy & Geosciences at Towson University, Maryland. *"I gave a talk at the event here (Towson Univ.) titled 'Science, Skepticism, and Global Warming', and am still walking upright. I pointed out how skepticism is central to the scientific enterprise and raised the question 'What if it's not CO2'?"*

Max S. Strozier is a chemist with 26 years in chemical lab analysis, and worked with the U.S. Department of Defense. *"Scientists across the globe are catching on - global warming is not real science. There is a sucker born every minute who believes in it, and Al Gore is playing the role of P.T. Barnum."*

Scott Sumner is a meteorologist from North Carolina. *"An increase in carbon dioxide (CO2) is not going to change the fact that EVERYTHING in weather and climate is cyclical. Nature is always trying to balance 'herself' with the current trend having been for less cold and snow during winter seasons and hotter and drier summers, but snow lovers do not fret."*

Brad Sussman is a meteorologist and past officer of the National Weather Association (NWA). *"Global warming has been happening on and off for millions of years. Millions of years when mankind wasn't driving around in SUVs and using coal for electric power! Believing that mankind is unequivocally responsible for global warming is the ultimate arrogance. Sorry to be humble, but we're not that special."*

Brian Sussman is a former member of the AMS Education Advisory Committee. *"Well fine. I'd rather be called a 'denier' than try to push a scheme that would make Karl Marx green with envy."*

Henrik Svensmark (M.S. and PhD in Physics) is well known for his theory of the effects of cosmic rays on cloud formation as a distinct and indirect cause of global warming. *"Fewer cosmic rays meant fewer clouds—and a warmer world."*

John Takeuchi is a Japanese author and writes for the Daily Republic. *"The atmosphere has periodic warming and cooling cycles. The sun is the primary source of energy impacting the earth's surface. That energy heats the land and the seas, which then warm the air above them. Water vapor and other gases in the atmosphere also affect temperature."*

John S. Theon is a retired NASA atmospheric scientist, and the recipient of many awards. *"I appreciate the opportunity to add my name to those who disagree that global warming is man-made."*

Sherwood Thoele is an analytical chemist and mathematician. *"I submit that there is no manmade global cooling/warming, that there is no study or research data that makes a good argument to that effect when carefully examined objectively and that the Earth has many different and wide ranging cycles that man cannot control, no matter how much he would like to."*

Mike Thompson is a former U.S. Navy meteorologist and TV broadcaster. *"It is easier to silence scientific dissent by utilizing the politics of personal destruction, than to actually debate them on the merits of their arguments. That should tell you something about the global warming debate... there is none right now... it's either you believe, or you are to be discredited."*

Wolfgang P. Thuene is a meteorologist and former forecaster for the German Weather Service. *"All temperature and weather observations indicate that the earth isn't like a greenhouse and that there is in reality no 'natural greenhouse effect' which could warm up the earth by its own emitted energy and cause by re-emission a 'global warming effect'.*

Frank Tipler is a mathematical physicist who has authored 58 peer-reviewed publications and five books. *"It is obvious that anthropogenic global warming is not science at all, because a scientific theory makes non-obvious predictions which are then compared with observations that the average person can check for himself."*

Earl F. Titcomb Jr. is a professional geologist and co-authored analyses of geological and seismological hazards. *"It is amazing to me, as a professional geologist, how many otherwise intelligent people have, as some may say, 'drunk the Al Gore Kool-Aid' concerning global climate change."*

Eduardo Tonni is an Argentinian award-winning paleontologist. *"The [global warming] scaremongering has its justification in the fact that it is something that generates funds."*

Tom Tripp is a lead author of the UN IPCC. *"We're not scientifically there yet. Despite what you may have heard in the media, there is nothing like a consensus of scientific opinion that this is a problem. Because there is natural variability in the weather, you cannot statistically know for another 150 years."*

Gary L. Troyer is a nuclear chemist and a Fellow Scientist at the Westinghouse Hanford Company. *"Scientists have long known that umbrellas do not cause rain. Similarly, there is significant historical evidence that warming causes increases in atmospheric carbon dioxide."*

Ralf Tscheuschner is a German theoretical physicist. *"as countless examples in history have shown, scientific consensus bears no resemblance whatsoever to scientific validity. Consensus is a political term, not a scientific one."*

Anastasios Tsonis is a mathematician and atmospheric scientist. *"Very often, when I talk to the public or the media about "global warming", they ask me if I believe in global warming. This unfortunate question is apparently very popular among scientists as well. And I say 'unfortunate' because when we are dealing with a scientific problem 'believing' has no place. In science we either 'prove' or 'disprove'.*

Noor Van Andel was a physicist and former head of research at Akzo Nobel. *"The warming by increased CO2 can only result from "increased back radiation" from the atmosphere to the surface, and for this the warming of the troposphere due to increased CO2 must be more than the surface warming. all models predict much more warming at 300 - 400 hPa compared to the surface warming trend. This is not observed."*

Fritz Vahrenholt is a German politician and energy executive with a doctorate in chemistry. He co-authored (with geologist Sebastian Lüning) the book, *Die kalte Sonne: warum die Klimakatastrophe nicht stattfindet* (The Cold Sun: Why the Climate Crisis Isn't Happening). The book proposes that climate change is driven by variations in solar activity.

William W. Vaughan is an ex-NASA atmospheric scientist with many awards. *"The cause of these global changes is fundamentally due to the Sun* and its effect on the Earth as it moves about in its orbit. Not from man-made activities... Also, I believe any governmental efforts to control global warming caused by nature will be essentially ineffective."

Jan Veizer is the Distinguished University Professor (emeritus) of Earth Sciences at the University of Ottawa. He has a Ph.D., in Isotope Geology, (Australian National University 1971), and a Ph.D., in Sedimentology and Structural Geology, (Slovak Academy of Sciences 1968). *"Empirical observations on all time scales point to celestial phenomena as the principal driver of climate, with greenhouse gases acting only as potential amplifiers."*

Frank Wachowski is a retired atmospheric scientist for the National Weather Service. *"Obviously, it appears that the Arctic is getting warmer and causing problems for polar bears and other animals. But are we doing it? We don't have a long-enough period to study yet... everything goes through cycles."*

Gary Walker is a professional geologist and member of the Canadian Society of Petroleum Geologists. *"I have done extensive research regarding the global warming debate and have come to the conclusion that while global warming is real (the earth has been warming since the end of the last ice age), the amount of this warming that has taken place in the last 150 years that has been caused by anthropogenic CO2 is very much uncertain."*

Haydon Walker is an Australian long-range weather forecaster and runs World Weather. *"I am disgusted with what we are putting into the atmosphere but I believe the climate change debate is too politically driven."*

Kevin Warwick is an award-winning Professor of Cybernetics at the University of Reading, England. *"Big thing here is – do we know what we are doing that is bringing about climate change? At present the answer to this is NO."*

Anthony Watts is an American blogger who runs the popular blog, Watts Up With That? and is the founder of the 'Surface Stations' project, documenting the condition of U.S. weather stations. *"The conclusion is inescapable: The U.S. temperature record is unreliable... Antarctic ice is above normal. And the global total amount of sea ice is above normal. So it's not disappearing any time soon."*

Gerd-Rainer Weber is a German scientist of Meteorology and Atmospheric Sciences. He is listed as a reviewer of IPCCC reports. *"Most of the extremist views about climate change have little or no scientific basis. The rational basis for extremist views about global warming may be a desire to push for political action on global warming."*

Leonard Weinstein worked 35 years at the NASA Langley Research Center. *"Any reasonable scientific analysis must conclude the basic theory wrong!!"*

William L. Wells is a chemist and chemical engineer and Adjunct Professor of Chemistry at Murray State University. *"One source indicates that China has plans to add 500 coal-fired plants in the next decade, while India is right behind with 200 plants on the drawing board. Restricting U.S. anthropogenic emissions, only a small part of the CO2 released into the environment, is a way of cutting off our economic noses to spite our faces."*

Bruce West is an Army physicist and active advocate for the Sun as the crucial factor in determining the cause of climate change. *"changes in the earth's average surface temperature are directly linked to ... the short-term statistical fluctuations in the Sun's irradiance and the longer-term solar cycles."*

David Whitehouse is the Science Editor at the online magazine, Global Warming Policy Forum. (https://www.thegwpf.com) The foundation is linked to Nigel Lawson's Global Warming Policy Foundation.

John Williams is an Australian researcher, author, and educator. *"There are 'strong and powerful counter-arguments' to the theories on global warming and carbon trading that are not being fully considered... There is no proof that carbon dioxide is causing or precedes global warming."*

Terry Wimberley is a Professor of Ecological Studies at Florida Gulf Coast University. *"At issue is how big of a problem is human produced CO2 emissions. Undoubtedly to some marginal degree - which scientists debate about - it is a problem, but is it the major cause of global warming? No."*

David Wojick has a Ph.D. in philosophy of science and mathematical logic from the University of Pittsburgh, and contributes to the website of ClimateChangeDebate.org.

Frederick Wolf is a professor of physics at Keene State College in New Hampshire. *"Several things have contributed to my skepticism about global warming being due to human causes. We all know that the atmosphere is a very complicated system. Also, after studying climate, I am aware that there are cycles of warm and cold periods of varying lengths which are still not completely understood."*

George T. Wolff is an atmospheric scientist, a former member of the EPA's Science Advisory Board, and served on a committee of the National Oceanic and Atmospheric Administration (NOAA). *"As an atmospheric scientist for over thirty years, engaged in studying and seeking solutions to environmental problems, I am appalled at the state of discord in the field of climate science... For too many in the field, critical thinking, the basis for all scientific inquiry, is not only absent, it is disdained."*

Robert Woock is a geophysicist and past president of the Southwest Louisiana Geophysical Society. *"I am a Geophysicist by education and practice with over thirty years in practice. Having studied the paleoclimate and environment for over thirty-five years I have come to some fundamental conclusions about our current conditions. The global warming debate is not over. I do not see any evidence in nature or data to suggest that we are in any anthropologic climate cycle."*

Tom Wysmuller worked as a meteorologist at New York University and for five years at NASA before, during, and after the moon landings. *"If mankind shuts down every coal and gas fired power plant, steel mill, and every auto, plane, train and ship and if we revert to a bare bones subsistence economy, with minimal fire and combustion products, we might be able to reduce CO2 growth from its average annual increase of 1.5ppm over the Keeling Curve era (since 1958) to about 0.5ppm."*

Lynwood Yarbrough is a biochemist and molecular biologist who ran a research lab and served as a consultant for the National Institutes of Health. *"I consider myself a scientific skeptic and want to be convinced by the data before I accept something as 'true' (see Freeman Dyson at edge.org on skepticism in science). As a biologist, I am aware of a number of cases in which science has been led in directions not based on hard evidence."*

Gregory Young is a physicist/neuroscientist currently engaged in experimental biophysical research. *"Let me assure you that we're (deniers) not in good humor, nor take it kindly to be slurred and ridiculed by taking the other side in this debate. And our numbers are still growing. Indeed,* we're angry that the vast majority of American Scientists will not be heard by the media."

William R. Young is a veteran meteorologist with 37 years of practical experience. *"We can all debate global warming and how much of an impact it has had and will have on our future weather, but not all change is the result of man."*

Miklós Zágoni is a Hungarian physicist and environmental researcher. *"Nature's regulatory instrument is water vapor: more carbon dioxide leads to less moisture in the air, keeping the overall GHG content in accord with the necessary balance conditions."*

Antonino Zichichi is an Italian nuclear physicist, from the University of Bologna, and was President of the World Federation of Scientists, a climate change denialist group. *"Man's action affects climate by no more than 10%. Ninety percent of climate change is ruled by natural phenomena whose future evolutions scientists today, as I said, do not know and cannot know."*

17.2... FURTHER READING; FOOD FOR THOUGHT

Sources for further reading... (in no particular order)

www.thegwpf.com

http://www.federationofscientists.org/index.php

www.ClimateDepot.com

Atmospheric Dynamics Group

Heliogenic.blogspot.com

https://carbon-sense.com/

https://plantsneedco2.org/

https://www.patreon.com/FFFA

https://climatechangedispatch.com

http://www.climatescienceinternational.org/

https://www.desmogblog.com

https://wattsupwiththat.com/

http://www.theclimatebet.com/2014/12/

https://www.therightclimatestuff.com/

About the Author

Ian Hall is an award-winning novelist and author of over 50 novels in many genres.

Born in Edinburgh, Scotland, he now lives in Kansas, USA.

He is a master story-teller, and hopes his love of history shines through in all his fiction works.

Learn more about Ian and his books at his website; ianhallauthor.com

You can get a free book and sign up for his newsletter at; Newsletter

Other Non-Fiction by Ian Hall

Ridiculously Comprehensive Dictionary of British Slang

Thousands of entries and definitions, rude words and all. The truly comprehensive guide to all British words and phrases; includes thousands of examples of Cockney Rhyming slang.

Churchills Secret Armies; Ungentlemanly Warfare

A look at the various departments, secret units, and skullduggery that British Prime Minister Winston Churchill set in motion in the first years of the Second World War.

WORLD WAR 2 SPY SCHOOL; Counter Espionage

A complete transcript of the entire manual of the SOE, dated 1943, that proved the birth of modern counter espionage. Newly formatted, edited and provided with the author's introduction.

OTHER NOVEL SERIES BY IAN HALL

The Jamie Leith Chronicles
Real life adventure; Scotland's attempt to colonize Panama in 1698

The Avenging Steel series.
Alternative Fiction; set in 1940's Edinburgh after the German invasion of Britain.

James Baird, a philosophy student at Edinburgh University becomes a reluctant member of the Scottish resistance movement against the occupying Nazis.

(6 books)

Caledonii; Birth of a Celtic Nation
Alternative/Speculative Historical Fiction; set in Scotland in 80AD as the Romans advance north.

With the aid of the druid movement, Calach, chief's son of clan Caledonii, must unite the bickering tribes against the new menace.

(5 books)

Star-Eater Chronicles
Ray-guns and rockets Science Fiction; set in the 25th century.

Seth Gingko is one of a thousand MacCollies Scouts sent out to map the galaxy. On reaching the furthermost star, and looks into the void beyond, he finds far more than he was bargaining for.

(8 books)

Space Academy Rebels
Teen/Young Adult Science Fiction; set in 2617.

The new intake of Space Academy would change the universe forever. As their paranormal abilities grow, they find their benefactors are not as squeaky-clean as they first thought.

(Trilogy)

Vampire High School.

Young Adult/Teen Horror; set in Arizona, USA.

When his best friend is murdered in front of his eyes, Lyman Bracks is mortified. When it happens in the middle of the school marching band, it's difficult to ignore.

(4 books, plus others)

A Connecticut Vampire.

Time-Travel Horror; set in Tudor England.

Vampire Francis is at the top of the food chain in modern USA. When he is transported back to 1500's England, he finds himself at the sharp end of a very different society.

(Trilogy)

ONE LAST THING

If you enjoyed this book, I'd be very grateful if you could post a short 5-star review on Amazon. Your support really does make a difference, and I read all the reviews so I can get your feedback and make this book even better.

Thanks again for your support.

Ian Hall

Printed in Great Britain
by Amazon

Suffolk Libraries

AMAZ 6/24

50 Walks for Ipswich-Woodbridge-Felixstowe and the East Suffolk Coast
With Nordic Walking Introduction

Dr I Pearson

Copyright © I Pearson 2024

I Pearson asserts the right under the UK Copyright, Designs and Patents Act 1988 to be identified as the author of this work.

Self-published via Kindle Direct Publishing.

All rights reserved. No part of this publication may be reproduced, converted to any other form by any means, or further distributed in any way without the prior permission of the author.

ISBN: 9798323159765
Imprint: Independently published

Extensive AI Use

This book was almost entirely written by an LLM AI, in this case Claude 3 Opus, in April 2024. It varies in depth throughout the book, with generally increasing detail as it progresses. I have not thoroughly checked it, so there will almost certainly be a few errors and omissions.

50 Walks for Ipswich-Woodbridge-Felixstowe and the East Suffolk Coast
With a Nordic Walking Introduction

For my beautiful wife Brigitta

Contents

Preface 7
Introduction to Nordic Walking 8
Equipment and Gear 10
Proper Technique and Form 12
Training and Progression 14
Nordic Walking for Specific Groups 16
Safety and Precautions 18
Nutrition and Hydration 20
Nordic Walking and Health 22
Nordic Walking and Other Related Sports 23
The Walks 25
Minsmere Walk 1 - 5.5 miles (8.9 km), Circular - Estimated Time: 3-4 hours 26
RSPB Minsmere Walk 2 - 3.5 miles (5.6 km) - Circular - Estimated Time: 2-3 hours 30
Dunwich Heath Walk 1 - 4 miles (6.4 km), Circular - Estimated Time: 2-3 hours 35
Dunwich Heath and Beach Walk 2 - 4.5 miles (7.2 km) - Circular - Estimated Time: 2-3 hours 39
Dunwich Heath Coastal Walk 3 - 6 miles (9.6 km) - Circular - Estimated Time: 3-4 hours 44
Dunwich Heath and Beach Walk 4 - 4 miles (6.4 km) - Circular - Estimated Time: 2-3 hours 48
Ickworth Park Walk - 3.5 miles (5.6 km), Circular - Estimated Time: 2-3 hours 53
Lavenham and the Lavenham Walk - 3 miles (4.8 km) - Circular - Estimated Time: 1.5-2 hours 58
Needham Lake Walk - 2 miles (3.2 km), Circular - Estimated Time: 1-1.5 hours 63
Nacton Circular Walk - 4 miles (6.4 km), Circular - Estimated Time: 2-3 hours 68
Nacton Shores Walk - 4 miles (6.4 km), Coastal - Estimated Time: 2-3 hours 73
Orwell Country Park Walk - 2 miles (3.2 km), Circular - Estimated Time: 1-1.5 hours 78
Orwell Walk - 12 miles (19.3 km), Circular - Estimated Time: 6-7 hours 83
Freston Wood Walk - 3 miles (4.8 km), Circular - Estimated Time: 1.5-2 hours 88
Pin Mill and Chelmondiston Walk - 5 miles (8 km), Circular - Estimated Time: 2.5-3 hours 93
Holbrook Circular Walk - 5 miles (8 km), Circular - Estimated Time: 2.5-3 hours 98
Shotley Peninsula Walk - 10 miles (16 km), Circular - Estimated Time: 5-6 hours 103
Shotley Gate Walk - 4 miles (6.4 km), Circular - Estimated Time: 2-2.5 hours 109
Butley Circular Walk - 5 miles (8 km), Circular - Estimated Time: 2.5-3 hours 114
Harkstead Circular Walk - 5 miles (8 km), Circular - Estimated Time: 2.5-3 hours 119
Harkstead and Erwarton Walk - 5 miles (8 km), Circular - Estimated Time: 2.5-3 hours 124

Tattingstone Wonder Walk - 3 miles (4.8 km), Circular - Estimated Time: 1.5-2 hours *130*

Alton Water Walk - 8 miles (12.9 km), Circular - Estimated Time: 4-5 hours *136*

East Bergholt Circular Walk - 6 miles (9.6 km), Circular - Estimated Time: 3-4 hours *141*

Ipswich Waterfront Walk - 2 miles (3.2 km), Non-circular - Estimated Time: 1-1.5 hours *147*

Ipswich Town Centre Walk - 2 miles (3.2 km), Non-circular - Estimated Time: 1-1.5 hours *152*

Freston Wood Walk - 3 miles (4.8 km), Circular - Estimated Time: 1.5-2 hours *157*

Kesgrave Woods Walk - 3 miles (4.8 km), Circular - Estimated Time: 1.5-2 hours *162*

Great Bealings Circular Walk - 6 miles (9.7 km), Circular - Estimated Time: 3-4 hours *168*

Rushmere Heath Walk - 2 miles (3.2 km), Circular - Estimated Time: 1-1.5 hours *173*

Newbourne Springs Walk - 5 miles (8 km), Circular - Estimated Time: 2.5-3 hours *179*

Waldringfield Circular Walk - 4 miles (6.4 km), Circular - Estimated Time: 2-2.5 hours *185*

Martlesham Heath Walk - 4 miles (6.4 km), Circular - Estimated Time: 2-2.5 hours *191*

Martlesham Creek Walk - 3 miles (4.8 km), Circular - Estimated Time: 1.5-2 hours *197*

Martlesham Circular Walk - 4 miles (6.4 km), Circular - Estimated Time: 2-2.5 hours *203*

Woodbridge Riverside Walk - 4 miles (6.4 km), Linear - Estimated Time: 2-2.5 hours *209*

River Deben Walk - 8 miles (12.9 km), Circular - Estimated Time: 4-5 hours *216*

Boulge Circular Walk - 4 miles (6.4 km), Circular - Estimated Time: 2-2.5 hours *222*

Hollesley Common Walk - 4 miles (6.4 km), Circular - Estimated Time: 2-2.5 hours *228*

Boyton and Hollesley Marshes Walk - 8 miles (12.9 km), Circular - Estimated Time: 4-5 hours *234*

Sutton Hoo Walk - 3 miles (4.8 km), Circular - Estimated Time: 1.5-2 hours *241*

Staverton Thicks Walk - 3 miles (4.8 km), Circular - Estimated Time: 1.5-2 hours *248*

Sutton Common Walk - 5 miles (8 km), Circular - Estimated Time: 2.5-3 hours *255*

Rendlesham Forest Walk - 5 miles (8 km), Circular - Estimated Time: 2.5-3 hours *261*

Bawdsey Quay Walk - 2 miles (3.2 km), Non-circular - Estimated Time: 1-1.5 hours *268*

Bawdsey Circular Walk - 5 miles (8 km), Circular - Estimated Time: 2.5-3 hours *275*

Bawdsey to Felixstowe Ferry Walk - 6 miles (9.7 km), Linear - Estimated Time: 3-4 hours *282*

Blaxhall Common Walk - 3 miles (4.8 km), Circular - Estimated Time: 1.5-2 hours *289*

Orford Ness Nature Reserve Walk - 5 miles (8 km), Circular - Estimated Time: 2.5-3 hours *295*

Orford Ness National Nature Reserve Walk - 5 miles (8 km), Linear - Estimated Time: 2.5-3 hours *302*

Shingle Street Walk - 3 miles (4.8 km), Circular - Estimated Time: 1.5-2 hours *309*

River Alde Walk - 10 miles (16 km), Circular - Estimated Time: 5-6 hours *316*

Bucklesham Circular Walk - 5 miles (8 km), Circular - Estimated Time: 2-2.5 hours *325*

Trimley St Martin Walk - 3 miles (4.8 km), Circular - Estimated Time: 1.5-2 hours *334*

Trimley Marshes Walk - 5 miles (8 km), Circular - Estimated Time: 2-3 hours *343*
Felixstowe Promenade North Walk - 2 miles (3.2 km) - Non-circular - Estimated Time: 1-1.5 hours *354*
Felixstowe Promenade South Walk - 2 miles (3.2 km) - Seafront - Estimated Time: 1-1.5 hours *365*
Felixstowe Ferry Walk - 3 miles (4.8 km), Circular - Estimated Time: 1.5-2 hours *372*
Felixstowe Hills Walk - 5 miles (8 km), Circular - Estimated Time: 2.5-3 hours *381*
Ipswich to Felixstowe Walk - 13 miles (20.9 km), Linear - Estimated Time: 6-7 hours....................*393*
Ipswich to Martlesham Heath Walk - 7 miles (11.3 km), Linear - Estimated Time: 3-4 hours.........*406*
Birds to Spot...*418*
Animals to Spot *421*
Insects to Spot *423*
Plants to Spot *425*
Sites of Special Historical Interest: *428*
Areas of Outstanding Beauty: *429*
Walks for Weeks..*430*
Epilogue *438*

Preface

In the enchanting county of Suffolk, nestled along the eastern coast of England, lie the charming towns of Ipswich, Woodbridge, and Felixstowe. This area is a true walker's paradise, boasting an array of breathtaking landscapes, captivating wildlife, and intriguing historical sites. To celebrate the natural and cultural wonders of this region, I have compiled a comprehensive guide to 50 walks that showcase the best of what these towns and their surroundings have to offer.

Each walk in this guide not only covers the unique features, views, and experiences that each has to offer but also delves into the fascinating historical context, the rare and captivating birds, insects, and other wildlife you may encounter, and the diverse array of plant life that adorns these paths. From the serene coastal routes of Shingle Street to the tranquil woodlands of Rendlesham Forest, there's a walk for everyone, whether you're a seasoned hiker or a casual stroller.

In addition to detailed descriptions of each walk, I have included information on the most notable features and views along the way, as well as a list of the bird species, rare insects, other wildlife, and unique plant life that you may encounter during your journey. These insights will not only help you appreciate the natural beauty of the area but also encourage you to keep an eye out for those special moments that make each walk memorable.

Moreover, I have delved into the rich history of the region, highlighting the fascinating stories and landmarks that have shaped these landscapes over the centuries. From ancient ruins and medieval churches to the tales of smugglers and seafarers, each walk is a journey through time, offering a glimpse into the enduring legacy of this captivating corner of England.

To ensure that your adventure is as seamless as possible, I have provided grid coordinates for the starting point of each walk and the location of the nearest car park. These coordinates will help you easily locate the beginning of each route and ensure that you're never far from a convenient parking spot. Additionally, I have included information on the facilities and accessibility of each walk, as well as seasonal highlights and safety considerations, so that you can plan your journey with confidence and ease.

As you embark on these walks, I hope that this guide becomes your trusted companion, revealing the hidden gems and enchanting landscapes of Ipswich, Woodbridge, Felixstowe, and their surrounding areas. With 50 walks to choose from, there's no shortage of opportunities to explore, appreciate, and fall in love with the natural wonders of this captivating region.

From the salt-sprayed cliffs of the coast to the gently rolling hills of the countryside, each walk is a celebration of the beauty, diversity, and resilience of the Suffolk landscape. Whether you're a birdwatcher, a history buff, a nature lover, or simply someone in search of a moment of peace and tranquility, these walks promise to inspire, delight, and invigorate you, leaving you with memories that will last a lifetime.

So lace up your walking boots, grab your binoculars and your sense of adventure, and let this guide be your passport to the wonders of the Suffolk countryside. As you explore these paths and trails, remember to tread lightly, leave no trace, and open your heart to the magic and mystery of the natural world. The beauty and the stories of this special place are waiting to be discovered, and all you have to do is take that first step.

Remember: don't get lost, don't drown, and don't get hit by a tree.

Have fun!

Introduction to Nordic Walking

What is Nordic Walking?

Nordic Walking is a full-body, low-impact fitness activity that combines the natural walking motion with the use of specially designed poles. The poles engage the upper body muscles, making it a more effective workout compared to regular walking. Nordic Walking is suitable for people of all ages and fitness levels, as it can be adapted to individual preferences and goals.

The concept of Nordic Walking involves using the poles to push off the ground with each stride, activating the arm, shoulder, chest, and back muscles. This additional upper body engagement leads to increased calorie burn, improved posture, and reduced stress on the joints compared to regular walking.

History and Origin of Nordic Walking

Nordic Walking has its roots in Finland, where it was developed in the 1930s as a summer training method for cross-country skiers. Finnish athletes and coaches discovered that using ski poles while walking helped maintain their fitness levels during the off-season.

The activity gained popularity in Europe during the 1980s and 1990s, particularly in Finland, Germany, and Austria. Manufacturers began producing specialized Nordic Walking poles, and the activity spread to other countries as a means of improving overall fitness and well-being. Today, Nordic Walking is recognized and practiced worldwide, with millions of enthusiasts in Europe, North America, Australia, and Asia. It has become a popular choice for individuals looking for a low-impact, full-body workout that can be enjoyed in various outdoor settings.

Benefits of Nordic Walking

Nordic Walking offers numerous benefits for both physical and mental health:

1. Improved cardiovascular fitness and endurance: The increased upper body engagement and overall intensity of Nordic Walking lead to improved heart and lung function, as well as increased stamina.

2. Full-body workout: Nordic Walking engages up to 90% of the body's muscles, providing a comprehensive workout that targets the arms, shoulders, chest, back, core, and legs.

3. Low-impact nature: The use of poles reduces the impact on joints, particularly the knees and hips, making it an ideal activity for individuals with joint issues, arthritis, or those recovering from injuries.

4. Suitable for various ages and fitness levels: Nordic Walking can be adapted to suit different fitness levels by adjusting the speed, terrain, and intensity of the workout. This makes it accessible to a wide range of individuals, from beginners to athletes.

5. Stress relief and mental well-being: Exercising outdoors in nature has been shown to reduce stress, improve mood, and increase overall mental well-being. Nordic Walking provides an opportunity to disconnect from daily stressors and enjoy the benefits of being in a natural environment.

6. Adaptability to various terrains and weather conditions: Nordic Walking can be practiced on various

surfaces, including pavement, trails, grass, and sand. The poles provide additional stability and support, making it possible to walk in different weather conditions and on uneven terrain.

The combination of these benefits makes Nordic Walking an attractive choice for those seeking a fun, effective, and low-impact way to improve their overall health and fitness.

Equipment and Gear

Nordic Walking Poles

Having the right equipment is crucial for a comfortable and effective Nordic walking experience. The most important piece of gear is a pair of high-quality Nordic walking poles, which are designed specifically for this activity. Nordic Walking poles are specifically designed to provide support, stability, and increased upper body engagement during the activity. There are three main types of poles:

1. Fixed-length poles: These poles have a set length and are suitable for individuals who consistently walk on similar terrains and have a preferred pole height.

2. Adjustable poles: These poles can be adjusted to different lengths, allowing for customization based on the user's height and the terrain. They are ideal for those who walk on varied surfaces or share poles with others.

3. Collapsible poles: These poles can be folded or collapsed for easy storage and transportation, making them convenient for travel or when storage space is limited.

Nordic Walking poles are typically made from lightweight materials such as aluminum, carbon fiber, or composite materials. The choice of material affects the price, weight, and durability of the poles.

When choosing poles, consider factors such as your height, the terrain you'll be walking on, and your personal preferences. Poles come in various materials, such as aluminum and carbon fiber, and may feature adjustable or fixed lengths. When selecting poles, it is essential to consider the proper size and adjustment. Adjustable poles are ideal for beginners, as they can be tailored to your height and shared among multiple users. Poles should be adjusted to a height that allows the elbow to be bent at a 90-degree angle when the pole tip is placed on the ground near the foot. This ensures proper form and maximizes the benefits of Nordic Walking.

Pole tips and baskets are another important consideration. Rubber tips are suitable for hard surfaces like pavement, while metal tips provide better traction on soft surfaces like grass or dirt. Baskets can be attached to the poles to prevent them from sinking into soft ground or snow.

Footwear

In addition to poles, invest in a pair of comfortable, supportive walking shoes with good traction. Opt for breathable, moisture-wicking clothing that allows for easy movement, and consider layering to accommodate changing weather conditions. Proper footwear is crucial for comfort, support, and injury prevention during Nordic Walking. The ideal shoe should provide the following:

1. Stability: A stable base to support the foot and prevent excessive pronation or supination.
2. Cushioning: Adequate cushioning to absorb shock and reduce impact on the joints.
3. Support: Proper arch support and heel counter to maintain proper foot alignment.

Different types of shoes can be suitable for Nordic Walking, depending on the terrain and individual preferences:

1. Walking shoes: These shoes are designed specifically for walking and offer a combination of stability, cushioning, and flexibility. They are suitable for most types of terrain.

2. Trail runners: These shoes provide additional traction and support for walking on uneven trails or off-road terrain. They often have more aggressive outsoles and reinforced uppers.
3. Hiking boots: For more challenging terrain or longer walks, hiking boots offer added support, stability, and protection. They are particularly suitable for rocky or steep trails.

In addition to shoes, choosing the right socks is important for comfort and blister prevention. Moisture-wicking socks made from materials like merino wool or synthetic blends help keep the feet dry and reduce friction.

Clothing

Dressing appropriately for Nordic Walking involves layering to accommodate different weather conditions and personal preferences. The key principles of clothing selection include:

1. Moisture-wicking fabrics: Choose base layers made from materials that wick sweat away from the skin to keep you dry and comfortable.
2. Breathable layers: Opt for breathable mid-layers and outer layers to allow for temperature regulation and prevent overheating.
3. Weather-appropriate gear: Consider the season and weather conditions when selecting clothing. Lightweight, UV-protective fabrics for summer; insulating layers for colder temperatures.
4. Comfort and mobility: Choose clothing that allows for a full range of motion and does not restrict movement.

Accessories

Several accessories can enhance comfort, safety, and enjoyment during Nordic Walking:

1. Gloves: Gloves provide comfort, grip, and blister prevention when using poles. Choose gloves with a comfortable fit and breathable materials.
2. Hydration systems: Staying hydrated is essential during Nordic Walking. Carry water using a water bottle, hydration belt, or hydration backpack, depending on the duration of your walk and personal preference.
3. Safety gear: When walking in low-light conditions or on roads, reflective clothing and accessories can improve visibility and safety. A headlamp or flashlight may also be necessary for walks in the dark.
4. Sun protection: Protect your skin from sun damage by wearing a hat, sunglasses, and sunscreen with an appropriate SPF rating.
5. A small backpack or hydration pack to carry water, snacks, and essentials,

By selecting the right equipment and gear, Nordic Walkers can ensure a safe, comfortable, and enjoyable experience while maximizing the benefits of the activity.

Proper Technique and Form

Basic Nordic Walking Technique

Mastering the proper technique is essential for maximizing the benefits of Nordic walking and preventing injury. The basic Nordic walking technique involves a natural, forward arm swing, coupled with a rolling foot motion and an opposite-leg-and-arm pattern.

The basic technique involves the following elements:

To begin, stand tall with your feet shoulder-width apart and your poles positioned at a 45-degree angle behind you, with the tips pointing diagonally backward. The poles should be placed opposite to the leading foot, so when the left foot is forward, the right pole is planted, and vice versa.

As you step forward with your left foot, swing your right arm forward to plant the pole tip near your foot. Push down and back on the pole as you walk, engaging your upper body muscles and propelling yourself forward.

Repeat this motion with the opposite arm and leg, establishing a rhythmic, alternating pattern. Keep your elbows straight but not locked, and maintain a slight bend in your knees to absorb impact. Focus on a natural, upright posture, with your core engaged and your gaze forward.

Arm swing and coordination with legs: Swing your arms naturally, keeping them close to your body. The arm swing should be coordinated with your leg movement, so your left arm is forward when your right foot is forward, and your right arm is forward when your left foot is forward.

Stride length and rhythm: Maintain a natural, comfortable stride length similar to your normal walking stride. Find a rhythmic pattern that synchronizes your arm and leg movements, allowing for a smooth and efficient motion.

As you become more comfortable with the basic technique, you can experiment with variations such as increasing your speed, lengthening your stride, or adding hills and uneven terrain to your route. Remember to listen to your body and start slowly, gradually increasing the intensity and duration of your Nordic walks as your fitness improves.

Common Mistakes and How to Avoid Them

Beginners often make some common mistakes when learning Nordic Walking technique. Being aware of these mistakes can help you avoid them and improve your form:

Improper pole placement and grip: Avoid planting the poles too far in front of your body or gripping them too tightly. Maintain a relaxed grip and place the poles at a 45-degree angle behind you.

Hunched posture and insufficient core engagement: Keep your body upright and engage your core muscles to maintain good posture. Avoid leaning forward or hunching your shoulders.

Overstriding and improper foot strike: Maintain a natural stride length and avoid overstriding, which can lead to a heel strike and increased impact on the joints. Aim for a mid-foot strike to promote a smooth and efficient gait.

Variations and Advanced Techniques

As you become more comfortable with the basic Nordic Walking technique, you can explore variations and advanced techniques to challenge yourself and enhance your workout:

Double poling: This technique involves planting both poles simultaneously and pushing off with both arms, engaging more of the upper body muscles. Double poling is useful for increasing intensity or navigating steep inclines.

Uphill and downhill techniques: When walking uphill, shorten your stride and use the poles to help propel you forward. On downhill sections, lengthen your stride and use the poles for balance and stability.

Nordic running and bounding: For a more intensive workout, incorporate short intervals of running or bounding while using the poles. This advanced technique can improve cardiovascular fitness and leg strength.

Warm-up and Cool-down Exercises

Proper warm-up and cool-down routines are important for preparing your body for exercise, reducing the risk of injury, and promoting recovery. Here are some recommended exercises:

Warm-up (5-10 minutes):
- Dynamic stretches: leg swings, arm circles, trunk rotations
- Light walking or marching in place
- Gradually increasing the intensity of your Nordic Walking

Cool-down (5-10 minutes):
- Gradually decreasing the intensity of your Nordic Walking
- Static stretches: quadriceps, hamstrings, calves, chest, and shoulders
- Deep breathing and relaxation exercises

Remember to listen to your body and progress gradually to avoid overexertion or injury. If you experience pain or discomfort, stop the activity and consult a healthcare professional.

Training and Progression

Getting Started with Nordic Walking

To get the most out of your Nordic walking experience, it's important to approach it with a structured training plan and a focus on progressive overload. This means starting with shorter, less challenging walks and gradually increasing the distance, intensity, and frequency over time.

Begin with flat, even surfaces such as a park, paved trail, or athletic track, aiming for 2-3 Nordic walks per week, each lasting 20-30 minutes. As you build endurance and confidence, slowly increase the duration of your walks, eventually working up to 45-60 minutes per session. Incorporate a variety of terrains and routes to keep your workouts engaging and challenging.

Start with shorter durations and gradually increase the length of your walks. A beginner might start with 10-15 minute sessions, 2-3 times per week, and gradually increase to 30-60 minute sessions, 3-5 times per week, as fitness improves.

In addition to your Nordic walking sessions, consider complementary exercises to improve your overall fitness and prevent muscle imbalances. Strength training exercises targeting your core, upper body, and lower body can help improve your Nordic walking performance and reduce the risk of injury.

Setting realistic goals and tracking your progress can help you stay motivated and celebrate your achievements. Consider using a fitness tracker, smartphone app, or journal to log your walks, distance, time, and any other relevant metrics.

Remember to listen to your body and allow for adequate rest and recovery between workouts. Stay hydrated, maintain a balanced diet, and prioritize quality sleep to support your fitness goals. By following a consistent, progressive training plan, you'll be well on your way to reaping the many benefits of Nordic walking.

Choosing the Right Intensity and Duration

To maximize the benefits of Nordic Walking, it is important to exercise at the right intensity and duration for your fitness level.

Here are some guidelines:

Target heart rate: Aim for a target heart rate of 50-85% of your maximum heart rate, depending on your fitness level and goals. You can estimate your maximum heart rate by subtracting your age from 220.

Perceived exertion: Use the "talk test" to gauge your intensity. You should be able to carry on a conversation while Nordic Walking, but not sing. If you are too breathless to talk, reduce your intensity.

Duration: Aim for at least 150 minutes of moderate-intensity aerobic activity or 75 minutes of vigorous-intensity activity per week, spread across several sessions.

Rest and recovery: Allow for adequate rest and recovery between Nordic Walking sessions to prevent overuse injuries and promote adaptation. Include at least one or two rest or low-intensity days per week.

Progressing and Increasing Difficulty

As your fitness improves, you can challenge yourself by increasing the intensity, duration, or difficulty of your Nordic Walking workouts:

Intervals and speed workouts: Incorporate short intervals of faster walking or Nordic running to boost cardiovascular fitness and calorie burn. For example, alternate 1-2 minutes of fast walking with 2-3 minutes of slower recovery walking.

Strength training exercises: Include bodyweight or resistance exercises to complement your Nordic Walking and improve overall strength. Examples include squats, lunges, push-ups, and core exercises.

Cross-training: Incorporate other low-impact activities like swimming, cycling, or yoga to prevent boredom, improve overall fitness, and reduce the risk of overuse injuries.

Sample Training Plans for Different Fitness Levels

Here are sample training plans for beginner, intermediate, and advanced Nordic Walkers:

Beginner:
- Week 1-2: 10-15 minutes, 2-3 times per week
- Week 3-4: 15-20 minutes, 3 times per week
- Week 5-6: 20-30 minutes, 3-4 times per week

Intermediate:
- Week 1-2: 30-40 minutes, 3-4 times per week
- Week 3-4: 40-50 minutes, 4 times per week, with one interval session
- Week 5-6: 50-60 minutes, 4-5 times per week, with one interval session and one hill session

Advanced:
- Week 1-2: 45-60 minutes, 4-5 times per week, with two interval sessions and one hill session
- Week 3-4: 60-75 minutes, 5 times per week, with two interval sessions, one hill session, and one long endurance session
- Week 5-6: 75-90 minutes, 5-6 times per week, with two interval sessions, one hill session, one long endurance session, and one cross-training session

Remember to adapt these plans to your individual goals, preferences, and schedule. Progress gradually and listen to your body to avoid overexertion or injury.

Nordic Walking for Specific Groups

Nordic Walking for Seniors

Nordic Walking is an excellent activity for seniors, as it provides numerous benefits for balance, stability, and bone density. The use of poles helps distribute weight evenly, reducing the impact on joints and providing additional support. This can be particularly beneficial for those with age-related conditions such as arthritis or osteoporosis.

When introducing Nordic Walking to seniors, it is important to start slowly and gradually increase intensity and duration. Seniors should focus on maintaining proper form and technique, and listen to their bodies to avoid overexertion. It may be necessary to adapt the technique or use shorter poles to accommodate individual needs and limitations.

Nordic Walking can also provide social and community benefits for seniors. Joining a Nordic Walking group or club can promote social interaction, motivation, and a sense of belonging. Many communities offer senior-specific Nordic Walking programs or events.

Nordic Walking for Weight Loss

Nordic Walking can be an effective tool for weight loss, as it burns more calories compared to regular walking. The increased upper body engagement and overall intensity of Nordic Walking lead to a higher calorie expenditure, making it a valuable addition to a weight loss program.

To maximize the weight loss benefits of Nordic Walking, consider incorporating high-intensity interval training (HIIT) sessions. Alternate short periods of fast walking or Nordic running with periods of slower recovery walking. This type of training can boost metabolism and improve cardiovascular fitness, leading to increased calorie burn even after the workout.

Combining Nordic Walking with a balanced, calorie-controlled diet is essential for sustainable weight loss. Focus on whole, nutrient-dense foods and maintain a modest calorie deficit to promote gradual weight loss while supporting overall health.

Nordic Walking for Athletes and Cross-training

Athletes can benefit from incorporating Nordic Walking into their training plans as a low-impact cross-training activity. Nordic Walking can help improve endurance, strength, and flexibility without adding excessive stress to the joints.

For runners, Nordic Walking can be a valuable tool for active recovery or injury prevention. The use of poles reduces the impact on the lower body while still providing a cardiovascular workout. This can be particularly beneficial for those recovering from running-related injuries or looking to maintain fitness while reducing mileage.

Athletes can also use Nordic Walking to enhance their overall conditioning and muscle balance. The engagement of upper body muscles can help improve posture, core stability, and arm strength, which can translate to better performance in their primary sport.

Nordic Walking during Pregnancy

Nordic Walking can be a safe and beneficial activity during pregnancy, provided that certain precautions and adaptations are made. Always consult with a healthcare provider before starting or continuing any exercise program during pregnancy.

Benefits of Nordic Walking during pregnancy include improved cardiovascular fitness, reduced back pain, and better postural support. The use of poles can help alleviate some of the pressure on the lower back and pelvis, which can be particularly beneficial as the pregnancy progresses.

When Nordic Walking during pregnancy, be mindful of the following safety guidelines:

1. Maintain a moderate intensity and avoid overexertion. Use the "talk test" to ensure you can carry on a conversation while walking.
2. Stay hydrated and avoid exercising in hot or humid conditions.
3. Wear supportive footwear and comfortable, breathable clothing.
4. Listen to your body and stop if you experience any pain, dizziness, or unusual symptoms.

As the pregnancy progresses, it may be necessary to adapt the Nordic Walking technique to accommodate changes in balance and centre of gravity. Shortening the stride length, using shorter poles, or walking on flatter terrain can help maintain stability and comfort.

Postpartum, Nordic Walking can be a gentle way to return to physical activity and regain fitness. Begin slowly and gradually increase intensity and duration, focusing on proper form and core engagement. It is important to allow adequate time for recovery and to follow the guidance of a healthcare provider before resuming exercise after childbirth.

Safety and Precautions

While Nordic walking is a low-impact, safe activity for most people, it's essential to take certain precautions to prevent injury and ensure a positive experience. Before starting any new exercise routine, consult with your healthcare provider to ensure that Nordic walking is appropriate for your individual needs and health status.

When Nordic walking, always be aware of your surroundings and follow local trail etiquette. Stay on designated paths, yield to other trail users, and be mindful of weather conditions and potential hazards such as uneven terrain, wet surfaces, or low-hanging branches.

Dress appropriately for the weather, wearing layers that can be easily added or removed as needed. Protect your skin with sunscreen, even on cloudy days, and wear a hat or sunglasses to shield your face from the sun.

Stay hydrated by carrying water with you and drinking regularly throughout your walk. If walking for extended periods or in hot weather, consider bringing electrolyte-rich snacks or beverages to replenish lost nutrients.

In case of emergency, carry a charged mobile phone and a small first-aid kit. Familiarize yourself with the area you'll be walking in and let someone know your planned route and expected return time.

Importance of Proper Form to Prevent Injuries

Maintaining proper Nordic Walking form is crucial for preventing injuries and maximizing the benefits of the activity. Improper technique can lead to undue stress on joints and muscles, increasing the risk of common Nordic Walking injuries such as:

Shoulder pain: Overuse or improper pole placement can lead to shoulder strain or rotator cuff injuries.
Elbow and wrist pain: Incorrect pole grip or overextension can cause strain on the elbow and wrist joints.
Knee pain: Overstriding or excessive impact can lead to knee pain or aggravate pre-existing knee conditions.

To minimize the risk of injuries, focus on the following techniques:

Maintain a relaxed, neutral grip on the poles, avoiding excessive tension in the hands and wrists.
Keep the elbows close to the body and avoid overextending the arms behind the body.
Plant the poles at a 45-degree angle behind the body, with the tips pointing diagonally backward.
Use a natural stride length and avoid overstriding, which can lead to a heel strike and increased impact on the joints.

Exercising in Various Weather Conditions

Nordic Walking can be enjoyed in a variety of weather conditions, but it is important to take appropriate precautions to ensure safety and comfort:

Heat and sun safety:
- Stay hydrated by drinking water before, during, and after your walk.
- Wear lightweight, breathable clothing and a hat to protect your head and face from the sun.
- Apply sunscreen with an appropriate SPF rating to exposed skin.
- Avoid Nordic Walking during the hottest parts of the day, and seek shade when possible.

Cold weather precautions:

- Dress in layers to allow for temperature regulation. Start with a moisture-wicking base layer, add insulating mid-layers, and finish with a wind- and water-resistant outer layer.
- Protect exposed skin from wind and cold with gloves, a hat, and a neck gaiter or scarf.
- Be aware of the wind chill factor and adjust your clothing and intensity accordingly.
- Watch for signs of hypothermia, such as shivering, confusion, or slurred speech.

Wet and slippery conditions:
- Choose footwear with good traction and consider using pole tips designed for improved grip on wet surfaces.
- Take shorter, more controlled steps to maintain balance and stability.
- Be aware of puddles, mud, or ice patches, and navigate around them when possible.
- Wear bright or reflective clothing to improve visibility in low-light conditions.

Nordic Walking Etiquette and Rules

When Nordic Walking in public spaces or on shared trails, it is important to follow proper etiquette and rules to ensure a safe and enjoyable experience for everyone:

Share the trail: Be aware of other trail users, such as walkers, runners, cyclists, and horses. Yield to slower-moving individuals and give a friendly warning when passing.

Stay to the right: When walking on a shared path or trail, stay to the right side to allow others to pass on the left.

Respect private property: Obey posted signs and avoid entering private property or restricted areas without permission.

Leave no trace: Practice responsible outdoor ethics by packing out any trash, avoiding damage to vegetation, and respecting wildlife.

Dealing with Common Nordic Walking Injuries

If you experience pain or discomfort during or after Nordic Walking, it is important to address the issue promptly to prevent further injury. Follow the RICE protocol for initial treatment:

Rest: Stop the activity and avoid putting weight or stress on the affected area.
Ice: Apply ice or a cold pack to the injured area for 15-20 minutes at a time, several times a day, to reduce pain and inflammation.
Compression: Use an elastic bandage or compression sleeve to provide support and minimize swelling.
Elevation: Elevate the affected limb above the level of the heart to reduce swelling and promote drainage.

If pain persists or worsens after 24-48 hours of self-treatment, seek medical attention. A healthcare professional can provide a proper diagnosis and recommend appropriate treatment, such as physical therapy or rehabilitation exercises.

To prevent future injuries, focus on gradual progression, proper form, and adequate rest and recovery between Nordic Walking sessions. Incorporating strength training and flexibility exercises can also help improve overall conditioning and reduce the risk of overuse injuries.

Nutrition and Hydration

Importance of Staying Hydrated during Nordic Walking

Proper hydration is essential for maintaining performance, regulating body temperature, and preventing heat-related illnesses during Nordic Walking. Dehydration can lead to symptoms such as:

Thirst
Dry mouth and lips
Fatigue
Dizziness or lightheadedness
Dark yellow urine

To stay hydrated, follow these guidelines:

Drink 17-20 ounces (500-590 ml) of water 2-3 hours before your Nordic Walking session.
Drink 7-10 ounces (200-300 ml) of water 10-20 minutes before starting your walk.
During your walk, drink 7-10 ounces (200-300 ml) of water every 10-20 minutes, depending on the intensity and weather conditions.
After your walk, drink 16-24 ounces (475-710 ml) of water for every pound (0.5 kg) of body weight lost during the session.

In addition to water, you can also consume sports drinks containing electrolytes for walks lasting longer than 60 minutes or in hot and humid conditions.

Pre- and Post-Workout Nutrition

Proper nutrition before and after Nordic Walking can help optimize performance, recovery, and overall health. Here are some general guidelines:

Pre-workout nutrition (1-3 hours before):
- Choose meals or snacks that are high in carbohydrates, moderate in protein, and low in fat and fiber to provide sustained energy without causing digestive discomfort.
- Examples include a banana with peanut butter, oatmeal with berries and nuts, or a turkey and vegetable wrap.
- For walks lasting less than 60 minutes, a small snack or a well-balanced meal consumed a few hours prior should be sufficient.

Post-workout nutrition (within 30-60 minutes):
- Consume a combination of carbohydrates and protein to replenish energy stores, promote muscle repair and growth, and support immune function.
- Aim for a ratio of 3:1 carbohydrates to protein.
- Examples include a smoothie with fruit, yogurt, and protein powder; a whole-grain sandwich with lean protein; or a stir-fry with rice and chicken.

Snacks and Meal Ideas for Longer Walks

For Nordic Walking sessions lasting more than 60-90 minutes, it may be necessary to consume snacks or small meals to maintain energy levels and prevent fatigue. Here are some portable and non-perishable options:

Energy bars or granola bars
Trail mix with dried fruit and nuts
Peanut butter or almond butter packets
Fresh or dried fruit, such as apples, bananas, or raisins
Whole-grain crackers or rice cakes
Electrolyte gels or chews

When planning snacks or meals for longer walks, focus on a balance of carbohydrates for energy, protein for muscle support, and healthy fats for satiety. Avoid foods high in fiber or fat, as they may cause digestive discomfort during exercise.

For endurance events or all-day Nordic Walking excursions, it's important to have a fueling strategy that includes regular snacks or small meals consumed every 30-60 minutes. Experiment with different foods and timing during training to determine what works best for your individual needs and preferences.

Remember that nutrition and hydration needs can vary depending on factors such as individual physiology, climate, and exercise intensity. Listen to your body and adjust your intake accordingly to maintain optimal performance and well-being during Nordic Walking.

Nordic Walking and Health

Cardiovascular Health

Nordic Walking provides significant benefits for cardiovascular health. The increased upper body engagement and overall intensity of Nordic Walking lead to improved heart and lung function, as well as increased circulation. Regular Nordic Walking can help:

Lower blood pressure
Improve cholesterol levels
Reduce the risk of heart disease, stroke, and type 2 diabetes
Enhance cardiovascular endurance and efficiency

Weight Management

Nordic Walking is an effective tool for weight management, as it increases calorie burn compared to regular walking. The additional upper body engagement and higher overall intensity lead to increased energy expenditure, making it easier to create a calorie deficit necessary for weight loss.

In addition to the direct calorie burn during Nordic Walking, the activity can also help boost metabolism, leading to increased calorie burn even at rest. This can be particularly beneficial for long-term weight maintenance.

Bone and Joint Health

As a weight-bearing exercise, Nordic Walking can help improve bone density and reduce the risk of osteoporosis. The gentle impact and resistance provided by the poles stimulate bone growth and strengthening, particularly in the spine, hips, and wrists.

Nordic Walking is also a low-impact activity, making it an excellent option for those with joint concerns or conditions such as arthritis. The use of poles helps distribute weight evenly and reduces the impact on the knees, hips, and ankles. This can help alleviate pain, stiffness, and inflammation in the joints.

Mental Health and Well-being

Engaging in regular Nordic Walking can have significant benefits for mental health and well-being. Exercise, particularly when done outdoors in nature, has been shown to:

Reduce stress and anxiety
Improve mood and self-esteem
Enhance cognitive function and memory
Promote better sleep quality

The rhythmic, meditative nature of Nordic Walking can also provide a sense of calm and relaxation, helping to reduce stress and promote overall well-being. Additionally, the social aspect of Nordic Walking, when done in groups or with friends, can provide a sense of connection and support that further enhances mental health benefits.

Nordic Walking and Other Related Sports

Comparison of Fitness Benefits

Nordic Walking shares some similarities with other outdoor activities, such as regular walking, hiking, and running. However, Nordic Walking offers unique benefits that set it apart:

Compared to regular walking, Nordic Walking engages more upper body muscles and provides a higher overall intensity, leading to increased calorie burn and cardiovascular benefits.
Compared to hiking, Nordic Walking is typically done on more even terrain and with a more consistent pace, making it a more accessible option for a wider range of fitness levels.
Compared to running, Nordic Walking is a lower-impact activity that places less stress on the joints while still providing a challenging cardiovascular workout.

Calorie Expenditure

The number of calories burned during Nordic Walking can vary depending on factors such as body weight, walking speed, terrain, and pole use. On average, a 150-pound (68 kg) person can burn approximately 400-500 calories per hour of Nordic Walking at a moderate pace.

Increasing the intensity, either by walking faster or incorporating intervals, can significantly increase calorie burn. For example, a 150-pound (68 kg) person can burn up to 600-700 calories per hour of vigorous Nordic Walking.

Intensity and Demands

The intensity of Nordic Walking can be easily adjusted to suit individual fitness levels and goals. Beginners can start with shorter durations and a slower pace, gradually increasing as their fitness improves. More advanced Nordic Walkers can incorporate intervals, hills, and longer distances to challenge themselves and continue making progress.

The use of poles and the increased upper body engagement in Nordic Walking can make it feel more demanding compared to regular walking, even at a similar pace. However, the low-impact nature of the activity makes it accessible and sustainable for a wide range of people, from beginners to athletes.

Precautions for Individuals with Illnesses

While Nordic Walking is generally a safe and beneficial activity, individuals with certain health conditions should take precautions and consult with their healthcare provider before starting a program. This includes:

Cardiovascular conditions, such as heart disease, hypertension, or peripheral artery disease
Respiratory conditions, such as asthma or chronic obstructive pulmonary disease (COPD)
Metabolic conditions, such as diabetes or thyroid disorders
Musculoskeletal conditions, such as arthritis, osteoporosis, or chronic pain

A healthcare provider can offer guidance on appropriate intensity levels, any necessary modifications to the Nordic Walking technique, and warning signs to watch for during exercise. They may also recommend a supervised cardiac rehabilitation program for those with cardiovascular conditions.

Emergency Situations and Response

Although rare, emergency situations can occur during Nordic Walking. It's important to be prepared and know how to respond:

Recognize signs and symptoms of distress, such as chest pain, shortness of breath, dizziness, or sudden weakness.
If you or someone else experiences these symptoms, stop the activity immediately and assess the situation.
If the symptoms persist or worsen, call emergency services (e.g., 911 in the United States) and follow their instructions.
If you are trained in CPR and the person is unresponsive and not breathing, begin CPR and continue until emergency help arrives.

To prepare for potential emergencies, consider taking a CPR and first aid certification course. Always carry a mobile phone and identification with emergency contact information during your Nordic Walking sessions. Let someone know your intended route and expected return time, especially if walking alone or in remote areas.

By understanding the health benefits, comparisons to other activities, precautions for specific conditions, and emergency response, you can make informed decisions and safely enjoy the many advantages of Nordic Walking as a fitness activity.

The Walks

I've written all the chapters to be fully self-contained, so you don't need to refer to other sections. The time estimates are for a leisurely stroll with a pet tortoise rather than a full-speed Nordic Walk, but when you've done one or two of them, you'll know how much faster you're likely to be.

AI makes mistakes

Almost all of the book was written by an AI Large Language Model, in this case Claude 3 Opus, in April 2024. It is generally pretty good, but it makes occasional mistakes and I haven't checked all the details. So if it tells you there is a toilet half way, and it turns out that is true but it's on the other side of the estuary, well, find a nice bush, hide behind it, and think about how safe you are from AI taking over the world.

Minsmere Walk 1 - 5.5 miles (8.9 km), Circular - Estimated Time: 3-4 hours

Starting Point Coordinates, Postcode: TM 473 672, Saxmundham IP17 3BY
Nearest Car Park, Postcode: Minsmere RSPB Visitor Centre, Sheepwash Ln, Saxmundham IP17 3BY

Embark on an unforgettable 5.5-mile (8.9 km) circular walk through the captivating RSPB Minsmere nature reserve, a hidden gem nestled along the picturesque Suffolk coast near Leiston. This walk is a true paradise for birdwatchers, nature enthusiasts, and anyone seeking to immerse themselves in the breathtaking beauty of the great outdoors.

As you step onto the trail, you'll find yourself transported into a world of diverse habitats, each one offering its own unique charm and wildlife encounters. From the enchanting woodland paths to the rugged heathland, the serene wetlands to the vibrant reedbeds, Minsmere is a true mosaic of natural wonders waiting to be explored.

Throughout your journey, keep your eyes peeled and your senses heightened, as you'll have the opportunity to spot some of the UK's rarest and most fascinating bird species. Listen for the haunting boom of the elusive bittern, a master of camouflage among the swaying reeds. Watch in awe as the majestic marsh harriers soar overhead, their distinctive silhouettes painting the sky. And delight in the charming antics of the bearded tits, flitting through the reedbeds like tiny feathered acrobats.

But the avian wonders are just the beginning of what Minsmere has to offer. As you traverse the coastal path, you'll be treated to breathtaking vistas of the glistening sea and the ever-changing tapestry of the Suffolk coastline. Feel the salty breeze on your face and the soft sand beneath your feet as you walk alongside the shore, perhaps even spotting the playful grey seals basking on the nearby beaches or bobbing curiously in the waves.

The Minsmere Walk is not just a physical journey, but a sensory experience that will leave you feeling rejuvenated and inspired. Immerse yourself in the sights, sounds, and scents of this extraordinary reserve, and let the worries of the world fade away as you connect with the raw beauty of nature.

Along the way, you'll encounter reminders of the area's rich history, such as the distinctive Scotts Hall, a historic building nestled within the reserve. You'll also have the chance to learn about the vital conservation work carried out by the RSPB, as they tirelessly strive to protect and preserve this unique ecosystem for generations to come.

As you make your way back to the visitor centre, taking in the final stretch of this incredible walk, you'll find yourself filled with a sense of accomplishment and a deeper appreciation for the natural world. The memories you've made and the wildlife encounters you've experienced will stay with you long after you've left the reserve, serving as a reminder of the precious beauty that exists just beyond our doorsteps.

So lace up your walking boots, grab your binoculars, and get ready to embark on an unforgettable adventure through the stunning landscapes of RSPB Minsmere. Whether you're a seasoned birdwatcher or simply a lover of the great outdoors, this walk promises to delight, inspire, and leave you with a renewed sense of wonder for the incredible natural heritage that surrounds us.

Detailed Route Directions:

1. Begin your adventure at the Minsmere RSPB Visitor Centre car park (Grid Reference: TM 473 672). From the car park, walk towards the visitor centre and pass through the wooden gate to the right of the building. Follow the path heading east (bearing 090°) towards the reserve entrance. (Estimated time: 5-10 minutes)

2. After approximately 200 meters, you'll pass through another wooden gate, officially entering the reserve. Immediately after the gate, take the right-hand fork in the path (bearing 135°), which leads you along a charming wooded area, gradually approaching the coast. (Estimated time: 10-15 minutes)

3. Continue along this path for about 500 meters until you reach a junction with a signpost. Turn left here (bearing 010°) and follow the footpath northward, hugging the stunning shoreline. You'll pass through a kissing gate and continue along the coastal path. (Estimated time: 15-20 minutes)

4. Immerse yourself in the beauty of the coastal path for approximately 1.5 miles (2.4 km), soaking in the awe-inspiring views of the sea and the surrounding habitats. Along this stretch, you'll encounter several benches, offering perfect spots to rest and take in the scenery. Keep an eye out for the iconic Minsmere Lookout, a raised platform providing panoramic views of the coast and the reserve. (Estimated time: 45-60 minutes)

5. Upon reaching the northern end of the reserve, marked by a wooden signpost, make a left turn (bearing 280°) and venture inland along the footpath, navigating through the enchanting reedbeds and wetlands. This section of the walk offers excellent opportunities for birdwatching, so have your binoculars ready! (Estimated time: 30-45 minutes)

6. Continue your journey along the footpath for about 1 mile (1.6 km) as it curves southwards (bearing 190°), guiding you through a tapestry of woodland and heathland habitats. Look out for the distinctive Scotts Hall, a historic building nestled within the reserve. (Estimated time: 30-45 minutes)

7. After approximately 1 more mile (1.6 km), you'll arrive at a junction with a signpost pointing towards the Island Mere Hide. Turn right here (bearing 230°) and follow the path for about 200 meters until you reach the hide, an exceptional spot for birdwatching and observing the local wildlife in their natural habitat. Spend some time here, as it offers a fantastic opportunity to see rare birds like bitterns and marsh harriers. (Estimated time: 30-45 minutes)

8. When you're ready to continue, retrace your steps back to the junction and turn right (bearing 320°), following the footpath towards the Minsmere RSPB Visitor Centre. After about 1 mile (1.6 km), you'll pass through a wooden gate, leaving the main part of the reserve. (Estimated time: 30-45 minutes)

9. From here, continue straight ahead (bearing 270°) for another 200 meters until you reach the visitor centre car park, where your circular walk comes to a satisfying conclusion. (Estimated time: 5-10 minutes)

Best Features and Views:

- Stunning coastal vistas that take your breath away, particularly from the Minsmere Lookout
- Diverse habitats, including enchanting woodland, rugged heathland, serene wetlands, and vibrant reedbeds
- Unparalleled birdwatching opportunities at the Island Mere Hide, with chances to spot rare species
- Historic Scotts Hall, adding a touch of heritage to the natural beauty of the reserve

Birds to Spot:

Keep your eyes peeled for the elegant avocets, elusive bitterns, majestic marsh harriers, charming bearded tits, and a wide variety of waders and wildfowl that call this reserve home. The Island Mere Hide is a particularly good spot for sightings. During spring and summer, listen for the distinctive booming call of the bittern echoing across the reedbeds.

Rare or Interesting Insects and Other Wildlife:

- Aerial acrobats: Marvel at the colourful butterflies fluttering by, the darting dragonflies, and the iridescent damselflies that add a touch of magic to the reserve.
- Graceful grazers: Keep an eye out for the majestic red deer that roam the reserve, particularly during the early morning or late evening hours.
- Coastal charmers: Spot the playful grey seals basking on the nearby beaches or bobbing in the waves, and delight in their curious nature.

Rare or Interesting Plants:

- Wetland wonders: Discover the unique wetland plants, such as the delicate bog bean with its fringed white flowers and the vibrant marsh cinquefoil with its sunny yellow blooms.
- Heathland hues: Admire the iconic heathland plants like the fragrant purple heather and the spiky yellow gorse, which paint the landscape in a stunning palette.
- Woodland wildflowers: Spot a variety of ferns and wildflowers in the woodland areas, from the dainty wood anemones to the striking foxgloves.

Facilities and Accessibility:

The Minsmere RSPB Visitor Centre offers a range of facilities to enhance your experience. The well-maintained toilets include accessible stalls and baby changing facilities. The cozy café serves a variety of refreshments and light meals, catering to different dietary requirements. The gift shop is filled with nature-inspired treasures, from books and souvenirs to binoculars and walking gear.

The nature reserve boasts a selection of accessible trails and hides, with some paths suitable for wheelchairs and mobility scooters. The visitor centre provides wheelchair hire for those who require it, subject to availability. The friendly staff at the visitor centre are always happy to offer expert advice, provide maps, and share wildlife guides to enrich your experience.

Please note that some sections of the footpath may be uneven or muddy, particularly after rainfall, so sturdy walking shoes or boots are highly recommended. The coastal path may be subject to erosion, so please stay on the designated path and follow any temporary diversion signs.

Seasonal Highlights:

The Minsmere reserve offers a dynamic and ever-changing landscape that showcases the best of each season:

- Spring: Witness the arrival of migrant birds, such as the wheatears and warblers, and enjoy the vibrant displays of wildflowers like the cowslips and bluebells. The dawn chorus of breeding birds fills the air with enchanting melodies.

- Summer: Delight in the sight of butterflies, dragonflies, and damselflies flitting amongst the wildflowers. Watch out for young birds fledging from their nests, and spot the iconic avocets with their striking black and white plumage.

- Autumn: Marvel at the gathering of thousands of wading birds and wildfowl, such as the black-tailed godwits and wigeons, as they prepare for migration or settle in for the winter. Admire the golden hues of the changing foliage in the woodland areas.

- Winter: Admire the large flocks of ducks, geese, and swans that seek refuge in the wetlands, including the majestic whooper swans. Keep an eye out for winter visitors like fieldfares and redwings, and enjoy the tranquility of the quieter months.

Difficulty Level and Safety Considerations:

This walk is rated as easy to moderate, with mostly flat terrain. Some sections of the path may be uneven or muddy, so appropriate footwear is essential. Be mindful of tidal conditions along the coast and avoid walking during high tide if possible.

Stay on the designated paths to minimize disturbance to the wildlife and follow any signs or instructions provided by the reserve staff. Remember to bring appropriate clothing for the weather conditions, as well as sunscreen, a hat, and insect repellent during the summer months. Carry enough water and snacks to keep you energized throughout the walk.

In case of emergencies, the visitor centre staff are trained in first aid and can provide assistance. Let someone know your planned route and expected return time before setting off on your walk.

Nearby Attractions:

- Dunwich Heath and Beach: Extend your exploration by visiting this stunning coastal landscape, featuring heathland and shingle beaches. It's perfect for a post-walk picnic or a refreshing swim.
- Leiston Abbey: Step back in time and discover the ruins of this once-thriving medieval monastery, founded in the 14th century by the Premonstratensian order.
- Long Shop Museum: For a taste of local industrial history, visit this museum in Leiston, which showcases the region's agricultural and engineering past.
- Suffolk Coast and Heaths AONB: Explore the breathtaking beauty and diverse wildlife of this Area of Outstanding Natural Beauty, which encompasses a variety of habitats, from estuaries to forests.

The Minsmere Walk offers a truly unforgettable experience for nature lovers and outdoor enthusiasts. With its diverse habitats, stunning coastal views, and incredible wildlife, this circular walk is sure to leave you with lasting memories and a deeper appreciation for the natural world. Embrace the opportunity to disconnect from the hustle and bustle of everyday life and immerse yourself in the serene beauty of the Suffolk coast.

RSPB Minsmere Walk 2 - 3.5 miles (5.6 km) - Circular - Estimated Time: 2-3 hours

Starting Point Coordinates, Postcode: TM 473 672, RSPB Minsmere IP16 4TU
Nearest Car Park, Postcode: RSPB Minsmere Visitor Centre car park, IP16 4TU

Embark on a captivating 5.5-mile (8.9 km) circular walk through the stunning RSPB Minsmere nature reserve, nestled along the picturesque Suffolk coast near Leiston. This walk is a paradise for birdwatchers and nature enthusiasts, offering an array of diverse habitats and an abundance of bird species to discover. As you traverse the trail, you'll be treated to breathtaking coastal views and the opportunity to spot fascinating wildlife, such as graceful deer and playful seals.

Along the coastal path, keep an eye out for the remnants of World War II defenses, including pill boxes and tank traps. These structures were part of the "Coastal Crust," a line of fortifications designed to protect the British coast from potential German invasions. The pill boxes served as small defensive forts, manned by soldiers to keep watch and defend the shoreline. Although no major battles took place here, the presence of these historical relics adds a layer of intrigue to the walk, reminding us of the area's wartime past.

As you walk through the wetlands and reedbeds, you'll have the chance to spot some of the UK's rarest bird species. The elusive bittern, known for its booming call that can carry for miles, has made a remarkable comeback at Minsmere. Thanks to the RSPB's conservation efforts, the bittern population has increased from just one booming male in 1997 to over 100 birds today. Listen carefully, and you might hear their distinctive call echoing across the reeds, particularly during the early morning or evening hours.

Another avian highlight is the graceful marsh harrier, a majestic bird of prey that has found refuge in Minsmere's expansive reedbeds. In the 1970s, the marsh harrier was on the brink of extinction in the UK, with only a single nesting pair remaining. However, dedicated conservation work has helped the species recover, and Minsmere now boasts one of the largest populations of marsh harriers in the country.

The Island Mere Hide, accessible via a combination of gravel paths and boardwalks, offers an exceptional opportunity to observe these fascinating birds and other wildlife. As you sit quietly in the hide, you may witness the incredible spectacle of hundreds of wading birds, such as avocets and black-tailed godwits, gathering to feed and rest during their migration. In the winter months, the mere becomes a haven for ducks, geese, and swans, including the regal whooper swans that travel from Iceland to escape the harsh Arctic winters.

Throughout the walk, take a moment to appreciate the vital conservation work carried out by the RSPB. Minsmere is a shining example of how dedicated efforts can restore and protect habitats, ensuring a future for countless species. The reserve's mosaic of habitats, including reedbeds, wetlands, heathland, and woodland, is carefully managed to provide optimal conditions for a wide range of flora and fauna.

As you conclude your walk and return to the visitor centre, reflect on the incredible diversity of life you've encountered and the stories of resilience and recovery that define Minsmere. This walk is more than just a physical journey; it is a testament to the power of nature and the importance of conservation in preserving our natural heritage for generations to come.

Route Directions:

1. Begin your journey at the RSPB Minsmere Visitor Centre car park (Grid Reference: TM 473 672). From the car park, follow the well-marked footpath leading towards the heart of the reserve, bearing 090°. (Estimated time: 10-15 minutes)

2. As you step into the reserve, take the left-hand fork in the path (bearing 315°), venturing through the enchanting woodland. Keep your senses attuned to the melodic calls and flitting movements of various woodland bird species, such as the striking woodpeckers, the agile nuthatches, and the spirited treecreepers. (Estimated time: 20-30 minutes)

3. Continue along the footpath, bearing 000°, as it guides you towards the captivating realm of reedbeds and wetlands. Here, pause to observe the graceful marsh harriers soaring overhead, the elusive bearded tits darting amongst the swaying reeds, and the majestic bitterns, whose booming calls resonate through the tranquil air. (Estimated time: 30-45 minutes)

4. From the reedbeds, follow the footpath (bearing 045°) as it leads you towards the coastal dunes, where you'll be greeted by awe-inspiring views of the glistening sea and the sprawling landscape that unfolds before you. (Estimated time: 20-30 minutes)

5. As you traverse the dunes, bearing 090°, keep your binoculars at the ready to spot the myriad coastal bird species that call this area home. Marvel at the elegant terns, the scurrying waders, and the wheeling gulls as they dance across the sky and the shoreline. (Estimated time: 30-45 minutes)

6. Complete your circular walk by following the footpath (bearing 180°) back to the RSPB Minsmere Visitor Centre car park, where your adventure began. (Estimated time: 20-30 minutes)

Best Features and Views:

- Immerse yourself in the diverse habitats of Minsmere, from the whispering woodland to the undulating coastal dunes.
- Unparalleled birdwatching opportunities, with the chance to spot a wide array of species in their natural environments. The best times for birdwatching are typically early morning or late afternoon, when birds are most active.
- Stunning coastal vistas that stretch out to the horizon, offering moments of serenity and connection with nature. The golden hour before sunset provides a particularly enchanting light for photography.

Birds to Spot:

- Woodland wonders: Woodpeckers, nuthatches, treecreepers, and a symphony of other woodland species.
- Reedbed residents: Marsh harriers, bearded tits, bitterns, and a host of wetland birds. The best time to hear the booming call of the bittern is during the early morning or evening hours in spring and early summer.
- Coastal companions: Terns, waders, gulls, and an ever-changing cast of avian visitors. During the migration seasons (spring and autumn), you can witness impressive gatherings of wading birds on the coastal mudflats.

Rare or Interesting Insects and Wildlife:

- Fluttering jewels: Discover a kaleidoscope of butterflies and dragonflies dancing amongst the reeds and wildflowers. Summer months are ideal for spotting these colourful insects.

- Majestic mammals: Keep an eye out for the graceful deer that roam the reserve's tranquil landscapes, particularly during the early morning or late evening hours.

Rare or Interesting Plants:

- Marshland marvels: Uncover the unique and adapted plants that thrive in the wetland habitats, such as the delicate marsh orchids and the vibrant purple loosestrife.
- Coastal flora: Explore the hardy and resilient vegetation that clings to the windswept dunes, including the distinctive marram grass and the colourful sea holly.
- Woodland wonders: Identify the diverse array of trees, shrubs, and wildflowers that paint the woodland canvas, from the stately oaks to the delicate bluebells that carpet the forest floor in spring.

Facilities and Accessibility:

The RSPB Minsmere Visitor Centre offers a range of facilities to enhance your experience. The well-maintained toilets are equipped with baby changing facilities and accessible stalls. The cozy café serves a variety of refreshments and light meals, with options for various dietary requirements, including vegetarian and gluten-free choices. The well-stocked shop offers a wide selection of nature-inspired gifts, books, and souvenirs, as well as essential items like binoculars and walking gear.

Several strategically placed birdwatching hides, some of which are wheelchair accessible, allow you to observe the reserve's avian residents up close. The Island Mere Hide, accessed via a combination of gravel paths and boardwalks, provides excellent views of the mere and its diverse birdlife. The visitor centre staff can provide the most up-to-date information on accessibility and assist with any specific requirements you may have.

To ensure a comfortable and enjoyable walk, it is recommended to wear sturdy walking shoes or boots that provide good support and traction, as the paths may be uneven or muddy in places. The visitor centre also offers a limited number of mobility scooters and wheelchairs for hire, subject to availability.

Seasonal Highlights:

The RSPB Minsmere reserve offers a dynamic and ever-changing landscape that showcases the best of each season:

- Spring: Witness the arrival of migratory birds, such as the elegant sandwich terns and the vibrant Mediterranean gulls. Delight in the blooming of delicate wildflowers, like the early purple orchids and the cowslips, and the awakening of the reserve's wildlife.

- Summer: Delight in the sight of breeding birds, such as the adorable avocets with their distinctive upturned beaks and the majestic marsh harriers soaring over the reedbeds. Admire the vibrant colours of the heathland, with the purple heather in full bloom, and the lush growth of the wetland vegetation.

- Autumn: Marvel at the spectacle of migratory birds gathering in preparation for their long journeys, including the striking black-tailed godwits and the elegant little egrets. Admire the golden hues of the changing foliage, as the leaves of the woodland trees turn to shades of amber and crimson.

- Winter: Discover the stark beauty of the wintering bird species, such as the regal whooper swans and the lively wigeons. Embrace the frosty landscapes, with the reedbeds and wetlands transformed into a winter wonderland, and enjoy the tranquility of the quieter months.

Difficulty Level and Safety Considerations:

This walk is classified as easy to moderate, depending on the specific paths you choose to follow. The terrain varies from well-maintained gravel paths and boardwalks to more uneven grass and sand tracks. As with any outdoor activity, it's essential to be mindful of the weather conditions and to dress appropriately. Bring layers to accommodate changes in temperature and a waterproof jacket in case of rain.

Stay on the designated paths to minimize disturbance to the delicate ecosystems and to ensure your safety. Be respectful of the wildlife by maintaining a safe distance and avoiding any actions that may cause stress or harm to the animals. If you encounter any livestock, such as grazing sheep or cattle, give them plenty of space and keep dogs on a lead.

Remember to carry water, snacks, and any necessary medications, and inform someone of your planned route and expected return time. In case of emergencies, the visitor centre staff are trained in first aid and can provide assistance.

Responsible Walking:

To minimize your impact on the environment and preserve the natural beauty of RSPB Minsmere, please follow these responsible walking practices:

- Stay on the designated paths to avoid disturbing sensitive habitats and wildlife.
- Respect wildlife by observing from a distance and avoiding any actions that may cause stress or harm to animals.
- Dispose of litter properly, either in the provided bins or by taking it home with you.
- Keep dogs on a lead and under control to prevent disturbance to wildlife.
- Support the conservation efforts of the RSPB by following their guidelines and considering a donation or membership.

By adopting these responsible walking practices, you can ensure that the RSPB Minsmere reserve remains a thriving haven for wildlife and a source of inspiration for generations of walkers and nature enthusiasts to come.

Nearby Attractions:

After your invigorating walk, take the time to explore some of the other fascinating attractions in the area:

- Leiston Abbey: Step back in time and discover the ruins of this once-thriving medieval monastery, founded in the 14th century by the Premonstratensian order. Learn about the abbey's history and its role in the local community.

- Dunwich Heath and Beach: Immerse yourself in the stunning coastal landscapes and vibrant heathland of this National Trust site. Explore the ruins of the lost city of Dunwich, which was once a thriving port before being claimed by the sea.

- Suffolk Coast and Heaths Area of Outstanding Natural Beauty: Explore the breathtaking beauty and diverse wildlife of this protected landscape, which encompasses a variety of habitats, from estuaries to forests. Discover the rich history and culture of the area, with its charming villages, ancient churches, and iconic landmarks like the Southwold Lighthouse.

- RSPB Minsmere Events: Check the RSPB Minsmere website or visit the visitor centre to find out about special events, guided walks, and family activities offered throughout the year. These events provide unique opportunities to learn about the reserve's wildlife, conservation efforts, and the fascinating stories behind its history.

Dunwich Heath Walk 1 - 4 miles (6.4 km), Circular - Estimated Time: 2-3 hours

Starting Point Coordinates, Postcode: TM 477 677, Saxmundham IP17 3DJ
Nearest Car Park, Postcode: Dunwich Heath National Trust car park, Minsmere Road, Dunwich, Saxmundham IP17 3DJ

Embark on an enchanting 4-mile (6.4 km) circular walk through the captivating landscapes of Dunwich Heath, a hidden gem nestled along the picturesque Suffolk coast near the charming village of Dunwich. This mesmerizing walk invites you to immerse yourself in the breathtaking beauty of the vibrant heathland, the allure of the stunning coastal scenery, and the opportunity to spot a diverse array of fascinating bird species, all while uncovering the intriguing history that has shaped this unique area.

As you step onto the trail, you'll find yourself transported into a world of natural wonders, where the vibrant hues of blooming heather and gorse paint the landscape in a mesmerizing tapestry of purple and gold. The fresh coastal breeze carries with it the invigorating scent of the sea, while the gentle rustling of the vegetation creates a soothing symphony that accompanies your every step.

Throughout your journey, keep your senses attuned to the abundant wildlife that calls Dunwich Heath home. Listen for the distinct calls of the lively Dartford warblers as they flit amongst the heather, their vibrant presence adding a spark of energy to the serene surroundings. Watch for the striking stonechats perched atop gorse bushes, their vibrant orange breasts catching the sunlight and creating a stunning contrast against the verdant backdrop.

As you make your way towards the coast, the landscape opens up to reveal breathtaking vistas of the glistening sea stretching out to the horizon. The soft sand beneath your feet and the rhythmic sound of the waves lapping against the shore create a sensory experience that invigorates the soul and calms the mind. Take a moment to pause and drink in the awe-inspiring beauty of the Suffolk coastline, with its towering cliffs and the endless expanse of the North Sea.

But the natural wonders of Dunwich Heath are not limited to its avian inhabitants and coastal charms. Keep an eye out for the elusive adders basking in the sun, their intricate patterns and graceful movements a testament to the delicate balance of the ecosystem. Watch for the playful antics of rabbits as they scamper amongst the heather and gorse, their presence a reminder of the thriving wildlife that flourishes in this unique habitat.

As you make your way back through the heathland, take a moment to appreciate the diverse array of grasses and wildflowers that sway in the coastal breeze. Each plant tells a story of resilience and adaptation, thriving in the challenging conditions of the heath and adding to the intricate tapestry of colours and textures that define this enchanting landscape.

The Dunwich Heath walk is not merely a physical journey, but an opportunity to reconnect with the raw beauty of nature and to find solace in the tranquility of the great outdoors. As you complete your circular route and return to the starting point, you'll find yourself filled with a sense of rejuvenation and a deeper appreciation for the precious natural heritage that surrounds us.

So lace up your walking boots, grab your binoculars, and prepare to embark on an unforgettable adventure through the captivating landscapes of Dunwich Heath. Whether you're a passionate birdwatcher, a nature enthusiast, or simply seeking a moment of serenity amidst the beauty of the Suffolk coast, this walk promises to delight, inspire, and leave you with memories that will linger long after you've left the heath behind.

Detailed Route Directions:

1. Begin your adventure at the Dunwich Heath National Trust car park (Grid Reference: TM 477 677). From the car park, follow the well-marked footpath heading south, bearing 180°, as it leads you into the heart of the heathland. (Estimated time: 10-15 minutes)

2. As you continue along the footpath, take a moment to admire the mesmerizing tapestry of colours created by the blooming heather and gorse, their vibrant hues painting the landscape in shades of purple and gold. (Estimated time: 15-20 minutes)

3. Upon reaching the southern edge of Dunwich Heath, turn left, bearing 270°, and follow the footpath as it meanders westward, running parallel to the captivating coastline. (Estimated time: 20-30 minutes)

4. After approximately 1 mile (1.6 km), the footpath will gently curve northward, bearing 000°, guiding you towards the inviting expanse of the beach. (Estimated time: 20-30 minutes)

5. As you arrive at the beach, turn left, bearing 270°, and immerse yourself in the sensory experience of walking along the shoreline. Feel the soft sand beneath your feet, listen to the soothing rhythm of the waves, and take in the awe-inspiring views of the glistening sea and the towering cliffs that rise above it. (Estimated time: 30-45 minutes)

6. Continue your leisurely stroll along the beach for about 1 mile (1.6 km) before turning left, bearing 000°, to head inland through the enchanting heathland once more. (Estimated time: 20-30 minutes)

7. Follow the footpath as it guides you back to the Dunwich Heath National Trust car park, bearing 045°, where your circular walk comes to a satisfying conclusion. (Estimated time: 15-20 minutes)

Best Features and Views:

- Marvel at the breathtaking expanse of Dunwich Heath, adorned with the vibrant colours of heather and gorse, creating a mesmerizing tapestry that shifts with the seasons.
- Immerse yourself in the stunning coastal views, with the glistening sea stretching out to the horizon and the majestic cliffs rising from the shore, offering a humbling reminder of the power and beauty of nature.
- Delight in the opportunity to observe a diverse array of birdlife in their natural habitats, from the lively Dartford warblers to the striking stonechats, each species adding its own unique character to the heathland symphony.

Birds to Spot:

- Flitting figments: Darting Dartford warblers: These lively little birds flit amongst the heather, their distinct calls filling the air with a joyful energy that perfectly complements the vibrant surroundings.
- Perching sentinels: Watch for these striking birds as they perch atop gorse bushes, their vibrant orange breasts catching the sunlight and creating a stunning contrast against the deep green of the vegetation.
- Melodic maestros: Listen for the enchanting songs of these ground-nesting birds, whose melodies enrich the heathland soundscape and add a touch of magic to your walk.
- Nocturnal serenaders: Enigmatic nightjars: As dusk falls, keep an ear out for the haunting churring calls of these elusive nocturnal birds, their mysterious presence adding an air of intrigue to the evening landscape.

Rare or Interesting Insects and Wildlife:

- Sunbathing serpents: Basking adders: Tread carefully and respectfully, as you may be lucky enough to spot these beautiful but venomous snakes basking in the sun, their intricate patterns and graceful movements a testament to the delicate balance of the ecosystem.
- Hopping habitants: Bounding rabbits: Watch for the playful antics of rabbits as they scamper amongst the heather and gorse, their presence a reminder of the thriving wildlife that flourishes in this unique habitat.
- Aerial artistry: Fluttering friends: Admire the delicate beauty of the various butterfly and moth species that call Dunwich Heath home, their colourful wings painting the air with ephemeral brushstrokes of nature's artistry.

Rare or Interesting Plants:

- Chromatic carpet: Heather and gorse: These iconic plants paint the heathland in a stunning palette of colours, from the deep purples of heather to the vibrant yellows of gorse, creating a living tapestry that shifts with the seasons.
- Swaying sentinels: Diverse grasses: Take a closer look at the variety of grasses that sway in the coastal breeze, each with its own unique texture and hue, their gentle movement adding a mesmerizing dimension to the landscape.
- Fragile flourishes: Wonderful wildflowers: Discover the delicate blooms of wildflowers that dot the heathland, their tender petals and vibrant colours adding splashes of life to the rugged terrain.

Facilities and Accessibility:

The Dunwich Heath National Trust car park offers essential facilities to enhance your visit, including well-maintained toilets for your comfort and convenience, as well as a charming café where you can enjoy refreshments and light meals, the perfect spot to relax and refuel before or after your walk.

The heathland paths are generally flat, making this walk accessible to most visitors, including those with moderate fitness levels. However, it's important to note that some areas may be uneven or muddy, particularly after rainfall, so it's essential to wear sturdy walking shoes or boots with good grip and support to ensure a comfortable and safe experience.

Seasonal Highlights:

The beauty of Dunwich Heath is ever-changing, with each season bringing its own unique charms and surprises, inviting you to return time and again to witness the landscape's captivating transformations:
- Spring and summer: Delight in the vibrant blooms of heather and gorse, their colours painting the heathland in a mesmerizing palette, while the lively activity of nesting birds fills the air with a joyous symphony of life.
- Autumn: Admire the changing colours of the heathland as it transitions into a rich tapestry of russet and gold, the warm hues of the fading foliage creating an enchanting and cozy atmosphere that invites contemplation and reflection.
- Winter: Embrace the stark beauty of the heathland, with its frosty mornings and the occasional dusting of snow, the monochromatic landscape offers a serene and meditative experience that allows you to connect with the raw essence of nature.

Difficulty Level and Safety Considerations:

This walk is classified as easy, with mostly flat terrain, making it suitable for walkers of all ages and abilities.

However, it's essential to stay on the designated paths to avoid disturbing the delicate ecosystem and to minimize the risk of encountering adders, which are venomous snakes native to the area. While the chances of an encounter are rare, it's crucial to remain vigilant and give these creatures a wide berth should you spot one.

As with any outdoor activity, it's important to be prepared for changing weather conditions. Bring appropriate clothing, including layers for warmth and a waterproof jacket to protect you from potential rain showers. The coastal breeze can be refreshing, but it can also make the temperature feel cooler than expected, so it's wise to be prepared.

Remember to carry an adequate supply of water and some energizing snacks to keep you hydrated and fuelled throughout your walk. It's also a good idea to pack any necessary medications and to inform someone of your intended route and expected return time, just as a precautionary measure.

Nearby Attractions:

After your invigorating walk, take the time to explore some of the other fascinating attractions that the area has to offer, each one providing a unique glimpse into the rich history and natural wonders of the Suffolk coast:

- Dunwich Museum: Delve into the captivating history of this once-thriving medieval port town and learn about the dramatic coastal erosion that has shaped the area, forever altering the landscape and the lives of those who called it home.
- Greyfriars' Abbey ruins: Step back in time and discover the haunting ruins of this 13th-century Franciscan friary, its crumbling walls and eerie atmosphere offering a glimpse into the spiritual and architectural heritage of the region.
- RSPB Minsmere nature reserve: Venture into one of the UK's premier birdwatching destinations, with its diverse habitats and abundant wildlife, offering a chance to immerse yourself in the wonders of nature and witness the incredible diversity of the Suffolk coast.

By exploring these nearby attractions, you'll gain a deeper appreciation for the rich tapestry of history, culture, and natural beauty that defines this extraordinary corner of England, making your visit to Dunwich Heath a truly unforgettable experience.

Dunwich Heath and Beach Walk 2 - 4.5 miles (7.2 km) - Circular - Estimated Time: 2-3 hours

Starting Point Coordinates, Postcode: TM 477 678, Dunwich IP17 3DJ
Nearest Car Park, Postcode: Dunwich Heath National Trust car park, Dunwich IP17 3DJ

Embark on an enchanting 4.5-mile (7.2 km) circular walk through the breathtaking landscapes of Dunwich Heath, a treasured National Trust site, and along the picturesque beach at Dunwich. This captivating walk invites you to explore a tapestry of diverse habitats, from the vibrant heathland to the lush woodland and the rolling coastal dunes, making it an ideal adventure for nature enthusiasts and those seeking a tranquil escape.

As you step onto the trail, you'll find yourself immersed in a world of natural wonders, where the vibrant hues of the heather and gorse create a stunning palette that paints the heathland in shades of purple and yellow. The gentle sea breeze carries with it the invigorating scent of the coast, while the rustling of the vegetation and the calls of the birds create a symphony of nature that accompanies your every step.

Throughout your journey, keep your senses attuned to the abundant wildlife that calls Dunwich Heath and Beach home. Listen for the distinct calls of the darting Dartford warblers as they flit amongst the heather, their lively presence adding a spark of energy to the serene surroundings. As dusk falls, the haunting churring songs of the enigmatic nightjars fill the air, creating an enchanting atmosphere that is both mysterious and captivating.

As you make your way through the lush woodland, the dappled sunlight filters through the canopy, casting a magical glow on the forest floor. The silver bark of the birch trees stands in stark contrast to the verdant foliage, creating a mesmerizing interplay of light and shadow. Take a moment to pause and appreciate the intricate details of the woodland ecosystem, from the delicate wildflowers to the scurrying creatures that call this habitat home.

Upon reaching the coastal path, the landscape transforms once again, revealing the awe-inspiring expanse of Dunwich Beach. The pebbled shores and the rolling dunes create a rugged beauty that is both humbling and invigorating. As you stroll along the shoreline, keep an eye out for the graceful terns, the spirited gulls, and the scurrying wading birds that frequent this coastal haven, each species adding its own unique character to the symphony of the sea.

As you make your way back through the heathland, take a moment to appreciate the hardy plants that thrive in this challenging environment. The heather and gorse, with their vibrant blooms, are a testament to the resilience and adaptability of nature. Keep an eye out for the basking adders, the slow-motion slow worms, and the fluttering butterflies that call this habitat home, each creature playing a vital role in the intricate web of life that defines Dunwich Heath.

The Dunwich Heath and Beach walk is not merely a physical journey, but an opportunity to reconnect with the raw beauty of nature and to find solace in the tranquility of the great outdoors. As you complete your circular route and return to the starting point, you'll find yourself filled with a sense of rejuvenation and a deeper appreciation for the precious natural heritage that surrounds us.

So lace up your walking boots, grab your binoculars, and prepare to embark on an unforgettable adventure through the captivating landscapes of Dunwich Heath and Beach. Whether you're a passionate birdwatcher, a nature enthusiast, or simply seeking a moment of serenity amidst the beauty of the Suffolk coast, this walk promises to delight, inspire, and leave you with memories that will linger long after you've left the heathland behind.

Route Directions:

1. Begin your journey at the Dunwich Heath National Trust car park (Grid Reference: TM 477 678). From the car park, follow the well-marked footpath, bearing 090°, as it leads you into the enchanting heathland. (Estimated time: 15-20 minutes)

2. As you traverse the heathland, take a moment to admire the stunning array of colours, from the deep purples of the heather to the vibrant yellows of the gorse. Keep your eyes and ears open for the darting Dartford warblers, the melodic songs of the nightjars, and the soaring woodlarks that call this habitat home. (Estimated time: 20-30 minutes)

3. Upon reaching the end of the heathland, turn left, bearing 000°, onto a footpath that guides you into a captivating wooded area. Immerse yourself in the sights, sounds, and scents of the woodland, as dappled sunlight filters through the canopy and the gentle rustling of leaves fills the air. (Estimated time: 20-30 minutes)

4. Continue along the woodland footpath until it joins the coastal path. Here, turn left, bearing 270°, and make your way towards the inviting expanse of Dunwich Beach. (Estimated time: 15-20 minutes)

5. As you step onto the shingle beach, pause to take in the awe-inspiring views of the North Sea stretching out before you. Stroll along the shoreline, feeling the satisfying crunch of the pebbles beneath your feet, and keep an eye out for the various bird species that frequent this coastal haven, from the graceful terns to the spirited gulls and the scurrying wading birds. (Estimated time: 30-45 minutes)

6. Continue your leisurely walk along the beach, bearing 180°, until you reach a footpath that leads you back into the captivating heathland. Turn left onto this path, bearing 090°, and begin your return journey. (Estimated time: 20-30 minutes)

7. Follow the footpath as it winds through the heathland, once again immersing you in the vibrant colours and diverse flora and fauna that make this habitat so special. Take your time to appreciate the intricate details of the landscape, from the delicate wildflowers to the basking adders and the fluttering butterflies. (Estimated time: 20-30 minutes)

8. As you near the end of your walk, you'll arrive back at the Dunwich Heath National Trust car park, where your circular adventure comes to a satisfying conclusion. (Estimated time: 5-10 minutes)

Best Features and Views:

- Lose yourself in the breathtaking beauty of Dunwich Heath, with its vibrant heather and gorse, creating a stunning tapestry of colours that shifts with the seasons, from the deep purples of summer to the rich golds of autumn.
- Discover the captivating coastal dunes and the expansive Dunwich Beach, with its pebbled shores and awe-inspiring views of the North Sea, where the rhythmic sound of the waves and the salty sea breeze create a sensory experience that invigorates the soul.
- Delight in the opportunity to observe a diverse array of bird species in their natural habitats, from the heathland specialists like the Dartford warblers and nightjars to the coastal visitors such as the terns, gulls, and wading birds, each adding their own unique character to the symphony of nature.

Birds to Spot:

- Feathered flashes: Darting Dartford warblers: Listen for their distinct calls and watch as they flit amongst the heather, their lively movements adding a spark of energy to the tranquil heathland.
- Twilight troubadours: Enigmatic nightjars: As dusk falls, keep an ear out for their haunting churring songs, a mesmerizing and mysterious sound that echoes through the evening air.
- Sky-borne serenades: Soaring woodlarks: Admire these ground-nesting birds as they take to the skies, their melodic songs filling the air with a joyful and uplifting tune.
- Shoreline sentinels: Coastal companions: Spot graceful terns, spirited gulls, and scurrying wading birds along the shoreline, each species perfectly adapted to the rhythms and challenges of life by the sea.

Rare or Interesting Insects and Wildlife:

- Serpentine sunbathers: Basking adders: Tread carefully and respectfully, as you may encounter these beautiful but venomous snakes soaking up the sun, their intricate patterns and graceful movements a testament to the complexity of nature.
- Legless wonders: Slow-motion slow worms: These legless lizards are a fascinating sight, often mistaken for snakes, but possessing their own unique charm and adaptations.
- Airborne artistry: Fluttering friends: Marvel at the colourful array of butterflies and dragonflies that dance amongst the heathland flowers, their delicate wings and aerial acrobatics adding a touch of magic to the landscape.

Rare or Interesting Plants:

- Chromatic carpet: Heather and gorse: These iconic plants paint the heathland in a stunning palette of purples and yellows, their vibrant blooms creating a living tapestry that shifts with the seasons and provides a vital habitat for countless species.
- Arboreal artistry: Silver-clad birches: Admire the striking silver bark of the birch trees that stand sentinel in the woodland, their slender trunks and delicate leaves creating a mesmerizing interplay of light and shadow.
- Resilient residents: Coastal flora: Discover the hardy plants that thrive in the challenging conditions of the coastal dunes, their adaptations and tenacity a testament to the incredible diversity and resilience of nature.

Facilities and Accessibility:

The Dunwich Heath Visitor Centre offers a range of facilities to enhance your experience, ensuring that your visit is both comfortable and enjoyable. Take advantage of the well-maintained toilets, perfect for a quick refresh before or after your walk. Stop by the cozy café to indulge in a selection of refreshments and light meals, the ideal spot to refuel and relax while soaking in the stunning views of the heathland. Don't forget to browse the shop, stocked with a delightful array of nature-inspired gifts and souvenirs, allowing you to take a piece of Dunwich Heath home with you.

This walk involves a mix of terrain, from the relatively flat heathland paths to the uneven shingle beach, offering a diverse and engaging experience for walkers. To ensure a comfortable and safe journey, it is highly recommended to wear sturdy walking shoes or boots with good ankle support and grip, allowing you to navigate the varying surfaces with ease and confidence.

Please note that some sections of the walk, particularly along the beach, may not be suitable for those with limited mobility or wheelchair users. The shingle beach can be challenging to traverse, and the uneven terrain may prove difficult for those with mobility concerns.

Seasonal Highlights:

The ever-changing nature of Dunwich Heath and Beach offers unique delights throughout the year, each season bringing its own special character and charm:

- Spring and summer: Witness the heathland burst into life with the blooming of heather and gorse, their vibrant colours painting the landscape in a stunning array of purples and yellows. Delight in the lively activity of breeding birds, as they busily build their nests and raise their young amidst the lush vegetation.
- Autumn: Admire the stunning autumnal hues as the heathland and woodland transform into a rich tapestry of golds, oranges, and reds, the changing foliage creating a warm and inviting atmosphere that invites contemplation and reflection.
- Winter: Embrace the stark beauty of the landscape, with its frosty mornings, windswept beaches, and the occasional sighting of overwintering birds. The monochromatic palette of the heathland in winter offers a serene and meditative experience, allowing you to connect with the raw essence of nature.

Difficulty Level and Safety Considerations:
This walk is classified as moderate due to the mixed terrain, including the shingle beach and some uneven footpaths. While the heathland paths are relatively flat, the beach section can be challenging, with the loose pebbles and shifting surface requiring extra effort and caution.

It's essential to stay on the designated trails to minimize erosion and protect the delicate ecosystems that thrive in this unique habitat. By following the marked paths, you not only ensure your own safety but also contribute to the conservation of this precious natural area.

Be prepared for changeable weather conditions by wearing layers and bringing a waterproof jacket. The coastal winds can be brisk, even on seemingly calm days, so a hat and gloves may also be necessary to keep you comfortable throughout your walk.

Remember to carry an adequate supply of water and some energizing snacks to keep you hydrated and fuelled during your journey. It's also wise to pack any necessary medications and to inform someone of your intended route and expected return time, just as a precautionary measure.

Nearby Attractions:

After your invigorating walk, take the opportunity to explore some of the other fascinating attractions that the area has to offer, each one providing a unique glimpse into the rich history and natural wonders of the Suffolk coast:

- Dunwich Museum: Delve into the rich history of this once-thriving medieval port town and learn about the dramatic coastal erosion that has shaped the area, forever altering the landscape and the lives of those who called it home.
- Greyfriars Priory: Step back in time and discover the haunting ruins of this 13th-century Franciscan priory, its crumbling walls and atmospheric surroundings offering a glimpse into the spiritual and architectural heritage of the region.
- RSPB Minsmere nature reserve: Venture into one of the UK's premier birdwatching destinations, with its diverse habitats and abundant wildlife, offering a chance to immerse yourself in the wonders of nature and witness the incredible diversity of the Suffolk coast.

By exploring these nearby attractions, you'll gain a deeper appreciation for the rich tapestry of history, culture, and natural beauty that defines this extraordinary corner of England, making your visit to Dunwich Heath and Beach a truly unforgettable experience.

Dunwich Heath Coastal Walk 3 - 6 miles (9.6 km) - Circular - Estimated Time: 3-4 hours

Starting Point Coordinates, Postcode: TM 477 678, Dunwich Heath IP17 3DJ
Nearest Car Park, Postcode: Dunwich Heath National Trust car park, IP17 3DJ

Immerse yourself in the captivating beauty of the Suffolk coast on this 6-mile (9.6 km) circular walk through the stunning landscapes of Dunwich Heath, a treasured National Trust site. This invigorating walk takes you on a journey through a tapestry of habitats, from the vibrant heathland to the lush woodland and the picturesque coastal cliffs, offering a perfect escape for nature enthusiasts and those seeking a tranquil adventure.

As you embark on this enchanting journey, you'll find yourself transported into a world of natural wonders, where the vibrant hues of the heather and gorse create a mesmerizing carpet that stretches out before you. The gentle sea breeze carries with it the invigorating scent of the coast, while the melodic calls of the birds and the rustling of the vegetation create a symphony of nature that accompanies your every step.

Venturing into the heart of the heathland, you'll be greeted by a sea of purple and yellow, as the blooming heather and gorse paint the landscape in a breathtaking display of colour. Take a moment to pause and appreciate the intricate beauty of these iconic plants, their delicate flowers swaying gently in the breeze, creating an enchanting dance that is sure to captivate your senses.

As you continue along the footpath, you'll find yourself immersed in the cool embrace of the woodland, where the dappled sunlight filters through the canopy, casting a magical glow on the forest floor. Listen for the distinctive calls of the woodpeckers, the enchanting songs of the warblers, and the hauntingly beautiful notes of the nightingales, each species adding its own unique voice to the woodland chorus.

Emerging from the trees, you'll be greeted by the awe-inspiring sight of the coastal cliffs, their rugged beauty a testament to the raw power of nature. Pause to take in the breathtaking panoramic views of the glistening sea, watching as the waves crash against the shore and the seabirds soar effortlessly on the currents of the wind.

Following the coastal path southward, you'll find yourself drawn to the picturesque beach at Dunwich, where the unspoiled shoreline stretches out before you, inviting you to take a moment to breathe in the salty sea air and listen to the soothing sound of the waves. If you're lucky, you may even spot some playful seals frolicking in the water or basking on the sand, their curious eyes watching you with a mix of interest and caution.

As you turn inland and make your way back through the enchanting heathland, take a final moment to reflect on the incredible beauty and diversity of the natural world that surrounds you. The sense of peace and tranquility that Dunwich Heath offers is a precious gift, one that will stay with you long after you've left this magical place behind.

Arriving back at the Dunwich Heath National Trust car park, your circular adventure comes to a close, but the memories of this stunning walk will remain etched in your mind, a testament to the power of nature to inspire, heal, and rejuvenate the soul.

Route Directions:

1. Begin your journey at the Dunwich Heath National Trust car park (Grid Reference: TM 477 678). From the

car park, follow the well-marked footpath, bearing 090°, into the heart of the heathland. (Estimated time: 20-30 minutes)

2. As you walk through the heathland, take a moment to admire the mesmerizing colours of the blooming heather and gorse, their vibrant hues creating a stunning natural painting that stretches out before you. (Estimated time: 30-40 minutes)

3. Continue along the footpath, bearing 045°, as it leads you into a captivating woodland area. Here, immerse yourself in the sights and sounds of the forest, listening for the melodic calls of woodpeckers, the enchanting songs of warblers, and the hauntingly beautiful notes of nightingales. (Estimated time: 30-40 minutes)

4. As you emerge from the woodland, bearing 000°, the path will guide you towards the breathtaking coastal cliffs. Pause to take in the awe-inspiring views of the glistening sea and the rugged coastline that stretches out to the horizon. (Estimated time: 20-30 minutes)

5. Follow the coastal path, bearing 180°, as it winds its way southward, leading you towards the picturesque beach at Dunwich. Along the way, keep your eyes peeled for the diverse array of coastal birds that call this area home, from the soaring seabirds to the scurrying waders. (Estimated time: 45-60 minutes)

6. Upon reaching Dunwich Beach, take a moment to appreciate the tranquil beauty of this unspoiled stretch of shoreline. If you're lucky, you may even spot some playful seals frolicking in the waves or basking on the sand. (Estimated time: 20-30 minutes)

7. When you're ready to continue, turn inland, bearing 270°, and follow the footpath as it leads you back through the enchanting heathland towards the car park. (Estimated time: 30-40 minutes)

8. As you near the end of your walk, take a final moment to reflect on the stunning natural beauty and the sense of peace and tranquility that Dunwich Heath offers, before arriving back at the Dunwich Heath National Trust car park, where your circular adventure comes to a close. (Estimated time: 10-15 minutes)

Best Features and Views:

- Lose yourself in the breathtaking tapestry of colours created by the heather and gorse that blanket Dunwich Heath, their vibrant hues shifting with the seasons, from the deep purples of summer to the golden yellows of autumn.
- Discover the awe-inspiring coastal cliffs, with their rugged beauty and panoramic views of the shimmering sea, where the power of the elements has carved a landscape of raw and untamed magnificence.
- Delight in the tranquil ambiance of Dunwich Beach, with its unspoiled shoreline and potential for seal sightings, a place where the stresses of modern life melt away, and you can reconnect with the simple joys of nature.

Birds to Spot:

- Feathered forest dwellers: Woodland wonders: Listen for the distinctive calls of woodpeckers, their rapid drumming echoing through the trees, the enchanting songs of warblers, their melodies weaving a magical tapestry of sound, and the hauntingly beautiful notes of nightingales, their voices a soulful reminder of the beauty and mystery of the natural world.
- Winged coastal wanderers: Coastal companions: Keep your eyes peeled for a variety of seabirds, from the graceful gulls soaring effortlessly on the sea breeze to the diving terns plunging into the waves with precision and grace, as well as the scurrying waders that forage along the shoreline, their delicate footprints etched in

the sand.

Rare or Interesting Insects and Wildlife:

- Serpentine sun worshippers: Basking adders: Tread carefully and respectfully, as you may encounter these beautiful but venomous snakes soaking up the sun, their intricate patterns and graceful movements a testament to the complexity and wonder of the natural world.
- Legless wonders: Slow-motion slow worms: These legless lizards are a fascinating sight, often mistaken for snakes, but possessing their own unique charm and adaptations that allow them to thrive in the heathland ecosystem.
- Aerial acrobats: Fluttering friends: Marvel at the colourful array of butterflies and dragonflies that dance amongst the heathland flowers, their delicate wings and agile movements a mesmerizing display of nature's artistry and grace.

Rare or Interesting Plants:

- Chromatic carpet: Heather and gorse: These iconic plants paint the heathland in a stunning palette of purples and yellows, their vibrant blooms creating a mesmerizing natural tapestry that shifts with the seasons and provides a vital habitat for a diverse array of species.
- Resilient coastal residents: Coastal flora: Discover the hardy and resilient plants that thrive in the challenging conditions of the coastal cliffs and dunes, their adaptations and tenacity a testament to the incredible strength and perseverance of nature in the face of adversity.
- Leafy forest treasures: Woodland wonders: Identify the diverse array of trees, shrubs, and wildflowers that make up the enchanting woodland ecosystem, from the majestic oaks and beeches to the delicate ferns and orchids, each species playing a vital role in the intricate web of life that thrives in the dappled shade of the forest.

Facilities and Accessibility:

The Dunwich Heath National Trust car park offers essential amenities to ensure a comfortable and enjoyable visit. Take advantage of the well-maintained toilets, perfect for a quick refresh before or after your walk. Stop by the charming café to indulge in a selection of refreshments and light meals, the ideal spot to relax and recharge while soaking in the stunning views of the heathland. The visitor centre also provides a wealth of information about the local area, its rich history, and the diverse wildlife that calls Dunwich Heath home, allowing you to deepen your understanding and appreciation of this remarkable landscape.

Please note that this walk involves some uneven terrain and steep coastal paths, which may present challenges for those with limited mobility or wheelchair users. To ensure a safe and enjoyable experience, it is highly recommended to wear sturdy walking shoes or boots with good ankle support and grip, allowing you to navigate the varying surfaces with confidence and stability.

Seasonal Highlights:

Dunwich Heath is a year-round destination, with each season offering its own unique charm and beauty, inviting you to return time and again to witness the ever-changing tapestry of colours, sounds, and sensations:

- Spring and summer: Witness the heathland erupting in a vibrant display of colour as the heather and gorse bloom, their delicate flowers painting the landscape in a breathtaking array of purples and yellows. Delight in the lively activity of nesting birds and buzzing insects, as they busily go about their lives amidst the lush

vegetation.
- Autumn: Admire the stunning autumnal hues as the heathland and woodland transform into a rich tapestry of golds, oranges, and reds, the changing foliage creating a warm and inviting atmosphere that invites contemplation and reflection. Enjoy the crisp coastal air, as it carries with it the promise of cozy evenings and heartwarming memories.
- Winter: Embrace the wild beauty of the wind-swept cliffs, the deserted beach, and the frosty heathland, as the stark and haunting landscapes offer a unique and unforgettable experience. Keep an eye out for overwintering birds seeking shelter along the coast, their presence a reminder of the resilience and adaptability of nature in the face of even the harshest conditions.

Difficulty Level and Safety Considerations:

This walk is classified as moderate, with some steep inclines and uneven terrain along the coastal path, making it suitable for walkers with a reasonable level of fitness and experience. It's essential to stay on the designated trails to minimize erosion and protect the delicate ecosystems that thrive in this unique habitat.

Be prepared for changeable weather conditions by wearing layers and bringing a waterproof jacket, as the coastal winds can be strong and unpredictable. A hat and gloves may also be necessary to keep you comfortable throughout your walk, especially during the cooler months.

Remember to carry an adequate supply of water and some energizing snacks to keep you hydrated and fuelled during your journey. It's also wise to pack any necessary medications and to inform someone of your intended route and expected return time, just as a precautionary measure.

Nearby Attractions:

After your invigorating walk, take the opportunity to explore some of the other fascinating attractions that the area has to offer, each one providing a unique glimpse into the rich history and natural wonders of the Suffolk coast:

- Dunwich Museum: Delve into the rich history of this once-thriving medieval port town and learn about the dramatic coastal erosion that has shaped the area, forever altering the landscape and the lives of those who called it home.
- Greyfriars Abbey ruins: Step back in time and discover the haunting ruins of this 13th-century Franciscan friary, its crumbling walls and atmospheric surroundings offering a glimpse into the spiritual and architectural heritage of the region.
- The historic village of Dunwich: Stroll through the charming streets of this ancient village, with its picturesque cottages and traditional pubs, and immerse yourself in the timeless beauty and tranquility of rural Suffolk life.

By exploring these nearby attractions, you'll gain a deeper appreciation for the rich tapestry of history, culture, and natural beauty that defines this extraordinary corner of England, making your visit to Dunwich Heath a truly unforgettable experience that will stay with you long after you've left the coast behind.

Dunwich Heath and Beach Walk 4 - 4 miles (6.4 km) - Circular - Estimated Time: 2-3 hours

Starting Point Coordinates, Postcode: TM 477 677, Dunwich IP17 3DJ
Nearest Car Park, Postcode: Dunwich Heath National Trust car park, IP17 3DJ

Embark on a captivating 4-mile (6.4 km) circular walk that takes you through the stunning heathland, enchanting woodlands, and along the sandy shores of Dunwich Heath and the Suffolk Coast National Nature Reserve. This walk offers a diverse array of habitats, making it a paradise for wildlife enthusiasts, birdwatchers, and anyone seeking to immerse themselves in the breathtaking beauty of the Suffolk coastline.

As you set foot on the trail, you'll find yourself transported into a world of natural wonders, where the vibrant hues of the heather and gorse create a mesmerizing tapestry that stretches out before you. The gentle sea breeze carries with it the invigorating scent of the coast, while the melodic calls of the birds and the rustling of the vegetation create a symphony of nature that accompanies your every step.

Traversing the heathland, you'll be greeted by a sea of purple and yellow, as the blooming heather and gorse paint the landscape in a breathtaking display of colour. Take a moment to pause and appreciate the intricate beauty of these iconic plants, their delicate flowers swaying gently in the breeze, creating an enchanting dance that is sure to captivate your senses.

As you make your way towards the beach, the landscape undergoes a stunning transformation, revealing the awe-inspiring expanse of the North Sea coastline. The sandy shores and the endless horizon create a sense of freedom and tranquility, inviting you to take a deep breath and let the stresses of everyday life melt away.

Strolling along the beach, you'll find yourself immersed in the fascinating world of coastal wildlife. Keep your eyes peeled for the graceful seabirds soaring overhead, the energetic terns diving into the waves in search of fish, and the lively waders scurrying along the shoreline, their delicate footprints etched in the sand.

As you venture inland and into the enchanting woodlands, the atmosphere shifts once more, enveloping you in a realm of dappled sunlight and the soothing sounds of nature. The melodic calls of various bird species, the playful antics of squirrels and rabbits, and the gentle rustling of leaves create a magical ambiance that is both calming and invigorating.

Throughout your journey, take a moment to appreciate the incredible diversity of landscapes and wildlife that you encounter. From the vibrant heathland to the sandy shores and the lush woodlands, each habitat offers a unique and captivating experience, showcasing the remarkable beauty and resilience of the natural world.

As you near the end of your walk and arrive back at the Dunwich Heath National Trust car park, you'll find yourself filled with a sense of accomplishment and a deeper connection to the stunning landscapes and wildlife of the Suffolk coast. This circular walk is more than just a physical journey; it is an opportunity to immerse yourself in the wonders of nature, to find solace in the tranquility of the great outdoors, and to create lasting memories that will stay with you long after you've left the heathland behind.

Route Directions:

1. Begin your adventure at the Dunwich Heath National Trust car park (Grid Reference: TM 477 677). From the car park, follow the well-marked footpath, bearing 180°, as it leads you southward into the heart of the

heathland. (Estimated time: 15-20 minutes)

2. As you traverse the heathland, take a moment to admire the stunning tapestry of colours that surrounds you, from the deep purples of the heather to the vibrant yellows of the gorse, painting the landscape in a mesmerizing palette that shifts with the seasons. (Estimated time: 20-30 minutes)

3. Continue along the footpath until you reach a T-junction. Here, turn right, bearing 270°, and follow the path as it guides you westward towards the inviting expanse of the beach. (Estimated time: 15-20 minutes)

4. Upon reaching the beach, turn right, bearing 000°, and begin your leisurely stroll along the sandy shoreline. Take in the awe-inspiring views of the North Sea stretching out before you, and feel the refreshing coastal breeze against your skin. (Estimated time: 30-40 minutes)

5. As you walk along the beach, keep your eyes peeled for the diverse array of bird species that call this coastal habitat home, from the graceful seabirds soaring overhead to the lively waders scurrying along the shore. (Estimated time: 30-40 minutes)

6. Continue your beach walk for approximately 1.5 miles (2.4 km) until you reach a footpath that leads inland. Turn right onto this path, bearing 045°, and begin your journey northward into the enchanting woodlands. (Estimated time: 20-30 minutes)

7. As you navigate through the woods, immerse yourself in the sights, sounds, and scents of this captivating habitat. Listen for the melodic calls of various bird species, spot the playful antics of squirrels and rabbits, and admire the dappled sunlight filtering through the leafy canopy above. (Estimated time: 20-30 minutes)

8. Continue along the woodland path until you reach a footpath intersection. Here, turn right, bearing 090°, and follow the path eastward, making your way back towards the Dunwich Heath car park. (Estimated time: 15-20 minutes)

9. As you near the end of your walk, take a moment to reflect on the incredible diversity of landscapes and wildlife you've encountered, from the vibrant heathland to the sandy shores and the enchanting woodlands. (Estimated time: 5-10 minutes)

10. Finally, arrive back at the Dunwich Heath National Trust car park, where your circular walk comes to a satisfying conclusion. (Estimated time: 1-5 minutes)

Best Features and Views:

- Lose yourself in the stunning beauty of Dunwich Heath, with its vibrant tapestry of heather and gorse, creating a mesmerizing landscape that shifts with the seasons, from the deep purples of summer to the golden yellows of autumn.
- Discover the awe-inspiring views of the North Sea coastline, with its sandy beaches and the endless expanse of the horizon, where the rhythmic sound of the waves and the salty sea breeze create a sensory experience that invigorates the soul.
- Explore the diverse habitats of the Suffolk Coast National Nature Reserve, from the lush woodlands to the coastal heathland, each offering a unique and captivating experience that showcases the remarkable beauty and resilience of the natural world.

Birds to Spot:

- Masters of the skies: Seabirds: Watch for the graceful gulls soaring overhead and the energetic terns diving into the waves, as they forage for fish along the coast, their aerial acrobatics a testament to their skill and adaptability.
- Shoreline sentinels: Waders: Observe the lively waders, such as oystercatchers, curlews, and sandpipers, as they scurry along the shoreline, probing the sand for hidden treasures, their distinctive calls and behaviours adding a lively soundtrack to your coastal walk.
- Feathered forest dwellers: Heathland and woodland birds: Listen for the melodic songs of various warblers, the distinctive calls of woodpeckers, and the cheerful chirps of tits and finches as you explore the heathland and woodland habitats, each species contributing to the rich tapestry of birdlife that thrives in these diverse landscapes.

Rare or Interesting Insects and Wildlife:

- Majestic grazers: Red deer: Keep an eye out for the majestic red deer, which can sometimes be spotted grazing in the heathland or seeking shelter in the woodland edges, their regal presence adding a touch of wilderness to the landscape.
- Hopping habitants: Rabbits: Watch for the playful antics of rabbits as they scamper through the heather and gorse, their fluffy tails bobbing as they go, their presence a reminder of the thriving wildlife that calls this habitat home.
- Airborne artistry: Butterflies and dragonflies: Admire the delicate beauty of the various butterfly species that flutter amongst the heathland flowers, their colourful wings painting the air with ephemeral brushstrokes, and the iridescent dragonflies that dart across the coastal ponds, their aerial prowess a marvel to behold.

Rare or Interesting Plants:

- Chromatic carpet: Heather and gorse: These iconic plants paint the heathland in a stunning palette of purples and yellows, creating a vibrant and ever-changing landscape that shifts with the seasons and provides a vital habitat for a diverse array of species.
- Fragile forest gems: Woodland wildflowers: Discover the delicate blooms of woodland wildflowers, such as bluebells, primroses, and wood anemones, as they carpet the forest floor in spring, their tender petals and vibrant colours adding a touch of magic to the dappled shade of the woodland.
- Resilient coastal residents: Coastal flora: Observe the hardy and resilient plants that thrive in the challenging conditions of the coastal dunes and salt marshes, such as sea holly, marram grass, and samphire, their adaptations and tenacity a testament to the incredible strength and perseverance of nature in the face of adversity.

Facilities and Accessibility:

The Dunwich Heath National Trust site offers a range of facilities to enhance your visit, ensuring that your experience is both comfortable and enjoyable. Take advantage of the well-maintained toilets, perfect for a quick refresh before or after your walk. Stop by the cozy café to indulge in a selection of refreshments and light meals, the ideal spot to relax and recharge while soaking in the stunning views of the heathland. Don't forget to visit the informative visitor centre, where you can learn more about the local area, its fascinating history, and the diverse wildlife that calls this remarkable landscape home.

The terrain on this walk is varied, with a mix of sandy beaches, heathland paths, and woodland trails, offering a diverse and engaging experience for walkers. Some sections may be uneven or muddy, particularly after rainfall, so it is highly recommended to wear sturdy walking shoes or boots with good grip and support, allowing you to navigate the varying surfaces with confidence and stability.

Seasonal Highlights:

The beauty of Dunwich Heath and the Suffolk Coast National Nature Reserve is ever-changing, with each season offering its own unique charms and surprises, inviting you to return time and again to witness the landscape's captivating transformations:

- Spring: Delight in the vibrant blooms of heather and gorse, their delicate flowers painting the heathland in a breathtaking array of purples and yellows. Enjoy the lively activity of nesting birds, as they busily build their nests and raise their young amidst the lush vegetation, and marvel at the emergence of delicate woodland wildflowers, their tender petals adding a touch of colour to the forest floor.
- Summer: Bask in the lush greenery of the heathland and woodland, as the vegetation reaches its peak of growth and vitality. Delight in the buzzing activity of insects, as they flit amongst the flowers and foliage, and enjoy the warm sandy beaches, perfect for a refreshing dip or a leisurely picnic by the shore.
- Autumn: Admire the stunning autumnal hues as the heathland and woodland transform into a rich tapestry of golds, oranges, and reds, the changing foliage creating a warm and inviting atmosphere that invites contemplation and reflection. Witness the migration of birds along the coast, as they gather in preparation for their long journeys to distant lands.
- Winter: Embrace the wild beauty of the windswept heathland, the deserted beaches, and the stark silhouettes of leafless trees, as the stark and haunting landscapes offer a unique and unforgettable experience. Keep an eye out for overwintering birds seeking shelter in the coastal habitats, their presence a reminder of the resilience and adaptability of nature in the face of even the harshest conditions.

Difficulty Level and Safety Considerations:

This walk is classified as easy to moderate, with mostly flat terrain and well-defined paths, making it suitable for walkers of all ages and abilities. However, some sections of the beach may be uneven or soft, and the woodland trails can be muddy or slippery after rain, so it's essential to take care and watch your step.

As with any outdoor activity, it's crucial to be prepared for changing weather conditions. Wear layers of clothing that can be easily adjusted to suit the temperature, and bring a waterproof jacket to protect you from potential rain showers. A hat and sunscreen are also recommended to shield you from the sun's rays, especially during the summer months.

Remember to carry an adequate supply of water and some energizing snacks to keep you hydrated and fuelled throughout your walk. It's also wise to pack any necessary medications and to inform someone of your intended route and expected return time, just as a precautionary measure.

Always keep to the designated paths to minimize your impact on the environment and to protect the delicate ecosystems that thrive in this unique habitat. Respect any wildlife you encounter by observing from a distance and avoiding any actions that may cause distress or harm. By following the countryside code and being a responsible walker, you can ensure that this beautiful landscape remains a haven for nature and a source of inspiration for generations to come.

Nearby Attractions:

After your invigorating walk, take the opportunity to explore some of the other fascinating attractions that the area has to offer, each one providing a unique glimpse into the rich history and natural wonders of the Suffolk coast:

- The coastal village of Walberswick: Stroll through the picturesque streets of this charming village, with its pretty cottages, traditional pubs, and stunning views of the River Blyth. Immerse yourself in the timeless beauty and tranquility of rural Suffolk life, and perhaps enjoy a well-deserved pint or a delicious meal at one of the friendly local establishments.
- The RSPB Minsmere nature reserve: Venture into one of the UK's premier birdwatching destinations, with its diverse habitats and abundant wildlife, including the iconic avocet and the elusive bittern. Explore the extensive network of trails and hides, and marvel at the incredible diversity of birdlife that thrives in this internationally important wetland site.
- The historic town of Southwold: Discover the timeless charm of this pretty seaside town, with its colourful beach huts, historic lighthouse, and bustling market square. Take a leisurely stroll along the promenade, enjoy a delicious ice cream or fish and chips, and soak in the nostalgic atmosphere of a traditional English coastal resort.

By exploring these nearby attractions, you'll gain a deeper appreciation for the rich tapestry of history, culture, and natural beauty that defines this extraordinary corner of England, making your visit to Dunwich Heath and Beach an unforgettable experience that will leave you with lasting memories and a renewed sense of connection to the great outdoors.

Ickworth Park Walk - 3.5 miles (5.6 km), Circular - Estimated Time: 2-3 hours

Starting Point Coordinates, Postcode: TL 816 605, Bury St Edmunds IP29 5QE
Nearest Car Park, Postcode: Ickworth House car park, Horringer, Bury St Edmunds IP29 5QE

Immerse yourself in the grandeur and tranquility of Ickworth Park on this delightful 3.5-mile (5.6 km) circular walk. Located near the historic town of Bury St Edmunds, this walk takes you through a landscape of stunning parkland, ancient woodland, and serene lakeside scenery, all set against the backdrop of the impressive Ickworth House, a magnificent neoclassical mansion with its iconic rotunda. This walk is perfect for those who appreciate the blend of natural beauty and architectural heritage, offering a peaceful escape from the hustle and bustle of everyday life.

As you embark on this enchanting journey, you'll find yourself transported into a world of timeless elegance and natural splendour. The grand facade of Ickworth House greets you as you set off, its impressive rotunda and stately columns a testament to the opulence and artistry of a bygone era. Take a moment to admire the architectural details and the sweeping views of the surrounding parkland, feeling the weight of history and the whisper of untold stories in the air around you.

Venturing into the parkland, you'll be embraced by a tapestry of lush green lawns, majestic ancient trees, and the gentle murmur of wildlife. The well-trodden paths guide you through this idyllic landscape, inviting you to slow down, breathe deep, and let the tranquility of nature wash over you. As you walk, keep an eye out for the playful antics of grey squirrels in the treetops, the darting flight of woodland birds, and the delicate blooms of wildflowers that add splashes of colour to the verdant canvas.

Continuing along the trail, you'll soon find yourself in the enchanting realm of the Italianate Garden, a horticultural masterpiece that showcases the elegance and refinement of classical design. Immerse yourself in the beauty of carefully manicured lawns, vibrant flower beds, and ornate statuary, each element a testament to the skill and artistry of the gardeners who have tended this oasis of tranquility through the generations.

As you leave the formal gardens behind and venture deeper into the woodland, the atmosphere shifts, enveloping you in a world of dappled sunlight, the soft crunch of leaves underfoot, and the gentle rustling of the breeze through the canopy. Let your senses be awakened by the earthy scents of the forest, the melodic birdsong that fills the air, and the occasional glimpse of wildlife that calls this woodland haven home.

Emerging from the trees, you'll be greeted by the serene beauty of the lake, its still waters reflecting the sky above and the lush greenery that surrounds it. Take a moment to pause and drink in the tranquil scene, watching as dragonflies skim across the surface and waterfowl glide gracefully by. This is a place of pure, unspoiled beauty, where the stresses of modern life seem to melt away, and the soul can find solace and renewal.

Continuing on your journey, you'll discover hidden gems nestled throughout the park, such as the charming Temple of Diana, a whimsical folly that adds a touch of romantic mystery to the landscape, and the bountiful Walled Garden, where the scents and colours of herbs, fruits, and vegetables mingle in a sensory feast.

As you make your way back towards Ickworth House, take a final moment to reflect on the incredible beauty and richness of the landscape you've explored. The timeless elegance of the architecture, the vibrant tapestry of the parkland and gardens, and the tranquility of the woodland and lake all combine to create an unforgettable experience that will stay with you long after you've left this enchanting corner of Suffolk behind.

Detailed Route Directions:

1. Begin your adventure at the Ickworth House car park (Grid Reference: TL 816 605). From the car park, make your way towards the entrance of Ickworth House, admiring the grandeur of the building as you approach. (Estimated time: 5-10 minutes)

2. At the entrance, take the footpath on the left, bearing 315°, and follow it alongside the impressive facade of Ickworth House. As you walk, take a moment to appreciate the architectural details and the beautiful views of the surrounding parkland. (Estimated time: 10-15 minutes)

3. Continue along the footpath, passing by the enchanting Stumpery and the elegant Italianate Garden, both of which are well worth a closer look. (Estimated time: 15-20 minutes)

4. At the end of the Italianate Garden, turn right, bearing 045°, and follow the footpath as it leads you into a delightful woodland area. Immerse yourself in the tranquility of the forest, listening to the gentle rustling of leaves and the melodic birdsong. (Estimated time: 20-30 minutes)

5. As you emerge from the woodland, continue straight ahead, bearing 090°, walking alongside the open parkland with the main road visible to your right. (Estimated time: 10-15 minutes)

6. When the footpath veers to the left, follow it, bearing 000°, as it takes you around the southern edge of the park. Along this stretch, you'll be treated to stunning views of the lake, its tranquil waters reflecting the surrounding landscape. (Estimated time: 20-30 minutes)

7. At the next intersection, turn left, bearing 270°, and follow the footpath along the western edge of the park. As you walk, keep an eye out for the Temple of Diana, a charming folly nestled amongst the trees. (Estimated time: 15-20 minutes)

8. Continue along the footpath until you reach the enchanting Walled Garden on your right. Take a moment to explore this horticultural gem, with its beautifully maintained flower beds, fruit trees, and vegetable patches. (Estimated time: 20-30 minutes)

9. After visiting the Walled Garden, follow the footpath back towards Ickworth House, bearing 135°, passing by the picturesque church along the way. (Estimated time: 15-20 minutes)

10. Finally, return to the Ickworth House car park, where your circular walk comes to a satisfying conclusion. (Estimated time: 5-10 minutes)

Best Features and Views:

- Stand in awe of the stunning architecture of Ickworth House, with its iconic rotunda and grand facade, a testament to the opulence and artistry of a bygone era, set against the backdrop of the beautiful parkland that surrounds it.
- Discover the tranquil beauty of the parkland and woodland, with their ancient trees, diverse flora, and peaceful atmosphere, where the stresses of modern life seem to melt away, and the soul can find solace and renewal.
- Delight in the elegance of the Italianate Garden, with its carefully manicured lawns, vibrant flower beds, and ornate statuary, each element a testament to the skill and artistry of the gardeners who have tended this oasis of tranquility through the generations.

- Uncover the charm of the Temple of Diana, a delightful folly tucked away in the western part of the park, adding a touch of romantic mystery to the landscape and inviting exploration and discovery.

Birds to Spot:

- Feathered forest dwellers: Woodland wonders: Listen for the distinctive drumming of woodpeckers, their rapid tapping echoing through the trees, the melodic songs of various thrushes, their haunting melodies a soundtrack to your woodland wanderings, and the cheerful chirps of tits and finches, their lively presence a constant companion on your journey.
- Winged guardians of the parkland: Parkland residents: Keep an eye out for the majestic presence of birds of prey, such as buzzards and kestrels, soaring over the open parkland in search of their next meal, their keen eyes and powerful talons a testament to their skill and adaptability in this wide-open landscape.

Rare or Interesting Insects and Wildlife:

- Nimble treetop acrobats: Grey squirrels: Watch for the playful antics of grey squirrels as they scamper through the woodland canopy and forage for acorns on the forest floor, their bushy tails and curious nature adding a touch of whimsy to your walk.
- Hopping heralds of the parkland: Rabbits: Spot the adorable rabbits hopping through the parkland grass, their white tails bobbing as they go, their presence a reminder of the abundance of life that thrives in this idyllic landscape.
- Winged wonders: Butterflies and moths: Admire the delicate beauty of the various butterfly and moth species that flutter amongst the wildflowers and woodland edges, their colourful wings and graceful movements a mesmerizing display of nature's artistry.

Rare or Interesting Plants:

- Sentinels of time: Ancient trees: Marvel at the majestic ancient trees that have stood sentinel in the park for centuries, their gnarled trunks and spreading canopies bearing witness to the passage of time and the countless stories they have silently observed.
- Botanical masterpieces: Ornamental plants: Delight in the vibrant colours and intricate designs of the ornamental plants and flowers that adorn the Italianate Garden, each bloom a work of art, carefully tended and arranged to create a stunning display of horticultural artistry.
- Floral gems of the wild: Wildflowers: Discover the delicate blooms of wildflowers that dot the woodland floor and the parkland margins, their tender petals and vibrant hues a testament to the resilience and beauty of nature, thriving in the untamed corners of this magnificent landscape.

Facilities and Accessibility:

Ickworth House offers a range of facilities to enhance your visit, ensuring that your experience is both comfortable and enjoyable. Take advantage of the well-appointed toilets, perfect for a quick refresh before or after your walk. Stop by the charming café to indulge in a selection of delicious refreshments and light meals, the ideal spot to relax and recharge while soaking in the stunning views of the parkland. Don't forget to browse the gift shop, stocked with unique souvenirs and local crafts, allowing you to take a piece of Ickworth magic home with you. For those with an appreciation for the arts, the on-site art gallery showcases an ever-changing collection of works by talented regional artists, adding a touch of cultural enrichment to your visit.

The parkland paths are generally well-maintained and relatively flat, making this walk accessible to most visitors, including those with moderate fitness levels. However, it's important to note that some areas may be uneven or muddy, particularly after periods of wet weather, so it's essential to wear sturdy walking shoes or boots with good grip and support to ensure a comfortable and safe experience.

Seasonal Highlights:

The beauty and character of Ickworth Park change with the seasons, each offering its own special charm and inviting you to return time and again to witness the landscape's captivating transformations:

- Spring: Delight in the vibrant blooms of spring flowers in the Italianate Garden, their delicate petals and fresh hues a celebration of new life and renewal. Enjoy the fresh green foliage of the woodland, as the trees awaken from their winter slumber, and revel in the lively birdsong that fills the air, a joyous symphony heralding the arrival of warmer days.
- Summer: Bask in the lush greenery of the parkland, the verdant lawns and majestic trees creating an oasis of tranquility beneath the warm summer sun. Seek out the dappled shade of the woodland, where the cool, leafy canopy offers a welcome respite from the heat, and admire the way the sunlight illuminates the stunning architecture of Ickworth House, casting long shadows and highlighting the intricate details of its facade.
- Autumn: Marvel at the breathtaking autumnal hues as the trees in the parkland and woodland transform into a rich tapestry of golds, oranges, and reds, the changing foliage creating a warm and inviting atmosphere that beckons you to linger and drink in the beauty of the season. Crunch through fallen leaves, breathe in the crisp autumn air, and savour the sense of tranquility that pervades the landscape during this time of transition.
- Winter: Embrace the tranquil beauty of the frost-covered landscape, the stark silhouettes of leafless trees etched against the pale winter sky, and the crisp, cold air that invigorates the senses and adds a touch of magic to your walk. Discover the subtle beauty of the winter parkland, with its muted colours and hushed atmosphere, and appreciate the way the season strips away the distractions, allowing you to focus on the essential elements of nature and the timeless elegance of Ickworth House.

Difficulty Level and Safety Considerations:

This walk is classified as easy, with mostly flat terrain and well-defined paths, making it suitable for walkers of all ages and abilities. However, as with any outdoor activity, it's essential to be prepared for changing weather conditions and to wear appropriate clothing and footwear to ensure your comfort and safety.

As you embark on your walk, be sure to carry water, snacks, and any necessary medications, keeping yourself hydrated and energized throughout your journey. It's also a good idea to inform someone of your intended route and expected return time, just as a precautionary measure.

Always stay on the designated paths to minimize your impact on the environment and to protect the delicate ecosystems that thrive within the park. Remember to respect any wildlife you encounter, observing from a distance and avoiding any actions that may cause distress or harm. By following the countryside code and being a responsible visitor, you can help ensure that this beautiful landscape remains a haven for nature and a source of enjoyment for generations to come.

Nearby Attractions:

After your invigorating walk, take the opportunity to explore some of the other fascinating attractions that the area has to offer, each one providing a unique glimpse into the rich history, culture, and natural wonders of the region:

- West Stow Anglo-Saxon Village: Step back in time and immerse yourself in the fascinating history of the Anglo-Saxons at this reconstructed village, where you can explore the thatched houses, witness traditional craft demonstrations, and participate in hands-on activities that bring the past to life.
- Bury St Edmunds Abbey and Gardens: Discover the stunning ruins of the ancient abbey, once one of the most important monasteries in England, and stroll through the beautiful gardens that now occupy the site, marvelling at the intricate floral displays and the tranquil atmosphere that pervades this historic oasis.
- Theatre Royal Bury St Edmunds: Experience the magic of live performance at this historic theatre, one of the oldest surviving Regency playhouses in Britain, with its elegant interior and diverse program of shows that cater to a wide range of tastes and interests.

By exploring these nearby attractions, you'll gain a deeper appreciation for the rich tapestry of history, culture, and natural beauty that defines this extraordinary corner of England, making your visit to Ickworth Park a truly unforgettable experience that will leave you with lasting memories and a renewed sense of connection to the world around you.

Lavenham and the Lavenham Walk - 3 miles (4.8 km) - Circular - Estimated Time: 1.5-2 hours

Starting Point Coordinates, Postcode: TL 913 490, Lavenham CO10 9QZ
Nearest Car Park, Postcode: Lavenham Market Place car park, CO10 9QZ

Step back in time and discover the timeless charm of Lavenham, one of England's finest medieval villages, on this delightful 3-mile (4.8 km) circular walk. This leisurely stroll takes you through a landscape of picturesque streets lined with historic timber-framed buildings, tranquil countryside, and along the banks of the gentle River Brett, offering a perfect blend of history, architecture, and natural beauty.

As you embark on this enchanting journey, you'll find yourself transported to a world where the past and present seamlessly intertwine. The picturesque streets of Lavenham, with their beautifully preserved medieval timber-framed buildings, create a captivating atmosphere that evokes a sense of wonder and nostalgia. Each building, with its unique character and history, tells a story of the village's rich heritage, inviting you to imagine the lives of those who walked these streets centuries ago.

Venturing beyond the village, the walk takes you into the heart of the tranquil countryside, where the gentle River Brett meanders through a tapestry of lush green fields and rolling hills. As you follow the footpath alongside the river, the hustle and bustle of modern life fades away, replaced by the soothing sounds of flowing water and the gentle rustling of leaves in the breeze. Take a moment to pause and appreciate the serene beauty of this natural oasis, letting the peacefulness wash over you and rejuvenate your senses.

Throughout your journey, you'll have the opportunity to spot a diverse array of wildlife and plants that thrive in this idyllic landscape. Listen for the melodic songs of skylarks as they soar above the fields, the cheerful chirps of yellowhammers in the hedgerows, and the distinctive calls of red kites soaring overhead. Along the riverbanks, keep an eye out for the graceful movements of swans, the darting flight of kingfishers, and the busy activity of wagtails, each species adding its own unique charm to the symphony of nature.

As you traverse the picturesque farmland, take note of the delicate blooms of wildflowers that dot the field margins and riverbanks, their vibrant colours and intricate petals a testament to the resilience and beauty of the natural world. Marvel at the majestic ancient trees that line the country lanes and stand sentinel in the fields, their gnarled trunks and spreading canopies bearing witness to the passage of time and the countless stories they have silently observed.

Returning to the heart of Lavenham, you'll once again find yourself immersed in the village's timeless charm. The historic Market Place, with its impressive Guildhall and fascinating Little Hall Museum, serves as a reminder of the rich history that permeates every corner of this remarkable destination. As you conclude your circular walk, take a moment to reflect on the beauty, tranquility, and resilience of both the natural world and the built environment that you've had the privilege to explore.

This Lavenham walk is more than just a physical journey; it is an opportunity to step back in time, to connect with the past, and to find solace and inspiration in the timeless beauty of the English countryside. Whether you're a history enthusiast, a nature lover, or simply seeking a moment of respite from the stresses of modern life, this walk promises to delight, enlighten, and rejuvenate your spirit.

Route Directions:

1. Begin your walk in Lavenham's Market Place (Grid Reference: TL 913 490). This historic square, surrounded by beautiful timber-framed buildings, offers ample parking and is the perfect starting point for your adventure. (Estimated time: 5-10 minutes)

2. From the Market Place, head east along the picturesque Water Street, passing by the impressive medieval Guildhall and the fascinating Little Hall Museum, both of which offer a glimpse into Lavenham's rich history. (Estimated time: 10-15 minutes)

3. At the end of Water Street, turn right onto Bridge Street and continue until you reach a footpath on your left, just past the charming bridge that crosses the River Brett. (Estimated time: 5-10 minutes)

4. Follow the footpath as it meanders alongside the tranquil River Brett, offering delightful views of the water and the surrounding countryside. Take a moment to appreciate the peaceful atmosphere and the gentle sounds of the flowing river. (Estimated time: 15-20 minutes)

5. Continue along the footpath for approximately 1 mile (1.6 km) until you reach a T-junction with a small lane. Here, turn left onto the lane, and then take the first right onto a public footpath that leads you through picturesque farmland. (Estimated time: 20-30 minutes)

6. As you follow the footpath through the fields, keep your eyes peeled for the various bird species and other wildlife that call this area home. Listen for the melodic songs of skylarks, the buzzing of bees, and the rustling of small mammals in the hedgerows. (Estimated time: 15-20 minutes)

7. At the end of the footpath, turn left onto a quiet country lane and follow it back towards Lavenham. As you approach the village, turn right onto Water Lane, which will lead you back to the historic Market Place. (Estimated time: 10-15 minutes)

8. Finally, return to Lavenham's Market Place, where your circular walk comes to a satisfying conclusion. Take a moment to reflect on the beauty and tranquility of the countryside you've just explored, and the timeless charm of the village itself. (Estimated time: 5-10 minutes)

Best Features and Views:

- Lose yourself in the picturesque streets of Lavenham, lined with beautifully preserved medieval timber-framed buildings, each with its own unique character and history, creating a captivating atmosphere that evokes a sense of wonder and nostalgia.
- Enjoy the tranquil beauty of the River Brett, as it meanders through the countryside, providing a peaceful backdrop to your walk and offering a serene oasis where the stresses of modern life fade away.
- Admire the bucolic charm of the surrounding farmland, with its patchwork of fields, hedgerows, and gently rolling hills, a timeless landscape that has inspired artists and writers for generations.

Birds to Spot:

- Melodic masters: Farmland friends: Listen for the melodic songs of skylarks as they soar above the fields, their joyful melodies a celebration of the wide-open spaces, the cheerful chirps of yellowhammers in the hedgerows, their vibrant plumage a flash of sunshine amidst the greenery, and the distinctive calls of red kites soaring overhead, their majestic presence a symbol of the resilience and adaptability of nature.
- Waterside wonders: Riverside residents: Keep an eye out for the graceful movements of swans, their elegant

forms gliding serenely across the surface of the River Brett, the darting flight of kingfishers, their iridescent blue and orange plumage a dazzling sight against the riverbank, and the busy activity of wagtails, their bobbing tails and lively personality a constant source of entertainment.

Rare or Interesting Insects and Wildlife:

- Winged wonders: Butterflies: Admire the delicate beauty of the various butterfly species that flutter amongst the wildflowers in the fields and along the river banks, such as the colourful peacock, with its striking eye-spot pattern, the vibrant red admiral, a master of aerial acrobatics, and the dainty common blue, its shimmering wings a reflection of the summer sky.
- Aerial acrobats: Dragonflies: Watch for the iridescent flashes of dragonflies as they dart across the water's surface, their vibrant colours catching the sunlight and their agile movements a testament to their skill and precision.
- Miniature marvels: Small mammals: Spot the signs of small mammals, such as field mice, voles, and shrews, as they scurry through the undergrowth and along the hedgerows, their tiny footprints and burrows a reminder of the intricate web of life that thrives in this pastoral paradise.

Rare or Interesting Plants:

- Floral treasures: Wildflowers: Discover the delicate blooms of wildflowers that dot the field margins and river banks, such as the vibrant poppies, their scarlet petals a symbol of remembrance and renewal, the nodding heads of cowslips, their gentle fragrance a herald of spring, and the frothy clouds of Queen Anne's lace, their intricate patterns a masterpiece of natural design.
- Sentinels of time: Ancient trees: Marvel at the majestic ancient trees that line the country lanes and stand sentinel in the fields, their gnarled trunks and spreading canopies a testament to their resilience and longevity, bearing witness to the passing of countless seasons and the ever-changing face of the landscape.
- Hedgerow havens: Hedgerow plants: Identify the diverse array of plants that make up the hedgerows, such as the prickly hawthorn, its delicate white blossoms a symbol of purity and hope, the fragrant elder, its creamy flowers and dark berries a staple of traditional medicine and cuisine, and the climbing tendrils of wild clematis, its star-like blooms a celestial adornment to the rural tapestry.

Facilities and Accessibility:

Lavenham offers a range of facilities to ensure a comfortable and enjoyable visit, catering to the needs of walkers and visitors of all ages and abilities. Take advantage of the conveniently located public toilets, perfect for a quick refresh before or after your walk. Indulge in a well-deserved break at one of the charming cafes, where you can savour a delicious cup of tea or coffee and a tempting selection of homemade cakes and pastries. For a more substantial meal or a refreshing pint, stop by one of the traditional pubs, each with its own unique character and a menu featuring hearty local fare. Don't forget to browse the independent shops, where you'll find an enticing array of local crafts, artisanal products, and one-of-a-kind souvenirs to remind you of your visit to this enchanting village.

The footpaths on this walk are generally well-maintained and relatively flat, making it accessible to most visitors, including those with moderate fitness levels. However, it's important to note that some sections may be uneven or muddy, particularly after periods of wet weather, so it's essential to wear sturdy walking shoes or boots with good grip and support to ensure a pleasant and safe experience.

Seasonal Highlights:

The character and beauty of Lavenham and its surrounding countryside change with the seasons, each offering its own special charm and inviting you to return time and again to witness the landscape's captivating transformations:

- Spring: Delight in the vibrant blooms of wildflowers in the fields and along the river banks, their delicate petals and fresh hues a celebration of new life and renewal. Enjoy the fresh green leaves of the trees, as they unfurl and fill the landscape with a sense of vitality and growth, and revel in the lively birdsong that fills the air, a joyous symphony heralding the arrival of warmer days.
- Summer: Bask in the lush greenery of the countryside, the sun-dappled fields and the gently swaying grasses creating an idyllic scene of pastoral tranquility. Seek out the cool shade of the trees, where the leafy canopy offers a welcome respite from the heat, and admire the way the golden sunlight illuminates the historic buildings of Lavenham, casting long shadows and highlighting the intricate details of their timber-framed facades.
- Autumn: Marvel at the breathtaking autumnal hues as the trees transform into a rich tapestry of golds, oranges, and reds, the changing foliage creating a warm and inviting atmosphere that beckons you to linger and drink in the beauty of the season. Crunch through fallen leaves, breathe in the crisp autumn air, and savour the sense of tranquility that pervades the landscape during this time of transition.
- Winter: Embrace the tranquil beauty of the frost-covered landscape, the stark silhouettes of leafless trees etched against the pale winter sky, and the crisp, cold air that invigorates the senses and adds a touch of magic to your walk. Discover the subtle beauty of the winter countryside, with its muted colours and hushed atmosphere, and appreciate the way the season strips away the distractions, allowing you to focus on the essential elements of the natural world and the timeless charm of Lavenham's historic heart.

Difficulty Level and Safety Considerations:

This walk is classified as easy, with mostly flat terrain and well-defined paths, making it suitable for walkers of all ages and abilities. However, as with any outdoor activity, it's essential to be prepared for changing weather conditions and to wear appropriate clothing and footwear to ensure your comfort and safety.

As you embark on your walk, be sure to carry water, snacks, and any necessary medications, keeping yourself hydrated and energized throughout your journey. It's also a good idea to inform someone of your intended route and expected return time, just as a precautionary measure.

Always stay on the designated paths to minimize your impact on the environment and to protect the delicate ecosystems that thrive in the countryside. Remember to respect any wildlife you encounter, observing from a distance and avoiding any actions that may cause distress or harm. If you find yourself in the company of livestock, such as grazing sheep or cattle, give them a wide berth and walk calmly and quietly to avoid disturbing them.

By following these simple guidelines and being a responsible and considerate walker, you can help ensure that the beautiful landscapes and historic charm of Lavenham and its surroundings remain a source of joy and inspiration for generations to come.

Nearby Attractions:

After your delightful walk, take the opportunity to explore some of the other fascinating attractions that the area has to offer, each one providing a unique glimpse into the rich history, culture, and natural beauty of the region:

- Lavenham Guildhall: Step back in time and immerse yourself in the fascinating history of Lavenham's medieval guildhall, where you can explore the beautiful timber-framed architecture, interact with engaging exhibits, and stroll through the tranquil gardens, gaining a deeper understanding of the village's prosperous past as a thriving centre of the wool trade.
- Little Hall Museum: Discover the charming rooms of this 14th-century hall house, each one filled with an enchanting collection of period furnishings, artworks, and artifacts that offer a intimate glimpse into the daily lives of Lavenham's residents through the centuries, from the humblest peasant to the wealthiest merchant.
- Long Melford and Kentwell Hall: Venture to the nearby village of Long Melford, where you can admire the impressive parish church, explore the historic high street lined with elegant Georgian buildings, and visit the stunning Kentwell Hall, a magnificent Tudor mansion set amidst beautifully landscaped gardens and grounds that transport you back to the grandeur of a bygone era.
- Kersey: Lose yourself in the picturesque charm of the village of Kersey, with its quaint ford, where the crystal-clear waters of a small stream cross the main street, its lovingly preserved historic buildings that seem frozen in time, and the peaceful countryside that envelops the village in a gentle embrace, offering a true escape from the modern world.

By exploring these nearby attractions, you'll gain a deeper appreciation for the rich tapestry of history, culture, and natural beauty that defines this extraordinary corner of Suffolk, making your visit to Lavenham and its surroundings an unforgettable experience that will leave you with lasting memories and a renewed sense of connection to England's captivating past and present.

Needham Lake Walk - 2 miles (3.2 km), Circular - Estimated Time: 1-1.5 hours

Starting Point Coordinates, Postcode: Needham Lake car park (Grid Reference: TM 089 463, Postcode: IP6 8NU).
Nearest Car Park, Postcode: Needham Lake car park (Postcode: IP6 8NU).

Embark on a delightful 2-mile (3.2 km) circular walk around the picturesque Needham Lake, a beloved spot for families and nature enthusiasts. This leisurely walk offers a perfect blend of tranquil woodland paths and serene lakeside trails, providing ample opportunities for birdwatching and immersing yourself in the beauty of the natural surroundings.

As you set out on this enchanting journey, you'll find yourself immediately captivated by the peaceful atmosphere that envelops Needham Lake. The shimmering waters of the lake, nestled amidst lush greenery and gently swaying reeds, create a scene of idyllic tranquility that soothes the soul and invigorates the senses. With each step along the shoreline, the stresses of everyday life melt away, replaced by a profound sense of connection to the natural world.

Venturing further along the trail, you'll soon find yourself immersed in the cool embrace of the woodland, where dappled sunlight filters through the leafy canopy, casting a magical glow upon the forest floor. The soft crunch of leaves underfoot and the gentle rustling of the breeze through the branches create a symphony of natural sounds that fill the air with an enchanting melody. Pause for a moment to breathe in the earthy scents of the woodland and let the tranquility wash over you, as you become one with the living tapestry of the forest.

As you continue your journey, the path will guide you to a charming bird hide, strategically placed to offer unparalleled views of the lake and its vibrant avian residents. Take a moment to pause and observe the diverse array of bird species that call Needham Lake home, from the vibrant flash of the kingfisher as it darts across the water's surface to the elegant grace of the great crested grebe as it glides effortlessly through the tranquil waters. The bird hide provides a unique opportunity to connect with the natural world and marvel at the intricate beauty of these feathered wonders.

Emerging from the woodland, you'll once again find yourself alongside the picturesque shores of Needham Lake. The lively atmosphere around the play park, where children's laughter mingles with the sounds of nature, adds a delightful counterpoint to the serene beauty of the surroundings. Take a moment to appreciate the joyful energy of families at play, as you reflect on the importance of preserving these precious natural spaces for future generations to enjoy.

As you complete your circular route and return to the starting point, you'll carry with you a renewed sense of peace and a deeper appreciation for the incredible beauty and diversity of the natural world. The Needham Lake Walk is more than just a physical journey; it is an opportunity to immerse yourself in the wonders of nature, to find solace in the tranquility of the landscape, and to create lasting memories that will stay with you long after you've left the shores of this enchanting lake behind.

Detailed Route Directions:

1. From the Needham Lake car park, set off along the path that skirts the edge of the lake, bearing 090° as you follow the shoreline in a clockwise direction. (Estimated time: 10-15 minutes)

2. Continue along the lakeside path, soaking in the peaceful atmosphere. As you walk, keep an eye out for the charming bird hide, where you can pause to observe the various avian species that call the lake home. You'll also pass several picturesque picnic areas, perfect for a mid-walk break. (Estimated time: 15-20 minutes)

3. As you approach the far end of the lake, the path will gently curve to the right, bearing 135°. Follow this path as it leads you into a delightful wooded area, where dappled sunlight filters through the leafy canopy. (Estimated time: 10-15 minutes)

4. Proceed through the woodland, immersing yourself in the tranquil surroundings. Listen for the melodic birdsong and the gentle rustling of leaves in the breeze. Eventually, you'll come to a footbridge that spans a small stream. (Estimated time: 10-15 minutes)

5. Cross the footbridge and immediately turn left, bearing 270°. This path will guide you back towards the shimmering waters of Needham Lake. (Estimated time: 5-10 minutes)

6. As you emerge from the woodland, you'll find yourself once again walking along the southern edge of the lake. Take a moment to appreciate the picturesque views and the lively atmosphere around the play park, where children's laughter mingles with the sounds of nature. (Estimated time: 15-20 minutes)

7. Continue along the lakeside path until you return to the Needham Lake car park, where your circular walk comes to a satisfying conclusion. (Estimated time: 5-10 minutes)

Best Features and Views:

- Lose yourself in the serene beauty of Needham Lake, with its tranquil waters that mirror the sky above and the lush greenery that surrounds its shores, creating an oasis of peace and tranquility amidst the bustling world beyond.
- Discover the enchanting woodland paths, where the soft dappled light and the gentle rustling of leaves create a magical atmosphere that transports you to a realm of natural wonder and inner peace.
- Delight in the strategically placed bird hide, a window into the vibrant world of Needham Lake's avian residents, where you can observe the intricate beauty and fascinating behaviours of a diverse array of bird species, from the majestic heron to the elusive kingfisher.

Historical Features of Interest:

Needham Lake has a fascinating history that speaks to the power of conservation and the importance of preserving natural spaces for future generations. Once a site of gravel extraction, the area has been transformed through dedicated efforts into the picturesque recreational area and wildlife haven that it is today, a testament to the resilience of nature and the positive impact of human stewardship.

Birds to Spot:

- Jewels of the lake: Kingfishers: Watch for the vibrant flash of blue and orange as these stunning birds dart along the water's edge, their swift and agile movements a mesmerizing display of natural prowess.
- Dancers on the water: Great Crested Grebes: Admire the elegant grace of these diving birds, with their distinctive crests and intricate courtship dances, a captivating sight that showcases the beauty and complexity of avian behaviour.
- Sentinels of the shore: Herons: Spot these majestic birds standing motionless at the water's edge, their keen

eyes and sharp bills poised to strike at unsuspecting fish, a living embodiment of patience and precision in the natural world.

Rare or Interesting Insects and Other Wildlife:

- Aerial acrobats: Dragonflies and Damselflies: During the warmer months, Needham Lake comes alive with the vibrant hues and graceful movements of dragonflies and damselflies. Their iridescent wings shimmer in the sunlight as they flit across the water's surface, a dazzling display of aerial artistry that adds a touch of magic to the lakeside landscape.

Rare or Interesting Plants:

- Sentinels of the shoreline: Reeds and Sedges: The swaying reeds and sedges that line the shores of Needham Lake are more than just a picturesque sight; they are a vital component of the lake's ecosystem, providing shelter and nourishment for a diverse array of wildlife, from nesting birds to tiny aquatic creatures.
- Floral tapestry: Woodland Wildflowers: As you traverse the woodland paths, keep an eye out for the delicate blooms of wildflowers that dot the forest floor, their vibrant colours and intricate petals a testament to the beauty and resilience of nature, thriving in the dappled shade of the trees.

Facilities and Accessibility:

Needham Lake offers a range of facilities to ensure a comfortable and enjoyable visit for walkers of all ages and abilities:

- Car park: Ample parking is available at Needham Lake, providing convenient access to the starting point of the walk and making it easy for visitors to begin their journey into the heart of nature.
- Public toilets: Well-maintained facilities are provided for visitor comfort, ensuring that walkers can refresh themselves before or after their invigorating lakeside stroll.
- Picnic areas: Several scenic spots along the route are perfect for enjoying a picnic and taking in the beautiful surroundings, allowing walkers to immerse themselves in the tranquil atmosphere and savour the simple pleasures of the great outdoors.
- Play park: A delightful play area is available for children to enjoy, making this walk ideal for families seeking to combine the joys of nature with the laughter and excitement of playtime.
- Nearby amenities: While there are no cafes directly on the route, the nearby town of Needham Market offers a selection of cozy establishments where you can grab a refreshment or a bite to eat, the perfect way to refuel after your invigorating walk.

The walk itself is mostly flat and follows well-maintained paths, making it accessible for most walkers, including those with moderate fitness levels. However, it's important to note that some sections may be uneven or muddy, particularly after rainfall, so sturdy footwear with good grip and support is recommended to ensure a safe and comfortable experience.

Seasonal Highlights:

Each season brings its own unique charm to the Needham Lake Walk, inviting you to return time and again to witness the ever-changing beauty of the landscape:

- Spring: Delight in the vibrant wildflowers that carpet the woodland floor, their delicate blooms a joyous celebration of new life and renewal. Revel in the lively birdsong that fills the air, a symphony of nature's

reawakening after the long winter slumber.
- Summer: Bask in the warm sunshine as you stroll along the lakeside paths, the lush greenery of the surrounding trees providing welcome shade and the buzzing activity of dragonflies and damselflies adding a touch of magic to the tranquil atmosphere. Enjoy a picnic by the water's edge, immersing yourself in the simple pleasures of the season.
- Autumn: Marvel at the stunning array of colours as the leaves change, painting the landscape in a breathtaking palette of gold, orange, and red. The crisp autumn air and the gentle rustling of fallen leaves underfoot create an enchanting atmosphere that invites contemplation and reflection.
- Winter: Embrace the tranquil beauty of the frosty landscape, the bare branches of the trees etched against the pale winter sky and the stillness of the lake's surface broken only by the occasional ripple of a passing bird. The crisp, fresh air invigorates the senses, making for an exhilarating and refreshing walk.

Difficulty Level and Safety Considerations:

This walk is classified as easy and is suitable for all ages and abilities, making it a perfect choice for families, nature enthusiasts, and those seeking a gentle and refreshing outdoor experience. The paths are mostly flat and well-maintained, ensuring a pleasant and comfortable journey for walkers of varying fitness levels.

However, as with any outdoor activity, it's essential to take certain precautions to ensure a safe and enjoyable experience. Do take care when walking along the lakeside paths, as they can become slippery when wet, particularly after rainfall. It's crucial to wear sturdy, comfortable footwear with good grip and support to minimize the risk of slips or falls.

Additionally, it's always a good idea to dress appropriately for the weather conditions, bringing layers that can be easily added or removed to maintain a comfortable temperature throughout your walk. Don't forget to carry water and some light snacks to keep you hydrated and energized, and always inform someone of your intended route and expected return time as a precautionary measure.

By following these simple guidelines and being mindful of your surroundings, you can ensure that your Needham Lake Walk is a safe, enjoyable, and truly unforgettable experience.

Nearby Attractions:

After your refreshing walk around Needham Lake, take the opportunity to explore some of the other fascinating attractions that the area has to offer, each one providing a unique glimpse into the rich history, culture, and natural beauty of the region:

- Needham Market: Step back in time as you visit this charming historic town, with its picturesque streets lined with quaint buildings and independent shops that offer a delightful array of local crafts and products. Stop by one of the welcoming pubs for a hearty meal or a refreshing pint, and immerse yourself in the warm and friendly atmosphere of this quintessential English town.
- Museum of East Anglian Life: Delve into the rich agricultural and industrial heritage of the region at this fascinating open-air museum in nearby Stowmarket. Explore the beautifully preserved historic buildings, interact with engaging exhibits, and gain a deeper understanding of the lives and livelihoods of the people who have shaped this unique corner of England.
- Suffolk Countryside: Venture further afield and lose yourself in the breathtaking beauty of the Suffolk countryside, a tapestry of gently rolling hills, picturesque villages, and enchanting rural landscapes that have inspired artists and writers for centuries. From the painted churches to the stunning coastline, there is an endless array of natural and cultural treasures waiting to be discovered.

By exploring these nearby attractions, you'll gain a deeper appreciation for the diverse and captivating character of this remarkable region, making your Needham Lake Walk the perfect starting point for a truly unforgettable journey through the heart of Suffolk.

Nacton Circular Walk - 4 miles (6.4 km), Circular - Estimated Time: 2-3 hours

Starting Point Coordinates, Postcode: St. Martin's Church, Nacton (Grid Reference: TM 196 422, Postcode: IP10 0HL).
Nearest Car Park, Postcode: On-street parking is available near St. Martin's Church in Nacton (Postcode: IP10 0HL).

Embark on a delightful 4-mile (6.4 km) circular walk that takes you through the charming village of Nacton and its idyllic surrounding countryside. As you traverse the picturesque landscape, you'll be treated to breathtaking views of the majestic River Orwell, expansive fields, and tranquil woodlands, creating a truly immersive and rejuvenating experience.

As you set out from the historic St. Martin's Church, you'll immediately find yourself enveloped by the tranquil beauty of the Suffolk countryside. The gentle rustling of leaves in the breeze and the melodic birdsong create a soothing soundtrack to your journey, inviting you to leave the stresses of everyday life behind and immerse yourself in the wonders of nature.

Venturing along the quiet country lanes and winding footpaths, you'll be greeted by a patchwork of vibrant fields, their colours shifting with the seasons, from the lush greens of spring to the golden hues of autumn. The expansive vistas stretch out before you, painting a picture of rural serenity that soothes the soul and invigorates the senses.

As you make your way towards the River Orwell, the landscape undergoes a stunning transformation. The glistening waters of the river come into view, their surface reflecting the ever-changing sky above and the lush greenery that lines the banks. Take a moment to pause and drink in the breathtaking scenery, watching as boats gently sail by and the occasional wader forages along the mudflats, their presence a testament to the rich biodiversity of this enchanting corner of Suffolk.

Continuing along the riverside path, you'll find yourself immersed in a world of natural wonders. The majestic trees that line the route create a cathedral-like canopy overhead, their branches reaching out to embrace you as you walk beneath them. The dappled sunlight that filters through the leaves casts an enchanting glow upon the path, creating an atmosphere of pure magic and tranquility.

As you venture back inland, the landscape once again shifts, revealing a tapestry of woodland and open fields. The peaceful seclusion of the woods envelops you, the soft crunch of leaves underfoot and the gentle rustling of wildlife in the undergrowth creating a sense of harmony and connection with the natural world. Emerge from the treeline and you'll be greeted by sweeping vistas of undulating fields, their colours and textures a testament to the timeless beauty of the Suffolk countryside.

Making your way back towards the village of Nacton, you'll have the opportunity to admire the charming cottages and historic buildings that line the streets, their traditional architecture and welcoming atmosphere a reminder of the rich heritage and community spirit that define this delightful corner of England.

As you return to St. Martin's Church, take a final moment to reflect on the incredible journey you've undertaken. The Nacton Circular Walk is more than just a physical experience; it is an opportunity to reconnect with the beauty and serenity of the natural world, to find peace and perspective amidst the chaos of modern life, and to create memories that will last a lifetime.

Detailed Route Directions:

1. From St. Martin's Church, head south along Church Road, bearing 180°. At the end of the road, turn left onto The Street, continuing your southward route. (Estimated time: 10-15 minutes)

2. As you stroll along The Street, keep an eye out for the footpath on the right, located just after the primary school. Turn right onto this path, bearing 270°, and follow it as it winds through picturesque fields and woodland. (Estimated time: 15-20 minutes)

3. Continue along the footpath, immersing yourself in the tranquil surroundings, until you reach the banks of the River Orwell. Here, turn right, bearing 000°, and follow the riverside path for a short distance, soaking in the stunning views of the water and the surrounding landscape. (Estimated time: 20-30 minutes)

4. After a brief stretch along the river, take the footpath on the left, bearing 270°, which leads you back inland. This path will guide you through more fields and woodland, offering a peaceful escape from the hustle and bustle of everyday life. (Estimated time: 20-30 minutes)

5. As you progress, you'll eventually come to a quiet country lane. Cross the lane and continue straight ahead, following the footpath as it traverses the beautiful countryside. (Estimated time: 15-20 minutes)

6. The footpath will lead you to another lane. Turn left, bearing 180°, and walk along the lane for a short distance before taking the footpath on the right, bearing 270°. This path will guide you back towards the village of Nacton. (Estimated time: 20-30 minutes)

7. As you approach the village, soak in the charming atmosphere and the picturesque views of the surrounding landscape. Make your way back to St. Martin's Church, where your circular walk comes to a satisfying conclusion. (Estimated time: 10-15 minutes)

Best Features and Views:

- Lose yourself in the stunning vistas of the River Orwell, its glistening waters and gently sailing boats creating a mesmerizing scene that invites contemplation and relaxation.
- Immerse yourself in the picturesque charm of the village of Nacton, with its quaint cottages, historic church, and friendly local atmosphere, a testament to the enduring beauty and community spirit of rural England.
- Discover the tranquil beauty of the expansive fields and woodlands that surround the village, their colours and textures shifting with the seasons, offering a peaceful escape into nature and a chance to reconnect with the simple joys of the great outdoors.

Historical Features of Interest:

Take a moment to admire the architectural beauty and rich heritage of St. Martin's Church in Nacton. This fascinating historical site has stood as a place of worship and community gathering for centuries, its walls bearing witness to the joys, sorrows, and triumphs of generations past. If possible, step inside the church to appreciate its serene ambiance and intricate details, a testament to the skill and devotion of the craftsmen who created this enduring monument to faith and fellowship.

Birds to Spot:

- Jewels of the river: Kingfishers: Watch for the brilliant flash of blue and orange as these stunning birds dart

along the river's edge, their swift and agile movements a mesmerizing display of natural beauty and precision.
- Sentinels of the shallows: Herons: Spot these majestic birds standing motionless in the shallows, their keen eyes and sharp bills poised to strike at unsuspecting prey, a living embodiment of patience and grace.
- Maestros of the mudflats: Waders: Observe the various species of wading birds, such as curlews and redshanks, as they forage in the mudflats along the river, their distinctive calls and elegant movements a captivating sight and sound.

Rare or Interesting Insects and Other Wildlife:

- Winged wonders: Butterflies: Admire the delicate beauty of the various butterfly species that flutter amongst the wildflowers in the fields and hedgerows, their colourful wings and graceful movements adding a touch of magic to the landscape.
- Aerial acrobats: Dragonflies: Watch for the iridescent shimmer of dragonflies as they dart across the river and the nearby ponds, their incredible aerial skills and vibrant colours a marvel to behold.
- Elusive enchantment: Otters: If you're lucky, you might catch a glimpse of these elusive and enchanting creatures playing along the riverbank, their sleek fur and playful antics a rare and wonderful sight.

Rare or Interesting Plants:

- Sentinels of time: Ancient trees: Marvel at the majestic oak, ash, and beech trees that have stood sentinel in the landscape for centuries, their gnarled trunks and spreading canopies a testament to the resilience and endurance of nature.
- Floral tapestries: Wildflowers: Delight in the vibrant tapestry of colours created by the diverse array of wildflowers that adorn the field margins and woodland edges, their delicate petals and intoxicating scents a feast for the senses.
- Carpets of blue: Bluebells: If you visit in late April or early May, you may be treated to the breathtaking sight of carpets of bluebells in the woodland areas, their delicate blooms and enchanting fragrance creating an unforgettable sensory experience.

Facilities and Accessibility:

While there are no public toilets directly along the route, you can find facilities at the nearby pub, The Red Lion, in Nacton, where you can also enjoy a refreshing drink or a delicious meal after your walk. The Red Lion offers a warm and welcoming atmosphere, perfect for a post-walk refreshment and a chance to reflect on the beauty and tranquility of the Nacton Circular Walk.

The walk itself is classified as easy to moderate, with some gentle hills and uneven terrain, making it accessible for most walkers with a reasonable level of fitness. However, it's essential to wear sturdy footwear with good grip, particularly after rainfall when the paths may be muddy or slippery, to ensure a safe and comfortable experience.

Seasonal Highlights:

The Nacton Circular Walk offers a unique and captivating experience in every season, each one revealing new facets of beauty and wonder in the Suffolk countryside:

- Spring: Witness the countryside bursting into life, with vibrant wildflowers painting the fields and hedgerows in a kaleidoscope of colours, lush green leaves unfurling on the trees, and the joyous birdsong of the breeding season filling the air with a symphony of new beginnings.

- Summer: Bask in the warm sunshine as you stroll along the footpaths, the buzzing of insects and the gentle rustling of the breeze through the foliage creating a soothing soundtrack to your journey. Seek out the dappled shade of the woodland paths, where the lush greenery provides a welcome respite from the heat.
- Autumn: Marvel at the stunning array of colours as the leaves change, transforming the landscape into a breathtaking tapestry of gold, orange, and red. The crisp autumn air and the gentle crunch of fallen leaves underfoot create an enchanting atmosphere that invites reflection and contemplation.
- Winter: Embrace the tranquil beauty of the frosty fields, the bare trees silhouetted against the pale winter sky, and the crisp, fresh air that invigorates the senses and adds a touch of magic to your walk. Discover the subtle beauty of the winter landscape, with its muted colours and hushed atmosphere, and appreciate the way the season strips away the distractions, allowing you to focus on the essential elements of nature.

Difficulty Level and Safety Considerations:

This walk is classified as easy to moderate, with some gentle hills and uneven terrain, making it suitable for most walkers with a reasonable level of fitness. However, as with any outdoor activity, it's essential to be prepared for changing weather conditions and to wear appropriate clothing and footwear to ensure your comfort and safety.

As you embark on your walk, be sure to carry water, snacks, and any necessary medications, keeping yourself hydrated and energized throughout your journey. It's also a good idea to bring a map or GPS device and to inform someone of your intended route and expected return time, just as a precautionary measure.

Remember to stay on the designated paths to minimize your impact on the environment and to respect the beautiful landscapes and wildlife you encounter along the way. Take care when walking on muddy or slippery surfaces, particularly after rainfall, and wear sturdy footwear with good grip to ensure a safe and enjoyable experience.

By following these simple guidelines and being mindful of your surroundings, you can fully immerse yourself in the beauty and tranquility of the Nacton Circular Walk, creating lasting memories and a deeper appreciation for the natural wonders of the Suffolk countryside.

Nearby Attractions:

After your invigorating walk, take the opportunity to explore some of the other fascinating attractions that the area has to offer, each one providing a unique glimpse into the rich history, culture, and natural beauty of the region:

- Orwell Country Park: Discover the stunning beauty of this nearby park, with its extensive network of trails winding through diverse habitats, from tranquil woodlands to open heathland. Enjoy breathtaking views of the River Orwell and keep an eye out for the abundant wildlife that calls this park home.
- Ipswich: Venture into the vibrant county town of Suffolk, where you can explore its rich history, from the ancient streets lined with timber-framed buildings to the impressive medieval churches and museums that showcase the town's fascinating past. Immerse yourself in the thriving waterfront area, with its bustling marinas, lively cafes, and cultural attractions.
- River Orwell: Embark on a boat trip along the picturesque River Orwell, taking in the stunning scenery and learning about the area's rich maritime heritage. Admire the impressive Orwell Bridge, spot the iconic Thames sailing barges, and discover the charming villages and historic landmarks that line the river's banks.

By exploring these nearby attractions, you'll gain a deeper appreciation for the diverse and captivating character of this remarkable region, making your Nacton Circular Walk the perfect starting point for a truly unforgettable journey through the heart of Suffolk.

Nacton Shores Walk - 4 miles (6.4 km), Coastal - Estimated Time: 2-3 hours

Starting Point Coordinates, Postcode: Orwell Country Park car park (Grid Reference: TM 204 415, Postcode: IP10 0JS).
Nearest Car Park, Postcode: Orwell Country Park car park (Postcode: IP10 0JS).

Embark on an exhilarating 4-mile (6.4 km) coastal walk along the picturesque Nacton Shores, where you'll be immersed in the captivating beauty of the Suffolk coastline. This walk takes you through an enchanting tapestry of habitats, from lush woodlands and serene marshes to the rugged beauty of the beach, offering a truly diverse and enriching experience for nature lovers and outdoor enthusiasts alike.

As you set out from the Orwell Country Park car park, you'll immediately find yourself enveloped by the tranquil beauty of the surrounding landscape. The well-marked path guides you towards the shoreline, the anticipation building with each step as you catch glimpses of the glistening waters of the River Orwell through the trees.

Descending towards the coast, the landscape opens up before you, revealing a breathtaking vista of the river stretching out to the horizon. The fresh sea breeze fills your lungs, and the gentle lapping of the waves against the shore creates a soothing soundtrack to your journey. Take a moment to pause and drink in the awe-inspiring scenery, feeling the stresses of everyday life melt away in the presence of such raw and unspoiled beauty.

As you turn eastward and follow the coastal path, you'll find yourself immersed in a captivating mix of habitats, each one offering its own unique charm and character. The lush woodlands envelop you in a world of dappled sunlight and the soft rustling of leaves, while the serene marshes invite you to pause and observe the intricate web of life that thrives in these wetland oases.

Continue along the path, your senses heightened by the intoxicating scents of the sea and the wildflowers that line the trail. Keep your eyes peeled for the diverse array of flora and fauna that call this area home, from the graceful wading birds foraging in the shallows to the colourful butterflies flitting amongst the blooms.

As you arrive at Nacton Shores, the rugged beauty of the beach unfolds before you, the golden sands stretching out like a ribbon along the coastline. The sound of the waves crashing against the shore fills the air, a powerful reminder of the untamed energy of the sea. Take time to explore this enchanting stretch of coastline, feeling the sand beneath your feet and the sun on your face, as you marvel at the endless expanse of the River Orwell.

When it's time to return, retrace your steps along the coastal path, once again immersing yourself in the captivating tapestry of habitats that make this walk so special. The ever-changing light and the shifting tides paint the landscape in a new light, offering fresh perspectives and new discoveries with each passing moment.

As you near the end of your journey and the woodland path leads you back to the car park, take a final moment to reflect on the incredible beauty and diversity of the natural world that you've experienced along the Nacton Shores Walk. This coastal adventure is more than just a physical journey; it is a chance to reconnect with the raw power and majesty of nature, to find solace in the tranquility of the landscape, and to create memories that will last a lifetime.

Detailed Route Directions:

1. From the car park at Orwell Country Park, follow the well-marked path, bearing 090°, as it leads you towards the shoreline. As you descend, take a moment to appreciate the stunning views of the River Orwell and the surrounding landscape. (Estimated time: 10-15 minutes)

2. Upon reaching the shoreline, turn right and head east, bearing 090°. The path will guide you through a captivating mix of woodland and marshy areas, each with its own unique charm and character. (Estimated time: 20-30 minutes)

3. Continue along the coastal path, immersing yourself in the sights, sounds, and scents of this incredible environment. Take your time to observe the diverse array of flora and fauna that call this area home. (Estimated time: 30-45 minutes)

4. As you progress, you'll find yourself at Nacton Shores, a beautiful stretch of beach that invites exploration. Spend some time here, enjoying the fresh sea breeze, the sound of the waves lapping against the shore, and the stunning views of the River Orwell. (Estimated time: 20-30 minutes)

5. When you're ready to continue, retrace your steps along the coastline, bearing 270°, back towards Orwell Country Park. Once again, take the opportunity to appreciate the diverse habitats and the breathtaking scenery as you walk. (Estimated time: 30-45 minutes)

6. Upon reaching the woodland near the starting point, follow the path as it leads you back to the car park at Orwell Country Park, where your coastal adventure comes to a satisfying conclusion. (Estimated time: 10-15 minutes)

Best Features and Views:

- Lose yourself in the stunning coastal vistas, the glistening waters of the River Orwell stretching out to the horizon, an endless expanse of blue that invites contemplation and wonder.
- Immerse yourself in the diverse habitats that make this walk so special, from the lush woodlands with their dappled sunlight and soft rustling of leaves to the serene marshes that teem with life, each one offering a unique and captivating experience.
- Delight in the opportunity to observe a wide variety of bird species, insects, and other wildlife in their natural environments, a chance to witness the incredible diversity and resilience of the natural world firsthand.

Historical Features of Interest:

As you walk along the coastline, keep an eye out for old shipwrecks and other remnants of the area's rich maritime history. These fascinating features offer a glimpse into the past, whispering stories of adventure, tragedy, and the enduring relationship between humans and the sea. Take a moment to imagine the lives of those who once sailed these waters and the tales that have been carried on the wind and waves for generations.

Birds to Spot:

- Elegant foragers: Wading birds: Look for the elegant avocets, with their slender, upturned bills and striking black and white plumage, the distinctive redshanks, their bright red legs a flash of colour against the mudflats, and other waders as they forage in the shallows and along the shore, their graceful movements a mesmerizing

display of natural adaptation.
- Masters of the skies: Seabirds: Watch for the graceful terns diving for fish, their sleek forms cutting through the air with precision and skill, the soaring gulls riding the air currents with effortless grace, and the cormorants perched on rocky outcrops, their wings spread wide to dry in the sun, each species a vital part of the coastal ecosystem.

Rare or Interesting Insects and Other Wildlife:

- Winged jewels: Butterflies: Admire the vibrant colours and delicate beauty of the various butterfly species that flutter amongst the wildflowers and along the woodland edges, their gossamer wings catching the light and adding a touch of magic to the landscape.
- Aerial acrobats: Dragonflies: Marvel at the iridescent shimmer of dragonflies as they dart across the marshes and ponds, their wings glinting in the sunlight like miniature stained-glass windows, a testament to the incredible diversity and beauty of the insect world.
- Elusive inhabitants: Mammals: Keep an eye out for signs of otters, water voles, and other mammals that make their home in the coastal environment, their presence a reminder of the delicate balance and interconnectedness of the ecosystems that thrive along the Nacton Shores.

Rare or Interesting Plants:

- Guardians of the forest: Woodland flora: Discover the diverse array of trees, shrubs, and wildflowers that thrive in the lush woodland areas, from the majestic oaks that tower overhead to the delicate bluebells that carpet the forest floor, each species playing a vital role in the intricate web of life that flourishes in the dappled shade.
- Wetland wonders: Marsh plants: Observe the specialized plants that have adapted to the unique conditions of the marshy areas, such as the slender reeds that sway in the breeze, the sturdy rushes that rise from the water's edge, and the graceful sedges that filter the sunlight, their presence a testament to the resilience and adaptability of nature.
- Coastal sentinels: Coastal vegetation: Admire the hardy plants that cling to the rugged coastline, such as the distinctive sea kale with its waxy, blue-green leaves, the delicate sea lavender that adds a splash of purple to the shoreline, and the tough marram grass that helps to stabilize the dunes, each species a vital component of the coastal ecosystem.

Facilities and Accessibility:

Orwell Country Park car park offers public toilets, ensuring a convenient and comfortable starting point for your coastal walk. Along the route, you'll find several picnic tables and benches strategically placed to offer scenic rest stops, perfect for taking a moment to catch your breath, enjoy a snack, and soak in the stunning views of the River Orwell and the surrounding landscape.

It's important to note that the Nacton Shores Walk includes some uneven terrain and may not be suitable for wheelchairs or pushchairs. To ensure a safe and enjoyable experience, it is highly recommended to wear sturdy walking shoes or boots that provide good support and traction, allowing you to navigate the varied surfaces, including sand, mud, and rocky areas, with confidence and ease.

Seasonal Highlights:

The Nacton Shores Walk offers a unique and captivating experience in every season, each one revealing new facets of beauty and wonder along the Suffolk coastline:

- Spring: Witness the explosion of colour as wildflowers carpet the woodlands and marshes, their delicate blooms and sweet fragrances a joyous celebration of new life and renewal. Delight in the lively birdsong of the breeding season, as the air fills with the melodies of courtship and the excited chirps of newly hatched chicks.
- Summer: Bask in the warm sunshine and the vibrant blooms of the coastal flowers, their petals painted in a kaleidoscope of colours that rival the beauty of the sea itself. Enjoy the buzzing activity of insects and other wildlife, as the coastline comes alive with the energy and vitality of the summer months.
- Autumn: Marvel at the stunning autumnal hues as the leaves change colour, transforming the woodlands into a breathtaking tapestry of gold, orange, and red. Embrace the crisp, invigorating air and the sense of tranquility that descends upon the landscape, inviting quiet reflection and a deeper connection with the natural world.
- Winter: Embrace the wild beauty of the wind-swept coast, the stark silhouettes of bare trees etched against the leaden sky, and the raw power of the waves as they crash upon the shore. Delight in the opportunity to spot overwintering birds, their presence a reminder of the resilience and adaptability of life in the face of even the harshest conditions.

Difficulty Level and Safety Considerations:

The Nacton Shores Walk is classified as moderate, with some uneven terrain and short, steep sections that may prove challenging for less experienced walkers. The paths can be muddy, slippery, or sandy in places, making sturdy footwear with good grip an essential item for a safe and enjoyable experience.

As you embark on your coastal adventure, it's crucial to be prepared for changeable weather conditions, especially along the exposed sections of the coastline where the wind and sea can be unpredictable. Bring appropriate clothing to protect you from the elements, as well as an ample supply of water and sunscreen to keep you hydrated and shielded from the sun's rays.

Always be mindful of the tides and avoid walking too close to the water's edge, as the shifting sands and powerful currents can pose a significant risk. Stick to the designated paths to minimize erosion and protect the delicate ecosystems that thrive along the Nacton Shores, ensuring that this beautiful landscape remains a haven for wildlife and a source of inspiration for generations to come.

Nearby Attractions:

After your invigorating walk along the Nacton Shores, take the opportunity to explore some of the other fascinating attractions that the area has to offer, each one providing a unique glimpse into the rich history, culture, and natural beauty of the Suffolk region:

- Ipswich: Discover the vibrant county town of Suffolk, where you can immerse yourself in its rich history, from the ancient streets lined with timber-framed buildings to the impressive museums that showcase the town's fascinating past. Explore the thriving waterfront area, with its bustling marinas, lively cafes, and cultural attractions that offer a window into contemporary life in this dynamic town.
- Felixstowe: Venture to the charming coastal town of Felixstowe, where you can stroll along the stunning seafront gardens, their colourful displays of flowers and meticulously manicured lawns a testament to the town's Victorian heritage. Visit the historic fort, a monument to the region's military past, and enjoy the lively atmosphere of the popular pier, with its amusements, cafes, and breathtaking views of the North Sea.
- Suffolk Coast and Heaths AONB: Immerse yourself in the breathtaking beauty of this Area of Outstanding Natural Beauty, a landscape of diverse habitats and stunning vistas that showcase the very best of the Suffolk coastline. Explore the picturesque villages that dot the countryside, their charming cottages and friendly pubs a welcome respite from the rugged beauty of the coast, and marvel at the abundance of wildlife that thrives in

this protected and cherished landscape.

By exploring these nearby attractions, you'll gain a deeper appreciation for the rich tapestry of history, culture, and natural beauty that defines this extraordinary corner of England, making your Nacton Shores Walk the perfect starting point for a truly unforgettable journey through the heart of Suffolk.

Orwell Country Park Walk - 2 miles (3.2 km), Circular - Estimated Time: 1-1.5 hours

Starting Point Coordinates, Postcode: Orwell Country Park car park, Ipswich (Grid Reference: TM 194 419, Postcode: IP10 0JS).
Nearest Car Park, Postcode: Orwell Country Park car park (Postcode: IP10 0JS).

Immerse yourself in the stunning beauty of Orwell Country Park on this delightful 2-mile (3.2 km) circular walk. This leisurely route takes you through an enchanting landscape of tranquil woodlands, open meadows, and along the banks of the magnificent River Orwell, offering breathtaking views and a chance to connect with nature. The well-maintained paths and mostly flat terrain make this walk accessible and enjoyable for all ages and abilities.

As you set out from the car park, the well-signposted path beckons you towards the River Orwell, the anticipation building with each step as you catch glimpses of the glistening water through the trees. Upon reaching the riverbank, the full splendour of the landscape unfolds before you, the river's gentle flow and the lush greenery creating a scene of tranquil beauty that soothes the soul and invigorates the senses.

Turning to follow the river's edge, you'll find yourself immersed in a world of natural wonders. The path meanders along the water's edge, offering ever-changing vistas of the river and the surrounding landscape. Keep your eyes peeled for the various bird species that call this area home, from the brilliant flash of the kingfisher darting along the riverbank to the majestic heron standing motionless in the shallows, each encounter a reminder of the incredible diversity of life that thrives in this enchanting corner of Suffolk.

As you continue along the trail, you'll soon find yourself passing beneath the impressive Orwell Bridge, its striking form a testament to human ingenuity and a fascinating contrast to the organic beauty of the park. Pause for a moment to appreciate the scale and grandeur of this iconic structure, the sound of the river flowing beneath it a constant reminder of the power and persistence of nature.

Leaving the bridge behind, the path gently curves inland, guiding you through a tapestry of picturesque woodland and open meadows. Immerse yourself in the tranquility of the forest, the dappled sunlight filtering through the canopy and the soft rustling of leaves in the breeze creating an atmosphere of pure serenity. Emerge from the treeline and you'll be greeted by the vibrant colours and gentle swaying of the meadow grasses, the wildflowers dotting the landscape like jewels in a crown.

As you navigate through these diverse habitats, take the time to observe the intricate details of the plants, trees, and wildflowers that adorn the landscape. Listen for the melodic birdsong that fills the air, each note a celebration of the vitality and resilience of the natural world. Breathe in the fresh, invigorating air and feel the stresses of everyday life melt away, replaced by a profound sense of connection and belonging to the earth beneath your feet.

The path eventually loops back towards the starting point, offering a chance to reflect on the beauty and tranquility of the journey you've just undertaken. As you return to the car park, carry with you the memories of the stunning vistas, the encounters with wildlife, and the moments of pure, unadulterated joy that come from immersing yourself in the wonders of nature.

The Orwell Country Park Walk is more than just a physical journey; it is an opportunity to reconnect with the natural world, to find peace and perspective amidst the chaos of modern life, and to discover the incredible beauty and diversity that exists just beyond our doorsteps. Whether you're a seasoned hiker or a casual walker, this trail promises to delight, inspire, and leave you with a renewed appreciation for the magnificent landscapes and rich ecological tapestry of Suffolk.

Detailed Route Directions:

1. From the car park, follow the well-signposted path, bearing 270°, as it leads you towards the River Orwell. Take a moment to appreciate the stunning views of the river and the lush greenery that surrounds you. (Estimated time: 5-10 minutes)

2. Upon reaching the riverbank, turn left and follow the path, bearing 180°, as it meanders along the water's edge. Soak in the tranquil atmosphere and keep an eye out for the various bird species that call this area home. (Estimated time: 15-20 minutes)

3. Continue along the river path, passing underneath the impressive Orwell Bridge. This iconic structure spans the river, providing a striking contrast to the natural beauty of the surrounding landscape. (Estimated time: 10-15 minutes)

4. After passing under the bridge, the path will gently curve inland, bearing 090°. Follow the trail as it leads you through picturesque woodland and open meadows, offering a chance to immerse yourself in the park's diverse habitats. (Estimated time: 20-30 minutes)

5. As you navigate through the woodland and meadows, take your time to observe the various plants, trees, and wildflowers that adorn the landscape. Listen for the melodic birdsong and the gentle rustling of leaves in the breeze. (Estimated time: 15-20 minutes)

6. The path will eventually loop back towards the starting point, guiding you once more to the car park at Orwell Country Park. As you complete the circular route, reflect on the beauty and tranquility of the natural world you've just experienced. (Estimated time: 5-10 minutes)

Best Features and Views:

- Lose yourself in the stunning vistas of the River Orwell, its glistening waters and gently sailing boats creating a mesmerizing scene that invites contemplation and wonder.
- Immerse yourself in the lush woodland and open meadows, each habitat offering a unique and captivating experience, from the dappled sunlight and soft rustling of leaves in the forest to the vibrant colours and gentle swaying of the meadow grasses.
- Marvel at the impressive Orwell Bridge, a striking feat of engineering that provides a fascinating contrast to the natural surroundings, its scale and grandeur a testament to human ingenuity and the enduring relationship between people and the landscape.

Historical Features of Interest:

As you walk through Orwell Country Park, keep an eye out for the informative panels that provide insights into the area's rich history. These fascinating glimpses into the past reveal how human activity has shaped the landscape over the centuries, from the ancient Bronze Age settlements to the industrial heritage of the 19th and 20th centuries. Take a moment to imagine the lives of those who have called this place home, and the stories that have been etched into the very fabric of the land itself.

Birds to Spot:

- Jewels of the riverbank: Kingfishers: Watch for the brilliant flash of blue and orange as these stunning birds dart along the river's edge, their swift and agile movements a mesmerizing display of natural beauty and precision.
- Sentinels of the shallows: Herons: Spot these majestic birds standing motionless in the shallows, their keen eyes and sharp bills poised to strike at unsuspecting prey, a living embodiment of patience and grace.
- Melodic masters: Songbirds: Listen for the enchanting melodies of various woodland birds, such as blackcaps, chiffchaffs, and wrens, their delightful notes filling the air with a joyous celebration of life and the changing seasons.

Rare or Interesting Insects and Other Wildlife:

- Aerial acrobats: Dragonflies and damselflies: Admire the iridescent shimmer of these fascinating insects as they dart across the ponds and along the river's edge, their delicate wings and graceful movements a testament to the incredible diversity and beauty of the natural world.
- Winged wonders: Butterflies: Delight in the vibrant colours and intricate patterns of the various butterfly species that flutter amongst the wildflowers and along the woodland edges, each one a living work of art that adds a touch of magic to the landscape.
- Elusive inhabitants: Mammals: Keep an eye out for signs of deer, foxes, and rabbits as they navigate through the woodland and meadows, their presence a reminder of the rich tapestry of life that thrives in the park's diverse habitats.

Rare or Interesting Plants:

- Guardians of the forest: Ancient woodland trees: Marvel at the majestic oaks, ashes, and beeches that have stood sentinel in the park for centuries, their gnarled trunks and spreading canopies a testament to the resilience and endurance of nature.
- Floral tapestries: Wildflowers: Discover the vibrant array of colours and delicate blooms that adorn the meadows and woodland edges, each species a vital thread in the intricate web of life that sustains the park's ecosystems.
- Sentinels of the riverbank: Riverside vegetation: Observe the specialized plants that thrive along the river's edge, such as the slender reeds, the sturdy rushes, and the graceful sedges, their presence a vital component of the aquatic habitat and a source of food and shelter for countless species.

Facilities and Accessibility:

The Orwell Country Park Walk is designed to be accessible for most visitors, with well-maintained paths and mostly flat terrain that make it a pleasure to navigate. However, it's important to note that some sections may be uneven or muddy, particularly after rainfall, so sturdy footwear with good grip and support is highly recommended to ensure a safe and comfortable experience.

While there are no toilets or cafes directly within the park, the nearby Strand Cafe at the Suffolk Food Hall offers a delightful spot to enjoy refreshments before or after your walk. This charming establishment provides a welcoming atmosphere and a tempting selection of locally sourced food and drink, the perfect way to refuel and relax after your invigorating journey through the park.

Seasonal Highlights:

Each season brings its own unique charm and character to the Orwell Country Park Walk, inviting you to return time and again to witness the landscape's captivating transformations:

- Spring: Witness the park bursting into life, with vibrant wildflowers painting the meadows in a kaleidoscope of colours, fresh green leaves unfurling on the trees, and the joyous birdsong of the breeding season filling the air with the melodies of new beginnings.
- Summer: Bask in the warm sunshine as you stroll along the riverbank, the buzzing of insects and the gentle rustling of the breeze through the foliage creating a soothing soundtrack to your journey. Seek out the dappled shade of the woodland paths, where the lush canopy provides a welcome respite from the heat.
- Autumn: Marvel at the stunning array of colours as the leaves change, transforming the landscape into a breathtaking tapestry of gold, orange, and red. Embrace the crisp, invigorating air and the sense of tranquility that descends upon the park, inviting quiet reflection and a deeper connection with the natural world.
- Winter: Embrace the tranquil beauty of the frosty ground, the bare trees silhouetted against the pale winter sky, and the crisp, fresh air that invigorates the senses and adds a touch of magic to your walk. Discover the subtle beauty of the winter landscape, with its muted colours and hushed atmosphere, and appreciate the way the season strips away the distractions, allowing you to focus on the essential elements of nature.

Difficulty Level and Safety Considerations:

This walk is classified as easy, with mostly flat terrain and well-maintained paths, making it suitable for walkers of all ages and abilities. However, as with any outdoor activity, it's essential to be prepared for changeable weather conditions and to wear appropriate clothing and footwear to ensure your comfort and safety.

As you embark on your walk, be sure to carry water, snacks, and any necessary medications, keeping yourself hydrated and energized throughout your journey. It's also a good idea to inform someone of your intended route and expected return time, just as a precautionary measure.

Remember to stay on the designated paths to minimize your impact on the environment and to respect the delicate ecosystems that thrive within the park. Take care when walking on muddy or slippery surfaces, particularly after rainfall, and wear sturdy footwear with good grip to ensure a safe and enjoyable experience.

By following these simple guidelines and the countryside code, you can fully immerse yourself in the beauty and tranquility of the Orwell Country Park Walk, creating lasting memories and a deeper appreciation for the natural wonders that surround us.

Nearby Attractions:

After your peaceful walk through Orwell Country Park, take the opportunity to explore some of the other fascinating attractions that the area has to offer, each one providing a unique glimpse into the rich history, culture, and natural beauty of the Suffolk region:

- Suffolk Food Hall: Visit this fantastic food and shopping destination, where you can sample an enticing array of local produce, indulge in a delicious meal, and browse the carefully curated selection of artisanal goods, each one a celebration of the region's culinary heritage and craftsmanship.
- Ipswich: Venture into the vibrant county town of Suffolk, where you can immerse yourself in its rich history, from the ancient streets lined with timber-framed buildings to the impressive museums that showcase the town's fascinating past. Explore the thriving waterfront area, with its bustling marinas, lively cafes, and cultural attractions that offer a window into contemporary life in this dynamic town.

- Christchurch Park: Discover another of Ipswich's beautiful green spaces, where you can stroll through the picturesque gardens surrounding the historic mansion, admire the tranquil ponds that reflect the changing skies, and lose yourself in the formal gardens that showcase the skill and artistry of the park's designers.

By exploring these nearby attractions, you'll gain a deeper appreciation for the diverse and captivating character of this remarkable region, making your Orwell Country Park Walk the perfect starting point for a truly unforgettable journey through the heart of Suffolk.

Orwell Walk - 12 miles (19.3 km), Circular - Estimated Time: 6-7 hours

Starting Point Coordinates, Postcode: Orwell Country Park Car Park, Nacton, Ipswich (Grid Reference: TM 205 414, Postcode: IP10 0AT).
Nearest Car Park, Postcode: Orwell Country Park Car Park, Nacton, Ipswich (Postcode: IP10 0AT).

Embark on an unforgettable 12-mile (19.3 km) circular walk along the picturesque River Orwell, immersing yourself in the breathtaking beauty of the Suffolk countryside. This challenging yet rewarding route offers stunning views of the river and the surrounding landscape, as well as the opportunity to discover rare birds, fascinating wildlife, and historical sites of interest. Perfect for experienced walkers seeking a day of adventure and exploration.

As you set out from the Orwell Country Park car park, the footpath beckons you into the tranquil surroundings of the park, the anticipation building with each step as you make your way towards the River Orwell. Upon reaching the riverbank, the full majesty of the landscape unfolds before you, the glistening waters and the gently swaying reeds creating a scene of serene beauty that soothes the soul and invigorates the senses.

Turning to follow the river southward, you'll find yourself immersed in a world of natural wonders. The path meanders along the water's edge, offering ever-changing vistas of the river and the surrounding countryside. Marvel at the graceful boats sailing by, their white sails catching the light and adding a touch of romance to the already enchanting scene.

As you continue along the trail, you'll soon find yourself enveloped by the cool shade of Bridge Wood, a charming wooded area nestled on the riverbank. Take a moment to pause and appreciate the dappled sunlight filtering through the leaves, the gentle rustling of the breeze in the branches, and the peaceful stillness that pervades this natural oasis.

Emerging from the woodland, the path continues southward, guiding you along the picturesque banks of the River Orwell. Keep your eyes peeled for the historic Woolverstone Marina, once a bustling Royal Navy training establishment, its presence a testament to the rich maritime heritage that has shaped this remarkable landscape.

Approaching Pin Mill, a delightful hamlet steeped in history and charm, you'll find yourself at a crossroads. Here, the path veers inland, leading you away from the river and towards the quaint village of Chelmondiston. Immerse yourself in the tranquil atmosphere of this traditional Suffolk village, admiring the charming cottages and the friendly locals as you make your way through its picturesque streets.

Leaving Chelmondiston behind, the footpath winds its way through the rolling fields and farmland of the Suffolk countryside, offering a chance to connect with the rural beauty and timeless character of this enchanting region. Breathe in the fresh country air, listen to the melodic birdsong, and savour the sense of peace and freedom that comes from exploring this unspoiled landscape on foot.

As you make your way back towards the River Orwell, the path once again joins the riverside, rewarding you with breathtaking views of the water and the surrounding landscape. The impressive Orwell Bridge looms on the horizon, its striking form a testament to human ingenuity and a fascinating contrast to the organic beauty of the river.

Ascending the short but steep incline to the top of the bridge, you'll be greeted by an awe-inspiring panorama of the River Orwell and the countryside stretching out before you. Take a moment to catch your breath and drink in the stunning vistas, feeling the exhilaration of standing high above the water and surveying the incredible landscape that you've traversed.

Descending back to the riverside path, the final leg of your journey takes you northward, following the meandering course of the river as it winds its way through the outskirts of Ipswich. Reflect on the incredible diversity of landscapes and experiences you've encountered along the way, from tranquil woodlands and picturesque villages to the raw power and beauty of the River Orwell itself.

As you arrive back at the Orwell Country Park car park, your circular walk comes to a satisfying close, the memories of your unforgettable adventure still fresh in your mind. The Orwell Walk is more than just a physical journey; it is a chance to immerse yourself in the rich history, stunning scenery, and vibrant wildlife of the Suffolk countryside, forging a deep connection with the natural world and creating memories that will last a lifetime.

Detailed Route Directions:

1. From the Orwell Country Park car park (Grid Reference: TM 205 414), follow the footpath, bearing 090°, as it leads you into the tranquil surroundings of the park. (Estimated time: 10-15 minutes)

2. Continue along the path as it descends towards the River Orwell. Upon reaching the riverbank, turn right and follow the path, bearing 180°, as it meanders along the water's edge, heading south. (Estimated time: 20-30 minutes)

3. After approximately 1.5 miles (2.4 km), you'll pass through Bridge Wood, a charming wooded area nestled on the riverbank. Take a moment to appreciate the dappled sunlight filtering through the trees and the peaceful atmosphere of this natural oasis. (Estimated time: 30-40 minutes)

4. Proceed south along the River Orwell for another 2 miles (3.2 km), taking in the stunning views of the water and the boats gently sailing by. You'll pass the historic Woolverstone Marina, once a Royal Navy training establishment, offering a glimpse into the area's rich maritime heritage. (Estimated time: 1-1.5 hours)

5. As you approach Pin Mill, a picturesque hamlet on the river's edge, look out for the footpath intersection marked by a signpost. Here, turn left, bearing 000°, and follow the path as it leads you inland, away from the river and towards the village of Chelmondiston. (Estimated time: 30-40 minutes)

6. Navigate through the charming streets of Chelmondiston, taking in the quaint cottages and the peaceful village atmosphere. At the end of the village, turn left onto the footpath, bearing 315°, which will guide you northward through the beautiful Suffolk countryside. (Estimated time: 30-40 minutes)

7. Continue along the footpath, passing through fields and farmland, until you reach a small bridge. Cross the bridge and then turn right onto the footpath, bearing 045°, which will lead you back towards the River Orwell. (Estimated time: 45-60 minutes)

8. Walk along the riverside path for approximately 2.5 miles (4 km), savouring the breathtaking views of the river and the surrounding landscape. As you approach the impressive Orwell Bridge, keep an eye out for the footpath on your right. (Estimated time: 1-1.5 hours)

9. Take the footpath on the right, bearing 135°, which will lead you up a short but steep incline to the top of the Orwell Bridge. Pause here to catch your breath and admire the panoramic views of the river and the

countryside stretching out before you. (Estimated time: 15-20 minutes)

10. After taking in the stunning vistas, carefully make your way back down to the riverside path and continue walking north, bearing 000°, for another 3 miles (4.8 km). You'll pass underneath the Orwell Bridge and through the outskirts of Ipswich, following the river as it winds its way through the landscape. (Estimated time: 1.5-2 hours)

11. Finally, you'll arrive back at the Orwell Country Park car park, where your circular walk comes to a satisfying conclusion. Take a moment to reflect on the incredible journey you've just completed and the unforgettable experiences you've had along the way. (Estimated time: 5-10 minutes)

Best Features and Views:

- Lose yourself in the stunning vistas of the River Orwell, its glistening waters and gently sailing boats creating a mesmerizing scene that invites contemplation and wonder, the ever-changing light and the play of shadows on the water's surface a testament to the raw beauty of the natural world.
- Immerse yourself in the tranquil charm of Bridge Wood, a verdant oasis nestled on the riverbank, where the dappled sunlight filtering through the leaves and the gentle rustling of the breeze in the branches create an atmosphere of pure serenity and peace.
- Marvel at the historic Woolverstone Marina, once a bustling hub of naval activity, its presence a poignant reminder of the rich maritime heritage that has shaped this remarkable landscape, the echoes of the past still palpable in the timeless beauty of the River Orwell.
- Stand in awe atop the impressive Orwell Bridge, the panoramic views of the river and the surrounding countryside stretching out before you, a breathtaking vista that showcases the incredible diversity and grandeur of the Suffolk landscape, from the winding course of the river to the rolling fields and picturesque villages that dot the horizon.

Historical Features of Interest:

Woolverstone Marina: Once a thriving Royal Navy training establishment, this historic site stands as a testament to the rich maritime heritage of the River Orwell. As you pass by the marina, take a moment to imagine the bustle of activity in days gone by, the young recruits learning the skills of seamanship and navigation, and the grand naval vessels that once graced these waters. The presence of Woolverstone Marina adds a layer of historical depth to the already captivating landscape, inviting you to ponder the stories and events that have shaped this remarkable corner of Suffolk.

Birds to Spot:

- Elegant avocets: These striking black and white waders, with their slender, upturned bills, are a true delight to spot along the riverbank. Watch as they wade through the shallows with a delicate grace, their presence a testament to the rich biodiversity of the River Orwell.
- Piping oystercatchers: Listen for the distinctive piping calls of these charismatic black and white birds as they forage for molluscs in the mudflats, their bright orange bills and lively behaviour adding a splash of colour and energy to the coastal landscape.
- Waders and wildfowl: Delight in the diverse array of waders and wildfowl that make their home along the River Orwell, from the elegant redshanks and curlews probing the mud for hidden treasures to the various species of ducks and geese floating serenely on the water's surface, each one a vital thread in the intricate tapestry of life that thrives in this unique habitat.

Rare or Interesting Insects and Other Wildlife:

- Shimmering dragonflies: Marvel at the iridescent colours and acrobatic flight of the various dragonfly species that dart across the water's surface and along the riverbank, their delicate wings catching the light and creating a mesmerizing display of natural artistry.
- Elusive water voles: Keep an eye out for signs of these charming and elusive mammals, such as their distinctive burrows along the riverbank and the gentle plopping sound as they dive into the water, their presence a heartening reminder of the resilience and adaptability of nature.
- Playful otters: If fortune favours you, you might catch a rare glimpse of these enigmatic creatures as they hunt for fish in the river or bask in the sun along the bank, their sleek fur and lively antics a source of endless fascination and delight.

Rare or Interesting Plants:

- Vibrant marsh marigolds: In spring, the riverbanks and damp meadows come alive with the vibrant yellow flowers of the marsh marigold, their cheerful blooms a joyous celebration of the season and a vital source of nectar for pollinating insects.
- Elegant yellow flag irises: Admire the stately beauty of these tall, elegant irises, their striking yellow flowers adding a regal touch to the riverbank and the surrounding wetlands, their presence a testament to the incredible diversity and resilience of the plant life that thrives in this unique habitat.

Facilities and Accessibility:

The Orwell Walk offers a range of facilities to ensure a comfortable and enjoyable experience for walkers. At Pin Mill, approximately halfway along the route, you'll find well-maintained public toilets and a delightful café, providing a welcome opportunity to take a break, refresh, and refuel before continuing your journey. The café's menu features a tempting selection of locally-sourced food and drink, the perfect way to savour the flavours of the Suffolk countryside while soaking in the charming atmosphere of this historic hamlet.

It's important to note that while the path is mostly flat, some sections may be muddy or uneven, particularly after periods of wet weather. This can make the walk challenging for wheelchair users and those with limited mobility. To ensure a safe and enjoyable experience, it is highly recommended to wear sturdy walking shoes or boots with good grip and support, allowing you to navigate the varied terrain with confidence and ease.

Seasonal Highlights:

The Orwell Walk offers a unique and captivating experience in every season, each one revealing new facets of beauty and wonder along the River Orwell:

- Spring and autumn: Witness the awe-inspiring spectacle of migrating wading birds, as they gather in vast numbers along the River Orwell to rest and refuel on their epic journeys to and from their distant breeding grounds. Marvel at the sight of thousands of birds wheeling through the sky, their calls filling the air with a cacophony of life and energy.
- Summer: Delight in the vibrant wildflowers that adorn the riverbanks and the surrounding meadows, their delicate blooms and sweet fragrances a joyous celebration of the season's warmth and vitality. Bask in the golden sunshine as you stroll along the picturesque river, the lush greenery and the gentle buzzing of insects creating an idyllic scene of rural tranquility.
- Winter: Embrace the wild beauty of the Suffolk coast in winter, as the River Orwell becomes a haven for large flocks of overwintering wildfowl, including a variety of ducks, geese, and swans. Admire the striking contrast of

their colourful plumage against the stark beauty of the frosty landscape, and marvel at their resilience in the face of the harsh Arctic conditions they have left behind.

Difficulty Level and Safety Considerations:

The Orwell Walk is classified as moderate, with some uneven terrain and muddy sections that can prove challenging for less experienced walkers. The route is best suited for those with a good level of fitness and comfortable with longer distances, as the 12-mile journey requires stamina and determination.

As you make your way along the River Orwell, it's crucial to be aware of the tidal conditions and to avoid walking too close to the water's edge, especially during high tide when the river can rise rapidly. Always stay on the designated path and exercise caution when navigating any steep or slippery sections, particularly near the Orwell Bridge.

To ensure a safe and enjoyable walk, be sure to wear sturdy, comfortable footwear with good grip and support, as well as clothing suitable for the changeable British weather. Carry an ample supply of water and energizing snacks to keep you hydrated and fuelled throughout your journey, and don't forget to pack any necessary medications or first aid supplies.

It's always a good idea to inform someone of your planned route and expected return time before setting off, and to carry a fully charged mobile phone in case of emergencies. By following these simple precautions and listening to your body's needs, you can fully immerse yourself in the beauty and tranquility of the Orwell Walk, creating memories that will last a lifetime.

Nearby Attractions:

After your invigorating walk along the River Orwell, take the opportunity to explore some of the other fascinating attractions that the area has to offer, each one providing a unique glimpse into the rich history, culture, and natural beauty of the Suffolk region:

- Jimmy's Farm & Wildlife Park: Step into a world of adventure and discovery at this popular family attraction, where you can get up close and personal with a delightful array of friendly farm animals, explore the beautiful gardens and woodland trails, and indulge in the delicious, locally-sourced food at the on-site restaurant.
- Ipswich Transport Museum: Embark on a fascinating journey through time at this unique museum, which houses an extensive collection of vehicles that showcase the rich transport heritage of Ipswich and the surrounding area. From horse-drawn carriages and vintage bicycles to classic buses and trams, each exhibit offers a glimpse into the past and the ways in which people have moved through this historic landscape over the centuries.
- Christchurch Park: Unwind and recharge in the tranquil surroundings of this beautiful park, located in the heart of Ipswich. Stroll through the picturesque gardens that surround the historic mansion, pause to admire the elegant Victorian fountains and the vibrant flower beds, and lose yourself in the peaceful atmosphere of this green oasis in the midst of the bustling town.

By exploring these nearby attractions, you'll gain a deeper appreciation for the diverse and captivating character of this remarkable region, making your Orwell Walk the perfect starting point for a truly unforgettable journey through the heart of Suffolk.

Freston Wood Walk - 3 miles (4.8 km), Circular - Estimated Time: 1.5-2 hours

Starting Point Coordinates, Postcode: The starting point for the Freston Wood Walk is at the Freston Wood car park (Grid Reference: TM 195 385, Postcode: IP9 1AF).
Nearest Car Park, Postcode: The nearest car park is the Freston Wood car park (Postcode: IP9 1AF).

Embark on a captivating 3-mile (4.8 km) circular walk through the enchanting Freston Wood, where the gentle embrace of nature invites you to leave behind the cares of the world and immerse yourself in a realm of tranquility and beauty. This easy to moderate route, with its well-maintained paths and breathtaking scenery, is perfect for those seeking a rejuvenating escape from the hustle and bustle of everyday life.

As you set out from the Freston Wood car park, a sense of anticipation builds with each step, the path beckoning you into the heart of the woodland. The moment you cross the threshold into this green sanctuary, you'll feel the stresses of daily life melt away, replaced by a profound sense of peace and connection with the natural world.

The trail winds through a tapestry of diverse habitats, each one revealing new facets of the woodland's charm. From the cool, mossy hollows where ferns unfurl their delicate fronds to the sun-dappled glades where wildflowers dance in the gentle breeze, every turn of the path brings fresh wonders to discover. Pause for a moment to listen to the melodic songs of birds flitting through the canopy, their joyful notes a celebration of the vibrant ecosystem that thrives in the heart of Freston Wood.

As you continue along the trail, the sights, sounds, and scents of the woodland engulf your senses, each one a delicate brush stroke in the masterpiece of nature. Take the time to examine the intricate details of the foliage, the patterns of bark on the ancient trees, and the tiny insects that bustle through the undergrowth, marvelling at the incredible complexity and resilience of the living world around you.

At the northern edge of the wood, the path curves gracefully, offering tantalizing glimpses of the surrounding countryside through the trees. Here, at the boundary between the sheltered world of the forest and the open expanse of fields and hedgerows, you may spot a shy deer grazing in the distance or hear the distant lowing of cattle, a reminder of the timeless rhythms of rural life that have shaped this landscape for generations.

As you loop back towards the starting point, the path guides you once more into the comforting embrace of the trees, their leaves whispering secrets in the gentle breeze. The play of light and shadow on the forest floor creates an ever-changing tapestry of patterns, each one a fleeting work of art that exists for a moment before dissolving back into the larger canvas of the woodland.

The Freston Wood Walk is more than just a physical journey; it is a chance to step out of the everyday and into a world of beauty, tranquility, and simple, profound truths. Whether you seek solace, inspiration, or simply a moment of quiet reflection, this enchanting woodland trail offers a sanctuary for the soul, a place where the cares of the world fade away, and the wisdom of the natural world speaks to the heart.

Detailed Route Directions:

1. From the Freston Wood car park, follow the path into the woodland, heading northeast (045°). Take a moment to appreciate the peaceful atmosphere and the lush greenery that surrounds you. (Estimated time: 5-10 minutes)

2. Continue along the path as it winds through the trees, bearing east (090°), and immerse yourself in the sights, sounds, and scents of this enchanting natural environment. Listen for the melodic birdsong and the gentle rustling of leaves in the breeze. (Estimated time: 20-30 minutes)

3. As you progress through the wood, take your time to observe the diverse array of plant life, from the majestic trees that tower overhead to the delicate wildflowers and ferns that carpet the forest floor. The path will gently turn to the southeast (135°). (Estimated time: 20-30 minutes)

4. At the northern edge of the wood, the path will curve to the south (180°). Follow the trail as it skirts along the boundary of the woodland, offering glimpses of the surrounding countryside. (Estimated time: 15-20 minutes)

5. The path will then turn to the southwest (225°), guiding you back into the heart of Freston Wood. As you walk, keep an eye out for the various woodland creatures that call this place home, from the playful squirrels to the shy deer. (Estimated time: 20-30 minutes)

6. Continue along the path, now heading west (270°), as it loops back towards the starting point. Take a moment to reflect on the beauty and tranquility of the woodland and the sense of peace and rejuvenation that spending time in nature can bring. (Estimated time: 15-20 minutes)

7. Finally, you'll arrive back at the Freston Wood car park, completing the circular walk. Before leaving, take one last look at the beautiful woodland and appreciate the opportunity to have experienced its wonders. (Estimated time: 5-10 minutes)

Best Features and Views:

- Immerse yourself in the serene and calming atmosphere of Freston Wood, a perfect retreat from the stresses of modern life. Let the gentle rustling of leaves, the melodic songs of birds, and the soft dappled sunlight filtering through the canopy transport you to a world of peace and tranquility.
- Marvel at the incredible diversity of plant life that thrives within the woodland. From the majestic oaks and beeches that have stood for centuries to the delicate wildflowers and ferns that carpet the forest floor, each species plays a vital role in the intricate web of life that defines this enchanting ecosystem.
- Delight in the opportunity to observe the woodland's fascinating wildlife in their natural habitat. Watch for the playful antics of squirrels as they scamper through the branches, listen for the distinctive drumming of woodpeckers, and keep an eye out for the timid deer that gracefully navigate the dappled shadows of the forest.

Historical Features of Interest:

While there are no specific historical features directly on the Freston Wood Walk, the woodland itself serves as a living testament to the enduring power and resilience of nature. As you wander along the trail, take a moment to consider the countless generations of plants, animals, and people that have called this place home, each leaving their mark on the landscape and contributing to the rich tapestry of history that underlies this beautiful woodland.

The presence of ancient trees, some of which may have stood for hundreds of years, offers a tangible connection to the past. These venerable giants have witnessed the passing of seasons, the changing of human cultures, and the ebb and flow of life itself, their gnarled trunks and spreading canopies a reminder of the timeless cycles that shape our world.

Additionally, the nearby Freston Tower, a fascinating 16th-century folly, stands as an intriguing historical feature just a short distance from the woodland. This six-story tower, with its commanding views of the surrounding countryside and the River Orwell, offers a glimpse into the lives and aspirations of those who once inhabited this landscape, adding an extra layer of depth and intrigue to your exploration of the area.

Birds to Spot:

- Agile nuthatches: Listen for the sharp, distinctive calls of these active birds as they climb headfirst down tree trunks, probing for insects in the bark crevices. Their sleek, blue-gray plumage and striking black eye stripe make them a delight to observe as they navigate the woodland with impressive dexterity.
- Melodious blackcaps: From April to September, keep an ear out for the rich, fluty songs of the blackcap, often referred to as the "northern nightingale" due to its beautiful voice. The male's striking black cap contrasts with its gray back and pale underparts, making it a memorable sight among the woodland foliage.
- Elusive treecreepers: These small, brown birds are master of camouflage, blending seamlessly with the bark of the trees they inhabit. Watch carefully as they spiral up tree trunks, meticulously probing the crevices for invertebrates, their long, curved bills perfectly adapted for extracting hidden prey from the nooks and crannies of the bark.

Rare or Interesting Insects and Other Wildlife:

- Enchanting butterflies: In the warmer months, the woodland edges and sunny glades come alive with the fluttering wings of various butterfly species. Look for the striking orange and brown patterning of the comma, the delicate beauty of the holly blue with its silvery-blue wings, and the unmistakable majesty of the purple emperor as it glides through the dappled sunlight.
- Industrious wood ants: As you walk through the woodland, keep an eye out for the impressive mounds of the wood ant colonies. These remarkable insects play a crucial role in the forest ecosystem, tirelessly foraging for food and maintaining the delicate balance of life within the woodland.
- Secretive badgers: Although elusive and primarily nocturnal, the presence of badgers can often be detected by the distinctive signs they leave behind. Look for the entrances to their underground setts, marked by large piles of soil and discarded bedding, and the well-worn paths they create through the undergrowth as they navigate their woodland domain.

Rare or Interesting Plants:

- Ethereal bluebells: In the spring, the woodland floor transforms into a mesmerizing carpet of bluebells, their delicate, nodding flowers casting a soft purple haze across the landscape. Take a moment to appreciate the ephemeral beauty of these enchanting blooms, a symbol of the woodland's enduring magic.
- Ancient trees: Stand in awe of the venerable oaks, beeches, and sweet chestnuts that have witnessed the passing of centuries. These ancient sentinels, with their gnarled bark and spreading canopies, serve as living monuments to the resilience and longevity of nature, their presence a humbling reminder of the timeless cycles that shape the woodland.
- Lush ferns: Discover the verdant beauty of the woodland's fern population, from the delicate fronds of the hart's tongue to the graceful arching stems of the male fern. These ancient plants, with their intricate and often prehistoric appearance, add a sense of primordial wonder to the forest floor, a testament to the enduring power of life in even the shadiest corners of the wood.

Facilities and Accessibility:

The Freston Wood Walk offers well-maintained paths that are generally accessible for most walkers. However, it is important to note that some sections may be uneven, rooty, or muddy, especially after periods of heavy rainfall. Sturdy walking shoes or boots with good grip are highly recommended to ensure a safe and comfortable experience.

While there are no toilets or other facilities directly on the walk route, the nearby village of Freston and the town of Ipswich offer a range of amenities for visitors, including public toilets, cozy pubs, and inviting cafes where you can refresh and refuel before or after your woodland adventure.

Due to the natural terrain and the presence of some uneven surfaces, the Freston Wood Walk may not be suitable for wheelchairs or pushchairs. Those with limited mobility or specific accessibility requirements should exercise caution and assess their ability to navigate the trail based on their individual needs and circumstances.

Seasonal Highlights:

The Freston Wood Walk offers a captivating and ever-changing experience throughout the year, with each season painting the landscape in its own unique hues and textures:

- Spring: Witness the woodland awakening from its winter slumber, as delicate green leaves unfurl on the branches and a vibrant carpet of bluebells, wood anemones, and primroses spreads across the forest floor. The air fills with the melodic songs of returning birds, and the first tentative buzzes of insects herald the arrival of warmer days ahead.
- Summer: Bask in the lush greenery of the woodland at the peak of its growth, the canopy providing a cool sanctuary from the heat of the summer sun. Observe the busy lives of the forest's inhabitants, from the industrious wood ants to the fluttering butterflies and the darting dragonflies that hover over sun-dappled glades.
- Autumn: Marvel at the breathtaking transformation of the woodland as the leaves shift from green to a stunning array of golds, oranges, and reds. The crisp, clean air and the satisfying crunch of fallen leaves underfoot create an invigorating atmosphere, perfect for a contemplative walk through the changing landscape.
- Winter: Embrace the stark beauty of the dormant woodland, with its bare branches etched against the cool, clear sky. Discover the subtle signs of life that persist even in the depths of winter, from the hardy evergreens that provide splashes of colour to the tracks of woodland creatures preserved in the frost or snow.

Difficulty Level and Safety Considerations:

The Freston Wood Walk is a relatively easy to moderate route, suitable for walkers of most fitness levels. The paths are generally well-maintained, but it is essential to remain aware of any uneven surfaces, exposed roots, or slippery sections, particularly after wet weather.

As with any outdoor activity, it is crucial to be prepared for changeable weather conditions. Ensure you wear appropriate footwear and clothing to suit the season, and carry sufficient water and snacks to keep yourself hydrated and energized throughout your walk.

Be mindful of the woodland's inhabitants, and remember to respect their habitat by staying on the designated paths and refraining from disturbing or feeding the wildlife. This not only helps to protect the delicate balance of the ecosystem but also ensures your own safety by minimizing the risk of unexpected encounters with wild animals.

It is always a good idea to inform someone of your planned route and expected return time before setting out, and to carry a charged mobile phone in case of emergencies. By following these simple precautions and exercising common sense, you can fully immerse yourself in the beauty and tranquility of Freston Wood, creating lasting memories and forging a deeper connection with the natural world.

Nearby Attractions:

After your rejuvenating walk through Freston Wood, take the opportunity to explore some of the other fascinating attractions that the surrounding area has to offer:

- Alton Water Park: Just a short drive from Freston Wood, Alton Water Park is a stunning 400-acre conservation area centred around a large reservoir. With its diverse range of habitats, including grasslands, woodlands, and wetlands, the park offers excellent opportunities for birdwatching, fishing, cycling, and water sports, making it a perfect destination for a full day of outdoor adventures.
- Jimmy's Farm and Wildlife Park: Experience the thrills and joys of a working farm at Jimmy's Farm and Wildlife Park, a unique attraction that combines agriculture, conservation, and education. Meet an array of friendly farm animals, explore the beautiful gardens and nature trails, and discover the exotic residents of the wildlife park, including meerkats, tapirs, and reindeer.
- Pin Mill: Take a step back in time as you visit the charming hamlet of Pin Mill, nestled on the banks of the River Orwell. This picturesque spot, with its historic cottages, traditional boats, and stunning views, has long been a source of inspiration for artists and writers. Enjoy a leisurely stroll along the river, stop for a pint in the 16th-century Butt and Oyster pub, and soak up the timeless atmosphere of this quintessential English village.

By combining your Freston Wood Walk with a visit to these captivating nearby attractions, you'll create an unforgettable itinerary that showcases the very best of Suffolk's natural beauty, rich history, and vibrant culture. Whether you're a nature enthusiast, a history buff, or simply seeking a memorable escape from the everyday, the Freston Wood area promises to delight and inspire you at every turn.

Pin Mill and Chelmondiston Walk - 5 miles (8 km), Circular - Estimated Time: 2.5-3 hours

Starting Point Coordinates, Postcode: Pin Mill car park (Grid Reference: TM 204 380, Postcode: IP9 1JW). Nearest Car Park, Postcode: Pin Mill car park (Postcode: IP9 1JW).

Embark on a delightful 5-mile (8 km) circular walk that takes you along the picturesque River Orwell and through the charming village of Chelmondiston. This walk offers a perfect blend of stunning river views, tranquil woodland paths, and the opportunity to explore the quaint village, immersing you in the beauty and history of the Suffolk countryside.

As you set out from the Pin Mill car park, the footpath beckons you alongside the River Orwell, the anticipation building with each step as you catch glimpses of the glistening water through the trees. The moment you emerge from the woodland, the full splendour of the river unfolds before you, the boats gently bobbing on the water and the distant horizon stretching out like a painted canvas.

Following the meandering path along the riverbank, you'll find yourself immersed in a world of tranquility and natural beauty. The soft lapping of the water against the shore and the gentle rustling of the leaves in the breeze create a soothing symphony that accompanies your every step. Take a moment to pause and drink in the breathtaking vistas, feeling the stresses of everyday life melt away in the presence of such serene beauty.

As you continue along the trail, the path guides you through enchanting woodland, the dappled sunlight filtering through the canopy and casting an ethereal glow upon the forest floor. The air is filled with the melodic songs of birds flitting through the branches, their joyful notes a celebration of the vibrant ecosystem that thrives in the heart of this green oasis. Breathe in the fresh, invigorating scent of the trees and let the tranquility of the woodland wash over you.

Emerging from the trees, you'll find yourself on the outskirts of the charming village of Chelmondiston. The path leads you into the heart of this picturesque community, where time seems to slow down and the cares of the modern world fade away. Stroll along the quaint streets lined with cottages that tell tales of a rich historical past, their colourful gardens and inviting doorways a testament to the warm and welcoming spirit of the villagers.

Take a moment to explore the village, discovering hidden gems at every turn. The historic church stands as a beacon of community life, its ancient stones echoing with the prayers and hymns of generations past. Nearby, a friendly local pub invites you to stop for a refreshing drink or a hearty meal, the perfect opportunity to immerse yourself in the convivial atmosphere and chat with the friendly residents who call this village home.

As you leave Chelmondiston behind, the path winds its way through the idyllic countryside once more, the fields stretching out like a patchwork quilt of vibrant greens and golden hues. The gentle undulations of the landscape create an ever-changing tableau, each vista more breathtaking than the last. Pause to admire the sweeping panoramas, the distant church spires and the lazy curl of smoke rising from farmhouse chimneys, a timeless scene that captures the essence of rural England.

The final leg of your journey brings you full circle, the path looping back towards Pin Mill and the starting point of your adventure. As you near the end of the walk, take a moment to reflect on the incredible beauty and diversity of the landscapes you've traversed, from the sparkling waters of the River Orwell to the charming streets of Chelmondiston and the tranquil embrace of the Suffolk countryside.

Back at the Pin Mill car park, your walk may have come to an end, but the memories of your experience will linger long after you've left this enchanting corner of England behind. The Pin Mill and Chelmondiston Walk is more than just a physical journey; it is a celebration of the simple joys and timeless beauty that can be found when we step off the beaten path and immerse ourselves in the wonders of the natural world.

Detailed Route Directions:

1. From the Pin Mill car park, set off along the footpath that runs alongside the River Orwell, bearing 000°. As you walk, take in the breathtaking views of the river and the boats gently bobbing on the water. (Estimated time: 20-30 minutes)

2. Continue along the path as it winds through enchanting woodland, the dappled sunlight filtering through the leaves. Listen for the melodic birdsong and the gentle rustling of the trees in the breeze. (Estimated time: 30-40 minutes)

3. As you approach Chelmondiston, the path will guide you into the heart of the village. Take a moment to explore the charming streets, admiring the picturesque cottages and the historic church. (Estimated time: 20-30 minutes)

4. Navigate through the village, following the roads and footpaths as they lead you onwards. Soak in the peaceful atmosphere and the friendly smiles of the locals you encounter along the way. (Estimated time: 20-30 minutes)

5. Upon leaving Chelmondiston, the path will bear 180°, guiding you southward. Immerse yourself once again in the tranquil surroundings of the countryside, with fields stretching out on either side. (Estimated time: 30-40 minutes)

6. The path will eventually loop back towards Pin Mill, bringing you full circle to the starting point at the car park. As you complete the walk, take a final moment to appreciate the beauty and serenity of the landscape you've just explored. (Estimated time: 20-30 minutes)

Best Features and Views:

- Lose yourself in the stunning vistas of the River Orwell, its glistening waters and gently sailing boats creating a mesmerizing scene that invites contemplation and wonder, the ever-changing light and the play of shadows on the water's surface a testament to the raw beauty of the natural world.
- Immerse yourself in the tranquil woodland paths, where the dappled sunlight filtering through the leaves and the gentle rustling of the breeze in the branches create an atmosphere of pure serenity and peace, a green oasis that soothes the soul and rejuvenates the spirit.
- Delight in the charm and character of the village of Chelmondiston, with its quaint cottages, historic church, and friendly local atmosphere, a timeless snapshot of rural English life that invites you to slow down, explore, and connect with the warm and welcoming community spirit.

Historical Features of Interest:

An undeniable highlight of the walk is the historic Pin Mill, a former tide mill dating back to the 16th century. This fascinating local landmark stands as a testament to the area's rich maritime heritage, its weathered walls and ancient timbers whispering stories of a bygone era. As you pass by this iconic structure, take a moment to imagine the lives of the millers and sailors who once called this place home, their tales forever woven into the fabric of the landscape.

Birds to Spot:

- Elegant egrets: These graceful white birds can often be spotted wading in the shallows or perched on the riverbank, their slender necks and delicate plumage a striking sight against the backdrop of the water and the reeds.
- Majestic herons: Watch for the imposing silhouette of a heron standing motionless in the water, its keen eyes fixed on the river's surface, waiting patiently for the perfect moment to strike at its unsuspecting prey.
- Diverse ducks: Observe the various species of ducks that make their home on the river, from the colourful and familiar mallards to the sleek and agile tufted ducks, each one a unique and fascinating character in the ever-changing tapestry of river life.

Rare or Interesting Insects and Other Wildlife:

- Dancing butterflies: In the warmer months, watch for the vibrant flashes of colour as butterflies dance amongst the trees and wildflowers, their delicate wings painted in a kaleidoscope of hues, from the regal purple of the peacock to the sunny yellow of the brimstone.
- Jewelled dragonflies: Marvel at the iridescent shimmer of dragonflies as they dart through the dappled sunlight, their gossamer wings catching the light like living gems, a mesmerizing display of aerial artistry that captivates the eye and the imagination.
- Woodland wonders: Keep an eye out for the playful antics of squirrels as they scamper through the trees, the bounding leaps of rabbits in the undergrowth, and the occasional glimpse of a deer in the distance, each encounter a reminder of the rich tapestry of life that thrives in the Suffolk countryside.

Rare or Interesting Plants:

- Seasonal wildflowers: In the spring and summer, delight in the vibrant blooms of wildflowers that carpet the fields and woodland edges, their delicate petals and sweet fragrances a joyous celebration of the season's bounty, from the cheerful daisies to the stately foxgloves.
- Enchanting bluebells: If you're fortunate enough to visit in late April or early May, you may be treated to the breathtaking sight of a sea of bluebells in the woodland areas, their delicate nodding heads creating a mesmerizing carpet of colour that stretches as far as the eye can see.
- Starry wood anemones: These dainty white flowers are a true harbinger of spring, dotting the forest floor with their star-like blooms, their presence a gentle reminder of the cyclical nature of life and the eternal renewal of the natural world.

Facilities and Accessibility:

While there are no public toilets directly on the route, you can find facilities at The Butt and Oyster, a charming pub near the starting point of the walk. This delightful establishment also offers a welcome opportunity to enjoy a well-earned refreshment after your adventure, with a tempting menu of locally-sourced food and drink that showcases the best of Suffolk's culinary heritage.

It's important to note that due to the uneven terrain and narrow paths, this walk is not suitable for wheelchairs or strollers. To ensure a safe and enjoyable experience, it is highly recommended to wear sturdy walking shoes with good grip, allowing you to navigate the trails with confidence and ease.

Seasonal Highlights:

Each season brings its own unique charm and character to the Pin Mill and Chelmondiston Walk, inviting you to return time and again to witness the ever-changing beauty of the Suffolk countryside:

- Spring: Witness the countryside bursting into life, with vibrant wildflowers painting the fields and hedgerows in a rainbow of colour, lush green foliage unfurling on the trees, and the joyous birdsong of the breeding season filling the air with the melodies of new beginnings.
- Summer: Bask in the warm sunshine as you stroll along the riverbank, the buzzing of insects and the gentle rustling of the breeze through the leaves creating a soothing soundtrack to your journey. Seek out the dappled shade of the woodland paths, where the lush canopy provides a welcome respite from the heat.
- Autumn: Marvel at the stunning array of colours as the leaves change, transforming the landscape into a breathtaking tapestry of gold, orange, and red. Embrace the crisp, invigorating air and the satisfying crunch of fallen leaves underfoot, as you savour the tranquil beauty of the countryside in its autumnal splendour.
- Winter: Embrace the tranquil beauty of the frosty riverbank, the bare trees etched against the pale winter sky, and the crisp, fresh air that invigorates the senses and adds a touch of magic to your walk. Discover the subtle signs of life that persist even in the coldest months, from the tracks of woodland creatures in the snow to the hardy evergreens that stand as beacons of resilience amidst the dormant landscape.

Difficulty Level and Safety Considerations:

This walk is classified as moderate, with some gentle inclines and uneven terrain that may prove challenging for less experienced walkers. The paths can be muddy or slippery, especially after rain, so it's essential to take care and wear appropriate footwear to ensure a safe and enjoyable experience.

As you make your way along the River Orwell, it's crucial to be cautious and keep a safe distance from the water's edge, as the tides and currents can be unpredictable. Always stay on the designated path and exercise caution when navigating any steep or slippery sections.

To ensure a comfortable and enjoyable walk, be sure to wear sturdy, comfortable footwear with good grip and support, as well as clothing suitable for the changeable British weather. Carry an ample supply of water and energizing snacks to keep you hydrated and fuelled throughout your journey, and don't forget to pack any necessary medications or first aid supplies.

It's always a good idea to inform someone of your planned route and expected return time before setting off, and to carry a fully charged mobile phone in case of emergencies. By following these simple precautions and the countryside code, you can fully immerse yourself in the beauty and tranquility of the Pin Mill and Chelmondiston Walk, creating memories that will last a lifetime.

Nearby Attractions:

After your invigorating walk, take the opportunity to explore some of the other fascinating attractions that the area has to offer, each one providing a unique glimpse into the rich history, culture, and natural beauty of the Suffolk region:

- Ipswich: Venture into the vibrant county town of Suffolk, where you can immerse yourself in its rich history, from the ancient streets lined with medieval buildings to the impressive waterfront that tells the story of the town's maritime past. Explore the excellent museums, galleries, and cultural attractions that showcase the diverse artistic and intellectual heritage of this dynamic community.
- Suffolk Coast & Heaths AONB: Discover the breathtaking beauty of this Area of Outstanding Natural Beauty, a

stunning landscape of diverse habitats and picturesque villages that showcases the very best of the Suffolk countryside. From the rugged coastline with its sweeping beaches and dramatic cliffs to the tranquil estuaries and the wildlife-rich wetlands, this protected area offers endless opportunities for exploration and adventure.

By exploring these nearby attractions, you'll gain a deeper appreciation for the diverse and captivating character of this remarkable region, making your Pin Mill and Chelmondiston Walk the perfect starting point for a truly unforgettable journey through the heart of Suffolk.

Holbrook Circular Walk - 5 miles (8 km), Circular - Estimated Time: 2.5-3 hours

Starting Point Coordinates, Postcode: Holbrook Village Hall, Ipswich (Grid Reference: TM 163 361, Postcode: IP9 2PZ).

Nearest Car Park, Postcode: Parking is available at Holbrook Village Hall (Postcode: IP9 2PZ).

Immerse yourself in the tranquil beauty of the Suffolk countryside on this delightful 5-mile (8 km) circular walk around the picturesque village of Holbrook and its surrounding landscape. This walk offers a perfect blend of charming village lanes, open fields, and peaceful riverside paths, providing a refreshing escape from the hustle and bustle of everyday life.

As you set out from the Holbrook Village Hall, the charming atmosphere of the village immediately envelops you, the quaint cottages and friendly locals creating a sense of warmth and community that sets the tone for your journey. The Street leads you southward, its gentle curve inviting you to explore the heart of this picturesque Suffolk village.

Leaving the village behind, you'll find yourself in the open countryside, the path guiding you through lush green fields that stretch out like a patchwork quilt of vibrant hues. The gentle breeze rustles through the hedgerows, carrying with it the sweet scent of wildflowers and the distant song of birds, a soothing symphony that accompanies your every step.

As you cross the bridge and turn onto the footpath, the landscape opens up before you, revealing a breathtaking vista of the River Stour, its glistening waters meandering through the countryside like a silver ribbon. Take a moment to pause and drink in the stunning views, feeling the stresses of everyday life melt away in the presence of such serene beauty.

Continuing along the riverside path, you'll find yourself immersed in a world of tranquility and natural wonder. The gentle lapping of the water against the shore and the soft rustling of the reeds create a soothing melody that fills the air, while the occasional splash of a fish or the distant cry of a waterbird adds a touch of excitement to the peaceful atmosphere.

As you make your way towards the historic Stutton Mill, take a moment to appreciate this charming reminder of the area's agricultural heritage. The weathered timbers and ancient stones of the mill seem to whisper stories of a bygone era, inviting you to imagine the lives of the farmers and millers who once called this place home.

Crossing the footbridge and joining the path on the other side of the river, you'll be greeted by a new perspective on the stunning countryside. The fields and hedgerows stretch out before you, their colours and textures creating an ever-changing tapestry that shifts with the seasons, from the lush greens of spring to the rich golds of autumn.

As you turn onto the country lane, the distant sound of birdsong grows louder, a joyful chorus that welcomes you back into the heart of the Suffolk countryside. The path leads you through the outskirts of Holbrook, where picturesque gardens and quaint cottages offer a charming glimpse into the daily lives of the village residents.

Finally, as you arrive back at the Holbrook Village Hall, take a moment to reflect on the incredible journey you've just completed. The Holbrook Circular Walk is more than just a physical excursion; it is a chance to reconnect with the beauty and simplicity of the natural world, to find peace and perspective amidst the chaos of modern life, and to create memories that will last a lifetime.

Detailed Route Directions:

1. From Holbrook Village Hall, head south along The Street, bearing 180°. Take in the charming atmosphere of the village as you walk, admiring the quaint cottages and the friendly locals going about their day. (Estimated time: 10-15 minutes)

2. At the end of The Street, turn left onto Mill Hill, bearing 090°. Follow the road as it leads you out of the village and towards the picturesque countryside. (Estimated time: 10-15 minutes)

3. After crossing the bridge, turn right onto a footpath, bearing 180°, which will guide you into the open fields. Immerse yourself in the tranquil surroundings, with the lush green grass stretching out before you and the gentle breeze rustling through the hedgerows. (Estimated time: 20-30 minutes)

4. Continue along the path as it winds through the fields, taking in the peaceful atmosphere and the distant views of the surrounding landscape. Eventually, you'll reach the historic Stutton Mill, a charming reminder of the area's agricultural heritage. (Estimated time: 30-40 minutes)

5. From Stutton Mill, continue along the path, bearing 090°, with the River Stour now on your right. Take a moment to appreciate the stunning views of the river, with its glistening waters and the boats gently sailing by. (Estimated time: 20-30 minutes)

6. Follow the riverside path, soaking in the tranquil ambiance, until you reach a footbridge. Cross the bridge, bearing 000°, and join the footpath on the other side of the river. (Estimated time: 10-15 minutes)

7. The path will now guide you back towards Holbrook, bearing 315°. As you walk, take in the beautiful countryside, with the fields and hedgerows stretching out on either side. (Estimated time: 30-40 minutes)

8. Upon reaching a country lane, turn left, bearing 270°, and follow the road for a short distance. Keep an eye out for the next footpath on your right, which will lead you back into the village. (Estimated time: 10-15 minutes)

9. Take the footpath on the right, bearing 000°, and follow it as it guides you through the outskirts of Holbrook. Admire the picturesque gardens and the quaint cottages as you make your way back to the heart of the village. (Estimated time: 15-20 minutes)

10. Finally, you'll arrive back at the Holbrook Village Hall, where your circular walk comes to a satisfying conclusion. Take a moment to reflect on the beauty and tranquility of the countryside you've just explored. (Estimated time: 5-10 minutes)

Best Features and Views:

- Lose yourself in the charming village of Holbrook, with its quaint cottages, friendly locals, and peaceful atmosphere, a perfect snapshot of traditional Suffolk life that invites you to slow down and appreciate the simple joys of rural living.
- Marvel at the stunning views of the River Stour, its glistening waters and gently sailing boats creating a mesmerizing scene that invites contemplation and wonder, the ever-changing light and the play of shadows on the water's surface a testament to the raw beauty of the natural world.

- Immerse yourself in the picturesque countryside, with its lush fields, ancient trees, and vibrant wildflowers, a living tapestry that shifts with the seasons and offers a breathtaking display of colour, texture, and life at every turn.

Historical Features of Interest:

The historic Stutton Mill stands as a captivating reminder of the area's rich agricultural heritage, its weathered timbers and ancient stones whispering stories of a bygone era. As you pass by this fascinating structure, take a moment to imagine the lives of the farmers and millers who once toiled here, their hard work and ingenuity shaping the landscape and the communities that called this place home. The presence of the mill adds a layer of depth and history to an already enchanting walk, inviting you to connect with the past and appreciate the enduring legacy of rural Suffolk.

Birds to Spot:

- Jewels of the river: Kingfishers: Watch for the brilliant flash of blue and orange as these stunning birds dart along the river's edge, their swift and agile movements a mesmerizing display of natural beauty and precision.
- Sentinels of the shallows: Herons: Spot these majestic birds standing motionless in the shallows, their keen eyes and sharp bills poised to strike at unsuspecting prey, a living embodiment of patience and grace.
- Elegant waders: Egrets: Look for the sleek white plumage of these graceful birds as they wade through the shallow waters or perch on the riverbank, their delicate features and refined movements a captivating sight against the backdrop of the river and the reeds.

Rare or Interesting Insects and Other Wildlife:

- Winged wonders: Butterflies: Admire the delicate beauty of the various butterfly species that flutter amongst the wildflowers in the meadows and along the hedgerows, their colourful wings and graceful movements adding a touch of magic to the countryside.
- Aerial acrobats: Dragonflies: Marvel at the iridescent colours and acrobatic flight of the dragonflies that dance above the river and the nearby ponds, their incredible aerial skills and vibrant hues a testament to the wonders of the insect world.
- Elusive enchantment: Otters: If you're incredibly lucky, you might catch a glimpse of these elusive and enchanting creatures playing along the riverbank or swimming in the water, their sleek fur and playful antics a rare and unforgettable sight.

Rare or Interesting Plants:

- Guardians of time: Ancient trees: Marvel at the majestic oaks, ashes, and beeches that have stood sentinel in the landscape for centuries, their gnarled branches reaching up to the sky and their mighty trunks bearing witness to the passage of countless seasons and the stories they have silently observed.
- Floral tapestries: Wildflowers: In the spring and summer, delight in the vibrant tapestry of colours created by the diverse wildflowers that adorn the meadows and woodland edges, their delicate petals and sweet fragrances a joyous celebration of nature's artistry and the resilience of life in the countryside.
- Wetland wonders: Reed beds: Observe the swaying reed beds along the river's edge, their tall, slender stems and feathery plumes creating a unique and captivating habitat that supports a diverse array of wildlife, from the tiniest insects to the most majestic birds.

Facilities and Accessibility:

While there are no public toilets directly along the route, you can find a delightful café in Holbrook village, the perfect spot to enjoy a refreshing drink or a light bite before or after your walk. This charming establishment offers a warm and welcoming atmosphere, with friendly staff and a tempting selection of locally-sourced food and drinks that showcase the best of Suffolk's culinary heritage.

The Holbrook Circular Walk is classified as easy to moderate, with some gentle hills and uneven terrain that may prove challenging for those with limited mobility. To ensure a safe and comfortable experience, it is highly recommended to wear sturdy footwear with good grip, particularly after rainfall when the paths may be muddy or slippery.

Seasonal Highlights:

The Holbrook Circular Walk offers a unique and captivating experience in every season, each one revealing new facets of beauty and wonder in the Suffolk countryside:

- Spring: Witness the countryside bursting into life, with vibrant wildflowers painting the meadows and hedgerows in a rainbow of colours, lush green leaves unfurling on the trees, and the joyous birdsong of the breeding season filling the air with the melodies of new beginnings.
- Summer: Bask in the warm sunshine as you stroll along the riverside paths, the buzzing of insects and the gentle rustling of the breeze through the foliage creating a soothing soundtrack to your journey. Seek out the dappled shade of the trees, where the lush canopy provides a welcome respite from the heat.
- Autumn: Marvel at the stunning array of colours as the leaves change, transforming the landscape into a breathtaking tapestry of gold, orange, and red. Embrace the crisp, invigorating air and the satisfying crunch of fallen leaves underfoot, as you savour the tranquil beauty of the countryside in its autumnal splendor.
- Winter: Embrace the tranquil beauty of the frosty fields, the bare trees etched against the pale winter sky, and the crisp, fresh air that invigorates the senses and adds a touch of magic to your walk. Discover the subtle signs of life that persist even in the coldest months, from the tracks of woodland creatures in the snow to the hardy evergreens that stand as beacons of resilience amidst the dormant landscape.

Difficulty Level and Safety Considerations:

This walk is classified as easy to moderate, with some gentle hills and uneven terrain, making it suitable for most walkers with a reasonable level of fitness. However, as with any outdoor activity, it's essential to be prepared for changeable weather conditions and to wear appropriate clothing and footwear to ensure your comfort and safety.

As you embark on your walk, be sure to carry water, snacks, and any necessary medications, keeping yourself hydrated and energized throughout your journey. It's also a good idea to bring a map or GPS device and to inform someone of your intended route and expected return time, just as a precautionary measure.

Remember to stay on the designated paths to minimize your impact on the environment and to protect the delicate ecosystems that thrive in the countryside. Take care when walking on muddy or slippery surfaces, particularly after rainfall, and wear sturdy footwear with good grip to ensure a safe and enjoyable experience.

By following these simple guidelines and being mindful of your surroundings, you can fully immerse yourself in the beauty and tranquility of the Holbrook Circular Walk, creating lasting memories and a deeper appreciation for the natural wonders of the Suffolk countryside.

Nearby Attractions:

After your invigorating walk, take the opportunity to explore some of the other fascinating attractions that the area has to offer, each one providing a unique glimpse into the rich history, culture, and natural beauty of the region:

- Royal Hospital School: Step back in time as you discover the rich history and stunning architecture of this prestigious boarding school, founded in the 18th century to educate the sons of Royal Navy personnel. Explore the impressive grounds, marvel at the grand buildings, and learn about the school's fascinating past and its enduring legacy in the community.
- Shotley Peninsula: Immerse yourself in the breathtaking beauty of this unique landscape, where the Orwell and Stour rivers meet to create a stunning tapestry of estuaries, mudflats, and salt marshes. Discover the incredible diversity of wildlife that thrives in this protected area, from the elegant wading birds and the playful seals to the rare plants and insects that call this special place home.
- Ipswich Waterfront: Venture into the vibrant heart of Ipswich, where the historic buildings and bustling marinas create a captivating blend of old and new. Explore the excellent selection of cafes, restaurants, and bars that line the waterfront, each one offering a unique taste of the town's culinary and cultural scene, and soak up the lively atmosphere of this dynamic and ever-evolving area.

By exploring these nearby attractions, you'll gain a deeper appreciation for the diverse and captivating character of this remarkable region, making your Holbrook Circular Walk the perfect starting point for a truly unforgettable journey through the heart of Suffolk.

Shotley Peninsula Walk - 10 miles (16 km), Circular - Estimated Time: 5-6 hours

Starting Point Coordinates, Postcode: Shotley Gate, near the Bristol Arms pub (Grid Reference: TM 235 339, Postcode: IP9 1QJ).
Nearest Car Park, Postcode: Shotley Gate, near the Bristol Arms pub (Postcode: IP9 1QJ).

Embark on an unforgettable 10-mile (16 km) circular walk that takes you through the stunning landscapes of the Shotley Peninsula, a hidden gem on the Suffolk coast. This walk offers a perfect blend of breathtaking river views, rolling farmland, and charming villages, providing a truly immersive experience for nature lovers and outdoor enthusiasts alike.

As you set out from Shotley Gate, the coastal path along the River Orwell beckons you forward, the anticipation building with each step as you catch glimpses of the sparkling water through the trees. The moment you emerge from the shelter of the village, the full majesty of the river unfolds before you, the boats gently bobbing on the waves and the distant horizon stretching out like a painted canvas.

Following the meandering path along the riverbank, you'll find yourself immersed in a world of tranquility and natural beauty. The soft lapping of the water against the shore and the gentle rustling of the leaves in the breeze create a soothing symphony that accompanies your every step. Take a moment to pause and drink in the stunning vistas, feeling the stresses of everyday life melt away in the presence of such serene beauty.

As you approach Shotley Marina, the bustling activity of this charming coastal community adds a lively counterpoint to the peaceful surroundings. The colourful boats and the friendly faces of the sailors and fishermen create a vibrant tableau that speaks to the enduring connection between the people of the peninsula and the water that surrounds them.

Turning inland, you'll soon find yourself in the heart of Shotley village, where the picturesque streets and historic buildings transport you back in time. The magnificent St. Mary's Church, with its soaring tower and intricate stonework, stands as a testament to the rich history and heritage of this special place, inviting you to step inside and marvel at the beauty of its ancient architecture.

Leaving the village behind, the path winds its way through rolling farmland, the fields stretching out like a patchwork quilt of vibrant greens and golden hues. The gentle undulations of the landscape create an ever-changing tableau, each vista more breathtaking than the last. Pause to admire the sweeping panoramas, the distant church spires, and the lazy curl of smoke rising from farmhouse chimneys, a timeless scene that captures the essence of rural England.

Joining the Arthur Ransome Trail, you'll find yourself walking in the footsteps of the famous author, the landscape that inspired his beloved stories unfolding before you in all its glory. The River Stour sparkles in the sunlight, its banks lined with graceful willows and reeds that sway in the breeze. Keep your eyes peeled for the darting flight of kingfishers and the stealthy movements of otters, each encounter a reminder of the rich tapestry of life that thrives along these ancient waterways.

As you approach the charming hamlet of Pin Mill, the sight of the historic buildings and the boats moored along the riverbank creates a scene of timeless beauty. The clinking of rigging and the distant laughter of sailors mingle with the cries of the gulls, a sensory feast that immerses you in the authentic atmosphere of this quintessential Suffolk destination.

The final leg of your journey takes you back towards Shotley Gate, the path winding through a landscape of breathtaking diversity and contrast. The vast expanse of the estuaries gives way to the intimate beauty of hidden coves and saltmarshes, each one a haven for rare and fascinating wildlife. As you walk, reflect on the incredible richness of the natural world and the way in which the land and the water have shaped the lives of the people who call this place home.

Arriving back at your starting point, the sense of accomplishment and wonder that fills your heart is a testament to the transformative power of this unforgettable journey. The Shotley Peninsula Walk is more than just a physical challenge; it is an opportunity to immerse yourself in the breathtaking beauty and fascinating history of one of Suffolk's most captivating landscapes, forging a deep and lasting connection with the natural world that will stay with you long after you've left this enchanting corner of England behind.

Detailed Route Directions:

1. From Shotley Gate, follow the coastal path along the River Orwell, bearing 045°. Take in the stunning views of the river and the boats gently bobbing on the water as you make your way towards Shotley Marina. (Estimated time: 30-40 minutes)

2. Continue along the riverbank, soaking in the peaceful atmosphere and the gentle breeze coming off the water. As you walk, keep an eye out for the various bird species that call this area home. (Estimated time: 45-60 minutes)

3. Upon reaching Shotley Marina, take a moment to admire the beautiful boats and the bustling activity of this charming coastal community. From here, head inland towards the village of Shotley, bearing 315°. (Estimated time: 20-30 minutes)

4. As you enter Shotley, take the time to explore the picturesque streets and the historic St. Mary's Church, a beautiful example of medieval architecture. After taking in the sights, continue through the village, bearing 000°. (Estimated time: 30-45 minutes)

5. Upon leaving Shotley, follow the footpath as it leads you towards Chelmondiston, bearing 045°. Immerse yourself in the rolling farmland, with fields stretching out on either side and the occasional copse of trees providing shade and shelter. (Estimated time: 1-1.5 hours)

6. As you approach Chelmondiston, join the Arthur Ransome Trail, a scenic route that runs alongside the River Stour. Follow the trail, bearing 090°, and take in the stunning views of the river and the surrounding countryside. (Estimated time: 1-1.5 hours)

7. Continue along the Arthur Ransome Trail until you reach the charming hamlet of Pin Mill, a popular destination for sailing enthusiasts. Take a moment to admire the historic buildings and the boats moored along the riverbank. (Estimated time: 30-45 minutes)

8. From Pin Mill, follow the footpath that leads back to Shotley Gate, bearing 225°. As you walk, reflect on the incredible landscapes and the rich history you've encountered along the way. (Estimated time: 1.5-2 hours)

9. Finally, you'll arrive back at Shotley Gate, where your circular walk comes to a satisfying conclusion. Take a final moment to appreciate the beauty and tranquility of the Shotley Peninsula before heading home. (Estimated time: 5-10 minutes)

Best Features and Views:

- Lose yourself in the stunning vistas of the River Orwell and the River Stour, their glistening waters and gently sailing boats creating a mesmerizing scene that invites contemplation and wonder, the ever-changing light and the play of shadows on the water's surface a testament to the raw beauty of the natural world.
- Immerse yourself in the timeless charm of the villages of Shotley and Chelmondiston, with their picturesque streets, historic churches, and friendly local atmosphere, each one a living snapshot of the rich cultural heritage and enduring community spirit that define this remarkable corner of Suffolk.
- Marvel at the sweeping panoramas of the rolling farmland and the peaceful countryside that surrounds the peninsula, the patchwork of fields and the distant church spires creating a landscape of breathtaking beauty and diversity that seems to stretch out forever, inviting you to lose yourself in the simple joys and profound truths of the natural world.

Historical Features of Interest:

- St. Mary's Church, Shotley: This magnificent medieval church stands as a testament to the rich history and heritage of the area, its soaring tower and intricate stonework a breathtaking example of the skill and devotion of the craftsmen who built it centuries ago. As you step inside, the serene atmosphere and the beautiful stained-glass windows invite you to pause and reflect on the generations of worshippers who have found solace and inspiration within these ancient walls.
- Pin Mill: This historic hamlet, with its charming buildings and bustling sailing community, offers a fascinating glimpse into the area's maritime past and its enduring connection to the sea. As you wander along the riverbank, the sight of the traditional sailing barges and the sound of the clinking rigging transport you back to a time when these waters were the lifeblood of the local economy, a reminder of the hardships and the triumphs of the generations who have called this place home.

Birds to Spot:

- Majestic marsh harriers: Watch for these magnificent birds of prey as they soar over the wetlands and marshes, their distinctive silhouettes and powerful wingbeats a breathtaking sight against the vast expanse of the sky. Marvel at their skill and grace as they hunt for prey, their keen eyes scanning the landscape for the slightest movement or sign of life.
- Elegant avocets: Delight in the sight of these striking black and white waders, their slender, upturned bills perfectly adapted for foraging in the mudflats and shallow waters along the rivers. Watch as they wade through the shallows with a delicate grace, their presence a testament to the incredible diversity and resilience of the birdlife that thrives in this unique habitat.
- Ghostly barn owls: If fortune favours you, you may catch a rare glimpse of these ethereal birds as they hunt over the fields and farmland, their silent flight and luminous white plumage a haunting and unforgettable sight in the gathering dusk. Listen for their eerie screeches and the soft rustle of their wings, a sound that has echoed across this landscape for countless generations.

Rare or Interesting Insects and Other Wildlife:

- Jewelled dragonflies and damselflies: Marvel at the iridescent colours and acrobatic flight of these ancient insects as they dart across the rivers and ponds, their gossamer wings catching the light like living gems. Watch as they hover and swoop, their aerial mastery a testament to the incredible adaptations and resilience of the natural world.
- Painted butterflies: Delight in the vibrant hues and intricate patterns of the various butterfly species that flutter amongst the wildflowers and along the hedgerows, each one a miniature work of art that adds a splash

of colour and movement to the landscape. From the bold red admirals to the delicate common blues, these enchanting creatures are a joy to behold and a vital part of the complex web of life that sustains the peninsula.
- Bounding hares: Keep your eyes peeled for these elusive and magnificent creatures as they race across the fields and farmland, their powerful hind legs propelling them at breathtaking speeds. Admire their grace and agility, and the way in which they seem to embody the wild and untamed spirit of the countryside, a reminder of the beauty and mystery that lies just beyond the edges of our everyday lives.

Rare or Interesting Plants:

- Hardy saltmarsh specialists: Discover the unique and fascinating plants that have adapted to thrive in the challenging conditions of the saltmarshes, their leaves and stems specially designed to cope with the high salinity and the constant ebb and flow of the tides. From the delicate sea lavender to the succulent sea purslane, these remarkable species are a testament to the incredible diversity and resilience of the plant kingdom.
- Vibrant wildflower meadows: In the spring and summer, lose yourself in the breathtaking tapestry of colours that adorns the meadows, verges, and woodland edges, each flower a tiny miracle of beauty and design. From the cheerful yellows of the cowslips to the delicate pinks of the cuckoo flowers, these ephemeral blooms are a joyous celebration of the cycle of life and the enduring wonder of the natural world.
- Living history hedgerows: Marvel at the gnarled and twisted branches of the ancient hedgerows that line the fields and lanes, each one a living museum of the countless generations of farmers and labourers who have shaped this landscape over the centuries. Look closer and you'll discover a hidden world of plant and animal life, from the delicate flowers of the dog rose to the darting movements of the tiny wrens and hedgehogs that make their homes in these green corridors.

Facilities and Accessibility:

The Shotley Peninsula Walk offers a range of facilities to ensure a comfortable and enjoyable experience for walkers. At Shotley Marina, approximately halfway along the route, you'll find well-maintained public toilets and a delightful café, providing a welcome opportunity to take a break, refresh, and refuel before continuing your journey. The café's menu features a tempting selection of locally-sourced food and drink, the perfect way to savour the flavours of the Suffolk countryside while soaking in the lively atmosphere of the marina.

It's important to note that while the walk is undeniably rewarding, it does include some hilly sections and uneven terrain that may prove challenging for those with limited mobility. To ensure a safe and enjoyable experience, it is essential to wear sturdy footwear with good grip, particularly after rainfall when the paths may be muddy or slippery. By taking these precautions and listening to your body's needs, you can fully immerse yourself in the stunning landscapes and rich history of the Shotley Peninsula, creating memories that will last a lifetime.

Seasonal Highlights:

The Shotley Peninsula Walk offers a unique and captivating experience in every season, each one revealing new facets of beauty and wonder in this enchanting corner of Suffolk:

- Spring: Witness the countryside bursting into life, with vibrant wildflowers painting the meadows and verges in a rainbow of colour, lush green leaves unfurling on the trees, and the joyous birdsong of the breeding season filling the air with the melodies of new beginnings. Feel the sun's warmth on your face and the fresh breeze in your hair, as you marvel at the incredible resilience and vitality of the natural world.
- Summer: Bask in the golden light and the heady scents of high summer, as you stroll along the riverside

paths and through the sun-dappled woodlands. Delight in the buzzing of insects and the gentle rustling of the breeze through the foliage, each sound a reminder of the pulsing energy and vitality that infuses the landscape during these long, lazy days.
- Autumn: Marvel at the breathtaking array of colours as the leaves turn, transforming the peninsula into a tapestry of gold, orange, and red that seems to stretch out forever. Savour the crisp, invigorating air and the satisfying crunch of fallen leaves underfoot, as you drink in the mellow beauty and poignant sense of change that characterizes this magical season.
- Winter: Embrace the wild beauty of the wind-swept rivers, the frosty fields, and the stark silhouettes of bare trees etched against the leaden sky. Wrap up warm and set out to discover the subtle signs of life that persist even in the depths of winter, from the hardy seabirds that ride the icy gusts to the delicate tracery of frost on the hedgerows, each one a reminder of the enduring wonder and mystery of the natural world.

Difficulty Level and Safety Considerations:

The Shotley Peninsula Walk is classified as moderate, with some hilly sections and uneven terrain that may prove challenging for less experienced walkers. The 10-mile distance also requires a good level of fitness and stamina, so it's important to be realistic about your abilities and to take regular breaks as needed.

As you make your way along the rivers and through the countryside, it's crucial to be prepared for changeable weather conditions, as the exposed coastal sections can be particularly windswept and chilly, even on seemingly mild days. Make sure to dress in layers and to bring a waterproof jacket, as well as plenty of water and energizing snacks to keep you going throughout the day.

The paths can be muddy or slippery, especially after heavy rain, so sturdy hiking boots with good grip are essential to ensure a safe and comfortable walk. It's also a good idea to carry a map and compass or GPS device, as well as a fully charged mobile phone, in case of any unexpected difficulties or emergencies.

By taking these sensible precautions and staying alert to your surroundings, you can fully immerse yourself in the stunning landscapes and fascinating history of the Shotley Peninsula, forging a deep and lasting connection with the natural world that will stay with you long after you've completed this unforgettable walk.

Nearby Attractions:

After your invigorating walk, take the opportunity to explore some of the other fascinating attractions that the area has to offer, each one providing a unique glimpse into the rich history, vibrant culture, and stunning natural beauty of the Suffolk coast and countryside:

- Ipswich: Venture into the vibrant county town of Suffolk, where you can immerse yourself in its rich history, from the ancient streets lined with timber-framed buildings to the impressive waterfront that tells the story of the town's maritime past. Explore the excellent museums, galleries, and cultural attractions that showcase the diverse artistic and intellectual heritage of this dynamic community.
- Harwich: Step back in time as you visit the historic port town of Harwich, with its impressive fortifications that have guarded the harbour for centuries. Delve into the town's fascinating maritime heritage at the excellent museums and historic sites, and sample the delicious seafood at the many excellent restaurants that line the quaint streets and bustling quayside.
- Suffolk Food Hall: Indulge your senses at this fantastic food and shopping destination, where you can sample an enticing array of local produce, from artisanal cheeses and cured meats to fresh fruits and vegetables straight from the surrounding fields. Enjoy a delicious meal at the on-site restaurant, which celebrates the flavours and traditions of Suffolk, and browse the fantastic selection of gifts and souvenirs that capture the essence of this remarkable region.

By exploring these nearby attractions, you'll gain a deeper appreciation for the diverse and captivating character of the Suffolk coast and countryside, making your Shotley Peninsula Walk the perfect starting point for a truly unforgettable journey through this enchanting corner of England. Whether you're a keen hiker, a nature lover, or simply someone who appreciates the beauty and tranquility of the great outdoors, this walk promises to delight, inspire, and leave you with memories that will last a lifetime.

As you set out on this incredible adventure, remember to tread lightly, respect the environment, and take the time to appreciate the countless wonders that await you at every turn. From the stunning vistas of the rivers and estuaries to the rich history and vibrant communities that have shaped this landscape over the centuries, the Shotley Peninsula Walk is a testament to the enduring beauty, resilience, and spirit of this unique and captivating corner of Suffolk.

So lace up your boots, grab your backpack, and set out to discover the magic and mystery of the Shotley Peninsula for yourself. Whether you're a seasoned walker or a curious newcomer, this unforgettable journey will open your eyes, lift your heart, and remind you of the incredible power of the natural world to inspire, heal, and transform us all.

Shotley Gate Walk - 4 miles (6.4 km), Circular - Estimated Time: 2-2.5 hours

Starting Point Coordinates, Postcode: Shotley Marina car park (Grid Reference: TM 236 339, Postcode: IP9 1QJ).
Nearest Car Park, Postcode: Shotley Marina car park (Postcode: IP9 1QJ).

Discover the hidden gems of Shotley Gate on this delightful 4-mile (6.4 km) circular walk that takes you along the stunning Rivers Orwell and Stour. Immerse yourself in the beauty of the Suffolk coastline, explore the charming village, and uncover the fascinating history of this unique corner of East Anglia.

As you set out from the Shotley Marina car park, the bustling atmosphere of the waterfront sets the stage for your adventure. The Bristol Arms pub, a local landmark that has welcomed visitors for centuries, beckons you with its timeless charm and the promise of a refreshing pint at the end of your journey.

Turning onto the riverside footpath, you'll immediately find yourself captivated by the breathtaking views of the River Orwell. The glistening waters stretch out before you, the gentle lapping of the waves and the distant cries of seabirds creating a soothing soundtrack to your walk. As you follow the path along the river's edge, the impressive Port of Felixstowe comes into view on the opposite bank, its towering cranes and bustling activity a reminder of the region's rich maritime heritage.

Continuing along the riverside, you'll soon reach the point where the River Orwell meets the River Stour, a confluence of two of Suffolk's most beautiful waterways. Here, the landscape opens up, revealing a stunning panorama of the estuary and the surrounding countryside. Take a moment to pause and drink in the view, feeling the gentle breeze on your face and the warm sun on your skin.

As you turn to follow the River Stour, keep your eyes peeled for the diverse array of bird species that call this area home. From the elegant avocets with their upturned bills to the majestic marsh harriers soaring overhead, each encounter is a reminder of the incredible biodiversity that thrives in this unique coastal habitat.

The path leads you past Shotley Marshes, a haven for wildlife and a Site of Special Scientific Interest. This fragile ecosystem, with its saltmarshes, mudflats, and reed beds, is a testament to the delicate balance of nature and the importance of conservation. Take a moment to appreciate the subtle beauty of this special place, from the vibrant purple of the sea lavender to the soft rustle of the reeds in the breeze.

As you leave the marshes behind and turn towards the village of Shotley Gate, you'll find yourself immersed in the rich history of this fascinating area. From its role in the defense of the realm to its thriving maritime heritage, every step brings a new story to light. The path meanders through the heart of the village, revealing charming cottages, historic buildings, and friendly locals going about their day.

Take the time to explore the hidden corners and secret gardens of Shotley Gate, each one a small marvel waiting to be discovered. Whether it's the colourful blooms of a cottage garden or the intricate brickwork of a Victorian terrace, every detail adds to the timeless charm of this special place.

As you make your way back to the starting point, the Shotley Marina car park, take a final moment to reflect on the beauty and tranquility of this remarkable corner of Suffolk. The Shotley Gate Walk is more than just a physical journey; it is a chance to connect with the natural world, to discover the hidden gems that lie just beyond the beaten path, and to create memories that will last a lifetime.

Detailed Route Directions:

1. From the car park, head southwest towards the picturesque Bristol Arms pub, a local landmark that has welcomed visitors for centuries. Take a moment to admire the historic building before turning right onto the riverside footpath, bearing 315°. (Estimated time: 5-10 minutes)

2. Follow the path as it winds along the River Orwell, offering breathtaking views of the water and the bustling Port of Felixstowe on the opposite bank. Listen to the gentle lapping of the waves and the distant cries of the seabirds as you walk. (Estimated time: 20-30 minutes)

3. Continue along the riverside path until you reach the point where the River Orwell meets the River Stour, a confluence of two of Suffolk's most beautiful waterways. Here, turn right and follow the path along the River Stour, bearing 045°. (Estimated time: 10-15 minutes)

4. As you walk along the River Stour, take in the stunning views of the estuary and the surrounding countryside. Keep an eye out for the various bird species that call this area home, from the elegant avocets to the majestic marsh harriers. (Estimated time: 20-30 minutes)

5. The path will lead you past Shotley Marshes, a haven for wildlife and a Site of Special Scientific Interest. Take a moment to appreciate the unique beauty of this fragile ecosystem, with its saltmarshes, mudflats, and reed beds. (Estimated time: 15-20 minutes)

6. After passing the marshes, turn right onto a footpath that heads back towards the village of Shotley Gate, bearing 135°. As you walk, reflect on the rich history of this area, from its role in the defense of the realm to its thriving maritime heritage. (Estimated time: 20-30 minutes)

7. Follow the path as it meanders through the village, taking in the charming cottages, historic buildings, and friendly locals. Discover the hidden corners and secret gardens that make Shotley Gate such a special place. (Estimated time: 15-20 minutes)

8. Finally, you'll arrive back at the Shotley Marina car park, where your circular walk comes to a satisfying conclusion. Take a final moment to appreciate the beauty and tranquility of this remarkable corner of Suffolk. (Estimated time: 5-10 minutes)

Best Features and Views:

- Lose yourself in the stunning vistas of the River Orwell and the River Stour, their glistening waters and gently sailing boats creating a mesmerizing scene that invites contemplation and wonder, the ever-changing light and the play of shadows on the water's surface a testament to the raw beauty of the natural world.
- Immerse yourself in the timeless charm of the village of Shotley Gate, with its picturesque cottages, historic buildings, and friendly local atmosphere, a living snapshot of the rich cultural heritage and enduring community spirit that define this remarkable corner of Suffolk.
- Marvel at the unique beauty of Shotley Marshes, a haven for wildlife and a Site of Special Scientific Interest, where the delicate balance of nature is on full display, from the vibrant colours of the saltmarsh plants to the intricate web of life that thrives in this fragile ecosystem.

Historical Features of Interest:

As you walk along the riverside, you'll pass by the site of HMS Ganges, a former Royal Navy training establishment that played a vital role in the defense of the realm. Take a moment to imagine the lives of the young recruits who once called this place home, their stories forever woven into the fabric of Shotley Gate's rich history. Discover the fascinating legacy of this iconic institution and its lasting impact on the local community, a testament to the enduring spirit of service and sacrifice that has shaped this special corner of Suffolk.

Birds to Spot:

- Elegant waders: Avocets: Look for these striking black and white waders, their slender, upturned bills perfectly adapted for foraging in the mudflats and shallow waters along the rivers, their presence a testament to the incredible diversity and resilience of the birdlife that thrives in this unique coastal habitat.
- Majestic hunters: Marsh harriers: Watch for these impressive birds of prey as they soar over the marshes, their distinctive silhouettes and powerful wingbeats a breathtaking sight against the vast expanse of the sky, their keen eyes scanning the landscape for the slightest movement or sign of prey.
- Scarlet sentinels: Redshanks: Listen for the distinctive piping calls of these charismatic waders as they patrol the water's edge, their bright red legs a flash of colour against the muted tones of the mudflats, their lively behaviour and alert nature a constant source of fascination and delight.

Rare or Interesting Insects and Other Wildlife:

- Jewelled acrobats: Dragonflies and damselflies: Marvel at the iridescent colours and aerial prowess of these ancient insects as they dart and hover above the water, their gossamer wings catching the light like living gems, their incredible agility and grace a testament to the wonders of the natural world.
- Fluttering kaleidoscope: Butterflies: Delight in the vibrant hues and intricate patterns of the various butterfly species that flutter amongst the wildflowers and along the path edges, each one a delicate masterpiece of nature's artistry, their ephemeral beauty a reminder of the fleeting magic of the moment.
- Elusive engineers: Water voles: Keep an eye out for signs of these charming and industrious mammals, their distinctive burrows and the gentle plopping sound as they dive into the water a tantalizing hint of their presence, their important role in shaping the wetland ecosystem a fascinating story of adaptation and resilience.

Rare or Interesting Plants:

- Purple haze: Sea lavender: In the summer months, lose yourself in the delicate beauty of the sea lavender, its soft purple flowers carpeting the saltmarshes in a haze of colour, its hardy nature a testament to the incredible adaptations of the plants that thrive in this challenging environment.
- Pastel pioneer: Sea aster: Discover the understated charm of the sea aster, its pale pink blooms a welcome splash of colour amidst the muted tones of the estuary, its tough, resilient nature a vital source of nectar for the pollinating insects that buzz and flutter along the coast.
- Salty delicacy: Marsh samphire: Keep an eye out for this intriguing edible plant, its fleshy green stems a familiar sight in the saltmarshes, its crisp, salty flavour a beloved delicacy that has been harvested and enjoyed for centuries, a living link to the traditional foods and flavours of the Suffolk coast.

Facilities and Accessibility:

The Shotley Gate Walk offers a range of facilities to ensure a comfortable and enjoyable experience for walkers. At the Shotley Marina, you'll find well-maintained public toilets and a delightful café, providing a welcome opportunity to take a break, refresh, and refuel before or after your walk. The café's menu features a tempting selection of locally-sourced food and drink, the perfect way to savour the flavours of the Suffolk countryside while soaking in the lively atmosphere of the waterfront.

It's important to note that some parts of the walk may be unsuitable for wheelchairs and pushchairs due to uneven terrain and narrow paths. To ensure a safe and enjoyable experience, it is highly recommended to wear sturdy footwear with good grip, particularly after rainfall when the paths may be muddy or slippery. By taking these precautions and being mindful of your surroundings, you can fully immerse yourself in the beauty and tranquility of the Shotley Gate Walk, creating lasting memories and a deeper connection with the natural world.

Seasonal Highlights:

The Shotley Gate Walk offers a unique and captivating experience in every season, each one revealing new facets of beauty and wonder in this enchanting corner of Suffolk:

- Spring: Witness the marshes bursting into life, with vibrant wildflowers painting the landscape in a rainbow of colour, lush green shoots emerging from the earth, and the joyous birdsong of the breeding season filling the air with the melodies of new beginnings and the promise of renewal.
- Summer: Bask in the warm sunshine and the heady scents of high summer, as you stroll along the riverside paths and through the village streets. Delight in the buzzing of insects and the gentle rustling of the foliage, each sound a reminder of the pulsing energy and vitality that infuses the landscape during these long, lazy days.
- Autumn: Marvel at the stunning array of colours as the leaves turn, transforming the countryside into a tapestry of gold, orange, and red that seems to stretch out forever. Savour the crisp, invigorating air and the satisfying crunch of fallen leaves underfoot, as you drink in the mellow beauty and poignant sense of change that characterizes this magical season.
- Winter: Embrace the wild beauty of the wind-swept rivers, the frosty marshes, and the stark silhouettes of bare trees etched against the leaden sky. Wrap up warm and set out to discover the subtle signs of life that persist even in the depths of winter, from the hardy seabirds that ride the icy gusts to the delicate tracery of frost on the hedgerows, each one a reminder of the enduring wonder and resilience of the natural world.

Difficulty Level and Safety Considerations:

The Shotley Gate Walk is classified as moderate, with some uneven terrain and muddy sections that may prove challenging for less experienced walkers. While the distance is relatively short at 4 miles, it's important to take your time and watch your step, particularly when navigating the narrow paths and the areas near the water's edge.

As you make your way along the rivers and through the village, be sure to stay on the designated paths and keep a safe distance from the water to avoid any accidents. It's also crucial to be prepared for changeable weather conditions, as the coastal climate can be unpredictable and the exposed sections of the walk can be particularly bracing on windy days.

To ensure a comfortable and enjoyable walk, be sure to wear sturdy, comfortable footwear with good grip and support, as well as clothing suitable for the season and the weather. Carry an ample supply of water and some energizing snacks to keep you hydrated and fuelled throughout your journey, and don't forget to pack any necessary medications or first aid supplies.

By following these simple precautions and the countryside code, you can fully immerse yourself in the beauty and fascination of the Shotley Gate Walk, forging a deep and lasting connection with the natural world and the rich history of this unique corner of Suffolk.

Nearby Attractions:

After your invigorating walk, take the opportunity to explore some of the other fascinating attractions that the area has to offer, each one providing a unique glimpse into the rich history, vibrant culture, and stunning natural beauty of the Suffolk coast and countryside:

- Ipswich: Venture into the vibrant county town of Suffolk, where you can immerse yourself in its rich history, from the ancient streets lined with timber-framed buildings to the impressive waterfront that tells the story of the town's maritime past. Explore the excellent museums, galleries, and cultural attractions that showcase the diverse artistic and intellectual heritage of this dynamic community.
- Suffolk Coast & Heaths AONB: Discover the breathtaking beauty of this Area of Outstanding Natural Beauty, a stunning landscape of diverse habitats and picturesque villages that showcases the very best of the Suffolk countryside. From the rugged coastline with its sweeping beaches and dramatic cliffs to the tranquil estuaries and the wildlife-rich wetlands, this protected area offers endless opportunities for exploration and adventure.
- Pin Mill: Step back in time as you visit this historic hamlet on the banks of the River Orwell, its charming buildings and bustling sailing community a testament to the enduring connection between the people of Suffolk and the water that shapes their lives. Enjoy a refreshing pint at the excellent pub, immerse yourself in the rich maritime heritage of the area, and soak up the timeless atmosphere of this quintessential Suffolk destination.

By exploring these nearby attractions, you'll gain a deeper appreciation for the diverse and captivating character of the Suffolk coast and countryside, making your Shotley Gate Walk the perfect starting point for a truly unforgettable journey through this enchanting corner of East Anglia. Whether you're a keen walker, a nature lover, or simply someone who appreciates the beauty and tranquility of the great outdoors, this area promises to delight, inspire, and leave you with memories that will last a lifetime.

Butley Circular Walk - 5 miles (8 km), Circular - Estimated Time: 2.5-3 hours

Starting Point Coordinates, Postcode: Village hall in Butley (Grid Reference: TM 365 498, Postcode: IP12 3PA). Nearest Car Park, Postcode: Limited on-street parking is available near the village hall in Butley (Postcode: IP12 3PA).

Step back in time and discover the tranquil beauty of the Suffolk countryside on this delightful 5-mile (8 km) circular walk around the charming village of Butley. Immerse yourself in the peaceful landscapes, explore the rich history, and uncover the hidden gems of this unspoiled corner of East Anglia.

As you set out from the village hall, the quiet country lane beckons you westward, the anticipation building with each step as you leave the gentle hum of village life behind. The picturesque surroundings envelop you in their embrace, the soft rustling of leaves and the melodic birdsong creating a symphony of natural sounds that soothes the soul and invigorates the senses.

Turning onto another lane, you'll find yourself winding through the heart of the Suffolk countryside, the patchwork of fields and hedgerows unfolding before you like a living tapestry. The gentle undulations of the landscape create an ever-changing tableau, each vista more breathtaking than the last. Pause to admire the sweeping panoramas, the distant church spires, and the lazy drift of clouds across the vast expanse of sky, a timeless scene that captures the essence of rural England.

As you step onto the footpath, the tranquil woodland welcomes you into its dappled shade, the sunlight filtering through the canopy and casting an enchanted glow upon the forest floor. Immerse yourself in the cool, green sanctuary, the air filled with the heady scent of earth and foliage, a balm for the senses that soothes and rejuvenates. Listen to the gentle rustling of leaves overhead and the soft crunch of twigs underfoot, each sound a reminder of the woodland's enduring presence and the countless stories it holds.

Emerging from the trees, you'll find yourself once again amidst the open farmland, the path stretching out before you like a ribbon of gold amidst the fields of green and gold. The wide, open skies seem to go on forever, a vast canvas painted in hues of blue and white, with the occasional bird soaring high above, a silent guardian watching over the landscape below.

As you make your way along the ancient hedgerows and wildflower meadows, take a moment to appreciate the incredible diversity of life that thrives in these precious habitats. The delicate blooms of cowslips and oxeye daisies sway gently in the breeze, their petals brushed with the soft hues of sunshine, while the vibrant flashes of butterflies and the iridescent shimmer of dragonflies add a touch of magic to the scene.

Approaching the village once more, the path guides you through the final stretch of picturesque farmland, the distant rooftops of Butley coming into view like a welcoming beacon. The sense of accomplishment and joy that fills your heart is a testament to the transformative power of this gentle countryside ramble, a reminder of the beauty and resilience of the natural world and the importance of preserving these special places for generations to come.

As you arrive back at the village hall, take a final moment to reflect on the incredible journey you've just completed. The Butley Circular Walk is more than just a physical expedition; it is a chance to connect with the land, to discover the hidden wonders that lie just beyond the hustle and bustle of modern life, and to find a sense of peace and belonging in the timeless beauty of the Suffolk countryside.

Detailed Route Directions:

1. From the village hall, head west along the quiet country lane, bearing 270°. Take in the picturesque surroundings and the gentle sounds of nature as you walk. (Estimated time: 10-15 minutes)

2. After about half a mile, turn left onto another lane, bearing 180°. Follow the lane as it winds through the peaceful countryside, passing by fields and hedgerows. (Estimated time: 20-25 minutes)

3. Continue along the lane until you reach a footpath on your right, bearing 270°. Turn onto the footpath and follow it as it leads you into a tranquil woodland, where dappled sunlight filters through the leaves and the air is filled with the sound of birdsong. (Estimated time: 15-20 minutes)

4. Emerge from the woodland onto another quiet lane, bearing 180°. Turn left onto the lane and follow it as it takes you deeper into the heart of the Suffolk countryside. (Estimated time: 20-25 minutes)

5. After a short distance, turn right onto a footpath, bearing 270°. The path will guide you through open farmland, offering stunning views of the surrounding landscape and the chance to spot local wildlife. (Estimated time: 25-30 minutes)

6. Continue along the footpath until you reach another lane, bearing 090°. Turn left onto the lane and follow it eastward, back towards the village of Butley. (Estimated time: 20-25 minutes)

7. As you approach the village, turn left onto a footpath, bearing 000°. This path will take you through more picturesque farmland, with ancient hedgerows and wildflower meadows. (Estimated time: 15-20 minutes)

8. Finally, you'll arrive back at the village hall in Butley, where your circular walk comes to a satisfying conclusion. Take a moment to reflect on the beauty and tranquility of the countryside you've just explored. (Estimated time: 5-10 minutes)

Best Features and Views:

- Lose yourself in the peaceful country lanes, their ancient hedgerows and vibrant wildflowers creating a living tapestry that seems to stretch out forever, a testament to the enduring beauty and resilience of the natural world.
- Immerse yourself in the tranquil woodland, its dappled sunlight and gentle birdsong a soothing balm for the senses, a green sanctuary that invites quiet reflection and a deeper connection with the rhythms of the earth.
- Marvel at the stunning views over the open farmland, the patchwork of fields, copses, and gentle hills creating a landscape of breathtaking beauty and diversity, a timeless scene that captures the very essence of rural England.

Historical Features of Interest:

As you explore the village of Butley, take a moment to seek out the ruins of Butley Priory, an Augustinian monastery that dates back to the 12th century. These ancient stones stand as a silent witness to the passage of time, their weathered surfaces and crumbling arches a poignant reminder of the lives and stories that have unfolded within these walls over the centuries. Imagine the whispered prayers of the monks who once called this place home, and the echoes of their footsteps that still seem to linger in the quiet corners of the ruins, a tangible link to the rich history and spiritual heritage of this special place.

Birds to Spot:

- Majestic hunters: Buzzards: Listen for the distinctive mewing calls of these impressive birds of prey as they soar overhead, their keen eyes scanning the landscape for the slightest movement or sign of prey, their powerful wings carrying them effortlessly through the vast expanse of sky.
- Vibrant songsters: Yellowhammers: Delight in the bright yellow plumage and cheerful melodies of these charming little birds, their joyful songs a celebration of the beauty and vitality of the Suffolk countryside, their presence a welcome splash of colour amidst the greens and golds of the fields and hedgerows.
- Woodland drummers: Woodpeckers: Keep an ear out for the rhythmic drumming sounds of woodpeckers as they excavate their nests in the trees, and try to spot the flash of green or black and white as they flit through the branches, their striking colours and lively behaviour a constant source of fascination and wonder.

Rare or Interesting Insects and Other Wildlife:

- Winged jewels: Butterflies: In the summer months, lose yourself in the vibrant colours and delicate beauty of the various butterfly species that flutter amongst the wildflowers and along the woodland edges, each one a living masterpiece of nature's artistry, their ephemeral presence a reminder of the fleeting magic of the moment.
- Aerial acrobats: Dragonflies: Watch in awe as the iridescent shimmer of dragonflies darts across ponds and streams, their gossamer wings catching the light like living gems, their incredible agility and grace a testament to the wonders of the natural world.
- Elusive inhabitants: Mammals: Keep an eye out for the telltale signs of deer, foxes, and badgers as you walk through the countryside, their tracks and trails a tantalizing hint of the secret lives that unfold in the fields and forests, their presence a reminder of the rich tapestry of life that thrives in these unspoiled landscapes.

Rare or Interesting Plants:

- Floral treasures: Wildflowers: In the spring and summer, delight in the delicate blooms of wildflowers that adorn the meadows and verges, from the sunny faces of cowslips and oxeye daisies to the blushing petals of red campion, each one a tiny miracle of beauty and resilience that adds a splash of colour to the green mantle of the countryside.
- Verdant sentinels: Ferns: Discover the lush green fronds of ferns that thrive in the shady nooks and crannies of the woodland, from the delicate tracery of maidenhair spleenwort to the stately plumes of royal fern, each one a living sculpture that speaks to the incredible diversity and adaptability of the plant kingdom.
- Autumnal wonders: Fungi: In the mellow days of autumn, keep an eye out for the strange and marvelous world of fungi that springs up from the forest floor, from the classic red and white spotted cap of the fly agaric to the bizarre and beautiful forms of the cauliflower fungus, each one a reminder of the hidden wonders that lie waiting to be discovered in the depths of the wood.

Facilities and Accessibility:

While there are no cafes or pubs directly on the route, the nearby town of Orford offers a range of options for those seeking refreshment and sustenance after their walk. From cozy tea rooms serving homemade cakes and sandwiches to traditional pubs with hearty meals and local ales, there's something to suit every taste and preference, the perfect way to refuel and reflect on the day's adventures.

The Butley Circular Walk is classified as moderately easy, with some gentle inclines and uneven terrain that may prove challenging for those with limited mobility. To ensure a safe and comfortable experience, it is highly recommended to wear sturdy footwear with good grip and support, particularly after rainfall when the paths may be muddy or slippery.

Seasonal Highlights:

The Butley Circular Walk offers a unique and captivating experience in every season, each one revealing new facets of beauty and wonder in this enchanting corner of Suffolk:

- Spring: Witness the countryside bursting into life, with vibrant wildflowers painting the meadows and verges in a rainbow of colour, lush green leaves unfurling on the trees, and the joyous birdsong of the breeding season filling the air with the melodies of new beginnings and the promise of renewal.
- Summer: Bask in the warm sunshine and the heady scents of high summer, as you stroll along the country lanes and through the dappled shade of the woodland. Delight in the buzzing of insects and the gentle rustling of the foliage, each sound a reminder of the pulsing energy and vitality that infuses the landscape during these long, lazy days.
- Autumn: Marvel at the stunning array of colours as the leaves turn, transforming the countryside into a breathtaking tapestry of gold, orange, and red that seems to stretch out forever. Discover the fascinating world of fungi that springs up from the forest floor, each strange and wonderful form a reminder of the hidden treasures that lie waiting to be discovered in the depths of the wood.
- Winter: Embrace the tranquil beauty of the frosty fields, the bare trees silhouetted against the pale winter sky, and the crisp, cold air that invigorates the senses and adds a touch of magic to your walk. Delight in the chance to spot winter visiting birds and other wildlife, their presence a reminder of the resilience and adaptability of the natural world in the face of even the harshest conditions.

Difficulty Level and Safety Considerations:

This walk is classified as easy to moderate, with some gentle inclines and uneven terrain that may prove challenging for less experienced walkers. While the distance is relatively short at 5 miles, it's important to take your time and watch your step, particularly when navigating the muddy or slippery sections of the path.

As you make your way through the countryside, be sure to stay on the designated footpaths and respect the land and the wildlife that calls it home. Remember to close any gates behind you and keep dogs on a lead, particularly when passing through fields with livestock.

To ensure a safe and enjoyable walk, be sure to wear sturdy, comfortable footwear with good grip and support, as well as clothing suitable for the season and the weather. Carry an ample supply of water and some energizing snacks to keep you hydrated and fuelled throughout your journey, and don't forget to pack any necessary medications or first aid supplies.

By following these simple precautions and the countryside code, you can fully immerse yourself in the beauty and tranquility of the Butley Circular Walk, creating lasting memories and a deeper appreciation for the natural wonders of the Suffolk countryside.

Nearby Attractions:

After your invigorating walk, take the opportunity to explore some of the other fascinating attractions that the area has to offer, each one providing a unique glimpse into the rich history, vibrant culture, and stunning natural beauty of the Suffolk coast and countryside:

- Orford Castle: Step back in time as you discover the impressive ruins of this 12th-century castle, its towering

keep and crumbling walls a testament to the power and influence of the medieval nobility. Climb to the top of the keep for breathtaking views over the Suffolk coast, and immerse yourself in the fascinating stories and legends that have shaped this iconic landmark over the centuries.
- RSPB Havergate Island: Embark on a boat trip to this remote and beautiful island nature reserve, where you can experience the raw beauty and untamed wilderness of the Suffolk coast at its most pristine. Marvel at the incredible diversity of birdlife that thrives in this unique habitat, from the elegant avocets and the majestic marsh harriers to the lively terns and the elusive bitterns, each one a living jewel in the crown of the natural world.
- Sutton Hoo: Uncover the incredible story of the ship burial of an Anglo-Saxon king at this world-famous archaeological site, where the echoes of a long-vanished world still seem to whisper through the ancient landscape. Discover the astonishing treasures that were unearthed here, from the intricate gold and garnet fittings of the royal regalia to the eerie beauty of the burial mask, each one a testament to the skill and artistry of the Anglo-Saxon craftsmen and the enduring power of the human spirit.

By exploring these nearby attractions, you'll gain a deeper appreciation for the diverse and captivating character of the Suffolk coast and countryside, making your Butley Circular Walk the perfect starting point for a truly unforgettable journey through this enchanting corner of East Anglia. Whether you're a keen historian, a nature lover, or simply someone who appreciates the beauty and tranquility of the great outdoors, this area promises to delight, inspire, and leave you with memories that will last a lifetime.

Harkstead Circular Walk - 5 miles (8 km), Circular - Estimated Time: 2.5-3 hours

Starting Point Coordinates, Postcode: Public car park at Harkstead Church (Grid Reference: TM 195 352, Postcode: IP9 1DR).
Nearest Car Park, Postcode: Public car park at Harkstead Church (Postcode: IP9 1DR).

Discover the picturesque village of Harkstead and its idyllic surrounding countryside on this delightful 5-mile (8 km) circular walk. With stunning views of the River Stour, charming country lanes, and an array of fascinating historic sites, this walk offers the perfect escape from the hustle and bustle of everyday life.

As you set out from the car park at Harkstead Church, the tranquil atmosphere of the village envelops you, the distinctive round tower of St. Mary's Church standing as a beacon of the community's rich history and heritage. Follow Church Lane southward, the gentle rustling of leaves and the distant birdsong creating a soothing soundtrack to your journey.

Turning onto the footpath, you'll find yourself immersed in the heart of the Suffolk countryside, the patchwork of fields and hedgerows stretching out before you like a living tapestry. The gentle undulations of the landscape create an ever-changing tableau, each vista more breathtaking than the last. Pause to admire the sweeping panoramas, the distant church spires, and the lazy drift of clouds across the vast expanse of sky, a timeless scene that captures the essence of rural England.

As you navigate through the wooded area, the dappled sunlight filtering through the canopy creates an enchanting play of light and shadow, the air filled with the earthy scent of leaf litter and the gentle rustling of wildlife in the undergrowth. Emerge from the trees and into the open fields once more, the path stretching out before you like an invitation to explore the wonders of this unspoiled landscape.

Upon reaching the footpath that leads towards the River Stour, a new chapter of your adventure begins. The river's glistening waters and gently swaying reeds create a scene of serene beauty, the soft lapping of the water against the shore a soothing melody that fills the air. Take a moment to breathe in the fresh, invigorating scent of the water and to appreciate the incredible diversity of life that thrives along the riverbank.

As you follow the path along the River Stour, the stunning views across the water will take your breath away. Watch as the light dances on the surface, the reflections of the clouds and the trees creating an ever-shifting kaleidoscope of colour and texture. Keep an eye out for the graceful movements of swans and the darting flight of kingfishers, each encounter a reminder of the rich tapestry of wildlife that calls this special place home.

Continuing through the picturesque countryside, the ancient hedgerows and wildflower meadows paint a picture of timeless beauty, their vibrant colours and delicate blooms a testament to the resilience and wonder of the natural world. Listen for the buzz of insects and the gentle rustling of the breeze through the grasses, each sound a reminder of the intricate web of life that thrives in these unspoiled habitats.

As you approach Harkstead village once more, the charming cottages and the inviting atmosphere of the community welcome you back, the sense of coming full circle filling you with a deep sense of accomplishment and connection to the land. The sight of Harkstead Church, standing proud and timeless against the backdrop of the Suffolk sky, serves as a fitting end to your incredible journey.

The Harkstead Circular Walk is more than just a physical experience; it is an opportunity to immerse yourself in the beauty, history, and natural wonders of this enchanting corner of East Anglia. As you reflect on the sights, sounds, and sensations of your adventure, you'll carry with you a renewed appreciation for the power of the landscape to inspire, heal, and transform, and a deeper understanding of the importance of preserving these precious places for generations to come.

Detailed Route Directions:

1. From the car park, head south along Church Lane, bearing 180°. Admire the beautiful church of St. Mary's as you pass by, with its distinctive round tower and centuries of history. (Estimated time: 5-10 minutes)

2. At the end of Church Lane, turn right onto a footpath, bearing 270°. Follow the path as it leads you through open fields, offering stunning views of the surrounding countryside. (Estimated time: 15-20 minutes)

3. Continue along the footpath, passing through several gates and stiles, until you reach a junction with another path. Keep straight ahead at this point, following the path as it winds through a wooded area. (Estimated time: 20-25 minutes)

4. Emerge from the woods and continue straight ahead, now with open fields on either side. After approximately 2 miles (3.2 km), you'll reach a footpath on your right that leads towards the River Stour. (Estimated time: 30-40 minutes)

5. Turn right onto this footpath, bearing 000°, and follow it as it takes you along the banks of the River Stour. Take a moment to appreciate the stunning views across the water and the peaceful atmosphere of this beautiful stretch of the river. (Estimated time: 20-30 minutes)

6. Continue along the riverside path for some time, enjoying the sights and sounds of the water and the wildlife that calls this area home. Eventually, the path will veer away from the river and head northwards, back towards Harkstead village. (Estimated time: 30-40 minutes)

7. Follow the path as it takes you through more picturesque countryside, with ancient hedgerows and wildflower meadows lining the route. Keep an eye out for the various bird species and other animals that make their home in these tranquil surroundings. (Estimated time: 20-30 minutes)

8. As you approach Harkstead village, you'll come to a junction with a road. Turn right here, bearing 090°, and follow the road as it leads you back into the heart of the village. (Estimated time: 10-15 minutes)

9. Continue along the road until you reach Harkstead Church once again, where your circular walk comes to a satisfying conclusion. Take a final moment to reflect on the beauty and history of this special corner of Suffolk. (Estimated time: 5-10 minutes)

Best Features and Views:

- Lose yourself in the stunning views of the River Stour, its glistening waters and gently sailing boats creating a mesmerizing scene that invites contemplation and wonder, the ever-changing light and the play of shadows on the water's surface a testament to the raw beauty of the natural world.
- Immerse yourself in the picturesque countryside, with its ancient hedgerows, wildflower meadows, and gently rolling hills, a living tapestry that shifts with the seasons and offers a breathtaking display of colour, texture, and life at every turn.
- Delight in the historic charm of Harkstead village, with its charming cottages, beautiful church, and sense of

timeless tranquility, a place where the past and present intertwine, inviting you to slow down, explore, and connect with the enduring spirit of this remarkable community.

Historical Features of Interest:

As you stand before the impressive edifice of Harkstead Church, dedicated to St. Mary, take a moment to appreciate the rich history and heritage that this fascinating site embodies. Dating back to the 12th century, the church's distinctive round tower and beautiful interior serve as a testament to the skill and devotion of the craftsmen who built it, and to the generations of worshippers who have found solace and inspiration within its walls. Imagine the countless prayers, hymns, and stories that have echoed through this sacred space over the centuries, a living link to the past that continues to shape the identity and character of the village to this day.

Birds to Spot:

- Jewels of the river: Kingfishers: Watch for the brilliant flash of blue and orange as these stunning birds dart along the river's edge, their swift and agile movements a mesmerizing display of natural beauty and precision.
- Acrobatic aviators: Lapwings: Listen for the distinctive peewit calls of these charismatic waders as they perform their acrobatic display flights over the fields and meadows, their black and white plumage and wispy crests a striking sight against the backdrop of the Suffolk sky.
- Silent hunters: Barn Owls: If fortune favours you, you may catch a rare glimpse of these ethereal birds as they hunt over the fields and hedgerows, their ghostly white plumage and silent flight a haunting and unforgettable sight in the gathering dusk.

Rare or Interesting Insects and Other Wildlife:

- Winged jewels: Butterflies: In the summer months, lose yourself in the vibrant colours and delicate beauty of the various butterfly species that flutter amongst the wildflowers and along the woodland edges, each one a living masterpiece of nature's artistry, their ephemeral presence a reminder of the fleeting magic of the moment.
- Aerial acrobats: Dragonflies: Watch in awe as the iridescent shimmer of dragonflies darts across the river and nearby ponds, their gossamer wings catching the light like living gems, their incredible agility and grace a testament to the wonders of the natural world.
- Elusive engineers: Water Voles: Keep an eye out for signs of these charming and industrious mammals along the riverbank, their distinctive burrows and the gentle plopping sound as they dive into the water a tantalizing hint of their presence, their important role in shaping the wetland ecosystem a fascinating story of adaptation and resilience.

Rare or Interesting Plants:

- Floral treasures: Wildflowers: In the spring and summer, delight in the delicate blooms of wildflowers that adorn the meadows and verges, from the sunny faces of cowslips and oxeye daisies to the blushing petals of red campion, each one a tiny miracle of beauty and resilience that adds a splash of colour to the green mantle of the countryside.
- Living history: Ancient Hedgerows: Marvel at the gnarled and twisted branches of the ancient hedgerows that line the fields and lanes, each one a living testament to the countless generations of farmers and laborers who have shaped this landscape over the centuries, their rich diversity of plant and animal life a microcosm of the countryside's enduring vitality.
- Wetland wonders: Riverside Plants: Discover the specialized plants that thrive along the banks of the River Stour, from the regal spires of purple loosestrife to the sunny blooms of yellow flag iris and the aromatic

leaves of water mint, each one a vital part of the intricate web of life that flourishes in this unique and precious habitat.

Facilities and Accessibility:

While there are no cafes or pubs directly on the route, the nearby villages of Holbrook and Shotley offer a range of options for those seeking refreshment and sustenance after their walk. From cozy tea rooms serving homemade cakes and sandwiches to traditional pubs with hearty meals and local ales, there's something to suit every taste and preference, the perfect way to refuel and reflect on the day's adventures.

The Harkstead Circular Walk is classified as moderately easy, with mostly flat terrain and some gentle inclines that may prove challenging for those with limited mobility. To ensure a safe and comfortable experience, it is highly recommended to wear sturdy footwear with good grip and support, particularly after rainfall when some sections of the route may be uneven or muddy.

Seasonal Highlights:

The Harkstead Circular Walk offers a unique and captivating experience in every season, each one revealing new facets of beauty and wonder in this enchanting corner of Suffolk:

- Spring: Witness the countryside bursting into life, with vibrant wildflowers painting the meadows and verges in a rainbow of colour, lush green leaves unfurling on the trees, and the joyous birdsong of the breeding season filling the air with the melodies of new beginnings and the promise of renewal.
- Summer: Bask in the warm sunshine and the heady scents of high summer, as you stroll along the country lanes and riverside paths, the buzzing of insects and the gentle rustling of the breeze through the foliage creating a soothing symphony that celebrates the fullness and vitality of life.
- Autumn: Marvel at the stunning array of colours as the leaves turn, transforming the landscape into a breathtaking tapestry of gold, orange, and red that seems to stretch out forever. Embrace the crisp, invigorating air and the sense of change and transformation that characterizes this season of mists and mellow fruitfulness.
- Winter: Embrace the tranquil beauty of the frosty fields, the bare trees etched against the pale winter sky, and the crisp, cold air that invigorates the senses and adds a touch of magic to your walk. Delight in the chance to spot winter visiting birds and other wildlife, their presence a reminder of the resilience and adaptability of the natural world in the face of even the harshest conditions.

Difficulty Level and Safety Considerations:

The Harkstead Circular Walk is classified as moderately easy, with mostly flat terrain and some gentle inclines that may prove challenging for less experienced walkers. While the distance is relatively short at 5 miles, it's important to take your time and watch your step, particularly when navigating any uneven or muddy sections of the path.

As you make your way through the countryside, be sure to stay on the designated footpaths and respect the land and the wildlife that calls it home. Remember to close any gates behind you and keep dogs on a lead, particularly when passing through fields with livestock.

To ensure a safe and enjoyable walk, be sure to wear sturdy, comfortable footwear with good grip and support, as well as clothing suitable for the season and the weather. Carry an ample supply of water and some energizing snacks to keep you hydrated and fuelled throughout your journey, and don't forget to pack any necessary medications or first aid supplies.

By following these simple precautions and the countryside code, you can fully immerse yourself in the beauty and tranquility of the Harkstead Circular Walk, creating lasting memories and a deeper appreciation for the natural wonders of the Suffolk countryside.

Nearby Attractions:

After your invigorating walk, take the opportunity to explore some of the other fascinating attractions that the area has to offer, each one providing a unique glimpse into the rich history, vibrant culture, and stunning natural beauty of the Suffolk coast and countryside:

- Pin Mill: Step back in time as you visit this historic hamlet on the banks of the River Orwell, its charming buildings and bustling sailing community a testament to the enduring connection between the people of Suffolk and the water that shapes their lives. Enjoy a refreshing pint at the excellent pub, immerse yourself in the rich maritime heritage of the area, and soak up the timeless atmosphere of this quintessential Suffolk destination.
- Jimmy's Farm: Embark on a fun-filled family day out at this popular attraction, where you can meet a delightful array of friendly animals, explore the beautiful nature trails, and indulge in the delicious, locally-sourced food at the on-site restaurant. With its commitment to conservation, education, and sustainability, Jimmy's Farm offers a unique and inspiring glimpse into the world of modern farming and the importance of connecting with the natural world.
- Alton Water Park: Discover the breathtaking beauty of this stunning reservoir and country park, a haven for wildlife and outdoor enthusiasts alike. From the tranquil waters that are perfect for fishing and watersports to the scenic trails that wind through the surrounding countryside, Alton Water Park offers endless opportunities for adventure, relaxation, and discovery, a true gem in the crown of the Suffolk landscape.

By exploring these nearby attractions, you'll gain a deeper appreciation for the diverse and captivating character of the Suffolk coast and countryside, making your Harkstead Circular Walk the perfect starting point for a truly unforgettable journey through this enchanting corner of East Anglia. Whether you're a keen walker, a nature lover, or simply someone who appreciates the beauty and tranquility of the great outdoors, this area promises to delight, inspire, and leave you with memories that will last a lifetime.

Harkstead and Erwarton Walk - 5 miles (8 km), Circular - Estimated Time: 2.5-3 hours

Starting Point Coordinates, Postcode: The starting point for the Harkstead and Erwarton Walk is at St. Mary's Church, Harkstead (Grid Reference: TM 182 343, Postcode: IP9 1DB).
Nearest Car Park, Postcode: On-street parking is available near St. Mary's Church in Harkstead (Postcode: IP9 1DB).

Embark on a captivating 5-mile (8 km) circular walk through the picturesque villages of Harkstead and Erwarton, where the gentle embrace of the Suffolk countryside invites you to leave behind the cares of the world and immerse yourself in a realm of tranquility and beauty. This easy to moderate route, with its quiet lanes, rolling fields, and stunning views of the River Stour, offers the perfect escape from the hustle and bustle of everyday life, allowing you to reconnect with nature and discover the hidden gems of this enchanting corner of East Anglia.

As you set out from the historic St. Mary's Church in Harkstead, the peaceful atmosphere of the village envelops you, the gentle rustling of leaves and the distant birdsong creating a soothing backdrop to your journey. Follow Church Lane southward, the path beckoning you into the heart of the Suffolk countryside. With each step, a tapestry of gently rolling hills, lush green pastures, and ancient woodlands unfolds before you, a testament to the enduring beauty and vitality of this remarkable landscape.

Turning onto the footpath, you'll find yourself immersed in a world of natural wonders, the soft contours of the land creating an ever-changing tableau that invites contemplation and wonder. Navigate through the fields and the small woodland, taking a moment to appreciate the intricate details of the world around you. From the delicate wildflowers swaying in the breeze to the majestic trees standing sentinel over the land, each element contributes to the rich tapestry of life that defines this special place.

As you emerge from the woodland and follow the quiet lane towards Erwarton, the village's timeless charm greets you like an old friend. This picturesque community, with its charming cottages and historic church, seems frozen in time, a reminder of the rich heritage and enduring spirit of rural Suffolk. Pause to admire the beautiful architecture of St. Mary's Church, its ancient stones and peaceful surroundings a testament to the generations of worshippers who have found solace and inspiration within its walls.

Leaving Erwarton behind, the path guides you northward, towards the glistening waters of the River Stour. As you approach the riverbank, a breathtaking vista unfolds before you, the estuary and the surrounding countryside painted in a mesmerizing palette of colours. Take a moment to drink in the view, feeling the gentle breeze on your face and the warm sun on your skin, a sensory experience that will stay with you long after your journey's end.

Following the path along the River Stour, you'll find yourself immersed in a world of natural wonders, the ebb and flow of the tide creating an ever-changing tapestry of light and colour on the water's surface. Keep an eye out for the diverse array of birdlife that calls this area home, from the graceful lapwings and the soaring skylarks to the brilliant flash of a kingfisher darting along the riverbank. Each encounter serves as a reminder of the incredible diversity and resilience of the natural world, a living testament to the delicate balance of life in this remarkable corner of East Anglia.

As you turn onto the quiet lane and make your way back towards Harkstead, take a moment to reflect on the incredible journey you've just completed. The fields and hedgerows that surround you are more than just a backdrop to your walk; they are a living tapestry, woven with the stories and traditions of the generations who have shaped this land. Each step you take is a connection to this rich history, a reminder of the enduring spirit and beauty of the Suffolk countryside.

Arriving back at St. Mary's Church, your starting point, you'll find yourself filled with a sense of accomplishment and a deeper appreciation for the simple joys and profound truths that can be found when we step away from the distractions of modern life and immerse ourselves in the timeless wonders of the natural world. The Harkstead and Erwarton Walk is more than just a physical journey; it is a celebration of the beauty, resilience, and enduring spirit of this enchanting corner of East Anglia, an invitation to reconnect with the land, with ourselves, and with the boundless wonders that surround us.

Detailed Route Directions:

1. From St. Mary's Church, head south on Church Lane, bearing 170°, and continue onto the footpath. (Estimated time: 5-10 minutes)

2. Follow the path as it bends to the right, bearing 230°, passing through fields and a small woodland. (Estimated time: 15-20 minutes)

3. Turn left onto a quiet lane, bearing 140°, and follow it to the village of Erwarton. (Estimated time: 20-25 minutes)

4. Walk through the village, bearing 80°, passing St. Mary's Church, and continue onto the footpath, heading north towards the River Stour. (Estimated time: 15-20 minutes)

5. Follow the path as it bends to the left, bearing 350°, offering views of the river. (Estimated time: 20-25 minutes)

6. Continue along the path, bearing 280°, until you reach a quiet lane, and turn left. (Estimated time: 15-20 minutes)

7. Follow the lane, bearing 200°, until you reach a junction, then turn right onto another footpath. (Estimated time: 10-15 minutes)

8. Continue along this path, bearing 150°, passing through more fields, until you reach Harkstead Church Lane. (Estimated time: 20-25 minutes)

9. Turn right and follow the lane, bearing 350°, back to the starting point at St. Mary's Church. (Estimated time: 10-15 minutes)

Best Features and Views:

- Immerse yourself in the picturesque Suffolk countryside, with its gently rolling hills, ancient woodlands, and quiet lanes, a living tapestry that showcases the timeless beauty and tranquility of this enchanting corner of East Anglia.
- Marvel at the stunning views of the River Stour, its glistening waters and ever-changing light creating a mesmerizing scene that invites contemplation and wonder, a perfect backdrop for a moment of peaceful reflection and connection with the natural world.
- Delight in the historic charm of the villages of Harkstead and Erwarton, their beautiful churches and picturesque cottages a testament to the rich heritage and enduring spirit of rural Suffolk, a living link to the generations who have called this special place home.

Historical Features of Interest:

As you pass by the historic St. Mary's Churches in both Harkstead and Erwarton, take a moment to appreciate the rich history and architectural beauty that these sacred spaces embody. With their ancient stones and peaceful surroundings, these churches serve as a tangible link to the generations of worshippers who have found solace, inspiration, and a sense of community within their walls.

Imagine the countless prayers, hymns, and stories that have echoed through these hallowed halls, a living testament to the enduring power of faith and tradition in the hearts of the people who call this special place home. These churches are more than just buildings; they are a reflection of the deep spiritual connections that have shaped the communities of rural Suffolk for centuries, a reminder of the shared values, hopes, and dreams that bind us together across time and space.

As you take a moment to appreciate the intricate details of the architecture, from the graceful arches and elegant stonework to the beautiful stained glass windows and time-worn gravestones, let your mind wander to the lives and experiences of those who have come before you. Each person who has passed through these doors has left their mark, contributing to the rich tapestry of history that defines this enchanting corner of East Anglia.

In the peaceful surroundings of the churchyards, with the gentle rustling of leaves and the distant birdsong as your backdrop, you may find yourself feeling a deep sense of connection to the past, a recognition of the enduring spirit and resilience that have shaped this land and its people for generations. This is the true gift of these historic churches, the opportunity to step out of the present moment and into a timeless realm of reflection, inspiration, and wonder.

Birds to Spot:

- Lapwings - These striking black and white waders are a delight to observe as they perform their acrobatic display flights over the fields and meadows. Known for their distinctive "peewit" calls and wispy crests, lapwings are a charming and iconic presence in the Suffolk countryside. Keep an eye out for their graceful gliding and impressive aerial manoeuvres, a testament to their adaptability and resilience in the face of changing landscapes and environmental pressures.
- Skylarks - As you walk through the fields and along the quiet lanes, listen for the joyous and unbroken streams of melody that announce the presence of these incredible songbirds. Skylarks are known for their ability to sing while hovering high above the ground, their music filling the air with a sense of exultation and freedom. Take a moment to scan the skies for a glimpse of these small brown birds, their songs a celebration of the beauty and vitality of the natural world.
- Kingfishers - If fortune favours you, you may catch a fleeting glimpse of these brilliant blue and orange birds as they dart along the banks of the River Stour. Kingfishers are known for their swift and agile movements, their compact bodies perfectly adapted for diving into the water to catch small fish and aquatic insects. The sight of a kingfisher perched on a branch or flashing past in a blur of vibrant colour is a rare and unforgettable treat, a reminder of the incredible diversity and beauty of the wildlife that calls this area home.

Rare or Interesting Insects and Other Wildlife:

- Butterflies - In the warmer months, the fields, woodlands, and riverbanks come alive with the vibrant colours and delicate beauty of various butterfly species. From the striking orange and brown patterning of the comma to the iridescent blues of the holly blue and the regal majesty of the purple emperor, these fluttering gems

add a touch of magic and wonder to the Suffolk countryside. Take a moment to appreciate their ephemeral beauty, a reminder of the fleeting nature of life and the importance of cherishing each moment.
- Dragonflies - As you walk along the River Stour and through the fields, keep an eye out for the shimmering, jewel-like presence of dragonflies darting through the air. These incredible aerial acrobats are known for their impressive agility and speed, their gossamer wings propelling them with incredible precision as they hunt for small insects. From the electric blue of the southern hawker to the emerald green of the emperor dragonfly, these creatures are a mesmerizing display of nature's artistry and adaptability.
- Small Mammals - The hedgerows, fields, and woodlands of the Harkstead and Erwarton Walk are home to a diverse array of small mammals, each one playing an important role in the delicate ecosystem of the Suffolk countryside. As you walk, keep an eye out for the telltale signs of these elusive creatures, from the scurrying of field mice and voles to the gentle rustling of a hedgehog foraging for food. You may also spot the distinctive tracks of a badger or the darting form of a weasel or stoat, a reminder of the rich tapestry of life that thrives just beyond our sight.

Rare or Interesting Plants:

- Ancient Oaks - As you pass through the woodlands and along the hedgerows, take a moment to appreciate the majestic presence of the ancient oak trees that have stood witness to the passing of centuries. These venerable giants, with their gnarled bark and spreading canopies, are a living testament to the enduring strength and resilience of the natural world. Imagine the countless generations of wildlife that have found shelter and sustenance in their branches, and the stories they could tell of the changing landscape and the people who have walked beneath their boughs.
- Bluebells - If your walk takes you through the woodlands in the spring, you may be fortunate enough to witness the breathtaking spectacle of vast carpets of bluebells stretching out beneath the trees. These delicate flowers, with their nodding heads and sweet fragrance, are a true celebration of the season's renewal and the magic of the natural world. Take a moment to drink in their beauty, and to appreciate the fleeting nature of their presence, a reminder of the importance of cherishing each moment and finding joy in the simple wonders that surround us.
- Wildflowers - Throughout your walk, keep an eye out for the diverse array of wildflowers that adorn the fields, hedgerows, and riverbanks of the Suffolk countryside. From the cheerful faces of buttercups and daisies to the delicate pinks of campion and the regal purples of foxgloves, each bloom is a tiny miracle of beauty and resilience, a testament to the incredible adaptability and diversity of the plant world. Take a moment to appreciate their colours, textures, and fragrances, and to reflect on the important role they play in supporting the intricate web of life that thrives in this special place.

Facilities and Accessibility:

The Harkstead and Erwarton Walk is a relatively easy route, with mostly flat terrain and well-maintained paths. However, it's important to note that some sections may be uneven or muddy, especially after periods of heavy rain. Sturdy walking shoes or boots with good grip are highly recommended to ensure a safe and comfortable experience.

While there are no public toilets or other facilities directly along the walk route, the nearby villages of Harkstead and Erwarton offer some amenities for those in need of a comfort break or refreshment. In Harkstead, the Bakers Arms pub is a welcoming spot for a drink or a bite to eat, while in Erwarton, the Wheatsheaf Inn offers a cozy atmosphere and traditional pub fare.

It's important to note that the walk may not be suitable for those with limited mobility or wheelchair users, as some sections of the path may be narrow, uneven, or include stiles. As always, it's a good idea to check the local conditions and plan accordingly, bringing water, snacks, and any necessary medications or sun protection to ensure a safe and enjoyable experience.

Seasonal Highlights:

The Harkstead and Erwarton Walk offers a unique and captivating experience in every season, each one revealing new facets of beauty and wonder in this enchanting corner of East Anglia:

- Spring: As the land awakens from its winter slumber, the fields and hedgerows come alive with the vibrant colours and sweet fragrances of wildflowers, from the sunny yellows of primroses and cowslips to the delicate pinks of cuckoo flowers and the regal purples of bluebells. The air is filled with the joyous songs of returning birds, and the fresh green leaves of the trees create a lush canopy overhead, a symbol of the renewal and hope that spring brings.
- Summer: In the warm, sun-drenched days of summer, the Suffolk countryside is at its most vibrant and alive. The fields are a patchwork of golden hues as the crops ripen, and the hedgerows are adorned with the delicate blooms of dog roses and honeysuckle. The riverbanks are a hive of activity, with dragonflies and damselflies darting through the air, and the gentle buzzing of bees and other insects creating a soothing background hum. This is a time to bask in the fullness of life, to savour the long days and the gentle pace of the countryside.
- Autumn: As the leaves begin to turn, the landscape is transformed into a breathtaking tapestry of rich, warm colours, from the fiery reds and oranges of the oaks and beeches to the soft golds of the fields and hedgerows. The crisp, clean air and the gentle crunch of fallen leaves underfoot create a sense of coziness and introspection, a time to reflect on the cycles of life and the beauty of change. This is a season of abundance and gratitude, a chance to savour the fruits of the land and to appreciate the simple pleasures of a walk in the countryside.
- Winter: While the fields may be dormant and the trees may be bare, the Harkstead and Erwarton Walk takes on a special kind of beauty in the winter months. The frosty mornings and the low, pale light create a sense of stillness and serenity, a time for quiet reflection and solitude. The stark outlines of the trees and hedgerows create a dramatic contrast against the wide, open skies, and the occasional sighting of a flock of overwintering birds adds a spark of life and colour to the muted landscape. This is a time to embrace the restorative power of nature, to find joy in the simple things, and to appreciate the enduring beauty of the Suffolk countryside.

Difficulty Level and Safety Considerations:

The Harkstead and Erwarton Walk is a moderate route, with mostly flat terrain and well-maintained paths. However, as with any outdoor activity, it's important to be prepared and to take necessary precautions to ensure a safe and enjoyable experience.

As you make your way through the countryside, be sure to stay on the designated footpaths and respect the land and the wildlife that calls it home. Remember to close any gates behind you and keep dogs on a lead, particularly when passing through fields with livestock. This not only ensures the safety of the animals but also helps to maintain good relationships with local farmers and landowners.

It's also important to be aware of the weather conditions and to dress appropriately for the season. In the summer months, be sure to wear lightweight, breathable clothing and to bring plenty of water and sun protection to stay hydrated and comfortable. In the cooler months, layers are key, as temperatures can vary throughout the day. A warm, waterproof jacket and sturdy boots are essential for staying dry and comfortable in case of rain or muddy conditions.

When walking along the country lanes, be mindful of traffic and stay alert for passing cars or farm vehicles. While these roads are generally quiet, it's always a good idea to stay visible and to step aside to let vehicles pass safely.

Finally, it's always a good idea to let someone know your planned route and expected return time before setting out and to carry a charged mobile phone in case of emergencies. By following these simple guidelines and using common sense, you can fully immerse yourself in the beauty and tranquility of the Harkstead and Erwarton Walk, creating lasting memories and a deeper connection with the natural world.

Nearby Attractions:

After your invigorating walk through the picturesque countryside of Harkstead and Erwarton, why not extend your adventure and explore some of the other fascinating attractions that this beautiful corner of East Anglia has to offer? Here are a few nearby destinations that are sure to capture your imagination:

- Pin Mill: Just a short drive from Harkstead, the charming hamlet of Pin Mill is a must-visit for anyone who loves the water. This picturesque spot on the banks of the River Orwell has a rich maritime history, with its traditional Thames sailing barges and its atmospheric 17th-century pub, The Butt and Oyster. Take a stroll along the river, watch the boats come and go, and soak up the timeless atmosphere of this enchanting place.
- Sutton Hoo: For a fascinating glimpse into the region's ancient past, head to the world-famous archaeological site of Sutton Hoo, just a short drive from Erwarton. This incredible site, which dates back to the early 7th century, is home to a series of burial mounds that have yielded some of the most spectacular Anglo-Saxon treasures ever discovered in Britain. Visit the exhibition hall to see replicas of the famous Sutton Hoo helmet and other priceless artifacts, and take a walk around the beautiful grounds to imagine the lives of the people who once called this place home.
- Jimmy's Farm and Wildlife Park: If you're looking for a fun and educational day out for the whole family, look no further than Jimmy's Farm and Wildlife Park. This popular attraction, which is just a short drive from Harkstead, offers a unique blend of farm animals, exotic creatures, and beautiful gardens. Meet the friendly pigs, sheep, and goats, marvel at the majestic birds of prey, and explore the nature trails and adventure play areas. With its delicious on-site restaurant and its packed program of events and activities, Jimmy's Farm is the perfect destination for a memorable day out in the Suffolk countryside.

Whether you're a history buff, a nature lover, or simply someone who appreciates the beauty and tranquility of rural England, the Harkstead and Erwarton Walk and its surrounding attractions offer a wealth of opportunities for discovery, relaxation, and inspiration. So why not make a day of it, or even a weekend, and immerse yourself in the rich culture, stunning landscapes, and warm hospitality of this very special corner of East Anglia? You're sure to leave with memories that will last a lifetime and a deeper appreciation for the simple joys and timeless beauty of the English countryside.

Tattingstone Wonder Walk - 3 miles (4.8 km), Circular - Estimated Time: 1.5-2 hours

Starting Point Coordinates, Postcode: The starting point for the Tattingstone Wonder Walk is at St. Mary's Church, Tattingstone (Grid Reference: TM 135 371, Postcode: IP9 2NA).
Nearest Car Park, Postcode: There is limited on-street parking available near St. Mary's Church (Postcode: IP9 2NA).

Embark on an enchanting 3-mile (4.8 km) circular walk through the picturesque Suffolk countryside, where you'll uncover the captivating Tattingstone Wonder, a unique architectural folly that adds an air of intrigue and whimsy to the landscape. This easy walk offers a perfect blend of history, natural beauty, and the chance to immerse yourself in the tranquil atmosphere of rural England.

As you set out from the historic St. Mary's Church in Tattingstone, the peaceful charm of the village envelops you, the gentle rustling of leaves and the distant birdsong creating a soothing backdrop to your journey. Follow Church Road westward, the path beckoning you away from the hustle and bustle of modern life and into the heart of the Suffolk countryside. With each step, a tapestry of gently rolling hills, lush green pastures, and ancient hedgerows unfolds before you, a testament to the enduring beauty and vitality of this remarkable landscape.

Turning onto the footpath, you'll find yourself immersed in a world of natural wonders, the soft contours of the land creating an ever-changing tableau that invites contemplation and wonder. Navigate through the fields, taking a moment to appreciate the simple beauty of the wildflowers swaying in the breeze, their delicate petals and vibrant colours a reminder of the resilience and charm of the natural world.

As you approach Tattingstone Wonder, the striking silhouette of this curious structure comes into view, its unusual design and mysterious purpose capturing your imagination. Built in the 18th century by local squire Edward White, this architectural folly resembles a church tower, its presence adding an element of eccentricity and surprise to the otherwise serene countryside. Pause to explore this fascinating landmark, pondering the stories and motivations behind its creation, and let your mind wander to the lives and aspirations of those who once called this place home.

Continuing along the path, you'll find yourself immersed in the timeless beauty of the Suffolk landscape, the vast expanse of arable fields stretching out before you like a patchwork quilt of earthy hues. The gentle whisper of the wind through the crops and the distant calls of farmland birds create a symphony of rural life, reminding you of the deep connection between human activity and the natural world.

As you turn eastward, the path guides you through a series of picturesque hedgerows, their gnarled branches and dense foliage providing shelter and sustenance for a diverse array of wildlife. Watch for the quick darting movements of rabbits and hares, their presence a charming reminder of the abundant life that thrives in these quiet corners of the countryside.

The final leg of your journey takes you back towards Tattingstone, the village church once again coming into view, its ancient stones and peaceful surroundings a welcome sight after your invigorating walk. As you reflect on the beauty and tranquility of the landscape you've just explored, you'll carry with you a sense of connection to the rich history and enduring spirit of rural Suffolk.

The Tattingstone Wonder Walk is a celebration of the simple joys and hidden treasures that can be found when we step off the beaten path and open our hearts to the wonders of the natural world. Whether you're a history enthusiast, a nature lover, or simply someone in search of a moment of peace and reflection, this enchanting circular route promises to delight and inspire, leaving you with memories that will last a lifetime.

Detailed Route Directions:

1. From St. Mary's Church, head west along Church Road, bearing 280°. (Estimated time: 5-10 minutes)

2. Turn right onto a footpath, bearing 350°, that leads across fields and past Tattingstone Wonder. (Estimated time: 15-20 minutes)

3. Continue north along the path, bearing 010°, turning right at a T-junction onto another footpath. (Estimated time: 20-25 minutes)

4. Follow this path eastward, bearing 080°, eventually turning south and continuing along a footpath that leads back to Church Road. (Estimated time: 25-30 minutes)

5. Turn right onto Church Road, bearing 190°, and return to the starting point at St. Mary's Church. (Estimated time: 10-15 minutes)

Best Features and Views:

- Discover the intriguing Tattingstone Wonder, a unique architectural folly that adds a touch of whimsy and mystery to the landscape, its unusual design and enigmatic purpose captivating the imagination and inviting exploration.
- Immerse yourself in the picturesque Suffolk countryside, with its gently rolling hills, lush green pastures, and vast expanses of arable fields, a living tapestry that showcases the timeless beauty and tranquility of rural England.
- Delight in the lovely views throughout the walk, from the charming wildflowers that adorn the fields to the distant horizons that seem to stretch out forever, each vista a reminder of the incredible diversity and resilience of the natural world.

Historical Features of Interest:

As you approach Tattingstone Wonder, take a moment to appreciate the fascinating history and eccentric character of this remarkable structure. Built in the 18th century by local squire Edward White, this architectural folly stands as a testament to the whimsical spirit and creative vision of a bygone era.

Its striking resemblance to a church tower adds an element of intrigue and surprise to the landscape, inviting you to ponder the stories and motivations behind its creation. Imagine the conversations and laughter that must have echoed through these fields when the folly was first unveiled, a playful addition to the timeless beauty of the Suffolk countryside.

The presence of Tattingstone Wonder serves as a reminder of the enduring human impulse to leave a mark on the landscape, to create something that captures the imagination and sparks curiosity in those who encounter it. As you explore this unique landmark, consider the ways in which it reflects the values, aspirations, and eccentricities of the society that built it, a window into the rich tapestry of history that underlies this enchanting corner of England.

In the peaceful surroundings of the fields and hedgerows, with the gentle rustling of leaves and the distant birdsong as your backdrop, you may find yourself feeling a deep sense of connection to the past, a recognition of the enduring spirit and resilience that have shaped this land and its people for generations. This is the true gift of Tattingstone Wonder, the opportunity to step out of the present moment and into a timeless realm of wonder, curiosity, and reflection.

Birds to Spot:

- Skylarks - As you walk through the open fields, listen for the joyous and unbroken streams of melody that announce the presence of these incredible songbirds. Skylarks are known for their ability to sing while hovering high above the ground, their music filling the air with a sense of exultation and freedom. Scan the skies for a glimpse of these small brown birds, their songs a celebration of the beauty and vitality of the arable landscape.
- Yellowhammers - These vibrant buntings are a delight to spot, their bright yellow heads and distinctive "a little bit of bread and no cheese" song bringing a splash of colour and charm to the hedgerows and field margins. Keep an eye out for their striking plumage and lively presence, a reminder of the rich diversity of birdlife that thrives in the Suffolk countryside.
- Linnets - Watch for the lively flocks of these small, streaky-brown finches as they forage among the stubble fields and hedgerows, their twittering calls and acrobatic flight adding a spark of energy and movement to the tranquil surroundings. These sociable birds are a characteristic sight in the arable landscapes of East Anglia, their presence a testament to the enduring connection between farming and wildlife.

Rare or Interesting Insects and Other Wildlife:

- Rabbits and Hares - As you navigate the field margins and pass through the ancient hedgerows, keep an eye out for the quick, bounding movements of these enchanting creatures. Rabbits and hares are a common sight in the Suffolk countryside, their presence a charming reminder of the abundant life that thrives in these quiet corners of the landscape. Watch for their characteristic ears and fluffy tails, and delight in the playful antics of these endearing mammals.
- Butterflies - In the warmer months, the fields and hedgerows come alive with the delicate flutter of butterfly wings, their colourful patterns and graceful flight adding a touch of magic to the walk. From the vibrant orange and brown of the comma to the intricate marbling of the marbled white, each species is a unique masterpiece of nature's artistry. Take a moment to appreciate their ephemeral beauty, a reminder of the fleeting wonders that surround us.
- Bees and Other Pollinators - As you walk through the wildflower-studded meadows and along the buzzing hedgerows, pay attention to the industrious activities of bees, hoverflies, and other pollinating insects. These tiny creatures play a vital role in maintaining the health and diversity of the ecosystem, their tireless work ensuring the continuation of life in all its forms. Listen for their gentle humming and watch for their purposeful movements, a testament to the intricate web of connections that sustains the natural world.

Rare or Interesting Plants:

- Arable Flora - Take a moment to appreciate the rich tapestry of colours and textures that make up the arable fields you pass, from the waving golden wheat to the vibrant red of poppies and the delicate blue of cornflowers. These wildflowers, once a common sight in agricultural landscapes, are now increasingly rare, their presence a reminder of the delicate balance between human activity and the natural world. Look for the subtle beauty of these humble plants, and consider the ways in which they contribute to the biodiversity and resilience of the countryside.
- Mixed Hedgerows - Marvel at the intricate web of branches, leaves, and flowers that make up the ancient hedgerows along your route, these living boundaries providing food, shelter, and corridors for a wide range of

wildlife. From the frothy white of hawthorn blossom in spring to the rich reds and purples of blackberries and sloes in autumn, each season brings its own unique palette of colours and flavours to these linear woodlands. Take the time to explore the hidden treasures of the hedgerows, and appreciate the vital role they play in the ecosystem of the Suffolk countryside.
- Wildflower Meadows - If your walk takes you past any areas of unimproved grassland or traditional hay meadows, keep an eye out for the stunning diversity of wildflowers that thrive in these precious habitats. From the nodding heads of cowslips and the sunshine-yellow of buttercups to the delicate pinks of ragged robin and the regal purples of orchids, each bloom is a tiny miracle of adaptation and survival. Take a moment to drink in the beauty and fragrance of these floral gems, and reflect on the importance of preserving these increasingly rare and vulnerable ecosystems for future generations.

Facilities and Accessibility:

The Tattingstone Wonder Walk is a relatively easy route, with mostly flat terrain and well-maintained paths. However, it's important to note that some sections may be uneven, muddy, or overgrown, especially after periods of heavy rain or during the growing season. Sturdy walking shoes or boots with good grip are highly recommended to ensure a safe and comfortable experience.

While there are no public toilets or other facilities directly on the walk route, the nearby village of Tattingstone does offer some amenities for visitors, including a charming pub, The White Horse, where you can enjoy a refreshing drink or a hearty meal before or after your walk.

It's important to note that the walk may not be suitable for those with limited mobility or wheelchair users, as some sections of the path may include stiles or kissing gates. As always, it's a good idea to check the local conditions and plan accordingly, bringing water, snacks, and any necessary medications or sun protection to ensure a safe and enjoyable experience.

Seasonal Highlights:

The Tattingstone Wonder Walk offers a unique and captivating experience in every season, each one revealing new facets of beauty and wonder in this enchanting corner of Suffolk:

- Spring: Witness the countryside bursting into life, with vibrant wildflowers painting the fields and hedgerows in a rainbow of colour, from the delicate pinks of cuckooflower to the sunshine-yellow of primroses and the regal purples of early purple orchids. The air is filled with the joyous songs of returning birds, and the fresh green leaves of the trees create a lush canopy overhead, a symbol of the renewal and hope that spring brings.
- Summer: Bask in the warm sunshine and the heady scents of high summer, as you stroll through the gently swaying fields of ripening wheat, barley, and oats, the golden hues of the crops creating a mesmerizing tapestry that stretches to the horizon. The hedgerows are alive with the buzzing of insects and the fluttering of butterflies, while the distant hum of farm machinery and the lazy buzzing of bees create a soothing soundtrack that celebrates the abundance and vitality of the season.
- Autumn: Marvel at the stunning transformation of the landscape as the leaves turn, painting the hedgerows and trees in a breathtaking palette of gold, orange, and red. The crisp, clean air and the gentle crunch of fallen leaves underfoot create a sense of coziness and introspection, a time to reflect on the cycles of life and the beauty of change. Watch for the busy antics of wildlife as they prepare for the winter ahead, from the industrious foraging of squirrels to the impressive murmurations of starlings that dance across the evening sky.
- Winter: Embrace the stark beauty of the dormant countryside, with its frosty fields, skeletal trees, and wide, open skies that seem to stretch on forever. The low, golden light of the winter sun casts long shadows across the landscape, creating a sense of drama and mystery that invites quiet contemplation and wonder. Listen for

the haunting cries of overwintering birds, from the lonely piping of a single redwing to the raucous chatter of a flock of fieldfares, and delight in the subtle signs of life that persist even in the depths of the coldest season.

Difficulty Level and Safety Considerations:

The Tattingstone Wonder Walk is a moderate route, with mostly flat terrain and well-maintained paths. However, as with any outdoor activity, it's important to be prepared and to take necessary precautions to ensure a safe and enjoyable experience.

As you make your way through the countryside, be sure to stay on the designated footpaths and respect the land and the wildlife that calls it home. Remember to close any gates behind you and keep dogs on a lead, particularly when passing through fields with livestock or during the ground-nesting bird season.

It's also important to be aware of the weather conditions and to dress appropriately for the season. In the summer months, be sure to wear lightweight, breathable clothing and to bring plenty of water and sun protection to stay hydrated and comfortable. In the cooler months, layers are key, as temperatures can vary throughout the day. A warm, waterproof jacket and sturdy boots are essential for staying dry and comfortable in case of rain or muddy conditions.

When walking along the quiet lanes, be mindful of any traffic and stay alert for passing cars or farm vehicles. While these roads are generally peaceful, it's always a good idea to stay visible and to step aside to let vehicles pass safely.

Finally, it's always a good idea to let someone know your planned route and expected return time before setting out, and to carry a charged mobile phone in case of emergencies. By following these simple guidelines and using common sense, you can fully immerse yourself in the beauty and tranquility of the Tattingstone Wonder Walk, creating lasting memories and a deeper connection with the natural world.

Nearby Attractions:

After your invigorating walk through the picturesque countryside surrounding Tattingstone, why not extend your adventure and explore some of the other fascinating attractions that this beautiful corner of Suffolk has to offer? Here are a few nearby destinations that are sure to capture your imagination:

- Alton Water Park: Just a short drive from Tattingstone, Alton Water Park is a stunning 400-acre conservation area centred around a large reservoir. With its diverse range of habitats, including grasslands, woodlands, and wetlands, the park offers excellent opportunities for birdwatching, fishing, cycling, and water sports, making it a perfect destination for a full day of outdoor adventures.
- Flatford Mill: Step into the world of John Constable, one of England's most beloved landscape painters, with a visit to the charming hamlet of Flatford. This picturesque spot, nestled in the heart of the Dedham Vale AONB, was a frequent subject of Constable's paintings, and it's easy to see why. With its timeless beauty, historic buildings, and tranquil river scenes, Flatford is a living testament to the enduring appeal of the English countryside.
- Jimmy's Farm and Wildlife Park: If you're looking for a fun and educational day out for the whole family, head to Jimmy's Farm and Wildlife Park, just a short drive from Tattingstone. This popular attraction offers a unique blend of farm animals, exotic creatures, and beautiful gardens, with plenty of opportunities for up-close encounters and hands-on experiences. With its delicious on-site restaurant and packed program of events and activities, Jimmy's Farm is the perfect destination for a memorable day out in the Suffolk countryside.

By combining your Tattingstone Wonder Walk with a visit to these captivating nearby attractions, you'll create an unforgettable itinerary that showcases the very best of Suffolk's natural beauty, rich history, and vibrant culture. Whether you're a nature enthusiast, a history buff, or simply seeking a memorable escape from the everyday, the Tattingstone area promises to delight and inspire you at every turn.

As you explore this enchanting corner of East Anglia, take a moment to reflect on the incredible diversity and resilience of the landscapes and communities that have shaped this region for centuries. From the ancient hedgerows and wildflower meadows to the historic churches and charming villages, each element of the countryside tells a story of the people, plants, and animals that have called this place home, weaving a rich tapestry of natural and cultural heritage that is truly unparalleled.

So why not make a day of it, or even a weekend, and immerse yourself in the timeless beauty and enduring spirit of the Suffolk countryside? With its stunning scenery, fascinating history, and warm, welcoming atmosphere, this special part of England is sure to leave you with memories that will last a lifetime, and a deeper appreciation for the simple joys and profound wonders that can be found when we step off the beaten path and open our hearts to the world around us.

Whether you're a seasoned rambler or a curious first-time visitor, the Tattingstone Wonder Walk and its surrounding attractions offer a wealth of opportunities for discovery, relaxation, and personal growth. So lace up your walking boots, pack a picnic, and set out on an adventure that will nourish your body, mind, and soul, leaving you with a renewed sense of connection to the land, to yourself, and to the boundless possibilities of the great outdoors.

Alton Water Walk - 8 miles (12.9 km), Circular - Estimated Time: 4-5 hours

Starting Point Coordinates, Postcode: The starting point for the Alton Water Walk is at the Alton Water Visitor Centre (Grid Reference: TM 163 364, Postcode: IP9 2RY).
Nearest Car Park, Postcode: There is a car park at the Alton Water Visitor Centre (Postcode: IP9 2RY).

Embark on an invigorating 8-mile (12.9 km) circular walk around the stunning Alton Water reservoir, a man-made wonder nestled in the heart of the Suffolk countryside. This diverse and engaging route takes you through a captivating mix of woodland, meadows, and along the picturesque shoreline, offering a truly immersive experience for nature lovers and outdoor enthusiasts alike.

As you set out from the Alton Water Visitor Centre, the well-signposted footpath beckons you along the reservoir's edge, the anticipation building with each step as you catch glimpses of the glistening water through the trees. The gentle lapping of the waves against the shore and the distant calls of waterfowl create a soothing soundtrack to your journey, inviting you to leave the stresses of everyday life behind and immerse yourself in the tranquility of the surrounding landscape.

Venturing into the lush woodlands that fringe the reservoir, you'll find yourself enveloped in a world of dappled sunlight and the soft rustling of leaves overhead. The earthy scent of the forest floor and the gentle creaking of the trees in the breeze create a sensory experience that reconnects you with the rhythms of the natural world. Keep your eyes peeled for the darting movements of woodland birds and the occasional flash of a bushy tail as squirrels scamper through the undergrowth.

Emerging from the cool embrace of the forest, the path guides you into the open expanse of wildflower-studded meadows, their vibrant colours and delicate blooms a testament to the incredible diversity of the local flora. The gentle hum of bees and the fluttering of butterflies add to the enchanting atmosphere, while the distant views of the reservoir and the surrounding countryside create a stunning backdrop that seems to stretch out forever.

As you continue along the route, you'll encounter several bridges that span the smaller inlets and streams feeding into the reservoir, each one offering a unique perspective on the water's edge and the wildlife that thrives in these sheltered corners. Pause for a moment to appreciate the reflections of the sky and the trees in the still surface of the water, a mirror that reveals the hidden depths and the ever-changing moods of the landscape.

The bird-watching hides scattered along the trail provide the perfect opportunity to observe the reservoir's feathered residents up close, from the majestic great crested grebes gliding across the water to the playful antics of the tufted ducks diving for food. Take a moment to sit quietly and let the sights and sounds of the birdlife wash over you, a reminder of the incredible diversity and resilience of the natural world.

As you make your way back towards the Visitor Centre, the sense of accomplishment and renewal that comes from completing this incredible walk fills your heart. The Alton Water Walk is more than just a physical journey; it is a chance to reconnect with the beauty and simplicity of the great outdoors, to find solace and inspiration in the timeless rhythms of the Suffolk countryside, and to create memories that will last a lifetime.

Detailed Route Directions:

1. From the Alton Water Visitor Centre, follow the signed footpath along the reservoir shoreline. (Estimated time: 30-45 minutes)

2. Continue along the path as it winds through woodlands and meadows, immersing yourself in the diverse habitats and the stunning scenery. (Estimated time: 1-1.5 hours)

3. Cross several bridges that span the smaller inlets and streams feeding into the reservoir, each one offering a unique perspective on the water's edge and the wildlife that thrives in these sheltered corners. (Estimated time: 30-45 minutes)

4. Pass by various bird-watching hides, taking the opportunity to observe the reservoir's feathered residents up close and appreciate the incredible diversity of the local birdlife. (Estimated time: 30-45 minutes)

5. Continue following the well-signposted route, with markers and occasional information boards guiding your way. (Estimated time: 1-1.5 hours)

6. After completing the circuit, you will arrive back at the Alton Water Visitor Centre, where your journey comes to a satisfying conclusion. (Estimated time: 30-45 minutes)

Best Features and Views:

- Immerse yourself in the stunning views of the Alton Water reservoir and the surrounding countryside, the glistening expanse of water and the patchwork of fields and woodlands creating a breathtaking panorama that invites contemplation and wonder.
- Discover the diverse habitats that fringe the reservoir, from the lush woodlands with their dappled sunlight and gentle rustling of leaves to the wildflower-studded meadows that burst with colour and life, each one a microcosm of the larger landscape waiting to be explored.
- Delight in the picturesque shoreline of the reservoir, the gentle lapping of the waves and the reflections of the sky and trees in the still surface of the water creating a mesmerizing scene that soothes the soul and invigorates the senses.

Historical Features of Interest:

While there are no significant historical sites directly along the walk, the Alton Water reservoir itself stands as a fascinating example of modern water engineering and the ingenuity of human endeavour. Constructed in the 1970s to meet the growing demand for water in the region, the reservoir is a testament to the delicate balance between the needs of society and the preservation of the natural environment. As you walk along its shores and through the surrounding landscapes, take a moment to reflect on the incredible feat of engineering that brought this vast body of water into being, and the countless lives and livelihoods that depend on its continued presence in the heart of the Suffolk countryside.

Birds to Spot:

- Elegant waterbirds: Great crested grebes: Watch for these majestic birds gliding across the surface of the reservoir, their striking black and white plumage and regal crests a breathtaking sight against the backdrop of the water and the sky.
- Diving dabblers: Tufted ducks: Delight in the playful antics of these charming ducks as they dive beneath the

surface in search of food, their distinctive tufted heads and contrasting black and white plumage a lively addition to the reservoir's avian community.
- Aerial acrobats: Various species of terns: Keep your eyes peeled for the graceful flight and precise diving skills of the different tern species that frequent the reservoir, from the common tern to the rarer black tern, each one a master of the skies and the water.

Rare or Interesting Insects and Other Wildlife:

- Jewelled aviators: Dragonflies and damselflies: In the summer months, marvel at the iridescent colours and acrobatic flight of these ancient insects as they dart and hover over the water's edge and the nearby vegetation, their presence a vibrant reminder of the reservoir's thriving ecosystem.
- Bounding bunnies: Rabbits: As you walk through the meadows and along the woodland edges, watch for the quick, darting movements of rabbits as they forage for food and play in the grass, their presence a charming addition to the pastoral landscape.
- Bushy-tailed foragers: Squirrels: Keep an eye out for the lively antics of squirrels as they scamper through the branches and along the forest floor, their russet coats and fluffy tails a delightful sight amidst the green and brown hues of the woodland.
- Majestic mammals: Occasional deer: If fortune favours you, you may catch a fleeting glimpse of a deer as it moves through the dappled shade of the forest or grazes in the open meadows, its graceful form and gentle presence a reminder of the untamed beauty that still exists in the Suffolk countryside.

Rare or Interesting Plants:

- Meadow marvels: Wildflowers: In the spring and summer, lose yourself in the vibrant tapestry of colours that adorns the meadows along the walk, from the cheerful faces of ox-eye daisies and buttercups to the delicate hues of cuckoo flowers and red clover, each bloom a tiny miracle of resilience and beauty.
- Woodland wonders: Bluebells and wood anemones: As you walk through the dappled shade of the forest, keep an eye out for the enchanting carpets of bluebells that appear in the spring, their delicate, nodding heads and sweet fragrance a true delight for the senses. Alongside them, the starry white blooms of wood anemones add a touch of ethereal magic to the forest floor.
- Shoreline specialists: Reeds and rushes: Along the water's edge, discover the towering stands of reeds and rushes that sway in the breeze, their slender stems and feathery plumes creating a natural tapestry that provides shelter and sustenance for a wide variety of wildlife, from nesting birds to tiny invertebrates.

Facilities and Accessibility:

The Alton Water Visitor Centre offers a range of facilities to ensure a comfortable and enjoyable experience for walkers, including well-maintained public toilets, a cozy café serving refreshments and light meals, and a small shop stocked with essential items and souvenirs. The picnic areas and children's playground provide the perfect spot for a relaxing break or a family-friendly adventure, while the knowledgeable staff at the Visitor Centre are always on hand to offer advice and information about the local area and the reservoir's rich ecosystem.

However, it's important to note that the Alton Water Walk itself is not suitable for wheelchairs or strollers due to the uneven terrain, narrow paths, and occasional steep inclines. To ensure a safe and comfortable walk, it is highly recommended to wear sturdy hiking shoes or boots with good grip and support, as well as weather-appropriate clothing that allows for ease of movement and protection from the elements.

Seasonal Highlights:

The Alton Water Walk offers a unique and captivating experience in every season, each one revealing new facets of beauty and wonder in this stunning corner of Suffolk:

- Spring: Witness the countryside bursting into life, with vibrant wildflowers painting the meadows in a rainbow of colours, fresh green leaves unfurling on the trees, and the joyous birdsong of the breeding season filling the air with the melodies of renewal and rebirth.
- Summer: Bask in the warm sunshine and the heady scents of high summer, as you stroll along the reservoir's edge and through the sun-dappled woodlands, the buzzing of insects and the gentle splashing of waterfowl creating a sensory symphony that celebrates the fullness and vitality of life.
- Autumn: Marvel at the breathtaking array of colours as the leaves turn, transforming the woodlands into a tapestry of gold, orange, and red that seems to set the landscape ablaze with the fiery hues of the season. Embrace the crisp, invigorating air and the satisfying crunch of fallen leaves underfoot, as you savour the poignant beauty and sense of change that autumn brings.
- Winter: Discover the tranquil beauty of the frosty shoreline, the bare branches of the trees etched against the pale winter sky, and the haunting cries of overwintering birds echoing across the still water, a time of quiet reflection and the chance to appreciate the stark, elemental power of the natural world in its winter slumber.

Difficulty Level and Safety Considerations:

The Alton Water Walk is classified as a moderate difficulty trail, with some gentle inclines, uneven terrain, and occasional steep sections that may prove challenging for less experienced walkers. The 8-mile distance also requires a good level of fitness and stamina, so it's important to pace yourself and take regular breaks as needed.

As you make your way around the reservoir, be sure to stay on the designated paths and be cautious when walking near the water's edge, as the ground can be slippery and the water can be deceptively deep. It's also crucial to respect the wildlife and the habitats that thrive along the trail, observing any signs or instructions regarding the protection of sensitive areas or nesting sites.

To ensure a safe and enjoyable walk, be sure to wear sturdy, comfortable footwear with good grip and support, as well as clothing suitable for the season and the weather conditions. Carry an ample supply of water and energy-rich snacks to keep you hydrated and fuelled throughout your journey, and don't forget to pack any necessary medications, sunscreen, or insect repellent, depending on the time of year.

By following these simple precautions and the principles of responsible outdoor recreation, you can fully immerse yourself in the beauty and wonder of the Alton Water Walk, forging a deep and lasting connection with the natural world and creating memories that will endure long after you've left the shores of this enchanting reservoir behind.

Nearby Attractions:

After your invigorating walk around Alton Water, take the opportunity to explore some of the other fascinating attractions that the local area has to offer, each one providing a unique glimpse into the rich history, vibrant culture, and stunning natural beauty of the Suffolk countryside:

- Hadleigh: Step back in time as you visit the charming market town of Hadleigh, with its picturesque streets lined with carefully preserved timber-framed buildings and the imposing grandeur of its medieval church. Explore the town's fascinating history at the local museum, sample the delicious local produce at the traditional shops and markets, and soak up the timeless atmosphere of this quintessential Suffolk community.

- Dedham Vale AONB: Venture into the heart of Constable Country, the breathtaking landscape that inspired one of England's most beloved painters, John Constable. Marvel at the gentle beauty of the Dedham Vale Area of Outstanding Natural Beauty, with its rolling hills, meandering rivers, and picturesque villages that seem to have stepped straight out of a painting. Walk in the footsteps of the great artist himself, discovering the scenes that he immortalized on canvas and gaining a deeper appreciation for the enduring allure of the English countryside.

- Jimmy's Farm and Wildlife Park: Embark on a fun-filled family day out at Jimmy's Farm and Wildlife Park, a working farm and wildlife sanctuary that offers a unique and hands-on experience for visitors of all ages. Meet the friendly animals, from the rare breeds of pigs and sheep to the exotic creatures of the tropical house, and learn about the importance of sustainable farming and conservation. With its scenic nature trails, engaging educational exhibits, and delicious on-site restaurant, Jimmy's Farm is a must-visit attraction for anyone looking to connect with the natural world and support the vital work of local farmers and conservationists.

By exploring these nearby attractions, you'll gain a deeper appreciation for the diverse and captivating character of the Suffolk countryside, making your Alton Water Walk the perfect starting point for a truly unforgettable journey through this enchanting corner of East Anglia. Whether you're a keen walker, a nature lover, or simply someone who appreciates the beauty and tranquility of the great outdoors, this area promises to delight, inspire, and leave you with memories that will last a lifetime.

East Bergholt Circular Walk - 6 miles (9.6 km), Circular - Estimated Time: 3-4 hours

Starting Point Coordinates, Postcode: The starting point for the East Bergholt Circular Walk is at St. Mary's Church, East Bergholt (Grid Reference: TM 070 344, Postcode: CO7 6TG).
Nearest Car Park, Postcode: On-street parking is available near St. Mary's Church in East Bergholt (Postcode: CO7 6TG).

Embark on an enchanting 6-mile (9.6 km) circular walk through the picturesque countryside surrounding the charming village of East Bergholt, a landscape that has captivated artists and nature lovers for centuries. This walk, steeped in history and natural beauty, offers a perfect escape from the hustle and bustle of everyday life, inviting you to immerse yourself in the tranquil scenery that inspired the renowned painter John Constable.

As you set out from the historic St. Mary's Church, with its unique bell cage, the footpath beckons you westward, leading you through a tapestry of lush meadows and ancient woodlands. With each step, you'll find yourself transported to a world of timeless beauty, where the gentle rustling of leaves and the distant birdsong create a soothing symphony that accompanies your journey.

Continuing along the path, you'll soon reach the former site of Old Hall, once a magnificent Tudor mansion that stood as a testament to the area's rich history. Though the hall itself is long gone, its presence can still be felt in the atmospheric surroundings, inviting you to imagine the lives and stories of those who once called this place home.

As you make your way through the meadows and woodlands, the landscape reveals its ever-changing charms. In the spring and summer, the fields come alive with a vibrant tapestry of wildflowers, their delicate blooms and sweet fragrances a delight for the senses. The autumn months paint the woods in a breathtaking array of golden hues, the crisp air and the soft crunch of fallen leaves underfoot creating an enchanting atmosphere that beckons you to linger and explore.

Upon reaching Flatford Mill, you'll find yourself standing in the very landscape that inspired John Constable's most iconic paintings. The picturesque mill and the gently flowing River Stour create a scene of idyllic beauty, the timeless charm of the English countryside captured in every brush stroke. Take a moment to appreciate the tranquil surroundings, imagining the artist at work, his keen eye and skilled hand translating the essence of this remarkable place onto canvas.

Crossing the River Stour at the famous Bridge Cottage, you'll embark on a delightful riverside path that leads you southward along the banks of the river. Here, the landscape opens up, revealing sweeping views of the water and the lush green fields beyond. Keep your eyes peeled for the flash of a kingfisher's vivid plumage or the graceful silhouettes of swans gliding along the river's surface, each encounter a reminder of the rich biodiversity that thrives in this unique habitat.

As you turn eastward and make your way along the valley, the path offers a chance to disconnect from the distractions of modern life and reconnect with the simple pleasures of the natural world. The rhythmic flowing of the river and the gentle breeze carrying the scent of the countryside create a sensory experience that soothes the soul and invigorates the spirit.

Passing through the hamlet of Dedham Vale, you'll find yourself immersed in a landscape that has remained largely unchanged for centuries. The thatched cottages, historic churches, and traditional farms stand as testaments to the enduring spirit of rural England, their weathered stones and timeless charm a reminder of the generations who have lived and worked in this beautiful corner of the country.

The final stretch of your journey takes you northward once more, through a series of tranquil pastures and along quiet country lanes, the pastoral landscape unfolding before you like a living painting. The fields and hedgerows stretch out to the horizon, their colours and textures shifting with the changing light, a testament to the enduring beauty and resilience of the East Anglian countryside.

As you return to the heart of East Bergholt, the iconic bell cage of St. Mary's Church comes into view once more, signalling the end of your remarkable journey. With a sense of accomplishment and renewal, you'll carry the memories of your walk with you, the sights, sounds, and sensations of this enchanting landscape forever etched in your mind and heart.

The East Bergholt Circular Walk is a true celebration of the natural and cultural heritage of this extraordinary corner of England. It is an invitation to slow down, to observe, and to marvel at the incredible beauty that lies just beyond our doorsteps. Whether you're a lover of art, a history enthusiast, or simply someone who appreciates the restorative power of a walk in the countryside, this route promises to delight, inspire, and leave you with a profound sense of connection to the landscape and its enduring spirit.

Detailed Route Directions:

1. From St. Mary's Church, head west along the footpath, bearing 280°, passing the former site of Old Hall. (Estimated time: 20-30 minutes)

2. Continue through meadows and woodlands, bearing 240°, immersing yourself in the tranquil beauty of the countryside. (Estimated time: 30-45 minutes)

3. Reach Flatford Mill, bearing 200°, a iconic landmark featured in John Constable's paintings. Take a moment to appreciate the picturesque surroundings. (Estimated time: 15-20 minutes)

4. Cross the River Stour at Bridge Cottage, bearing 160°, and follow the riverside path south, enjoying the views of the water and the lush green fields. (Estimated time: 30-45 minutes)

5. Turn east, bearing 90°, and walk along the valley, keeping an eye out for kingfishers and swans. (Estimated time: 30-45 minutes)

6. Pass through the hamlet of Dedham Vale, bearing 60°, admiring the historic architecture and timeless charm of rural England. (Estimated time: 20-30 minutes)

7. Turn north, bearing 340°, and walk through tranquil pastures and along quiet country lanes, the bell cage of St. Mary's Church guiding you back to East Bergholt. (Estimated time: 30-45 minutes)

8. Complete the walk at St. Mary's Church, bearing 020°, taking a moment to reflect on the beauty and inspiration you've encountered along the way. (Estimated time: 5-10 minutes)

Best Features and Views:

- Immerse yourself in the picturesque countryside that inspired John Constable, the landscape forever immortalized in his iconic paintings, a living testament to the enduring beauty of the East Anglian landscape.
- Delight in the beautiful views of meadows, woodlands, and riverside paths, each one a unique and

captivating scene that showcases the incredible diversity and charm of the natural world.
- Marvel at the timeless allure of Flatford Mill and Bridge Cottage, landmarks that have captured the hearts and imaginations of artists and visitors alike, their weathered walls and tranquil setting a symbol of the rich cultural heritage of the region.

Historical Features of Interest:

As you set out from St. Mary's Church, take a moment to appreciate the unique bell cage that stands in the churchyard. This remarkable structure, dating back to the 16th century, was built to house the church bells when the tower was deemed too weak to support them. The bell cage stands as a testament to the ingenuity and adaptability of the local community, a symbol of the enduring spirit that has shaped East Bergholt over the centuries.

Continuing along the path, you'll soon reach the former site of Old Hall, a magnificent Tudor mansion that once stood as a symbol of wealth and power in the region. Built in the 16th century by the powerful Daundy family, Old Hall was a grand residence that played host to many notable figures over the years, including Queen Elizabeth I. Though the hall itself was demolished in the 18th century, its legacy lives on in the atmospheric surroundings, inviting you to imagine the lives and stories of those who once called this place home.

Upon reaching Flatford Mill, you'll find yourself standing in the footsteps of John Constable, the celebrated artist whose paintings captured the very essence of this remarkable landscape. The mill, which dates back to the 18th century, was a frequent subject of Constable's work, its picturesque setting and timeless charm providing endless inspiration for the artist. As you gaze upon the weathered walls and the gentle flow of the River Stour, you'll see the world through Constable's eyes, appreciating the incredible skill and vision that allowed him to translate the beauty of this place onto canvas.

Crossing the river at Bridge Cottage, you'll have the opportunity to visit another iconic landmark that features prominently in Constable's paintings. This charming cottage, which dates back to the 16th century, served as a crossing point for the River Stour and was home to the bridge warden who collected tolls from passing river traffic. Today, the cottage stands as a museum and a testament to the rich history and cultural heritage of the Dedham Vale, offering a fascinating glimpse into the lives of the people who have shaped this remarkable landscape over the centuries.

As you make your way through the hamlet of Dedham Vale, take a moment to appreciate the historic architecture and timeless charm of the thatched cottages, traditional farms, and ancient churches that dot the landscape. These buildings, some of which date back hundreds of years, stand as testaments to the enduring spirit of rural England, their weathered stones and rustic beauty a reminder of the generations who have lived and worked in this extraordinary corner of the country.

Birds to Spot:

- Kingfishers: As you walk along the River Stour, keep your eyes peeled for the brilliant flash of blue and orange that signals the presence of a kingfisher. These stunning birds, known for their incredible diving abilities, are a true delight to behold, their vibrant plumage and swift movements a testament to the remarkable beauty and adaptability of nature.
- Swans: The River Stour is home to a thriving population of mute swans, the elegant white birds that have captured the hearts and imaginations of people for centuries. As you stroll along the riverside path, watch for the graceful silhouettes of these majestic creatures gliding along the water's surface, their serene presence a reminder of the tranquility and beauty of the natural world.

- Woodland Birds: As you make your way through the ancient woodlands that line the route, keep an ear out for the melodic songs and lively calls of the many bird species that make their home among the trees. From the cheerful chirping of robins and wrens to the soft cooing of wood pigeons and the lively chattering of finches, each voice contributes to the enchanting symphony of the forest, a celebration of the vibrant life that thrives in these peaceful surroundings.

Rare or Interesting Insects and Other Wildlife:

- Dragonflies: In the summer months, the riverbanks and wetlands along the route come alive with the vibrant colours and acrobatic movements of dragonflies. These incredible insects, with their iridescent wings and striking patterns, are a true marvel of nature, their aerial displays a mesmerizing spectacle that captures the imagination and ignites a sense of wonder.
- Butterflies: The meadows and hedgerows of the East Bergholt Circular Walk are home to a stunning array of butterfly species, each one a delicate work of art that flutters and dances among the wildflowers. From the vibrant orange and black of the peacock butterfly to the delicate blues and silvers of the common blue, these enchanting creatures add a touch of magic and beauty to the landscape, their ephemeral presence a reminder of the fleeting nature of life and the importance of cherishing each moment.
- Otters: While they are elusive and rarely seen, the River Stour is home to a healthy population of otters, the playful and intelligent mammals that have captured the hearts of nature lovers around the world. As you walk along the riverbank, keep an eye out for the tell-tale signs of otter activity, from the webbed footprints in the mud to the sleek, sinuous shapes that occasionally break the surface of the water, a thrilling reminder of the incredible diversity of life that thrives in this remarkable corner of England.

Rare or Interesting Plants:

- Wildflower Meadows: In the spring and summer months, the meadows along the East Bergholt Circular Walk explode into a vibrant tapestry of colour, as countless species of wildflowers burst into bloom. From the cheerful yellows of buttercups and cowslips to the delicate pinks of cuckoo flowers and the regal purples of orchids, each blossom is a tiny miracle of nature, a testament to the incredible resilience and adaptability of the plant kingdom.
- Ancient Woodland Flora: As you walk through the ancient woodlands that line the route, take a moment to appreciate the lush green understory of ferns, mosses, and woodland plants that thrive in the dappled shade of the trees. From the delicate fronds of hart's tongue fern to the glossy leaves of wild garlic and the nodding bells of wood anemone, each species plays a vital role in the complex ecosystem of the forest, a reminder of the intricate web of life that sustains us all.
- Hedgerow Herbs: The hedgerows that crisscross the landscape are not only beautiful to behold but also home to a fascinating array of medicinal and culinary herbs that have been used by people for centuries. As you walk along the field margins, keep an eye out for the delicate white flowers of cow parsley, the pungent leaves of wild garlic, and the cheerful yellow blooms of St. John's wort, each one a reminder of the deep connection between human culture and the natural world.

Facilities and Accessibility:

The East Bergholt Circular Walk offers a range of facilities to ensure a comfortable and enjoyable experience for walkers. Along the route, you'll find benches and picnic areas where you can take a break, enjoy a snack, and soak in the stunning views of the surrounding countryside. At Flatford Mill, approximately halfway through the walk, you'll also find public toilets and a charming tea room, providing a welcome opportunity to refresh and refuel before continuing your journey.

While the walk is relatively easy, with mostly flat terrain and well-maintained paths, it's important to note that some sections may be muddy or uneven, particularly after periods of heavy rain. As such, it's highly recommended to wear sturdy walking shoes or boots with good grip and support to ensure a safe and comfortable experience.

It's also worth noting that some sections of the walk may not be suitable for wheelchairs or pushchairs due to the presence of stiles and kissing gates. However, much of the route is accessible to a wide range of abilities, and with a little planning and preparation, everyone can enjoy the beauty and tranquility of this remarkable landscape.

Seasonal Highlights:

The East Bergholt Circular Walk offers a unique and captivating experience in every season, each one revealing new facets of beauty and wonder in this enchanting corner of England:

- Spring: As the land awakens from its winter slumber, the meadows and woodlands along the route burst into life, with vibrant wildflowers painting the landscape in a kaleidoscope of colour. From the delicate pinks of cuckoo flowers to the cheerful yellows of primroses and the nodding bells of bluebells, each bloom is a celebration of the season's renewal and the promise of new beginnings.
- Summer: In the warm, sun-drenched days of summer, the countryside is at its most lush and verdant, with the gentle greens of the fields and trees creating a soothing backdrop to your walk. The riverbanks are alive with the vibrant colours of dragonflies and the gentle buzzing of bees, while the meadows are a sea of waving grasses and wildflowers, each one a tiny miracle of nature's abundance and resilience.
- Autumn: As the leaves begin to turn, the landscape is transformed into a breathtaking tapestry of gold, orange, and red, the crisp air and the soft crunch of fallen leaves underfoot creating a sense of magic and enchantment. The hedgerows are heavy with the rich hues of ripe berries and the sounds of foraging birds, a reminder of the incredible bounty that nature provides in even the most unassuming of places.
- Winter: While the fields and trees may be bare, the East Bergholt Circular Walk takes on a special kind of beauty in the winter months, with the stark outlines of the landscape etched against the pale, watery light of the sun. The frosty mornings and the misty afternoons create a sense of mystery and tranquility, inviting you to slow down and savour the quiet wonders of the season, from the delicate tracery of ice on a leaf to the distant calls of overwintering birds.

Difficulty Level and Safety Considerations:

The East Bergholt Circular Walk is a relatively easy route, suitable for walkers of most fitness levels. The terrain is mostly flat, with well-maintained paths and clear waymarking throughout. However, as with any outdoor activity, it's important to be prepared and to take necessary precautions to ensure a safe and enjoyable experience.

As you set out on your walk, be sure to wear appropriate footwear and clothing for the season and the weather conditions. Sturdy walking shoes or boots with good grip and support are essential, particularly in wet or muddy conditions. It's also a good idea to bring a waterproof jacket and layers of warm clothing, even in the summer months, as the weather can be changeable and unpredictable.

While the route is generally safe and well-maintained, it's important to stay aware of your surroundings and to follow the countryside code at all times. This means staying on designated paths, closing gates behind you, and respecting the land and the wildlife that calls it home. If you're walking with a dog, be sure to keep it on a lead where necessary and to clean up after it to help preserve the natural beauty of the area.

It's also a good idea to carry a map and compass, or a GPS device, to help you navigate the route and stay on track. While the waymarking is generally clear and easy to follow, having a backup navigation aid can provide peace of mind and help you avoid getting lost or disoriented.

Finally, it's always a good idea to let someone know your planned route and expected return time before setting out, and to carry a fully charged mobile phone in case of emergencies. By taking these simple precautions and using common sense, you can fully immerse yourself in the beauty and tranquility of the East Bergholt Circular Walk, creating lasting memories and a deeper appreciation for the natural world.

Nearby Attractions:

After your invigorating walk, take the opportunity to explore some of the other fascinating attractions that the local area has to offer, each one providing a unique glimpse into the rich history, vibrant culture, and stunning natural beauty of the Dedham Vale and the surrounding region:

1. Flatford Mill (National Trust): Delve deeper into the life and work of John Constable with a visit to Flatford Mill, the iconic landmark that inspired so many of his most famous paintings. Explore the beautifully preserved buildings, including the mill itself, Bridge Cottage, and Willy Lott's House, and learn about the fascinating history of the area through interactive exhibits and guided tours.

2. Dedham Village: Take a stroll through the charming village of Dedham, just a short distance from East Bergholt, and discover the timeless beauty and rich cultural heritage of this quintessential English countryside. Visit the stunning 15th-century church of St. Mary the Virgin, browse the independent shops and galleries that line the high street, and enjoy a delicious meal or a refreshing drink at one of the many pubs and restaurants that serve locally sourced produce.

3. The Munnings Art Museum: For art lovers, a visit to the Munnings Art Museum in Dedham is a must. Housed in the former home and studio of Sir Alfred Munnings, one of England's most celebrated equestrian artists, the museum showcases a stunning collection of his paintings, drawings, and sculptures, as well as works by other notable artists of the period. The beautiful grounds and gardens surrounding the museum are also well worth a visit, offering a tranquil retreat from the hustle and bustle of modern life.

4. The Beth Chatto Gardens: Just a short drive from East Bergholt, the Beth Chatto Gardens are a horticultural paradise, showcasing an incredible diversity of plants from around the world. Created by the renowned plantswoman Beth Chatto, the gardens are a testament to her lifelong passion for plants and her pioneering approach to ecological gardening. With its stunning displays of flowers, foliage, and texture, the Beth Chatto Gardens are a must-visit for anyone with an interest in horticulture, nature, or simply the beauty of the natural world.

By combining your East Bergholt Circular Walk with visits to these and other nearby attractions, you'll create an unforgettable itinerary that showcases the very best of the Dedham Vale and beyond. Whether you're a lover of art, history, nature, or simply the great outdoors, this enchanting corner of England promises to delight and inspire you at every turn, leaving you with memories that will last a lifetime.

Ipswich Waterfront Walk - 2 miles (3.2 km), Non-circular - Estimated Time: 1-1.5 hours

Starting Point Coordinates, Postcode: The starting point is at Ipswich Railway Station (Grid Reference: TM 159 444, Postcode: IP2 8AL).
Nearest Car Park, Postcode: The nearest car park is the Ipswich Station Car Park (Postcode: IP2 8AL).

Embark on a delightful 2-mile (3.2 km) non-circular walk along Ipswich's revitalized waterfront, where the charm of the past seamlessly blends with the vibrant energy of the present. This easy stroll offers a fascinating journey through time, inviting you to discover the rich history and stunning views that make this iconic destination a true gem in the heart of Suffolk.

As you set out from the bustling Ipswich Railway Station, the anticipation builds with each step as you make your way towards the town centre along Princes Street. The lively atmosphere of the urban landscape gradually gives way to a sense of tranquility as you approach the River Orwell, its glistening waters beckoning you to explore the wonders that await along its banks.

Crossing the bridge over the river, you'll find yourself immersed in a world of nautical beauty and timeless elegance. Turn left onto the waterfront promenade, and allow yourself to be captivated by the stunning vistas that unfold before you. The marina stretches out like a sea of masts and sails, the gentle bobbing of the boats creating a soothing rhythm that seems to echo the ebb and flow of life itself.

As you continue along the promenade, the rich tapestry of Ipswich's maritime heritage reveals itself in the form of historic buildings that line the waterfront. Old warehouses and custom houses stand as silent witnesses to the countless stories and adventures that have unfolded here over the centuries, their weathered facades and intricate architectural details a testament to the skill and craftsmanship of a bygone era.

Among these architectural treasures, the Old Neptune Inn holds a special place in the hearts of locals and visitors alike. Dating back to the 16th century, this iconic pub has been a gathering place for seafarers, merchants, and travelers for generations, its cozy interior and welcoming atmosphere a haven of warmth and conviviality amidst the ever-changing tides of time.

As you make your way past these historic landmarks, the waterfront promenade seamlessly blends the old with the new, the sleek lines and modern aesthetics of the contemporary apartments and developments creating a striking contrast against the backdrop of the ancient buildings. This harmonious fusion of past and present serves as a powerful reminder of the enduring spirit of Ipswich, a town that has embraced change and innovation while never losing sight of its rich cultural heritage.

Throughout your walk, take a moment to pause and appreciate the vibrant ecosystem that thrives along the waterfront. The calls of seagulls and the gentle splashing of waterfowl add a lively soundtrack to your journey, while the occasional flutter of a butterfly or the darting of a dragonfly serves as a reminder of the incredible diversity of life that exists even in the heart of the urban landscape.

As you reach the far end of the waterfront, take a moment to drink in the panoramic views of the river and the surrounding countryside, the vast expanse of sky and water creating a sense of freedom and possibility that seems to stretch out to the horizon. Here, at the edge of the town and the brink of the unknown, you'll find yourself filled with a renewed sense of wonder and a deep appreciation for the incredible beauty and resilience of the natural world.

Turning around to retrace your steps, you'll see the waterfront promenade in a whole new light, the familiar sights and sounds now imbued with a deeper sense of meaning and connection. Each step brings you closer to your starting point, the railway station once again coming into view, a symbol of the journeys that have brought you here and the countless adventures that still lie ahead.

As you complete your walk and step back into the hustle and bustle of everyday life, you'll carry with you a newfound appreciation for the incredible history, beauty, and vibrancy of Ipswich's waterfront. This short but memorable stroll is a testament to the enduring power of place, a reminder that even in the midst of the modern world, there are still pockets of magic and wonder waiting to be discovered by those who take the time to look.

Detailed Route Directions:

1. From Ipswich Railway Station, walk towards the town centre along Princes Street. (Estimated time: 5-10 minutes)

2. Cross the bridge over the River Orwell and turn left onto the waterfront promenade. (Estimated time: 5-10 minutes)

3. Continue along the promenade, passing the marina, historic buildings, and modern apartments. Take your time to appreciate the views and soak in the atmosphere. (Estimated time: 20-30 minutes)

4. At the far end of the waterfront, pause to enjoy the panoramic views of the river and the surrounding countryside. (Estimated time: 5-10 minutes)

5. Turn around and retrace your steps back along the waterfront promenade, seeing the sights in a new light. (Estimated time: 20-30 minutes)

6. Cross the bridge over the River Orwell once again and make your way back to Ipswich Railway Station, where your walk comes to an end. (Estimated time: 5-10 minutes)

Best Features and Views:

- Immerse yourself in the stunning views of the marina, with its sea of masts and sails creating a mesmerizing tableau that captures the essence of Ipswich's maritime heritage.
- Marvel at the historic buildings that line the waterfront, their weathered facades and intricate architectural details a testament to the rich history and enduring spirit of this iconic destination.
- Delight in the harmonious blend of old and new, as the sleek lines and modern aesthetics of the contemporary apartments and developments create a striking contrast against the backdrop of the ancient buildings.

Historical Features of Interest:

As you walk along the waterfront, take a moment to appreciate the Old Neptune Inn, a 16th-century pub that has been a beloved gathering place for generations of seafarers, merchants, and travelers. Step inside this iconic establishment and allow yourself to be transported back in time, the cozy interior and welcoming atmosphere a testament to the enduring power of community and shared experience. Imagine the countless stories and adventures that have been shared within these walls, the laughter and the tears, the triumphs and the struggles, all woven together into the rich tapestry of Ipswich's maritime heritage.

Birds to Spot:

- Graceful gliders: Seagulls: Watch as these iconic coastal birds soar effortlessly above the waterfront, their distinctive cries and elegant flight a constant reminder of the ever-present connection between land and sea.
- Waterfowl wonders: Ducks and geese: Keep an eye out for the various species of waterfowl that make their home along the River Orwell, from the stately swans to the lively ducks and geese, each one a charming addition to the vibrant ecosystem of the waterfront.

Rare or Interesting Insects and Other Wildlife:

- Aerial acrobats: Dragonflies: If you're lucky, you may catch a glimpse of these stunning insects darting and hovering over the water's edge, their iridescent wings and incredible agility a marvel of the natural world.
- Fluttering gems: Butterflies: While the urban environment may not be a haven for wildlife, keep an eye out for the occasional butterfly flitting among the plants and flowers along the promenade, a delicate reminder of the beauty and resilience of nature.

Rare or Interesting Plants:

Due to the urban nature of the walk, there may be few rare or interesting plants to discover. However, take a moment to appreciate the carefully tended planters and flower beds that adorn the waterfront, their vibrant blooms and lush greenery a welcome splash of colour and life amidst the built environment. These small oases of nature serve as a reminder of the importance of green spaces and the vital role they play in creating a more harmonious and sustainable urban landscape.

Facilities and Accessibility:

The Ipswich Waterfront Walk offers a range of facilities to ensure a comfortable and enjoyable experience for all visitors. Along the waterfront, you'll find well-maintained public toilets, perfect for a quick comfort break during your stroll. If you're feeling peckish or in need of refreshment, the numerous cafes and restaurants that line the promenade provide a delightful selection of food and drink options, from cozy coffee shops to stylish bistros serving up the best of local cuisine.

For those arriving by train, the Ipswich Railway Station itself offers a variety of amenities, including restrooms, shops, and eateries, making it a convenient starting point for your waterfront adventure.

The walk is designed to be accessible for all ages and abilities, with mostly flat and paved surfaces that make for easy and comfortable navigation. Whether you're a seasoned walker or simply looking for a leisurely stroll, the Ipswich Waterfront Walk provides a welcoming and inclusive environment that caters to the needs of all visitors.

Seasonal Highlights:

The Ipswich Waterfront Walk offers a unique and captivating experience throughout the year, each season bringing its own special charm and character to this dynamic and ever-changing destination:

- Spring: As the weather warms and the days grow longer, the waterfront comes alive with the vibrant colours of spring flowers and the lively atmosphere of outdoor cafes and events, a perfect time to enjoy the fresh air and the awakening energy of the new season.

- Summer: During the summer months, the waterfront buzzes with activity, from lively festivals and outdoor performances to the joyful laughter of children playing along the promenade. Bask in the warm sunshine and soak up the convivial atmosphere, as the marina becomes a hub of social interaction and shared experiences.
- Autumn: As the leaves begin to turn and the air takes on a crisp, invigorating quality, the waterfront transforms into a stunning tableau of golden hues and soft, diffused light. Take a moment to appreciate the changing of the seasons and the subtle beauty of this transitional time, as the town prepares for the cozy delights of the colder months ahead.
- Winter: While the waterfront may be quieter during the winter season, it holds a special charm all its own, with the twinkling lights of the marina reflecting off the still waters and the historic buildings taking on a magical, frosty appearance. Bundle up and enjoy a bracing walk along the promenade, stopping to warm up with a hot cocoa or a hearty meal at one of the inviting eateries along the way.

Difficulty Level and Safety Considerations:

The Ipswich Waterfront Walk is classified as an easy walk, suitable for all ages and abilities. The mostly flat and paved surfaces make for a comfortable and accessible experience, with few challenges or obstacles to navigate. However, as with any outdoor activity, it's essential to take basic precautions to ensure a safe and enjoyable journey.

Be sure to wear comfortable, supportive footwear with good grip, especially in wet or slippery conditions. Stay aware of your surroundings and keep a safe distance from the water's edge, particularly with children or those with limited mobility.

In case of inclement weather, be prepared with appropriate clothing, such as a waterproof jacket or umbrella, to stay dry and comfortable throughout your walk. During the warmer months, don't forget to apply sunscreen, wear a hat, and stay hydrated to protect yourself from the sun's rays.

By following these simple guidelines and using common sense, you can fully immerse yourself in the beauty and wonder of the Ipswich Waterfront Walk, creating lasting memories and a deeper appreciation for this incredible destination.

Nearby Attractions:

After your invigorating walk along the waterfront, take the opportunity to explore some of the other fascinating attractions that Ipswich has to offer, each one providing a unique glimpse into the rich history, vibrant culture, and diverse interests of this dynamic town:

- Ipswich Transport Museum: Step back in time and discover the fascinating world of transportation at this unique museum, home to an extensive collection of vehicles that span over 100 years of history. From vintage cars and classic buses to historic trams and bicycles, this engaging attraction offers a hands-on and immersive experience for visitors of all ages.
- Christchurch Park: Escape the hustle and bustle of the town centre and lose yourself in the tranquil beauty of Christchurch Park, a stunning green oasis that offers a welcome respite from the demands of modern life. Stroll through the beautifully landscaped gardens, admire the historic Christchurch Mansion, and enjoy a picnic or a game of frisbee on the lush, rolling lawns.
- Ipswich Museum: Immerse yourself in the rich history and diverse natural world of Ipswich and beyond at this fascinating museum, home to an eclectic collection of artifacts and exhibits that span millions of years. From ancient fossils and Roman treasures to Victorian curiosities and contemporary art, this engaging attraction offers a window into the incredible tapestry of human and natural history that has shaped the region.

- Town Centre: No visit to Ipswich would be complete without exploring the vibrant town centre, a hub of shopping, dining, and entertainment that caters to every taste and interest. Browse the local boutiques and independent shops, sample the diverse culinary delights on offer, and soak up the lively atmosphere of this historic market town, where past and present collide in a celebration of community and culture.

By exploring these nearby attractions and the countless other hidden gems that Ipswich has to offer, you'll gain a deeper appreciation for the unique character and enduring appeal of this fascinating town, making your waterfront walk just the beginning of an unforgettable journey through the heart of Suffolk.

Ipswich Town Centre Walk - 2 miles (3.2 km), Non-circular - Estimated Time: 1-1.5 hours

Starting Point Coordinates, Postcode: The starting point for the Ipswich Town Centre Walk is at the Ipswich Tourist Information Centre (Grid Reference: TM 163 444, Postcode: IP1 1DH).
Nearest Car Park, Postcode: The nearest car park is the NCP Ipswich Black Horse Lane (Postcode: IP1 2EF).

Embark on a captivating 2-mile (3.2 km) walk through the historic heart of Ipswich, where centuries of history and architectural beauty intertwine to create a truly unforgettable experience. This easy, non-circular route is perfect for visitors of all ages and fitness levels, offering a fascinating glimpse into the rich cultural heritage and vibrant modern character of one of England's oldest towns.

As you set out from the Ipswich Tourist Information Centre, the anticipation builds as you make your way along St. Stephen's Lane, the bustling energy of the town centre enveloping you in its warm embrace. The striking façade of the Buttermarket shopping centre stands as a testament to the seamless blend of old and new that defines Ipswich, its contemporary design complementing the ancient streets and buildings that surround it.

Turning onto Tavern Street and Westgate Street, you'll find yourself immersed in a world of architectural wonders, each building and monument a silent witness to the countless stories and events that have shaped this remarkable town over the centuries. Follow the signs to St. Mary le Tower, a stunning medieval church whose soaring spire and intricate stained glass windows serve as a powerful reminder of the deep spiritual and artistic traditions that have long defined Ipswich.

As you continue along Northgate Street, the impressive façade of the Ipswich Museum and Art Gallery comes into view, its grand Victorian architecture a fitting showcase for the incredible collections housed within. From ancient fossils and Roman treasures to contemporary art and local history, this beloved institution offers a fascinating journey through time and space, inviting you to explore the rich tapestry of human experience that has shaped this corner of England.

Turning onto Elm Street and St. Margaret's Street, you'll soon find yourself standing before the historic St. Margaret's Church, a beautiful example of Gothic architecture that has served as a place of worship and community gathering for over 700 years. Take a moment to appreciate the intricate stone carvings and the peaceful atmosphere of the churchyard, a tranquil oasis amidst the bustle of the town centre.

As you make your way down St. Helen's Street and onto Grimwade Street, the impressive bulk of Christchurch Mansion comes into view, its grand Tudor façade and sprawling gardens a testament to the wealth and power of the local gentry who once called this magnificent estate home. Step inside and marvel at the beautifully preserved interiors, from the ornate plasterwork and fine art collections to the sumptuous furnishings and historic artifacts that offer a tantalizing glimpse into the lives of the aristocracy.

Emerging from the mansion, you'll find yourself in the tranquil surroundings of Christchurch Park, a lush green oasis that serves as a welcome respite from the demands of urban life. Stroll along the winding paths and immerse yourself in the vibrant colours and fragrances of the carefully tended flower beds, or simply sit and watch the world go by, enjoying the gentle birdsong and the soft rustling of leaves in the breeze.

As you wend your way back to the town centre along Soane Street, take a moment to reflect on the incredible journey you've just undertaken, the sights and sounds of Ipswich's rich history and vibrant modern culture forever etched in your memory. This short but deeply rewarding walk is a testament to the enduring power of place, a reminder that even in the heart of a bustling town, there is always something new and wonderful waiting to be discovered.

Detailed Route Directions:

1. From the Ipswich Tourist Information Centre, head east along St. Stephen's Lane, passing the Buttermarket shopping centre. (Estimated time: 5 minutes)

2. Turn left onto Tavern Street and continue onto Westgate Street. (Estimated time: 5 minutes)

3. Follow the signs for St. Mary le Tower, a beautiful medieval church with stunning stained-glass windows. Take a moment to admire the architecture and peaceful atmosphere. (Estimated time: 10 minutes)

4. Continue along Northgate Street, passing the impressive Ipswich Museum and Art Gallery. (Estimated time: 5 minutes)

5. Turn left onto Elm Street and then right onto St. Margaret's Street, where you can find the historic St. Margaret's Church. Appreciate the Gothic architecture and tranquil churchyard. (Estimated time: 10 minutes)

6. Walk down St. Helen's Street and turn left onto Grimwade Street to visit the magnificent Christchurch Mansion and Park. Explore the grand interiors and beautiful gardens. (Estimated time: 20-30 minutes)

7. From the park, head back to the town centre along Soane Street, reflecting on the rich history and culture you've experienced along the way. (Estimated time: 10 minutes)

Best Features and Views:

- Immerse yourself in the architectural wonders of Ipswich, from the stunning medieval churches to the grand Victorian buildings, each one a testament to the rich history and enduring beauty of this remarkable town.
- Discover the vibrant modern culture of Ipswich, with its bustling shopping centres, lively streets, and welcoming atmosphere, a perfect blend of old and new that captures the essence of contemporary urban life.
- Delight in the tranquil oasis of Christchurch Park, with its lush green spaces, colourful flower beds, and peaceful walking paths, a welcome respite from the demands of the town centre and a chance to reconnect with the natural world.

Historical Features of Interest:

As you explore the historic heart of Ipswich, take a moment to appreciate the incredible depth and diversity of the town's cultural heritage, embodied in the many landmarks and monuments you'll encounter along the way.

At St. Mary le Tower, marvel at the soaring spire and intricate stained-glass windows, a powerful expression of the medieval craftsmanship and spiritual devotion that have long defined this sacred space. Imagine the countless generations of worshippers who have found solace and inspiration within these ancient walls, their prayers and hymns echoing through the centuries.

Step inside the Ipswich Museum and Art Gallery and lose yourself in the vast collections of art, archaeology, and natural history, each exhibit a window into a different aspect of the human experience. From the fossilized remains of prehistoric creatures to the exquisite paintings and sculptures of local artists, this beloved institution offers a fascinating journey through time and space.

At St. Margaret's Church, take a moment to appreciate the intricate stone carvings and Gothic architecture, a testament to the skill and devotion of the medieval craftsmen who created this enduring monument to faith and community. Reflect on the countless baptisms, weddings, and funerals that have taken place here over the centuries, each one a thread in the rich tapestry of Ipswich's history.

And at Christchurch Mansion, step back in time to the grand era of the Tudor gentry, marvelling at the opulent interiors and fine art collections that offer a tantalizing glimpse into the lives of the wealthy and powerful. From the ornate plasterwork and sumptuous furnishings to the historic artifacts and family portraits, this magnificent estate is a living testament to the enduring legacy of Ipswich's aristocratic past.

Birds to Spot:

- Urban dwellers: Common species: As you walk through the town centre, keep an eye out for the ubiquitous pigeons, starlings, and house sparrows that thrive in the urban environment, their lively presence a reminder of the resilience and adaptability of nature.
- Parkland visitors: Christchurch Park: In the tranquil green spaces of Christchurch Park, you may spot a variety of more timid bird species, such as robins, blackbirds, and blue tits, their colourful plumage and melodic songs a delightful addition to the peaceful atmosphere of the gardens.

Rare or Interesting Insects and Other Wildlife:

- Scurrying foragers: Squirrels: As you stroll through Christchurch Park, watch for the playful antics of the resident squirrels as they scamper among the trees and across the lawns, their bushy tails and curious nature a charming sight in the heart of the town.
- Miniature marvels: Various insects: Keep an eye out for the many species of insects that call the park home, from the delicate butterflies and bees that flit among the flower beds to the fascinating beetles and bugs that crawl along the bark of the ancient trees.

Rare or Interesting Plants:

- Seasonal displays: Christchurch Park: Throughout the year, the carefully tended flower beds and gardens of Christchurch Park burst into a vibrant display of colour and fragrance, from the cheerful daffodils and tulips of spring to the lush roses and lavender of summer, each season bringing its own unique charm and beauty to this beloved green space.
- Ancient guardians: Trees: As you wander through the park, take a moment to appreciate the majestic trees that have stood watch over Ipswich for generations, their gnarled trunks and spreading canopies a testament to the enduring power and resilience of nature in the heart of the town.

Facilities and Accessibility:

The Ipswich Town Centre Walk offers a range of facilities and amenities to ensure a comfortable and enjoyable experience for all visitors. Public toilets can be found at the Buttermarket shopping centre and in Christchurch Park, providing convenient locations for a quick comfort break during your walk.

For those looking to refuel or relax, the numerous cafes, restaurants, and shops along the route offer a tempting array of refreshments and retail therapy options. Whether you're in the mood for a light snack, a satisfying meal, or a bit of souvenir shopping, you'll find plenty of opportunities to indulge your tastes and interests along the way.

The walk is generally flat and wheelchair accessible, making it a welcoming and inclusive experience for people of all abilities. However, it's worth noting that some of the older streets may have cobblestones or uneven surfaces, so those with mobility concerns may need to exercise caution in certain areas. If in doubt, don't hesitate to ask a local for advice or assistance – the friendly people of Ipswich are always happy to help!

Seasonal Highlights:

The Ipswich Town Centre Walk offers a unique and engaging experience throughout the year, with each season bringing its own special charm and character to this dynamic and ever-evolving urban landscape:

- Spring: As the days grow longer and the weather turns milder, the town centre comes alive with the vibrant colours and fragrances of spring flowers, from the cheerful daffodils and tulips in Christchurch Park to the delicate cherry blossoms lining the streets. Enjoy the fresh air and the renewed energy of the season as you explore the bustling shops and cafes.
- Summer: In the warm, sunny months of summer, Ipswich truly shines, with outdoor events, festivals, and performances adding an extra layer of excitement and culture to the town centre. Take advantage of the pleasant weather to enjoy a leisurely lunch at a sidewalk cafe, or simply bask in the lively atmosphere of the streets and parks, soaking up the sights and sounds of summer in the city.
- Autumn: As the leaves begin to turn and the air takes on a crisp, invigorating quality, Ipswich transforms into a stunning tableau of rich, warm hues, from the golden hues of the trees in Christchurch Park to the inviting glow of the shop windows along the streets. Enjoy a brisk walk through the town centre, stopping to savour a hot coffee or a hearty meal at one of the many cozy pubs or restaurants along the way.
- Winter: While the weather may be chilly, the town centre takes on a special charm in the winter months, with twinkling holiday lights, festive decorations, and a sense of community and warmth that defies the cold. Bundle up and enjoy a brisk walk through the frosty streets, stopping to admire the historic architecture and shop windows decked out in their winter finery. And be sure to warm up with a hot cocoa or a mulled wine at one of the inviting cafes or pubs along the route.

Difficulty Level and Safety Considerations:

The Ipswich Town Centre Walk is a relatively easy and straightforward route, suitable for people of all ages and fitness levels. The mostly flat terrain and well-maintained sidewalks and paths make for a comfortable and accessible experience, with few challenges or obstacles to navigate.

However, as with any urban walk, it's important to stay alert and aware of your surroundings, particularly when crossing busy streets or navigating crowded areas. Be sure to use designated crosswalks and follow traffic signals, and keep an eye out for vehicles and bicycles as you make your way through the town centre.

If you're walking with children or those with mobility concerns, be sure to supervise them closely and provide assistance as needed, particularly in areas with uneven surfaces or cobblestones. And don't hesitate to take breaks or adjust your pace as needed – the beauty of a self-guided walk is that you can go at your own speed and enjoy the experience on your own terms.

By following these simple precautions and using common sense, you can fully immerse yourself in the rich history, vibrant culture, and stunning architecture of Ipswich, creating lasting memories and a deeper appreciation for this fascinating and endlessly rewarding town.

Nearby Attractions:

After your invigorating walk through the historic heart of Ipswich, take the opportunity to explore some of the other fascinating attractions that this dynamic and diverse town has to offer:

- Ipswich Transport Museum: Step back in time and discover the fascinating world of transportation at this unique museum, home to an extensive collection of vehicles that span over 100 years of history. From vintage cars and classic buses to historic trams and fire engines, this hands-on museum offers an engaging and educational experience for visitors of all ages.
- Holywells Park: Escape the hustle and bustle of the town centre and immerse yourself in the tranquil beauty of Holywells Park, a stunning 67-acre green space that offers a peaceful retreat from the demands of urban life. Stroll through the beautifully landscaped gardens, explore the historic Stable Block and Orangery, and enjoy the playground, sports facilities, and wildlife that make this park a beloved destination for locals and visitors alike.
- Ipswich Waterfront: Just a short walk from the town centre, the vibrant and revitalized Ipswich Waterfront offers a perfect blend of history, culture, and modern amenities. Admire the stunning architecture of the historic buildings and the sleek lines of the contemporary developments, sample the delicious cuisine at the many restaurants and cafes, and soak up the lively atmosphere of this bustling and dynamic area.
- Christchurch Park: If you haven't already explored the full extent of Christchurch Park during your town centre walk, be sure to return and discover the many hidden gems and delights that this beloved green space has to offer. From the beautifully preserved Tudor mansion to the charming gardens, ancient trees, and wildlife, this park is a true oasis of natural beauty and history in the heart of Ipswich.

By venturing beyond the town centre and discovering these and the many other attractions that Ipswich has to offer, you'll gain a richer and more complete understanding of the unique character and enduring appeal of this fascinating and ever-evolving town. Whether you're a history buff, a nature lover, or simply a curious explorer, Ipswich promises to surprise, delight, and inspire you at every turn.

Freston Wood Walk - 3 miles (4.8 km), Circular - Estimated Time: 1.5-2 hours

Starting Point Coordinates, Postcode: The starting point for the Freston Wood Walk is at the Freston Wood car park (Grid Reference: TM 195 385, Postcode: IP9 1AF).
Nearest Car Park, Postcode: The nearest car park is the Freston Wood car park (Postcode: IP9 1AF).

Immerse yourself in the tranquil beauty of Freston Wood on this delightful 3-mile (4.8 km) circular walk. This gentle route takes you through a peaceful and scenic woodland, offering a chance to escape the hustle and bustle of everyday life and connect with nature. With well-maintained paths and easy to moderate terrain, this walk is perfect for those seeking a relaxing and rejuvenating experience.

As you set out from the Freston Wood car park, the path beckons you into the heart of the woodland, the anticipation building with each step as you leave the cares of the world behind. The moment you enter the cool, green embrace of the forest, you'll feel a sense of peace and tranquility wash over you, the stresses of daily life melting away in the presence of nature's soothing touch.

Following the winding path deeper into the wood, you'll find yourself immersed in a world of gentle beauty and quiet wonder. The soft rustle of leaves overhead and the dappled sunlight filtering through the canopy create an enchanting atmosphere, inviting you to slow down, breathe deeply, and savour the simple joys of being surrounded by the living, breathing landscape.

As you continue along the trail, the sights, sounds, and scents of the woodland engulf your senses, each one a delicate brush stroke in the masterpiece of nature. The air is filled with the melodic songs of birds flitting through the branches, their joyful notes a celebration of the vibrant ecosystem that thrives in the heart of Freston Wood. Pause for a moment to listen, and you may hear the gentle tapping of a woodpecker or the distant cry of a circling buzzard, each sound a reminder of the rich tapestry of life that flourishes in this green oasis.

The path winds on, guiding you through a diverse array of habitats and microclimates, each one revealing new facets of the woodland's beauty. From the cool, mossy hollows where ferns unfurl their delicate fronds to the sun-dappled glades where wildflowers dance in the gentle breeze, every step brings fresh wonders to discover. Take the time to examine the intricate details of the foliage, the patterns of bark on the ancient trees, and the tiny insects that bustle through the undergrowth, marvelling at the incredible complexity and resilience of the natural world.

As you approach the northern edge of the wood, the path curves gracefully, offering tantalizing glimpses of the surrounding countryside through the trees. Here, at the boundary between the sheltered world of the forest and the open expanse of fields and hedgerows, you may spot a shy deer grazing in the distance or hear the distant lowing of cattle, a reminder of the timeless rhythms of rural life that have shaped this landscape for generations.

Turning back towards the heart of Freston Wood, the path leads you once more into the comforting embrace of the trees, their leaves whispering secrets in the gentle breeze. The play of light and shadow on the forest floor creates an ever-changing tapestry of patterns, each one a fleeting work of art that exists for a moment before dissolving back into the larger canvas of the woodland.

As you near the end of your journey, the path gradually loops back towards the starting point, each step bringing you closer to the world you left behind. Yet, even as you emerge from the trees and catch sight of the car park, you carry with you the sense of peace, wonder, and connection that only time spent in nature can provide.

The Freston Wood Walk is more than just a physical journey; it is a chance to step out of the everyday and into a world of beauty, tranquility, and simple, profound truths. Whether you seek solace, inspiration, or simply a moment of quiet reflection, this gentle woodland trail offers a sanctuary for the soul, a place where the cares of the world fade away, and the wisdom of the natural world speaks to the heart.

Detailed Route Directions:

1. From the Freston Wood car park, follow the path into the woodland, heading northeast (045°). Take a moment to appreciate the peaceful atmosphere and the lush greenery that surrounds you. (Estimated time: 5-10 minutes)

2. Continue along the path as it winds through the trees, bearing east (090°), and immerse yourself in the sights, sounds, and scents of this enchanting natural environment. Listen for the melodic birdsong and the gentle rustling of leaves in the breeze. (Estimated time: 20-30 minutes)

3. As you progress through the wood, take your time to observe the diverse array of plant life, from the majestic trees that tower overhead to the delicate wildflowers and ferns that carpet the forest floor. The path will gently turn to the southeast (135°). (Estimated time: 20-30 minutes)

4. At the northern edge of the wood, the path will curve to the south (180°). Follow the trail as it skirts along the boundary of the woodland, offering glimpses of the surrounding countryside. (Estimated time: 15-20 minutes)

5. The path will then turn to the southwest (225°), guiding you back into the heart of Freston Wood. As you walk, keep an eye out for the various woodland creatures that call this place home, from the playful squirrels to the shy deer. (Estimated time: 20-30 minutes)

6. Continue along the path, now heading west (270°), as it loops back towards the starting point. Take a moment to reflect on the beauty and tranquility of the woodland and the sense of peace and rejuvenation that spending time in nature can bring. (Estimated time: 15-20 minutes)

7. Finally, you'll arrive back at the Freston Wood car park, completing the circular walk. Before leaving, take one last look at the beautiful woodland and appreciate the opportunity to have experienced its wonders. (Estimated time: 5-10 minutes)

Best Features and Views:

- Immerse yourself in the serene and calming atmosphere of Freston Wood, a perfect escape from the stresses of daily life, where the gentle rustling of leaves and the melodic songs of birds create a soothing symphony that calms the mind and rejuvenates the spirit.
- Marvel at the incredible diversity of plant life that thrives within the woodland, from the majestic oaks and beeches that tower overhead to the delicate wildflowers and ferns that carpet the forest floor, each species playing a vital role in the intricate web of life that defines this enchanting ecosystem.
- Delight in the chance to observe various woodland creatures in their natural habitat, from the playful antics of squirrels darting through the branches to the shy grace of deer grazing in the dappled sunlight, each encounter a reminder of the rich tapestry of life that flourishes in the heart of the forest.

Historical Features of Interest:

The woodland itself serves as a living testament to the enduring beauty and historical significance of the natural landscape. Freston Wood, like many ancient woodlands in the UK, has a rich history that dates back centuries. These woodlands were once an integral part of the rural economy, providing timber for construction, fuel for heating and cooking, and a variety of other resources essential to the lives of local communities.

In the past, Freston Wood would have been managed through traditional techniques such as coppicing and pollarding, which involved the periodic cutting of trees to encourage new growth and maintain a healthy, diverse ecosystem. These practices, passed down through generations, helped to shape the character of the woodland and ensure its longevity for future generations to enjoy.

As you walk through the wood, keep an eye out for signs of these historical management techniques, such as the distinctive multi-stemmed growth of coppiced trees or the gnarled, knobbly branches of pollarded oaks. These living monuments to the past serve as a reminder of the deep connection between human activity and the natural world, and the importance of preserving these ancient woodlands for generations to come.

While exploring the woodland, you may also come across remnants of old boundary markers, such as ancient hedgerows or earthen banks, which offer a glimpse into the historical divisions of the landscape and the ways in which people have shaped and defined the countryside over the centuries.

By immersing yourself in the timeless beauty of Freston Wood and understanding its historical significance, you'll gain a deeper appreciation for the complex web of natural and human history that has shaped this enchanting landscape, and the vital role that ancient woodlands play in preserving our shared heritage and securing a sustainable future for all.

Birds to Spot:

- Agile nuthatches: Listen for the sharp, distinctive calls of these active birds as they climb headfirst down tree trunks, probing for insects in the bark crevices. Their sleek, blue-gray plumage and striking black eye stripe make them a delight to observe as they navigate the woodland with impressive dexterity.
- Melodious blackcaps: From April to September, keep an ear out for the rich, fluty songs of the blackcap, often referred to as the "northern nightingale" due to its beautiful voice. The male's striking black cap contrasts with its gray back and pale underparts, making it a memorable sight among the woodland foliage.
- Elusive treecreepers: These small, brown birds are masters of camouflage, blending seamlessly with the bark of the trees they inhabit. Watch carefully as they spiral up tree trunks, meticulously probing the crevices for invertebrates, their long, curved bills perfectly adapted for extracting hidden prey from the nooks and crannies of the bark.

Rare or Interesting Insects and Other Wildlife:

- Enchanting butterflies: In the warmer months, the woodland edges and sunny glades come alive with the fluttering wings of various butterfly species. Look for the striking orange and brown patterning of the comma, the delicate beauty of the holly blue with its silvery-blue wings, and the unmistakable majesty of the purple emperor as it glides through the dappled sunlight.
- Industrious wood ants: As you walk through the woodland, keep an eye out for the impressive mounds of the wood ant colonies. These remarkable insects play a crucial role in the forest ecosystem, tirelessly foraging for food and maintaining the delicate balance of life within the woodland.
- Secretive badgers: Although elusive and primarily nocturnal, the presence of badgers can often be detected

by the distinctive signs they leave behind. Look for the entrances to their underground setts, marked by large piles of soil and discarded bedding, and the well-worn paths they create through the undergrowth as they navigate their woodland domain.

Rare or Interesting Plants:

- Ethereal bluebells: In the spring, the woodland floor transforms into a mesmerizing carpet of bluebells, their delicate, nodding flowers casting a soft purple haze across the landscape. Take a moment to appreciate the ephemeral beauty of these enchanting blooms, a symbol of the woodland's enduring magic.
- Ancient trees: Stand in awe of the venerable oaks, beeches, and sweet chestnuts that have witnessed the passing of centuries. These ancient sentinels, with their gnarled bark and spreading canopies, serve as living monuments to the resilience and longevity of nature, their presence a humbling reminder of the timeless cycles that shape the woodland.
- Lush ferns: Discover the verdant beauty of the woodland's fern population, from the delicate fronds of the hart's tongue to the graceful arching stems of the male fern. These ancient plants, with their intricate and often prehistoric appearance, add a sense of primordial wonder to the forest floor, a testament to the enduring power of life in even the shadiest corners of the wood.

Facilities and Accessibility:

The Freston Wood Walk offers well-maintained paths that are generally accessible for most walkers. However, it is important to note that some sections may be uneven, rooty, or muddy, especially after periods of heavy rainfall. Sturdy walking shoes or boots with good grip are highly recommended to ensure a safe and comfortable experience.

While there are no public toilets or other facilities directly on the walk route, the nearby village of Freston and the town of Ipswich offer a range of amenities for visitors, including public toilets, cozy pubs, and inviting cafes where you can refresh and refuel before or after your woodland adventure.

Due to the natural terrain and the presence of some uneven surfaces, the Freston Wood Walk may not be suitable for wheelchairs or pushchairs. Those with limited mobility or specific accessibility requirements should exercise caution and assess their ability to navigate the trail based on their individual needs and circumstances.

Seasonal Highlights:

The Freston Wood Walk offers a captivating and ever-changing experience throughout the year, with each season painting the landscape in its own unique hues and textures:

- Spring: Witness the woodland awakening from its winter slumber, as delicate green leaves unfurl on the branches and a vibrant carpet of bluebells, wood anemones, and primroses spreads across the forest floor. The air fills with the melodic songs of returning birds, and the first tentative buzzes of insects herald the arrival of warmer days ahead.
- Summer: Bask in the lush greenery of the woodland at the peak of its growth, the canopy providing a cool sanctuary from the heat of the summer sun. Observe the busy lives of the forest's inhabitants, from the industrious wood ants to the fluttering butterflies and the darting dragonflies that hover over sun-dappled glades.
- Autumn: Marvel at the breathtaking transformation of the woodland as the leaves shift from green to a stunning array of golds, oranges, and reds. The crisp, clean air and the satisfying crunch of fallen leaves

underfoot create an invigorating atmosphere, perfect for a contemplative walk through the changing landscape.
- Winter: Embrace the stark beauty of the dormant woodland, with its bare branches etched against the cool, clear sky. Discover the subtle signs of life that persist even in the depths of winter, from the hardy evergreens that provide splashes of colour to the tracks of woodland creatures preserved in the frost or snow.

Difficulty Level and Safety Considerations:

The Freston Wood Walk is a relatively easy to moderate route, suitable for walkers of most fitness levels. The paths are generally well-maintained, but it is essential to remain aware of any uneven surfaces, exposed roots, or slippery sections, particularly after wet weather.

As with any outdoor activity, it is crucial to be prepared for changeable weather conditions. Ensure you wear appropriate footwear and clothing to suit the season, and carry sufficient water and snacks to keep yourself hydrated and energized throughout your walk.

Be mindful of the woodland's inhabitants, and remember to respect their habitat by staying on the designated paths and refraining from disturbing or feeding the wildlife. This not only helps to protect the delicate balance of the ecosystem but also ensures your own safety by minimizing the risk of unexpected encounters with wild animals.

It is always a good idea to inform someone of your planned route and expected return time before setting out, and to carry a charged mobile phone in case of emergencies. By following these simple precautions and exercising common sense, you can fully immerse yourself in the beauty and tranquility of Freston Wood, creating lasting memories and forging a deeper connection with the natural world.

Nearby Attractions:

After your rejuvenating walk through Freston Wood, take the opportunity to explore some of the other fascinating attractions that the surrounding area has to offer:

- Alton Water Park: Just a short drive from Freston Wood, Alton Water Park is a stunning 400-acre conservation area centred around a large reservoir. With its diverse range of habitats, including grasslands, woodlands, and wetlands, the park offers excellent opportunities for birdwatching, fishing, cycling, and water sports, making it a perfect destination for a full day of outdoor adventures.
- Jimmy's Farm and Wildlife Park: Experience the thrills and joys of a working farm at Jimmy's Farm and Wildlife Park, a unique attraction that combines agriculture, conservation, and education. Meet an array of friendly farm animals, explore the beautiful gardens and nature trails, and discover the exotic residents of the wildlife park, including meerkats, tapirs, and reindeer.
- Pin Mill: Take a step back in time as you visit the charming hamlet of Pin Mill, nestled on the banks of the River Orwell. This picturesque spot, with its historic cottages, traditional boats, and stunning views, has long been a source of inspiration for artists and writers. Enjoy a leisurely stroll along the river, stop for a pint in the 16th-century Butt and Oyster pub, and soak up the timeless atmosphere of this quintessential English village.

By combining your Freston Wood Walk with a visit to these captivating nearby attractions, you'll create an unforgettable itinerary that showcases the very best of Suffolk's natural beauty, rich history, and vibrant culture. Whether you're a nature enthusiast, a history buff, or simply seeking a memorable escape from the everyday, the Freston Wood area promises to delight and inspire you at every turn.

Kesgrave Woods Walk - 3 miles (4.8 km), Circular - Estimated Time: 1.5-2 hours

Starting Point Coordinates, Postcode: The starting point for the Kesgrave Woods Walk is at the entrance to Kesgrave Woods on Bell Lane, Kesgrave (Grid Reference: TM 223 449, Postcode: IP5 1JF).
Nearest Car Park, Postcode: Parking is available on nearby residential streets, but please park considerately. The nearest car park is at the Tesco Superstore on Ropes Drive (Postcode: IP5 3RX), approximately a 15-minute walk from the starting point.

Embark on a delightful 3-mile (4.8 km) circular walk through the tranquil and picturesque Kesgrave Woods, a hidden gem nestled in the heart of Suffolk. This gentle route offers the perfect opportunity to escape the hustle and bustle of everyday life and immerse yourself in the beauty and serenity of nature.

As you make your way from the entrance to Kesgrave Woods on Bell Lane, the path beckons you into a world of lush greenery and dappled sunlight. The soft crunch of leaves underfoot and the gentle rustling of the breeze through the trees create a soothing symphony that accompanies you on your journey, inviting you to slow down and savour the simple pleasures of the woodland.

The well-maintained footpath winds its way through the heart of the forest, guiding you past towering oaks, graceful beeches, and slender birches. As you navigate the twists and turns of the trail, keep your eyes peeled for the rich array of wildlife that calls Kesgrave Woods home. From the striking black and white plumage of a great spotted woodpecker to the acrobatic antics of a nuthatch scaling a tree trunk, each encounter serves as a reminder of the incredible diversity and resilience of nature.

As you continue along the path, crossing charming wooden bridges that span babbling brooks and meandering streams, take a moment to breathe in the fresh, invigorating air and let the stresses of modern life melt away. The woodland's lush, tranquil atmosphere has a remarkable ability to soothe the soul and rejuvenate the spirit, making this walk the perfect antidote to the demands of the 21st century.

In the spring, the forest floor transforms into a mesmerizing carpet of colour, as delicate bluebells and vibrant wild garlic bloom in a majestic sea that fills the air with the sweet scents of spring. The seasonal shift serves as a powerful reminder of the cyclical nature of life and the enduring beauty of the natural world.

As you approach the halfway point of your journey, the path begins to loop back towards the starting point, offering a new perspective on the woodland landscape. The gentle inclines and uneven terrain provide a satisfying physical challenge, while the ever-changing play of light and shadow through the canopy creates a captivating visual spectacle that enchants and inspires.

Upon returning to the entrance of Kesgrave Woods, you'll feel a sense of accomplishment and renewal, your mind clear and your spirit lifted by the simple yet profound experience of connecting with nature. This walk serves as a testament to the enduring power of the great outdoors, a reminder that even in the midst of our busy, technology-driven lives, there is always a place of peace and beauty waiting to be discovered.

The Kesgrave Woods Walk is a celebration of the timeless beauty and restorative power of the natural world, an invitation to step away from the distractions and pressures of modern life and rediscover the simple joys of exploration and connection. Whether you're a seasoned hiker or simply a lover of nature, this enchanting circular route promises to delight and inspire, leaving you with memories that will last a lifetime.

Detailed Route Directions:

1. From the entrance to Kesgrave Woods on Bell Lane, head north into the woodland, bearing 000°. (Estimated time: 5-10 minutes)

2. Follow the main footpath as it winds through the trees, bearing 045°, crossing a small wooden bridge over a stream. (Estimated time: 15-20 minutes)

3. Continue along the path, now bearing 090°, as it gently inclines and weaves through the heart of the woodland. (Estimated time: 20-30 minutes)

4. The path will eventually curve to the left, bearing 315°, leading you through a particularly scenic section of the forest. (Estimated time: 25-35 minutes)

5. Cross another small bridge, bearing 270°, and follow the path as it begins to loop back towards the starting point. (Estimated time: 20-30 minutes)

6. The path will turn to the right, bearing 135°, guiding you through the final stretch of the woodland. (Estimated time: 15-20 minutes)

7. Arrive back at the entrance to Kesgrave Woods on Bell Lane, bearing 180°, where your journey began. (Estimated time: 5-10 minutes)

Best Features and Views:

- Immerse yourself in the tranquil and picturesque woodland setting of Kesgrave Woods, a peaceful haven that offers a much-needed escape from the stresses and distractions of modern life.
- Marvel at the diverse array of trees that call this forest home, from majestic oaks and graceful beeches to slender birches and sturdy pines, each one a testament to the enduring strength and beauty of nature.
- Delight in the opportunity to spot local wildlife, from the striking plumage of a great spotted woodpecker to the acrobatic antics of a nuthatch, each encounter a reminder of the incredible diversity and resilience of the natural world.

Historical Features of Interest:

While Kesgrave Woods does not boast any significant historical features along the walk, the forest itself serves as a living testament to the enduring legacy of the Suffolk landscape. These ancient woodlands have stood as silent witnesses to the passage of time, their roots reaching deep into the soil and their branches stretching towards the sky, a constant presence in a world of change.

As you walk beneath the leafy canopy, take a moment to consider the generations of people who have found solace, inspiration, and connection within these very woods. From the earliest inhabitants who relied on the forest for sustenance and shelter to the modern-day visitors who seek respite from the demands of urban life, Kesgrave Woods has long served as a sanctuary for the human spirit.

The act of walking through this timeless landscape, treading the same paths that countless others have travelled before you, creates a profound sense of connection to the past and the natural world. Each step becomes a meditation on the enduring power of nature and the importance of preserving these precious green spaces for future generations to enjoy and cherish.

In the stillness of the forest, with the gentle rustling of leaves and the distant birdsong as your backdrop, you may find yourself feeling a deep sense of belonging, a recognition of the inherent value of the world around you. This is the true gift of Kesgrave Woods, the opportunity to step out of the present moment and into a realm of timeless beauty and quiet reverence.

Birds to Spot:

- Great Spotted Woodpeckers: Listen for the distinctive drumming sound of these striking black and white birds as they hammer their beaks against tree trunks in search of insects. With a flash of red on their heads and a bold pattern of black and white on their wings, great spotted woodpeckers are a captivating sight in the woodland canopy.
- Nuthatches: Keep your eyes peeled for these agile little birds as they scurry up and down tree trunks, often head-first, probing for insects and seeds in the bark. Nuthatches are easily recognizable by their striking blue-grey backs, peachy undersides, and the black stripe that runs through their eyes, giving them a bandit-like appearance.
- Treecreepers: These small, brown birds are masters of camouflage, blending seamlessly with the bark of the trees they call home. Watch carefully as they spiral up tree trunks, methodically searching for insects and spiders in the crevices of the bark. Treecreepers' long, curved bills and stiff tails are perfectly adapted for their unique foraging style, making them a fascinating species to observe in the woodland setting.

Rare or Interesting Insects and Other Wildlife:

- Butterflies: During the warmer months, Kesgrave Woods comes alive with the delicate flutter of butterfly wings. From the vibrant orange and brown hues of the comma to the iridescent purple sheen of the purple hairstreak, these enchanting creatures add a splash of colour and beauty to the woodland scene. Keep an eye out for other species such as the speckled wood, peacock, and brimstone, each one a unique and captivating sight.
- Dragonflies: As you walk along the woodland's streams and damp areas, look for the iridescent shimmer of dragonflies darting through the air. These ancient insects, with their impressive aerial skills and striking colours, have long capture the human imagination. Species such as the southern hawker, with its vibrant green and blue hues, and the common darter, with its rich red colouration, are a delight to behold as they patrol their woodland territories.
- Slow Worms: While not actually worms at all, these legless lizards are a fascinating and often overlooked resident of Kesgrave Woods. Slow worms, with their smooth, glossy scales and gentle demeanor, can often be spotted basking in sunny patches or hiding beneath logs and stones. As you walk through the forest, keep an eye out for these secretive creatures, and take a moment to appreciate their unique place in the woodland ecosystem.

Rare or Interesting Plants:

- Bluebells: In the spring, the forest floor of Kesgrave Woods transforms into a mesmerizing carpet of blue, as thousands of delicate bluebells bloom in a spectacular display. The sight and scent of these enchanting flowers have long captivated the hearts and minds of nature lovers, and a walk through the bluebell-strewn woodland is an unforgettable experience.
- Wild Garlic: Alongside the bluebells, the pungent aroma of wild garlic fills the air in springtime. Also known as ramsons, these delicate white flowers and broad green leaves are a favourite of foragers, who prize the plant for its flavourful and nutritious properties. As you walk through the woods, look for patches of wild garlic lighting up the forest floor, and take a moment to appreciate their fragrance and beauty.
- Ferns: Kesgrave Woods is home to a diverse array of fern species, from the tall and stately bracken to the

delicate and lacy hart's tongue. These ancient plants, with their intricate fronds and vibrant green hues, add a touch of primeval magic to the woodland setting, reminding us of the long and complex history of life on Earth. Take a moment to examine the varied shapes and textures of the ferns you encounter, and marvel at their resilience and adaptability in the face of changing conditions.

Facilities and Accessibility:

The Kesgrave Woods Walk is a relatively easy route, suitable for walkers of most fitness levels. The paths are generally well-maintained, but it is important to note that some sections may be uneven, muddy, or include exposed tree roots, particularly after periods of heavy rainfall. Sturdy walking shoes or boots with good grip are highly recommended to ensure a safe and comfortable experience.

It's important to note that there are no facilities directly within Kesgrave Woods, so walkers should plan accordingly. This means bringing sufficient water, snacks, and any necessary medications or sun protection to ensure a comfortable and enjoyable walk.

Due to the natural terrain and the presence of some uneven surfaces, the walk may not be suitable for wheelchairs, mobility scooters, or pushchairs. Those with limited mobility or specific accessibility requirements should exercise caution and assess their ability to navigate the trail based on their individual needs and circumstances.

The nearest public toilets and other amenities can be found in the nearby town of Kesgrave, which is just a short drive or walk from the woodland entrance. Here, walkers can find a range of facilities, including cafes, shops, and pubs, perfect for a pre- or post-walk refreshment.

Seasonal Highlights:

The Kesgrave Woods Walk offers a unique and captivating experience in every season, each one revealing new facets of beauty and wonder within this enchanting woodland:

- Spring: As winter's chill gives way to the warmth and vitality of spring, Kesgrave Woods bursts into life with an explosion of colour and fragrance. Bluebells and wild garlic carpet the forest floor in a mesmerizing display, while the fresh green leaves of the trees create a lush canopy overhead. The air is filled with the joyful songs of returning birds, and the woodland is abuzz with the energy of new beginnings.
- Summer: In the height of summer, Kesgrave Woods provides a welcome respite from the heat and bustle of daily life. The dense foliage offers a cool, green sanctuary, dappled with golden sunlight filtering through the leaves. The woodland is alive with the hum of insects, the flutter of butterflies, and the gentle rustling of small mammals in the undergrowth. This is a time to savour the fullness and abundance of nature at its peak.
- Autumn: As the days grow shorter and the nights become crisp, Kesgrave Woods transforms into a breathtaking tapestry of gold, orange, and red. The autumn leaves create a warm and inviting atmosphere, perfect for a contemplative walk through the woods. The earthy scents of mushrooms and decaying leaves fill the air, a reminder of the ever-turning cycle of life and death in the natural world.
- Winter: While the woods may be stripped bare of their leaves, the stark beauty of the winter landscape offers its own unique charm. Frosty mornings and the occasional dusting of snow create a hushed and magical atmosphere, the bare branches of the trees etched against the pale sky. This is a time for quiet reflection and solitude, for embracing the stillness and serenity of the forest in its dormant state.

Difficulty Level and Safety Considerations:

The Kesgrave Woods Walk is classified as an easy to moderate trail, with mostly level terrain and well-defined paths. However, walkers should be aware of potential trip hazards, such as exposed tree roots, and take care when navigating muddy or slippery sections, especially after heavy rain.

As with any outdoor activity, it's essential to be prepared for changeable weather conditions. Ensure you wear appropriate footwear and clothing for the season, and always carry a waterproof jacket or umbrella, even on seemingly clear days.

While the woodland is generally a safe and welcoming environment, it's important to stay alert and aware of your surroundings. Stick to the designated paths, and avoid venturing off-trail to minimize the risk of getting lost or disturbing sensitive wildlife habitats.

If walking alone, it's a good idea to let someone know your intended route and estimated return time. Carry a fully charged mobile phone in case of emergencies, and familiarize yourself with the nearest exit points and landmarks within the woodland.

By following these simple guidelines and exercising common sense, walkers can fully immerse themselves in the beauty and tranquility of Kesgrave Woods, enjoying a safe and rewarding experience that nourishes both body and soul.

Nearby Attractions:

After your invigorating walk through the tranquil beauty of Kesgrave Woods, why not explore some of the other fascinating attractions that this charming corner of Suffolk has to offer? The nearby town of Kesgrave and the surrounding area are home to a wealth of historical sites, scenic landscapes, and local amenities that are sure to appeal to visitors of all interests.

- Kesgrave Town: Just a short walk from the woodland entrance, the town of Kesgrave offers a range of shops, cafes, and pubs where you can grab a refreshment or a bite to eat. Take a stroll through the town's pleasant streets, admiring the mix of traditional and modern architecture, and soak up the friendly, community atmosphere.
- The Suffolk Coast: A short drive from Kesgrave, the stunning Suffolk Coast beckons with its miles of sandy beaches, picturesque seaside towns, and rich maritime history. Explore the charming coastal villages of Aldeburgh and Southwold, sample locally caught seafood, and take in the breathtaking views of the North Sea.
- Ipswich: Just a few miles down the road, the vibrant county town of Ipswich offers a wealth of cultural and historical attractions. Visit the iconic Christchurch Mansion, a stunning Tudor house set in a beautiful park, or explore the Ipswich Museum, with its fascinating exhibits on local history and natural science. The town's thriving waterfront area, with its bustling marina and array of restaurants and bars, is also well worth a visit.
- Woodbridge: Another nearby town that merits exploration is the charming market town of Woodbridge. Nestled on the banks of the River Deben, Woodbridge boasts a rich history and a picturesque town centre, lined with independent shops, galleries, and eateries. The town is also home to the fascinating Tide Mill Living Museum, which tells the story of one of the country's last working tide mills.

By combining your Kesgrave Woods Walk with a visit to these nearby attractions, you'll have the opportunity to experience the very best of what this beautiful corner of Suffolk has to offer. From the tranquil beauty of the woodland to the rich history and vibrant culture of the surrounding towns and villages, this is a region that truly offers something for everyone.

So why not make a day of it, or even plan a weekend getaway? With its stunning landscapes, friendly communities, and wealth of things to see and do, the Kesgrave area promises to delight and inspire visitors of all ages and interests. Whether you're a nature lover, a history buff, or simply someone in search of a relaxing escape from the hustle and bustle of daily life, you'll find plenty to enjoy in this enchanting corner of East Anglia.

As you explore the region, take a moment to reflect on the deep sense of connection and belonging that comes from immersing yourself in the natural world. The tranquil beauty of Kesgrave Woods and the surrounding landscapes serves as a powerful reminder of the importance of preserving and cherishing these precious green spaces, not just for our own well-being, but for the countless species that call them home.

In a world that often feels increasingly disconnected and fast-paced, the simple act of taking a walk through the woods, breathing in the fresh air, and listening to the gentle sounds of nature can be a profoundly restorative and grounding experience. It's a chance to reconnect with the rhythms of the earth, to quiet the noise of our busy minds, and to rediscover a sense of wonder and appreciation for the world around us.

So whether you're a local resident or a visitor from further afield, be sure to add the Kesgrave Woods Walk and the surrounding attractions to your must-visit list. With its stunning scenery, rich history, and warm, welcoming atmosphere, this is a destination that promises to leave you feeling refreshed, inspired, and more deeply connected to the natural world and your place within it.

As you set out on your adventures, remember to tread lightly, to leave no trace, and to approach each encounter with an open heart and a curious mind. By doing so, you'll not only ensure that these beautiful landscapes remain pristine and unspoiled for generations to come, but you'll also open yourself up to the countless opportunities for growth, discovery, and transformation that nature has to offer.

In the words of the great naturalist John Muir, "In every walk with nature, one receives far more than he seeks." So step out into the beauty of Kesgrave Woods and the surrounding countryside, and allow yourself to be touched by the magic and wonder of the natural world. You never know what gifts and insights await you on the trail ahead.

Great Bealings Circular Walk - 6 miles (9.7 km), Circular - Estimated Time: 3-4 hours

Starting Point Coordinates, Postcode: The starting point for the Great Bealings Circular Walk is at St. Mary's Church, Great Bealings (Grid Reference: TM 242 487, Postcode: IP13 6NY).
Nearest Car Park, Postcode: There is limited on-street parking available near St. Mary's Church (Postcode: IP13 6NY).

Embark on a delightful 6-mile (9.7 km) circular walk through the charming village of Great Bealings and the picturesque Suffolk countryside, where a tapestry of woodland, arable fields, and meadows awaits. This moderate route offers a perfect blend of tranquility, natural beauty, and the opportunity to immerse yourself in the rich history and stunning landscapes of this enchanting corner of East Anglia.

As you set out from the historic St. Mary's Church, a sense of anticipation builds as you follow Church Road eastward, the gentle rustling of leaves and the distant birdsong a soothing soundtrack to your journey. Turn left onto Lodge Road, the path leading you towards a hidden world of natural wonders, before veering right onto a footpath that guides you into the cool embrace of a wooded area.

Immerse yourself in the dappled sunlight and the soft earthy scents of the woodland, the trees standing tall like silent guardians, their branches reaching out to create a living canopy above your head. As you continue eastward, the path eventually turns south, leading you along the edge of a vast field, the golden hues of the crops swaying gently in the breeze, a mesmerizing display of the bounty and beauty of the Suffolk countryside.

At the southern end of the field, turn left and follow the path as it skirts the field's edge, offering breathtaking views of the surrounding landscape. Here, the patchwork of colours and textures creates an ever-changing tableau, from the vibrant greens of the meadows to the rich, earthy tones of the freshly tilled soil, each one a testament to the enduring connection between the land and the people who have shaped it for generations.

As you continue along the path, eventually turning westward, take a moment to appreciate the intricate details of the hedgerows that line your route, their gnarled branches and dense foliage providing shelter and sustenance for a diverse array of wildlife. Keep your eyes peeled for the quick, darting movements of rabbits and hares, their presence a charming reminder of the abundant life that thrives in these quiet corners of the countryside.

The path then turns northward, guiding you back towards Lodge Road, where you'll turn right and retrace your steps for a short distance. As you approach the final leg of your journey, turn left onto Church Road, the iconic tower of St. Mary's Church once again coming into view, its weathered stone and timeless beauty a fitting symbol of the enduring spirit and rich history of Great Bealings.

As you complete the circular route and arrive back at your starting point, take a moment to reflect on the incredible journey you've just undertaken. The Great Bealings Circular Walk is more than just a physical excursion; it is a celebration of the simple joys and profound truths that can be found when we step away from the distractions of modern life and immerse ourselves in the timeless beauty of the natural world.

Whether you're a local resident or a visitor from afar, this enchanting walk promises to delight and inspire, offering a unique perspective on the stunning landscapes, rich heritage, and warm community spirit that define this special corner of Suffolk. So lace up your walking boots, grab your camera, and set out on an adventure that will nourish your soul and leave you with memories to last a lifetime.

Detailed Route Directions:

1. From St. Mary's Church, head east along Church Road, bearing 090°. (Estimated time: 10-15 minutes)

2. Turn left onto Lodge Road, bearing 000°, then right onto a footpath that leads into a wooded area. (Estimated time: 15-20 minutes)

3. Continue eastward, bearing 090°, eventually turning south and following a footpath that skirts the edge of a large field. (Estimated time: 30-40 minutes)

4. Turn left at the southern end of the field, bearing 180°, continuing along the edge of the field until you reach a footpath that leads westward. (Estimated time: 30-40 minutes)

5. Follow this path, bearing 270°, eventually turning north and returning to Lodge Road. (Estimated time: 30-40 minutes)

6. Turn right onto Lodge Road, bearing 000°, then left onto Church Road. (Estimated time: 15-20 minutes)

7. Return to the starting point at St. Mary's Church, bearing 270°. (Estimated time: 10-15 minutes)

Best Features and Views:

- Immerse yourself in the tranquil beauty of the Suffolk countryside, with its patchwork of arable fields, lush meadows, and ancient woodland, each one a unique and captivating display of nature's artistry.
- Delight in the stunning views that unfold as you navigate the gently rolling landscape, from the sweeping vistas of golden cropland to the intimate glimpses of wildlife-rich hedgerows, each one a reminder of the incredible diversity and resilience of the natural world.
- Discover the rich history and timeless charm of Great Bealings, with its picturesque village streets, historic church, and warm, welcoming community spirit, a testament to the enduring character and proud heritage of this special corner of East Anglia.

Historical Features of Interest:

As you set out on your walk, take a moment to appreciate the historic St. Mary's Church, a beautiful example of a medieval parish church that has stood at the heart of the Great Bealings community for centuries. The church's elegant tower and weathered stone walls speak to the skill and devotion of the craftsmen who built it, while the peaceful churchyard and well-tended graves offer a poignant reminder of the generations who have worshipped and found solace within its walls.

Step inside the church to discover a treasure trove of history, from the intricate stained glass windows and beautifully carved wooden pews to the ornate stone font and the evocative memorial plaques that line the walls. Each detail tells a story of faith, community, and the enduring human spirit, inviting you to imagine the lives and experiences of those who have called this special place home.

As you continue on your walk, keep an eye out for other signs of Great Bealings' rich history, from the charming thatched cottages and traditional farmhouses that dot the village to the ancient hedgerows and field boundaries that have shaped the landscape for generations. These tangible links to the past serve as a reminder of the deep roots and proud heritage that define this remarkable corner of Suffolk, offering a glimpse into the lives and stories of the people who have shaped this land over the centuries.

Birds to Spot:

- Skylarks: Listen for the joyous, unbroken song of these remarkable birds as they soar high above the fields and meadows, their melodic trills and whistles a celebration of the wide-open spaces and boundless freedom of the Suffolk countryside.
- Yellowhammers: Keep an eye out for the bright yellow plumage and distinctive "a-little-bit-of-bread-and-no-cheese" song of these charming buntings, their presence a welcome splash of colour and sound amidst the muted tones of the arable landscape.
- Linnets: Watch for the lively flocks of these small, streaky-brown finches as they flit among the hedgerows and field margins, their twittering calls and acrobatic flight a delightful addition to the natural soundtrack of your walk.

Rare or Interesting Insects and Other Wildlife:

- Rabbits and Hares: As you navigate the fields and woodland edges, be on the lookout for the quick, darting movements of these enchanting creatures, their soft fur and twitching noses a charming sight amidst the tranquil countryside.
- Butterflies: In the warmer months, the hedgerows and meadows come alive with the delicate flutter of butterfly wings, from the vibrant orange-and-brown of the Gatekeeper to the iridescent blue of the Common Blue, each one a mesmerizing display of nature's artistry.
- Dragonflies: Near ponds and streams, keep your eyes peeled for the iridescent shimmer and acrobatic flight of these ancient insects, their gossamer wings and jewel-like colours a captivating sight against the verdant backdrop of the Suffolk landscape.

Rare or Interesting Plants:

- Hedgerow Flowers: As you walk along the field margins and through the woodland, take a moment to appreciate the delicate blooms that adorn the hedgerows, from the frothy white of cow parsley to the vibrant pink of red campion, each one a testament to the incredible diversity and resilience of the plant kingdom.
- Ancient Woodland Indicators: In the wooded areas, look out for plants that hint at the ancient origins of these precious habitats, such as the delicate white stars of wood anemone, the lush green fronds of hart's tongue fern, and the pungent aroma of wild garlic, each one a living link to the primeval forests that once blanketed this land.
- Arable Wildflowers: Among the crops and at the field edges, keep your eyes peeled for the fleeting beauty of arable wildflowers, such as the vibrant blue of cornflowers, the delicate pink of corn cockle, and the sunny yellow of corn marigolds, their presence a reminder of the vital role that agriculture plays in shaping and sustaining the biodiversity of the Suffolk countryside.

Facilities and Accessibility:

The Great Bealings Circular Walk is a moderate route, with mostly level terrain and a mix of footpaths, bridleways, and quiet country lanes. However, it's essential to note that some sections of the walk may be uneven, muddy, or overgrown, particularly after periods of wet weather or during the growing season. As such, sturdy walking boots with good ankle support and grip are highly recommended to ensure a safe and comfortable experience.

While there are no public toilets or other facilities directly along the route, the village of Great Bealings does offer some amenities for walkers, including a charming pub, The Admirals Head, which serves refreshing drinks and delicious meals made with locally-sourced ingredients. This welcoming establishment provides an ideal spot to rest, refuel, and soak up the friendly atmosphere of the village before continuing your walk.

It's important to be aware that some parts of the walk may not be suitable for those with limited mobility or for pushchairs, due to the presence of stiles, kissing gates, and uneven ground. However, with careful planning and a willingness to take your time, most walkers will find this route to be an enjoyable and rewarding experience.

Seasonal Highlights:

The Great Bealings Circular Walk offers a wealth of delights and surprises throughout the year, with each season revealing new facets of beauty and wonder in this enchanting corner of Suffolk:

- Spring: As the countryside awakens from its winter slumber, the hedgerows and woodland burst into life with a profusion of delicate blooms, from the nodding heads of bluebells to the starry-white of wild garlic. The air is filled with the joyous songs of returning birds, and the fresh green of new leaves creates a vibrant canopy overhead, a celebration of the season's promise and renewal.
- Summer: In the long, languid days of summer, the landscape is at its most lush and verdant, the fields swaying with ripening crops and the meadows alive with the hum of insects. Wildflowers paint the hedgerows and verges in a kaleidoscope of colour, while the warm sunshine and gentle breezes create an idyllic atmosphere for a leisurely walk through the heart of the Suffolk countryside.
- Autumn: As the days grow shorter and the nights turn crisp, the Great Bealings landscape is transformed into a breathtaking tapestry of gold, russet, and amber, the turning leaves creating a fire of colour that warms the soul. The hedgerows are laden with the bounty of the harvest, from plump blackberries to glossy sloes, while the distant sounds of migrating birds and the earthy scent of fallen leaves create a poignant sense of the turning seasons.
- Winter: While the fields may be bare and the trees stripped of their leaves, the Great Bealings Circular Walk takes on a stark, elemental beauty in the winter months, the frosty ground and low sun creating a haunting, otherworldly atmosphere. Wrap up warm and enjoy the crunch of frozen leaves underfoot, the puffs of breath hanging in the chill air, and the profound sense of peace and solitude that comes from walking through a slumbering landscape, awaiting the first stirrings of spring.

Difficulty Level and Safety Considerations:

The Great Bealings Circular Walk is a moderate route, suitable for walkers of average fitness and mobility. While the terrain is mostly level, there are some sections of the walk that may be uneven, slippery, or overgrown, particularly after wet weather or during the summer months. As such, it's essential to wear appropriate footwear, such as sturdy walking boots with good grip and ankle support, and to take care when navigating these areas.

Walkers should also be aware of the presence of livestock in some of the fields along the route, particularly during the spring and summer months. It's important to follow the countryside code, keeping dogs on a lead and closing any gates behind you to ensure the safety of both animals and walkers alike.

As with any outdoor activity, it's crucial to be prepared for changeable weather conditions, so be sure to bring warm, waterproof layers, even on seemingly mild days. It's also a good idea to carry a map and compass or GPS device, as well as a charged mobile phone and a small first aid kit, in case of emergencies.

By following these simple precautions and using common sense, walkers can fully immerse themselves in the beauty and tranquility of the Great Bealings Circular Walk, creating lasting memories and a deeper appreciation for the natural wonders and rich heritage of this remarkable corner of Suffolk.

Nearby Attractions:

After your invigorating walk through the picturesque countryside of Great Bealings, why not extend your adventure and explore some of the other fascinating attractions that this beautiful corner of Suffolk has to offer? The surrounding area is home to a wealth of historic sites, charming villages, and stunning natural landscapes that are sure to captivate and inspire visitors of all ages and interests.

- Woodbridge: Just a short drive from Great Bealings lies the historic market town of Woodbridge, nestled on the banks of the River Deben. This charming town boasts a rich maritime heritage, a picturesque waterfront, and a thriving arts and culture scene. Explore the town's narrow streets and ancient buildings, visit the fascinating Tide Mill Living Museum, or simply relax and watch the boats go by from one of the many cafes and pubs that line the quayside.
- Sutton Hoo: For a truly unforgettable experience, head to the world-famous archaeological site of Sutton Hoo, just a few miles from Great Bealings. This incredible site, which dates back to the 7th century, is home to a series of ancient burial mounds that have yielded some of the most spectacular Anglo-Saxon treasures ever discovered in Britain. Take a guided tour of the site, marvel at the replicas of the famous Sutton Hoo helmet and other priceless artifacts, and immerse yourself in the fascinating history of this iconic landmark.
- Rendlesham Forest: Nature lovers and outdoor enthusiasts will delight in the stunning beauty of Rendlesham Forest, a vast expanse of ancient woodland and heathland that lies just a short distance from Great Bealings. With its miles of walking and cycling trails, its abundant wildlife, and its mysterious UFO legacy, Rendlesham Forest offers a truly unique and unforgettable experience for visitors of all ages and interests. Whether you're hoping to spot a rare bird, uncover the secrets of the forest's extraterrestrial past, or simply enjoy a peaceful walk in the dappled shade of the trees, this enchanting woodland is not to be missed.

By combining your Great Bealings Circular Walk with a visit to these and other nearby attractions, you'll have the opportunity to experience the very best of what this enchanting corner of Suffolk has to offer. From the timeless beauty of the countryside to the rich history and vibrant culture of the surrounding towns and villages, this is a region that truly offers something for everyone.

So why not make a day of it, or even plan a weekend getaway, and immerse yourself in the warmth, charm, and enduring allure of the Suffolk countryside? With its stunning landscapes, fascinating heritage, and welcoming communities, this special part of East Anglia promises to leave you with memories that will last a lifetime, and a deeper appreciation for the simple joys and profound wonders that can be found when we step off the beaten path and explore the world around us.

Rushmere Heath Walk - 2 miles (3.2 km), Circular - Estimated Time: 1-1.5 hours

Starting Point Coordinates, Postcode: The starting point for the Rushmere Heath Walk is at the Rushmere Heath car park (Grid Reference: TM 199 439, Postcode: IP4 3PD).
Nearest Car Park, Postcode: The nearest car park is at Rushmere Heath (Postcode: IP4 3PD).

Embark on a delightful 2-mile (3.2 km) circular walk through the enchanting landscape of Rushmere Heath, a precious remnant of the once-extensive heathlands that blanketed this corner of Suffolk. This easy route, with its mostly flat terrain and well-defined paths, offers the perfect opportunity for walkers of all ages and abilities to immerse themselves in the beauty and tranquility of this rare and captivating habitat.

As you set out from the Rushmere Heath car park, a sense of anticipation builds as the path leads you into a world of gently undulating heathland, where the soft hues of heather and gorse create a breathtaking tapestry that stretches out before you. With each step, you'll find yourself drawn deeper into the heart of this unique and fragile ecosystem, where the interplay of light and shadow, colour and texture, creates an ever-changing panorama that soothes the soul and invigorates the senses.

Follow the path as it winds through the heathland, keeping the well-manicured greens and fairways of the neighboring golf course to your right, a striking contrast to the untamed beauty of the heath itself. As you walk, take a moment to appreciate the incredible diversity of life that thrives in this seemingly sparse landscape, from the delicate blooms of ling and bell heather to the darting forms of lizards and the haunting calls of rare birds that make their home among the gorse and bracken.

Immerse yourself in the gentle rustling of the breeze through the heather and the distant hum of bees as they flit from flower to flower, each sound a reminder of the intricate web of life that sustains this precious habitat. As you continue along the path, you'll pass through pockets of woodland, where the cool shade of the trees offers a welcome respite from the sun and the chance to spot woodland birds and mammals that find shelter and sustenance among the branches.

At the far side of the heath, the path reaches a junction, inviting you to turn right and follow the trail as it loops back towards your starting point. As you navigate this section of the walk, take the time to reflect on the rich history and cultural significance of Rushmere Heath, a landscape that has been shaped by the interplay of human activity and natural processes for centuries. From its ancient origins as a site of prehistoric settlement to its more recent history as a treasured commons and a haven for rare and endangered species, this remarkable place stands as a testament to the enduring relationship between people and the natural world.

As you near the end of your walk and the car park comes into view once more, you'll carry with you a renewed sense of appreciation for the beauty and fragility of our heathland habitats, and a deeper understanding of the vital role they play in supporting biodiversity and enriching our lives. The Rushmere Heath Walk may be short in distance, but it is long on inspiration and insight, offering a precious opportunity to reconnect with nature and to rediscover the simple joys and profound truths that can be found when we step off the beaten path and open our hearts to the wonders that surround us.

Detailed Route Directions:

1. From the Rushmere Heath car park, head southeast, bearing 120°, following the footpath that leads into the heart of the heathland. (Estimated time: 5-10 minutes)

2. Continue along the path, bearing 100°, keeping the golf course to your right as you walk through the diverse and captivating heathland landscape. (Estimated time: 15-20 minutes)

3. As you reach the far side of the heath, you'll come to a junction. Turn right here, bearing 200°, following the path as it begins to loop back towards your starting point. (Estimated time: 10-15 minutes)

4. Follow the path as it winds through the heathland, bearing 280°, taking in the ever-changing panorama of colours and textures that define this unique habitat. (Estimated time: 15-20 minutes)

5. Continue along the path, now bearing 340°, as it passes through pockets of woodland and open heathland, each one revealing new facets of the landscape's beauty and diversity. (Estimated time: 10-15 minutes)

6. As you near the end of the walk, the path will guide you back to the Rushmere Heath car park, bearing 020°, where your journey began. (Estimated time: 5-10 minutes)

Best Features and Views:

- Immerse yourself in the captivating beauty of Rushmere Heath, with its gently undulating landscape of heather, gorse, and bracken, a rare and precious remnant of the once-extensive heathlands that defined this corner of Suffolk.
- Delight in the incredible diversity of life that thrives in this unique habitat, from the delicate blooms of ling and bell heather to the darting forms of lizards and the haunting calls of rare birds, each one a testament to the resilience and adaptability of nature.
- Enjoy the striking contrast between the untamed beauty of the heathland and the well-manicured greens and fairways of the neighbouring golf course, a reminder of the complex interplay between human activity and the natural world.

Historical Features of Interest:

Rushmere Heath has a rich and fascinating history that spans centuries, serving as a vital resource and a cherished commons for the local community. In prehistoric times, the heath was an important site of early human settlement, with archaeological evidence suggesting that people have lived and worked in this landscape for thousands of years.

Throughout the medieval period and beyond, Rushmere Heath served as a crucial source of fuel, building materials, and grazing land for local people, its resources carefully managed and maintained through a system of common rights and traditional practices. The heath also played a significant role in the social and cultural life of the community, serving as a gathering place for festivals, fairs, and other celebrations that brought people together and reinforced the bonds of kinship and belonging.

In more recent times, Rushmere Heath has been recognized as a vital haven for rare and endangered species, its unique mix of heathland, grassland, and woodland providing a sanctuary for plants and animals that have been pushed to the brink of extinction elsewhere. Today, the heath is carefully managed by local conservation organizations and volunteers, who work tirelessly to protect and preserve this precious landscape for future generations.

As you walk through the heath, take a moment to reflect on the countless generations of people who have shaped and been shaped by this remarkable place, from the ancient inhabitants who first set foot on its soil to the dedicated conservationists who work to safeguard its future. Each step you take is a connection to this rich and enduring history, a reminder of the timeless rhythms and cycles that have defined this landscape for centuries.

Birds to Spot:

- Dartford Warblers: These elusive and charismatic birds are a true specialty of heathland habitats, their distinctive song and secretive behaviour making them a prized sighting for birdwatchers. Listen for their scratchy, insect-like call, and watch for their small, dark forms as they flit among the gorse and heather, a testament to their remarkable adaptability and resilience.
- Nightjars: As dusk falls over the heathland, keep an ear out for the haunting, churring song of the nightjar, a mysterious and enigmatic bird that comes alive in the twilight hours. With their cryptic plumage and silent, moth-like flight, nightjars are a true wonder of the heathland, a reminder of the magic and mystery that can be found in even the most unassuming of places.
- Woodlarks: These delightful songbirds are a characteristic sound of the heathland, their melodious, fluting calls and distinctive display flights a joy to behold. Look for their rusty-brown plumage and streaked breasts as they forage among the heather and grassland, a charming and welcome presence in this unique and precious habitat.

Rare or Interesting Insects and Other Wildlife:

- Heathland Lizards: As you walk through the heath, keep an eye out for the quick, darting movements of the common lizard and the sand lizard, two of the UK's native reptile species that thrive in the warm, dry conditions of the heathland. These fascinating creatures are a vital part of the heathland ecosystem, their presence a sign of the health and vitality of this rare and precious habitat.
- Silver-studded Blue Butterflies: In the summer months, the heathland comes alive with the delicate beauty of the silver-studded blue butterfly, a rare and endangered species that depends on the specific conditions found in heathland habitats. Watch for the shimmering blue wings of the males as they flit among the heather and gorse, a mesmerizing sight that captures the ephemeral magic of the natural world.
- Raft Spiders: For those with a keen eye and a fascination for the smaller wonders of the heathland, the raft spider is a true marvel. These impressive arachnids, which can grow up to 7 cm in diameter, are found in the wet, boggy areas of the heath, where they hunt for prey on the surface of the water. Their intricate webs and striking patterns are a testament to the incredible diversity and adaptability of life in this unique and precious habitat.

Rare or Interesting Plants:

- Heather: The undisputed queen of the heathland, heather is the defining plant of this rare and precious habitat. From the delicate pink bells of ling heather to the rich, dark hues of bell heather, these hardy and adaptable plants create a stunning tapestry of colour and texture that blankets the heath in an ever-changing display of beauty and vitality.
- Gorse: With its vibrant yellow flowers and prickly, evergreen leaves, gorse is another iconic plant of the heathland, its distinctive aroma and vibrant hues a welcome sight throughout the year. This tough and resilient shrub provides shelter and sustenance for a wide range of heathland wildlife, from nesting birds to basking reptiles, its presence a vital component of the heathland ecosystem.
- Sundews: For those with a keen eye and a love of the unusual, the sundews of Rushmere Heath are a true wonder. These small, carnivorous plants, with their glistening, sticky leaves and delicate white flowers, are a fascinating example of the incredible adaptations that plants have evolved to thrive in the harsh conditions of the heathland. Look for them in the wetter, boggier areas of the heath, where they lie in wait for unsuspecting insects to fall prey to their trap-like leaves.

Facilities and Accessibility:

The Rushmere Heath Walk is a relatively easy and accessible route, with mostly flat terrain and well-defined paths suitable for walkers of all ages and abilities. However, it's important to note that some sections of the heath may be uneven, muddy, or overgrown, particularly after periods of wet weather or during the height of the growing season. As such, sturdy footwear with good grip and support is recommended to ensure a safe and comfortable walk.

While there are no public toilets or other facilities directly on the route, the nearby town of Ipswich offers a range of amenities for visitors, including cafes, shops, and public restrooms. The town centre is just a short drive or bus ride from the heath, making it easy to combine your walk with a visit to this historic and vibrant community.

It's worth noting that some parts of the walk may not be suitable for wheelchair users or those with limited mobility, due to the presence of uneven ground, gates, and other obstacles. However, much of the heath is relatively level and accessible, and with a bit of advance planning and preparation, most visitors should be able to enjoy the beauty and tranquility of this special place.

Seasonal Highlights:

The Rushmere Heath Walk offers a unique and captivating experience in every season, each one revealing new facets of beauty and wonder in this rare and precious habitat:

- Spring: As the days lengthen and the weather warms, the heathland comes alive with the vibrant colours and sweet scents of spring. The gorse bursts into bloom, its bright yellow flowers a beacon of hope and renewal, while the first tender shoots of heather emerge from the soil, their delicate hues a promise of the beauty yet to come. The air is filled with the song of returning birds, and the heathland hums with the energy of new life and growth.
- Summer: In the long, lazy days of summer, the heathland is at its most glorious, a sea of purple and pink as the heather reaches its peak of bloom. The air is heavy with the drone of bees and the chirping of grasshoppers, while butterflies flit from flower to flower in a mesmerizing dance of colour and grace. This is a time to bask in the warmth of the sun, to breathe in the heady scent of the heather, and to marvel at the incredible diversity and resilience of the heathland ecosystem.
- Autumn: As the days grow shorter and the nights turn crisp, the heathland takes on a new character, its colours shifting from the rich purples of summer to the warm golds and russets of autumn. The heather begins to fade, its blooms giving way to the rich, nutty scent of ripening seed heads, while the gorse blazes with a final burst of yellow before the winter sets in. This is a time of change and transformation, a reminder of the constant cycle of life and death that defines the natural world.
- Winter: While the heathland may seem bleak and lifeless in the depths of winter, there is still much to discover and appreciate in this quiet and contemplative season. The bare branches of the gorse and the muted hues of the heather create a stark and haunting beauty, while the low sun casts long shadows across the frosty ground. Listen for the gentle rustling of small mammals foraging among the leaf litter, and watch for the graceful silhouettes of deer as they pick their way through the heath, a reminder of the enduring spirit of the wild that persists even in the coldest and darkest of days.

Difficulty Level and Safety Considerations:

The Rushmere Heath Walk is a relatively easy and straightforward route, suitable for walkers of all ages and fitness levels. With mostly flat terrain and well-defined paths, this walk offers a gentle and accessible introduction to the beauty and diversity of the heathland habitat.

However, as with any outdoor activity, it's important to take certain precautions and to be aware of potential hazards. Some sections of the heath may be uneven, muddy, or slippery, particularly after periods of wet weather, so sturdy footwear with good grip is essential. It's also a good idea to wear long trousers and sleeves to protect against ticks and other biting insects, which can be prevalent in heathland habitats.

Walkers should also be mindful of the presence of wildlife on the heath, particularly adders, which are the UK's only venomous snake species. While adders are generally shy and non-aggressive, they may bite if disturbed or threatened, so it's important to stay alert and to give them plenty of space if encountered.

As with any walk in a natural area, it's always a good idea to carry a map and compass, as well as a fully charged mobile phone and a small first aid kit, in case of emergencies. It's also important to follow the countryside code, sticking to designated paths, closing gates behind you, and taking your litter home to help preserve the beauty and integrity of this special place.

By following these simple guidelines and using common sense, walkers can fully immerse themselves in the magic and wonder of Rushmere Heath, enjoying a safe and rewarding experience that nourishes both body and soul.

Nearby Attractions:

After your invigorating walk through the captivating landscape of Rushmere Heath, why not extend your adventure and explore some of the other fascinating attractions that this beautiful corner of Suffolk has to offer? The surrounding area is home to a wealth of historic sites, charming villages, and stunning natural wonders that are sure to captivate and inspire visitors of all ages and interests.

- Ipswich: Just a stone's throw from Rushmere Heath lies the vibrant and historic town of Ipswich, one of England's oldest continuously inhabited settlements. With its rich maritime heritage, stunning architecture, and thriving cultural scene, Ipswich offers a fascinating glimpse into the past and present of this unique corner of East Anglia. Explore the town's narrow streets and ancient buildings, visit the world-class museums and galleries, or simply relax and watch the world go by from one of the many cafes and restaurants that line the picturesque waterfront.
- Suffolk Coast & Heaths AONB: For those seeking a deeper immersion in the natural beauty of the region, the Suffolk Coast & Heaths Area of Outstanding Natural Beauty is a must-visit destination. This stunning landscape, which stretches along the coast from the River Stour to the edge of Ipswich, encompasses a diverse range of habitats, from sandy beaches and saltmarshes to ancient woodlands and wildflower meadows. Whether you're a keen birdwatcher, a seasoned hiker, or simply a lover of breathtaking scenery, this incredible area has something to offer every visitor.
- Dedham Vale: No visit to this corner of Suffolk would be complete without a trip to the iconic Dedham Vale, the landscape that inspired many of John Constable's most famous paintings. This picturesque valley, which straddles the border between Suffolk and Essex, is a living testament to the enduring beauty and timeless charm of the English countryside. Wander through the quaint villages and rolling hills that Constable so lovingly depicted, visit the museum dedicated to his life and work, or simply sit and soak up the peaceful atmosphere of this enchanting place, a world away from the stresses and strains of modern life.

By combining your Rushmere Heath Walk with a visit to these and other nearby attractions, you'll have the opportunity to experience the very best of what this enchanting corner of Suffolk has to offer. From the timeless beauty of the countryside to the rich history and vibrant culture of the region's towns and villages, this is an area that truly offers something for everyone.

So why not make a day of it, or even plan a longer stay, and immerse yourself in the warmth, charm, and enduring allure of the Suffolk landscape? With its stunning scenery, fascinating heritage, and welcoming communities, this special part of East Anglia promises to leave you with memories that will last a lifetime, and a deeper appreciation for the simple joys and profound wonders that can be found when we step off the beaten path and explore the world around us.

As you set out on your adventures, remember to tread lightly, to leave no trace, and to approach each encounter with an open heart and a curious mind. By doing so, you'll not only ensure that these beautiful landscapes remain pristine and unspoiled for generations to come, but you'll also open yourself up to the countless opportunities for growth, discovery, and transformation that nature has to offer.

In the words of the great naturalist and conservationist John Muir, "In every walk with nature, one receives far more than he seeks." So step out into the magic and wonder of Rushmere Heath and beyond, and allow yourself to be touched by the beauty, resilience, and enduring spirit of this remarkable corner of the world. Who knows what surprises and delights await you on the path ahead?

Newbourne Springs Walk - 5 miles (8 km), Circular - Estimated Time: 2.5-3 hours

Starting Point Coordinates, Postcode: The starting point for the Newbourne Springs Walk is at the car park of Newbourne Village Hall (Grid Reference: TM 277 463, Postcode: IP12 4NP).
Nearest Car Park, Postcode: The nearest car park is at the Newbourne Village Hall (Postcode: IP12 4NP).

Embark on a captivating 5-mile (8 km) circular walk through the enchanting landscape of Newbourne Springs Nature Reserve, where a tapestry of woodland, wetland, and heathland habitats awaits. This moderately challenging route offers a perfect escape from the hustle and bustle of everyday life, inviting you to immerse yourself in the tranquility and beauty of the Suffolk countryside.

As you set out from the car park at Newbourne Village Hall, a sense of anticipation builds as you follow the footpath southward, the gentle rustling of leaves and the distant birdsong a soothing soundtrack to your journey. With each step, you'll find yourself drawn deeper into the heart of the reserve, where the interplay of light and shadow, colour and texture, creates an ever-changing panorama that soothes the soul and invigorates the senses.

Navigating the waymarked trails through the reserve, you'll encounter a rich mosaic of habitats, each one revealing new facets of the landscape's diversity and charm. From the cool, dappled shade of the woodland to the lush, verdant wetlands, every turn of the path brings fresh wonders to discover. Take a moment to appreciate the incredible variety of plant and animal life that thrives in this special place, from the delicate blooms of orchids and marsh marigolds to the darting forms of dragonflies and the haunting calls of rare birds.

As you continue along the footpath, heading southward, you'll cross a picturesque railway bridge, a reminder of the area's rich history and the enduring connection between human activity and the natural world. From here, the path leads you towards the charming village of Waldringfield, where you'll turn left onto a footpath that follows the meandering course of the River Deben.

Immerse yourself in the stunning views of the riverscape, the sparkling waters and gently swaying reeds a mesmerizing sight that invites contemplation and wonder. Keep an eye out for the vibrant flash of a kingfisher or the graceful silhouette of a heron, each encounter a reminder of the precious and fragile beauty of the natural world.

As you follow the river's path back towards Newbourne, the landscape opens up into a patchwork of fields and woodland edges, the golden hues of ripening crops and the soft greens of the hedgerows a testament to the enduring rhythm of the seasons. Here, in the quiet spaces between the cultivated land and the untamed wilderness, you may spot the telltale signs of the reserve's more elusive residents, from the distinctive tracks of a water vole to the delicate wings of a white admiral butterfly.

Nearing the end of your journey, the footpath guides you once more through the tranquil embrace of the woodland, the ancient trees standing as silent guardians of this timeless landscape. As you emerge from the dappled shade and catch sight of the Newbourne Village Hall car park, you'll carry with you a renewed sense of connection to the natural world and a deepened appreciation for the beauty and resilience of the Suffolk countryside.

The Newbourne Springs Walk is a testament to the enduring power of nature to inspire, heal, and transform, offering a precious opportunity to step beyond the boundaries of our everyday lives and immerse ourselves in the wonder and majesty of the living world. Whether you're a seasoned hiker or simply a lover of the great outdoors, this remarkable journey promises to leave you with memories that will last a lifetime.

Detailed Route Directions:

1. From the Newbourne Village Hall car park, head south, bearing 170°, along the footpath to reach the entrance of the Newbourne Springs Nature Reserve. (Estimated time: 10-15 minutes)

2. Follow the waymarked trails through the reserve, bearing 200°, taking in the stunning woodland and wetland habitats. (Estimated time: 30-40 minutes)

3. Continue along the footpath to the south, bearing 190°, crossing over the railway bridge and heading towards the village of Waldringfield. (Estimated time: 20-30 minutes)

4. Turn left, bearing 90°, onto the footpath that runs alongside the River Deben, which will lead you back towards Newbourne. (Estimated time: 40-50 minutes)

5. Follow the footpath, bearing 20°, as it meanders through the fields and along the edge of the woodland. (Estimated time: 30-40 minutes)

6. Continue along the path, now bearing 350°, as it guides you through the final stretch of the reserve. (Estimated time: 20-30 minutes)

7. Arrive back at the Newbourne Village Hall car park, bearing 010°, where your journey began. (Estimated time: 5-10 minutes)

Best Features and Views:

- Discover the incredible diversity of habitats within the Newbourne Springs Nature Reserve, from the cool, dappled shade of the woodland to the lush, verdant wetlands, each one a testament to the rich tapestry of life that thrives in this special place.
- Delight in the stunning views of the River Deben, its sparkling waters and gently swaying reeds a mesmerizing sight that invites contemplation and wonder, a perfect backdrop for a moment of peaceful reflection.
- Immerse yourself in the tranquil beauty of the Suffolk countryside, with its patchwork of fields, ancient woodlands, and hedgerows, a living tapestry that showcases the timeless charm and enduring resilience of this remarkable landscape.

Historical Features of Interest:

While there are no specific historical sites or artifacts directly along the Newbourne Springs Walk, the area surrounding the reserve has a rich and fascinating history that spans centuries. The landscape bears the imprints of human activity dating back to the Roman, Saxon, and Medieval periods, a testament to the enduring relationship between people and the natural world.

As you walk through the fields and along the riverbank, take a moment to imagine the countless generations of farmers, fishermen, and local communities who have shaped and been shaped by this land over the centuries. From the ancient trackways and field boundaries to the distant ruins of long-abandoned settlements, each feature of the landscape tells a story of the lives and experiences of those who came before.

The presence of the railway bridge, which you'll cross during your walk, serves as a more recent reminder of the ways in which human ingenuity and industry have transformed the countryside. The arrival of the railway in the 19th century brought new opportunities and challenges to the area, connecting rural communities to the wider world and forever changing the face of the landscape.

As you immerse yourself in the tranquil beauty of the reserve, let your mind wander to the rich tapestry of human history that underlies this enchanting corner of Suffolk. Each step you take is a connection to the countless stories and experiences that have unfolded here, a reminder of the deep and enduring bond between people and the natural world.

Birds to Spot:

- Kingfishers: Keep your eyes peeled for the vibrant flash of blue and orange as these stunning birds dart along the riverbank, their swift and agile flight a marvel to behold. With a bit of patience and a keen eye, you may be lucky enough to spot a kingfisher perched on a branch, its iridescent plumage gleaming in the sunlight.
- Woodpeckers: Listen for the distinctive drumming sound of these striking birds as they tap their beaks against the trunks of trees in search of insects. The reserve is home to several species of woodpecker, including the green woodpecker with its distinctive red crown and the great spotted woodpecker with its bold black and white plumage.
- Warblers: From the melodious notes of the blackcap to the lively chirping of the chiffchaff, the reserve is a haven for a variety of warbler species. These small, often elusive birds are a delight to observe as they flit through the foliage, their intricate songs and calls a joyous celebration of the natural world.

Rare or Interesting Insects and Other Wildlife:

- Water Voles: As you walk along the riverbank and through the wetland areas of the reserve, keep an eye out for the telltale signs of water vole activity, from the distinctive burrows and feeding platforms to the occasional glimpse of these endearing creatures swimming in the water. The presence of water voles is a testament to the health and vitality of the reserve's aquatic ecosystems.
- Norfolk Hawker Dragonflies: In the summer months, the wetlands of Newbourne Springs come alive with the vibrant colours and acrobatic flight of the Norfolk hawker dragonfly. These impressive insects, with their emerald green eyes and striking brown and green patterning, are a true spectacle to behold as they dart and hover over the water's surface.
- White Admiral Butterflies: As you navigate the woodland paths, keep your eyes peeled for the delicate beauty of the white admiral butterfly, a rare and enchanting species that thrives in the dappled shade of the forest. With its striking white bands against a dark brown background, this elegant butterfly is a mesmerizing sight as it flutters among the trees.

Rare or Interesting Plants:

- Orchids: In the spring and summer, the reserve comes alive with the delicate blooms of various orchid species, from the vibrant purple of the early purple orchid to the subtle pinks and whites of the common spotted orchid. These enchanting flowers, with their intricate shapes and patterns, are a testament to the incredible diversity and beauty of the plant kingdom.
- Marsh Marigolds: As you explore the wetland areas of the reserve, you'll be treated to the vibrant yellow blooms of marsh marigolds, their cheerful faces a welcome sight amidst the lush green of the surrounding vegetation. These hardy plants, which thrive in damp and boggy conditions, are an important source of nectar for a variety of insects and pollinators.
- Bluebells: In the spring, the woodland floor of Newbourne Springs transforms into a mesmerizing carpet of blue, as thousands of delicate bluebells bloom in a spectacular display. The sight and scent of these

enchanting flowers have long captivated the hearts and minds of nature lovers, and a walk through the bluebell-strewn woodland is an unforgettable experience.

Facilities and Accessibility:

The Newbourne Springs Walk is a moderately challenging route, with some uneven terrain and muddy sections that may prove difficult for those with limited mobility. While the paths are generally well-maintained, it's essential to wear sturdy walking boots or shoes with good grip and ankle support to ensure a safe and comfortable experience.

It's important to note that there are no public toilets or other facilities directly along the walk route, so walkers should plan accordingly. The nearest amenities can be found in the nearby villages of Newbourne and Waldringfield, where you'll find a selection of pubs, cafes, and small shops.

Due to the nature of the terrain and the presence of stiles and kissing gates, the walk is not suitable for wheelchairs or pushchairs. Those with limited mobility or specific accessibility requirements should exercise caution and assess their ability to navigate the route based on their individual needs and circumstances.

Seasonal Highlights:

The Newbourne Springs Walk offers a unique and captivating experience in every season, each one revealing new facets of beauty and wonder within this enchanting corner of Suffolk:

- Spring: As the days grow longer and the sun's warmth returns, the reserve bursts into life with an explosion of colour and sound. The woodland floor is carpeted with a sea of bluebells, their delicate blooms and sweet fragrance a joyous celebration of the season's renewal. The air is filled with the melodic songs of returning birds, and the first butterflies and dragonflies begin to emerge, their vibrant wings a welcome sight after the long winter months.
- Summer: In the height of summer, the reserve is at its most lush and verdant, the trees in full leaf and the wetlands alive with the buzz of insects and the gentle croaking of frogs. The meadows are a riot of wildflowers, from the delicate pinks of ragged robin to the sunny yellows of buttercups, while the riverbanks are lined with the striking purple blooms of purple loosestrife. This is a time to bask in the abundance of life, to savour the warm sun on your face and the heady scents of the blooming landscape.
- Autumn: As the days shorten and the nights grow crisp, the reserve takes on a new character, its colours shifting from the vibrant greens of summer to the rich golds, reds, and browns of autumn. The woodland is a tapestry of falling leaves, their rustling a gentle accompaniment to the sounds of foraging birds and mammals. The wetlands are alive with the calls of migrating wildfowl, while the hedgerows are laden with the bounty of ripening berries and fruits, a feast for both wildlife and human foragers alike.
- Winter: While the reserve may seem quieter and more subdued in the depths of winter, there is still much to discover and enjoy in this season of stillness and contemplation. The bare branches of the trees create a stark and beautiful silhouette against the pale winter sky, while the frosty ground crunches underfoot, a satisfying sound in the crisp, cold air. Keep an eye out for the tracks and signs of the reserve's winter residents, from the delicate footprints of a fox to the distinctive heart-shaped prints of a roe deer, a reminder of the enduring cycle of life that persists even in the darkest and coldest of months.

Difficulty Level and Safety Considerations:

The Newbourne Springs Walk is a moderately challenging route, with some uneven terrain, muddy sections, and a total distance of 5 miles (8 km). While the walk is suitable for most people with a reasonable level of fitness, it's essential to be prepared for the physical demands of the journey and to take necessary precautions to ensure a safe and enjoyable experience.

As you set out on your walk, be sure to wear appropriate footwear, such as sturdy hiking boots or walking shoes with good grip and ankle support. The paths can be slippery and uneven in places, particularly after wet weather, so take care to watch your step and avoid any hazards.

It's also important to be aware of the presence of livestock in some of the fields along the route, particularly during the spring and summer months. Remember to follow the countryside code, keeping dogs on a lead when necessary and closing any gates behind you to ensure the safety of both animals and walkers.

As with any outdoor activity, it's crucial to be prepared for changeable weather conditions. Bring appropriate clothing and layers to suit the season, as well as a waterproof jacket and trousers in case of rain. It's also a good idea to carry a map and compass or GPS device, as well as a charged mobile phone and a small first aid kit, in case of emergencies.

By following these simple guidelines and using common sense, walkers can fully immerse themselves in the beauty and tranquility of the Newbourne Springs Walk, enjoying a safe and rewarding experience that nourishes both body and soul.

Nearby Attractions:

After your invigorating walk through the enchanting landscape of Newbourne Springs, why not extend your adventure and explore some of the other fascinating attractions that this beautiful corner of Suffolk has to offer? The surrounding area is home to a wealth of natural wonders, historic sites, and charming towns and villages that are sure to captivate and inspire visitors of all ages and interests.

- Suffolk Coast & Heaths AONB: Just a short distance from Newbourne Springs lies the breathtaking Suffolk Coast & Heaths Area of Outstanding Natural Beauty, a stunning landscape of sandy beaches, windswept heaths, and tranquil estuaries. Whether you're a keen birdwatcher, a seasoned hiker, or simply a lover of wide-open spaces and stunning coastal vistas, this exceptional area has something to offer everyone. Explore the many walking and cycling trails, visit the charming seaside towns of Aldeburgh and Southwold, or simply sit and soak up the peace and beauty of this remarkable corner of England.
- Woodbridge: For a taste of Suffolk's rich history and vibrant culture, be sure to visit the nearby town of Woodbridge, a charming market town nestled on the banks of the River Deben. With its picturesque waterfront, independent shops and cafes, and fascinating museums and galleries, Woodbridge is the perfect destination for a day of exploration and relaxation. Take a stroll along the river, visit the historic Tide Mill Living Museum, or simply sit and watch the boats go by from one of the many riverside pubs and restaurants.
- Sutton Hoo: Just a short drive from Newbourne Springs lies the world-famous archaeological site of Sutton Hoo, a fascinating glimpse into the rich history and culture of Anglo-Saxon England. This incredible site, which dates back to the 7th century, is home to a series of ancient burial mounds that have yielded some of the most spectacular treasures ever discovered in Britain, including the iconic Sutton Hoo helmet and a wealth of gold and silver artifacts. Take a guided tour of the site, explore the award-winning museum and visitor centre, and immerse yourself in the captivating story of this remarkable place and the people who once called it home.

By combining your Newbourne Springs Walk with a visit to these and other nearby attractions, you'll have the opportunity to experience the very best of what this enchanting corner of Suffolk has to offer. From the timeless beauty of the countryside to the rich history and vibrant culture of the region's towns and villages, this is an area that truly offers something for everyone.

As you explore the wonders of the Suffolk landscape, take a moment to reflect on the deep sense of connection and belonging that comes from immersing yourself in the natural world. The tranquil beauty of Newbourne Springs and the surrounding areas serves as a powerful reminder of the importance of preserving and cherishing these precious wild spaces, not just for our own well-being, but for the countless species that call them home.

In a world that often feels increasingly disconnected and fast-paced, the simple act of taking a walk through the countryside, breathing in the fresh air, and listening to the gentle sounds of nature can be a profoundly restorative and grounding experience. It's a chance to reconnect with the rhythms of the earth, to quiet the noise of our busy minds, and to rediscover a sense of wonder and appreciation for the world around us.

So whether you're a local resident or a visitor from further afield, be sure to add the Newbourne Springs Walk and the surrounding attractions to your must-visit list. With its stunning scenery, fascinating history, and warm, welcoming atmosphere, this is a destination that promises to leave you feeling refreshed, inspired, and more deeply connected to the natural world and your place within it.

As you set out on your adventures, remember to tread lightly, to leave no trace, and to approach each encounter with an open heart and a curious mind. By doing so, you'll not only ensure that these beautiful landscapes remain pristine and unspoiled for generations to come, but you'll also open yourself up to the countless opportunities for growth, discovery, and transformation that nature has to offer.

In the words of the great naturalist and conservationist Rachel Carson, "Those who contemplate the beauty of the earth find reserves of strength that will endure as long as life lasts." So step out into the wonder and majesty of the Suffolk countryside, and allow yourself to be touched by the power and resilience of the natural world. You never know what insights and revelations await you on the path ahead, but one thing is certain – you'll emerge from your journey with a renewed sense of awe, gratitude, and connection to the incredible web of life that surrounds and sustains us all.

Waldringfield Circular Walk - 4 miles (6.4 km), Circular - Estimated Time: 2-2.5 hours

Starting Point Coordinates, Postcode: The starting point for the Waldringfield Circular Walk is at the Waldringfield village green (Grid Reference: TM 285 465, Postcode: IP12 4QZ).
Nearest Car Park, Postcode: Limited on-street parking is available near Waldringfield village green (Postcode: IP12 4QZ).

Embark on a delightful 4-mile (6.4 km) circular walk through the charming village of Waldringfield and its enchanting surroundings, where the tranquil beauty of the Suffolk countryside and the picturesque River Deben intertwine to create a truly captivating landscape. This easy to moderate route offers a perfect blend of riverside paths, lush meadows, and serene woodlands, inviting you to immerse yourself in the timeless charm and natural wonders of this idyllic corner of East Anglia.

As you set out from the historic village green, a sense of anticipation builds as you follow the footpath southward, the gentle rustling of leaves and the distant calls of wading birds a soothing soundtrack to your journey. With each step, you'll find yourself drawn closer to the glistening waters of the River Deben, a hidden gem that has long been a haven for sailors, shipbuilders, and lovers of the great outdoors.

Upon reaching the riverbank, take a moment to breathe in the fresh, invigorating air and soak up the stunning views of the estuary, its sparkling waters and gently swaying reeds a mesmerizing sight that invites contemplation and wonder. As you follow the path westward, keep an eye out for the elegant silhouettes of curlews and redshanks, their distinctive calls and graceful movements a reminder of the rich biodiversity that thrives along the river's edge.

Continuing along the path, you'll soon find yourself on the outskirts of Waldringfield, where the village's nautical heritage and timeless charm become even more apparent. From the quaint thatched cottages to the historic boathouses that line the shore, every aspect of the village speaks to its deep connection with the water and the enduring spirit of the local community.

As you turn northward, the landscape opens up into a patchwork of lush meadows and tranquil woodlands, the vibrant greens and earthy hues a testament to the incredible fertility and diversity of the Suffolk countryside. Immerse yourself in the peaceful surroundings, listening for the gentle babbling of a small stream as you cross it, and take a moment to appreciate the intricate beauty of the wildflowers and grasses that sway in the gentle breeze.

Following the edge of the woods, you'll be treated to dappled sunlight filtering through the canopy, the cool shade a welcome respite on warmer days. Here, the air is filled with the soft rustling of leaves and the melodic songs of woodland birds, each sound a reminder of the intricate web of life that thrives in these quiet corners of the landscape.

As you turn eastward, the path guides you through more picturesque meadows, their vibrant colours and textures a feast for the senses. In the spring and summer months, these fields come alive with a dazzling array of wildflowers, from the delicate pinks of cuckoo flowers to the sunny yellows of buttercups, each bloom a tiny miracle of nature's artistry.

Nearing the end of your journey, you'll once again catch sight of Waldringfield village green, its historic buildings and friendly atmosphere a welcome sight after your invigorating walk. As you complete the circular route, take a moment to reflect on the beauty and tranquility of the landscapes you've explored, and the sense of connection and well-being that comes from immersing yourself in the great outdoors.

The Waldringfield Circular Walk is a celebration of the simple joys and timeless beauty that can be found in the Suffolk countryside, offering a precious opportunity to slow down, unwind, and reconnect with the natural world. Whether you're a local resident or a visitor from afar, this enchanting journey promises to leave you with lasting memories and a deepened appreciation for the incredible landscapes and communities that make this corner of England so special.

Detailed Route Directions:

1. From the Waldringfield village green, head south, bearing 180°, along the footpath that leads to the River Deben. (Estimated time: 10-15 minutes)

2. Upon reaching the riverbank, turn left and follow the path westward, bearing 270°, taking in the stunning views of the estuary and the rich birdlife that inhabits its shores. (Estimated time: 20-30 minutes)

3. As you reach the outskirts of Waldringfield, turn right and head north, bearing 000°, passing through picturesque meadows and woodlands. (Estimated time: 25-35 minutes)

4. Cross the small stream and continue along the edge of the woods, bearing 020°, immersing yourself in the peaceful surroundings and the gentle sounds of nature. (Estimated time: 20-30 minutes)

5. Turn right and head east, bearing 090°, traversing more beautiful meadows and wildflower-filled fields. (Estimated time: 25-35 minutes)

6. Finally, make your way back to the Waldringfield village green, bearing 135°, where your circular walk comes to an end. (Estimated time: 10-15 minutes)

Best Features and Views:

- Marvel at the stunning views of the River Deben, its sparkling waters and gently swaying reeds creating an enchanting backdrop for your walk, and offering a glimpse into the rich maritime history of the area.
- Delight in the picturesque meadows and wildflower-filled fields, their vibrant colours and delicate blooms a testament to the incredible diversity and beauty of the Suffolk countryside.
- Immerse yourself in the tranquil woodlands, where dappled sunlight, soft rustling leaves, and the melodic songs of birds create a serene and rejuvenating atmosphere, perfect for a moment of quiet reflection.

Historical Features of Interest:

As you explore the charming village of Waldringfield, take a moment to appreciate its rich maritime heritage and the enduring legacy of the shipbuilders, sailors, and river communities that have shaped this landscape for centuries. The village's picturesque riverside setting and historic buildings serve as a testament to the deep connection between the people of Waldringfield and the waters that have sustained them for generations.

One of the most striking examples of this nautical heritage can be found in the town's boathouses and yards, where skilled craftsmen have been building and repairing vessels for centuries. These workshops, with their weathered timber frames and salt-worn doors, are a living link to the past, offering a fascinating glimpse into the traditional techniques and materials that have been used to create the boats that have plied the River Deben and beyond.

As you walk along the riverbank, imagine the bustling scenes of yesteryear, when the shores would have been lined with boats of all shapes and sizes, from small fishing skiffs to majestic sailing ships. The air would have been filled with the sounds of hammering and sawing, the shouts of the workers, and the gentle lapping of the water against the hulls, a symphony of industry and innovation that has echoed through the ages.

Today, while the boatyards may be quieter and the pace of life more relaxed, the spirit of Waldringfield's maritime past can still be felt in every corner of the village. From the carefully preserved cottages and public houses to the memorial benches and weathervanes that adorn the waterfront, each element of the landscape tells a story of the people and traditions that have shaped this special place.

As you immerse yourself in the timeless beauty of Waldringfield and its surroundings, take a moment to reflect on the countless generations of men and women who have lived and worked here, their lives and labours intertwined with the rhythms of the river and the changing seasons. In doing so, you'll gain a deeper appreciation for the rich tapestry of history and culture that underlies this enchanting corner of Suffolk, and the enduring spirit of resilience and adaptability that has sustained its communities through the centuries.

Birds to Spot:

- Curlews: Listen for the haunting, bubbling calls of these distinctive wading birds as they probe the mudflats and saltmarshes along the River Deben, their long, curved bills perfectly adapted for extracting hidden prey from the soft sediment. With their mottled brown plumage and elegant, sweeping flight, curlews are a captivating sight and a true icon of the Suffolk coastline.
- Redshanks: Keep an eye out for these striking waders, with their bright orange-red legs and bold black-and-white wing patterns, as they forage along the river's edge or take to the air in a flurry of rapid wingbeats. The evocative "teu-hoo" call of the redshank is a familiar sound in the estuaries and wetlands of East Anglia, and a reminder of the rich biodiversity that thrives in these special habitats.
- Avocets: If you're lucky, you may catch a glimpse of these elegant black-and-white waders, with their slender, upturned bills and long, bluish legs, as they grace the shallow waters and mudflats of the River Deben. Once a rare sight in the UK, avocets have made a remarkable comeback in recent years thanks to dedicated conservation efforts, and are now a celebrated symbol of the resilience and beauty of our coastal birdlife.

Rare or Interesting Insects and Other Wildlife:

- Butterflies: In the spring and summer months, the meadows and hedgerows around Waldringfield come alive with the vibrant colours and delicate flutter of butterfly wings. From the striking orange-and-brown of the comma to the iridescent blues of the common blue and the holly blue, these enchanting insects are a delight to behold as they dance among the wildflowers and bask in the warm sunshine.
- Dragonflies: As you walk along the riverbank and through the damper areas of the meadows, keep an eye out for the iridescent shimmer and acrobatic flight of dragonflies and damselflies. These ancient insects, with their intricate wing patterns and jewel-like colours, are a fascinating and beautiful presence in the Suffolk countryside, and a reminder of the incredible diversity of life that thrives in our wetland habitats.
- Otters: Although elusive and often difficult to spot, otters have made a remarkable resurgence along the River Deben in recent years, thanks to improvements in water quality and habitat conservation. As you walk along the riverbank, look out for the tell-tale signs of otter activity, from the distinctive webbed footprints in the mud to the occasional glimpse of a sleek, furry head breaking the surface of the water, a thrilling reminder of the wild and untamed beauty of this special landscape.

Rare or Interesting Plants:

- Wildflower Meadows: In the spring and summer months, the meadows around Waldringfield burst into a dazzling display of colour and diversity, as countless species of wildflowers bloom in a glorious celebration of life and renewal. From the delicate pinks of cuckoo flowers and the soft purples of lady's smock to the vibrant yellows of cowslips and the deep blues of meadow cranesbill, each bloom is a tiny miracle of nature, a testament to the incredible resilience and adaptability of our native flora.
- Saltmarsh Plants: Along the edges of the River Deben, where the fresh water of the river meets the salt water of the sea, a unique and fascinating community of plants thrives in the challenging conditions of the saltmarsh. Look out for the fleshy leaves and pale pink flowers of sea thrift, the delicate white blooms of scurvy grass, and the striking purple spikes of sea lavender, each species a marvel of adaptation and survival in this harsh but beautiful environment.
- Woodland Flowers: As you walk through the dappled shade of the woodlands, keep an eye out for the delicate beauty of the woodland flowers that carpet the forest floor in the spring and early summer. From the nodding heads of bluebells and the starry white blooms of wood anemones to the delicate pinks of wild garlic and the cheerful yellows of primroses, these enchanting flowers are a reminder of the magic and wonder that can be found in even the quietest corners of the Suffolk countryside.

Facilities and Accessibility:

The Waldringfield Circular Walk is a relatively easy to moderate route, with mostly level terrain and well-maintained paths. However, it's important to note that some sections of the walk may be uneven, muddy, or slippery, particularly after periods of wet weather or during the winter months. As such, sturdy walking shoes or boots with good grip and support are highly recommended to ensure a safe and comfortable experience.

While there are no public toilets directly along the route, the village of Waldringfield does offer some amenities for walkers, including a charming pub, The Maybush, which serves delicious food and refreshing drinks. This welcoming establishment provides an ideal spot to rest, refuel, and soak up the friendly atmosphere of the village before continuing your walk.

It's worth noting that some parts of the walk may not be suitable for wheelchairs or pushchairs, due to the presence of kissing gates, uneven ground, and narrow paths. However, with careful planning and a willingness to take your time, most walkers will find this route to be an enjoyable and rewarding experience.

Seasonal Highlights:

The Waldringfield Circular Walk offers a wealth of delights and surprises throughout the year, with each season revealing new facets of beauty and wonder in this enchanting corner of Suffolk:

- Spring: As the land awakens from its winter slumber, the countryside around Waldringfield bursts into life, with vibrant greens, delicate blossoms, and the joyous songs of returning birds filling the air. The riverbanks and meadows are adorned with a colourful array of wildflowers, from the cheery yellows of lesser celandine to the soft blues of forget-me-nots, while the woodlands are carpeted with a mesmerizing sea of bluebells, their delicate scent and nodding heads a true celebration of the season.
- Summer: In the long, lazy days of summer, the landscape is at its most verdant and lush, with the meadows and hedgerows alive with the hum of insects and the gentle rustle of the warm breeze through the grasses. The riverbanks are a hive of activity, with wading birds, dragonflies, and damselflies all making the most of the abundant food and shelter, while the woodlands provide a cool, green sanctuary from the heat of the sun, the dappled light filtering through the leaves and creating an enchanting play of light and shadow.
- Autumn: As the days grow shorter and the air turns crisp, the countryside takes on a new character, with the

vibrant greens of summer giving way to a stunning palette of golds, oranges, and reds. The woodlands are a riot of colour, with the leaves of the trees turning to a fiery display of autumnal hues, while the hedgerows are laden with the rich bounty of fruits and berries, providing a vital food source for wildlife as they prepare for the winter ahead. The light takes on a softer, more mellow quality, and there is a sense of calm and tranquility that pervades the landscape, inviting quiet reflection and contemplation.
- Winter: While the countryside may be at its most stark and subdued in the depths of winter, there is still much to appreciate and enjoy on a crisp, clear day. The bare branches of the trees stand out in sharp relief against the pale sky, their intricate patterns and textures revealed in all their glory, while the frosty ground crunches underfoot, a satisfying sound in the stillness of the air. The river takes on a new character, with the low sun glinting off the surface and casting long shadows across the water, and there is a sense of peace and solitude that is both invigorating and deeply restorative.

Difficulty Level and Safety Considerations:

The Waldringfield Circular Walk is a relatively easy to moderate route, with mostly level terrain and well-defined paths. However, walkers should be aware of a few potential hazards and take necessary precautions to ensure a safe and enjoyable experience.

As with any outdoor activity, it's important to be prepared for changeable weather conditions. Even on seemingly mild days, it's a good idea to bring along a waterproof jacket and extra layers, as the wind and rain can pick up unexpectedly, particularly along the exposed riverbank. Sturdy, comfortable footwear with good grip is also essential, as some sections of the walk may be uneven, muddy, or slippery, especially after wet weather.

Walkers should also be mindful of the tides when walking along the River Deben. While the path is generally well above the high-water mark, it's always a good idea to check the tide times before setting out and to be aware of any areas where the path may become submerged or difficult to navigate at high tide.

It's also important to respect the wildlife and habitats along the route, sticking to the designated paths and leaving no trace of your visit. This not only helps to protect the delicate ecosystems of the area but also ensures that future generations will be able to enjoy the same beautiful landscapes and diverse flora and fauna that make this walk so special.

By following these simple guidelines and using common sense, walkers of all ages and abilities can immerse themselves in the beauty and tranquility of the Waldringfield Circular Walk, creating memories that will last a lifetime and forging a deeper connection with the natural world and the rich heritage of this remarkable corner of Suffolk.

Nearby Attractions:

After your invigorating walk around the picturesque village of Waldringfield and its stunning surroundings, why not take the opportunity to explore some of the other fascinating attractions that this beautiful part of Suffolk has to offer? The area is rich in history, culture, and natural wonders, with something to appeal to every interest and taste.

Just a short drive from Waldringfield is the charming market town of Woodbridge, with its picturesque riverfront, independent shops, and fascinating museums. Take a stroll along the River Deben, admiring the sailing boats and historic buildings, before popping into one of the many excellent cafes, pubs, or restaurants for a well-earned refreshment. Don't miss the chance to visit the town's iconic Tide Mill, one of the oldest working tide mills in the country, and a testament to the ingenuity and resilience of past generations.

For those with an interest in history and archaeology, the National Trust's Sutton Hoo is an absolute must-see. This world-famous site, located just a few miles from Waldringfield, is home to a remarkable Anglo-Saxon burial ground, where a treasure trove of gold and silver artifacts was discovered in 1939. Take a guided tour of the site, explore the fascinating exhibition hall, and marvel at the skill and craftsmanship of the ancient artisans who created these incredible objects.

Of course, no visit to this part of Suffolk would be complete without exploring the stunning Suffolk Coast and Heaths Area of Outstanding Natural Beauty. This vast, unspoiled landscape, stretching from the River Stour in the south to the bustling port town of Lowestoft in the north, is a paradise for nature lovers, walkers, and anyone seeking a sense of peace and tranquility. With its wide, sandy beaches, secluded coves, and sweeping vistas of sea and sky, this is a place where the stresses and strains of modern life simply melt away, leaving you refreshed, rejuvenated, and reconnected with the natural world.

Whether you're a history buff, a foodie, a nature lover, or simply someone in search of a little relaxation and escape, the area around Waldringfield truly has something for everyone. So why not make a day of it, or even plan a longer stay, and immerse yourself in the rich culture, stunning scenery, and warm hospitality of this very special corner of East Anglia?

As you explore the many delights and hidden gems of the Suffolk countryside and coast, take a moment to reflect on the incredible diversity and resilience of the natural world, and the ways in which it has shaped and been shaped by the human communities that have called this place home for countless generations. From the ancient burial mounds of Sutton Hoo to the timeless beauty of the River Deben, every aspect of the landscape tells a story of the enduring relationship between people and place, and the countless ways in which we are all connected to the living world around us.

So whether you're a local resident rediscovering the joys of your own backyard, or a visitor from further afield, seeking new horizons and fresh perspectives, the Waldringfield Circular Walk and its surroundings offer a wealth of opportunities for discovery, adventure, and personal growth. By opening yourself up to the beauty and wonder of the natural world, and the rich tapestry of human history and culture that is woven through it, you'll come away with a deeper appreciation for the incredible diversity and resilience of life in all its forms, and a renewed sense of your own place within the great web of being.

In the words of the renowned naturalist and conservationist, Sir David Attenborough, "No one will protect what they don't care about, and no one will care about what they have never experienced." By immersing yourself in the stunning landscapes, fascinating history, and vibrant communities of this special corner of Suffolk, you'll not only create lasting memories and forge deeper connections with the world around you, but you'll also be playing a vital role in ensuring that these precious places and experiences are protected and cherished for generations to come.

So lace up your walking boots, grab your camera and your sense of adventure, and set out on a journey of discovery that will nourish your body, mind, and soul. The Waldringfield Circular Walk and the many wonders of the Suffolk countryside and coast are waiting to be explored, and there's never been a better time to start your adventure than now. Happy trails!

Martlesham Heath Walk - 4 miles (6.4 km), Circular - Estimated Time: 2-2.5 hours

Starting Point Coordinates, Postcode: The starting point for the Martlesham Heath Walk is at the Martlesham Heath Village Green (Grid Reference: TM 258 470, Postcode: IP12 4RF).
Nearest Car Park, Postcode: There is on-street parking available near the Village Green (Postcode: IP12 4RF).

Embark on a delightful 4-mile (6.4 km) circular walk through the enchanting landscape of Martlesham Heath, where the rugged beauty of heathland and the tranquil charm of woodland intertwine to create a captivating tapestry of nature. This easy route offers the perfect opportunity to immerse yourself in the rich biodiversity and stunning vistas of the Suffolk countryside, providing a much-needed escape from the hustle and bustle of everyday life.

As you set out from the picturesque Martlesham Heath Village Green, a sense of anticipation builds as you follow the footpath westward, the gentle rustling of the breeze through the foliage and the distant birdsong a soothing soundtrack to your journey. With each step, you'll find yourself drawn deeper into a world of natural wonders, where the vibrant hues of heather and gorse paint the landscape in a mesmerizing palette of purples, pinks, and golds.

Upon reaching the edge of the heathland, turn left and allow the well-trodden path to guide you along the fringes of this precious habitat, where the open expanse of the heath gives way to the cool, dappled shade of the surrounding woodlands. As you walk, take a moment to appreciate the incredible diversity of life that thrives in this unique environment, from the elusive nightjars and woodlarks that haunt the heathland to the myriad butterflies and dragonflies that dance among the wildflowers.

Continuing along the path, you'll soon find yourself immersed in a world of ever-changing vistas, the rugged beauty of the heathland juxtaposed with the soft, rolling contours of the nearby countryside. Here, the wide-open skies and far-reaching views invite a sense of freedom and possibility, offering a welcome respite from the confines of modern life and a chance to reconnect with the simple pleasures of the great outdoors.

As you navigate the gentle undulations of the heath, the path will eventually lead you to a junction, where you'll turn left and venture into the heart of the heathland itself. Embrace the opportunity to explore this rare and precious habitat up close, marvelling at the intricate interplay of light and shadow, texture and colour that defines this extraordinary landscape.

Crossing the heathland, the path will curve gracefully to the right, guiding you through a small yet enchanting patch of woodland, where the cool, earthy scent of leaf litter and the soft crunch of twigs underfoot create a sense of intimacy and seclusion. Here, in the quiet spaces between the trees, you may catch a glimpse of the woodland's more elusive residents, from the majestic deer that roam the undergrowth to the tiny, iridescent beetles that scurry among the roots and fallen branches.

Emerging from the woodland, you'll arrive at another junction, where you'll turn right and head southward along a well-defined path that winds its way through the dappled shade of the trees. As you walk, take the time to observe the subtle changes in the forest's character, from the lush, verdant growth of spring and summer to the rich, burnished hues of autumn and the stark, sculptural beauty of winter.

Eventually, the path will lead you back to the edge of the heathland, where you'll once again find yourself immersed in the wild, untamed beauty of this iconic habitat. Follow the path eastward, drinking in the sweeping vistas and the ever-changing play of light and shadow across the landscape, until you finally arrive back at your starting point, the Martlesham Heath Village Green.

As you complete your circular journey, take a moment to reflect on the extraordinary natural heritage and timeless allure of Martlesham Heath, a landscape that has captivated the hearts and minds of generations. This walk, though easy in terms of terrain and distance, offers a profound and enriching experience, one that invites us to slow down, to observe, and to marvel at the incredible diversity and resilience of the natural world.

Whether you're a seasoned rambler or a curious first-time visitor, the Martlesham Heath Walk promises to delight and inspire, offering a welcome reminder of the beauty, tranquility, and enduring importance of our precious heathland habitats. So why not lace up your walking boots, grab your binoculars, and set out on an adventure that will nourish your body, mind, and soul, leaving you with memories to cherish and a renewed sense of connection to the living world around you?

Detailed Route Directions:

1. From the Martlesham Heath Village Green, head west along the footpath, bearing 270°. (Estimated time: 10-15 minutes)

2. Upon reaching the edge of the heathland, turn left and follow the path as it skirts the edge of the heathland and woodlands, bearing 200°. (Estimated time: 25-35 minutes)

3. Continue along the path, following the edge of the heath, bearing 240°, until you reach a junction. (Estimated time: 20-30 minutes)

4. At the junction, turn left, bearing 170°, and cross over the heathland. (Estimated time: 15-20 minutes)

5. Follow the path as it curves right, bearing 280°, and passes through a small patch of woodland. (Estimated time: 10-15 minutes)

6. At the next junction, turn right, heading south along a well-defined path through the woodland, bearing 190°. (Estimated time: 25-35 minutes)

7. The path will eventually lead you back to the edge of the heathland. Follow the path along the edge of the heathland, heading east, bearing 80°, until you reach the starting point at the Village Green. (Estimated time: 20-30 minutes)

Best Features and Views:

- Immerse yourself in the rugged beauty of Martlesham Heath, where the vibrant hues of heather and gorse create a mesmerizing tapestry that shifts and changes with the seasons, offering a stunning display of nature's artistry.
- Delight in the sweeping vistas and far-reaching views of the surrounding countryside, where the wide-open skies and gently rolling hills invite a sense of freedom, tranquility, and connection to the natural world.
- Discover the rich biodiversity of the heathland and woodland habitats, from the haunting calls of nightjars and woodlarks to the delicate beauty of butterflies and wildflowers, each a testament to the incredible resilience and adaptability of life in these unique environments.

Historical Features of Interest:

Although there are no significant historical features directly along the Martlesham Heath Walk, the landscape itself serves as a living testament to the rich natural and cultural heritage of the Suffolk countryside. The heathland, with its distinctive flora and fauna, is a rare and precious habitat that has been shaped by centuries of human activity, from the grazing of livestock to the harvesting of heather and gorse for fuel and fodder.

In the past, heathlands like Martlesham Heath were an integral part of the rural economy, providing a vital resource for local communities and supporting a way of life that was deeply connected to the rhythms and cycles of the natural world. Today, these habitats are recognized as important sites for biodiversity conservation, with many species of plants and animals depending on the unique conditions found in heathland environments.

As you walk through the heath and woodland, take a moment to imagine the generations of people who have lived and worked in this landscape, their lives and livelihoods intertwined with the ebb and flow of the seasons and the ever-changing character of the land. From the ancient peoples who first cleared the forest to create open grazing land to the modern-day conservationists who work to protect and restore these precious habitats, each has played a role in shaping the Martlesham Heath we see today.

In the quiet solitude of the heath, with the wind whispering through the heather and the distant calls of birds echoing across the landscape, you may find yourself feeling a deep sense of connection to the natural world and the countless stories and experiences that have unfolded here over the centuries. This is the true gift of Martlesham Heath, the opportunity to step out of the present moment and into a timeless realm of beauty, resilience, and enduring wonder.

Birds to Spot:

- Nightjars: As dusk falls over the heathland, listen for the distinctive "churring" call of these elusive birds, their cryptic plumage and silent, moth-like flight making them a challenge to spot. Nightjars are a true specialty of heathland habitats, their presence a sign of the rich diversity of insects and other prey that thrive in these unique environments.
- Woodlarks: Keep an ear out for the melodious, fluting song of these small, brown birds as they soar above the heathland, their distinctive display flights a joyous celebration of the wide-open spaces and the freedom of the skies. Woodlarks are a characteristic species of heathland and open country, their presence a testament to the importance of these habitats for birds and other wildlife.
- Stonechats: Watch for the bold, striking plumage of these lively little birds as they perch atop gorse bushes and heather, their sharp "chack" calls and restless, flicking tails a constant presence on the heath. Stonechats are a resilient and adaptable species, their ability to thrive in the harsh conditions of the heathland a reminder of the incredible tenacity and resourcefulness of nature.

Rare or Interesting Insects and Other Wildlife:

- Butterflies: In the spring and summer months, the heathland comes alive with the vibrant colours and delicate flutter of butterfly wings, from the rich oranges and browns of the small copper to the iridescent blues and purples of the common blue. These enchanting insects are a vital part of the heathland ecosystem, their presence a sign of the health and diversity of the wildflower communities that sustain them.
- Dragonflies: As you walk along the paths and through the woodland, keep an eye out for the iridescent shimmer and acrobatic flight of dragonflies and damselflies, their jewel-like colours and intricate wing patterns a marvel of nature's artistry. From the large and powerful emperor dragonfly to the delicate and graceful demoiselle damselflies, these ancient insects are a fascinating and beautiful presence in the heathland and

woodland habitats.
- Deer: For those with a keen eye and a quiet step, the woodland edges and dense thickets of the heath may offer a glimpse of the secretive deer that call this landscape home. From the majestic red deer to the smaller and more elusive roe deer, these graceful creatures are a reminder of the rich tapestry of life that thrives in the diverse habitats of Martlesham Heath, their presence a testament to the enduring wildness and beauty of the natural world.

Rare or Interesting Plants:

- Heather: The undisputed queen of the heathland, heather is the defining plant of this rare and precious habitat, its delicate purple and pink flowers creating a stunning tapestry that stretches across the landscape. From the deep, rich hues of bell heather to the paler, more subtle shades of ling heather, these hardy and resilient plants are a vital source of food and shelter for a wide range of insects, birds, and other wildlife.
- Gorse: With its vibrant yellow flowers and sharp, prickly stems, gorse is another iconic plant of the heathland, its distinctive coconut-like scent and year-round blooms a welcome sight in even the harshest of conditions. Gorse provides a vital source of shelter and nesting sites for birds and small mammals, its dense, thorny thickets a safe haven from predators and the elements.
- Wildflowers: As you walk through the heath and woodland, take a moment to appreciate the delicate beauty and incredible diversity of the wildflowers that thrive in these unique habitats. From the cheerful yellow blooms of tormentil and the delicate pinks of common centaury to the striking purples of heath milkwort and the frothy whites of heath bedstraw, each species is a tiny miracle of adaptation and resilience, a testament to the rich and complex web of life that sustains the heathland ecosystem.

Facilities and Accessibility:

The Martlesham Heath Walk is a relatively easy route, with mostly level terrain and well-maintained paths suitable for walkers of all ages and abilities. However, it's important to note that some sections of the heath and woodland may be uneven, muddy, or slippery, particularly after periods of wet weather or during the winter months. As such, sturdy walking shoes or boots with good grip and support are highly recommended to ensure a safe and comfortable experience.

While there are no public toilets or other facilities directly along the walk route, the nearby village of Martlesham Heath does offer some amenities for visitors, including a small selection of shops, cafes, and pubs where you can grab a refreshment or a bite to eat before or after your walk.

It's worth noting that some parts of the walk may not be suitable for wheelchairs or pushchairs due to the presence of kissing gates, steps, and uneven ground. However, much of the route is relatively level and accessible, and with a bit of advance planning and preparation, most walkers should be able to enjoy the beauty and tranquility of this special landscape.

Seasonal Highlights:

The Martlesham Heath Walk offers a unique and captivating experience in every season, with each time of year revealing new facets of beauty and wonder in this enchanting corner of Suffolk:

- Spring: As the days lengthen and the sun's warmth returns, the heathland bursts into life with an explosion of colour and sound. The gorse is a blaze of vibrant yellow, its coconut-scented blooms a beacon for bees and other pollinators, while the first tentative shoots of heather emerge from the soil, their delicate hues a promise of the glory to come. The air is filled with the joyous songs of skylarks and the churring calls of

nightjars, the wings of countless butterflies and damselflies shimmering in the clear spring light.
- Summer: In the long, lazy days of high summer, the heathland reaches the peak of its beauty, the purple and pink hues of the heather stretching out in a mesmerizing carpet that seems to shimmer and dance in the heat. The air is heavy with the drone of bees and the chirping of crickets, while the woodland edges are alive with the fluttering of butterfly wings and the gentle rustle of leaves. This is a time to bask in the abundance of life, to savour the heady scents and the vibrant colours of the heathland in full bloom.
- Autumn: As the days grow shorter and the nights turn crisp, the heath takes on a new character, its colours shifting from the rich purples of summer to the warm golds and russets of autumn. The heather fades to a muted brown, its flowers giving way to the tiny, bead-like seed heads that will ensure its survival through the winter months. The woodland edges are a riot of colour, with the leaves of the trees turning to a blaze of oranges, reds, and yellows before their final descent to the forest floor. This is a time of change and transformation, a reminder of the cyclical nature of life and the endless adaptability of the natural world.
- Winter: While the heathland may seem stark and lifeless in the depths of winter, there is a haunting beauty to the landscape in its dormant state. The bare branches of the gorse and the skeletal stems of the heather stand out in sharp relief against the pale, watery light of the winter sun, while the frosty ground crunches underfoot, a satisfying sound in the crisp, cold air. The woodland edges are a study in subtlety, with the muted greys and browns of the trees creating a sense of quiet stillness and introspection. Yet even in this season of rest and renewal, there are signs of life all around, from the tracks of deer and foxes in the snow to the twittering of flocks of overwintering birds in the branches above.

Difficulty Level and Safety Considerations:

The Martlesham Heath Walk is classified as an easy route, suitable for walkers of all ages and fitness levels. The terrain is mostly level and the paths are generally well-maintained, making for a comfortable and enjoyable experience. However, as with any outdoor activity, it's important to be aware of potential hazards and to take necessary precautions to ensure a safe and rewarding walk.

One of the main considerations when walking on heathland is the presence of uneven ground, hidden holes, and slippery surfaces, particularly after rain or during the winter months. It's essential to wear sturdy, comfortable footwear with good grip and ankle support, and to watch your step when navigating these areas.

Another factor to keep in mind is the exposed nature of the heathland, which can leave walkers vulnerable to the elements, particularly on hot, sunny days or in strong winds. Be sure to check the weather forecast before setting out and to dress appropriately for the conditions, with layers that can be easily added or removed as needed.

It's also important to stay on the designated paths and to avoid disturbing the delicate heathland vegetation or the wildlife that calls this habitat home. This not only helps to protect the fragile ecosystem of the heath but also ensures that you don't inadvertently stumble into any hidden hazards or get lost in the more remote areas of the landscape.

As with any walk in a natural area, it's always a good idea to carry a map and compass or a GPS device, as well as a fully charged mobile phone and a small first aid kit, just in case of emergencies. It's also wise to let someone know your intended route and estimated return time before setting out, so that they can raise the alarm if you fail to return as planned.

By following these simple guidelines and using common sense, walkers of all abilities can fully immerse themselves in the beauty and tranquility of the Martlesham Heath Walk, enjoying a safe and rewarding experience that nourishes both body and soul.

Nearby Attractions:

After your invigorating walk through the captivating landscape of Martlesham Heath, why not take the opportunity to explore some of the other fascinating attractions that this beautiful corner of Suffolk has to offer? The surrounding area is rich in history, culture, and natural wonders, with something to appeal to every interest and taste.

Just a short drive from Martlesham Heath is the charming market town of Woodbridge, nestled on the picturesque banks of the River Deben. This historic town boasts a wealth of independent shops, cafes, and restaurants, as well as a vibrant arts and cultural scene. Take a stroll along the riverside, admiring the beautiful architecture and the bustling marina, or visit one of the many museums and galleries that showcase the rich heritage and creativity of the region.

For those with a passion for history and archaeology, the world-famous Sutton Hoo site is an absolute must-see. Located just a few miles from Woodbridge, this incredible site is home to a remarkable Anglo-Saxon burial ground, where a treasure trove of gold and silver artifacts was unearthed in 1939. Take a guided tour of the site, marvel at the exquisite craftsmanship of the Sutton Hoo helmet and other precious objects, and immerse yourself in the fascinating story of this ancient kingdom and the people who once called it home.

Of course, no visit to this part of Suffolk would be complete without exploring the stunning natural beauty of the River Deben and the surrounding countryside. Whether you're a keen birdwatcher, a passionate conservationist, or simply someone who loves to get out and explore the great outdoors, this area offers endless opportunities for discovery and adventure. From the tranquil waters of the estuary to the rolling hills and ancient woodlands of the Deben Valley, every turn of the path reveals new wonders and delights, inviting you to connect with the timeless rhythms of the natural world.

As you explore the many attractions and hidden gems of this enchanting corner of East Anglia, take a moment to reflect on the incredible richness and diversity of the landscapes and communities that have shaped this region over countless generations. From the ancient peoples who first settled on the shores of the Deben to the modern-day artists, makers, and innovators who continue to draw inspiration from the beauty and vitality of the Suffolk countryside, each has left their mark on this special place, weaving a tapestry of stories and experiences that is as varied and fascinating as the land itself.

So why not make a day of it, or even plan a longer stay, and immerse yourself in the warm hospitality, stunning scenery, and enduring charm of the Suffolk heartland? Whether you're a local resident rediscovering the joys of your own backyard or a visitor from further afield, seeking new horizons and fresh perspectives, the Martlesham Heath Walk and its surroundings offer a wealth of opportunities for exploration, relaxation, and personal growth.

As you set out on your adventures, remember to tread lightly, to leave no trace, and to approach each encounter with an open heart and a curious mind. By doing so, you'll not only deepen your own understanding and appreciation of the natural world and its many wonders, but you'll also play a vital role in ensuring that these precious landscapes and habitats are protected and cherished for generations to come.

In the words of the great naturalist and conservationist, John Muir, "In every walk with nature, one receives far more than he seeks." So step out into the beauty and diversity of the Suffolk countryside, and allow yourself to be transformed by the simple yet profound act of putting one foot in front of the other. Who knows what surprises and revelations await you on the trail ahead, but one thing is certain – you'll emerge from your journey with a renewed sense of wonder, gratitude, and connection to the incredible web of life that surrounds and sustains us all. Happy trails!

Martlesham Creek Walk - 3 miles (4.8 km), Circular - Estimated Time: 1.5-2 hours

Starting Point Coordinates, Postcode: The starting point for the Martlesham Creek Walk is at the car park on Sandy Lane, Martlesham (Grid Reference: TM 257 469, Postcode: IP12 4SD).
Nearest Car Park, Postcode: The nearest car park is the starting point car park on Sandy Lane (Postcode: IP12 4SD).

Embark on a delightful 3-mile (4.8 km) circular walk along the picturesque Martlesham Creek, where the tranquil waters and lush countryside intertwine to create a captivating landscape that soothes the soul and invigorates the senses. This moderate route offers a perfect escape from the hustle and bustle of everyday life, inviting you to immerse yourself in the natural beauty and rich biodiversity of the Suffolk coast and heaths.

As you set out from the car park on Sandy Lane, a sense of anticipation builds as you follow the footpath eastward, the gentle lapping of the creek's waters and the distant calls of wading birds a soothing soundtrack to your journey. With each step, you'll find yourself drawn deeper into a world of ever-changing vistas and hidden wonders, where the ebb and flow of the tide and the play of light on the water create an enchanting tableau that shifts and transforms with every passing moment.

Keep your eyes peeled for the myriad bird species that call this area home, from the majestic herons and elegant egrets to the lively redshanks and oystercatchers that forage along the mudflats and saltmarshes. The creek is a haven for birdwatchers and nature enthusiasts alike, its rich tapestry of habitats providing a vital refuge for both resident and migratory species throughout the year.

As you continue along the footpath, hugging the creek's edge, take a moment to appreciate the subtle beauty and incredible diversity of the plant life that thrives in this unique environment. From the delicate blooms of sea lavender and the vibrant green shoots of samphire to the sturdy stems of cord-grass and the intricately patterned leaves of sea purslane, each species is a testament to the remarkable adaptability and resilience of nature in the face of the elements.

Immerse yourself in the peaceful atmosphere of the creek, the soft breeze carrying the scent of salt and the gentle rustling of the reeds a soothing balm for the senses. Here, in the quiet spaces between land and water, you may catch a glimpse of the creek's more elusive inhabitants, from the darting forms of small fish and crustaceans to the sleek, sinuous shapes of otters and water voles that make their home in the banks and channels.

As you approach the footbridge that spans the creek, take a moment to pause and drink in the stunning panorama that unfolds before you, the vast expanse of sky and water a humbling reminder of the raw power and beauty of the natural world. Cross the bridge, the change in perspective offering new insights and discoveries, before following the path as it turns north and then west, guiding you back towards your starting point.

Along the way, keep an eye out for the vibrant flashes of colour and iridescent wings of the many dragonflies and damselflies that hover and dart among the reeds and rushes, their acrobatic flights and intricate patterns a mesmerizing display of nature's artistry. These ancient insects, with their otherworldly beauty and remarkable adaptations, are a fascinating window into the complex web of life that thrives in the wetland habitats of Martlesham Creek.

As you near the end of your walk, the car park on Sandy Lane once again comes into view, the sense of accomplishment and renewal that comes from immersing oneself in the natural world a fitting reward for your efforts. Take a moment to reflect on the incredible journey you've just undertaken, the sights, sounds, and sensations of the creek forever etched in your memory, a reminder of the enduring power and resilience of the living world that surrounds and sustains us.

The Martlesham Creek Walk is a celebration of the simple joys and profound truths that can be found when we step off the beaten path and open our hearts to the wonders of the natural world. Whether you're a seasoned hiker or a curious first-time explorer, this enchanting circular route promises to delight and inspire, offering a much-needed respite from the stresses and distractions of modern life and a chance to reconnect with the timeless rhythms and cycles of the earth.

Detailed Route Directions:

1. From the car park on Sandy Lane, head east along the footpath, bearing 090°, following the creek's edge. (Estimated time: 20-30 minutes)

2. Continue along the footpath, keeping an eye out for various bird species as you go. The path will continue to follow the creek, bearing 100°. (Estimated time: 25-35 minutes)

3. Upon reaching the footbridge, cross over the creek, bearing 000°. (Estimated time: 5-10 minutes)

4. Once across the footbridge, follow the path as it turns north, bearing 350°, and then west, bearing 280°, heading back towards the starting point. (Estimated time: 30-40 minutes)

5. Continue along the path, now bearing 260°, as it leads you back to the car park on Sandy Lane, where your circular walk began. (Estimated time: 15-20 minutes)

Best Features and Views:

- Immerse yourself in the tranquil beauty of Martlesham Creek, its sparkling waters and gently swaying reeds creating a mesmerizing backdrop for your walk, inviting moments of peaceful reflection and quiet contemplation.
- Delight in the stunning views of the surrounding countryside, where the lush greens of the marshes and meadows meet the vast, open skies, offering a humbling reminder of the raw power and majesty of the natural world.
- Marvel at the incredible diversity of bird species that call this area home, from the graceful herons and egrets to the lively waders and songbirds, each one a vibrant thread in the rich tapestry of life that defines this unique and precious habitat.

Historical Features of Interest:

Although there are no specific historical landmarks or artifacts directly along the Martlesham Creek Walk, the landscape itself is a living testament to the rich natural and cultural heritage of the Suffolk coast and heaths. The creek and its surrounding wetlands have played a vital role in the lives of the local communities for centuries, providing a source of food, water, and transportation, as well as a place of beauty, solace, and inspiration.

As you walk along the footpath, imagine the countless generations of people who have lived and worked in harmony with this dynamic landscape, their lives shaped by the rhythms of the tides and the cycles of the seasons. From the ancient peoples who first settled along the shores of the creek, harvesting the bounty of the saltmarshes and mudflats, to the modern-day conservationists and nature enthusiasts who work tirelessly to protect and preserve this fragile ecosystem, each has left their mark on the land, contributing to the rich and varied history of this special place.

The presence of the creek itself, with its ever-shifting channels and constantly evolving shoreline, serves as a powerful reminder of the dynamic nature of the natural world, and the ways in which the forces of water, wind, and time have shaped and reshaped the landscape over countless millennia. As you cross the footbridge and gaze out over the vast expanse of the creek, consider the countless storms and tidal surges that have swept through this area, the ebb and flow of the water sculpting the banks and sandbars into new and unexpected forms, a testament to the enduring resilience and adaptability of the coastal environment.

In the quiet moments of your walk, as the sounds of the outside world fade away and the beauty of the creek envelops you, you may find yourself feeling a deep sense of connection to the natural world and the countless stories and experiences that have unfolded here over the ages. This is the true gift of the Martlesham Creek Walk, the opportunity to step out of the present moment and into a timeless realm of wonder, reflection, and discovery, where the boundary between the human and the natural world dissolves, and we are reminded of our place within the greater web of life.

Birds to Spot:

- Little Egrets: Keep an eye out for the striking white plumage and elegant, elongated neck of these small herons as they wade through the shallows of the creek, their quick, darting movements a testament to their remarkable agility and precision as they hunt for small fish and invertebrates.
- Redshanks: Listen for the distinctive "teu-hoo" call of these vibrant waders as they forage along the mudflats and saltmarshes, their bright orange-red legs and bold black-and-white wing patterns a lively addition to the ever-changing tapestry of the creek's avian life.
- Reed Warblers: As you walk along the footpath, keep an ear out for the lively chattering and complex songs of these small, brown birds as they flit among the reeds and rushes, their melodies a joyous celebration of the vitality and resilience of the wetland habitats they call home.

Rare or Interesting Insects and Other Wildlife:

- Hairy Dragonfly: In the late spring and early summer, watch for the striking black-and-yellow patterned wings and the dark, hairy thorax of this impressive dragonfly species as it patrols the edges of the creek, its powerful flight and aggressive demeanour a captivating display of nature's raw energy and vitality.
- Water Vole: If you're lucky, you may catch a glimpse of these elusive and enchanting mammals as they swim along the edges of the creek or forage among the reeds and rushes, their small, rounded ears and furry bodies a charming sight in the dappled light of the wetland habitat.
- Common Darter: As you walk along the footpath, keep an eye out for the vibrant red bodies and distinctive black wing spots of these abundant dragonflies as they bask in the warm sunshine or dart across the surface of the water, their aerial acrobatics a mesmerizing display of speed, agility, and grace.

Rare or Interesting Plants:

- Sea Lavender: In the late summer and early autumn, delight in the delicate purple blooms and slender stems of this beautiful saltmarsh plant, its flowers adding a splash of colour and charm to the otherwise muted tones of the coastal vegetation.

- Golden Samphire: Discover the vibrant yellow flowers and succulent, segmented leaves of this edible wild plant as it grows in dense clusters along the upper reaches of the saltmarsh, its tangy, crisp flavour a beloved ingredient in many traditional coastal dishes.
- Cord-grass: Marvel at the tall, slender stems and gracefully arching leaves of this resilient saltmarsh grass, its extensive root systems playing a vital role in stabilizing the creek's banks and providing shelter and nutrients for a wide range of marsh-dwelling creatures.

Facilities and Accessibility:

The Martlesham Creek Walk is a moderate route that includes some uneven terrain and potentially muddy sections, making it unsuitable for wheelchairs and pushchairs. It's essential to wear sturdy, comfortable footwear with good grip and support to ensure a safe and enjoyable experience.

While there are no facilities directly along the walk route, the nearby village of Martlesham does offer some amenities for visitors, including a small selection of shops, pubs, and cafes where you can grab a refreshment or a bite to eat before or after your walk.

It's important to note that there are no public toilets or other facilities along the route, so plan accordingly and make sure to bring enough water, snacks, and any necessary medications or sun protection to ensure a comfortable and safe journey.

As always, it's a good idea to check the local tide times and weather conditions before setting out, as the creek's water levels and the condition of the footpath can be affected by tidal changes and rainfall. By being prepared and taking necessary precautions, walkers of all ages and abilities can immerse themselves in the beauty and tranquility of the Martlesham Creek Walk, forging a deeper connection with the natural world and creating memories that will last a lifetime.

Seasonal Highlights:

The Martlesham Creek Walk offers a unique and captivating experience in every season, with each time of year revealing new facets of beauty and wonder in this dynamic coastal landscape:

- Spring: As the land awakens from its winter slumber, the creek and its surroundings burst into life, with vibrant green shoots emerging from the saltmarshes and a symphony of birdsong filling the air. The arrival of migratory birds, such as the elegant little terns and the lively whimbrels, adds an extra layer of excitement and energy to the already bustling wetland scene.
- Summer: In the long, languid days of summer, the creek is alive with the buzzing of insects and the gentle lapping of the water against the shore. The saltmarshes are a riot of colour, with the delicate blooms of sea lavender and the vibrant yellow of golden samphire creating a stunning contrast against the lush greens of the cord-grass and the muted tones of the mudflats. This is a time of abundance and vitality, with the creek's many inhabitants making the most of the warm sunshine and the rich bounty of food.
- Autumn: As the days grow shorter and the nights turn crisp, the creek takes on a new character, with the fading light casting long shadows across the water and the first hints of golden hues appearing in the saltmarsh vegetation. The departure of the summer visitors and the arrival of overwintering birds, such as the majestic brent geese and the lively dunlins, mark a shift in the rhythm of life along the creek, a reminder of the endless cycle of the seasons.
- Winter: While the creek may seem still and silent in the depths of winter, there is a stark, elemental beauty to the landscape in this season of rest and renewal. The bare branches of the trees and the frost-covered reeds create a haunting, otherworldly atmosphere, while the low sun casts a pale, ethereal light across the water and the mudflats. Wrap up warm and enjoy the crunch of frozen ground beneath your feet, the puffs of

breath hanging in the chill air, and the profound sense of peace and solitude that comes from walking in a landscape stripped back to its essential elements.

Difficulty Level and Safety Considerations:

The Martlesham Creek Walk is a moderate route that includes some uneven terrain and potentially muddy sections, so a reasonable level of fitness and mobility is required. While the walk is not overly strenuous, it's essential to take necessary precautions and be prepared for the conditions to ensure a safe and enjoyable experience.

One of the main considerations when walking along the creek is the potential for slippery or unstable ground, particularly in areas where the footpath is close to the water's edge or crosses muddy sections of the saltmarsh. It's crucial to wear sturdy, non-slip footwear with good ankle support and to watch your step at all times, especially after rainfall or during high tide.

Another factor to keep in mind is the exposed nature of the coastal environment, which can leave walkers vulnerable to the elements, particularly on windy or rainy days. Be sure to check the weather forecast before setting out and to dress appropriately for the conditions, with layers that can be easily added or removed as needed, as well as a waterproof jacket and trousers in case of sudden downpours.

It's also important to be aware of the tidal nature of the creek and to check the local tide times before embarking on your walk. While the footpath is generally well above the high-water mark, some sections may become submerged or difficult to navigate during exceptionally high tides or storm surges, so it's always better to err on the side of caution.

As with any walk in a natural area, it's a good idea to carry a map and compass or a GPS device, as well as a fully charged mobile phone and a small first aid kit, in case of emergencies. It's also wise to let someone know your intended route and estimated return time before setting out, so that they can raise the alarm if you fail to return as planned.

By following these simple guidelines and using common sense, walkers of all ages and abilities can safely enjoy the beauty and tranquility of the Martlesham Creek Walk, immersing themselves in the wonders of the natural world and forging a deeper connection with the incredible coastal landscapes of Suffolk.

Nearby Attractions:

After your refreshing walk along the picturesque Martlesham Creek, why not explore some of the other fascinating attractions that this corner of Suffolk has to offer? The surrounding area is rich in history, culture, and natural beauty, with plenty of opportunities for discovery and adventure.

For aviation enthusiasts, a visit to the Suffolk Aviation Heritage Museum is a must. Located just a short drive from Martlesham, this fascinating museum is dedicated to preserving the rich aviation history of the region, with a particular focus on the role of the Royal Air Force and the United States Army Air Forces during World War II. Explore the museum's collection of vintage aircraft, artifacts, and photographs, and learn about the brave men and women who served in the skies over East Anglia.

If you're in the mood for a bit of luxury and pampering, head to the nearby Kesgrave Hall, a stunning Grade II listed mansion set in 38 acres of beautiful grounds. This elegant hotel and restaurant offers a range of indulgent spa treatments, as well as delicious seasonal cuisine crafted from the finest local ingredients. Whether you're looking for a romantic getaway, a family celebration, or simply a chance to unwind and recharge, Kesgrave Hall is the perfect destination.

For those who want to delve deeper into the history and culture of the region, the charming town of Woodbridge is just a short drive away. This historic market town, nestled on the banks of the River Deben, is a treasure trove of ancient buildings, independent shops, and cultural attractions. Take a stroll along the picturesque Thoroughfare, lined with timber-framed houses and quaint tea rooms, or visit the town's iconic Tide Mill, one of the oldest working tide mills in the country. Don't miss the chance to explore the Woodbridge Museum, which tells the fascinating story of the town's maritime heritage and its role in the development of the wider area.

Of course, no visit to this part of Suffolk would be complete without experiencing the incredible natural beauty of the coast and countryside. The nearby Suffolk Coast & Heaths Area of Outstanding Natural Beauty is a paradise for walkers, birdwatchers, and anyone who loves the great outdoors. With its miles of unspoiled beaches, tranquil estuaries, and wild, open heaths, this stunning landscape is home to an incredible diversity of wildlife and offers endless opportunities for exploration and adventure.

Whether you're a keen hiker, a history buff, or simply someone who appreciates the finer things in life, the area around Martlesham Creek has something for everyone. So why not make a day of it, or even plan a longer stay, and immerse yourself in the rich culture, stunning scenery, and warm hospitality of this enchanting corner of East Anglia?

As you explore the many attractions and hidden gems of the Suffolk countryside and coast, take a moment to reflect on the incredible resilience and adaptability of the natural world, and the ways in which it has shaped and been shaped by the human communities that have called this place home for countless generations. From the ancient creek-dwellers who first navigated the tidal waters of Martlesham to the modern-day conservationists and nature-lovers who work to protect and preserve its fragile ecosystems, each has played a part in the ongoing story of this remarkable landscape.

So whether you're a lifelong resident rediscovering the joys of your own backyard, or a first-time visitor eager to explore new horizons, the Martlesham Creek Walk and its surroundings offer a wealth of opportunities for learning, growth, and connection. By immersing yourself in the beauty and complexity of the natural world, and opening yourself up to the stories and experiences of those who have gone before, you'll gain a deeper appreciation for the incredible web of life that surrounds and sustains us all.

As you set out on your adventures, remember to tread lightly, to leave no trace, and to approach each encounter with an open heart and a curious mind. By doing so, you'll not only enrich your own understanding and enjoyment of the world around you, but you'll also help to ensure that these precious landscapes and habitats are preserved and cherished for generations to come.

In the words of the great conservationist and author, Rachel Carson, "There is something infinitely healing in the repeated refrains of nature – the assurance that dawn comes after night, and spring after winter." So embrace the healing power of the natural world, and let the beauty and resilience of the Suffolk coast and countryside be your guide and your inspiration, as you navigate the ever-changing landscape of your own life's journey. Happy exploring!

Martlesham Circular Walk - 4 miles (6.4 km), Circular - Estimated Time: 2-2.5 hours

Starting Point Coordinates, Postcode: The starting point for the Martlesham Circular Walk is at St. Michael and All Angels Church in Martlesham (Grid Reference: TM 262 471, Postcode: IP12 4RW).
Nearest Car Park, Postcode: On-street parking is available near St. Michael and All Angels Church in Martlesham (Postcode: IP12 4RW).

Embark on an enchanting 4-mile (6.4 km) circular walk through the delightful village of Martlesham and its picturesque surroundings, where the tranquil charm of the Suffolk countryside and the rich tapestry of local history intertwine to create a truly captivating experience. This easy to moderate route offers a perfect blend of quiet lanes, woodland paths, and rolling fields, inviting you to immerse yourself in the timeless beauty and gentle rhythms of rural life.

As you set out from the historic St. Michael and All Angels Church, take a moment to appreciate the elegant architecture and peaceful atmosphere of this ancient place of worship, its weathered stones and soaring tower a testament to the enduring faith and craftsmanship of generations past. With each step along Church Lane, feel the stresses of modern life melt away, replaced by a sense of connection to the land and its stories.

Turning left onto The Street, the heart of Martlesham village, you'll find yourself transported to a world of quaint cottages, colourful gardens, and friendly faces, the essence of English country life distilled into a few picturesque streets. Pause to admire the traditional thatched roofs and half-timbered facades, each one a living link to the rich heritage and vibrant community spirit that define this special corner of Suffolk.

Just before reaching the inviting Red Lion Pub, a footpath on the left beckons you into the lush embrace of the surrounding countryside. Follow the well-trodden trail as it winds through sun-dappled fields and ancient woodlands, the soft rustling of leaves and the gentle chirping of birds a soothing soundtrack to your journey. Immerse yourself in the ever-changing tableau of colours and textures, from the vibrant wildflowers that paint the meadows to the gnarled trunks and sprawling canopies of the majestic oaks that stand sentinel over the landscape.

As you navigate the quiet country lanes that crisscross the area, take a moment to breathe in the fresh, clean air and savour the simple pleasures of a walk in nature. Let your mind wander to the countless generations of farmers, laborers, and villagers who have shaped this land over the centuries, their stories etched into every hedgerow and field boundary, a living testament to the enduring connection between people and place.

Passing through sun-drenched farmland and dappled woodland, the footpath eventually brings you to another tranquil lane, where you'll turn left and continue your exploration of the Martlesham countryside. Here, the landscape opens up before you, revealing sweeping vistas of gently rolling hills, patchwork fields, and distant church spires, a bucolic scene that seems to have sprung from the pages of a storybook.

As you loop back towards Martlesham, the footpath guides you through a final stretch of woodland and farmland, the changing light and shifting shadows creating an enchanting play of colour and texture that captivates the eye and soothes the soul. With each step, feel the worries of the world fall away, replaced by a profound sense of peace, gratitude, and wonder at the beauty and resilience of the natural world.

Finally, the towers of St. Michael and All Angels Church come into view once more, signaling the end of your circular journey. Take a moment to reflect on the sights, sounds, and sensations of your walk, the memories of this special place forever etched in your mind and heart. The Martlesham Circular Walk is more than just a physical journey; it is a celebration of the timeless beauty, rich history, and enduring spirit of the Suffolk countryside, an invitation to reconnect with the land, with ourselves, and with the simple joys of a life well-lived.

Detailed Route Directions:

1. From St. Michael and All Angels Church, head south on Church Lane, bearing 180°. (Estimated time: 5 minutes)

2. Turn left onto The Street, bearing 90°, and continue along the road. (Estimated time: 10 minutes)

3. Just before the Red Lion Pub, take the footpath on the left, bearing 45°, and follow it through fields and woodland. (Estimated time: 25-30 minutes)

4. Upon reaching a quiet country lane, turn left, bearing 0°, and walk for a short distance. (Estimated time: 5 minutes)

5. Take the footpath on the right, bearing 90°, and walk through more fields. (Estimated time: 20-25 minutes)

6. Turn left onto another quiet country lane, bearing 0°, and follow the lane until you reach a junction. (Estimated time: 15 minutes)

7. At the junction, turn right, bearing 135°, and continue along the road for a short distance. (Estimated time: 5 minutes)

8. Take the footpath on the left, bearing 45°, leading through farmland and woodland. (Estimated time: 25-30 minutes)

9. When the footpath meets another country lane, turn left, bearing 315°, and continue back to St. Michael and All Angels Church, where the walk ends. (Estimated time: 15-20 minutes)

Best Features and Views:

- Immerse yourself in the picturesque charm of Martlesham village, with its quaint cottages, colourful gardens, and friendly atmosphere, a quintessential example of English country life at its finest.
- Delight in the peaceful countryside surrounding the village, from the sun-dappled fields and ancient woodlands to the gently rolling hills and patchwork farmland, each vista a testament to the enduring beauty of the Suffolk landscape.
- Discover the rich history and elegant architecture of St. Michael and All Angels Church, a beautiful historical site that serves as a living link to the generations of worshippers who have found solace and inspiration within its walls.

Historical Features of Interest:

As you begin your walk at the historic St. Michael and All Angels Church, take a moment to appreciate the rich history and architectural beauty of this ancient place of worship. The church, which dates back to the 13th century, has served as a spiritual and social centre for the Martlesham community for over 700 years, its weathered stones and graceful arches a testament to the skill and devotion of the medieval craftsmen who built it.

Step inside the church to discover a treasure trove of historical features, from the intricately carved wooden pews and the stunning stained-glass windows to the ornate stone font and the evocative memorial plaques that line the walls. Each element tells a story of faith, community, and the enduring human spirit, inviting you to imagine the countless generations of worshippers who have found comfort, joy, and inspiration within these hallowed walls.

As you explore the churchyard, with its ancient yew trees and weathered gravestones, reflect on the lives and legacies of those who have come before, their stories woven into the very fabric of the landscape. From the humble farmers and labourers who toiled in the fields to the wealthy landowners and merchants who shaped the course of local history, each has left an indelible mark on the village and its surroundings.

Throughout your walk, keep an eye out for other signs of Martlesham's rich heritage, from the traditional thatched cottages and half-timbered houses that line the streets to the old field boundaries and woodland paths that have been trodden by countless feet over the centuries. These tangible links to the past serve as a reminder of the deep roots and enduring spirit that define this special corner of Suffolk, offering a glimpse into the lives and dreams of the people who have called this place home.

As you immerse yourself in the timeless beauty of the Martlesham countryside, let your mind wander to the rich tapestry of history that underlies every step of your journey. From the ancient Saxons who first settled in the area to the modern-day residents who continue to shape its character and identity, each generation has added its own unique thread to the ever-evolving story of this remarkable community.

Birds to Spot:

- Great Spotted Woodpeckers: Keep your eyes peeled for the striking black-and-white plumage and bright red underbelly of these charming birds as they flit among the trees, their distinctive drumming a familiar sound in the woodlands around Martlesham.
- Goldfinches: Listen for the sweet, tinkling songs of these colourful finches as they gather in small flocks along the hedgerows and field edges, their vibrant gold, red, and black plumage a delight to behold against the verdant backdrop of the countryside.
- Common Buzzards: Scan the skies for the broad wings and mottled brown plumage of these majestic raptors as they soar effortlessly over the fields and woodlands, their plaintive mewing calls a haunting presence in the tranquil landscape.

Rare or Interesting Insects and Other Wildlife:

- Marbled White Butterflies: In the summer months, the meadows and grasslands around Martlesham come alive with the delicate black-and-white checkered wings of these beautiful butterflies, their graceful flight and intricate patterning a mesmerizing sight to behold.
- Hazel Dormice: Though elusive and seldom seen, these enchanting little mammals make their home in the ancient woodlands and hedgerows of the area, their golden-brown fur and large, black eyes a charming reminder of the rich diversity of life that thrives in these quiet corners of the countryside.
- Stoats: As you walk along the field margins and woodland edges, keep an eye out for the sleek, agile form of these small but feisty predators, their chestnut-brown fur and distinctive black-tipped tails a fleeting but unforgettable sight in the dappled sunlight of the Suffolk countryside.

Rare or Interesting Plants:

- Bluebells: In the spring, the woodlands around Martlesham are transformed into a mesmerizing sea of blue, as countless bluebells burst into bloom beneath the dappled canopy of the trees. Take a moment to breathe in their delicate fragrance and marvel at the ephemeral beauty of these enchanting wildflowers.
- Wild Garlic: Follow your nose to the pungent aroma of wild garlic as you walk through the shaded woodland paths, the delicate white flowers and broad green leaves of this flavourful plant a welcome sight in the early spring months.
- Oxeye Daisies: In the summer, the meadows and field margins are adorned with the cheerful white and yellow blooms of oxeye daisies, their simple yet striking beauty a testament to the resilience and adaptability of nature in the face of an ever-changing landscape.

Facilities and Accessibility:

The Martlesham Circular Walk is a relatively easy route, with mostly level terrain and a mix of quiet lanes, footpaths, and tracks. However, it's important to note that some sections of the walk may be uneven, muddy, or overgrown, particularly after periods of wet weather or during the height of summer. As such, sturdy walking shoes or boots with good grip and support are highly recommended to ensure a safe and comfortable experience.

While there are no public toilets or other facilities directly along the route, the village of Martlesham does offer some amenities for walkers, including the welcoming Red Lion Pub, where you can enjoy a refreshing drink or a hearty meal before or after your walk. Please be aware that opening hours and availability may vary, so it's always a good idea to check ahead or plan accordingly.

It's also worth noting that some parts of the walk may not be suitable for those with limited mobility or for pushchairs, due to the presence of stiles, kissing gates, and uneven ground. However, with careful planning and a willingness to take your time, most walkers will find this route to be an enjoyable and rewarding experience.

Seasonal Highlights:

The Martlesham Circular Walk offers a unique and captivating experience in every season, with each time of year revealing new facets of beauty and wonder in this enchanting corner of Suffolk:

- Spring: As the land awakens from its winter slumber, the woodlands and hedgerows around Martlesham burst into life, with delicate wildflowers like bluebells and wild garlic carpeting the ground in a mesmerizing display of colour and fragrance. The air is filled with the joyous songs of nesting birds, and the fresh green leaves of the trees create a lush canopy overhead, a symbol of the renewal and hope that spring brings.
- Summer: In the long, sun-drenched days of summer, the countryside is alive with the buzzing of insects and the gentle rustling of the warm breeze through the fields of ripening wheat and barley. The meadows are a riot of colour, with oxeye daisies, poppies, and other wildflowers creating a vibrant tapestry that stretches to the horizon, while the cool shade of the woodlands offers a welcome respite from the heat of the day.
- Autumn: As the days grow shorter and the nights turn crisp, the landscape around Martlesham is transformed into a breathtaking palette of gold, orange, and red, the turning leaves of the trees creating a fiery display that warms the soul. The hedgerows are laden with the rich bounty of blackberries, sloes, and other wild fruits, while the distant sounds of migrating geese and the earthy scent of fallen leaves create a poignant sense of the turning seasons.
- Winter: While the fields may be fallow and the trees bare, the Martlesham Circular Walk takes on a stark, ethereal beauty in the depths of winter. The frosty ground and the low, pale light create a sense of quiet

introspection, while the occasional sighting of a red kite or a flock of fieldfares adds a spark of life and colour to the muted landscape. This is a time for brisk walks and cozy firesides, for savouring the simple pleasures and the warm hospitality of village life.

Difficulty Level and Safety Considerations:

The Martlesham Circular Walk is a relatively easy route, suitable for walkers of most ages and fitness levels. However, as with any outdoor activity, it's essential to be prepared for the conditions and to take necessary precautions to ensure a safe and enjoyable experience.

As you navigate the quiet lanes and footpaths around the village, be sure to stay alert to any traffic, particularly on the short stretches of road where there may not be a dedicated pedestrian path. Wear bright or reflective clothing to ensure that you're visible to drivers, and always walk facing oncoming traffic where possible.

When walking through fields with livestock, particularly during the spring and summer months, it's important to give the animals a wide berth and to keep dogs on a lead. Remember that even the friendliest-looking cattle can become agitated or territorial, especially when protecting their young, so it's always better to err on the side of caution.

As you make your way along the woodland paths and tracks, be mindful of any exposed roots, overhanging branches, or slippery surfaces, particularly after rainfall or during the autumn and winter months when fallen leaves can obscure potential hazards. Wear sturdy, comfortable footwear with good grip and support, and take your time to navigate any challenging sections.

It's also a good idea to carry a map and compass or GPS device, as well as a fully charged mobile phone and a small first aid kit, in case of emergencies. Let someone know your intended route and estimated return time before setting off, and be sure to carry enough water and snacks to keep you hydrated and energized throughout your walk.

By following these simple guidelines and exercising common sense, walkers of all abilities can fully immerse themselves in the beauty and tranquility of the Martlesham Circular Walk, forging a deeper connection with the natural world and creating memories that will last a lifetime.

Nearby Attractions:

After your invigorating walk through the picturesque countryside around Martlesham, why not extend your adventure and explore some of the other fascinating attractions that this beautiful corner of Suffolk has to offer? The surrounding area is rich in history, culture, and natural wonders, with something to appeal to every interest and taste.

Just a short drive from Martlesham is the stunning expanse of Martlesham Heath, a designated Site of Special Scientific Interest and a haven for rare wildlife and plants. This unique landscape, which was once home to a bustling airfield during World War II, now offers miles of walking and cycling trails, as well as a fascinating insight into the area's rich aviation heritage. Explore the atmospheric ruins of the old control tower, marvel at the vibrant heathland flowers, and keep an eye out for the elusive woodlarks and nightjars that make their home among the gorse and bracken.

For those with a passion for history and architecture, the nearby town of Woodbridge is an absolute must-visit. This charming market town, nestled on the banks of the River Deben, boasts an array of historic buildings, independent shops, and cultural attractions that are sure to delight and inspire. Take a stroll along the picturesque Thoroughfare, lined with timber-framed houses and quirky boutiques, or visit the town's iconic Tide Mill, one of the oldest working tide mills in the country. Don't miss the chance to explore the Woodbridge Museum, which tells the fascinating story of the town's maritime heritage and its role in the development of the wider region.

Of course, no visit to this part of Suffolk would be complete without experiencing the incredible natural beauty of the River Deben and the surrounding countryside. This tranquil waterway, which winds its way through the heart of the Suffolk Coast and Heaths Area of Outstanding Natural Beauty, offers endless opportunities for walking, cycling, birdwatching, and simply immersing yourself in the timeless rhythms of the natural world. Take a leisurely boat trip along the river, marvelling at the stunning scenery and the abundant wildlife, or pack a picnic and find a quiet spot along the banks to while away a lazy afternoon in the warm embrace of the Suffolk sun.

As you explore the many attractions and hidden gems of the Martlesham area, take a moment to reflect on the incredible richness and diversity of the landscapes and communities that have shaped this corner of England over countless generations. From the ancient woodlands and wildflower meadows to the historic churches and thatched cottages, every aspect of the region tells a story of the enduring relationship between people and place, and the countless ways in which we are all connected to the living world around us.

So whether you're a seasoned rambler or a curious first-time visitor, be sure to add the Martlesham Circular Walk and its surrounding attractions to your must-visit list. With its stunning scenery, fascinating heritage, and warm, welcoming atmosphere, this enchanting corner of Suffolk promises to delight and inspire you at every turn, leaving you with memories that will last a lifetime and a deeper appreciation for the simple joys and profound wonders of the English countryside.

As you set out on your adventures, remember to tread lightly, to leave no trace, and to approach each encounter with an open heart and a curious mind. By doing so, you'll not only enrich your own understanding and enjoyment of the natural world, but you'll also help to ensure that these precious landscapes and habitats are preserved and cherished for generations to come.

In the words of the great poet and naturalist, William Wordsworth, "Nature never did betray the heart that loved her." So let the beauty and wisdom of the Suffolk countryside be your guide and your inspiration, as you embark on a journey of discovery that will nourish your body, mind, and soul, and leave you with a renewed sense of wonder and connection to the incredible web of life that surrounds and sustains us all.

Happy wandering!

Woodbridge Riverside Walk - 4 miles (6.4 km), Linear - Estimated Time: 2-2.5 hours

Starting Point Coordinates, Postcode: The starting point for the Woodbridge Riverside Walk is at Woodbridge Railway Station (Grid Reference: TM 270 490, Postcode: IP12 4AU).
Nearest Car Park, Postcode: The nearest car park is the Station Road Car Park in Woodbridge (Postcode: IP12 4AU).

Embark on a delightful 4-mile (6.4 km) linear walk along the enchanting River Deben in Woodbridge, where the picturesque charm of the Suffolk countryside and the rich maritime history of this ancient market town intertwine to create a truly captivating experience. This easy route, suitable for all ages, offers stunning views of the river, the historic Tide Mill, and the surrounding landscape, inviting you to immerse yourself in the timeless beauty and gentle rhythms of life on the water's edge.

As you set out from Woodbridge Railway Station, a sense of anticipation builds as you cross the railway bridge and turn onto Quay Street, the gateway to the town's vibrant riverside community. With each step, feel the stresses of modern life melt away, replaced by a growing sense of connection to the land, the water, and the countless generations who have lived and worked along the banks of the River Deben.

Your first stop is the iconic Woodbridge Tide Mill, a fascinating 12th-century water mill that stands as a testament to the ingenuity and resilience of the local people. Take a moment to explore this beautifully preserved piece of living history, marvelling at the ancient machinery and the skill of the millers who have kept the wheels turning for over 800 years. As you watch the ebb and flow of the tide and listen to the gentle creaking of the wooden gears, you'll gain a deeper appreciation for the enduring connection between the town and the river that has shaped its destiny.

Leaving the Tide Mill behind, follow the riverside path as it winds along the banks of the Deben, revealing stunning vistas of the sparkling water and the lush green marshlands beyond. Breathe in the fresh, salt-tinged air and let your mind wander to the countless sailors, fishermen, and merchants who have navigated these waters over the centuries, their stories and adventures woven into the very fabric of the landscape.

As you pass by bustling boatyards, inviting pubs, and charming restaurants, take a moment to soak up the lively atmosphere and the friendly spirit of the riverside community. Watch the colourful sails of the boats bobbing gently on the water, and listen to the laughter and chatter of the people enjoying a meal or a drink by the river's edge. This is a place where the past and the present mingle seamlessly, creating a timeless tapestry of life and culture that is uniquely Woodbridge.

Continuing along the path, you'll soon leave the town behind and enter a world of open countryside, where the river meanders lazily through a patchwork of fields, woods, and wetlands. Here, the pace of life slows to a gentle rhythm, and the sounds of civilization fade away, replaced by the soothing melody of birdsong and the whisper of the wind through the reeds.

As you walk, keep your eyes peeled for the diverse array of wildlife that calls the River Deben home. From the elegant avocets and the lively redshanks that wade through the shallows, to the iridescent dragonflies and the elusive water voles that dart among the reed beds, each encounter is a reminder of the incredible biodiversity and ecological importance of this precious waterway.

Eventually, you'll reach the Wilford Bridge area, where you can pause to take in the stunning panorama of the river and the surrounding landscape. This is a place of quiet contemplation and deep connection to the natural world, where the timeless beauty of the Suffolk countryside and the enduring spirit of the River Deben come together in a symphony of colour, light, and life.

From here, you have the choice to either retrace your steps back to Woodbridge, savouring the views and the experiences of your outward journey, or to arrange for transport back to your starting point, carrying with you the memories and the insights of this unforgettable walk.

The Woodbridge Riverside Walk is more than just a physical journey; it is a celebration of the rich history, vibrant culture, and stunning natural beauty that define this enchanting corner of Suffolk. Whether you're a lover of the outdoors, a history enthusiast, or simply someone in search of a moment of peace and tranquility, this walk promises to delight and inspire you, leaving you with a deeper appreciation for the simple joys and profound wonders of life on the river's edge.

Detailed Route Directions:

1. From Woodbridge Railway Station, head east along Station Road, bearing 090°. (Estimated time: 5 minutes)

2. Cross the railway bridge and turn left onto Quay Street, bearing 000°. (Estimated time: 5 minutes)

3. Continue along Quay Street until you reach the historic Woodbridge Tide Mill. (Estimated time: 10 minutes)

4. Follow the riverside path along the River Deben, bearing 045°, passing boatyards, pubs, and restaurants. (Estimated time: 30 minutes)

5. Continue along the path as it takes you past Woodbridge Cruising Club, bearing 090°. (Estimated time: 20 minutes)

6. The path then leads you through open countryside, bearing 060°, with views of the river and marshlands. (Estimated time: 45 minutes)

7. Eventually, you'll reach the Wilford Bridge area. From here, you can either turn back and retrace your steps to Woodbridge, bearing 240°, or arrange for transport back to the starting point. (Estimated time: 60 minutes if returning on foot)

Best Features and Views:

- Marvel at the stunning views of the River Deben, its sparkling waters and gently flowing tides creating an ever-changing tableau of light and colour that captivates the eye and soothes the soul.
- Discover the historic Woodbridge Tide Mill, a 12th-century marvel of engineering and a testament to the ingenuity and resilience of the local people, its ancient walls and turning wheels a living link to the town's rich maritime heritage.
- Immerse yourself in the picturesque charm of the Suffolk countryside, with its lush green marshlands, rolling fields, and tranquil woodlands, a patchwork of natural beauty that stretches out as far as the eye can see.

Historical Features of Interest:

As you embark on your walk along the River Deben, take a moment to appreciate the rich history and cultural significance of the Woodbridge Tide Mill, a fascinating 12th-century water mill that stands as a testament to the town's enduring relationship with the river and the sea.

One of the oldest working tide mills in the country, this remarkable structure has been harnessing the power of the tides to grind wheat into flour for over 800 years, a living example of the ingenuity and resourcefulness of the people who have called this place home for generations. Step inside the mill and marvel at the ancient machinery, the wooden gears and cogs still turning with the ebb and flow of the tide, just as they have done for centuries.

Imagine the lives of the millers who have toiled within these walls, their days and nights governed by the rhythms of the river and the turning of the seasons. From the earliest days of the mill's existence, when the flour it produced was a vital staple for the local community, to the present day, when it stands as a beloved landmark and a symbol of Woodbridge's rich heritage, the Tide Mill has been a constant presence in the life of the town, a reminder of the deep and enduring connection between the people and the water that sustains them.

As you explore the mill and its surroundings, take a moment to reflect on the countless stories and experiences that have unfolded here over the centuries, from the daily labours of the millers and the comings and goings of the boats that once lined the quayside, to the laughter and chatter of the visitors who now flock to this special place to learn about its fascinating history and to soak up its unique atmosphere.

Throughout your walk, keep an eye out for other signs of Woodbridge's rich maritime heritage, from the bustling boatyards and the elegant sailing vessels that still ply the river's waters, to the historic pubs and warehouses that line the quayside, each one a reminder of the town's long and storied relationship with the sea.

In the tranquil beauty of the river and the timeless charm of the Tide Mill, you'll find a powerful sense of connection to the past, a feeling of being part of something much greater than yourself, a link in the long chain of human history that stretches back through the ages and connects us all to the enduring rhythms and cycles of the natural world.

Birds to Spot:

- Avocets: Keep your eyes peeled for these elegant black-and-white waders, with their slender, upturned bills and long, bluish legs, as they grace the mudflats and shallows of the River Deben. Once a rare sight in the UK, avocets have made a remarkable comeback in recent years, and are now a treasured symbol of the Suffolk coast and its incredible birdlife.
- Redshanks: Listen out for the distinctive "teu-hoo" call of these lively waders as they forage along the river's edge, their bright orange-red legs and bold black-and-white wing patterns a striking sight against the muted tones of the mudflats. With their energetic movements and charismatic presence, redshanks are a true delight to observe, and a reminder of the rich diversity of life that thrives along the River Deben.
- Curlews: As you walk along the river, keep an ear out for the haunting, bubbling call of these majestic waders, their long, down-curved bills and mottled brown plumage perfectly adapted for probing the mud and sand in search of tasty morsels. With their evocative calls and striking appearance, curlews are a true icon of the Suffolk coast, and a symbol of the wild and untamed beauty of the River Deben and its surrounding marshlands.

Rare or Interesting Insects and Other Wildlife:

- Dragonflies: In the summer months, the riverbanks and reed beds come alive with the vibrant colours and acrobatic movements of a variety of dragonfly species, from the large and powerful Emperor Dragonfly to the more delicate and agile Banded Demoiselle. These enchanting insects, with their iridescent wings and jewel-

like eyes, are a true wonder of the natural world, and a delight to observe as you walk along the River Deben.
- Water Voles: As you make your way along the river, keep your eyes peeled for these elusive and enchanting mammals, often known as "water rats" due to their resemblance to their land-dwelling cousins. With their silky brown fur, rounded faces, and endearing habits, water voles are a true delight to spot, and a reminder of the incredible diversity of life that thrives in and around the River Deben.
- Marsh Harriers: Scan the skies over the reed beds and marshlands for the distinctive silhouette of these majestic birds of prey, their broad wings and forked tails a striking sight against the vast East Anglian sky. As you watch them soaring and diving in search of prey, marvel at their incredible aerial skills and their perfect adaptation to life in the wetland habitats of the Suffolk coast.

Rare or Interesting Plants:

- Reed Beds: As you walk along the River Deben, take a moment to appreciate the vast expanses of reed beds that line its banks, their tall, slender stems swaying gently in the breeze and providing a vital habitat for a wide range of wildlife. From the bearded tits and reed warblers that nest among their stalks to the water voles and otters that find shelter and sustenance in their dense growth, the reed beds are a true wonder of the Suffolk countryside, and a testament to the incredible resilience and adaptability of nature.
- Sea Lavender: In the summer months, the saltmarshes along the River Deben are transformed into a sea of delicate purple flowers, as the sea lavender bursts into bloom. This hardy and adaptable plant, with its slender stems and clusters of tiny, lavender-coloured blooms, is a true icon of the Suffolk coast, and a reminder of the unique and precious habitats that thrive in the transition zone between land and sea.
- Marsh Samphire: Keep an eye out for the succulent, green stems of this edible wild plant as you walk along the riverbank, its distinctive shape and texture a familiar sight to anyone who has explored the coastal marshes of East Anglia. Also known as glasswort, marsh samphire is a beloved delicacy among foragers and food enthusiasts, its crisp, salty flavour a perfect complement to fish and other seafood dishes, and a true taste of the wild and untamed beauty of the Suffolk coast.

Facilities and Accessibility:

The Woodbridge Riverside Walk is a relatively easy and accessible route, with mostly flat terrain and well-maintained paths suitable for walkers of all ages and abilities. However, it's important to note that some sections of the walk, particularly those closest to the river's edge, may be uneven or muddy, especially after periods of wet weather or high tide. As such, sturdy walking shoes or boots with good grip are highly recommended to ensure a safe and comfortable experience.

Along the route, you'll find a range of facilities and amenities to enhance your enjoyment of the walk. In Woodbridge itself, there are plenty of public toilets, cafes, pubs, and restaurants where you can take a break, grab a refreshment, or enjoy a delicious meal featuring the best of Suffolk's local produce and seafood.

As you make your way along the river, you'll also pass several benches and picnic areas, perfect for taking a moment to rest, soak up the stunning views, or enjoy an al fresco lunch while watching the boats go by. These facilities are well-maintained and easily accessible, making them a great option for walkers of all abilities.

It's worth noting that while much of the walk is suitable for wheelchairs and pushchairs, some sections may be more challenging due to the presence of steps, narrow paths, or uneven terrain. However, with a little advance planning and a willingness to take your time, most visitors should be able to enjoy the majority of the route and the beautiful scenery it has to offer.

Seasonal Highlights:

The Woodbridge Riverside Walk offers a unique and enchanting experience in every season, with each time of year revealing new facets of beauty and wonder along the banks of the River Deben:

- Spring: As the land awakens from its winter slumber, the riverbanks and marshlands burst into life, with vibrant green reeds and a profusion of delicate wildflowers creating a stunning display of colour and vitality. The air is filled with the joyous songs of returning birds, from the melodious warbles of the reed warblers to the haunting cries of the curlews, while the first butterflies and dragonflies of the year flutter and dart among the sunlit glades.
- Summer: In the long, lazy days of high summer, the River Deben is at its most inviting, its sparkling waters and gently swaying reed beds a cool and tranquil oasis from the heat of the sun. The marshlands are a riot of colour, with the purple haze of sea lavender and the vibrant pinks and yellows of the saltmarsh flowers creating a stunning tapestry that stretches out to the horizon. This is a time of abundance and joy, with the river and its surroundings teeming with life and energy.
- Autumn: As the leaves begin to turn and the nights grow longer, the Woodbridge Riverside Walk takes on a new character, its colours shifting from the lush greens of summer to the rich golds, reds, and browns of autumn. The reed beds rustle with the sounds of migrating birds, while the river's surface is painted with the reflected hues of the changing foliage. This is a time of quiet contemplation and gentle melancholy, a chance to savour the fleeting beauty of the season and to reflect on the cycles of life and death that shape the natural world.
- Winter: While the riverbanks may be stark and bare in the depths of winter, there is a haunting beauty to the landscape in this season of stillness and solitude. The skeletal forms of the reeds and the frosted leaves of the saltmarsh plants create an ethereal, otherworldly atmosphere, while the low sun casts long shadows across the water and the distant calls of overwintering birds echo through the chill air. Wrap up warm and embrace the bracing wind, and you'll discover a new appreciation for the raw power and elemental majesty of the Suffolk coast in winter.

Difficulty Level and Safety Considerations:

The Woodbridge Riverside Walk is generally considered an easy route, suitable for walkers of all ages and abilities. The terrain is mostly flat, with well-maintained paths and clear signage throughout, making it a great option for families, casual walkers, and those looking for a gentle and relaxing stroll along the beautiful River Deben.

However, as with any outdoor activity, it's important to be aware of potential hazards and to take necessary precautions to ensure a safe and enjoyable experience. One of the main considerations when walking along the river is the risk of slips and falls, particularly on the sections of path closest to the water's edge, where the ground may be uneven, muddy, or slippery, especially after rain or at high tide.

To mitigate this risk, it's essential to wear sturdy, non-slip footwear with good ankle support, and to take extra care when navigating these areas. Walking poles can also be a useful aid for maintaining balance and stability on uneven ground, particularly for those with mobility issues or concerns about falls.

It's also important to be mindful of the tides and weather conditions when planning your walk, as the water levels and the condition of the paths can change rapidly throughout the day. Check the local tide times and weather forecast before setting out, and be prepared to adjust your route or turn back if necessary to avoid any areas that may become flooded or inaccessible.

Another factor to consider is the exposed nature of some sections of the walk, which can leave you vulnerable to the sun, wind, and rain, particularly during the summer months or in changeable weather. Be sure to wear appropriate clothing and sun protection, and to carry plenty of water and snacks to keep you hydrated and energized throughout your journey.

As with any walk in a nature area, it's also a good idea to be aware of any potential wildlife hazards, such as ticks, stinging insects, or nesting birds, and to take appropriate precautions to avoid disturbing or damaging the delicate ecosystems along the river's edge. By staying on designated paths, keeping dogs on a lead, and taking your litter home with you, you can help to protect and preserve the unique and precious habitats of the River Deben for generations to come.

Despite these considerations, the Woodbridge Riverside Walk remains a safe, accessible, and immensely rewarding experience for the vast majority of visitors, offering a chance to connect with the stunning natural beauty, rich history, and vibrant culture of this iconic Suffolk town and its surrounding landscape. By following a few simple guidelines and using common sense, walkers of all ages and abilities can enjoy the many benefits of this unforgettable route, from the fresh air and exercise to the mental and spiritual refreshment that comes from spending time in nature.

Nearby Attractions:

After your walk along the picturesque River Deben, why not take some time to explore the many other attractions and activities that Woodbridge and the surrounding area have to offer? From fascinating historical sites and museums to stunning natural landscapes and outdoor adventures, there's something for everyone in this charming corner of Suffolk.

One of the top attractions in Woodbridge itself is the town's iconic Tide Mill, a beautifully preserved example of a working watermill that dates back to the 12th century. Take a guided tour of the mill to learn about its fascinating history and the ingenious engineering that allowed it to harness the power of the tides to grind wheat into flour for centuries. Afterward, be sure to browse the gift shop for unique souvenirs and locally produced crafts, or enjoy a delicious meal at the mill's charming riverside cafe.

For those interested in exploring the town's rich maritime heritage, a visit to the Woodbridge Museum is a must. Housed in a beautiful Georgian building on the Market Hill, the museum offers a fascinating glimpse into the history of Woodbridge and its relationship with the River Deben, from the earliest days of Saxon settlement to the present day. With exhibits on everything from shipbuilding and fishing to the town's role in defending the East Coast during World War II, the museum is a treasure trove of local knowledge and a great way to deepen your appreciation for this special place.

Just a short drive from Woodbridge, the National Trust's Sutton Hoo is one of the most important archaeological sites in Britain, and a fascinating destination for history buffs and curious visitors alike. This ancient Anglo-Saxon burial ground, discovered in 1939, yielded an astonishing array of treasures, including the iconic Sutton Hoo helmet and a magnificent ship burial that shed new light on the culture and beliefs of early medieval England. Today, visitors can explore the site's award-winning exhibition centre, take a guided tour of the burial mounds, and walk in the footsteps of the ancient kings and warriors who once called this place home.

For those looking to immerse themselves in the stunning natural beauty of the Suffolk coast and countryside, the nearby Area of Outstanding Natural Beauty is a paradise waiting to be discovered. With miles of pristine beaches, wild heathlands, and tranquil estuaries, this protected landscape is home to an incredible diversity of wildlife and offers endless opportunities for walking, cycling, birdwatching, and simply soaking up the peace and tranquility of one of England's most beautiful regions.

Whether you're a nature lover, a history buff, or simply someone in search of a relaxing and rejuvenating escape from the hustle and bustle of modern life, Woodbridge and its surroundings have something truly special to offer. So why not make a day of it, or even plan a longer stay, and discover for yourself the many charms and delights of this enchanting corner of Suffolk? With its stunning scenery, fascinating heritage, and warm, welcoming community, Woodbridge is a destination that will capture your heart and leave you with memories to last a lifetime.

River Deben Walk - 8 miles (12.9 km), Circular - Estimated Time: 4-5 hours

Starting Point Coordinates, Postcode: The starting point for the River Deben Walk is at Woodbridge Tide Mill, Quayside, Woodbridge (Grid Reference: TM 274 490, Postcode: IP12 1BG).
Nearest Car Park, Postcode: The nearest car park is the Station Road Car Park in Woodbridge (Postcode: IP12 4AU).

Embark on an enchanting 8-mile (12.9 km) circular walk along the picturesque River Deben, where the gentle flow of the water and the tranquil beauty of the Suffolk countryside intertwine to create a truly captivating experience. This moderate route offers a perfect blend of riverside trails, charming villages, and historic sites, inviting you to immerse yourself in the rich natural and cultural heritage of this stunning corner of East Anglia.

As you set out from the iconic Woodbridge Tide Mill, a beautifully restored 18th-century marvel of engineering, feel the anticipation build as you follow the riverside path southward, the gentle lapping of the water and the distant cries of wading birds a soothing soundtrack to your journey. With each step, allow the stresses of modern life to melt away, replaced by a growing sense of connection to the land, the river, and the countless generations who have walked this path before you.

Passing through the quaint village of Melton, with its charming cottages and friendly locals, you'll soon find yourself at Wilford Bridge, a picturesque crossing point that offers stunning views up and down the River Deben. Pause for a moment to take in the vista, the play of light on the water and the gentle swaying of the reeds a mesmerizing sight that invites quiet contemplation and reflection.

Crossing the bridge, follow the path along the opposite bank of the river, the landscape opening up before you in a patchwork of riverside meadows, rolling fields, and distant woodland. Here, the tranquility of the countryside envelops you, the only sounds the rustling of the wind through the grasses and the occasional burst of birdsong from the hedgerows and copses that line your route.

As you continue northward, the villages of Martlesham and Waldringfield emerge from the landscape, their ancient churches and timeworn houses a testament to the enduring spirit of the communities that have thrived along the River Deben for centuries. Take a moment to explore these charming settlements, perhaps stopping for a refreshing pint or a hearty meal at one of the welcoming local pubs, before setting off once more along the river's edge.

At Waldringfield, the path turns inland for a short distance, offering a change of scenery and a chance to appreciate the beauty of the Suffolk countryside from a different perspective. Here, amidst the gently rolling hills and the patchwork fields, you may catch a glimpse of the region's iconic wildlife, from the majestic red deer to the elusive barn owl, each a reminder of the incredible diversity and resilience of the natural world.

Looping back towards the river, you'll once again find yourself immersed in the tranquil beauty of the Deben, the path winding through a landscape of ever-changing vistas and hidden delights. As you near the end of your journey, the historic Woodbridge Tide Mill comes into view once more, its weathered walls and turning waterwheel a fitting symbol of the town's rich maritime heritage and the enduring relationship between the river and the people who call this place home.

As you complete your circular route, take a moment to reflect on the incredible journey you've just undertaken, the sights, sounds, and sensations of the River Deben forever etched in your memory. This walk is more than just a physical challenge; it is a celebration of the beauty, history, and resilience of the Suffolk countryside, an opportunity to reconnect with the natural world and to find solace and inspiration in the timeless flow of the river.

Detailed Route Directions:

1. From Woodbridge Tide Mill, head south along the riverside path, bearing 170°. (Estimated time: 30 minutes)

2. Pass through the village of Melton, continuing along the river, bearing 190°. (Estimated time: 45 minutes)

3. Reach Wilford Bridge and cross the river, bearing 280°. (Estimated time: 1 hour)

4. Follow the path on the opposite bank of the river, heading north, bearing 350°. (Estimated time: 1 hour 15 minutes)

5. Pass through the villages of Martlesham and Waldringfield, bearing 020°. (Estimated time: 1 hour 30 minutes)

6. At Waldringfield, head inland for a short distance, bearing 290°. (Estimated time: 20 minutes)

7. Loop back towards the river, following the path, bearing 100°. (Estimated time: 30 minutes)

8. Continue along the riverside path, heading back to Woodbridge Tide Mill, bearing 010°. (Estimated time: 1 hour)

Best Features and Views:

- Immerse yourself in the tranquil beauty of the River Deben, its gently flowing waters and picturesque banks creating an enchanting backdrop for your walk, inviting moments of reflection and connection with the natural world.
- Discover the charming villages of Melton, Martlesham, and Waldringfield, their ancient churches, timeworn houses, and friendly locals offering a glimpse into the rich history and enduring spirit of the communities along the river.
- Delight in the stunning views of the Suffolk countryside, from the riverside meadows and rolling fields to the distant woodlands and wide-open skies, each vista a testament to the incredible diversity and beauty of this corner of East Anglia.

Historical Features of Interest:

As you embark on your walk along the River Deben, take a moment to appreciate the historic Woodbridge Tide Mill, a beautifully restored 18th-century marvel of engineering that stands as a testament to the town's rich maritime heritage. This fascinating site, one of the oldest working tide mills in the country, offers a unique glimpse into the ingenuity and resourcefulness of past generations, who harnessed the power of the tides to grind wheat into flour for the local community.

Step inside the mill to explore its intricate machinery and learn about the daily lives of the millers who once toiled here, their work governed by the rhythms of the river and the turning of the seasons. From the creaking of the wooden gears to the gentle sloshing of the water, every sight and sound within these ancient walls tells a story of the enduring connection between the people of Woodbridge and the River Deben.

As you continue your walk, keep an eye out for the signs directing you to the world-famous Sutton Hoo, an extraordinary archaeological site just a short detour from the riverside path. This awe-inspiring location, discovered in 1939, is home to a remarkable Anglo-Saxon burial ground, where a treasure trove of priceless artifacts was unearthed, including the iconic Sutton Hoo helmet and a magnificent ship burial.

Take the time to visit the site's award-winning exhibition centre, where you can marvel at the exquisite craftsmanship of the unearthed treasures and learn about the fascinating history of the Anglo-Saxon kingdom that once thrived in this corner of Suffolk. As you walk in the footsteps of the ancient kings and warriors who were laid to rest here, you'll gain a deeper appreciation for the rich cultural heritage that underlies the beautiful landscapes you've been exploring.

Throughout your journey along the River Deben, you'll encounter countless other reminders of the region's storied past, from the grand estates and manor houses that dot the countryside to the humble cottages and farmsteads that have sheltered generations of hard-working families. Each of these historic sites offers a unique window into the lives and experiences of those who came before, inviting you to imagine the joys, sorrows, and daily struggles that have shaped this landscape over the centuries.

As you near the end of your circular route, take a moment to reflect on the ways in which the history of the River Deben and its surrounding communities continues to influence and inspire us today. From the enduring legacy of the Woodbridge Tide Mill to the groundbreaking discoveries at Sutton Hoo, these sites serve as powerful reminders of the ingenuity, resilience, and creativity of the human spirit, and the countless ways in which we are all connected to the land, the water, and the generations who have gone before us.

Birds to Spot:

- Kingfishers: Keep your eyes peeled for the brilliant blue and orange flash of these elusive birds as they dart along the riverbanks, their swift, agile flight and sharp, pointed bills perfectly adapted for catching small fish and aquatic insects.
- Herons: Scan the shallows and riverbanks for the stately silhouettes of gray herons, their long legs and sharp beaks poised to strike at unwary fish or frogs. These majestic birds, with their impressive wingspans and haunting calls, are a iconic presence along the River Deben.
- Egrets: Watch for the graceful, white forms of little egrets as they stalk through the shallows or perch in the riverside trees, their elegant plumage and slender, black bills a striking contrast to the lush green vegetation. These beautiful birds, once a rare sight in the UK, are now a delightful addition to the River Deben's avian community.

Rare or Interesting Insects and Other Wildlife:

- Dragonflies: In the summer months, the riverbanks and nearby meadows come alive with the vibrant colours and acrobatic flight of numerous dragonfly species, from the large, powerful emperor dragonfly to the more delicate and iridescent banded demoiselle. These ancient insects, with their intricate wing patterns and jewel-like eyes, are a fascinating and beautiful presence along the River Deben.
- Otters: Though elusive and often difficult to spot, otters have made a remarkable comeback along the River Deben in recent years, thanks to conservation efforts and improvements in water quality. As you walk along the riverbanks, keep an eye out for the telltale signs of otter activity, such as webbed footprints in the mud or the remains of fish and crustaceans on the shore.
- Water Voles: As you explore the riverbanks and nearby ditches, watch for the charming, round faces and furry bodies of water voles, often referred to as "water rats" due to their resemblance to their land-dwelling cousins. These endearing mammals, with their large eyes and distinctive chomping habits, are a vital part of

the River Deben's ecosystem and a delight to observe in their natural habitat.

Rare or Interesting Plants:

- Yellow Flag Iris: In late spring and early summer, the riverbanks and wetlands along the Deben come alive with the vibrant yellow blooms of the yellow flag iris, its tall, sword-like leaves and showy flowers a striking presence amidst the lush green vegetation.
- Marsh Marigold: Look for the glossy, heart-shaped leaves and brilliant golden flowers of marsh marigolds in the wet meadows and marshes near the river, their cheerful blooms a welcome sight in the early spring months.
- Purple Loosestrife: From midsummer to early autumn, the riverbanks and damp meadows are adorned with the tall, purple spikes of purple loosestrife, its showy flowers attracting a wide array of butterflies, bees, and other pollinators.

Facilities and Accessibility:

The River Deben Walk is a moderate route with some uneven terrain and occasional inclines, making it less suitable for those with limited mobility or fitness. However, the path is generally well-maintained and clearly signposted, with several benches and rest areas along the way for those who need to take a break or enjoy a picnic with a view.

In terms of facilities, the charming town of Woodbridge, at the start and end of the walk, offers a range of amenities, including public toilets, cafes, pubs, and restaurants. There are also several pubs and inns along the route in the villages of Melton, Martlesham, and Waldringfield, providing welcome refreshment and a chance to sample the local cuisine and ales.

It's important to note that some sections of the walk, particularly those closest to the river, may be prone to flooding during high tides or after heavy rainfall. It's always a good idea to check the local tide times and weather forecast before setting out, and to wear appropriate footwear and clothing for the conditions.

Additionally, while the route is generally well-signposted, it's recommended to carry a map and compass or GPS device, as well as a fully charged mobile phone, in case of emergencies or unexpected detours.

By taking these precautions and being mindful of the terrain and weather conditions, walkers of moderate fitness and ability can fully enjoy the beauty and tranquillity of the River Deben Walk, immersing themselves in the stunning landscapes and fascinating history of this enchanting corner of Suffolk.

Seasonal Highlights:

The River Deben Walk offers a unique and captivating experience in every season, with each time of year revealing new facets of beauty and wonder along this enchanting waterway:

- Spring: As the land awakens from its winter slumber, the riverbanks and nearby meadows burst into life, with vibrant wildflowers like marsh marigolds and yellow flag irises painting the landscape in a palette of gold and purple. The air is filled with the joyous songs of nesting birds, and the fresh green leaves of the trees create a lush canopy overhead, a symbol of renewal and growth.
- Summer: In the long, lazy days of summer, the River Deben is at its most inviting, with the warm sun sparkling on the water's surface and the gentle breeze carrying the scent of wildflowers and ripening fruit. The riverbanks are alive with the hum of insects and the fluttering of butterflies, while the nearby fields and

meadows are a riot of colour, with purple loosestrife, ox-eye daisies, and other summer blooms creating a breathtaking tapestry of life.
- Autumn: As the leaves begin to turn and the days grow shorter, the River Deben takes on a new character, its waters reflecting the fiery hues of the changing foliage and the mellow light of the autumn sun. The hedgerows and copses are laden with the bounty of the season, from shiny chestnuts and plump blackberries to the bright red hips of the wild roses, while the calls of migrating birds fill the air with a poignant sense of departure and change.
- Winter: Though the riverbanks may be bare and the fields fallow, the River Deben Walk is no less beautiful in the depths of winter. The stark outlines of the trees and the frosty glitter of the water create a haunting, almost ethereal atmosphere, while the distant calls of overwintering birds and the occasional flash of a kingfisher's bright plumage add a spark of life and colour to the subdued landscape. Wrap up warm and embrace the invigorating chill of the winter air, and you'll find a profound sense of peace and solitude in this season of rest and reflection.

Difficulty Level and Safety Considerations:

The River Deben Walk is a moderate route, suitable for walkers of average fitness and ability. The terrain is varied, with a mix of well-maintained footpaths, riverside trails, and country lanes, as well as some uneven ground and occasional inclines. As such, it's essential to wear sturdy, comfortable walking shoes or boots with good grip and ankle support, and to be prepared for muddy or slippery conditions, especially after rain or during the winter months.

One of the main safety considerations when walking along the River Deben is the risk of flooding or tidal surges, particularly in the sections of the route closest to the water's edge. It's crucial to check the local tide times and weather forecast before setting out, and to be aware of any areas where the path may become submerged or difficult to navigate at high tide. If in doubt, it's always best to err on the side of caution and to seek an alternative route or turn back if necessary.

Another factor to keep in mind is the exposure to the elements, especially on the more open stretches of the walk where there is little shelter from the sun, wind, or rain. Be sure to wear appropriate clothing for the season and the weather conditions, with layers that can be easily added or removed, as well as a waterproof jacket and sunscreen as needed.

It's also important to stay hydrated and energized throughout the walk, so be sure to carry plenty of water and snacks, and to take regular breaks to rest and enjoy the scenery. If you have any medical conditions or allergies, it's a good idea to inform someone of your route and estimated return time, and to carry any necessary medications or first aid supplies.

As with any outdoor activity, it's essential to follow the countryside code and to respect the environment and the local community. This means staying on designated footpaths, closing gates behind you, keeping dogs under control, and taking your litter home with you. By doing so, you'll help to protect the delicate ecosystems and preserve the natural beauty of the River Deben for generations to come.

Despite these considerations, the River Deben Walk remains a safe and immensely rewarding experience for the vast majority of walkers, offering a unique opportunity to connect with the stunning landscapes, rich history, and vibrant wildlife of this iconic Suffolk river. By following these simple guidelines and using common sense, you can fully immerse yourself in the beauty and tranquillity of the Deben Valley, creating memories that will last a lifetime.

Nearby Attractions:

After your invigorating walk along the River Deben , why not take some time to explore the many other attractions and delights that this beautiful corner of Suffolk has to offer? From fascinating historical sites and museums to stunning coastal landscapes and charming market towns, there's something to capture the imagination of every visitor.

One of the most iconic attractions in the area is the National Trust's Sutton Hoo, an extraordinary archaeological site just a short distance from the River Deben. This world-famous location, discovered in 1939, is home to a remarkable Anglo-Saxon burial ground, where a treasure trove of priceless artifacts was unearthed, including the iconic Sutton Hoo helmet and a magnificent ship burial. Take a guided tour of the site, explore the award-winning exhibition centre, and marvel at the incredible craftsmanship and artistry of the ancient people who once called this place home.

For those interested in the rich maritime heritage of the Suffolk coast, a visit to the charming town of Woodbridge is a must. This historic port, nestled on the banks of the River Deben, has been a centre of shipbuilding, trade, and seafaring for centuries, and its picturesque streets and waterfront are steeped in the tales and traditions of generations of sailors and merchants. Take a stroll along the Quayside, admire the beautifully preserved Tide Mill, or visit the town's excellent museum to learn more about the fascinating history of this thriving community.

Nature lovers and outdoor enthusiasts will be spoilt for choice in this stunning corner of East Anglia, with a wealth of reserves, parks, and protected landscapes to explore. The nearby Suffolk Coast and Heaths Area of Outstanding Natural Beauty is a particular highlight, with its miles of unspoiled beaches, wild heathlands, and tranquil estuaries offering endless opportunities for walking, cycling, birdwatching, and simply immersing yourself in the breathtaking beauty of the natural world.

For a taste of the region's delicious local produce and culinary traditions, be sure to visit one of the many farmers' markets, food festivals, and artisanal producers that can be found throughout the area. From the famous Suffolk cyder and beer to the delicate flavours of locally caught seafood and the hearty goodness of traditional farmhouse cheeses and cured meats, there's a wealth of gastronomic delights to discover and savour.

As you explore the many attractions and hidden gems of the Suffolk countryside and coast, take a moment to reflect on the deep sense of history, community, and connection to the land that pervades every aspect of life in this special corner of England. From the ancient burial mounds and medieval churches to the timeless rhythms of the tides and the turning of the seasons, every sight and sound speaks to the enduring spirit and resilience of the people who have called this place home for generations.

So whether you're a history buff, a foodie, a nature lover, or simply someone in search of a little peace and quiet in a beautiful setting, the River Deben Walk and its surroundings offer a wealth of opportunities for discovery, relaxation, and enjoyment. Why not extend your stay and take the time to truly immerse yourself in the charm, beauty, and warm hospitality of this enchanting region? With so much to see and do, you're sure to leave with memories that will last a lifetime and a deep appreciation for the unique character and heritage of the Suffolk landscape.

Boulge Circular Walk - 4 miles (6.4 km), Circular - Estimated Time: 2-2.5 hours

Starting Point Coordinates, Postcode: The starting point for the Boulge Circular Walk is at St. Michael's Church in Boulge (Grid Reference: TM 246 544, Postcode: IP13 6NR).
Nearest Car Park, Postcode: On-street parking is available near St. Michael's Church in Boulge (Postcode: IP13 6NR).

Embark on a peaceful 4-mile (6.4 km) circular walk through the charming village of Boulge and its picturesque surroundings, where the tranquil beauty of the Suffolk countryside and the rich tapestry of local history intertwine to create a truly enchanting experience. This easy to moderate route offers a perfect blend of woodland paths, quiet country lanes, and serene farmland, inviting you to immerse yourself in the timeless charm and gentle rhythms of rural life.

As you set out from the historic St. Michael's Church, take a moment to appreciate the elegant architecture and peaceful atmosphere of this ancient place of worship, its weathered stones and soaring tower a testament to the enduring faith and craftsmanship of generations past. With each step along the footpath beside the churchyard, feel the stresses of modern life melt away, replaced by a sense of connection to the land and its stories.

Cross the small bridge and enter the cool embrace of the woodland, where the soft rustling of leaves and the gentle chirping of birds create a soothing soundtrack to your journey. Follow the winding path as it leads you through the dappled shade of the trees, the air filled with the earthy scents of moss and leaf litter, a reminder of the endless cycles of growth and decay that shape the natural world.

Emerging from the woodland, the path opens up into peaceful farmland, where the gentle undulations of the fields and the distant horizons invite a sense of openness and possibility. As you navigate the well-trodden trails that crisscross this patchwork landscape, take a moment to appreciate the simple beauty of the wildflowers that sway in the breeze, their delicate petals and vibrant colours a testament to the resilience and adaptability of nature.

Continuing along the footpath, you'll soon find yourself on a quiet country lane, where the occasional passing of a tractor or the distant lowing of cattle are the only sounds that break the tranquil silence. Take a left and follow the lane for a short distance, immersing yourself in the bucolic charm of the Suffolk countryside, before taking a right onto another footpath that leads you back into the fields.

As you walk through the gently rolling landscape, keep an eye out for the varied birdlife that calls this area home, from the soaring buzzards and graceful skylarks to the colourful yellowhammers that flit among the hedgerows. Each encounter serves as a reminder of the incredible diversity and beauty of the natural world, and the importance of preserving these precious habitats for generations to come.

Crossing a small stream and entering another stretch of woodland, the path winds through a world of dappled light and shadow, the ancient trees standing as silent guardians of the secrets and stories of the land. Take a moment to run your hand along the gnarled bark of an old oak, feeling the weight of history and the connection to the countless lives that have passed beneath its boughs.

Emerging once more onto a quiet country lane, you'll pass by the impressive Boulge Hall, a grand Georgian mansion that stands as a testament to the wealth and status of the local gentry who once called this place home. Imagine the elegant parties and lively social gatherings that must have taken place within its walls, the laughter and music drifting out across the tranquil countryside.

As you turn onto the final footpath that leads back towards St. Michael's Church, take a moment to reflect on the journey you have just completed, the sights and sounds of the Boulge landscape forever etched in your memory. This circular walk is more than just a physical route; it is a celebration of the enduring beauty, rich history, and peaceful character of the Suffolk countryside, an invitation to slow down, to breathe deeply, and to reconnect with the simple joys and profound truths of life in the slow lane.

Detailed Route Directions:

1. From St. Michael's Church, head west along the footpath alongside the churchyard, bearing 280°. (Estimated time: 5-10 minutes)

2. Cross a small bridge and continue through the woodland, following the path as it bends to the right, bearing 350°. (Estimated time: 10-15 minutes)

3. As the path opens up into farmland, continue along the footpath, crossing several fields, bearing 290°. (Estimated time: 25-30 minutes)

4. Upon reaching a quiet country lane, turn left and follow the lane for a short distance, bearing 230°. (Estimated time: 5-10 minutes)

5. Take the footpath on the right, walking through more fields, bearing 320°. (Estimated time: 20-25 minutes)

6. Cross a small stream and continue through woodland, bearing 010°. (Estimated time: 15-20 minutes)

7. Join another quiet country lane, turn left, and follow the lane, passing by Boulge Hall, until you reach a junction, bearing 140°. (Estimated time: 25-30 minutes)

8. Turn right onto another footpath, continuing along the edge of the fields, bearing 080°. (Estimated time: 20-25 minutes)

9. The path will veer left, bearing 010°, leading back towards St. Michael's Church, where the walk ends. (Estimated time: 10-15 minutes)

Best Features and Views:

- Immerse yourself in the tranquil charm of the Suffolk countryside, with its gently rolling hills, patchwork fields, and distant horizons, a landscape that invites contemplation and a sense of connection to the natural world.
- Discover the enchanting woodland paths that wind through the dappled shade of ancient trees, their rustling leaves and chirping birds creating a soothing symphony that calms the mind and soothes the soul.
- Delight in the peaceful farmland that surrounds the village of Boulge, where the simple beauty of wildflowers, the gentle murmur of streams, and the occasional sight of grazing cattle create a timeless tableau of rural life.

Historical Features of Interest:

As you begin your walk at the historic St. Michael's Church, take a moment to appreciate the rich history and architectural beauty of this ancient place of worship. The church, which dates back to the 13th century, has served as a spiritual and social centre for the Boulge community for over 700 years, its weathered stones and graceful lines a testament to the skill and devotion of the medieval craftsmen who built it.

Step inside the church to discover a treasure trove of historical features, from the intricate stained-glass windows and beautifully carved wooden pews to the ornate stone font and the evocative memorial plaques that line the walls. Each element tells a story of faith, community, and the enduring human spirit, inviting you to imagine the countless generations of worshippers who have found comfort, joy, and inspiration within these hallowed walls.

As you explore the churchyard, with its ancient yew trees and weathered gravestones, reflect on the lives and legacies of those who have come before, their stories woven into the very fabric of the landscape. From the humble farmers and labourers who toiled in the fields to the wealthy landowners and merchants who shaped the course of local history, each has left an indelible mark on the village and its surroundings.

Throughout your walk, keep an eye out for other signs of Boulge's rich heritage, from the traditional timber-framed cottages and red-brick farmhouses that dot the countryside to the old field boundaries and woodland paths that have been trodden by countless feet over the centuries. These tangible links to the past serve as a reminder of the deep roots and enduring spirit that define this special corner of Suffolk, offering a glimpse into the lives and dreams of the people who have called this place home.

As you pass by the impressive Boulge Hall, a grand Georgian mansion set amidst the peaceful countryside, imagine the lives of the wealthy families who once resided within its walls. Picture the elegant soirées and lively hunting parties that would have filled the house with music, laughter, and the clinking of glasses, a world away from the daily toil of the farmers and labourers who worked the surrounding land.

In the tranquil beauty of the Boulge landscape, with the gentle rustling of leaves and the distant birdsong as your backdrop, you may find yourself feeling a deep sense of connection to the past, a recognition of the timeless rhythms and enduring values that have shaped this land and its people for generations. This is the true gift of the Boulge Circular Walk, the opportunity to step out of the present moment and into a world of history, beauty, and quiet reflection, where the boundaries between past and present dissolve, and the essence of rural England is revealed in all its glory.

Birds to Spot:

- Yellowhammers: Listen for the cheerful, repetitive song of these colourful buntings as you walk along the field edges and hedgerows, their bright yellow heads and chestnut-streaked backs a welcome sight among the greens and browns of the countryside.
- Skylarks: Keep your eyes and ears open for these unassuming brown birds as they soar high above the fields, their enchanting, melodious songs filling the air with a sense of joy and freedom. Skylarks are a true emblem of the open countryside, their presence a testament to the enduring beauty and vitality of the rural landscape.
- Buzzards: Scan the skies for the broad, rounded wings and distinctive mewing calls of these majestic birds of prey, often seen soaring effortlessly over the fields and woodlands of Boulge. As you watch them ride the thermals, marvel at their incredible aerial skills and their perfect adaptation to life in the Suffolk countryside.

Rare or Interesting Insects and Other Wildlife:

- Butterflies: In the spring and summer months, the fields, hedgerows, and woodland edges of Boulge come alive with the delicate flutter of butterfly wings. From the vibrant orange-and-brown of the comma to the iridescent blues of the holly blue and the common blue, these enchanting insects add a splash of colour and

movement to the tranquil landscape.
- Dragonflies: As you cross the small streams and pass by the damp, grassy areas along your walk, keep an eye out for the iridescent shimmer and acrobatic flight of dragonflies and damselflies. These ancient insects, with their intricate wing patterns and jewel-like colours, are a fascinating and beautiful presence in the Suffolk countryside.
- Small Mammals: Throughout your walk, stay alert for signs of the various small mammals that make their homes in the fields, hedgerows, and woodlands of Boulge. From the darting movements of field mice and voles to the occasional glimpse of a shy rabbit or a curious stoat, each encounter offers a reminder of the rich tapestry of life that thrives in this peaceful corner of England.

Rare or Interesting Plants:

- Ancient Oaks: As you walk through the woodlands and along the field boundaries of Boulge, take a moment to appreciate the majestic presence of the ancient oak trees that have stood watch over this landscape for centuries. These venerable giants, with their gnarled bark and spreading canopies, are a living link to the past, their enduring strength and beauty a testament to the resilience of nature.
- Bluebells: In the spring, the woodland floors of Boulge are transformed into a mesmerizing carpet of blue, as thousands of delicate bluebells bloom beneath the dappled shade of the trees. Take a moment to breathe in their sweet fragrance and marvel at the ephemeral beauty of these enchanting wildflowers, a true celebration of the season's renewal.
- Wildflower Meadows: Throughout the summer months, the fields and grasslands of Boulge are dotted with a vibrant array of wildflowers, from the cheery yellows of buttercups and cowslips to the delicate pinks of cuckoo flowers and the regal purples of knapweed. These diverse and colourful blooms not only add to the visual appeal of the landscape but also provide vital nectar sources for pollinators and other wildlife.

Facilities and Accessibility:

The Boulge Circular Walk is a relatively easy route, with mostly level terrain and a mix of well-maintained footpaths, bridleways, and quiet country lanes. However, it's important to note that some sections of the walk may be uneven, muddy, or overgrown, particularly after periods of wet weather or during the height of summer. As such, sturdy walking shoes or boots with good grip and support are highly recommended to ensure a safe and comfortable experience.

While there are no public toilets or other facilities directly along the route, the nearby village of Boulge does have a few amenities for walkers, including a charming parish church where you can take a moment to rest and appreciate the peaceful surroundings.

It's worth noting that some parts of the walk may not be suitable for those with limited mobility or for pushchairs, due to the presence of stiles, kissing gates, and uneven ground. However, with careful planning and a willingness to take your time, most walkers will find this route to be an enjoyable and rewarding experience.

Seasonal Highlights:

The Boulge Circular Walk offers a unique and captivating experience in every season, with each time of year revealing new facets of beauty and wonder in this enchanting corner of Suffolk:

- Spring: As the land awakens from its winter slumber, the woodlands and hedgerows of Boulge burst into life, with delicate wildflowers like bluebells and wood anemones carpeting the ground in a mesmerizing display of

colour and fragrance. The air is filled with the joyous songs of nesting birds, and the fresh green leaves of the trees create a lush canopy overhead, a symbol of the renewal and hope that spring brings.
- Summer: In the long, sun-drenched days of summer, the countryside around Boulge is at its most vibrant and alive. The fields are a patchwork of golden hues as the crops ripen, and the wildflower meadows are a riot of colour, with butterflies and bees flitting from bloom to bloom. The woodlands provide a cool, green sanctuary from the heat of the day, the dappled light filtering through the leaves and creating an enchanting play of light and shadow.
- Autumn: As the days grow shorter and the nights turn crisp, the Boulge landscape is transformed into a breathtaking tapestry of gold, orange, and red, the turning leaves of the trees creating a warm and inviting atmosphere, perfect for a contemplative walk through the countryside. The hedgerows are laden with the rich bounty of berries and nuts, while the distant sounds of migrating birds and the earthy scent of fallen leaves create a poignant sense of the turning seasons.
- Winter: While the fields may be bare and the trees stripped of their leaves, the Boulge Circular Walk takes on a stark, elemental beauty in the winter months, the frosty ground and the pale, watery light creating a sense of quiet introspection and solitude. The bare branches of the trees stand out in sharp relief against the sky, their intricate patterns and textures revealed in all their glory, while the occasional sighting of a red fox or a flock of fieldfares adds a spark of life and colour to the subdued landscape.

Difficulty Level and Safety Considerations:

The Boulge Circular Walk is a relatively easy route, suitable for walkers of most ages and fitness levels. With mostly level terrain and well-maintained paths, this walk offers a gentle and accessible introduction to the beauty and tranquility of the Suffolk countryside.

However, as with any outdoor activity, it's important to take certain precautions and to be aware of potential hazards. Some sections of the walk may be uneven, muddy, or slippery, particularly after periods of wet weather, so sturdy footwear with good grip is essential. It's also a good idea to wear long trousers and sleeves to protect against ticks and other biting insects, which can be prevalent in rural areas.

Walkers should also be mindful of the presence of livestock in some of the fields along the route, particularly during the spring and summer months. It's important to follow the countryside code, keeping dogs on a lead where necessary and closing any gates behind you to ensure the safety of both animals and walkers.

As with any walk in a rural area, it's always a good idea to carry a map and compass or GPS device, as well as a fully charged mobile phone and a small first aid kit, in case of emergencies. It's also wise to inform someone of your planned route and expected return time before setting out.

By following these simple guidelines and using common sense, walkers can fully immerse themselves in the beauty and peacefulness of the Boulge Circular Walk, enjoying a safe and rewarding experience that nourishes both body and soul.

Nearby Attractions:

After your invigorating walk through the picturesque countryside surrounding Boulge, why not extend your adventure and explore some of the other fascinating attractions that this beautiful corner of Suffolk has to offer? The area is rich in history, culture, and natural beauty, with plenty of opportunities for discovery and relaxation.

Just a short drive from Boulge lies the historic market town of Woodbridge, nestled on the banks of the River Deben. This charming town boasts a wealth of independent shops, cafes, and restaurants, as well as a vibrant arts and culture scene. Take a stroll along the picturesque Thoroughfare, lined with timber-framed buildings and quirky boutiques, or visit the town's iconic Tide Mill, one of the oldest working tide mills in the country. Don't miss the chance to explore the Woodbridge Museum, which tells the fascinating story of the town's maritime heritage and its role in the development of the wider region.

For those with a passion for history and archaeology, the nearby Sutton Hoo is an absolute must-see. This world-famous site, located just a few miles from Boulge, is home to a remarkable Anglo-Saxon burial ground, where a treasure trove of gold and silver artifacts was unearthed in 1939. Take a guided tour of the site, marvel at the exquisite craftsmanship of the Sutton Hoo helmet and other precious objects, and immerse yourself in the captivating story of this ancient kingdom and the people who once called it home.

Nature lovers and outdoor enthusiasts will be spoilt for choice in this stunning corner of East Anglia, with a wealth of reserves, parks, and protected landscapes to explore. The nearby Suffolk Coast and Heaths Area of Outstanding Natural Beauty is a particular highlight, with its miles of unspoiled beaches, wild heathlands, and tranquil estuaries offering endless opportunities for walking, cycling, birdwatching, and simply soaking up the beauty of the natural world.

For a taste of the region's rich agricultural heritage and delicious local produce, be sure to visit one of the many farm shops, farmers' markets, and artisanal food producers that can be found throughout the area. From the famous Suffolk cyder and beer to the delicate flavours of locally grown fruits and vegetables, there's a wealth of culinary delights to discover and savour.

As you explore the many attractions and hidden gems of the Suffolk countryside, take a moment to reflect on the deep sense of history, community, and connection to the land that pervades every aspect of life in this special corner of England. From the ancient churches and thatched cottages to the timeless rhythms of the seasons and the enduring beauty of the landscape, every sight and sound speaks to the rich tapestry of human experience that has shaped this region over countless generations.

So whether you're a history buff, a nature lover, a foodie, or simply someone in search of a little peace and quiet in a beautiful setting, the Boulge Circular Walk and its surroundings offer a wealth of opportunities for exploration, inspiration, and rejuvenation. Why not make a day of it, or even plan a longer stay, and immerse yourself in the warmth, charm, and gentle pace of life that makes this part of Suffolk so special?

As you set out on your adventures, remember to tread lightly, to respect the local environment and communities, and to approach each new experience with an open heart and a curious mind. By doing so, you'll not only deepen your own understanding and appreciation of the world around you, but you'll also help to ensure that the beauty, diversity, and rich cultural heritage of this enchanting region can be enjoyed by generations to come.

In the words of the great naturalist and writer, Edward Abbey, "Wilderness is not a luxury but a necessity of the human spirit." So embrace the wild beauty and timeless charm of the Suffolk countryside, and let the Boulge Circular Walk be your guide and your inspiration as you embark on a journey of discovery that will nourish your soul and renew your sense of wonder at the incredible world we inhabit.

Happy exploring!

Hollesley Common Walk - 4 miles (6.4 km), Circular - Estimated Time: 2-2.5 hours

Starting Point Coordinates, Postcode: The starting point for the Hollesley Common Walk is at the Hollesley Common car park (Grid Reference: TM 345 449, Postcode: IP12 3NA).
Nearest Car Park, Postcode: The nearest car park is the Hollesley Common car park (Postcode: IP12 3NA).

Embark on an enchanting 4-mile (6.4 km) circular walk through the diverse and captivating landscape of Hollesley Common, where the rugged beauty of heathland, the tranquil charm of woodland, and the lush expanse of grassland intertwine to create a truly unique and inspiring natural haven. This easy to moderate route offers a perfect opportunity for walkers of most abilities to immerse themselves in the rich tapestry of Suffolk's countryside, discovering the fascinating flora and fauna that thrive in this special corner of England.

As you set out from the Hollesley Common car park, feel the anticipation build as you follow the well-marked footpath westward, venturing into the heart of the heathland. With each step, the stresses of modern life melt away, replaced by a growing sense of connection with the natural world and its myriad wonders.

The heathland sprawls before you, a patchwork of vibrant colours and textures, from the delicate pinks and purples of the heather to the vibrant yellow of the gorse. Take a moment to breathe in the heady scent of these hardy plants, their fragrance carried on the gentle breeze that rustles through the landscape. As you walk, keep your eyes peeled for the diverse array of plant life that flourishes in this unique habitat, each species a testament to the incredible resilience and adaptability of nature.

After approximately a mile, the path guides you into a tranquil woodland, where the dappled sunlight filters through the canopy of leaves, creating an enchanting play of light and shadow on the forest floor. Immerse yourself in the peaceful atmosphere, listening to the gentle rustling of the trees and the melodic birdsong that fills the air. The woodland is a haven for various bird species, so keep your ears attuned to the different calls and songs, and your eyes sharp for any flashes of colour amidst the foliage.

As you navigate through the woodland, the path gradually turns southward, leading you through a landscape that seems to shift and change with every step. Marvel at the intricate ecosystem that thrives beneath the trees, from the delicate ferns and mosses to the scurrying insects and foraging small mammals. Each creature plays a vital role in the complex web of life that sustains this enchanting woodland.

Emerging from the cool embrace of the trees, you'll find yourself in an expanse of lush grassland, where the path meanders southeast, guiding you back towards your starting point. Take a moment to appreciate the open vistas and the feeling of space and freedom that comes with walking through this verdant landscape. The grassland is home to a myriad of wildflowers and grasses, their colours and textures creating a mesmerizing tapestry that sways gently in the breeze.

As you make your way through the grassland, keep an eye out for the various insects and reptiles that thrive in this habitat. From the iridescent shimmer of dragonflies darting above the grass to the slender form of a common lizard basking on a sun-warmed rock, each encounter serves as a reminder of the astonishing diversity of life that can be found in even the most unassuming of places.

The final stretch of your walk brings you back to the Hollesley Common car park, where your journey began. As you complete the circular route, take a moment to reflect on the beauty, tranquillity, and natural wonders you've experienced along the way. The Hollesley Common Walk is more than just a physical journey; it is a celebration of the incredible diversity and resilience of Suffolk's countryside, an opportunity to reconnect with the Earth and to find solace and inspiration in the timeless rhythms of the natural world.

Detailed Route Directions:

1. From the Hollesley Common car park, follow the marked footpath into the heathland, heading west (bearing 270°). (Estimated time: 20-25 minutes)

2. Continue along the path through the heathland for approximately 1 mile, keeping an eye out for the diverse plant life. (Estimated time: 20-25 minutes)

3. As the path enters a wooded area, continue through the woods, following the path as it turns south (bearing 180°). (Estimated time: 25-30 minutes)

4. Exit the woods and encounter the grasslands, continuing to follow the path southeast (bearing 135°). (Estimated time: 25-30 minutes)

5. The path will lead you back towards the Hollesley Common car park, completing the circular route. (Estimated time: 20-25 minutes)

Best Features and Views:

- Immerse yourself in the captivating beauty of Hollesley Common, where the rugged charm of heathland, the tranquil allure of woodland, and the lush expanse of grassland create a diverse and inspiring natural tapestry.
- Marvel at the incredible array of plant life that thrives in the heathland, from the delicate heather and vibrant gorse to the myriad of other hardy species that have adapted to this unique environment.
- Delight in the peaceful atmosphere of the woodland, where dappled sunlight, gentle birdsong, and the soft rustling of leaves create a soothing symphony that calms the mind and soothes the soul.

Historical Features of Interest:

While there are no significant historical features directly along the Hollesley Common Walk, the surrounding area is rich in history and cultural heritage, offering fascinating insights into the long and storied past of this enchanting corner of Suffolk.

Just a short drive from Hollesley Common, the world-renowned archaeological site of Sutton Hoo awaits, promising a captivating journey through the mists of time. Discovered in 1939, this ancient Anglo-Saxon burial ground has yielded some of the most remarkable and valuable treasures ever found in Britain, including the iconic Sutton Hoo helmet and a magnificent ship burial that has shed new light on the culture, beliefs, and artistic achievements of the early medieval period.

Visitors to Sutton Hoo can explore the atmospheric burial mounds, marvel at the exquisite craftsmanship of the excavated artifacts in the on-site museum, and immerse themselves in the fascinating story of the wealthy and powerful Anglo-Saxon kingdom that once held sway over this landscape. Through guided tours, interactive exhibits, and thought-provoking displays, the site brings the distant past vividly to life, inviting reflection on the enduring legacy of those who came before us and the ways in which their lives and experiences continue to shape our understanding of the world today.

Another nearby historical gem is the imposing Orford Castle, a magnificent 12th-century fortress that stands as a testament to the turbulent history and strategic importance of the Suffolk coast. Built by King Henry II to protect the region from invasion and to assert his authority over his rebellious barons, the castle has borne witness to centuries of conflict, intrigue, and political upheaval, its mighty walls and soaring keep a silent monument to the courage, ambition, and ingenuity of the medieval mind.

Visitors can explore the castle's fascinating architecture, from the intricate stonework of the great hall to the dizzying heights of the keep, and learn about the daily lives of the lords, ladies, and soldiers who once called this place home. With its commanding views over the River Ore and the surrounding countryside, Orford Castle offers a unique and unforgettable perspective on the rich history and enduring beauty of the Suffolk landscape.

As you walk through the tranquil beauty of Hollesley Common, let your mind wander to the countless generations of people who have shaped and been shaped by this land, from the ancient hunters and gatherers who first set foot on these shores to the farmers, fishermen, and artisans who have carried on the traditions and crafts of their forebears. Each step you take is a connection to this rich and varied past, a reminder of the enduring power of place to inspire, to challenge, and to transform the human spirit.

Birds to Spot:

- Nightjars: As dusk falls over the heathland, listen for the distinctive "churring" call of these elusive birds, their cryptic plumage and silent, moth-like flight making them a mesmerizing presence in the twilight landscape.
- Woodlarks: Keep your ears attuned to the melodious, fluting song of these ground-nesting birds, often heard before they are seen, as they soar above the heathland and woodland edges in a joyous display of aerial mastery.
- Dartford Warblers: Watch for these small, lively warblers as they flit among the gorse and heather, their distinctive reddish-brown plumage and long, slightly cocked tails a charming sight in the vibrant tapestry of the heathland.

Rare or Interesting Insects and Other Wildlife:

- Dragonflies and Damselflies: Marvel at the iridescent shimmer and acrobatic flight of these ancient insects as they dart and hover above the grassland and along the woodland edges, their presence a testament to the health and vitality of the wetland habitats that sustain them.
- Butterflies: From the striking orange and brown hues of the grayling to the delicate beauty of the silver-studded blue, the heathland and grassland of Hollesley Common are home to a fascinating array of butterfly species, each one a delightful addition to the ever-changing kaleidoscope of colour and movement.
- Reptiles: Keep your eyes peeled for the slender form of a common lizard basking on a sun-warmed rock, or the distinctive diamond-shaped pattern of an adder as it slithers through the undergrowth, each encounter a reminder of the incredible diversity of life that thrives in these unique habitats.

Rare or Interesting Plants:

- Heather: In late summer and early autumn, the heathland of Hollesley Common is transformed into a sea of delicate pink and purple blooms, as the heather bursts into flower, creating a breathtaking display that stretches as far as the eye can see.
- Gorse: With its vibrant yellow flowers and prickly, evergreen stems, gorse is a defining feature of the heathland landscape, its coconut-scented blooms and dense, impenetrable thickets providing food and shelter for a wide range of insects, birds, and small mammals.
- Sundews: Look closely in the damper areas of the heathland, and you may spot the delicate, glistening leaves of these tiny carnivorous plants, their sticky tentacles poised to trap unsuspecting insects and supplement

their nutrient-poor diet.

Facilities and Accessibility:

The Hollesley Common Walk is a relatively easy to moderate route, with mostly level terrain and well-defined paths. However, it's important to note that some sections of the heathland and woodland may be uneven, muddy, or overgrown, particularly after periods of wet weather or during the height of summer. As such, sturdy walking shoes or boots with good grip and support are highly recommended to ensure a safe and comfortable experience.

While there are no facilities directly at the starting point or along the route, it's essential to come prepared with enough water, snacks, and any necessary medications or sun protection to ensure a comfortable and enjoyable walk. It's also a good idea to make use of facilities and stock up on refreshments before starting the walk, as there are no amenities available on the common itself.

It's worth noting that some parts of the walk may not be suitable for those with limited mobility or for pushchairs, due to the presence of uneven ground, tree roots, and narrow paths. However, with careful planning and a willingness to take your time, most walkers will find this route to be a rewarding and manageable experience.

Seasonal Highlights:

The Hollesley Common Walk offers a unique and captivating experience in every season, with each time of year revealing new facets of beauty and wonder in this enchanting corner of Suffolk:

- Spring: As the land awakens from its winter slumber, the heathland and grassland burst into life, with vibrant green shoots emerging from the earth and a profusion of delicate wildflowers painting the landscape in a rainbow of colours. The air is filled with the joyous songs of breeding birds, and the first butterflies and dragonflies of the year take to the wing, heralding the return of warmer days and the promise of new beginnings.
- Summer: In the long, languid days of high summer, the heathland is a riot of colour and sound, with the purple haze of heather, the golden glow of gorse, and the constant hum of insects creating a mesmerizing tapestry that seems to shimmer and dance in the heat. The woodland edges are alive with the fluttering of butterfly wings and the darting flight of dragonflies, while the cool shade beneath the trees offers a welcome respite from the sun's intense rays.
- Autumn: As the days grow shorter and the nights turn crisp, the heath takes on a new character, its colours shifting from the vibrant purples and yellows of summer to the rich, burnished hues of copper, bronze, and gold. The air is filled with the earthy scent of decaying vegetation and the gentle rustling of fallen leaves, while the mournful cries of migrating birds and the busy foraging of small mammals add a poignant sense of the turning seasons and the never-ending cycle of life and death.
- Winter: Though the heath may seem bleak and lifeless in the depths of winter, there is a stark, elemental beauty to the landscape in this season of rest and renewal. The bare branches of the trees stand out in sharp relief against the pale, watery sky, while the frost-kissed heather and gorse create an ethereal, almost otherworldly atmosphere that invites quiet contemplation and introspection. Wrap up warm and embrace the bracing chill of the winter air, and you may be rewarded with the sight of a majestic hen harrier soaring low over the heath, or the delicate tracks of a red deer picked out in the snow, a reminder of the enduring wonder and resilience of the natural world.

Difficulty Level and Safety Considerations:

The Hollesley Common Walk is classified as an easy to moderate route, with mostly level terrain and well-defined paths. However, as with any outdoor activity, it's essential to be prepared for changeable weather conditions and to take necessary precautions to ensure a safe and enjoyable experience.

As you navigate the heathland and woodland trails, be sure to stay on the designated paths and to avoid disturbing any wildlife or damaging the delicate habitats that sustain them. Keep an eye out for uneven ground, exposed tree roots, and overhanging branches, particularly in the wooded areas, and take care when crossing any muddy or waterlogged sections, especially after heavy rain.

It's also important to be aware of the presence of livestock on the common, particularly during the grazing season. While the animals are generally accustomed to walkers and pose no threat, it's always advisable to give them a wide berth and to keep dogs on a lead to avoid any potential conflicts or disturbance.

When walking in the heathland areas, be mindful of the risk of tick bites, particularly during the warmer months. These small, blood-sucking arachnids can carry Lyme disease, a potentially serious bacterial infection that can cause a range of symptoms if left untreated. To minimize the risk of bites, wear long-sleeved shirts and trousers, tuck your socks into your boots, and use an insect repellent containing DEET. After your walk, be sure to check yourself, your clothing, and your pets thoroughly for any attached ticks, and remove them promptly and safely using a tick removal tool or fine-tipped tweezers.

As with any walk in a remote or unfamiliar area, it's always a good idea to carry a map and compass or GPS device, as well as a fully charged mobile phone and a small first aid kit, in case of emergencies. Let someone know your intended route and estimated return time before setting out, and be sure to carry enough water and snacks to keep you hydrated and energized throughout your walk.

By following these simple guidelines and using common sense, walkers of all ages and abilities can safely enjoy the beauty and tranquillity of the Hollesley Common Walk, immersing themselves in the wonders of the Suffolk countryside and forging a deeper connection with the natural world around them.

Nearby Attractions:

After your invigorating walk through the captivating landscape of Hollesley Common, why not extend your adventure and explore some of the other fascinating attractions that this beautiful corner of Suffolk has to offer? From world-renowned historical sites to picturesque coastal towns and stunning nature reserves, there's something to capture the imagination and delight the senses of every visitor.

Just a short drive from Hollesley Common, the iconic Sutton Hoo archaeological site awaits, offering a unique and unforgettable window into the rich history and culture of Anglo-Saxon England. Discover the incredible story of the ship burial and the priceless treasures that were unearthed here, and marvel at the exquisite craftsmanship and artistry of the Sutton Hoo helmet and other artifacts, which have shed new light on the sophistication and creativity of this ancient civilization.

For those who love the sea and the timeless charm of traditional coastal towns, the nearby villages of Aldeburgh and Orford are a must-visit. Stroll along the picturesque beaches, lined with colourful fishing boats and charming seafront cottages, and sample some of the freshest and most delicious seafood you'll ever taste, caught daily by the skilled local fishermen. Browse the independent shops and galleries that line the quaint high streets, and soak up the warm, welcoming atmosphere of these friendly and unspoiled communities, where the spirit of the Suffolk coast is alive and well.

Nature enthusiasts and birdwatchers will be in their element at the nearby RSPB Minsmere nature reserve, a vast and pristine wetland habitat that is home to an incredible array of wildlife, from the elusive bittern and the majestic marsh harrier to the playful otter and the graceful avocet. With its miles of trails, hides, and viewing platforms, Minsmere offers unparalleled opportunities to observe and photograph the stunning diversity of the Suffolk coastline, and to connect with the rhythms and wonders of the natural world.

For a taste of the region's rich agricultural heritage and thriving food culture, be sure to visit one of the many farmers' markets, farm shops, and artisanal producers that can be found throughout the area. Sample the creamy, tangy flavours of the famous Suffolk cheeses, savour the succulent meats and freshly harvested vegetables that are the hallmark of the county's traditional cuisine, and wash it all down with a refreshing pint of locally brewed ale or a crisp, fragrant glass of Suffolk cider.

As you explore the many attractions and hidden gems of this enchanting region, take a moment to reflect on the deep sense of history, community, and connection to the land that pervades every aspect of life in Suffolk. From the ancient burial mounds and medieval castles to the timeless beauty of the heathlands and the enduring traditions of the coastal towns, every sight and sound tells a story of the remarkable people and places that have shaped this unique and captivating corner of England.

So whether you're a history buff, a nature lover, a foodie, or simply someone in search of a little peace and tranquillity in a beautiful and unspoiled setting, the Hollesley Common Walk and its surroundings offer a wealth of opportunities for discovery, relaxation, and inspiration. Why not linger a while longer and immerse yourself in the subtle charms and hidden delights of this very special part of the world, and let the magic of Suffolk work its spell on your heart and soul?

As you set out on your adventures, remember to tread lightly, to respect the local environment and communities, and to approach each new experience with an open mind and a sense of wonder. By doing so, you'll not only enrich your own understanding and appreciation of the world around you, but you'll also help to preserve and celebrate the priceless natural and cultural heritage of this remarkable region for generations to come.

In the words of the great naturalist and conservationist, Sir David Attenborough, "It seems to me that the natural world is the greatest source of excitement; the greatest source of visual beauty; the greatest source of intellectual interest. It is the greatest source of so much in life that makes life worth living." So embrace the wild beauty and enduring spirit of the Suffolk countryside, and let the Hollesley Common Walk be your gateway to a lifetime of exploration, discovery, and joy in the wonders of the natural world.

Happy wanderings!

Boyton and Hollesley Marshes Walk - 8 miles (12.9 km), Circular - Estimated Time: 4-5 hours

Starting Point Coordinates, Postcode: The starting point for the Boyton and Hollesley Marshes Walk is at the Boyton Marshes RSPB car park (Grid Reference: TM 387 479, Postcode: IP12 3NE).
Nearest Car Park, Postcode: The nearest car park is the Boyton Marshes RSPB car park (Postcode: IP12 3NE).

Embark on an enchanting 8-mile (12.9 km) circular walk through the breathtaking landscapes of Boyton and Hollesley Marshes, a jewel in the crown of the Suffolk Coast and Heaths Area of Outstanding Natural Beauty. This captivating route takes you on a journey through a tapestry of diverse habitats, from the serene beauty of the marshland to the lush tranquility of the woodland and the rugged charm of the heathland, offering unparalleled opportunities for birdwatching, wildlife spotting, and immersing yourself in the natural wonders of the Suffolk coast.

As you set out from the Boyton Marshes RSPB car park, feel the anticipation build as you head eastward along the sea wall, the vast expanse of the marshes stretching out before you in a mesmerizing palette of greens, blues, and golds. With each step, allow the stresses of modern life to melt away, replaced by a growing sense of connection with the landscape and the myriad species that call this unique ecosystem home.

The path winds along the edge of the marshes, the gentle lapping of the water and the distant cries of the birds a soothing soundtrack to your journey. Keep your eyes peeled for the graceful silhouettes of avocets and lapwings, their striking black-and-white plumage a stark contrast to the muted tones of the marsh. Listen for the haunting calls of the marsh harriers as they soar overhead, their keen eyes scanning the reed beds for signs of prey.

As you continue along the sea wall, the landscape begins to shift and change, the vast openness of the marshes giving way to the cool, green embrace of the coastal woodland. Here, the air is filled with the gentle rustling of leaves and the melodic songs of woodland birds, a soothing balm for the soul. Take a moment to breathe in the earthy scents of moss and leaf litter, and to marvel at the intricate patterns of light and shadow that dance across the forest floor.

Emerging from the woodland, you'll find yourself in the rugged, wild beauty of the heathland, where the vibrant purples and pinks of the heather create a stunning carpet that stretches out to the horizon. This unique habitat is home to a fascinating array of plants and animals, from the delicate blooms of the bog asphodel to the elusive Dartford warbler, each one a testament to the incredible adaptability and resilience of nature.

As you cross the road at the northern end of the marshes and continue along the footpath, take a moment to reflect on the rich history and cultural heritage of this remarkable landscape. For centuries, the Suffolk coast has been shaped by the interplay of human activity and natural processes, from the ancient salt-making industry that once thrived in the marshes to the modern-day efforts to conserve and protect this fragile ecosystem for future generations.

Rejoining the sea wall at the southern end of Boyton Marshes, you'll be treated to a stunning panorama of the coast, the vast expanse of the North Sea stretching out to the horizon in a shimmering blue haze. Take a moment to drink in the raw beauty and power of this elemental landscape, and to feel the deep sense of connection and belonging that comes from immersing yourself in the natural world.

As you complete the circular route and return to the Boyton Marshes RSPB car park, take a moment to reflect on the incredible journey you've just undertaken. The Boyton and Hollesley Marshes Walk is more than just a physical challenge; it is a celebration of the breathtaking diversity and enduring beauty of the Suffolk coast, a chance to reconnect with nature and to find solace, inspiration, and joy in the simple act of putting one foot in front of the other.

Detailed Route Directions:

1. From the Boyton Marshes RSPB car park, head east along the sea wall, following the edge of Boyton Marshes, bearing 090°. (Estimated time: 1 hour)

2. Continue along the sea wall until you reach Hollesley Bay, keeping the marshes on your left, bearing 045°. (Estimated time: 1 hour)

3. At Hollesley Bay, turn inland and follow the path through Hollesley Marshes, passing through woodland and heathland, bearing 315°. (Estimated time: 1 hour 15 minutes)

4. Cross the road at the northern end of the marshes and continue along the footpath, bearing 225°. (Estimated time: 45 minutes)

5. Rejoin the sea wall at the southern end of Boyton Marshes, bearing 135°. (Estimated time: 1 hour)

6. Follow the sea wall back to the Boyton Marshes RSPB car park to complete the walk, bearing 270°. (Estimated time: 1 hour)

Best Features and Views:

- Immerse yourself in the serene beauty of Boyton and Hollesley Marshes, where the vast expanse of the wetlands creates a breathtaking panorama of colour, texture, and life, inviting quiet contemplation and a deep sense of connection with the natural world.
- Delight in the lush tranquility of the coastal woodland, where the cool green canopy and the gentle rustling of leaves create a soothing oasis of calm, a perfect spot for a moment of reflection or a picnic in the dappled shade.
- Marvel at the rugged, wild beauty of the heathland, where the vibrant purples and pinks of the heather create a stunning carpet that stretches out to the horizon, a testament to the incredible diversity and resilience of the Suffolk coast.

Historical Features of Interest:

While there are no specific historical sites directly along the route, the Boyton and Hollesley Marshes Walk takes you through a landscape that is steeped in history and cultural heritage, offering a fascinating glimpse into the complex relationship between humans and the natural world on the Suffolk coast.

For centuries, the marshes and wetlands of this area have been a vital resource for local communities, providing a rich array of products and services that have sustained generations of people. One of the most significant industries in the region was salt-making, which thrived in the medieval period and played a crucial role in the economy and social structure of the coastal villages.

The process of salt-making involved the careful management of the marshes, with networks of ditches and sluices used to control the flow of seawater into shallow evaporation pans. The resulting salt crystals were then harvested and traded widely, providing a valuable commodity that was essential for preserving food and enhancing flavour in a time before refrigeration.

As you walk along the sea wall and through the marshes, keep an eye out for the remnants of this ancient industry, from the subtle undulations in the landscape that mark the outlines of long-abandoned salt pans to the occasional fragments of clay pipes and pottery that hint at the lives and labours of the salt-makers who once toiled here.

Beyond the salt-making industry, the marshes and wetlands of the Suffolk coast have also played a vital role in the region's agriculture and fishing heritage. For generations, local farmers have grazed their cattle on the lush pastures of the marshes, while fishermen have harvested the bountiful waters of the estuaries and the open sea, creating a way of life that is deeply rooted in the rhythms and cycles of the natural world.

Today, the Boyton and Hollesley Marshes are recognized as a vital haven for wildlife and biodiversity, with conservation organizations like the RSPB working tirelessly to protect and preserve this fragile ecosystem for future generations. As you walk through this stunning landscape, take a moment to reflect on the countless generations of people who have shaped and been shaped by this place, and to appreciate the deep sense of history and continuity that underlies every step of your journey.

Birds to Spot:

- Avocets: Keep your eyes peeled for these elegant black-and-white waders, with their slender upturned bills and long bluish legs, as they forage in the shallows of the marshes. Once a rare sight in the UK, avocets have made a remarkable comeback in recent years, thanks to dedicated conservation efforts, and are now a beloved symbol of the Suffolk coast.
- Marsh Harriers: Watch for the distinctive silhouette of these majestic birds of prey as they soar over the reed beds and marshes, their rich brown plumage and broad wings a striking sight against the vast East Anglian sky. Listen for their haunting cries and marvel at their incredible aerial skills as they hunt for small mammals and birds among the dense vegetation.
- Bearded Tits: These charming little birds, with their striking black face masks and rusty-orange plumage, are a true delight to spot among the reed beds of the marshes. Listen for their distinctive "pinging" calls and watch for the flash of their long tails as they flit among the swaying stems, a joyful presence in this unique and precious habitat.

Rare or Interesting Insects and Other Wildlife:

- Dragonflies and Damselflies: In the summer months, the marshes and wetlands come alive with the vibrant colours and acrobatic flight of numerous dragonfly and damselfly species, from the large and powerful emperor dragonfly to the delicate and iridescent banded demoiselle. These ancient insects, with their intricate wing patterns and jewel-like eyes, are a fascinating and beautiful presence in the Suffolk countryside.
- Water Voles: As you walk along the edges of the ditches and waterways that crisscross the marshes, keep an eye out for the furry brown bodies and rounded faces of water voles, often affectionately referred to as "water rats" due to their resemblance to their terrestrial cousins. These charming mammals are a vital part of the wetland ecosystem and a delight to spot as they swim and forage among the reeds.
- Otters: Though elusive and often difficult to spot, otters have made a remarkable resurgence in the rivers and wetlands of the Suffolk coast in recent years, thanks to improved water quality and conservation efforts. As you explore the marshes and waterways, look out for the tell-tale signs of otter activity, from the webbed footprints in the mud to the sleek, sinuous shapes that occasionally break the surface of the water, a thrilling reminder of the wild beauty and resilience of this special landscape.

Rare or Interesting Plants:

- Reed Beds: As you walk through the marshes, take a moment to appreciate the vast expanses of reed beds that stretch out before you, their tall, slender stems swaying gently in the breeze. These dense stands of common reed (Phragmites australis) provide a vital habitat for a wide range of birds, insects, and other wildlife, and play a crucial role in the ecology and biodiversity of the wetland ecosystem.
- Sea Lavender: In the late summer and early autumn, the saltmarshes of the Suffolk coast are transformed into a stunning carpet of delicate purple flowers, as the sea lavender (Limonium vulgare) bursts into bloom. This hardy and resilient plant, with its slender stems and clusters of tiny, lavender-coloured blooms, is a true icon of the coastal landscape, and a vital source of nectar for bees and other pollinators.
- Bog Asphodel: As you walk through the damper areas of the heathland, look out for the striking yellow spikes of bog asphodel (Narthecium ossifragum), a rare and beautiful plant that thrives in the acidic, nutrient-poor soils of the Suffolk coast. With its star-shaped flowers and slender, grass-like leaves, this exquisite wildflower is a true gem of the heathland flora, and a reminder of the incredible diversity and adaptability of the natural world.

Facilities and Accessibility:

The Boyton and Hollesley Marshes Walk is a moderate route that offers a mix of flat, easy terrain along the sea wall and some more challenging sections through the marshes and heathland. While much of the walk is on well-maintained paths and tracks, there are some areas that can be muddy, wet, or uneven, particularly after rain or during the winter months.

As such, it is essential to wear sturdy, waterproof walking boots with good ankle support and to dress appropriately for the weather conditions, with layers that can be easily added or removed as needed. It's also a good idea to bring a waterproof jacket and trousers, as well as a hat and sunscreen to protect against the elements.

It's important to note that there are no public toilets, cafes, or other facilities directly along the route, so it is crucial to come prepared with plenty of water, snacks, and any necessary medications or first aid supplies. The nearest amenities can be found in the nearby villages of Boyton and Hollesley, where you can find a selection of pubs, cafes, and small shops.

While the majority of the walk is relatively accessible, with gates and stiles providing access through fences and field boundaries, some sections may not be suitable for those with limited mobility or for pushchairs, due to the uneven terrain and the presence of steps and narrow paths. However, with careful planning and a willingness to take your time, most walkers will find this route to be a rewarding and enjoyable experience.

Seasonal Highlights:

The Boyton and Hollesley Marshes Walk offers a unique and captivating experience in every season, with each time of year revealing new facets of beauty and wonder in this extraordinary coastal landscape:

- Spring: As the land awakens from its winter slumber, the marshes and wetlands burst into life, with the fresh green shoots of the reeds emerging from the water and the first wildflowers of the year beginning to bloom. The air is filled with the joyous songs of breeding birds, from the melodious warbles of the sedge warblers to the booming calls of the bitterns, while the heathland is ablaze with the vibrant yellows of the gorse and the delicate pinks of the spring squill.
- Summer: In the long, warm days of summer, the marshes are alive with the buzzing of insects and the constant activity of countless bird species, from the graceful flights of the marsh harriers to the busy foraging

of the avocets and lapwings. The heathland is a riot of colour, with the purple haze of the heather and the golden blooms of the bog asphodel creating a stunning tapestry that stretches out to the horizon, while the sea lavender transforms the saltmarshes into a mesmerizing carpet of delicate purple flowers.
- Autumn: As the leaves begin to turn and the days grow shorter, the marshes take on a new character, with the fading light casting long shadows across the water and the distant cries of the overwintering birds filling the air. The heathland is ablaze with the rich, warm hues of the turning foliage, while the fruit and berries of the coastal woodlands provide a vital source of food for the wildlife preparing for the long winter ahead.
- Winter: Though the landscape may seem stark and empty in the depths of winter, there is a haunting beauty to the marshes and wetlands in this season of rest and renewal. The bare branches of the trees and the frost-covered reeds create an ethereal, almost otherworldly atmosphere, while the low sun casts a pale, pink light across the water and the distant calls of the wading birds echo through the stillness. Wrap up warm and embrace the invigorating chill of the coastal winds, and you'll discover a profound sense of peace and solitude in this wild and elemental landscape.

Difficulty Level and Safety Considerations:

The Boyton and Hollesley Marshes Walk is a moderate route that offers a mix of flat, easy terrain along the sea wall and some more challenging sections through the marshes and heathland. While much of the walk is on well-maintained paths and tracks, it's important to be aware of the potential hazards and to take necessary precautions to ensure a safe and enjoyable experience.

One of the main challenges of walking in the marshes and wetlands is the risk of getting stuck or sinking in the soft, muddy ground, particularly after heavy rain or during the winter months. It's essential to wear sturdy, waterproof walking boots with good ankle support and to take care when navigating through these areas, sticking to the designated paths and avoiding any patches of standing water or deep mud.

Another factor to consider is the exposed nature of the coastal landscape, which can leave you vulnerable to the elements, particularly on windy or rainy days. Make sure to check the weather forecast before setting out and to dress appropriately for the conditions, with warm, waterproof layers and a hat and gloves if necessary.

It's also important to be mindful of the tides when walking along the sea wall or near the water's edge, as the water levels can rise quickly and unexpectedly, particularly during spring tides or stormy weather. Always check the tide times before you set out and be prepared to adjust your route if necessary to avoid any areas that may become submerged or difficult to navigate.

As with any walk in a remote or unfamiliar area, it's a good idea to carry a map and compass or GPS device, as well as a fully charged mobile phone and a small first aid kit, in case of emergencies . Let someone know your intended route and estimated return time before you set off, and be sure to carry enough water, snacks, and any necessary medications to keep you safe and comfortable throughout your walk.

Finally, it's crucial to follow the Countryside Code and to respect the delicate ecosystems and wildlife habitats that you'll be exploring along the way. This means sticking to the designated paths and tracks, keeping dogs on a lead, and taking all your litter home with you to help preserve the natural beauty and integrity of this precious landscape for generations to come.

By following these simple guidelines and using common sense, walkers of all ages and abilities can safely enjoy the stunning scenery, rich history, and incredible biodiversity of the Boyton and Hollesley Marshes Walk, immersing themselves in the timeless rhythms and enduring wonders of the Suffolk coast.

Nearby Attractions:

After your invigorating walk through the captivating landscapes of Boyton and Hollesley Marshes, why not extend your adventure and explore some of the other fascinating attractions that this beautiful corner of Suffolk has to offer? From picturesque seaside towns to world-renowned nature reserves and cultural landmarks, there's something to capture the imagination and delight the senses of every visitor.

Just a short drive from the marshes, the charming coastal town of Aldeburgh awaits, with its colourful seafront cottages, independent shops and galleries, and famous fish and chip shops. Take a stroll along the pebble beach, admiring the iconic scallop sculpture by local artist Maggi Hambling, or visit the Aldeburgh Museum to learn about the town's rich maritime history and its connections to the composer Benjamin Britten.

For nature lovers and birdwatchers, the nearby RSPB Minsmere reserve is an absolute must-see. This stunning coastal sanctuary, with its mosaic of reed beds, lagoons, and woodland, is home to an astonishing variety of wildlife, from the majestic marsh harriers and bitterns to the playful otters and water voles. With its network of trails, hides, and visitor centre, Minsmere offers unparalleled opportunities to explore and appreciate the incredible diversity of the Suffolk coast.

Culture enthusiasts will delight in a visit to the world-famous Snape Maltings, a restored Victorian maltings complex that now serves as a hub for music, art, and creativity. Attend a concert or recital in the stunning Concert Hall, browse the independent shops and galleries showcasing the work of local artists and craftspeople, or simply soak up the unique atmosphere of this cultural landmark, set amidst the breathtaking beauty of the Suffolk countryside.

For a taste of the region's rich agricultural heritage and delicious local produce, be sure to visit one of the many farmers' markets, farm shops, and artisanal food producers that can be found throughout the area. Sample the creamy, tangy cheeses, the succulent meats and seafood, and the crisp, refreshing ciders and beers that are the hallmark of Suffolk's thriving food and drink scene, and take home a piece of this special place to savour long after your walk has ended.

As you explore the many attractions and hidden gems of the Suffolk coast and countryside, take a moment to reflect on the deep sense of history, community, and connection to the land that pervades every aspect of life in this enchanting region. From the ancient monasteries and castles to the vibrant seaside towns and the timeless beauty of the marshes and heaths, every sight and sound tells a story of the remarkable people and places that have shaped this unique corner of England.

So whether you're a nature enthusiast, a history buff, a foodie, or simply someone in search of a little beauty and tranquillity in a fast-paced world, the Boyton and Hollesley Marshes Walk and its surroundings offer a wealth of opportunities for discovery, relaxation, and inspiration. Why not linger a while and immerse yourself in the magic and wonder of the Suffolk coast, and let the gentle rhythms and enduring spirit of this special place work their spell on your heart and soul?

As you set out on your adventures, remember to tread lightly, to respect the local environment and communities, and to approach each new encounter with an open mind and a sense of curiosity. By doing so, you'll not only enrich your own experience and understanding of the world around you, but you'll also help to preserve and celebrate the priceless natural and cultural heritage of this remarkable region for generations to come.

In the words of the renowned naturalist and broadcaster, Sir David Attenborough, "No one will protect what they don't care about; and no one will care about what they have never experienced." So embrace the wild beauty and enduring charm of the Suffolk coast, and let the Boyton and Hollesley Marshes Walk be your gateway to a lifetime of discovery, wonder, and love for the incredible diversity and resilience of the natural world.

Happy walking!

Sutton Hoo Walk - 3 miles (4.8 km), Circular - Estimated Time: 1.5-2 hours

Starting Point Coordinates, Postcode: The starting point for the Sutton Hoo Walk is at the Sutton Hoo Visitor Centre car park (Grid Reference: TM 288 487, Postcode: IP12 3DJ).
Nearest Car Park, Postcode: The nearest car park is the main car park at Sutton Hoo Visitor Centre (Grid Reference: TM 288 487, Postcode: IP12 3DJ).

Embark on a captivating 3-mile (4.8 km) circular walk around the world-renowned Sutton Hoo burial mounds, where the fascinating history of Anglo-Saxon England and the breathtaking beauty of the Suffolk countryside intertwine to create an unforgettable experience. This easy to moderate route offers a perfect blend of historical intrigue, stunning riverside views, and diverse wildlife, making it an ideal choice for history buffs, nature lovers, and those seeking a delightful escape from the hustle and bustle of everyday life.

As you set out from the Sutton Hoo Visitor Centre car park, a sense of anticipation builds as you follow the footpath, the burial mounds gradually coming into view across the expansive grassy field. With each step, you'll find yourself drawn deeper into the captivating story of this ancient site, where the incredible treasures and ship burial of an Anglo-Saxon king were discovered in 1939, forever changing our understanding of early medieval England.

Take your time exploring the burial mounds, pausing to read the informative panels that shed light on the site's rich history and the painstaking archaeological work that uncovered its secrets. As you stand atop the mounds, let your imagination transport you back to a time of great halls, epic battles, and the rise of powerful kingdoms, the lives and legends of the Anglo-Saxons etched into the very landscape that surrounds you.

Continuing along the footpath, head south towards the River Deben, the trail winding through a picturesque tapestry of fields and hedgerows. As you near the river, the landscape opens up to reveal stunning vistas of the glistening water and the lush green marshes beyond, a timeless scene that has inspired countless artists and writers throughout the centuries.

Follow the path along the riverbank, immersing yourself in the tranquil beauty of this iconic Suffolk waterway. Pause to rest on a bench beneath a majestic oak tree, taking in the serene views and listening to the gentle lapping of the water and the distant calls of wading birds. This is a moment to slow down, to breathe deeply, and to connect with the natural world, letting the stresses of modern life melt away in the embrace of the countryside.

As you continue along the riverside path, keep your eyes peeled for the abundance of wildlife that calls this area home. From the darting flights of iridescent dragonflies to the vibrant flashes of kingfishers, every moment brings a new opportunity for discovery and wonder. In the wooded areas, listen for the drumming of woodpeckers and the melodious songs of warblers, while on the riverbanks, watch for the telltale signs of otters and water voles, elusive creatures that thrive in this pristine habitat.

Turning away from the river, follow the footpath back towards Sutton Hoo, the trail leading you through a small woodland alive with the delicate blooms of bluebells in the spring and the earthy scents of wild mushrooms in the autumn. Here, amidst the dappled shade of ancient trees, you'll find a moment of quiet reflection, a chance to ponder the countless generations who have walked this same path, each leaving their mark on the ever-evolving story of the landscape.

As you emerge from the woods and catch sight of the burial mounds once more, you'll carry with you a newfound appreciation for the rich history and enduring beauty of Sutton Hoo. This walk is a testament to the power of place, a celebration of the ways in which the natural world and human heritage are forever intertwined, offering us a profound sense of connection to the past, present, and future.

Whether you're a seasoned hiker or a curious first-time visitor, the Sutton Hoo Walk promises to delight and inspire, inviting you to step into a world of wonder, where the echoes of ancient kings and the timeless rhythms of nature converge to create an unforgettable experience. So lace up your walking boots, grab your binoculars, and set out on a journey of discovery that will nourish your mind, body, and soul, leaving you with memories to cherish for a lifetime.

Detailed Route Directions:

1. From the Sutton Hoo Visitor Centre car park, locate the footpath just to the right of the main entrance and follow it for approximately 300 meters, bearing 150°. (Estimated time: 5 minutes)

2. At the first intersection, turn left, bearing 60°, towards the burial mounds visible in the large grassy field ahead. (Estimated time: 5 minutes)

3. Explore the burial mounds, taking the time to read the informative panels and appreciate the site's historical significance. After visiting the mounds, continue along the footpath, heading south towards the River Deben, bearing 170°. (Estimated time: 20-30 minutes)

4. Upon reaching the river, turn right and follow the footpath along the riverside, bearing 230°, passing through the wooded area known as Tranmer House. Take a moment to enjoy the lovely views of the river and surrounding countryside from the bench near the large oak tree. (Estimated time: 20-25 minutes)

5. Continue along the riverside path for about 0.9 miles (1.4 km), bearing 280°, until you reach a footpath intersection marked by a wooden signpost. (Estimated time: 25-30 minutes)

6. At the signpost, turn left, bearing 350°, and follow the footpath back towards the Sutton Hoo Visitor Centre. (Estimated time: 20-25 minutes)

7. As you approach the visitor centre, navigate through the small woodland area, bearing 60°, noting the diverse plant life. The path will then open up to a grassy field with views of the burial mounds in the distance. (Estimated time: 10-15 minutes)

8. Complete the circular walk by arriving back at the Sutton Hoo Visitor Centre car park, bearing 90°. (Estimated time: 5 minutes)

Best Features and Views:

- Immerse yourself in the captivating history of the Anglo-Saxon burial mounds at Sutton Hoo, where the discovery of incredible treasures and a magnificent ship burial has shed light on the rich culture and power of early medieval England.
- Delight in the stunning views of the River Deben, its glistening waters and lush green marshes creating a timeless and inspiring landscape that has captivated artists and writers for centuries.
- Embrace the tranquility and beauty of the Suffolk countryside, from the picturesque tapestry of fields and hedgerows to the peaceful woodland paths, each offering a chance to reconnect with nature and find solace in the great outdoors.

Historical Features of Interest:

The Sutton Hoo burial grounds, discovered in 1939, represent one of the most significant archaeological finds in British history, offering an unparalleled glimpse into the world of the Anglo-Saxons and the power and wealth of their kings. The site, dating back to the 6th and 7th centuries AD, contains a range of burial mounds, with the most impressive being the ship burial of an Anglo-Saxon king, believed to be Rædwald of East Anglia.

The ship burial, excavated from the largest mound, revealed an astonishing array of treasures, including a magnificent helmet, a sword, and a sceptre, all crafted with exquisite skill and adorned with precious materials such as gold, garnet, and enamel. These artifacts, now on display at the British Museum, provide a tantalizing window into the sophistication and artistry of Anglo-Saxon culture, as well as the far-reaching trade networks and cultural exchanges that shaped this fascinating period of history.

As you explore the burial mounds and learn about the meticulous archaeological work that uncovered their secrets, you'll gain a deeper appreciation for the ingenuity and dedication of the archaeologists and historians who have worked tirelessly to piece together the story of Sutton Hoo. From the painstaking excavation process to the cutting-edge scientific analysis of the artifacts, the study of this remarkable site has revolutionized our understanding of the Anglo-Saxon world and its enduring legacy.

Beyond the burial mounds themselves, the landscape of Sutton Hoo is steeped in history, with evidence of human occupation stretching back thousands of years. As you walk along the River Deben, imagine the countless generations of people who have navigated its waters, from the early hunter-gatherers who fished and foraged along its banks to the medieval merchants who plied their trade on its tides.

The nearby Tranmer House, a former manor house that now serves as a visitor centre, adds another layer of history to the site. Built in the early 20th century, the house was once home to Edith Pretty, the landowner who commissioned the excavation of the Sutton Hoo mounds, setting in motion the incredible chain of events that would lead to the discovery of the ship burial and its treasures.

As you immerse yourself in the rich history of Sutton Hoo, take a moment to reflect on the enduring power of archaeology and the ways in which the study of the past can enrich and transform our understanding of the present. The story of Sutton Hoo is not just a tale of kings and treasures, but a testament to the resilience, creativity, and ingenuity of the human spirit, a reminder that even the most distant and enigmatic chapters of our history have the power to inspire and enlighten us today.

Birds to Spot:

- Green Woodpecker: Listen for the distinctive laughing call and look for the bright green plumage of this striking woodland bird as it forages for ants on the forest floor or clings to the trunks of trees.
- Marsh Harrier: Scan the skies over the River Deben for the distinctive silhouette of this majestic bird of prey, its broad wings and long tail a familiar sight above the marshes and wetlands of the Suffolk coast.
- Kingfisher: Keep your eyes peeled for the electric blue flash of a kingfisher darting along the riverbank, its vibrant colours and quick, agile flight a true delight to behold.

Rare or Interesting Insects and Other Wildlife:

- Hairy Dragonfly: In the summer months, look for the impressive black and yellow-patterned wings and the hairy thorax of this striking dragonfly species as it patrols the riverbanks and nearby ponds.
- Otter: Although elusive and often difficult to spot, otters have been known to frequent the River Deben and its tributaries. Keep an eye out for their sleek, sinuous forms playing in the water or the telltale signs of their

presence, such as webbed footprints in the mud or the remains of fish on the riverbank.
- Purple Emperor: If you're lucky, you may catch a glimpse of this majestic butterfly species in the woodland areas along the walk, its iridescent purple and white wings a stunning sight as it glides through the dappled sunlight of the forest canopy.

Rare or Interesting Plants:

- Wild Garlic: In the spring, the woodland floors of Sutton Hoo are carpeted with the delicate white flowers and vibrant green leaves of wild garlic, their pungent aroma filling the air and adding a delightful sensory dimension to your walk.
- Bluebells: From April to May, the woodlands come alive with the enchanting sight and sweet fragrance of bluebells, their delicate purple blooms creating a mesmerizing sea of colour beneath the trees.
- Yellow Flag Iris: In the summer months, look for the striking yellow flowers of the yellow flag iris along the banks of the River Deben, their vibrant petals and sword-like leaves adding a splash of colour to the lush green vegetation of the marshes.

Facilities and Accessibility:

The Sutton Hoo Visitor Centre offers a range of facilities to enhance your experience, including a café serving delicious locally-sourced food and drinks, a gift shop stocked with unique souvenirs and books related to the site's history, and well-maintained restrooms. The visitor centre itself is fully accessible for wheelchair users, with step-free access and accessible toilets available.

The Sutton Hoo Walk is generally considered an easy to moderate route, with mostly level terrain and well-defined footpaths. However, it's important to note that some sections of the trail may be uneven, muddy, or slightly overgrown, particularly after periods of wet weather or during the summer months. As such, sturdy walking shoes or boots with good grip and ankle support are highly recommended to ensure a safe and comfortable walk.

While much of the route is accessible to a wide range of abilities, some parts of the walk, particularly the sections closest to the River Deben, may not be suitable for wheelchairs or pushchairs due to the presence of gates, steps, and narrow or uneven paths. Visitors with limited mobility are advised to check with the visitor centre staff for the most up-to-date information on accessibility and alternative routes.

For those who prefer a more leisurely pace or require frequent rest stops, there are plenty of benches and picnic spots along the route, offering the perfect opportunity to take a break, enjoy the stunning views, and soak up the peaceful atmosphere of the Suffolk countryside.

Seasonal Highlights:

The Sutton Hoo Walk offers a unique and captivating experience in every season, with each time of year revealing new facets of beauty and wonder in this fascinating corner of East Anglia:

- Spring: As the land awakens from its winter slumber, the woodlands and hedgerows burst into life with a vibrant display of colourful wildflowers, from the delicate pinks of cuckoo flower to the cheerful yellows of primroses and the enchanting purples of bluebells. The air is filled with the melodious songs of returning migrant birds, such as chiffchaffs and blackcaps, while the fresh green leaves of the trees create a lush canopy overhead, dappling the forest floor with a mesmerizing play of light and shadow.
- Summer: In the long, lazy days of summer, the landscape is alive with the hum of insects and the gentle

rustling of the warm breeze through the grasslands and reedbeds. The riverbanks are adorned with the striking yellow blooms of flag iris, while the nearby meadows are a sea of waving grasses and wildflowers, attracting a myriad of butterflies, bees, and other pollinators. This is a time to bask in the sun's warmth, to seek out the cool shade of the woodland trails, and to delight in the vibrant colours and sounds of the Suffolk countryside at its most lush and bountiful.

- Autumn: As the leaves begin to turn and the days grow shorter, the Sutton Hoo landscape is transformed into a breathtaking tapestry of gold, orange, and red, the changing foliage creating a warm and inviting atmosphere that beckons you to explore. The woodland floors are carpeted with a crunchy layer of fallen leaves, while the hedgerows and trees are laden with a bountiful harvest of berries, nuts, and seeds, providing a vital food source for the birds and mammals preparing for the winter ahead. This is a time for quiet reflection and gentle exploration, for savouring the earthy scents and mellow light of the season, and for marvelling at the cyclical beauty of the natural world.

- Winter: Though the landscape may be stark and stripped back to its essentials, the Sutton Hoo Walk takes on a unique beauty and character in the winter months, with the frosty ground and the low, slanting light creating an almost ethereal atmosphere. The bare branches of the trees stand out in sharp relief against the pale sky, their intricate patterns and textures revealed in exquisite detail, while the distant calls of overwintering birds, such as redwings and fieldfares, add a haunting soundtrack to the otherwise silent landscape. Wrap up warm and embrace the invigorating chill of the winter air, and you'll discover a profound sense of peace, clarity, and connection with the enduring spirit of this ancient and fascinating place.

Safety Considerations:

When embarking on the Sutton Hoo Walk, it's essential to keep in mind a few key safety considerations to ensure a pleasant and incident-free experience. While the route is generally well-maintained and easy to navigate, there are a few potential hazards to be aware of, particularly in relation to the terrain and weather conditions.

As with any outdoor activity, it's crucial to wear appropriate footwear and clothing for the conditions. Sturdy, comfortable walking shoes or boots with good grip and ankle support are a must, especially on the sections of the trail that may be uneven, muddy, or slippery. In wet or cold weather, make sure to bring warm, waterproof layers to protect against the elements, as well as a hat and gloves if necessary.

When walking along the River Deben, exercise caution and stay back from the water's edge, particularly after heavy rainfall or during high tides, when the banks may be unstable or the water levels higher than usual. Children and pets should be closely supervised at all times, and it's advisable to keep dogs on a lead to prevent them from disturbing wildlife or straying into dangerous areas.

It's also important to be mindful of any livestock you may encounter along the route, particularly if you are walking through fields or meadows where sheep or cattle are grazing. Give the animals plenty of space, avoid making sudden movements or loud noises that may startle them, and follow any specific instructions or signage provided by the landowners or the visitor centre staff.

In terms of navigation, the Sutton Hoo Walk is generally well-signposted and easy to follow, but it's always a good idea to bring a map and compass or a GPS device, just in case you need to double-check your route or find your way back to the visitor centre. Make sure to familiarize yourself with the trail map before setting out, and take note of any key landmarks or junctions that will help you stay on track.

As with any outdoor pursuit, it's also crucial to be prepared for emergencies or unexpected situations. Carry a fully charged mobile phone with you at all times, and make sure to let someone know your intended route and estimated return time before setting off. In the event of an accident or injury, dial 999 and follow the instructions of the emergency services, making sure to provide as much detail as possible about your location and the nature of the incident.

By following these simple safety guidelines and using common sense, walkers of all ages and abilities can fully immerse themselves in the beauty, history, and natural wonders of the Sutton Hoo Walk, creating lasting memories and forging a deeper connection with this extraordinary corner of the Suffolk countryside.

Nearby Attractions:

After your invigorating walk around the Sutton Hoo burial grounds, why not extend your visit and explore some of the other fascinating attractions that this beautiful part of Suffolk has to offer? From charming historic towns to stunning coastal landscapes and world-class cultural destinations, there's something to capture the imagination and delight the senses of every visitor.

Just a short drive from Sutton Hoo, the picturesque town of Woodbridge awaits, with its quaint streets, independent shops, and welcoming pubs and cafes. Take a stroll along the River Deben, admiring the beautiful boats and the historic buildings that line its banks, or visit the town's famous Tide Mill, one of the oldest working tide mills in the country. Discover the town's rich maritime heritage at the Woodbridge Museum, or simply soak up the relaxed, friendly atmosphere of this charming Suffolk gem.

For a truly unforgettable cultural experience, head to the world-renowned Snape Maltings, a stunning collection of Victorian malthouses that have been transformed into a vibrant arts complex, home to the famous Aldeburgh Festival of Music. Attend a concert or recital in the breathtaking Concert Hall, explore the art galleries and shops showcasing the work of local artists and craftspeople, or simply enjoy a leisurely walk around the beautiful grounds, taking in the views of the River Alde and the surrounding countryside.

Nature lovers and birdwatchers will be in their element at the nearby RSPB Minsmere nature reserve, one of the UK's premier wildlife-watching destinations. With its diverse mosaic of reedbeds, wetlands, and coastal habitats, Minsmere is home to an astonishing array of bird species, from the elusive bittern and the majestic marsh harrier to the colourful avocet and the comical puffin. Follow the network of trails and boardwalks that crisscross the reserve, visit the hides and viewing platforms that offer unparalleled views of the wildlife, and immerse yourself in the sights, sounds, and rhythms of the natural world.

For a taste of the region's rich history and archaeology, be sure to visit the nearby town of Rendlesham, once the royal seat of the Anglo-Saxon kings of East Anglia. Explore the fascinating remains of the royal palace complex, learn about the incredible finds that have been discovered here in recent years, and gain a deeper appreciation for the wealth, power, and sophistication of the early medieval kingdoms that flourished in this part of England.

As you explore the many attractions and hidden gems of the Suffolk coast and countryside, take a moment to reflect on the deep sense of place and belonging that pervades every aspect of life in this special corner of the world. From the ancient burial mounds and the timeless beauty of the landscape to the vibrant creativity and warm hospitality of the local communities, every experience offers a chance to connect with the enduring spirit and character of this remarkable region.

So whether you're a history buff, a nature lover, a culture vulture, or simply someone in search of a memorable and enriching day out, the Sutton Hoo Walk and its surroundings offer a wealth of opportunities for discovery, inspiration, and sheer enjoyment. Why not linger a while longer and immerse yourself in the magic and wonder of the Suffolk countryside, and let the echoes of the past and the beauty of the present combine to create an unforgettable experience that will stay with you long after you've left?

As you set out on your adventures, remember to tread lightly, to respect the local environment and communities, and to approach each new encounter with an open mind and a curious heart. By doing so, you'll not only deepen your own understanding and appreciation of this extraordinary place and its people, but you'll also help to ensure that the natural and cultural treasures of the Suffolk coast and countryside can continue to be enjoyed and cherished by generations to come.

Happy exploring!

Staverton Thicks Walk - 3 miles (4.8 km), Circular - Estimated Time: 1.5-2 hours

Starting Point Coordinates, Postcode: The starting point for the Staverton Thicks Walk is at the layby on Wantisden Road (Grid Reference: TM 393 547, Postcode: IP12 3DJ).
Nearest Car Park, Postcode: The nearest car park is the layby on Wantisden Road (Postcode: IP12 3DJ).

Embark on an enchanting 3-mile (4.8 km) circular walk through the ancient woodland of Staverton Thicks, where the majestic oaks and hollies stand as silent guardians of a forest untouched by time. This easy to moderate route offers a perfect escape from the hustle and bustle of modern life, inviting you to immerse yourself in the tranquil beauty and rich biodiversity of this unique and captivating landscape.

As you set out from the layby on Wantisden Road, feel the anticipation build as you step into the cool, green embrace of the woodland, the well-trodden path leading you deeper into a world of dappled sunlight and gentle rustling leaves. With each step, the stresses of everyday life melt away, replaced by a growing sense of peace and connection with the natural world.

The path winds westward through the heart of Staverton Thicks, guiding you past towering oaks and hollies, their gnarled trunks and sprawling canopies a testament to the centuries of growth and change they have witnessed. Take a moment to run your hand along the rough bark, feeling the weight of history and the enduring strength of these ancient trees, each one a living monument to the resilience and wonder of nature.

As you continue along the trail, keep your eyes and ears attuned to the sights and sounds of the woodland, from the melodic songs of the birds flitting through the branches to the delicate rustling of small mammals foraging in the undergrowth. Each encounter serves as a reminder of the incredible diversity of life that thrives in this special place, a complex web of interactions and adaptations that have evolved over countless generations.

Emerging into a sun-dappled clearing, pause to take in the peaceful surroundings, the open space offering a chance to catch your breath and marvel at the intricate tapestry of light and shadow that dances across the forest floor. Here, amidst the wildflowers and the gentle hum of insects, you may find a moment of quiet reflection, a chance to ponder the timeless cycles of growth and decay that shape the woodland ecosystem.

Turning northward, follow the path as it leads you through the clearing and back into the cool, green sanctuary of the trees, the trail winding through a landscape of ever-changing textures and hues. From the soft, spongy moss that carpets the forest floor to the vibrant splashes of colour provided by the woodland flowers, every step reveals new wonders and delights, inviting a sense of curiosity and appreciation for the natural world.

As you continue westward, the path gradually curves back towards the east, guiding you through a final stretch of ancient woodland before depositing you once more at the starting point on Wantisden Road. Take a moment to look back at the forest you've just explored, the sense of accomplishment and renewal that comes from immersing yourself in nature's embrace a fitting reward for your efforts.

The Staverton Thicks Walk is more than just a physical journey; it is a celebration of the enduring beauty and resilience of our ancient woodlands, a reminder of the profound connection that exists between humans and the natural world. Whether you're a seasoned hiker or a casual wanderer, this enchanting route promises to delight and inspire, offering a chance to step back in time and rediscover the simple joys and timeless truths that can be found in the heart of the forest.

Detailed Route Directions:

1. From the layby on Wantisden Road, head west into Staverton Thicks, following the well-trodden path, bearing 270°. (Estimated time: 15-20 minutes)

2. Continue walking westward through the woodland, admiring the ancient oak and holly trees, maintaining your bearing of 270°. (Estimated time: 20-25 minutes)

3. Upon reaching a clearing, turn north, bearing 360°, and follow the path through the clearing. (Estimated time: 10-15 minutes)

4. Re-enter the woodland on the western side of the clearing, bearing 270°, and continue walking through the trees. (Estimated time: 20-25 minutes)

5. The path will eventually turn east, bearing 90°. Follow this path as it leads you back towards the starting point. (Estimated time: 25-30 minutes)

6. Arrive back at the layby on Wantisden Road, where your circular walk began. (Estimated time: 5-10 minutes)

Best Features and Views:

- Immerse yourself in the timeless beauty of Staverton Thicks, an ancient woodland that has stood as a silent witness to the passing of centuries, its majestic oaks and hollies a testament to the enduring resilience of nature.
- Delight in the peaceful atmosphere of the forest, the dappled sunlight and the gentle rustling of leaves creating a soothing symphony that calms the mind and soothes the soul.
- Discover the rich biodiversity of the woodland ecosystem, from the melodic songs of the birds to the delicate wildflowers and the fascinating array of insects and mammals that call this special place home.

Historical Features of Interest:

As you walk through the ancient woodland of Staverton Thicks, take a moment to consider the rich and fascinating history that underlies this remarkable landscape. Believed to have been continuously wooded since the end of the last Ice Age, some 12,000 years ago, Staverton Thicks stands as a rare and precious remnant of the primeval forests that once blanketed much of the British Isles.

The majestic oaks and hollies that tower overhead, their gnarled trunks and sprawling canopies a testament to their great age, are living links to a distant past, their roots reaching deep into the soil and their branches stretching towards the sky. Some of these venerable trees may be hundreds, if not thousands, of years old, having witnessed the rise and fall of countless generations of plants, animals, and humans.

As you navigate the winding trails and sun-dappled glades of the woodland, imagine the countless feet that have trodden these same paths over the centuries, from the early hunter-gatherers who sought shelter and sustenance in the forest's depths to the medieval lords and ladies who may have hunted deer and wild boar in the tangled undergrowth.

Indeed, it is thought that Staverton Thicks may have once served as a medieval deer park, a carefully managed landscape where the nobility would have pursued their quarry on horseback, the sound of baying hounds and the blast of hunting horns echoing through the trees. Today, the woodland is a peaceful and tranquil place, a far cry from the pageantry and excitement of the medieval hunt, but the echoes of this ancient tradition can still be felt in the quiet corners and hidden glades of the forest.

As you immerse yourself in the timeless beauty of Staverton Thicks, take a moment to reflect on the enduring power of nature to shape and inspire the human experience. From the earliest days of our species, when the forest was a place of mystery, danger, and enchantment, to the present day, when we seek solace and renewal in the embrace of the trees, the woodland has played a vital role in our physical, emotional, and spiritual well-being.

By walking in the footsteps of those who came before us, and by opening ourselves up to the wonder and wisdom of the natural world, we can gain a deeper appreciation for the rich tapestry of history and ecology that defines this extraordinary place, and a renewed sense of our own place within the great web of life.

Birds to Spot:

- Tawny Owl: As you walk through the shadowy depths of the woodland, listen for the haunting, quavering hoot of the tawny owl, a master of camouflage whose mottled brown plumage blends seamlessly with the bark of the trees. These nocturnal hunters are a key part of the forest ecosystem, helping to keep populations of rodents and other small mammals in check.
- Great Spotted Woodpecker: Keep an eye out for the striking black-and-white plumage and the distinctive red undertail of the great spotted woodpecker, a lively and industrious bird that can often be heard drumming on dead wood in search of insects. These charismatic birds are a common sight in Staverton Thicks, their presence a sign of the health and diversity of the woodland habitat.
- Nuthatch: Listen for the sharp, ringing calls of the nuthatch, a small but feisty bird with a slate-blue back, a rusty-orange breast, and a distinctive black eye stripe. These agile climbers are often seen scurrying up and down tree trunks, probing for insects and seeds in the bark crevices, their presence adding a lively and engaging element to the forest soundscape.

Rare or Interesting Insects and Other Wildlife:

- Purple Emperor: In the summer months, keep your eyes peeled for the magnificent purple emperor butterfly, a large and stunning insect with iridescent purple wings and a powerful, soaring flight. These elusive butterflies are a rare and cherished sight in the woodland glades, their presence a testament to the health and diversity of the forest ecosystem.
- Stag Beetle: As you walk through the dappled shade of the trees, look out for the impressive stag beetle, a large and distinctive insect with fearsome-looking mandibles and a shiny black carapace. These fascinating creatures, which can grow up to 8 cm in length, play a vital role in the decomposition of dead wood, helping to recycle nutrients back into the soil.
- Fallow Deer: If you're lucky, you may catch a glimpse of the graceful fallow deer, a beautiful and elusive mammal with a rich chestnut coat and striking white spots. These elegant creatures, which have roamed the forests of Britain since the medieval period, are a reminder of the ancient traditions of hunting and land management that have shaped the landscape of Staverton Thicks over the centuries.

Rare or Interesting Plants:

- Ancient Oaks: Take a moment to marvel at the majestic ancient oaks that tower over the woodland floor, their gnarled trunks and sprawling canopies a testament to their great age and enduring resilience. These

venerable trees, some of which may be hundreds or even thousands of years old, are a vital part of the forest ecosystem, providing shelter and sustenance for countless species of plants, animals, and fungi.
- Holly: As you walk through the dappled shade of the trees, keep an eye out for the glossy green leaves and bright red berries of the holly, a distinctive and beloved tree that has long been associated with winter celebrations and festivities. In Staverton Thicks, the holly grows alongside the oak in a unique and enchanting forest community, its presence adding a touch of colour and cheer to the woodland scene.
- Bluebells: If you visit Staverton Thicks in the spring, you may be lucky enough to witness the breathtaking spectacle of a bluebell carpet, a shimmering sea of purple and blue that stretches out beneath the trees as far as the eye can see. These delicate and aromatic flowers, which bloom for just a few short weeks each year, are a true celebration of the woodland's enduring beauty and vitality.

Facilities and Accessibility:

The Staverton Thicks Walk is an easy to moderate route that is suitable for most walkers with a reasonable level of fitness. The terrain is mostly flat and even, with well-defined paths that wind through the heart of the woodland, making for a comfortable and enjoyable walking experience.

However, it is important to note that the forest floor can be uneven and slippery in places, particularly after rain or during the wetter months of the year. As such, it is highly recommended that walkers wear sturdy, supportive footwear with good grip and ankle support, to ensure a safe and stable footing on the trails.

At present, there are no facilities such as public toilets, cafes, or visitor centres directly at the starting point or along the route of the walk. Walkers are advised to bring their own water, snacks, and any necessary medications or first aid supplies, and to make use of facilities in nearby towns or villages before embarking on the walk.

It is also worth noting that, while the paths are generally wide and accessible, the woodland environment may not be suitable for those with limited mobility or pushchairs, due to the presence of tree roots, branches, and other natural obstacles. Walkers with specific accessibility requirements are encouraged to contact local authorities or visit the website of the Suffolk Coast & Heaths Area of Outstanding Natural Beauty for more detailed information and guidance.

Despite these limitations, the Staverton Thicks Walk remains a highly rewarding and enjoyable experience for those who are able to undertake it, offering a chance to immerse oneself in the timeless beauty and tranquility of one of England's most enchanting ancient woodlands.

Seasonal Highlights:

The Staverton Thicks Walk offers a unique and captivating experience in every season, with each time of year revealing new facets of beauty and wonder in this enchanting ancient woodland:

- Spring: As the winter chill gives way to the warmth and vitality of spring, Staverton Thicks bursts into life with an explosion of colour and sound. The forest floor is carpeted with a mesmerizing sea of bluebells, their delicate purple and blue blooms creating a fragrant and enchanting tableau beneath the trees. The air is filled with the joyous songs of nesting birds, and the fresh green leaves of the oaks and hollies create a lush canopy overhead, dappling the woodland floor with a play of light and shadow.
- Summer: In the long, languid days of summer, the woodland is a cool and tranquil oasis, the dense foliage providing a welcome respite from the heat and bustle of the outside world. The forest is alive with the hum of insects and the gentle rustling of the breeze through the leaves, while the dappled sunlight creates an ever-changing tapestry of light and colour on the woodland floor. This is a time to bask in the fullness and

abundance of nature, to savour the rich scents and sounds of the forest at the height of its growth and vitality.
- Autumn: As the days grow shorter and the nights turn crisp, Staverton Thicks takes on a new character, its colours shifting from the lush greens of summer to the rich, warm hues of gold, orange, and red. The woodland floor is a carpet of fallen leaves, their earthy scent mingling with the cool, damp air to create a heady and evocative atmosphere. Fungi of all shapes and sizes sprout from the decaying wood and leaf litter, adding a touch of mystery and otherworldliness to the scene, while the soft light filtering through the thinning canopy creates an enchanting play of shadow and illumination.
- Winter: Though the trees may be bare and the ground frozen, Staverton Thicks takes on a stark and haunting beauty in the depths of winter, the skeletal forms of the oaks and hollies etched against the pale, watery light of the low sun. The woodland is a place of stillness and quiet contemplation, the only sounds the distant calls of winter birds and the crunch of frost beneath one's feet. Yet even in this season of dormancy and rest, there are signs of life all around, from the delicate tracery of lichen on the bark to the fresh green shoots of early spring flowers pushing up through the leaf litter, a reminder of the eternal cycle of death and rebirth that lies at the heart of the forest's enduring magic.

Difficulty Level and Safety Considerations:

The Staverton Thicks Walk is considered an easy to moderate route, suitable for walkers of most ages and abilities. The terrain is generally flat and even, with well-maintained paths that wind through the heart of the woodland, making for a comfortable and enjoyable walking experience.

However, as with any outdoor activity, it is essential to be aware of potential hazards and to take necessary precautions to ensure a safe and responsible journey. One of the main considerations when walking in an ancient woodland like Staverton Thicks is the risk of tripping or falling due to uneven ground, exposed tree roots, or slippery surfaces, particularly after rain or during the wetter months.

To mitigate this risk, walkers are advised to wear sturdy, supportive footwear with good grip and ankle support, and to take their time when navigating any challenging or uneven sections of the trail. It is also a good idea to carry a hiking pole or walking stick for added stability and balance, particularly for those with mobility issues or concerns.

Another factor to keep in mind is the possibility of falling branches or debris from the ancient trees that tower overhead. While the risk of injury from falling objects is relatively low, it is important to stay alert and aware of one's surroundings, particularly during windy or stormy weather when the risk may be higher.

Walkers should also be mindful of the presence of wildlife in the woodland, from the more familiar species such as deer and squirrels to the less common and potentially hazardous creatures such as adders or hornets. While the vast majority of wildlife encounters in Staverton Thicks are harmless and enriching, it is always wise to give animals a respectful distance and to avoid disturbing or feeding them.

As with any walk in a rural or remote area, it is also essential to carry a fully charged mobile phone and a small first aid kit in case of emergencies, as well as to inform someone of your intended route and estimated return time before setting off. By following these simple guidelines and exercising common sense and caution, walkers of all ages and abilities can safely immerse themselves in the timeless beauty and tranquility of Staverton Thicks, forging a deeper connection with the natural world and creating memories that will last a lifetime.

Nearby Attractions:

After your invigorating walk through the enchanting ancient woodland of Staverton Thicks, why not take the opportunity to explore some of the other fascinating attractions that this beautiful corner of Suffolk has to offer? From stunning coastal landscapes to charming historic villages and world-class nature reserves, there is something to capture the imagination and delight the senses of every visitor.

Just a short drive from Staverton Thicks lies the magnificent Suffolk Coast and Heaths Area of Outstanding Natural Beauty, a breathtaking expanse of unspoilt coastline, heathland, and forest that stretches from the River Stour in the south to the bustling port of Lowestoft in the north. This nationally protected landscape is a paradise for walkers, birdwatchers, and nature lovers of all ages, with miles of scenic trails, pristine beaches, and abundant wildlife to discover and enjoy.

One of the jewels in the crown of the Suffolk Coast and Heaths AONB is the picturesque village of Orford, a charming and historic settlement nestled on the banks of the River Ore. With its quaint cottages, independent shops, and welcoming pubs, Orford is the perfect place to relax and recharge after a day of exploring the great outdoors. Be sure to visit the impressive Orford Castle, a 12th-century keep that offers stunning views over the surrounding countryside and coast, and learn about the village's rich maritime history and its enduring connection to the sea.

For those with a passion for wildlife and conservation, the nearby RSPB Minsmere nature reserve is an absolute must-visit. This internationally renowned site, which covers over 2,500 acres of wetland, woodland, and coastal habitat, is home to an astonishing array of bird species, from the elusive bittern and the majestic marsh harrier to the colourful avocet and the comical puffin. With its network of trails, hides, and visitor facilities, Minsmere offers unparalleled opportunities to observe and learn about the incredible diversity of the natural world, and to connect with the beauty and resilience of the Suffolk landscape.

And for those seeking a deeper understanding of the region's rich cultural heritage, the charming market town of Woodbridge is just a short journey away. With its handsome Georgian architecture, vibrant arts scene, and excellent museums and galleries, Woodbridge is a treasure trove of history and creativity, offering a fascinating glimpse into the lives and traditions of the people who have called this corner of Suffolk home for generations.

As you explore the many attractions and hidden gems of the Suffolk countryside and coast, take a moment to reflect on the deep sense of place and belonging that pervades every aspect of life in this special part of the world. From the ancient woodlands and the timeless rhythms of the sea to the warm hospitality and enduring spirit of the local communities, every experience is an invitation to connect with the essence of what makes Suffolk such a unique and captivating destination.

So whether you're a keen rambler, a history buff, a wildlife enthusiast, or simply someone in search of a little peace and inspiration in a beautiful setting, the Staverton Thicks Walk and its surroundings offer a wealth of opportunities for discovery, relaxation, and personal enrichment. Why not linger a while and immerse yourself in the magic and wonder of the Suffolk landscape, and let the beauty and diversity of the natural world work its restorative charm on your mind, body, and soul?

As you set out on your adventures, remember to tread lightly, to respect the local environment and communities, and to approach each new encounter with an open heart and a curious mind. By doing so, you'll not only deepen your own understanding and appreciation of the world around you, but you'll also help to ensure that the precious natural and cultural heritage of this remarkable region can be enjoyed and cherished by generations to come.

In the words of the great naturalist and conservationist, John Muir, "In every walk with nature, one receives far more than he seeks." So step out into the beauty and diversity of the Suffolk countryside, and let the timeless magic of Staverton Thicks and its surroundings be your guide and your inspiration as you embark on a journey of discovery that will nourish your spirit and enrich your life in countless ways.

Happy trails!

Sutton Common Walk - 5 miles (8 km), Circular - Estimated Time: 2.5-3 hours

Starting Point Coordinates, Postcode: The starting point for the Sutton Common Walk is at Sutton Memorial Hall, Woodbridge (Grid Reference: TM 301 479, Postcode: IP12 3JQ).
Nearest Car Park, Postcode: Parking is available at Sutton Memorial Hall (Postcode: IP12 3JQ).

Embark on a delightful 5-mile (8 km) circular walk through the serene landscape of Sutton Common, where the tranquil beauty of the Suffolk countryside and the rich tapestry of rural life intertwine to create an unforgettable experience. This easy to moderate route offers a perfect blend of scenic woodlands, open fields, and peaceful country paths, inviting you to immerse yourself in the gentle rhythms and timeless charm of this enchanting corner of East Anglia.

As you set out from the historic Sutton Memorial Hall, feel the anticipation build as you follow the footpath westward, the hall's weathered bricks and elegant lines a fitting starting point for your journey into the heart of the Suffolk countryside. With each step, the stresses of modern life begin to melt away, replaced by a growing sense of peace and connection with the natural world that surrounds you.

The path winds through a patchwork of fields, the lush greens and golden hues of the crops swaying gently in the breeze, a mesmerizing display of the bounty and beauty of the rural landscape. Take a moment to breathe in the fresh, clean air and to listen to the melodic birdsong that fills the sky, each note a joyous celebration of the abundance and diversity of life that thrives in these peaceful surroundings.

As you continue along the trail, you'll soon find yourself immersed in the cool, green embrace of Sutton Common's woodlands, where the dappled sunlight filters through the canopy of leaves and the soft rustling of the wind in the branches creates a soothing natural symphony. Let your senses be filled with the earthy scents of moss and leaf litter, and feel the gentle crunch of twigs and acorns beneath your feet as you navigate the winding paths that lead you deeper into this enchanting realm.

Emerging from the woods, you'll find yourself once again in the open expanse of the fields, the vast East Anglian sky stretching out above you in an endless blue canopy. Here, amidst the gently undulating landscape, you may catch a glimpse of the area's rich wildlife, from the graceful curves of a soaring buzzard to the darting form of a hare bounding through the long grass, each encounter a reminder of the incredible resilience and adaptability of the creatures that call this place home.

As you turn onto another quiet country lane, take a moment to appreciate the timeless beauty of the traditional hedgerows that line the route, their gnarled branches and dense foliage a living testament to the centuries of human stewardship that have shaped and nurtured this landscape. Imagine the countless generations of farmers and labourers who have walked these same paths, their lives and stories woven into the very fabric of the land itself.

Continuing through more woodlands and fields, the path eventually guides you back towards Sutton Memorial Hall, the starting point of your journey and a welcome sight after your invigorating walk. As you complete the circular route, take a moment to reflect on the beauty and tranquillity of the landscapes you've encountered, and the sense of peace and connection that comes from immersing yourself in the simple pleasures of the natural world.

The Sutton Common Walk is a celebration of the enduring charm and rich heritage of the Suffolk countryside, offering a chance to step back from the demands of modern life and to reconnect with the timeless rhythms and gentle beauty of the rural landscape. Whether you're a seasoned rambler or a curious first-time visitor, this enchanting route promises to delight and inspire, leaving you with memories that will last a lifetime.

Detailed Route Directions:

1. From Sutton Memorial Hall, head west on the footpath that runs alongside the hall, bearing 270°. (Estimated time: 10 minutes)

2. Follow the path through fields, maintaining your westward heading (270°), until you reach a quiet country lane. (Estimated time: 25 minutes)

3. Turn left onto the lane, bearing 180°, and walk for a short distance. (Estimated time: 5 minutes)

4. Take the footpath on the right, bearing 270°, which leads into Sutton Common. (Estimated time: 10 minutes)

5. Wander through the woodlands, following the path as it meanders through the trees, bearing approximately 315°. (Estimated time: 30 minutes)

6. Continue across the fields, keeping a general northward heading (360°), eventually reaching another country lane. (Estimated time: 25 minutes)

7. Turn left onto the lane, bearing 270°, and walk along it for a short distance. (Estimated time: 5 minutes)

8. Take the footpath on the right, bearing 360°, and follow it through more woodlands and fields. (Estimated time: 35 minutes)

9. The path will eventually lead you back to Sutton Memorial Hall, bearing 90°, completing the circular walk. (Estimated time: 15 minutes)

Best Features and Views:

- Immerse yourself in the tranquil atmosphere of Sutton Common's woodlands, where the dappled sunlight, gentle rustling of leaves, and melodic birdsong create a soothing natural symphony that calms the mind and soothes the soul.
- Delight in the picturesque countryside views as you traverse the open fields, the vast East Anglian sky stretching out above you and the gently undulating landscape showcasing the timeless beauty of the rural setting.
- Discover the rich tapestry of rural life that is woven into the very fabric of the landscape, from the traditional hedgerows and ancient trees to the distant sight of historic farmhouses and church spires, each element a testament to the enduring connection between the land and its people.

Historical Features of Interest:

Although there are no specific historical landmarks along the Sutton Common Walk, the landscape itself is imbued with a deep sense of rural history and a connection to the generations of people who have shaped and been shaped by this corner of Suffolk. As you walk through the fields and woodlands, you'll be following in the footsteps of countless farmers, labourers, and local residents who have called this place home for centuries.

The traditional hedgerows that line the route are a particularly evocative reminder of the area's agricultural heritage. These living boundaries, some of which may be hundreds of years old, were originally planted to enclose fields, mark property lines, and provide shelter for livestock. Over time, they have become an integral part of the Suffolk countryside, supporting a rich diversity of plant and animal life and serving as corridors for wildlife to move through the landscape.

As you pass by the ancient trees that dot the woodlands and fields, take a moment to consider the stories they could tell if they could speak. These venerable giants have stood witness to the passing of generations, their gnarled trunks and spreading canopies a testament to the enduring strength and resilience of the natural world in the face of change and adversity.

Even the quiet country lanes that wind through the landscape are themselves a part of the area's rich history, their narrow, twisting routes a reminder of the days when the movement of people and goods was a slower, more localized affair. Imagine the horse-drawn carts and early motor vehicles that once traveled these same roads, carrying the produce of the fields to market and bringing news and supplies to the scattered farmsteads and villages.

As you immerse yourself in the timeless beauty of the Sutton Common landscape, let your mind wander to the lives and experiences of those who have called this place home over the centuries. From the earliest settlers who cleared the forests and tilled the soil to the modern-day farmers and conservationists who work to preserve and protect this precious heritage, each has left their mark on the land, contributing to the rich tapestry of history that underlies every step of your journey.

Birds to Spot:

- Green Woodpecker: Listen for the distinctive laughing call and look for the bright green plumage of this striking woodland bird as it forages for ants on the forest floor or clings to the trunks of trees.
- Tawny Owl: As you walk through the shadowy depths of the woodlands, keep an ear out for the haunting, drawn-out hoots of the tawny owl, a master of camouflage whose mottled brown plumage blends seamlessly with the bark of the trees.
- Treecreeper: These small, unobtrusive birds are often heard before they are seen, their high-pitched, thin calls betraying their presence as they spiral up the trunks of trees in search of insects. Look for their mottled brown backs and white undersides as they flit from tree to tree, always on the move.

Rare or Interesting Insects and Other Wildlife:

- Purple Emperor Butterfly: In the summer months, keep your eyes peeled for the majestic purple emperor butterfly, a large and stunning insect with iridescent purple wings and a powerful, soaring flight. These elusive butterflies are a rare and cherished sight in the woodlands of Sutton Common.
- Roe Deer: As you walk through the fields and woodland edges, you may be lucky enough to spot the graceful form of a roe deer, a native species known for its elegant appearance and shy, elusive nature. Look for their reddish-brown summer coats or grayish-brown winter coats, and listen for the gentle rustling of the undergrowth as they move through their territories.
- Hazel Dormouse: Although incredibly rare and difficult to spot, the hazel dormouse is known to inhabit the ancient woodlands and hedgerows of the Suffolk countryside. These enchanting little creatures, with their golden-brown fur, large black eyes, and long, furry tails, are a protected species and a symbol of the rich biodiversity that thrives in the area's natural habitats.

Rare or Interesting Plants:

- Bluebells: In the spring, the woodlands of Sutton Common are transformed by the appearance of vast carpets of bluebells, their delicate, nodding flowers creating a mesmerizing sea of blue beneath the trees. Take a moment to breathe in their sweet, heady fragrance and to appreciate the ephemeral beauty of these enchanting wildflowers.
- Wild Garlic: As you walk through the woods in the spring and early summer, you may catch the pungent aroma of wild garlic wafting on the breeze. Look for the delicate white flowers and bright green leaves of this flavourful plant, which is a favourite of foragers and chefs alike.
- Ancient Trees: Throughout your walk, keep an eye out for the venerable old trees that stand sentinel over the landscape, their gnarled trunks and spreading canopies a testament to their great age and enduring presence. From mighty oaks to graceful beeches, these ancient specimens are a vital part of the woodland ecosystem and a living link to the area's rich natural and cultural heritage.

Facilities and Accessibility:

The Sutton Common Walk is a relatively easy route, with mostly gentle terrain and a mix of well-maintained footpaths and quiet country lanes. However, it's important to note that some sections of the walk may be uneven, muddy, or overgrown, particularly after periods of wet weather or during the height of summer. As such, sturdy walking shoes or boots with good grip and support are highly recommended to ensure a safe and comfortable experience.

While there are no public toilets or cafes directly along the route, the nearby village of Sutton does offer some amenities for walkers, including a charming pub, The Cherry Tree Inn, which serves refreshing drinks and hearty meals made with locally-sourced ingredients. This welcoming establishment provides an ideal spot to rest and refuel before or after your walk, and to soak up the friendly atmosphere of the local community.

It's worth noting that some parts of the walk may not be suitable for those with limited mobility or for pushchairs, due to the presence of stiles, kissing gates, and uneven ground. However, with careful planning and a willingness to take your time, most walkers will find this route to be a manageable and enjoyable experience.

Seasonal Highlights:

The Sutton Common Walk offers a wealth of delights and surprises throughout the year, with each season revealing new facets of beauty and wonder in this enchanting corner of Suffolk:

- Spring: As the land awakens from its winter slumber, the woodlands and hedgerows burst into life with an explosion of colour and sound. Bluebells carpet the forest floor in a mesmerizing display of blue and purple, while the cheerful songs of nesting birds fill the air with the promise of new beginnings. The fields are a vibrant green, dotted with the bright yellows of cowslips and primroses, and the fresh scent of wild garlic wafts on the breeze.
- Summer: In the long, lazy days of summer, the landscape is at its most lush and verdant, the fields swaying with ripening crops and the woodlands providing a cool, green sanctuary from the heat of the sun. Butterflies dance among the wildflowers, while dragonflies skim the surface of ponds and streams. The air is heavy with the drowsy hum of insects, and the evenings are filled with the soft hooting of tawny owls and the gentle rustling of nocturnal creatures.
- Autumn: As the leaves begin to turn and the days grow shorter, the Sutton Common Walk takes on a new character, the rich hues of gold, orange, and red transforming the woodlands into a breathtaking tapestry of colour. Fungi of all shapes and sizes emerge from the leaf litter, while the hedgerows are heavy with the

bounty of berries and nuts. The crisp, clean air and the soft crunch of fallen leaves underfoot make for an invigorating walk, the perfect way to embrace the changing of the seasons.
- Winter: Though the landscape may be stark and stripped back to its essentials, the Sutton Common Walk is no less beautiful in the depths of winter. The bare branches of the trees create intricate patterns against the pale sky, while the frosty fields and silent woods are imbued with a sense of timeless peace and solitude. Wrap up warm and enjoy the crunch of frozen ground beneath your feet, keeping an eye out for the tracks of deer and foxes in the snow, and the hardy birds that brave the cold to forage for food.

Difficulty Level and Safety Considerations:

The Sutton Common Walk is a relatively easy route, suitable for walkers of most ages and fitness levels. The terrain is mostly gentle, with a mix of well-maintained footpaths and quiet country lanes, making for a pleasant and accessible walking experience.

However, as with any outdoor activity, it's important to be aware of potential hazards and to take necessary precautions to ensure a safe and enjoyable walk. One of the main considerations is the possibility of uneven or slippery ground, particularly in the wooded areas or after periods of wet weather. Walkers are advised to wear sturdy, supportive footwear with good grip, and to take extra care when navigating any muddy or uneven sections of the path.

Another factor to keep in mind is the need to respect the countryside and the wildlife that calls it home. This means sticking to the designated footpaths, closing any gates behind you, and taking your litter home with you to help preserve the natural beauty of the area. If you encounter any livestock along the route, give them plenty of space and avoid disturbing them, especially during the spring and summer months when young animals may be present.

It's also a good idea to check the weather forecast before setting out and to dress appropriately for the conditions. In the summer months, be sure to carry plenty of water and sun protection, while in the cooler months, layers are key to staying comfortable and warm. And as with any walk in a rural area, it's always wise to carry a map and compass or GPS device, as well as a fully charged mobile phone and a small first aid kit, in case of emergencies.

By following these simple guidelines and using common sense, walkers of all abilities can fully immerse themselves in the beauty and tranquility of the Sutton Common Walk, forging a deeper connection with the natural world and the rich history of the Suffolk countryside.

Nearby Attractions:

After your invigorating walk through the picturesque landscape of Sutton Common, why not extend your adventure and explore some of the other fascinating attractions that this beautiful corner of Suffolk has to offer? From ancient forests to charming coastal towns and world-class cultural destinations, there's something to capture the imagination and delight the senses of every visitor.

Just a short drive from Sutton Common lies the enchanting Rendlesham Forest, a vast and atmospheric woodland that is steeped in history and legend. This former royal hunting ground is now a haven for wildlife and outdoor enthusiasts, with miles of walking and cycling trails that wind through the trees and open up to stunning vistas of the surrounding countryside. Keep an eye out for the elusive red deer that roam the forest, and discover the mysterious UFO trail that commemorates the famous incident of 1980, when strange lights were reported in the night sky above the trees.

For those who love the sea and the timeless charm of the English coast, the nearby town of Aldeburgh is a must-visit destination. This picturesque seaside resort, with its colourful beach huts, historic high street, and world-famous fish and chips, is the perfect place to spend a lazy afternoon, soaking up the sun and the salty air. Take a stroll along the shingle beach, admiring the striking sculpture of the Scallop by local artist Maggi Hambling, or visit the Aldeburgh Museum to learn about the town's rich maritime history and its connections to the composer Benjamin Britten.

Culture vultures and art lovers will find plenty to inspire and delight them in the nearby town of Snape, home to the world-renowned Snape Maltings concert hall and arts complex. Housed in a stunning collection of converted Victorian malthouses, this cultural hub hosts a diverse program of music, theatre, and dance performances throughout the year, as well as exhibitions, workshops, and festivals that celebrate the creativity and innovation of the region. Take a tour of the historic buildings, shop for unique gifts and souvenirs in the independent boutiques, or simply soak up the atmosphere of this vibrant and dynamic destination.

For a taste of Suffolk's rich agricultural heritage and delicious local produce, be sure to visit one of the many farmers' markets, farm shops, and artisanal food producers that can be found throughout the area. From the creamy, tangy cheeses of the Suffolk Farmhouse Cheeses company to the succulent meats and fresh vegetables of the Suffolk Food Hall, there's a wealth of culinary delights to discover and savour, each one a celebration of the county's thriving food and drink scene.

As you explore the many attractions and hidden gems of the Suffolk countryside and coast, take a moment to reflect on the deep sense of place and belonging that pervades every aspect of life in this special corner of England. From the ancient woodlands and wildflower meadows to the historic towns and villages that have shaped the region's character over centuries, every experience is an invitation to connect with the enduring spirit of Suffolk, and to find your own place within its rich and varied story.

So whether you're a keen rambler, a history buff, a foodie, or simply someone in search of a little beauty and tranquility in a fast-paced world, the Sutton Common Walk and its surroundings offer a wealth of opportunities for discovery, relaxation, and inspiration. Why not linger a while and immerse yourself in the magic and wonder of this enchanting landscape, and let the gentle rhythms and timeless charm of Suffolk work their spell on your mind, body, and soul?

As you set out on your adventures, remember to tread lightly, to respect the local environment and communities, and to approach each new encounter with an open heart and a curious mind. By doing so, you'll not only enrich your own experience and understanding of the world around you, but you'll also help to preserve and celebrate the priceless natural and cultural heritage of this remarkable region for generations to come.

In the words of the great nature writer Robert Macfarlane, "Paths are the habits of a landscape. They are acts of consensual making. It's hard to create a footpath on your own...Paths connect. This is their first duty and their chief reason for being. They relate places in a literal sense, and by extension they relate people."

So let the paths of the Sutton Common Walk be your guide and your companion as you explore the beauty and diversity of the Suffolk countryside, connecting you not only to the landscapes and communities of the present, but also to the countless generations who have walked these ways before you, each leaving their own indelible mark on the story of this ancient and fascinating land.

Happy wanderings!

Rendlesham Forest Walk - 5 miles (8 km), Circular - Estimated Time: 2.5-3 hours

Starting Point Coordinates, Postcode: The starting point for the Rendlesham Forest Walk is at the Rendlesham Forest car park (Grid Reference: TM 353 473, Postcode: IP12 3NF).
Nearest Car Park, Postcode: The nearest car park is at the Rendlesham Forest car park (Postcode: IP12 3NF).

Embark on an enchanting 5-mile (8 km) circular walk through the captivating and enigmatic Rendlesham Forest, where the tranquil beauty of nature intertwines with the allure of the unknown. This easy to moderate route offers a perfect blend of well-maintained paths, gentle inclines, and the intriguing Rendlesham Forest UFO trail, making it an ideal choice for families, nature enthusiasts, and those seeking a fascinating outdoor adventure.

As you set out from the Rendlesham Forest car park, feel the anticipation build as you follow the signposted trails into the heart of the forest. With each step, the stresses of modern life melt away, replaced by a growing sense of peace and connection with the natural world that surrounds you.

The path winds through a mesmerizing tapestry of coniferous and deciduous woodland, the contrasting textures and hues of the trees creating an enchanting play of light and shadow. Take a moment to breathe in the fresh, pine-scented air and to listen to the soothing sounds of birdsong and rustling leaves, allowing the forest's tranquil atmosphere to envelop you.

As you continue along the trail, keep an eye out for the thought-provoking Rendlesham Forest UFO trail markers, which provide fascinating insights into the mysterious events that unfolded here in December 1980. Immerse yourself in the captivating story of the UFO incident, one of the UK's most famous and well-documented cases, and ponder the questions it raises about our place in the universe and the possibility of life beyond our world.

The well-maintained paths guide you through the forest's diverse habitats, from the cool, green depths of the coniferous plantations to the dappled sunlight and vibrant understory of the deciduous woodlands. Along the way, keep your eyes peeled for the rich array of wildlife that calls Rendlesham Forest home, from the darting forms of woodland birds and the fluttering wings of butterflies to the graceful silhouettes of deer and the playful antics of squirrels.

As you navigate the gentle inclines and winding trails, take the time to appreciate the forest's seasonal highlights. In spring, the woodland floor is awash with a breathtaking display of bluebells and other delicate wildflowers, their vibrant colours and sweet fragrance a true feast for the senses. Autumn brings a spectacular show of fiery hues as the leaves turn to gold, orange, and red, while winter casts a magical, frosty spell over the landscape, transforming it into a silent, ethereal wonderland.

Throughout your walk, embrace the opportunity to disconnect from the distractions of daily life and to reconnect with the simple joys and profound truths that can be found in nature. Whether you're seeking solace, inspiration, or a chance to bond with loved ones, the Rendlesham Forest Walk offers a haven of peace and beauty that will nourish your mind, body, and soul.

As you loop back to the car park after 5 miles, take a moment to reflect on the incredible journey you've just experienced. The Rendlesham Forest Walk is more than just a physical adventure; it is a reminder of the awe-inspiring power of nature, the enduring mysteries of the universe, and the importance of preserving and cherishing our precious green spaces for generations to come.

Detailed Route Directions:

1. From the Rendlesham Forest car park, follow the signposted trails into the forest, bearing approximately 45°. (Estimated time: 10-15 minutes)

2. Continue along the well-maintained paths, passing through a mix of coniferous and deciduous woodland, maintaining a general heading of 45°. (Estimated time: 30-40 minutes)

3. Keep an eye out for the Rendlesham Forest UFO trail markers, which provide information about the famous UFO incident. Stop to read the information boards and immerse yourself in the intriguing story. (Estimated time: 10-15 minutes)

4. Proceed through the forest, following the paths as they wind through the diverse habitats, with a general heading of 90°. (Estimated time: 45-60 minutes)

5. Continue to follow the trails, now heading roughly 135°, as they loop back towards the car park. (Estimated time: 30-40 minutes)

6. Arrive back at the Rendlesham Forest car park, where your circular walk began. (Estimated time: 5-10 minutes)

Best Features and Views:

- Immerse yourself in the tranquil beauty of Rendlesham Forest, with its enchanting mix of coniferous and deciduous woodland, each habitat offering a unique and captivating atmosphere that soothes the soul and invigorates the senses.
- Embark on a thought-provoking journey along the Rendlesham Forest UFO trail, delving into the intriguing story of the famous UFO incident and pondering the mysteries of the universe and our place within it.
- Delight in the opportunity to disconnect from the stresses of modern life and reconnect with the simple joys and profound truths that can be found in nature, as you walk along the well-maintained paths and gentle inclines of this beautiful forest.

Historical Features of Interest:

The Rendlesham Forest UFO incident, which occurred in December 1980, is one of the most famous and well-documented UFO sightings in the UK, and has become an integral part of the forest's rich history and cultural heritage. The incident involved a series of reported sightings of unexplained lights and a supposed landing of a craft in the forest, near the nearby RAF Woodbridge airbase.

On the night of December 26th, 1980, military personnel from the airbase, including Deputy Base Commander Lieutenant Colonel Charles Halt, reported seeing strange lights descending into the forest. Upon investigation, they discovered a triangular pattern of indentations in the ground, as well as burn marks and broken branches on nearby trees. Radiation levels at the site were also reportedly higher than normal.

Two nights later, on December 28th, Lt. Col. Halt and a team of servicemen ventured back into the forest to further investigate the incident. They reportedly observed more unexplained lights, which seemed to move through the trees and sky, and even recorded their observations on audio tape, which has since become known as the "Halt Tape."

The Rendlesham Forest incident quickly gained notoriety and has been the subject of numerous investigations, books, documentaries, and television shows over the years. While skeptics have offered various explanations for the sightings, including misidentified celestial objects, lighthouse beams, and pranks, many people continue to believe that the incident represents genuine evidence of extraterrestrial visitation.

As you walk along the Rendlesham Forest UFO trail, you'll encounter a series of information boards that provide a detailed account of the incident, including eyewitness testimonies, official reports, and various theories and explanations. These boards offer a fascinating insight into the enduring mystery of the incident and its impact on popular culture and the public imagination.

Beyond the UFO incident, Rendlesham Forest itself has a rich and varied history, with evidence of human activity dating back thousands of years. The area was once part of a vast royal hunting forest, and has played a significant role in the region's economy and way of life for centuries, providing timber, fuel, and other resources for local communities.

During World War II, the forest was home to a number of military installations, including RAF Woodbridge and the nearby Bentwaters airbase, which played a crucial role in the Allied war effort. The legacy of this military presence can still be seen in the forest today, with remnants of wartime buildings and infrastructure scattered throughout the woodland.

As you immerse yourself in the beauty and tranquility of Rendlesham Forest, take a moment to reflect on the many layers of history and human experience that have shaped this remarkable landscape over the centuries. From the ancient peoples who first settled in the area to the modern-day visitors who come to explore its mysteries and wonders, each has left their mark on the forest, contributing to the rich tapestry of stories and legends that define this enchanting corner of Suffolk.

Birds to Spot:

- Green Woodpecker: Listen for the distinctive laughing call and look for the bright green plumage of this striking woodland bird as it forages for ants on the forest floor or clings to the trunks of trees.
- Nuthatch: Keep your eyes peeled for these agile birds as they scurry up and down tree trunks and branches, often head-first, in search of insects and seeds. Look for their distinct blue-gray backs, rusty-orange underparts, and black eye stripes.
- Treecreeper: These small, brown birds are well-camouflaged against the bark of trees, making them a challenge to spot. Listen for their high-pitched, thin calls and watch for their unique behaviour of spiralling up tree trunks in search of insects, using their stiff tails for support.

Rare or Interesting Insects and Other Wildlife:

- Purple Emperor Butterfly: In the summer months, keep an eye out for these magnificent butterflies as they soar through the woodland canopy. The males have stunning iridescent purple wings, while the females are a more subtle brown colour with white bands.
- Roe Deer: As you walk through the forest, you may catch a glimpse of these graceful deer, known for their reddish-brown summer coats and distinctive white rump patches. Be quiet and patient, as they are often shy and elusive.
- Eurasian Red Squirrel: Although rare and threatened by the introduced grey squirrel, Rendlesham Forest is one of the few remaining strongholds for the native red squirrel in England. Keep your eyes peeled for these charming, bushy-tailed rodents as they scamper through the trees and forage on the forest floor.

Rare or Interesting Plants:

- Bluebells: In the spring, parts of Rendlesham Forest are carpeted with a stunning display of native English bluebells, their delicate, nodding flowers creating a mesmerizing sea of blue beneath the trees. Remember to stick to the paths to avoid trampling these beautiful and fragile wildflowers.
- Wild Garlic: Also known as ramsons, wild garlic is a pungent and flavourful woodland plant that blooms in the spring. Look for its broad, green leaves and clusters of small, white, star-shaped flowers, and take a moment to enjoy its distinctive aroma.
- Wood Anemone: These delicate, white, or pale pink flowers are a harbinger of spring in Rendlesham Forest, often appearing before the trees have fully leafed out. Look for their slender stems and deeply divided leaves on the forest floor, and appreciate their ephemeral beauty.

Facilities and Accessibility:

The Rendlesham Forest Walk offers a range of facilities to ensure a comfortable and enjoyable experience for visitors. At the Rendlesham Forest car park, you'll find public toilets and a café, perfect for a pre-walk snack or a post-walk refreshment. The car park also serves as an information hub, with maps, leaflets, and helpful staff on hand to answer any questions you may have about the forest and its many attractions.

The paths throughout the forest are well-maintained and suitable for strollers, making it an ideal destination for families with young children. However, some sections of the trail may have uneven terrain or gentle inclines, which could pose a challenge for wheelchair users. It's always a good idea to check with the local forest authorities or visitor centre for the most up-to-date information on accessibility and trail conditions.

While dogs are welcome in Rendlesham Forest, it's essential to keep them under close control and to clean up after them to help preserve the delicate ecosystems and protect the wildlife that calls the forest home. Be mindful of any seasonal restrictions or guidance, particularly during the bird nesting season or when livestock may be present in nearby fields.

For those seeking a more immersive experience, Rendlesham Forest offers a range of additional facilities and activities, including picnic areas, barbecue sites, and dedicated cycling trails. The nearby Tangham Campsite provides an opportunity for overnight stays, allowing visitors to fully embrace the peace and tranquility of the forest and to wake up to the enchanting sights and sounds of nature all around them.

Seasonal Highlights:

Rendlesham Forest is a year-round destination that offers a unique and captivating experience in every season, with each month revealing new facets of beauty and wonder in this enchanting woodland:

- Spring: As the forest awakens from its winter slumber, the woodland floor comes alive with a breathtaking display of wildflowers, from the delicate wood anemones and the vibrant yellow of lesser celandine to the enchanting carpets of bluebells that stretch out beneath the trees. The fresh green leaves of the deciduous trees create a lush canopy overhead, while the air is filled with the joyous songs of returning migrant birds, such as chiffchaffs, blackcaps, and willow warblers.
- Summer: In the height of summer, Rendlesham Forest is a cool and tranquil oasis, offering a welcome respite from the heat and bustle of the outside world. The trees are in full leaf, creating a dense green canopy that filters the sunlight and dapples the forest floor with shifting patterns of light and shade. The air is alive with the hum of insects, from the gentle buzzing of bees to the colourful flashes of butterflies, such as the majestic purple emperor and the delicate silver-washed fritillary.
- Autumn: As the days grow shorter and the nights turn crisp, Rendlesham Forest is transformed into a

breathtaking tapestry of gold, orange, and red, the changing leaves creating a fiery display that is a feast for the senses. The forest floor is carpeted with a crunchy layer of fallen leaves, while the cool, clear air is filled with the earthy scents of autumn and the distant calls of migrating birds. This is a time for quiet reflection and gentle exploration, for savouring the last warm days of the year and marvelling at the cyclical beauty of the natural world.
- Winter: Though the trees may be bare and the ground hard with frost, Rendlesham Forest takes on a stark and ethereal beauty in the depths of winter. The skeletal forms of the deciduous trees stand out in sharp relief against the pale, watery light, while the evergreens provide a welcome splash of colour and a reminder of the enduring resilience of nature. Wrap up warm and listen for the haunting calls of overwintering birds, such as redwings and fieldfares, or the soft hooting of tawny owls in the twilight. On crisp, clear days, the low sun casts long shadows across the forest floor, creating an enchanting play of light and shade that is a photographer's dream.

Difficulty Level and Safety Considerations:

The Rendlesham Forest Walk is a relatively easy to moderate route, suitable for walkers of most ages and fitness levels. The paths are well-maintained and clearly signposted, with gentle inclines and occasional uneven sections that may require a bit more care and attention, particularly in wet or icy conditions.

As with any outdoor activity, it's essential to come prepared with appropriate footwear, clothing, and equipment for the conditions. Sturdy, comfortable walking shoes or boots with good grip are a must, especially on the forest trails where roots, rocks, and muddy patches can create potential tripping hazards. In wet weather, waterproof clothing and extra layers are advisable, while in the summer months, sun protection and insect repellent are recommended.

When exploring the forest, it's crucial to stay on the designated paths and trails to minimize your impact on the delicate ecosystems and to avoid disturbing the wildlife that calls the woodland home. Be mindful of any seasonal restrictions or guidance, such as during the bird nesting season or when forestry operations are taking place, and always follow the countryside code, taking your litter home with you and leaving no trace of your visit.

It's also a good idea to familiarize yourself with the trail map and to carry a compass or GPS device, as well as a fully charged mobile phone, in case of emergencies. Let someone know your intended route and expected return time, and be sure to check the weather forecast before setting out, as conditions can change quickly in the forest environment.

By following these simple guidelines and using common sense, walkers of all abilities can safely enjoy the beauty, tranquility, and fascinating history of the Rendlesham Forest Walk, creating unforgettable memories and forging a deeper connection with the natural world and the enduring mysteries of the universe.

Nearby Attractions:

After your exciting walk through the enchanting and mysterious Rendlesham Forest, why not explore some of the other fascinating attractions that this beautiful corner of Suffolk has to offer? From ancient archaeological sites to charming coastal towns and world-class nature reserves, there's something to capture the imagination and delight the senses of every visitor.

Just a short drive from Rendlesham Forest lies the world-famous Sutton Hoo, an Anglo-Saxon burial site of immense historical and cultural significance. Here, in 1939, archaeologists unearthed an astonishing treasure trove of artifacts, including an ornate ship burial believed to belong to a 7th-century East Anglian king. Visit the exhibition hall to marvel at the exquisite craftsmanship of the Sutton Hoo helmet and other priceless objects, and take a guided tour of the atmospheric burial mounds , immersing yourself in the fascinating story of this ancient kingdom and its enduring legacy.

For a taste of Suffolk's picturesque coastal charm, head to the nearby town of Woodbridge, nestled on the banks of the beautiful River Deben. Stroll along the historic Thoroughfare, lined with independent shops, cafes, and galleries, and visit the town's iconic Tide Mill, one of the oldest working tide mills in the country. Take a leisurely boat trip along the river, soaking up the stunning views of the surrounding countryside, or simply relax and watch the world go by from one of the many welcoming pubs and restaurants that line the quayside.

Nature enthusiasts and birdwatchers will be in their element at the nearby RSPB Minsmere nature reserve, a sprawling coastal sanctuary that is home to an incredible diversity of wildlife. Explore the reed beds, wetlands, and woodlands that make up this internationally important habitat, and keep your eyes peeled for some of the reserve's star species, such as the elusive bittern, the majestic marsh harrier, and the charming avocet. With its network of trails, hides, and visitor facilities, Minsmere offers unparalleled opportunities to connect with the natural world and to marvel at the breathtaking beauty of the Suffolk coast.

For a deeper dive into the region's rich maritime heritage and unique cultural identity, be sure to visit the Suffolk Coast & Heaths Area of Outstanding Natural Beauty, a stunning landscape of unspoiled beaches, heathland, and estuaries that stretches from the River Stour in the south to the bustling port of Lowestoft in the north. Walk along the windswept shores, explore the charming seaside towns and villages, and discover the area's fascinating history, from the ancient Roman settlement at Dunwich to the iconic lighthouse at Orford Ness.

As you explore the many attractions and hidden gems of the Suffolk countryside and coast, take a moment to reflect on the deep sense of place and belonging that pervades every aspect of life in this special corner of England. From the mysterious depths of Rendlesham Forest to the timeless beauty of the seaside, every sight, sound, and experience is an invitation to connect with the enduring spirit of the land and its people, and to find your own place within the rich tapestry of stories and legends that define this extraordinary region.

So whether you're a history buff, a nature lover, a UFO enthusiast, or simply someone in search of a little magic and wonder in a world that can sometimes feel all too ordinary, the Rendlesham Forest Walk and its surroundings offer a wealth of opportunities for discovery, inspiration, and sheer delight. Why not linger a while and immerse yourself in the enchanting atmosphere of this unique and captivating landscape, and let the mysteries and marvels of Suffolk work their spell on your heart and imagination?

As you set out on your adventures, remember to approach each new experience with an open mind and a sense of curiosity, ready to be surprised and amazed by the incredible richness and diversity of the world around you. And as you walk in the footsteps of the countless generations who have called this place home, from the ancient Anglo-Saxons to the modern-day military witnesses of the Rendlesham Forest incident, know that you are part of an enduring story that stretches back through the mists of time and into the farthest reaches of the universe.

In the words of the great science fiction writer Arthur C. Clarke, "Two possibilities exist: either we are alone in the Universe, or we are not. Both are equally terrifying." But perhaps, as you stand beneath the starry skies of Rendlesham Forest, you may find a third possibility: that we are never truly alone, but part of an infinite and awe-inspiring cosmos, forever connected by the mysteries and wonders that we have yet to discover.

So let the magic and majesty of Rendlesham Forest be your guide and your inspiration, as you embark on a journey of exploration that will expand your horizons, enrich your understanding, and fill your heart with a profound sense of the beauty, the strangeness, and the endless possibilities of the universe we call home.

Happy trails, and may the mysteries of Rendlesham Forest forever captivate your imagination!

Bawdsey Quay Walk - 2 miles (3.2 km), Non-circular - Estimated Time: 1-1.5 hours

Starting Point Coordinates, Postcode: The starting point for the Bawdsey Quay Walk is at the Bawdsey Quay car park (Grid Reference: TM 349 402, Postcode: IP12 3AU).
Nearest Car Park, Postcode: The nearest car park is at the Bawdsey Quay (Postcode: IP12 3AU).

Embark on a delightful 2-mile (3.2 km) non-circular walk along the picturesque Bawdsey Quay, where the gentle lapping of the Deben Estuary and the rich history of the area combine to create a truly enchanting experience. This easy walk, suitable for all ages and abilities, offers a perfect opportunity to immerse yourself in the stunning coastal scenery and fascinating heritage of this unique corner of Suffolk.

As you set out from the Bawdsey Quay car park, take a moment to appreciate the tranquil atmosphere and the fresh sea breeze that greets you. Follow the well-maintained path leading to the waterfront, your senses awakening to the sights, sounds, and scents of the estuary.

Along the way, you'll pass the impressive Bawdsey Manor, a historic building with a captivating story. During World War II, this very site was home to the world's first operational radar station, playing a crucial role in the defense of the nation. The Radar Memorial, situated nearby, commemorates this significant milestone in military technology, offering a poignant reminder of the sacrifices and ingenuity of those who served here.

As you continue along the quayside, take the time to absorb the breathtaking views across the Deben Estuary. Watch the boats bobbing gently on the sparkling water, from the colourful sails of recreational craft to the purposeful movements of fishing vessels. The ever-changing light and the distant calls of seabirds create an enchanting tableau that invites quiet contemplation and a deep sense of connection with the natural world.

The coastal path, flanked by the diverse and resilient vegetation of the saltmarsh, guides you further along the shoreline. Keep an eye out for the fascinating plants that thrive in this unique habitat, such as the succulent sea purslane, the delicate sea lavender, and the edible samphire. These hardy species, adapted to the challenges of the coastal environment, are a testament to the remarkable adaptability and tenacity of life in the face of the elements.

As you walk, the Deben Estuary reveals itself as a haven for a wide array of wildlife. Birdwatchers will delight in the opportunity to spot various species, from the distinctive black-and-white oystercatchers probing the mudflats with their long, red bills to the elegant avocets gracefully sweeping their upturned beaks through the shallows. The estuary is also home to a variety of fish and the occasional seal, their presence a reminder of the rich biodiversity that flourishes in this special place.

Upon reaching the Bawdsey Ferry, you have the option to cross the river to Felixstowe Ferry, extending your adventure and exploring the charming coastal villages on the other side. However, if you choose to conclude your walk here, simply retrace your steps along the coastal path, savouring the views and the tranquility of the estuary once more as you make your way back to the Bawdsey Quay car park.

The Bawdsey Quay Walk is a celebration of the simple joys and profound beauty that can be found in the gentle exploration of our coastal heritage. Whether you're a lover of history, a keen birdwatcher, or simply someone in search of a peaceful escape from the hustle and bustle of modern life, this enchanting walk promises to delight and inspire, leaving you with memories that will linger long after the last echoes of the estuary have faded away.

Detailed Route Directions:

1. From the Bawdsey Quay car park, head towards the waterfront, following the clearly marked path. (Estimated time: 2-3 minutes)

2. Pass by the historic Bawdsey Manor on your left and the Radar Memorial on your right as you continue along the path. (Estimated time: 5-7 minutes)

3. Walk along the quayside, taking in the stunning views across the Deben Estuary. Watch the boats go by and enjoy the peaceful atmosphere. (Estimated time: 15-20 minutes)

4. Continue following the coastal path, passing through the diverse coastal vegetation and keeping an eye out for the various bird species that call the estuary home. (Estimated time: 25-30 minutes)

5. Upon reaching the Bawdsey Ferry, decide whether to cross the river to Felixstowe Ferry or conclude your walk. If concluding, turn around and retrace your steps along the coastal path. (Estimated time: 2-3 minutes)

6. Make your way back along the quayside, once again enjoying the views and the tranquil ambiance of the estuary. (Estimated time: 25-30 minutes)

7. Pass by the Radar Memorial and Bawdsey Manor as you return to the Bawdsey Quay car park, where your walk began. (Estimated time: 5-7 minutes)

Best Features and Views:

- Discover the historic significance of Bawdsey Manor, once home to the world's first operational radar station during World War II, and pay your respects at the poignant Radar Memorial, which commemorates this groundbreaking achievement in military technology.
- Immerse yourself in the breathtaking views across the Deben Estuary, with its sparkling waters, bobbing boats, and ever-changing light, a scene that invites quiet contemplation and a deep appreciation for the beauty of the natural world.
- Delight in the diverse coastal vegetation along the walk, from the succulent sea purslane to the delicate sea lavender, each plant a testament to the remarkable adaptability and resilience of life in the face of the elements.

Historical Features of Interest:

As you embark on the Bawdsey Quay Walk, you'll find yourself immersed in a landscape steeped in history, with the Bawdsey Manor and the Radar Memorial serving as powerful reminders of the area's significant role in shaping the course of World War II and the development of modern military technology.

Bawdsey Manor, the impressive building you'll pass early on in your walk, was the site of the world's first operational radar station, established in 1936 as part of a top-secret project known as "Chain Home." This groundbreaking early warning system, which used radio waves to detect incoming enemy aircraft, played a crucial role in the Battle of Britain, giving the RAF precious minutes to scramble their fighters and intercept the Luftwaffe's attacks.

The scientists and engineers who worked at Bawdsey Manor, led by the visionary physicist Robert Watson-Watt, were pioneers in the field of radar technology, their innovations and dedication laying the foundations for the sophisticated detection and tracking systems that are now an integral part of modern defense and aviation.

As you pause to reflect at the Radar Memorial, take a moment to imagine the buzz of activity and the weight of responsibility that must have permeated the atmosphere at Bawdsey Manor during those pivotal years. The men and women who served here, often working long hours under intense pressure, were the unsung heroes of the war effort, their contributions to the defense of the nation and the ultimate victory over Nazi Germany immeasurable.

Beyond its wartime significance, the Bawdsey Quay area has a rich and varied history that spans centuries, from the ancient Romans who first recognized the strategic importance of this coastal location to the medieval monks who established a ferry crossing here, connecting the communities on either side of the Deben Estuary.

In the 19th century, Bawdsey Quay was a thriving centre of maritime trade, with ships and barges loading and unloading their cargoes of coal, grain, and other goods at the bustling wharves and warehouses that lined the shore. The remnants of this industrial past can still be seen in the weathered brickwork and timeworn structures that dot the landscape, silent witnesses to the ebb and flow of history.

As you walk along the coastal path, take a moment to imagine the countless generations of fishermen, farmers, and seafarers who have made their living from the bounty of the estuary and the surrounding countryside, their lives and livelihoods inextricably linked to the rhythms of the tides and the turning of the seasons.

Today, the Bawdsey Quay Walk offers a chance to connect with this rich tapestry of history and to gain a deeper appreciation for the enduring spirit and resilience of the people who have called this place home. By immersing yourself in the tranquil beauty of the estuary and the fascinating stories that have shaped this landscape, you'll come away with a renewed sense of wonder and respect for the generations who have gone before us, and a deeper understanding of the complex interplay between human endeavour and the natural world.

Birds to Spot:

- Oystercatchers: Keep an eye out for these striking black-and-white waders, with their distinctive long, red bills, as they probe the mudflats and shoreline for shellfish and other invertebrates. Their loud, piping calls and charismatic behaviour make them a favourite among birdwatchers.
- Avocets: These elegant waders, with their slender, upturned bills and stunning black-and-white plumage, are a delight to observe as they sweep their beaks through the shallow water, searching for small crustaceans and insects. The Deben Estuary is a prime location for spotting these graceful birds.
- Curlews: Listen for the haunting, bubbling calls of these large, long-billed waders as they forage along the tideline or in the nearby saltmarshes. With their mottled brown plumage and down-curved beaks, curlews are a captivating sight and a reminder of the rich biodiversity of the estuary.

Rare or Interesting Insects and Other Wildlife:

- Common Seals: As you walk along the coastal path, keep your eyes peeled for the distinctive round heads and inquisitive faces of common seals bobbing in the water or hauled out on the mudflats. These charismatic mammals are a delight to spot and a testament to the ecological importance of the Deben Estuary.
- Shore Crabs: Peer into the shallows and rockpools along the shoreline, and you may catch a glimpse of these hardy crustaceans scuttling about, their mottled green and brown carapaces providing excellent camouflage against the seabed. Shore crabs play a vital role in the estuarine food web, both as predators and prey.

- Bumblebees: As you pass through the coastal vegetation, keep an eye out for the vibrant colours and gentle buzzing of bumblebees as they flit from flower to flower, collecting nectar and pollen. These essential pollinators are a sign of a healthy ecosystem and a reminder of the interconnectedness of life in the coastal environment.

Rare or Interesting Plants:

- Sea Purslane: This succulent, salt-tolerant plant, with its small, fleshy leaves and tiny, pink flowers, is a characteristic species of the saltmarsh habitat. Sea purslane is not only an important food source for various coastal animals but also plays a crucial role in stabilizing the shoreline and protecting against erosion.
- Sea Lavender: In the summer months, the saltmarsh comes alive with the delicate purple blooms of sea lavender, their slender stems swaying gently in the coastal breeze. This beautiful and resilient plant is adapted to the challenging conditions of the tidal zone, its presence a testament to the remarkable diversity of life in the estuary.
- Samphire: Also known as sea asparagus, this edible plant, with its distinctive fleshy, green stems, is a prized delicacy among foragers and chefs alike. Samphire thrives in the salty, nutrient-rich soils of the saltmarsh, its presence a reminder of the deep connection between the land and the sea, and the bounty that the coastal environment provides.

Facilities and Accessibility:

The Bawdsey Quay Walk is a relatively easy and accessible route, with flat terrain and well-maintained paths that make it suitable for walkers of all ages and abilities. The car park at Bawdsey Quay serves as the starting point for the walk, and offers ample space for vehicles, as well as public toilets for the convenience of visitors.

For those looking to take a break or enjoy a refreshment during their walk, there is a seasonal café that operates at Bawdsey Quay during the summer months. Here, you can sit back and relax with a cup of tea or coffee, perhaps indulging in a delicious homemade cake or a light snack, while taking in the stunning views across the Deben Estuary.

The coastal path itself is generally wide and well-maintained, with a surface that is mostly even and free from obstacles. This makes it accessible for people with reduced mobility, such as those using wheelchairs or mobility scooters, although it's always a good idea to check the current conditions before setting out, as some sections may be slightly uneven or muddy after periods of wet weather.

For families with young children or those with limited time, the Bawdsey Quay Walk is an ideal choice, as it offers a short yet rewarding experience that can be easily completed in an hour or two. The gentle pace and stunning scenery make it a perfect opportunity to spend quality time together, learning about the fascinating history and ecology of the area while enjoying the fresh coastal air and the tranquility of the estuary.

Whether you're a seasoned rambler or a casual walker, the Bawdsey Quay Walk promises to be a memorable and enjoyable experience, thanks to its combination of easy accessibility, excellent facilities, and the captivating beauty of the Suffolk coast. So why not gather your loved ones, pack a picnic, and set out on a gentle adventure that will leave you feeling refreshed, inspired, and more connected to the natural world around you?

Seasonal Highlights:

The Bawdsey Quay Walk offers a unique and enchanting experience in every season, with each time of year revealing new facets of beauty and wonder in this captivating corner of the Suffolk coast:

- Spring: As the days grow longer and the sun's warmth returns, the coastal path comes alive with the vibrant colours and delicate fragrances of spring wildflowers. The saltmarsh is dotted with the cheerful blooms of sea pink and sea aster, while the hedgerows and grasslands burst into life with the yellows, purples, and whites of cowslips, violets, and stitchworts. The air is filled with the joyous songs of breeding birds, and the estuary is a hive of activity as migratory species return to feed and nest along the shoreline.
- Summer: In the height of summer, the Deben Estuary is a shimmering expanse of blue, the gentle lapping of the waves and the distant cries of seabirds creating a soothing natural symphony. The coastal vegetation is lush and verdant, with the succulent leaves of sea purslane and the delicate purple blooms of sea lavender adding splashes of colour to the saltmarsh. Butterflies flit among the wildflowers, while dragonflies dart and hover over the water's edge, their iridescent wings catching the sunlight in a mesmerizing display.
- Autumn: As the leaves begin to turn and the days grow shorter, the Bawdsey Quay Walk takes on a new character, the mellow light and the crisp, clean air creating a sense of tranquility and reflection. The estuary becomes a hub of activity as migratory waders, such as curlews, redshanks, and dunlins, arrive to feed on the rich mudflats, their evocative calls and striking plumage a delight for birdwatchers. The coastal plants take on the warm, burnished hues of autumn, the seed heads and berries providing a vital food source for wildlife as they prepare for the winter ahead.
- Winter: Though the landscape may be stark and the weather bracing, the Bawdsey Quay Walk is no less beautiful in the depths of winter. The frosty mornings and the low, slanting light create an almost ethereal atmosphere, the pale colours of the sky and the sea blending seamlessly with the muted tones of the saltmarsh and the shoreline. Flocks of overwintering wildfowl, such as wigeons, teals, and pintails, gather on the estuary, their presence a reminder of the enduring cycle of life and the importance of these coastal habitats for the survival of countless species.

Difficulty Level and Safety Considerations:

The Bawdsey Quay Walk is a relatively easy and straightforward route, with flat terrain and well-maintained paths that make it suitable for walkers of all ages and abilities. However, as with any outdoor activity, it's essential to be aware of potential hazards and to take necessary precautions to ensure a safe and enjoyable experience.

One of the main considerations when walking along the coast is the changeable nature of the weather and the tides. The coastal path, while generally firm and stable, can become slippery or muddy in wet conditions, so it's crucial to wear sturdy, non-slip footwear with good tread and ankle support.

It's also a good idea to check the local tide times before setting out, as some sections of the path may be prone to flooding or may become inaccessible during high tides. If in doubt, it's always best to err on the side of caution and to plan your walk accordingly, allowing plenty of time to return safely to your starting point.

Another factor to keep in mind is the exposure to the elements, particularly on windy or sunny days. The coastal path offers little in the way of shelter, so it's important to come prepared with appropriate clothing and accessories, such as a windproof jacket, a hat, and sunscreen, to protect yourself from the sun, wind, and rain.

It's also worth noting that while the Deben Estuary is generally a safe and welcoming environment, it's essential to respect the wildlife and the habitats that make this area so special. This means keeping a respectful distance from any animals you may encounter, such as seals or birds, and avoiding disturbing them or damaging their habitats. It's also important to keep dogs under close control, particularly during the breeding season, to minimize any potential impact on the delicate ecosystems of the coast.

As with any walk in a rural or coastal area, it's always a good idea to carry a fully charged mobile phone, a map or GPS device, and a small first aid kit, just in case of emergencies. Let someone know your intended route and estimated return time, and don't hesitate to seek help or turn back if you feel unsure or uncomfortable at any point.

By following these simple guidelines and using common sense, walkers of all ages and abilities can safely enjoy the beauty, tranquility, and fascinating history of the Bawdsey Quay Walk, creating lasting memories and forging a deeper connection with the natural world and the rich heritage of the Suffolk coast.

Nearby Attractions:

After your delightful walk along the Bawdsey Quay, why not take the opportunity to explore some of the other fascinating attractions that this beautiful corner of Suffolk has to offer? From charming historic towns to stunning coastal landscapes and world-class nature reserves, there's something to capture the imagination and delight the senses of every visitor.

Just a short drive from Bawdsey Quay lies the picturesque town of Woodbridge, nestled on the banks of the River Deben. This historic market town boasts a rich maritime heritage, with a delightful mix of independent shops, galleries, and restaurants lining its quaint streets and lanes. Take a stroll along the riverside, admiring the beautiful boats and the historic tide mill, or visit the town's excellent museum to learn more about the area's fascinating past and its enduring connection to the sea.

For those with a passion for history and culture, the nearby Sutton Hoo is an absolute must-see. This world-famous archaeological site, located just a few miles from Bawdsey Quay, is home to a remarkable Anglo-Saxon burial ground, where a treasure trove of priceless artifacts was unearthed in 1939. Take a guided tour of the site, marvel at the exquisite craftsmanship of the Sutton Hoo helmet and other precious objects, and immerse yourself in the captivating story of this ancient kingdom and its enduring legacy.

Nature lovers and birdwatchers will be spoilt for choice in this stunning corner of the Suffolk Coast and Heaths Area of Outstanding Natural Beauty. The nearby RSPB Hollesley Marshes reserve is a haven for a wide variety of wetland and coastal birds, from the majestic marsh harriers and the elusive bitterns to the colourful avocets and the lively redshanks. With its network of trails, hides, and viewing points, this beautiful reserve offers unparalleled opportunities to observe and appreciate the incredible diversity of the Suffolk coastline.

For a taste of the region's rich agricultural heritage and delicious local produce, be sure to visit one of the many farmers' markets, farm shops, and artisanal food producers that can be found throughout the area. Sample the creamy, tangy cheeses, the succulent meats and fresh vegetables, and the crisp, refreshing ciders and beers that are the hallmark of Suffolk's thriving food and drink scene, and take home a piece of this special place to savour long after your visit has ended.

As you explore the many attractions and hidden gems of the Suffolk coast and countryside, take a moment to reflect on the deep sense of history, community, and connection to the land that pervades every aspect of life in this enchanting region. From the ancient coastal defenses and the historic quaysides to the timeless beauty of the estuaries and the saltmarshes, every sight and sound tells a story of the remarkable people and places that have shaped this unique and captivating corner of England.

So whether you're a keen rambler, a history buff, a wildlife enthusiast, or simply someone in search of a little peace and inspiration in a beautiful setting, the Bawdsey Quay Walk and its surroundings offer a wealth of opportunities for discovery, relaxation, and enjoyment. Why not linger a while and immerse yourself in the magic and wonder of the Suffolk coast, and let the gentle rhythms and enduring charm of this special place work their spell on your heart and soul?

As you set out on your adventures, remember to tread lightly, to respect the local environment and communities, and to approach each new encounter with an open mind and a curious heart. By doing so, you'll not only enrich your own understanding and appreciation of the world around you, but you'll also help to ensure that the precious natural and cultural heritage of this remarkable region can be enjoyed and cherished by generations to come.

In the words of Rachel Carson, "Those who contemplate the beauty of the earth find reserves of strength that will endure as long as life lasts." So let the beauty and the strength of the Suffolk coast be your guide and your inspiration, as you embark on a journey of discovery that will nourish your body, mind, and spirit, and leave you with memories that will last a lifetime. Happy exploring!

Bawdsey Circular Walk - 5 miles (8 km), Circular - Estimated Time: 2.5-3 hours

Starting Point Coordinates, Postcode: The starting point for the Bawdsey Circular Walk is at the Bawdsey Quay car park (Grid Reference: TM 343 401, Postcode: IP12 3AS).
Nearest Car Park, Postcode: The nearest car park is the Bawdsey Quay car park (Postcode: IP12 3AS).

Embark on a captivating 5-mile (8 km) circular walk that takes you on a journey through the stunning coastal and rural landscapes surrounding the charming village of Bawdsey. This moderately challenging route offers a perfect blend of breathtaking views, peaceful woodland paths, and quiet country lanes, inviting you to immerse yourself in the natural beauty and rich history of this enchanting corner of Suffolk.

As you set out from the Bawdsey Quay car park, feel the anticipation build as you head north along the coastal path, the glistening waters of the River Deben stretching out to your right. Take a moment to breathe in the fresh, salty air and to soak up the magnificent views of the Suffolk coastline, the distant horizon melting into the endless blue of the sky.

Continue along the coastal path for approximately 1.5 miles, your senses alive with the sights, sounds, and scents of the sea. Watch for the graceful swooping of terns and the distinctive black-and-white plumage of oystercatchers as they forage along the shoreline, their presence a testament to the rich biodiversity of this unique coastal habitat.

As you reach the footpath leading inland to the west, bid farewell to the river and turn your steps towards the lush green countryside that awaits. Follow the path as it winds through gently rolling fields and dappled woodland, the soft rustling of leaves and the melodic birdsong a soothing accompaniment to your journey.

Emerging from the woodland, you'll find yourself on a quiet country lane, the epitome of rural tranquility. Turn south and let the lane guide you through a patchwork of fields and hedgerows, the vibrant colours and textures of the landscape a feast for the senses. Keep an eye out for the majestic silhouettes of marsh harriers soaring overhead or the fleeting glimpse of a kestrel hovering in search of prey, their presence a reminder of the delicate balance and resilience of the natural world.

As you continue along the lane, take a moment to appreciate the timeless beauty of the Suffolk countryside, the ancient oaks and wildflower-strewn verges a living connection to the generations who have walked this way before. The distant sight of Bawdsey Manor, with its striking architecture and historical significance, serves as a poignant reminder of the area's rich heritage and its role in shaping the course of history.

Turning east onto another footpath, you'll begin your return journey to Bawdsey, the village's charming cottages and friendly community a welcoming sight after your invigorating walk. As you approach the heart of the village, turn south and follow the footpath that leads you back to the Bawdsey Quay and your starting point, the River Deben once again coming into view, its sparkling waters a constant companion on your coastal adventure.

As you complete your circular walk, take a moment to reflect on the incredible natural beauty and fascinating history that you've encountered along the way. The Bawdsey Circular Walk is more than just a physical journey; it is a celebration of the enduring spirit and resilience of the Suffolk coast and countryside, a testament to the power of nature to inspire, heal, and transform.

Whether you're a seasoned rambler or a curious explorer, this captivating route promises to delight and inspire, offering a unique perspective on the landscapes and communities that make this corner of England so special. So lace up your walking boots, grab your camera, and set out on an adventure that will leave you with memories to cherish and a deeper appreciation for the simple joys and profound wonders of the great outdoors.

Detailed Route Directions:

1. From the Bawdsey Quay car park, head north along the coastal path, keeping the River Deben on your right. (Estimated time: 30-40 minutes)

2. Continue along the coastal path for approximately 1.5 miles until you reach a footpath leading inland to the west. (Estimated time: 30-40 minutes)

3. Follow the footpath through fields and woodland. (Estimated time: 20-30 minutes)

4. Upon reaching a quiet country lane, turn south and follow the lane as it winds through the countryside. (Estimated time: 30-40 minutes)

5. Turn east onto another footpath, leading back towards Bawdsey. (Estimated time: 20-30 minutes)

6. As you approach the village, turn south and follow a footpath that takes you back to Bawdsey Quay and the starting point. (Estimated time: 20-30 minutes)

Best Features and Views:

- Marvel at the stunning views of the River Deben and the Suffolk coastline, the endless blue of the sky and sea a breathtaking backdrop to your coastal adventure.
- Immerse yourself in the peaceful beauty of the countryside, with its gently rolling fields, ancient woodlands, and wildflower-strewn verges, a patchwork of colours and textures that delight the senses.
- Delight in the rich birdlife that calls this area home, from the graceful terns and oystercatchers of the coast to the majestic marsh harriers and kestrels of the countryside, each species a vital thread in the intricate web of the local ecosystem.

Historical Features of Interest:

As you embark on the Bawdsey Circular Walk, you'll find yourself immersed in a landscape steeped in history, with the striking presence of Bawdsey Manor serving as a powerful reminder of the area's significant role in shaping the course of World War II and the development of radar technology.

Bawdsey Manor, a magnificent Grade II listed building, was requisitioned by the Air Ministry in 1936 to serve as a top-secret research station for the development of the world's first operational radar system. Under the leadership of pioneering scientist Robert Watson-Watt, a team of brilliant minds worked tirelessly to create a groundbreaking early warning system that would prove crucial in the Battle of Britain and the ultimate victory of the Allied forces.

As you catch sight of the Manor's impressive façade and sprawling grounds, take a moment to imagine the hushed conversations and intense focus of the scientists and engineers who toiled within its walls, their groundbreaking work shrouded in secrecy and their achievements known only to a select few. The building itself stands as a testament to the ingenuity, determination, and sacrifice of these unsung heroes, whose efforts helped to turn the tide of the war and usher in a new era of technological innovation.

The legacy of Bawdsey Manor and its role in the development of radar technology is celebrated at the nearby Bawdsey Radar Transmitter Block Museum, a fascinating attraction that offers visitors a glimpse into the world of early radar and the incredible story of the men and women who made it possible. Through exhibits, artifacts, and interactive displays, the museum brings to life the drama and significance of this pivotal moment in history, inviting reflection on the enduring impact of scientific discovery and human perseverance.

As you continue your walk, keep an eye out for other traces of Bawdsey's wartime past, from the old pillboxes and defensive structures that dot the coastline to the distant silhouette of the Martello towers that once guarded the mouth of the River Deben. Each of these silent sentinels holds a story of courage, ingenuity, and resilience, a testament to the enduring spirit of a nation and a community that refused to surrender in the face of seemingly insurmountable odds.

Throughout your journey, let the landscape be your guide, the gentle undulations of the fields and the winding paths of the woodland inviting contemplation on the generations of people who have shaped and been shaped by this remarkable corner of Suffolk. From the ancient tribes who first settled along the river's banks to the farmers and fishermen who have sustained local communities for centuries, each has left their mark on the land, contributing to the rich tapestry of history that underlies every step of your walk.

As you near the end of your circular route and the village of Bawdsey comes into view once more, take a moment to reflect on the extraordinary stories and experiences that you've encountered along the way. The Bawdsey Circular Walk is more than just a physical journey through a beautiful landscape; it is an invitation to connect with the past, to gain a deeper understanding of the forces that have shaped our world, and to find inspiration in the enduring spirit of innovation, courage, and community that defines this special place.

Birds to Spot:

- Avocets: These elegant black-and-white waders, with their distinctive upturned bills, are a delight to observe along the mudflats and saltmarshes of the River Deben. Once a rare sight in the UK, avocets have made a remarkable comeback thanks to dedicated conservation efforts and are now a symbol of hope and resilience in the face of environmental challenges.
- Marsh Harriers: As you walk through the open countryside, keep your eyes to the skies for the unmistakable silhouette of the marsh harrier, a majestic bird of prey with a russet-brown plumage and a distinctive white crown. These skilled hunters are a key predator in the coastal ecosystem, their presence a sign of the health and diversity of the local wetlands and marshes.
- Redshanks: Listen for the evocative piping calls of these lively waders as they forage along the river's edge, their bright orange legs and distinctive red bills making them easy to spot among the mud and seaweed. Redshanks are a common sight along the Suffolk coast, their presence a reminder of the vital importance of intertidal habitats for the survival of countless species.

Rare or Interesting Insects and Other Wildlife:

- Common Seals: As you walk along the coastal path, keep your eyes peeled for the curious faces and sleek, spotted coats of common seals basking on the mudflats or bobbing in the waters of the River Deben. These charismatic mammals are a beloved feature of the Suffolk coastline, their playful antics and endearing expressions a source of joy and wonder for visitors and locals alike.
- Brown Hares: In the fields and meadows of the Suffolk countryside, you may be lucky enough to catch a glimpse of the elusive brown hare, a magnificent creature known for its long, powerful legs and its incredible speed and agility. Hares have been a part of the English landscape for centuries, their presence a symbol of

the wild beauty and untamed spirit of the rural heartland.
- Emperor Dragonflies: In the summer months, the wetlands and marshes along the River Deben come alive with the vibrant colours and acrobatic flight of emperor dragonflies, the largest and most impressive of Britain's dragonfly species. With their striking blue and green bodies and their powerful, darting movements, these incredible insects are a true wonder of the natural world, a marvel of adaptation and evolutionary design.

Rare or Interesting Plants:

- Sea Kale: As you walk along the coastal path, look out for the distinctive blue-green leaves and delicate white flowers of sea kale, a rare and beautiful plant that thrives in the challenging conditions of the shingle beaches and sand dunes. This hardy perennial, which is both edible and medicinal, is a vital stabilizer of the coastal ecosystem, its deep roots helping to bind the soil and prevent erosion.
- Marsh Orchids: In the wet meadows and marshlands that fringe the River Deben, you may be fortunate enough to spot the exquisite blooms of marsh orchids, their slender stems and intricate, purple-pink flowers a mesmerizing sight amidst the lush green vegetation. These enchanting wildflowers, which are increasingly rare due to habitat loss and degradation, are a poignant reminder of the fragility and beauty of our wetland ecosystems.
- Yellow Horned-poppy: Along the shingle beaches and coastal paths of the Suffolk coast, keep your eyes peeled for the striking yellow flowers and distinctive, horn-shaped seed pods of the yellow horned-poppy, a rare and fascinating plant that has adapted to the harsh and unforgiving conditions of the shoreline. This tenacious species, with its deep tap roots and its ability to tolerate salt spray and drought, is a true survivor, a testament to the incredible resilience and adaptability of nature in the face of adversity.

Facilities and Accessibility:

The Bawdsey Circular Walk is a moderately challenging route that offers a mix of coastal paths, woodland trails, and quiet country lanes, making it suitable for walkers with a reasonable level of fitness and mobility. The terrain is varied, with some uneven surfaces, gentle inclines, and potentially muddy sections, particularly after periods of wet weather.

At the starting point of the walk, the Bawdsey Quay car park provides ample space for parking, as well as public toilets for the convenience of visitors. During the summer months, a charming seasonal cafe operates at the Quay, offering a range of light refreshments, hot and cold drinks, and locally-sourced snacks and treats, perfect for fueling up before your walk or rewarding yourself upon your return.

While the majority of the route is accessible and well-maintained, it's important to note that some sections of the path, particularly along the coast and through the woodland, may be narrow, uneven, or slippery at times. As such, sturdy walking shoes or boots with good grip and ankle support are highly recommended to ensure a safe and comfortable walking experience.

For those with limited mobility or specific accessibility requirements, it's always a good idea to consult with local authorities or visit the Suffolk Coast & Heaths Area of Outstanding Natural Beauty website for the most up-to-date information on trail conditions, potential obstacles, and alternative routes.

Despite these considerations, the Bawdsey Circular Walk remains a highly rewarding and enjoyable experience for the majority of walkers, offering a fantastic opportunity to explore the diverse landscapes, rich history, and abundant wildlife of this enchanting corner of the Suffolk coast and countryside. With its stunning views, peaceful atmosphere, and fascinating stories, this walk is sure to leave you feeling refreshed, inspired, and more deeply connected to the natural world around you.

Seasonal Highlights:

The Bawdsey Circular Walk is a year-round destination that offers a unique and captivating experience in every season, with each month revealing new facets of beauty and wonder in this remarkable corner of Suffolk.

- Spring: As the land awakens from its winter slumber, the coastal paths and countryside trails come alive with the vibrant colours and sweet scents of spring wildflowers. The hedgerows and meadows are adorned with the delicate blooms of primroses, violets, and cowslips, while the woodland floor is carpeted with a mesmerizing sea of bluebells, their nodding heads and heady fragrance a true feast for the senses. The air is filled with the joyous songs of breeding birds, and the river's edge is a hive of activity as migratory waders and wildfowl return to feed and nest along the mudflats and saltmarshes.
- Summer: In the warm, sun-drenched months of summer, the Suffolk coast is at its most inviting, the sparkling waters of the River Deben and the distant horizon of the sea creating a stunning backdrop to your walk. The coastal vegetation is lush and verdant, with the distinctive blue-green leaves of sea kale and the delicate pink blooms of marsh orchids adding splashes of colour to the shingle beaches and wetland habitats. Butterflies flit among the wildflowers, while dragonflies and damselflies dart and hover over the river's edge, their iridescent wings catching the light in a mesmerizing display of aerial acrobatics.
- Autumn: As the leaves begin to turn and the days grow shorter, the Bawdsey Circular Walk takes on a new character, the golden light and the crisp, clear air creating a sense of magic and enchantment in the landscape. The woodland paths are strewn with a colourful carpet of fallen leaves, their rustling underfoot a satisfying accompaniment to the gentle sounds of birdsong and the distant murmur of the sea. Fungi of all shapes and sizes emerge from the forest floor and the decaying tree stumps, their bizarre forms and vivid hues a reminder of the cycle of life and death that underpins the natural world.
- Winter: Though the coastal winds may be bracing and the countryside stark and unadorned, the Bawdsey Circular Walk retains a austere beauty and a sense of wild, untamed energy even in the depths of winter. The bare branches of the trees stand out in sharp relief against the pale, watery light of the low sun, while the frosty ground and the distant cries of overwintering birds create a haunting, almost ethereal atmosphere. Wrap up warm and embrace the invigorating chill, letting the windswept beauty of the Suffolk coast fill your lungs and clear your mind, a bracing tonic for the soul in this season of quiet introspection.

Difficulty Level and Safety Considerations:

The Bawdsey Circular Walk is a moderately challenging route that requires a reasonable level of fitness and mobility to complete. While the terrain is varied and the paths generally well-maintained, there are some sections of the walk that may prove more demanding, particularly for those with limited experience or specific accessibility needs.

One of the main considerations when embarking on this walk is the potential for uneven, slippery, or muddy ground, especially along the coastal path and through the woodland areas. The trails can be narrow and rocky in places, with exposed tree roots, hidden obstacles, and occasional steep inclines that may pose a challenge for some walkers. As such, it is essential to wear sturdy, supportive footwear with good grip and ankle support, and to take extra care when navigating these more demanding sections of the route.

Another factor to keep in mind is the changeable nature of the British weather, particularly along the exposed coastline and open countryside. The wind, rain, and sun can all take their toll over the course of a 5-mile walk, so it's crucial to come prepared with appropriate clothing and gear for the conditions. Layering is key, with a waterproof jacket, hat, and gloves essential for staying warm and dry in inclement weather, while sun protection and plenty of water are a must for hot, sunny days.

It's also worth noting that while the majority of the walk is well-signposted and easy to follow, there are a few sections where the path may be less clear or may require a bit of navigation. As such, it's always a good idea to carry a map and compass or GPS device, as well as a fully-charged mobile phone and a small first aid kit, just in case of emergencies.

When walking along the coastal path, it's important to be aware of the tides and to check the local tide times before setting out, as some sections of the trail may become submerged or inaccessible during high water. Stick to the designated paths and heed any warning signs or barriers, as the cliffs and mudflats can be treacherous and unstable, particularly after heavy rain or stormy weather.

As with any outdoor activity, it's also crucial to respect the natural environment and the wildlife that calls this area home. Keep a safe distance from any animals you may encounter, such as seals or birds, and avoid disturbing their habitats or leaving behind any litter or waste. By following the countryside code and taking a responsible, mindful approach to your walk, you can help to preserve the beauty and integrity of this special place for generations to come.

Despite these challenges and considerations, the Bawdsey Circular Walk remains a highly rewarding and enjoyable experience for the vast majority of walkers, offering a unique opportunity to explore the stunning landscapes, fascinating history, and abundant wildlife of the Suffolk coast and countryside. With a bit of preparation, common sense, and a sense of adventure, this walk promises to be a highlight of any visit to this enchanting corner of England.

Nearby Attractions:

After your invigorating walk around the stunning coastal and rural landscapes of Bawdsey, why not take the opportunity to explore some of the other fascinating attractions that this beautiful corner of Suffolk has to offer? From ancient historical sites to charming seaside towns and world-class cultural destinations, there is something to capture the imagination and delight the senses of every visitor.

Just a stone's throw from Bawdsey Quay, the impressive Bawdsey Manor and the Bawdsey Radar Transmitter Block Museum offer a unique and compelling glimpse into the area's pivotal role in the development of radar technology during World War II. Take a guided tour of the manor house and learn about the groundbreaking work of the scientists and engineers who helped to change the course of history, or explore the hands-on exhibits and interactive displays at the museum, which bring the drama and significance of this incredible story to life.

For those seeking a deeper connection with the natural world, the nearby Suffolk Coast and Heaths Area of Outstanding Natural Beauty is a true paradise, with its miles of unspoiled beaches, tranquil estuaries, and wild, windswept heaths. Take a stroll along the iconic Shingle Street, a remote and hauntingly beautiful stretch of coastline that has inspired artists and writers for generations, or explore the lush wetlands and reedbeds of the Butley River, home to an astonishing array of birdlife and rare plant species.

No visit to this part of Suffolk would be complete without a trip to the charming seaside town of Aldeburgh, a beloved destination for music lovers, foodies, and anyone in search of a quintessential English coastal experience. Wander along the picturesque high street, lined with colourful cottages and independent shops, or take a bracing walk along the pebble beach, breathing in the fresh sea air and marvelling at the sculptural beauty of the Scallop, a striking tribute to the town's most famous resident, the composer Benjamin Britten.

For a taste of the region's rich maritime heritage and vibrant cultural scene, be sure to visit the nearby town of

Woodbridge, nestled on the banks of the River Deben. Explore the fascinating collections of the Woodbridge Tide Mill Museum, which tells the story of the town's ancient tidal mill and its vital role in the local economy, or take in a performance at the Riverside Theatre, a thriving arts venue that hosts a diverse program of music, drama, and comedy throughout the year.

As you explore the many attractions and hidden gems of this enchanting corner of Suffolk, take a moment to reflect on the deep sense of place and belonging that pervades every aspect of life in this special part of the world. From the rugged beauty of the coastline to the gentle rolling hills of the countryside, every sight and sound is a reminder of the enduring connection between people and the land, and the countless generations who have shaped and been shaped by this remarkable landscape.

So whether you're a history buff, a nature lover, a culture vulture, or simply someone in search of a little peace and tranquility in a beautiful setting, Bawdsey and its surroundings offer a wealth of opportunities for discovery, relaxation, and inspiration. Why not linger a while and immerse yourself in the magic and wonder of the Suffolk coast and countryside, and let the timeless beauty and rich heritage of this extraordinary place work its spell on your mind, body, and soul?

As you set out on your adventures, remember to tread lightly, to respect the local environment and communities, and to approach each new experience with an open heart and a curious mind. By doing so, you'll not only enrich your own understanding and appreciation of the world around you, but you'll also help to ensure that the precious natural and cultural treasures of this remarkable region can be cherished and enjoyed by generations to come.

Happy exploring!

Bawdsey to Felixstowe Ferry Walk - 6 miles (9.7 km), Linear - Estimated Time: 3-4 hours

Starting Point Coordinates, Postcode: The starting point for the Bawdsey to Felixstowe Ferry Walk is at Bawdsey Quay (Grid Reference: TM 344 401, Postcode: IP12 3AS).
Nearest Car Park, Postcode: The nearest car park is the Bawdsey Quay car park (Postcode: IP12 3AS).

Embark on a captivating 6-mile (9.7 km) linear walk along the breathtaking Suffolk coastline, connecting the charming villages of Bawdsey and Felixstowe Ferry. This delightful route offers a perfect blend of stunning coastal views, tranquil estuary landscapes, and peaceful countryside, inviting you to immerse yourself in the natural beauty and rich history of this enchanting corner of East Anglia.

As you set out from the picturesque Bawdsey Quay, feel the anticipation build as you head north along the coast, following the well-trodden footpath that runs parallel to the glistening waters of the River Deben. With each step, the stresses of modern life begin to melt away, replaced by a growing sense of peace and connection with the magnificent landscape that surrounds you.

The path winds its way along the shoreline, offering unparalleled views of the shimmering estuary and the lush green countryside beyond. Take a moment to breathe in the fresh, salty air and to marvel at the ever-changing play of light on the water, the distant horizon seeming to beckon you onwards to new adventures and undiscovered delights.

As you continue along the route, you'll pass through the charming village of Ramsholt, where the historic Ramsholt Arms pub offers a welcome respite for weary walkers. Take a moment to step inside this cozy establishment, with its traditional decor and friendly atmosphere, and enjoy a refreshing drink or a hearty meal made with the finest local ingredients.

Leaving Ramsholt behind, the path continues northward, hugging the riverbank and offering increasingly spectacular views of the coastal landscape. Keep your eyes peeled for the abundant birdlife that calls this area home, from the graceful terns and the distinctive avocets to the lively oystercatchers and the majestic herons that stalk the mudflats in search of their prey.

As you near the mouth of the River Deben at Felixstowe Ferry, the landscape opens up before you, revealing a stunning panorama of the vast, open sea and the rugged beauty of the Suffolk coastline. Take a moment to drink in the awe-inspiring view, feeling the power and majesty of the natural world wash over you in a moment of pure, unadulterated joy.

At the end of your walk, you have the option to take the charming foot ferry across the river to explore the delightful seaside town of Felixstowe, with its sandy beaches, Victorian gardens, and lively promenade. Alternatively, for those who wish to return to Bawdsey, simply retrace your steps along the coastal path, savouring the stunning views and the tranquil atmosphere of the estuary once more.

As you make your way back to your starting point, take a moment to reflect on the incredible journey you've just undertaken. The Bawdsey to Felixstowe Ferry Walk is more than just a physical excursion; it is a celebration of the timeless beauty and enduring spirit of the Suffolk coast, a chance to reconnect with the natural world and to find peace, inspiration, and renewal in the simple act of putting one foot in front of the other.

Whether you're a seasoned hiker or a casual walker, this unforgettable route promises to delight and inspire, offering a unique perspective on the landscapes, history, and wildlife of this remarkable corner of England. So lace up your boots, grab your camera, and set out on an adventure that will stay with you long after the last echoes of the waves have faded away.

Detailed Route Directions:

1. From Bawdsey Quay, head north along the coast, following the footpath that runs parallel to the River Deben. (Estimated time: 45-60 minutes)

2. Continue along the path, passing through the village of Ramsholt. Look out for the historic Ramsholt Arms pub, which makes for a great refreshment stop. (Estimated time: 60-75 minutes)

3. Proceed northward along the riverbank, enjoying the stunning coastal views and keeping an eye out for various bird species. (Estimated time: 75-90 minutes)

4. As you approach the mouth of the River Deben at Felixstowe Ferry, take a moment to appreciate the breathtaking panorama of the open sea and the Suffolk coastline. (Estimated time: 30-45 minutes)

5. To explore Felixstowe, take the foot ferry across the river. Alternatively, to return to Bawdsey, retrace your steps along the coastal path. (Estimated time: 60-75 minutes)

Best Features and Views:

- Immerse yourself in the stunning coastal views, from the glistening waters of the River Deben to the rugged beauty of the Suffolk shoreline, each vista offering a unique and captivating perspective on this enchanting landscape.
- Delight in the tranquil estuary landscapes, where the ever-changing interplay of water, land, and sky creates a mesmerizing tableau that soothes the soul and invigorates the senses.
- Discover the peaceful countryside that surrounds the coastal path, from the lush green fields to the gently rolling hills, each one a testament to the enduring beauty and richness of the East Anglian landscape.

Historical Features of Interest:

As you set out on your walk from Bawdsey Quay, take a moment to consider the fascinating history that has shaped this unique corner of the Suffolk coast. Just a short distance from the starting point, the impressive Bawdsey Manor stands as a testament to the area's pivotal role in the development of radar technology during World War II.

Requisitioned by the Air Ministry in 1936, Bawdsey Manor became the site of the world's first operational radar station, where a team of pioneering scientists and engineers worked tirelessly to create a groundbreaking early warning system that would prove crucial in the Battle of Britain and the ultimate victory of the Allied forces.

As you pass by the manor, pause to imagine the incredible feats of ingenuity and perseverance that took place within its walls, and the countless lives that were saved as a result of the breakthroughs made here. The story of Bawdsey Manor is a powerful reminder of the transformative impact of scientific discovery and the enduring spirit of innovation that has shaped our world in countless ways.

Continuing along the coastal path, you'll find yourself walking in the footsteps of the many generations who have called this place home, from the ancient tribes who first settled along the banks of the River Deben to the fishermen, farmers, and seafarers who have plied their trades here for centuries. Each passing mile reveals new insights into the rich tapestry of history that underlies this timeless landscape, inviting you to contemplate the enduring connections between people, place, and the natural world.

As you near the end of your walk at Felixstowe Ferry, take a moment to consider the many chapters of human experience that have unfolded along this stretch of coastline, from the bustling maritime trade of the Victorian era to the strategic importance of the port during the World Wars. The story of this place is one of constant change and adaptation, a testament to the resilience and ingenuity of the human spirit in the face of ever-shifting circumstances.

Whether you're marvelling at the technological wonders of Bawdsey Manor or simply savouring the timeless beauty of the Suffolk coast, the Bawdsey to Felixstowe Ferry Walk offers a unique and compelling window into the history and heritage of this remarkable region. So step back in time, let your imagination soar, and allow yourself to be captivated by the enduring magic and mystery of this unforgettable corner of England.

Birds to Spot:

- Avocets: These elegant black-and-white waders, with their distinctive upturned bills, are a true delight to spot along the mudflats and estuaries of the Suffolk coast. Once a rare sight in the UK, avocets have made a remarkable comeback in recent years, thanks to dedicated conservation efforts, and are now a symbol of hope and resilience for the natural world.
- Little Terns: Keep your eyes peeled for these diminutive seabirds as they dart and dive over the waves, their quick, agile movements and sharp calls a lively counterpoint to the gentle rhythm of the tides. Little terns are one of the UK's rarest breeding seabirds, and the Suffolk coast is a vital stronghold for these enchanting creatures.
- Marsh Harriers: As you walk along the coastal path, scan the skies for the distinctive silhouette of the marsh harrier, with its broad wings, long tail, and characteristic V-shaped flight pattern. These majestic birds of prey are a captivating sight as they soar and glide over the reedbeds and marshes, their presence a sign of the health and vitality of the coastal ecosystem.

Rare or Interesting Insects and Other Wildlife:

- Essex Skipper Butterfly: In the summer months, the coastal grasslands and meadows come alive with the delicate, golden-brown wings of the Essex skipper butterfly, a charming and elusive species that is found almost exclusively in the southeast of England. Keep your eyes peeled for these tiny, fast-flying insects as they flit among the wildflowers and bask in the warm sunshine.
- Brown Hare: As you walk through the peaceful countryside surrounding the coastal path, you may be lucky enough to catch a glimpse of the elusive brown hare, a magnificent creature known for its long, powerful legs and its incredible speed and agility. Hares have been a beloved feature of the Suffolk landscape for centuries, their presence a symbol of the enduring wildness and beauty of the rural heartland.
- Common Seal: The waters of the River Deben and the surrounding coastline are home to a thriving population of common seals, these playful and inquisitive mammals a joy to spot as they bask on the mudflats or bob in the waves. Keep your eyes peeled for their sleek, spotted coats and their charming, expressive faces, a reminder of the incredible diversity of life that flourishes along the Suffolk shore.

Rare or Interesting Plants:

- Sea Lavender: In the late summer months, the saltmarshes and mudflats of the Suffolk coast are transformed

into a stunning carpet of delicate purple flowers, as the sea lavender bursts into bloom. This hardy and resilient plant is a true icon of the coastal landscape, its ability to thrive in the challenging conditions of the tidal zone a testament to the remarkable adaptability of nature.
- Samphire: Also known as sea asparagus, this succulent, salty-flavoured plant is a beloved delicacy among foragers and food lovers, its crisp, green stems a familiar sight in the marshes and estuaries of the Suffolk coast. Samphire is not only a delicious and nutritious ingredient but also plays a vital role in the coastal ecosystem, helping to stabilize the mudflats and providing food and shelter for a wide range of wildlife.
- Yellow Horned-poppy: As you walk along the shingle beaches and coastal paths, keep an eye out for the striking yellow flowers and long, curved seed pods of the yellow horned-poppy, a rare and fascinating plant that has adapted to the harsh and unforgiving conditions of the shoreline. With its deep tap roots and its ability to tolerate salt spray and drought, this tenacious species is a true survivor, a living testament to the incredible resilience of the natural world.

Facilities and Accessibility:

The Bawdsey to Felixstowe Ferry Walk is a relatively easy and accessible route, with mostly flat terrain and well-maintained paths that make it suitable for walkers of all ages and abilities. However, it's important to note that some sections of the walk, particularly along the coastal path, may be uneven, muddy, or slippery, especially after periods of wet weather or during high tide.

At the starting point in Bawdsey Quay, you'll find a range of facilities to ensure a comfortable and enjoyable walking experience, including a public car park, toilets, and a charming cafe that serves delicious refreshments and light meals. The cafe is the perfect spot to fuel up before your walk or to relax and unwind after a day of exploration and adventure.

As you make your way along the route, you'll pass through the delightful village of Ramsholt, where the historic Ramsholt Arms pub offers a warm welcome and a range of tasty food and drinks. This cozy, traditional inn is a great place to stop for a bite to eat or a refreshing pint, and to soak up the friendly, laid-back atmosphere of the Suffolk countryside.

While the majority of the walk is relatively easy and accessible, it's worth noting that some sections of the coastal path may be narrow, uneven, or exposed, and may not be suitable for those with limited mobility or specific access requirements. It's always a good idea to check the local conditions and weather forecast before setting out, and to wear sturdy, comfortable footwear with good grip and support.

At the end of your walk, you have the option to take the foot ferry across the River Deben to the charming seaside town of Felixstowe, where you'll find a range of amenities and attractions, including public toilets, cafes, and shops. Alternatively, if you choose to retrace your steps back to Bawdsey Quay, you can enjoy the stunning coastal views and the tranquil atmosphere of the estuary once more, before returning to your starting point.

Overall, the Bawdsey to Felixstowe Ferry Walk offers a fantastic opportunity to explore the natural beauty, rich history, and unique character of the Suffolk coast, with a range of facilities and amenities that make it accessible and enjoyable for walkers of all ages and abilities. Whether you're a seasoned hiker or a curious first-time visitor, this unforgettable route promises to delight and inspire, leaving you with memories that will last a lifetime.

Seasonal Highlights:

The Bawdsey to Felixstowe Ferry Walk is a year-round destination that offers a unique and captivating experience in every season, with each month revealing new facets of beauty and wonder along this remarkable stretch of the Suffolk coastline.

- Spring: As the days grow longer and the sun's warmth returns, the coastal path comes alive with the vibrant colours and sweet scents of spring wildflowers. The saltmarshes and meadows are adorned with a carpet of sea lavender and samphire, while the hedgerows and verges burst into life with the cheerful blooms of primroses, violets, and forget-me-nots. The air is filled with the joyous songs of skylarks and the gentle humming of bees, and the estuaries and mudflats are a hive of activity as migratory birds arrive to feed and breed.
- Summer: In the long, languid days of summer, the Suffolk coast is at its most inviting, the sparkling waters of the River Deben and the shimmering horizon of the sea creating an irresistible backdrop for a leisurely walk. The coastal vegetation is lush and verdant, with the distinctive yellow flowers of the horned poppy and the delicate pink blooms of the sea bindweed adding splashes of colour to the shingle beaches and dunes. Butterflies dance among the wildflowers, while dragonflies and damselflies skim the surface of the water, their iridescent wings catching the sunlight in a mesmerizing display.
- Autumn: As the leaves begin to turn and the days grow shorter, the Bawdsey to Felixstowe Ferry Walk takes on a new character, the golden light and the crisp, clean air creating a sense of tranquility and reflection. The hedgerows and thickets are laden with the rich bounty of blackberries, sloes, and rosehips, while the mudflats and saltmarshes are alive with the sounds of thousands of overwintering birds, from the haunting cries of curlews and redshanks to the raucous cackling of Brent geese. This is a time to savour the subtle beauty and quiet majesty of the coastal landscape, as nature prepares for the challenges of the winter ahead.
- Winter: Though the winds may be bracing and the skies leaden, the Suffolk coast remains a place of raw, elemental beauty even in the depths of winter. The muted colours of the landscape, from the steely grey of the sea to the pale gold of the dormant grasses, create a stunning contrast with the vibrant red berries of the hawthorn and the glossy green leaves of the holly. Wrap up warm and embrace the invigorating chill, reveling in the sense of solitude and the opportunity to witness the untamed power of the natural world in all its wild, untamed glory.

Difficulty Level and Safety Considerations:

The Bawdsey to Felixstowe Ferry Walk is a relatively easy and straightforward route, suitable for walkers of most ages and fitness levels. The terrain is mostly flat and the paths are generally well-maintained, making for a comfortable and enjoyable walking experience.

However, as with any coastal walk, it's important to be aware of potential hazards and to take necessary precautions to ensure a safe and responsible journey. One of the main considerations when walking along the Suffolk coast is the changeable nature of the weather and the tides, which can have a significant impact on the accessibility and condition of the paths.

One of the main considerations when walking along the Suffolk coast is the changeable nature of the weather and the tides, which can have a significant impact on the accessibility and condition of the paths. The coastal path, while generally firm and stable, can become slippery or muddy in wet conditions, so it's crucial to wear sturdy, non-slip footwear with good tread and ankle support.

It's also a good idea to check the local tide times before setting out, as some sections of the path may be prone to flooding or may become inaccessible during high tides. If in doubt, it's always best to err on the side of caution and to plan your walk accordingly, allowing plenty of time to return safely to your starting point.

Another factor to keep in mind is the exposure to the elements, particularly on windy or sunny days. The coastal path offers little in the way of shelter, so it's important to come prepared with appropriate clothing and accessories, such as a windproof jacket, a hat, and sunscreen, to protect yourself from the sun, wind, and rain.

It's also worth noting that while the River Deben and the surrounding coastline are generally safe and welcoming, it's essential to respect the water and to be aware of any potential hazards, such as strong currents or deep mud. Always keep a safe distance from the water's edge, and avoid swimming or wading in unfamiliar areas, particularly during high tides or rough weather.

As with any walk in a rural or coastal area, it's always a good idea to carry a fully charged mobile phone, a map or GPS device, and a small first aid kit, just in case of emergencies. Let someone know your intended route and estimated return time, and don't hesitate to seek help or turn back if you feel unsure or uncomfortable at any point.

By following these simple guidelines and using common sense, walkers of all ages and abilities can safely enjoy the beauty, tranquility, and fascinating history of the Bawdsey to Felixstowe Ferry Walk, creating lasting memories and forging a deeper connection with the natural world and the rich heritage of the Suffolk coast.

Nearby Attractions:

After your invigorating walk along the stunning Suffolk coastline, why not take the opportunity to explore some of the other fascinating attractions that this beautiful corner of East Anglia has to offer? From charming seaside towns to world-class nature reserves and historic monuments, there's something to capture the imagination and delight the senses of every visitor.

Just a short distance from the starting point of your walk, the impressive Bawdsey Manor and the Bawdsey Radar Museum offer a unique and compelling glimpse into the area's pivotal role in the development of radar technology during World War II. Discover the story of the brilliant scientists and engineers who worked tirelessly to create a system that would change the course of history, and marvel at the incredible ingenuity and determination that characterized this groundbreaking endeavor.

For those with a love of the great outdoors, the nearby RSPB Havergate Island nature reserve is an absolute must-visit. Accessible only by boat, this remote and wild island is a haven for a remarkable variety of birdlife, from the majestic marsh harriers and the elusive bitterns to the lively avocets and the comical puffins. With its pristine saltmarshes, shingle beaches, and brackish lagoons, Havergate Island offers a unique and unforgettable experience of the raw beauty and diversity of the Suffolk coast.

If you're looking to immerse yourself in the rich maritime heritage and vibrant cultural scene of the region, be sure to spend some time exploring the charming seaside towns of Felixstowe and Woodbridge. In Felixstowe, you can stroll along the elegant promenade, take in the stunning views from the historic seafront gardens, or simply relax and unwind on the sandy beaches that stretch for miles along the coast. And in Woodbridge, you'll find a picturesque waterfront lined with boats and barges, a bustling high street filled with independent shops and cafes, and a wealth of fascinating museums and galleries that showcase the town's long and storied history.

For a taste of the region's delicious local produce and traditional crafts, why not visit one of the many farmers' markets, food festivals, and artisan workshops that can be found throughout the Suffolk countryside? From the creamy, tangy cheeses and the succulent, locally-reared meats to the crisp, refreshing ciders and the handcrafted pottery and textiles, there's a wealth of culinary and creative delights to discover and enjoy, each one a celebration of the county's rich agricultural and artistic heritage.

As you explore the many attractions and hidden gems of this enchanting corner of England, take a moment to reflect on the deep sense of place and belonging that pervades every aspect of life along the Suffolk coast. From the ancient fishing villages and the historic seaside towns to the wild, untamed beauty of the estuaries and the marshes, every sight and sound tells a story of the enduring connection between people and the land, and the countless generations who have shaped and been shaped by this extraordinary landscape.

So whether you're a keen rambler, a wildlife enthusiast, a history buff, or simply someone in search of a little adventure and inspiration, the Bawdsey to Felixstowe Ferry Walk and its surroundings offer a wealth of opportunities for discovery, relaxation, and personal enrichment. Why not linger a while and immerse yourself in the magic and wonder of the Suffolk coast, and let the timeless rhythms and enduring charm of this special place work their spell on your heart and soul?

As you set out on your adventures, remember to tread lightly, to respect the local environment and communities, and to approach each new experience with an open mind and a curious heart. By doing so, you'll not only enrich your own understanding and appreciation of the world around you, but you'll also help to ensure that the precious natural and cultural treasures of this remarkable region can be enjoyed and cherished by generations to come.

In the words of the great naturalist and conservationist, Sir David Attenborough, "No one will protect what they don't care about, and no one will care about what they have never experienced." So step out into the beauty and diversity of the Suffolk coast, and let the wonder and magic of this extraordinary place be your guide and your inspiration, as you embark on a journey of discovery that will nourish your mind, body, and spirit, and leave you with memories that will last a lifetime.

Happy wandering!

Blaxhall Common Walk - 3 miles (4.8 km), Circular - Estimated Time: 1.5-2 hours

Starting Point Coordinates, Postcode: The starting point for the Blaxhall Common Walk is at the entrance to Blaxhall Common on the B1069 road (Grid Reference: TM 365 568, Postcode: IP12 2DY).
Nearest Car Park, Postcode: There is a small layby on the B1069 road near the entrance to the common (Postcode: IP12 2DY).

Embark on a delightful 3-mile (4.8 km) circular walk through the tranquil and picturesque Blaxhall Common, a hidden gem nestled within the stunning Suffolk Coast and Heaths Area of Outstanding Natural Beauty. This relatively easy route, suitable for all ages, offers a perfect escape from the hustle and bustle of everyday life, inviting you to immerse yourself in the beauty and serenity of the heathland and woodland habitats that define this enchanting landscape.

As you set out from the entrance to Blaxhall Common, feel the anticipation build as you follow the footpath into the heart of this peaceful oasis. With each step, the stresses and distractions of the modern world begin to melt away, replaced by a growing sense of connection with the natural world and its myriad wonders.

The path winds its way through a captivating tapestry of heathland and woodland, the contrasting textures and hues of the landscape creating an enchanting interplay of light and shadow. Take a moment to breathe in the fresh, invigorating air, filled with the subtle scents of heather, gorse, and wildflowers, and let the gentle rustling of leaves and the distant birdsong soothe your soul.

As you navigate the trails that branch off from the main path, keep your eyes and ears attuned to the rich diversity of plant and animal life that thrives in this unique ecosystem. From the delicate blooms of the heathland wildflowers to the majestic oaks and beeches that tower overhead, each species plays a vital role in the intricate web of life that sustains this precious habitat.

Listen for the distinctive calls of rare and uncommon birds, such as the haunting churr of the nightjar, the melodious song of the woodlark, or the lively twitter of the Dartford warbler. These elusive creatures, perfectly adapted to the challenges and opportunities of the heathland environment, are a testament to the resilience and adaptability of nature in the face of an ever-changing world.

As you continue to explore the common, marvel at the vibrant array of insects and other wildlife that call this place home. From the iridescent flutter of dragonflies and the delicate dance of butterflies to the playful antics of rabbits darting through the undergrowth, each encounter serves as a reminder of the incredible diversity and vitality of the natural world.

The walk's circular route eventually loops back towards the starting point, offering a chance to reflect on the beauty and tranquility of the landscape you've just traversed. As you near the end of your journey, take a moment to appreciate the profound sense of peace and rejuvenation that comes from immersing oneself in the timeless rhythms and simple pleasures of the great outdoors.

The Blaxhall Common Walk is more than just a physical journey; it is a celebration of the enduring power of nature to inspire, heal, and transform. Whether you're a seasoned hiker or a curious first-time visitor, this enchanting route promises to delight and uplift, leaving you with memories that will last a lifetime and a deepened appreciation for the extraordinary wonders that lie just beyond our doorsteps.

Detailed Route Directions:

1. From the entrance to Blaxhall Common on the B1069 road, follow the footpath leading into the common, bearing approximately 45°. (Estimated time: 5-10 minutes)

2. Continue along the main path as it winds through the heathland and woodland habitats, maintaining a general heading of 45°. (Estimated time: 20-30 minutes)

3. As you reach a junction with several smaller trails branching off, keep following the main path, now bearing approximately 90°. (Estimated time: 15-20 minutes)

4. The path will eventually curve towards the north, bearing roughly 0°. Continue along this section of the walk, enjoying the peaceful surroundings. (Estimated time: 20-30 minutes)

5. As the path begins to loop back towards the starting point, maintain a general heading of 270°. (Estimated time: 15-20 minutes)

6. Finally, the path will lead you back to the entrance of Blaxhall Common on the B1069 road, where your walk began. (Estimated time: 5-10 minutes)

Best Features and Views:

- Immerse yourself in the tranquil beauty of Blaxhall Common, with its enchanting mix of heathland and woodland habitats, each offering a unique and captivating atmosphere that soothes the soul and invigorates the senses.
- Delight in the diverse array of plant life that thrives in this special place, from the delicate blooms of heathland wildflowers to the majestic oaks and beeches that tower overhead, each species a vital thread in the rich tapestry of the common's ecosystem.
- Marvel at the incredible variety of wildlife that calls Blaxhall Common home, from the rare and elusive birds of the heathland to the vibrant insects and playful mammals that bring the landscape to life, each encounter a reminder of the breathtaking diversity and resilience of the natural world.

Historical Features of Interest:

While there are no specific historical sites or artifacts directly along the route, the Blaxhall Common Walk offers a unique opportunity to connect with the rich cultural heritage and enduring traditions of the Suffolk countryside. As you navigate the peaceful trails and immerse yourself in the timeless beauty of the heathland and woodland, you'll be following in the footsteps of countless generations who have lived, worked, and found solace in this enchanting landscape.

The heathland itself is a product of centuries of human activity, shaped by the age-old practices of grazing, cutting, and burning that have maintained this unique habitat for countless generations. These traditional land management techniques, passed down from one generation to the next, have created a landscape that is both stunningly beautiful and ecologically invaluable, supporting a wealth of rare and specialized species that have adapted to thrive in the challenging conditions of the heath.

As you walk through the common, imagine the lives of the shepherds, turf-cutters, and charcoal burners who once made their living from the bounty of the heathland, their labors woven into the very fabric of the landscape. From the distant past to the present day, Blaxhall Common has served as a vital resource and a cherished sanctuary for the local community, its enduring beauty and tranquility a testament to the deep connection between people and place that lies at the heart of rural life.

In the stillness of the common, with the wind whispering through the heather and the birdsong echoing across the landscape, you may find yourself feeling a profound sense of kinship with the generations who have walked this way before, their stories and experiences etched into every contour and feature of the land. This is the true gift of the Blaxhall Common Walk, the chance to step out of the present moment and into a world of timeless beauty, where the boundary between past and present dissolves, and the essence of the Suffolk countryside is revealed in all its glory.

Birds to Spot:

- Nightjars: As dusk falls over the heathland, listen for the distinctive churring call of these elusive birds, their cryptic plumage and silent, moth-like flight making them a mesmerizing presence in the gathering twilight.
- Woodlarks: Keep an ear out for the melodious, fluting song of these small, brown birds as they soar above the heathland and woodland edges, their aerial displays a joyous celebration of the wide-open spaces and the freedom of the skies.
- Dartford Warblers: Watch for these lively, long-tailed warblers as they flit among the gorse and heather, their distinctive reddish-brown plumage and insect-like call a charming addition to the heathland chorus.

Rare or Interesting Insects and Other Wildlife:

- Silver-studded Blue Butterflies: In the summer months, the heathland comes alive with the iridescent blue wings of these rare and beautiful butterflies, their delicate markings and graceful flight a true wonder to behold.
- Common Lizards: As you walk through the sunny glades and open areas of the common, keep your eyes peeled for the quick, darting movements of these small, agile reptiles, their mottled brown and green scales providing perfect camouflage against the heathland vegetation.
- Rabbits: The heathland and woodland edges of Blaxhall Common are home to a thriving population of rabbits, their playful antics and fluffy white tails a delightful sight as they scamper through the undergrowth and bask in the warm sunshine.

Rare or Interesting Plants:

- Heather: The undisputed queen of the heathland, heather is the defining plant of this rare and precious habitat, its delicate purple and pink flowers creating a stunning carpet that stretches out across the landscape in late summer and early autumn.
- Gorse: With its vibrant yellow flowers and prickly, evergreen stems, gorse is another iconic plant of the heathland, its coconut-scented blooms and dense, thorny thickets providing shelter and sustenance for a wide range of insects, birds, and mammals.
- Sheepsbit: Look closely among the short, rabbit-grazed turf of the heathland, and you may spot the delicate blue flowers of sheepsbit, a tiny, low-growing plant that thrives in the nutrient-poor soils and exposed conditions of the common, its presence a testament to the incredible adaptability and resilience of nature.

Facilities and Accessibility:

The Blaxhall Common Walk is a relatively easy and accessible route, with mostly level terrain and well-defined paths suitable for walkers of all ages and abilities. However, it's important to note that some sections of the common may be uneven, muddy, or overgrown, particularly during the wetter months or after periods of heavy rain. As such, sturdy walking shoes or boots with good grip are highly recommended to ensure a safe and comfortable experience.

While there are no facilities such as toilets or cafes directly on the common, the nearby village of Blaxhall offers a charming pub, The Ship Inn, where you can enjoy a refreshing drink or a hearty meal before or after your walk. The village also has a small shop and a church, providing a glimpse into the traditional rural life of the Suffolk countryside.

It's worth noting that Blaxhall Common is a protected wildlife habitat, and as such, it's essential to follow the designated paths and to respect the delicate ecosystems and sensitive species that make this place so special. Please keep dogs on a lead, refrain from littering or disturbing the vegetation, and take only photographs, leaving nothing but footprints behind.

By treating the common with care and consideration, visitors of all abilities can enjoy the beauty, tranquility, and fascinating natural heritage of this enchanting corner of the Suffolk Coast and Heaths AONB, creating memories that will last a lifetime and helping to preserve this precious landscape for generations to come.

Seasonal Highlights:

The Blaxhall Common Walk offers a unique and captivating experience in every season, with each time of year revealing new facets of beauty and wonder within this enchanting heathland and woodland landscape:

- Spring: As the days lengthen and the first warm rays of sun caress the land, the common bursts into life with an explosion of colour and vitality. The gorse is a blaze of vibrant yellow, its coconut-scented blooms filling the air with their heady fragrance, while the delicate pinks and purples of the heather begin to emerge from the wiry stems. The birdsong reaches a crescendo as the breeding season gets underway, and the first butterflies and bees take to the wing, flitting among the wildflowers and basking in the gentle spring light.
- Summer: In the long, languid days of high summer, the heathland is at its most glorious, the purple and pink hues of the heather stretching out in a mesmerizing carpet that seems to shimmer and dance in the heat. The air is heavy with the drone of insects, the buzzing of bees and the chirping of grasshoppers creating a soothing symphony that perfectly captures the essence of the season. The woodland edges are a cool and shady refuge, the dappled light filtering through the canopy and painting the forest floor in shifting patterns of gold and green.
- Autumn: As the days grow shorter and the nights turn crisp, the common takes on a new character, its colours shifting from the rich purples of summer to the warm golds and russets of autumn. The heather fades to a muted brown, its flowers giving way to the tiny, bead-like seed heads that will ensure its survival through the winter months. Fungi of all shapes and sizes emerge from the woodland floor and the decaying stumps, their bizarre forms and lurid hues adding a touch of magic and mystery to the landscape. The distant calls of migrating birds and the rustle of small mammals preparing for the cold months ahead fill the air with a sense of change and anticipation.
- Winter: Though the heathland may seem stark and lifeless in the depths of winter, there is a haunting beauty to the landscape in its dormant state. The bare stems of the heather and gorse stand out in sharp relief against the frost-kissed earth, while the skeletal forms of the trees etch intricate patterns against the pale, watery sky. The stillness is broken only by the occasional gust of wind or the distant call of a raven, the common's wild inhabitants hunkered down and conserving their energy for the challenges of the season. Yet even in this time of rest and renewal, there are signs of life all around, from the tracks of a fox or rabbit in the snow to the hardy evergreens that add a splash of colour to the muted palette of the heath.

Difficulty Level and Safety Considerations:

The Blaxhall Common Walk is a relatively easy route, suitable for walkers of most ages and fitness levels. The terrain is mostly flat and the paths are generally well-maintained, making for a comfortable and enjoyable walking experience. However, as with any outdoor activity, it's essential to be aware of potential hazards and to take necessary precautions to ensure a safe and responsible journey.

One of the main considerations when walking on heathland is the risk of tripping or stumbling on the uneven ground, particularly in areas where the vegetation is dense or the paths are less well-defined. It's crucial to wear sturdy, comfortable footwear with good grip and ankle support, and to watch your step when navigating these more challenging sections of the route.

Another factor to keep in mind is the changeable nature of the British weather, which can range from warm and sunny to cold, wet, and windy, often within the space of a single walk. Be sure to check the forecast before setting out and to dress appropriately for the conditions, with layers that can be easily added or removed as needed, as well as a waterproof jacket and sturdy boots.

It's also important to stay hydrated and energized throughout your walk, so be sure to bring plenty of water and some healthy snacks to keep you going. If you have any medical conditions or allergies, it's a good idea to carry any necessary medications or first aid supplies, just in case.

When exploring the heathland and woodland habitats of Blaxhall Common, it's essential to respect the delicate ecosystems and sensitive species that make this place so special. Stick to the designated paths, keep dogs on a lead, and avoid disturbing the vegetation or wildlife. Remember that this is a protected area, and that every visitor has a responsibility to help preserve its beauty and integrity for generations to come.

By following these simple guidelines and using common sense, walkers of all abilities can safely immerse themselves in the tranquility, diversity, and natural wonders of the Blaxhall Common Walk, forging a deeper connection with the landscape and creating memories that will last a lifetime.

Nearby Attractions:

After your invigorating walk through the enchanting heathland and woodland of Blaxhall Common, why not extend your adventure and explore some of the other fascinating attractions that this beautiful corner of Suffolk has to offer? From charming coastal towns to world-class nature reserves and cultural landmarks, there's something to capture the imagination and delight the senses of every visitor.

Just a short drive from Blaxhall Common, the picturesque seaside town of Aldeburgh awaits, with its colourful beach huts, historic high street, and famous fish and chip shops. Take a stroll along the shingle beach, breathe in the fresh sea air, and enjoy a delicious meal of locally-caught seafood while watching the boats bobbing gently on the waves. Be sure to visit the iconic Scallop sculpture by local artist Maggi Hambling, a striking tribute to the town's fishing heritage and a beloved landmark of the Suffolk coast.

For those with a passion for nature and birdwatching, the nearby RSPB Minsmere reserve is an absolute must-see. This internationally-renowned wetland and coastal sanctuary is home to an astonishing variety of wildlife, from the majestic avocets and bitterns to the playful otters and the elusive bearded tits. With its network of trails, hides, and visitor facilities, Minsmere offers unparalleled opportunities to observe and learn about the incredible diversity of the Suffolk countryside, and to experience the raw beauty and power of the natural world in all its glory.

Culture vultures and art lovers will find plenty to inspire and delight them in the nearby town of Snape, home to the world-famous Snape Maltings concert hall and arts complex. Housed in a stunning collection of converted Victorian malthouses, this cultural hub hosts a diverse program of music, theater, and dance performances throughout the year, as well as exhibitions, workshops, and festivals that celebrate the creativity and innovation of the region. Take a tour of the historic buildings, browse the independent shops and galleries, and soak up the unique atmosphere of this vibrant and dynamic destination.

For a taste of Suffolk's rich agricultural heritage and delicious local produce, be sure to visit the nearby Orford Farmers' Market, held every Saturday morning in the picturesque village of Orford. Here, you'll find an enticing array of fresh, seasonal fruits and vegetables, artisan cheeses and breads, locally-reared meats, and handmade crafts and gifts, all showcasing the best of the county's thriving food and farming scene. Enjoy a cup of freshly-brewed coffee and a homemade pastry while chatting with the friendly stallholders, and take home a piece of Suffolk's culinary and cultural bounty to savour long after your visit.

As you explore the many attractions and hidden gems of this enchanting corner of East Anglia, take a moment to reflect on the deep sense of place and belonging that pervades every aspect of life in the Suffolk countryside. From the ancient heathlands and wildflower meadows to the historic towns and villages that have shaped the region's character over centuries, every sight and sound tells a story of the enduring connection between people and the land, and the countless generations who have found solace, inspiration, and joy in the beauty of the natural world.

So whether you're a keen rambler, a nature enthusiast, a foodie, or simply someone in search of a little peace and tranquility in a beautiful setting, Blaxhall Common and its surroundings offer a wealth of opportunities for discovery, relaxation, and personal enrichment. Why not linger a while and immerse yourself in the magic and wonder of the Suffolk countryside, and let the gentle rhythms and timeless charm of this special place work their spell on your mind, body, and soul?

As you set out on your adventures, remember to tread lightly, to respect the local environment and communities, and to approach each new experience with an open heart and a curious mind. By doing so, you'll not only deepen your own understanding and appreciation of the world around you, but you'll also help to ensure that the precious natural and cultural treasures of this remarkable region can be enjoyed and cherished by generations to come.

In the words of the great naturalist and conservationist, John Muir, "In every walk with nature, one receives far more than he seeks." So step out into the beauty and diversity of the Suffolk countryside, and let the wonder and wisdom of the natural world be your guide and your inspiration, as you embark on a journey of discovery that will nourish your spirit, enrich your life, and leave you with memories that will last a lifetime.

Happy trails!

Orford Ness Nature Reserve Walk - 5 miles (8 km), Circular - Estimated Time: 2.5-3 hours

Starting Point Coordinates, Postcode: The starting point for the Orford Ness Nature Reserve Walk is at Orford Quay (Grid Reference: TM 421 496, Postcode: IP12 2NU).
Nearest Car Park, Postcode: The nearest car park is the Orford Quay car park (Postcode: IP12 2NU).

Embark on an unforgettable 5-mile (8 km) circular walk through the breathtaking and mysterious landscapes of Orford Ness, a remote and wild corner of the Suffolk coast that has been shaped by the forces of nature and the echoes of human history. This unique National Trust nature reserve, once a top-secret military testing site, offers a rare opportunity to explore a hauntingly beautiful and ecologically diverse wilderness, where the interplay of land, sea, and sky creates a truly awe-inspiring experience.

Your journey begins at the picturesque Orford Quay, where you'll board the National Trust ferry for a short but memorable voyage across the glistening waters of the River Ore. As you watch the bustling quayside recede into the distance, feel the anticipation build as you approach the enigmatic shores of Orford Ness, a shingle spit that stretches for miles along the Suffolk coast, its stark beauty and sense of isolation a world away from the trappings of modern life.

Once on the Ness, follow the well-marked footpath as it winds through a tapestry of shingle ridges, saltmarshes, and grazing marshes, each habitat a haven for a remarkable array of flora and fauna. As you walk, take a moment to appreciate the subtle textures and hues of the shingle beneath your feet, a constantly shifting mosaic of pebbles and shells that tells the story of countless centuries of coastal change and adaptation.

Pause to marvel at the resilience and ingenuity of the plants that thrive in this challenging environment, from the robust sea kale with its waxy blue-green leaves to the delicate sea pea with its vibrant purple flowers. These hardy pioneers are a testament to the incredible adaptability of nature, their deep roots and salt-tolerant leaves a perfect match for the harsh conditions of the coastal fringe.

As you continue along the path, keep your eyes peeled for the rich birdlife that calls Orford Ness home, from the elegant avocets and lapwings that grace the mudflats to the terns and gulls that wheel and dive over the open water. The Ness is a vital sanctuary for these feathered wonders, its remote location and diverse habitats providing a safe haven for breeding, feeding, and resting during their epic migrations.

But Orford Ness is not just a place of natural wonders; it is also a site steeped in history and intrigue, its eerie landscapes bearing the scars of a secretive past. As you pass by the crumbling remnants of military buildings and testing facilities, let your imagination wander to the clandestine experiments and classified operations that once took place here, a chilling reminder of the darker chapters of human ingenuity and the long shadow of conflict.

One of the most iconic landmarks on your walk is the Orford Ness Lighthouse, a striking red and white structure that has stood watch over the coastline since the 18th century. Take a moment to appreciate the graceful lines and timeless beauty of this historic beacon, its presence a comforting constant in a landscape that is forever shifting and changing with the tides and the seasons.

As you turn inland and follow the path through the lush grazing marshes, keep a watchful eye out for the brown hares that bound through the tussocky grass and the seals that bask on the muddy banks of the tidal creeks. These charismatic creatures, along with the myriad butterflies and dragonflies that flit and dart among the reeds, are a joyous celebration of the incredible diversity of life that flourishes in this unique and precious habitat.

Completing the circular route, you'll arrive back at the ferry landing point, your senses invigorated and your spirit uplifted by the raw beauty and untamed energy of Orford Ness. As you board the ferry for your return journey to Orford Quay, take a moment to reflect on the extraordinary landscape you've just explored, a place where the forces of nature and the hand of humankind have intertwined to create a truly unforgettable experience.

The Orford Ness Nature Reserve Walk is a journey of discovery and revelation, an invitation to step beyond the ordinary and to immerse yourself in a world of wonder, mystery, and endless possibility. Whether you're a nature lover, a history buff, or simply a curious adventurer, this remarkable corner of the Suffolk coast promises to delight, inspire, and challenge you in equal measure, leaving you with memories that will last a lifetime.

Detailed Route Directions:

1. Begin your walk at Orford Quay (Grid Reference: TM 421 496). Board the National Trust ferry to Orford Ness, which operates between April and October, weather permitting. Please note that a fee applies for the ferry. (Estimated time: 10-15 minutes)

2. Upon arriving at Orford Ness, follow the clearly marked footpath as it leads you through the captivating shingle and saltmarsh habitats. Take your time to observe and appreciate the diverse flora and fauna that thrive in this unique environment. (Estimated time: 45-60 minutes)

3. Continue along the footpath, passing by the intriguing remnants of military testing sites that hint at the Ness's secretive past. Keep an eye out for the iconic Orford Ness Lighthouse, a striking red and white structure that serves as a landmark for your walk. (Estimated time: 30-45 minutes)

4. As you reach the lighthouse, turn inland and follow the footpath as it guides you through the lush grazing marshes. Stay alert for the various bird species that call this habitat home, such as the elegant avocets, the lively lapwings, and the graceful redshanks. (Estimated time: 45-60 minutes)

5. Complete the circular walk by making your way back to the ferry landing point on Orford Ness. Take a moment to reflect on the extraordinary landscapes and wildlife you've encountered during your journey. (Estimated time: 30-45 minutes)

6. Board the ferry for your return trip to Orford Quay, bringing your unforgettable Orford Ness Nature Reserve Walk to a close. (Estimated time: 10-15 minutes)

Best Features and Views:

- Marvel at the iconic Orford Ness Lighthouse, a striking red and white structure that has stood watch over the coastline since the 18th century, its timeless beauty and historic significance a testament to the enduring human presence on this wild and remote spit of land.
- Discover the haunting remnants of military testing sites scattered across the Ness, their crumbling buildings and rusting structures a chilling reminder of the secretive experiments and classified operations that once took place here, a darker chapter in the story of human ingenuity and the long shadow of conflict.
- Immerse yourself in the stunning diversity of habitats that define Orford Ness, from the stark beauty of the shingle ridges to the lush vitality of the saltmarshes and grazing marshes, each one a unique and precious

ecosystem that supports an astonishing array of plants, birds, and other wildlife.

Birds to Spot:

- Avocets: Delight in the grace and elegance of these striking black and white waders, their slender, upturned bills perfectly adapted for foraging in the shallow waters and mudflats of Orford Ness. A true emblem of conservation success, the avocet's presence here is a testament to the incredible resilience and adaptability of nature.
- Lapwings: Watch for the acrobatic aerial displays and distinctive "peewit" calls of these charismatic waders, their iridescent green and purple plumage and wispy crests a mesmerizing sight against the open skies of the Ness. Lapwings are a key species of the grazing marshes, their presence a vital indicator of the health and diversity of this precious habitat.
- Redshanks: Listen for the evocative piping calls and look for the bright orange legs of these lively waders as they probe the mud and shallow water for prey, their energetic movements and bold patterning a delight to behold. Redshanks are a common sight on Orford Ness, their presence a reassuring constant in the ever-changing coastal landscape.

Rare or Interesting Insects and Other Wildlife:

- Brown Hares: Keep your eyes peeled for the bounding forms and tawny coats of these elusive mammals as they navigate the tussocky grass and open spaces of the grazing marshes, their long, black-tipped ears and powerful hind legs a marvel of natural adaptation. Brown hares are an iconic species of the British countryside, their presence on Orford Ness a testament to the wild and untamed character of this unique landscape.
- Seals: Scan the muddy banks and tidal creeks for the sleek, dappled forms of seals basking in the sun or bobbing in the water, their inquisitive faces and playful demeanour a charming sight for any wildlife lover. Both common and grey seals can be spotted around Orford Ness, their presence a reminder of the rich marine life that thrives just offshore and the vital connection between land and sea in this coastal wilderness.
- Butterflies and Dragonflies: Marvel at the vibrant colours and intricate patterns of the many butterfly and dragonfly species that flit and dart among the wildflowers and reeds of Orford Ness, their delicate beauty and aerial agility a mesmerizing sight to behold. From the striking orange and brown of the wall brown butterfly to the iridescent blues and greens of the emperor dragonfly, these ephemeral creatures are a vital part of the complex web of life that defines this exceptional nature reserve.

Rare or Interesting Plants:

- Sea Kale: Discover the robust, blue-green leaves and delicate white flowers of this hardy coastal plant, its thick, waxy foliage perfectly adapted to the harsh conditions of the shingle ridges. Sea kale is not only a vital stabilizer of the shifting pebbles but also an important food source for a variety of insects and birds, its presence a keystone of the shingle ecosystem.
- Sea Pea: Delight in the vibrant purple blooms and tendrils of this low-growing, scrambling plant as it cascades over the shingle and adds a splash of colour to the muted tones of the coast. Sea pea is a rare and specialized species that thrives in the nutrient-poor soils of Orford Ness, its ability to fix nitrogen from the air a remarkable adaptation that allows it to flourish where other plants struggle.
- Saltmarsh and Grazing Marsh Plants: Immerse yourself in the lush greens and muted browns of the saltmarsh and grazing marsh vegetation, from the swaying stems of sea lavender and the fleshy leaves of glasswort to the tufted grasses and delicate rushes that carpet the damp ground. These diverse and resilient plant communities are the foundation of the marsh ecosystem, their complex root systems and dense growth providing food, shelter, and breeding sites for an astonishing array of wildlife.

Facilities and Accessibility:

The Orford Ness Nature Reserve Walk offers a unique and immersive experience for visitors, but it is important to note that the remote and wild character of the site means that facilities are limited and the terrain can be challenging.

At the starting point of your walk, the charming village of Orford provides a range of amenities to ensure a comfortable and enjoyable visit. The Orford Quay car park offers ample space for vehicles, while the nearby public toilets and inviting café are perfect for a quick comfort break or a refreshing snack before embarking on your adventure.

Once on Orford Ness itself, visitors should be prepared for a more rugged and isolated environment. The nature reserve is only accessible by the National Trust ferry, which operates between April and October, weather permitting. It is essential to check the ferry times and availability before planning your visit, as well as any potential restrictions or closures due to conservation work or unsafe conditions.

The walk itself involves traversing a variety of challenging terrains, from the uneven and shifting shingle ridges to the muddy and slippery saltmarshes. Sturdy walking shoes or boots with good ankle support and grip are an absolute must to ensure a safe and comfortable experience, and visitors should be prepared for sudden changes in weather conditions, particularly wind and rain.

While the footpaths on Orford Ness are generally well-maintained and clearly marked, the remote nature of the site means that they may not be suitable for all abilities. The uneven ground, narrow boardwalks, and lack of handrails or benches may prove challenging for those with mobility issues or limited stamina, and the exposed, open landscape can be disorienting for some visitors.

Despite these challenges, the Orford Ness Nature Reserve Walk remains a highly rewarding and enriching experience for those who are able to undertake it. By preparing adequately, following the guidance of the National Trust staff, and embracing the wild and untamed character of the landscape, visitors can immerse themselves in the unique beauty, history, and ecology of this extraordinary corner of the Suffolk coast.

Seasonal Highlights:

The Orford Ness Nature Reserve Walk offers a dynamic and ever-changing experience throughout the year, with each season revealing new facets of beauty and wonder in this exceptional coastal wilderness:

- Spring: As the days lengthen and the first warm rays of sun kiss the shingle, the Ness bursts into life with an explosion of colour and activity. The saltmarshes and grazing marshes are adorned with a vibrant carpet of wildflowers, from the delicate pink of thrift to the cheery yellow of buttercups, while the air is filled with the joyous songs of skylarks and the gentle humming of bees. Along the shoreline, the first migratory birds begin to arrive, with wheatears and sandpipers joining the resident waders in their search for food among the rocks and seaweed.
- Summer: In the long, languid days of summer, Orford Ness is at its most lively and inviting, with the shimmering waters of the North Sea and the distant haze of the horizon creating an irresistible backdrop for adventure and discovery. The shingle is alive with the buzzing of insects and the scurrying of lizards, while the saltmarshes are a riot of colour, with sea lavender and sea aster adding a splash of purple and gold to the green expanse. Butterflies and dragonflies flit among the wildflowers, and the haunting cries of terns and gulls echo across the vast, open skies.
- Autumn: As the leaves begin to turn and the nights grow longer, Orford Ness takes on a new character, with

the mellow light and the distant cries of overwintering birds creating a sense of peace and solitude. The shingle is strewn with the muted gray of driftwood and the rich brown of seaweed, while the grazing marshes are a patchwork of russet and gold, the fading colours of the grasses and reeds a poignant reminder of the passing of the seasons. Flocks of geese and ducks arrive from the north, their raucous calls and lively movements bringing a spark of energy to the tranquil landscape.

- Winter: Though the wind may be bracing and the skies leaden, Orford Ness is no less beautiful in the depths of winter. The stark outlines of the lighthouse and the military buildings stand out against the pale, watery light, while the shingle and the saltmarshes are etched with the delicate tracery of frost and the softening blanket of snow. Wrap up warm and embrace the invigorating chill of the coastal air, marvelling at the raw power of the elements and the enduring strength of the natural world. Keep an eye out for the hardy birds that brave the cold, from the tiny dunlins and sanderlings that scurry along the tideline to the majestic birds of prey that soar overhead, their presence a reminder of the untamed and unconquerable spirit of this wild and wonderful place.

Difficulty Level and Safety Considerations:

The Orford Ness Nature Reserve Walk is a moderate trek that requires a reasonable level of fitness and mobility to complete safely and comfortably. While the distance of 5 miles (8 km) may not seem overly challenging, the unique and varied terrain of the Ness presents a range of potential hazards and difficulties that walkers should be aware of before setting out.

One of the most significant challenges of the walk is the uneven and shifting nature of the shingle beach, which makes up a large portion of the route. The loose, rounded pebbles can be difficult to navigate, placing strain on the ankles and knees and increasing the risk of slips, trips, and falls. Sturdy walking boots with good ankle support and a robust tread are essential for maintaining balance and stability on this challenging surface.

The exposed and open character of the Ness is another factor that walkers should take into account when planning their visit. With little in the way of shelter or shade, the landscape can be subject to strong winds, driving rain, and intense sunlight, all of which can take their toll over the course of a 5-mile walk. It is crucial to check the weather forecast before setting out and to dress appropriately for the conditions, with layers that can be easily added or removed as needed, as well as a water proof jacket, hat, and sunscreen for protection against the elements.

The tidal nature of the saltmarshes and creeks that crisscross the Ness is another important safety consideration. While the footpaths are generally well-marked and maintained, there is always a risk of becoming disoriented or stranded by the rising water, particularly in poor visibility or changing weather conditions. It is essential to check the tide times before embarking on the walk and to allow plenty of time to complete the route and return to the ferry landing point before the water levels become dangerously high.

Walkers should also be mindful of the military history of Orford Ness and the potential hazards posed by the remnants of its former life as a testing site. While the majority of the buildings and structures are now derelict and off-limits to visitors, there may be hidden dangers such as unstable foundations, broken glass, or rusting metal that could cause injury if not approached with caution. It is vital to stick to the designated footpaths and to heed any warning signs or barriers that may be in place to ensure your safety and the preservation of the site.

Despite these challenges and considerations, the Orford Ness Nature Reserve Walk remains a highly rewarding and enriching experience for those who are prepared and respectful of the unique character of the landscape. By taking sensible precautions, following the guidance of the National Trust staff, and allowing plenty of time to enjoy the scenery and wildlife at a relaxed pace, walkers of moderate fitness and ability can safely immerse themselves in the rugged beauty and fascinating history of this extraordinary corner of the Suffolk coast.

As with any outdoor activity, it is always a good idea to carry a fully charged mobile phone, a map or GPS device, and a small first aid kit, just in case of emergencies. Let someone know your intended route and estimated return time, and don't hesitate to turn back or seek assistance if you feel uncomfortable or unsure at any point.

By embracing the wild and untamed spirit of Orford Ness while also respecting its power and fragility, you can forge a deeper connection with the natural world and the human stories that have shaped this incredible landscape over the centuries. So step out onto the shingle with confidence and curiosity, and let the wind, the waves, and the wide-open skies be your guide on this unforgettable coastal adventure.

Nearby Attractions:

After your bracing walk through the wild and captivating landscapes of Orford Ness, why not take the opportunity to explore some of the other fascinating attractions that this charming corner of Suffolk has to offer? From ancient castles and picturesque villages to world-class cultural destinations and mouthwatering culinary experiences, there is something to delight and inspire every visitor to this enchanting region.

Just a stone's throw from Orford Quay, the imposing bulk of Orford Castle rises majestically above the rooftops, its towering keep and massive walls a testament to the power and prestige of the medieval lords who once held sway over this strategic corner of the Suffolk coast. Built in the 12th century by Henry II to guard the entrance to the River Ore, the castle has stood witness to centuries of conflict, intrigue, and change, its storied walls whispering tales of battles, sieges, and secret plots. Today, visitors can explore the castle's atmospheric chambers and winding staircases, marvelling at the ingenuity of its Norman builders and imagining the lives of the soldiers, servants, and nobles who once called this incredible fortress home.

For a deeper dive into the rich history and culture of the area, be sure to visit the Orford Museum, a charming little gem that offers a fascinating glimpse into the lives and livelihoods of the people who have shaped this unique corner of the Suffolk coast over the centuries. From the Bronze Age settlers who first made their homes along the river's edge to the fishermen, farmers, and sailors who have plied their trades here for generations, the museum's collection of artifacts, photographs, and personal stories paints a vivid picture of the enduring human spirit and the unbreakable bonds between people and place.

No visit to Orford would be complete without a leisurely stroll through the picturesque streets and lanes of the village itself, a delightful jumble of colourful cottages, quirky shops, and inviting pubs that exudes an irresistible charm and character. Wander along the quayside, watching the boats bob gently on the glittering waters of the River Ore, or sample some of the delicious local produce at one of the many excellent eateries, from freshly caught seafood and succulent marsh-grazed beef to artisanal cheeses and homemade cakes.

For a true taste of the Suffolk good life, head to the world-renowned Pump Street Bakery, a haven of slow-fermented sourdough, handcrafted pastries, and mouthwatering chocolate that has garnered a cult following among foodies and locals alike. Inhale the heady scent of freshly baked bread, savour the flaky, buttery perfection of a handmade croissant, and let the rich, complex flavours of single-origin chocolate transport you to a state of pure bliss.

As you explore the many delights and hidden gems of Orford and its surrounding landscape, take a moment to reflect on the incredible resilience, creativity, and adaptability of the people who have called this special place home over the millennia. From the ancient tribes who first navigated the tidal waters of the Ore to the visionary conservationists and local communities who work tirelessly to preserve and protect the fragile ecosystems of Orford Ness, each generation has left its mark on the land, weaving a rich tapestry of human experience that is as enduring and inspiring as the wild beauty of the coast itself.

So whether you're a history buff, a nature lover, a foodie, or simply someone in search of a little escape and adventure in a breathtakingly beautiful setting, Orford and its surroundings offer a wealth of opportunities for discovery, relaxation, and unforgettable memories. Let the timeless rhythms of the tides and the echoes of the past be your guide, as you embark on a journey of exploration that will nourish your mind, body, and soul, and leave you with a profound sense of connection to the people and places that make this enchanting corner of England so truly special.

Orford Ness National Nature Reserve Walk - 5 miles (8 km), Linear - Estimated Time: 2.5-3 hours

Starting Point Coordinates, Postcode: The starting point for the Orford Ness National Nature Reserve Walk is at Orford Quay (Grid Reference: TM 422 496, Postcode: IP12 2NU).
Nearest Car Park, Postcode: The nearest car park is the Orford Quay car park (Postcode: IP12 2NU).

Embark on an unforgettable 5-mile (8 km) linear walk through the hauntingly beautiful and ecologically significant landscapes of the Orford Ness National Nature Reserve, a remote and wild spit of land on the Suffolk coast that has been shaped by the forces of nature and the echoes of human history. This unique walk offers a rare opportunity to explore one of Europe's largest and most important vegetated shingle habitats, where the interplay of land, sea, and sky creates a truly awe-inspiring and thought-provoking experience.

Your journey begins at the picturesque Orford Quay, where you'll board the National Trust ferry for a short but memorable crossing to the mysterious shores of Orford Ness. As you leave the bustling quayside behind and watch the mainland recede into the distance, feel the anticipation build as you approach this otherworldly landscape, a place where the boundaries between the natural and the human, the ancient and the modern, seem to blur and dissolve.

Once on the Ness, follow the well-marked footpath as it leads you through a mesmerizing tapestry of shingle ridges, saltmarshes, and coastal lagoons, each habitat a testament to the incredible resilience and adaptability of the species that call this place home. Marvel at the subtle beauty of the sea kale and yellow-horned poppy that cling tenaciously to the shifting pebbles, their delicate blooms and striking foliage a vibrant contrast to the muted tones of the shingle.

As you walk, keep your eyes and ears attuned to the rich birdlife that thrives in this unique environment, from the haunting cries of the oystercatchers and the elegant grace of the avocets to the lively chatter of the terns and the occasional glimpse of a majestic marsh harrier soaring overhead. The Ness is a haven for both resident and migratory species, its remote location and diverse habitats providing a vital sanctuary for countless birds throughout the year.

But the Orford Ness National Nature Reserve is not just a place of natural wonders; it is also a site steeped in mystery and intrigue, its eerie landscapes bearing the scars of a long and secretive military history. As you make your way along the coast, you'll pass by the crumbling remnants of old testing facilities and bunkers, their weathered concrete and rusting metal a stark reminder of the clandestine experiments and operations that once took place here, hidden from public view.

Take a moment to contemplate the juxtaposition of these man-made structures against the untamed beauty of the surrounding landscape, a poignant symbol of the complex and often fraught relationship between humanity and the natural world. Let your imagination wander to the stories and secrets that these silent ruins might hold, a testament to the ingenuity, ambition, and folly of our species in the face of the awesome power of nature.

As you near the end of the spit, pause to drink in the breathtaking views of the open sea and the distant coastline, the vast expanse of the sky and the ceaseless motion of the waves a humbling reminder of our own small place in the grand scheme of things. Here, at the edge of the land, where the elements converge in a ceaseless dance of wind, water, and stone, you may find a moment of profound connection and clarity, a sense of being part of something much greater than yourself.

Finally, retrace your steps along the shingle back to the ferry landing point, your mind and heart filled with the sights, sounds, and sensations of this extraordinary place. As you board the ferry for your return journey to Orford Quay, take a moment to reflect on the incredible diversity and resilience of the life that flourishes in this challenging environment, and the enduring impact of human activity on even the most remote and isolated corners of our planet.

The Orford Ness National Nature Reserve walk is a journey of discovery and contemplation, an invitation to step beyond the ordinary and to immerse yourself in a landscape that is at once beautiful, haunting, and deeply thought-provoking. Whether you are a nature lover, a history buff, or simply someone in search of a unique and unforgettable experience, this walk promises to leave you with memories and insights that will stay with you long after you have left the shingle behind.

Detailed Route Directions:

1. Begin your journey at Orford Quay (Grid Reference: TM 422 496), where you will need to take the National Trust ferry to Orford Ness. The ferry operates seasonally, and tickets can be purchased on arrival. Be sure to check the National Trust website for current ferry times and prices. (Estimated time: 10-15 minutes)

2. Upon disembarking the ferry at Orford Ness, follow the clearly marked footpath as it leads you through the fascinating shingle habitat. As you walk, take the time to observe and appreciate the unique plant life that thrives in this challenging environment, such as the yellow-horned poppy and sea kale. Keep your eyes peeled for the diverse array of bird species that call the Ness home, including avocets, oystercatchers, and terns. (Estimated time: 45-60 minutes)

3. Continue along the footpath, which will guide you past the intriguing remnants of Orford Ness's military past. Take a moment to contemplate the old testing facilities and bunkers, their weathered structures standing in stark contrast to the wild beauty of the surrounding landscape. (Estimated time: 30-45 minutes)

4. As you reach the end of the spit, pause to soak in the breathtaking views of the sea and the distant coastline. Allow yourself a moment of quiet reflection, marvelling at the vastness of the sky and the ceaseless motion of the waves. (Estimated time: 15-30 minutes)

5. To complete your walk, simply retrace your steps along the shingle footpath, making your way back to the ferry landing point. Take your time to enjoy the unique landscape and wildlife once more, reflecting on the profound experience of exploring this remarkable corner of the Suffolk coast. (Estimated time: 60-75 minutes)

6. Board the ferry for your return trip to Orford Quay, bringing your unforgettable journey through the Orford Ness National Nature Reserve to a close. (Estimated time: 10-15 minutes)

Best Features and Views:

- Immerse yourself in the haunting beauty of the unique shingle habitat, where the subtle colours and textures of the pebbles create an otherworldly landscape that seems to stretch to the horizon, inviting quiet contemplation and a deep sense of connection to the raw power of nature.
- Marvel at the incredible diversity of birdlife that finds sanctuary in the remote and varied habitats of Orford Ness, from the elegant wading of the avocets in the shallow waters to the lively aerial acrobatics of the terns as they dive for fish in the coastal lagoons.
- Discover the enigmatic remnants of the Ness's military history, the crumbling bunkers and testing facilities standing as eerie reminders of the secret experiments and operations that once took place here, a thought-provoking contrast to the timeless rhythms of the natural world that surrounds them.

Birds to Spot:

- Avocets: Delight in the graceful presence of these striking black-and-white waders, their slender, upturned bills perfectly adapted for sweeping through the shallow waters and mudflats in search of small invertebrates. As a symbol of conservation success and resilience, the avocet's return to Orford Ness is a heartening reminder of nature's capacity for recovery and renewal.
- Oystercatchers: Listen for the distinctive piping calls and look for the bold black-and-white plumage of these charismatic shorebirds as they forage along the tideline, their bright orange bills a vivid contrast to the muted tones of the shingle. Oystercatchers are a lively and endearing presence on the Ness, their interactions and behaviours a source of endless fascination for birdwatchers and nature lovers alike.
- Terns: Watch for the quick, darting flight and the sharp, pointed wings of the various tern species that make their summer homes on Orford Ness, from the diminutive little tern to the more robust Sandwich and common terns. These elegant seabirds are a delight to observe as they hover over the water, their aerial agility and keen eyesight a marvel of evolutionary adaptation.

Rare or Interesting Insects and Other Wildlife:

- Grayling Butterflies: On sunny summer days, look for the cryptic gray-brown wings and distinctive orange eyespots of the grayling butterfly as it basks on the warm shingle or flutters low over the sparse vegetation. This well-camouflaged species is perfectly adapted to the harsh conditions of the coastal environment, its mottled colouration and quick, erratic flight making it a challenge to spot and a thrill to observe.
- Rare Moths: As dusk falls, keep an eye out for the nocturnal fluttering of some of the UK's rarest and most specialized moth species, such as the white-line dart and Archer's dart, which find a precious foothold in the unique habitats of Orford Ness. These secretive and often overlooked insects are a testament to the incredible diversity and complexity of life that exists even in the most seemingly barren and inhospitable landscapes.
- Common Seals: Scan the waters of the Ore estuary and the North Sea for the sleek, dappled forms of common seals as they swim and dive in search of fish, their playful antics and curious nature a charming sight for visitors to the Ness. Though more elusive than their larger grey seal cousins, common seals are a vital part of the marine ecosystem here, their presence a sign of the healthy and productive waters that sustain the abundant birdlife and other wildlife of the reserve.

Rare or Interesting Plants:

- Yellow-horned Poppy: Marvel at the striking golden blooms and intricately-shaped seed pods of this hardy coastal specialist, its deep taproot and waxy, blue-green leaves perfectly adapted to the shifting, nutrient-poor shingle. The yellow-horned poppy is a key pioneer species on the Ness, its ability to colonize and stabilize the bare, exposed gravel a vital first step in the long, slow process of ecological succession that ultimately gives rise to the reserve's rich mosaic of habitats.
- Sea Kale: Discover the thick, fleshy leaves and delicate white flowers of this edible coastal plant, its robust, branching stems and deep, penetrating roots an effective anchor against the wind and waves. Sea kale is not only a valuable stabilizer of the shingle but also an important food source for a variety of insects and birds, its presence a keystone of the intricate web of life that defines this unique and fragile ecosystem.
- Sea Pea: Delight in the vibrant purplish-pink blooms and scrambling, tendriled growth of this attractive and resilient legume, its ability to fix atmospheric nitrogen and thrive in the harsh, salty conditions of the coastal fringe a remarkable adaptation that allows it to flourish where few other plants can survive. The sea pea's dense mats of foliage and extensive root system play a vital role in binding and enriching the meager soil of the shingle, paving the way for the establishment of more diverse and complex plant communities over time.

Facilities and Accessibility:

The Orford Ness National Nature Reserve walk is a unique and immersive experience that offers visitors a chance to explore one of the UK's most unusual and ecologically significant coastal landscapes. However, it is important to note that the remote and challenging nature of the site means that facilities are limited and accessibility can be an issue for some visitors.

At the starting point of your journey, the charming village of Orford provides a range of amenities to help you prepare for your visit to the Ness. The Orford Quay car park has ample space for vehicles, while the nearby public toilets and well-stocked village stores offer a convenient place to take care of any last-minute needs or stock up on supplies for your walk.

To reach Orford Ness itself, visitors must take a short ferry ride from the quay, which operates on a seasonal basis and is subject to weather conditions and tidal restrictions. The ferry is run by the National Trust, and tickets can be purchased on the day from the small kiosk at the quay. It is essential to check the ferry times and availability before planning your visit, as well as any potential closures or limitations due to the sensitive nature of the reserve.

Once on the Ness, visitors should be prepared for a rugged and challenging environment, with limited shelter and no facilities beyond the basic landing stage and visitor hut. The 5-mile walk takes place entirely on the loose, uneven shingle, which can be difficult and tiring to navigate, particularly in hot or windy weather. Sturdy walking shoes or boots with good ankle support and a robust tread are an absolute must, and visitors should bring plenty of water, snacks, and sun protection to ensure a safe and comfortable experience.

Due to the nature of the terrain and the lack of infrastructure, the Orford Ness walk is not currently suitable for wheelchair users or those with mobility impairments. The uneven, shifting surface of the shingle and the absence of dedicated paths or boardwalks can make progress slow and arduous, even for experienced walkers. Visitors with limited mobility may find it easier to explore the nearby Orford Quay and village, which offer a range of more accessible opportunities to enjoy the local history, culture, and scenic views.

For those who are able to undertake the walk, however, the rewards are truly unparalleled. The chance to immerse oneself in the wild, elemental beauty of the Ness, to discover its fascinating natural and human histories, and to connect with a landscape that has been shaped by the forces of nature and the hands of humans over countless generations, is an experience that will stay with you long after you have left the shingle behind.

By following the guidance and recommendations of the National Trust staff, respecting the fragile ecosystems and sensitive wildlife of the reserve, and coming prepared for the physical demands of the walk, visitors can safely and responsibly enjoy the unique wonders and timeless allure of the Orford Ness National Nature Reserve, a place like no other on the Suffolk coast or indeed in the whole of the UK.

Seasonal Highlights:

The Orford Ness National Nature Reserve walk offers a captivating and ever-changing experience throughout the year, with each season revealing new facets of beauty, wonder, and ecological significance in this extraordinary coastal wilderness:

- Spring: As the days lengthen and the first warmth of the sun caresses the shingle, the Ness comes alive with a burst of colour and activity. The bright yellow blooms of the horned poppy and the delicate white flowers of the sea kale paint the pebbles with a vibrant palette, while the arrival of the first migratory birds fills the air

with a symphony of song and wing-beats. Oystercatchers and ringed plovers stake out their territories on the open shingle, their lively courtship displays and nest-building efforts a heartening sign of nature's enduring cycles of renewal and regeneration.
- Summer: In the long, languid days of high summer, the Ness is at its most lively and inviting, the shimmering blues of the sea and sky a mesmerizing backdrop to the ceaseless movement of wildlife on the wing and on the shore. The terns and gulls wheel and dive over the glittering waters, their raucous calls and flashing wings a vivid expression of the season's energy and abundance, while the shingle echoes with the tiny patter of countless unseen feet, as rare moths and other nocturnal creatures emerge to find mates and lay their eggs among the sparse, sun-baked vegetation.
- Autumn: As the light begins to soften and the leaves on the distant trees take on the fiery hues of the turning season, the Ness assumes a new character, the fading greens and purples of the sea kale and sea pea giving way to the warm, burnished tones of the shingle and the driftwood that washes up along the shore. The southward migration of waders and wildfowl brings a new cast of avian visitors to the reserve, their evocative calls and restless energy a poignant reminder of the great journeys and transformations that mark the passage of the seasons.
- Winter: Though the winds may be sharp and the skies leaden, the Ness holds a stark, elemental beauty in the depths of winter, the monochromatic palette of the shingle and the churning grey of the sea and sky creating a haunting, almost otherworldly atmosphere. Flocks of overwintering ducks and geese shelter in the bays and saltmarshes, their lively chatter and sudden flights a welcome spark of life in the dormant landscape, while the hardiest of the resident birds, like the turnstones and sanderlings, pick their way among the frost-rimmed pebbles, their tenacity and resilience a humbling reminder of the enduring power of nature to adapt and survive in even the harshest of conditions.

Difficulty Level and Safety Considerations:

The Orford Ness National Nature Reserve walk is a moderate to challenging excursion that requires a good level of physical fitness, stamina, and surefootedness to complete safely and enjoyably. While the distance of 5 miles (8 km) may not seem overly daunting, the unique and demanding terrain of the shingle spit presents a range of potential hazards and difficulties that walkers should be keenly aware of before setting out.

The most significant challenge of the walk is the loose, uneven, and constantly shifting surface of the shingle, which makes up the entirety of the route. The rounded pebbles and coarse gravel can be extremely tiring and uncomfortable to walk on for extended periods, placing considerable strain on the ankles, knees, hips, and lower back. The risk of slips, trips, and falls is high, particularly in wet or windy conditions, or when fatigue begins to set in towards the end of the walk.

To mitigate these risks, it is essential to wear sturdy, well-fitting walking boots with good ankle support, a thick, cushioned sole, and a durable, high-grip tread. Walking poles can also be helpful for maintaining balance and stability on the unstable surface, as well as reducing the impact on joints and muscles over the course of the walk.

Another key consideration is the exposed, open nature of the Ness, which offers little to no shelter from the elements. The site is fully exposed to the strong winds, driving rain, and intense sunlight that can sweep in from the North Sea at any time of year, making for challenging and potentially dangerous conditions if walkers are not adequately prepared.

It is vital to check the weather forecast before setting out and to dress in layers that can be easily adjusted to accommodate changes in temperature, wind chill, or precipitation. A waterproof and windproof outer layer, as well as a hat, gloves, and sunscreen, are essential items for any visit to the Ness, regardless of the season.

The remote and isolated character of the reserve also means that help may be far away in case of an emergency or accident. It is crucial to carry a fully charged mobile phone, a map or GPS device, and a small first aid kit, as well as plenty of water and high-energy snacks to maintain hydration and stamina throughout the walk.

Walkers should also be mindful of the tides and the risk of becoming cut off or stranded by rising water levels, particularly at the far end of the spit where the shingle narrows and the sea can encroach rapidly. It is essential to check the tide times and to allow ample time to complete the walk and return to the ferry pick-up point before the water becomes dangerously high.

Lastly, visitors should be aware of the potential hazards posed by the abandoned military structures and debris that litter the Ness, relics of its former life as a secret testing facility. While the majority of these sites are clearly marked and fenced off, there may be hidden dangers such as unstable foundations, sharp edges, or contaminated soil that could cause injury or illness if disturbed or encountered.

Walkers should stay on the designated paths at all times, heed any warning signs or barriers, and refrain from entering or tampering with any of the derelict buildings or equipment. By treating the Ness with respect and caution, visitors can minimize the risks to themselves and others, and help to preserve this unique and fragile landscape for future generations to enjoy and learn from.

Despite these challenges and considerations, the Orford Ness National Nature Reserve walk remains a deeply rewarding and enriching experience for those who are prepared, informed, and motivated to undertake it. With proper planning, equipment, and mindset, walkers of moderate to advanced ability can safely immerse themselves in the wild, untamed beauty of the Ness, and gain a profound appreciation for the incredible forces of nature and history that have shaped this remarkable corner of the Suffolk coast.

Nearby Attractions:

After your bracing and thought-provoking walk through the wild landscapes of Orford Ness, why not take some time to explore the many other fascinating attractions and destinations that this charming corner of Suffolk has to offer? From ancient castles and picturesque villages to world-class nature reserves and cultural landmarks, there is something to delight and inspire every visitor to this enchanting region.

Just a short ferry ride from the Ness, the historic village of Orford awaits, its quaint streets and cozy pubs a welcome respite after a long day on the shingle. Be sure to visit the imposing bulk of Orford Castle, a magnificent 12th-century keep that offers breathtaking views over the village and the surrounding countryside from its lofty battlements. Explore the castle's atmospheric chambers and passageways, learn about its turbulent history and strategic importance, and imagine the lives of the knights, lords, and ladies who once called this incredible fortress home.

For a fascinating glimpse into the rich history and culture of the local area, head to the Orford Museum, a small but engaging collection of artifacts, photographs, and stories that paint a vivid picture of life on the Suffolk coast over the centuries. From prehistoric flint tools and Roman pottery to vintage fishing gear and wartime memorabilia, the exhibits offer a unique and personal perspective on the people and events that have shaped this remarkable community.

Nature lovers and birdwatchers will not want to miss the chance to visit the nearby RSPB Havergate Island nature reserve, a pristine wilderness of mudflats, saltmarshes, and lagoons that is accessible only by boat. This remote and unspoiled island is a haven for a wide variety of rare and migratory birds, including avocets, spoonbills, and marsh harriers, as well as countless species of wading birds and wildfowl. Take a guided tour with a knowledgeable RSPB warden, marvel at the stunning views and the incredible diversity of birdlife, and feel the stresses and distractions of modern life melt away in the timeless rhythms of the tides and the turning seasons.

For a truly unforgettable cultural experience, make the short journey to the world-renowned Snape Maltings concert hall and arts complex, a stunning collection of converted Victorian industrial buildings that now host an eclectic program of music, theatre, and visual arts throughout the year. Attend a performance by the world-class Aldeburgh Festival Orchestra, explore the galleries and exhibitions showcasing the work of local and international artists, or simply enjoy a relaxing walk through the beautiful grounds and gardens, soaking up the creative energy and natural beauty of this incredible site.

Of course, no visit to the Suffolk coast would be complete without sampling some of the region's famous culinary delights, from the succulent Orford-caught lobsters and oysters to the creamy, tangy cheeses and hearty, locally-reared meats that grace the menus of the area's many excellent pubs, restaurants, and cafes. Be sure to wash it all down with a pint or two of the county's renowned craft beers and ciders, or a refreshing glass of crisp, aromatic Aspall's Suffolk Cyder, a true taste of the region's rich agricultural heritage.

As you explore the many delights and hidden gems of this enchanting corner of East Anglia, take a moment to reflect on the deep and enduring connection between the land, the sea, and the people who have shaped and been shaped by this extraordinary landscape over the centuries. From the ancient Britons who first settled along the coast to the visionary conservationists and local communities who work tirelessly to protect and preserve its fragile ecosystems today, each generation has left its mark on the fabric of the Suffolk countryside, weaving a rich tapestry of history, culture, and natural wonder that is truly unparalleled.

So whether you're a history buff, a nature lover, a foodie, or simply someone in search of a little adventure and inspiration in a breathtakingly beautiful setting, Orford and its surroundings offer a wealth of opportunities for discovery, relaxation, and personal enrichment. Let the wild beauty of the Ness and the timeless charm of the Suffolk coast be your guide and your muse, as you embark on a journey of exploration that will nourish your mind, body, and soul, and leave you with memories to cherish for a lifetime.

Shingle Street Walk - 3 miles (4.8 km), Circular - Estimated Time: 1.5-2 hours

Starting Point Coordinates, Postcode: The starting point for the Shingle Street Walk is at the small car park at Shingle Street (Grid Reference: TM 364 470, Postcode: IP12 3BE).
Nearest Car Park, Postcode: The nearest car park is the Shingle Street car park (Postcode: IP12 3BE).

Embark on a captivating 3-mile (4.8 km) circular walk along the unique and dramatic coastline of Shingle Street, where the wild beauty of the North Sea meets the haunting remoteness of the Suffolk shore. This easy route, with its mostly flat terrain and pebble-strewn paths, offers a perfect escape from the hustle and bustle of modern life, inviting you to immerse yourself in the timeless rhythms of the tide and the secrets of a landscape shaped by the forces of nature and the echoes of human history.

As you set out from the small car park at Shingle Street, feel the crunch of the shingle beneath your feet and the bracing tang of the sea air on your face, the vast expanse of the ocean stretching out before you in a shimmering blue haze. Follow the shoreline eastward, picking your way along the shifting and uneven surface of the beach, each step a reminder of the incredible power and resilience of the natural world.

Take a moment to pause and marvel at the unique coastal flora that clings tenaciously to the storm-battered shingle, from the robust, waxy-leaved sea kale to the delicate, purple-flowered sea pea. These hardy pioneers are a testament to the incredible adaptability of life in the face of the harshest conditions, their deep roots and salt-resistant leaves a perfect match for the challenges of the coastal environment.

As you continue along the beach, keep an eye out for the rich diversity of birdlife that calls this special place home, from the graceful terns that wheel and dive over the breaking waves to the elusive skylarks that soar and sing above the windswept expanse of the shore. The marshes that fringe the coastline are a haven for wading birds like avocets and redshanks, their distinctive calls and striking plumage a welcome splash of colour and sound in the muted tones of the landscape.

Upon reaching the imposing bulk of the Martello Tower, a relic of the Napoleonic Wars that stands guard over the lonely coastline, turn inland and make your way towards the serene tranquility of the marshes. Here, amidst the waving grasses and the glittering channels of water, you'll find a world of quiet beauty and hidden wonders, where the gentle rhythms of the seasons and the ebb and flow of the tides create an ever-changing tapestry of light, colour, and texture.

As you follow the path alongside the marshes, keep your senses attuned to the sights, sounds, and sensations of this unique and precious habitat, from the darting iridescence of dragonflies to the soft rustle of unseen creatures in the undergrowth. Each step brings new discoveries and fresh perspectives, inviting you to slow down, to breathe deeply, and to reconnect with the simple joys and profound truths of the natural world.

Looping back towards the car park at Shingle Street, take a moment to reflect on the extraordinary journey you've just completed, the wild beauty and haunting remoteness of the coast forever etched in your mind and heart. This walk is a celebration of the enduring power of place, a reminder of the countless stories and secrets that lie hidden in the contours of the land and the whispers of the wind and waves.

Whether you're a lover of nature, a history buff, or simply someone in search of a moment of solace and stillness in a world that often feels too fast and too loud, the Shingle Street Walk promises to refresh and restore, to challenge and inspire. So let the timeless magic of the Suffolk shore be your guide and your muse, as you embark on a journey of discovery that will leave you with memories to cherish and a deeper appreciation for the incredible diversity and resilience of the natural world.

Detailed Route Directions:

1. From the car park at Shingle Street, head eastward along the shingle beach, bearing 90°. Take your time to navigate the uneven surface of the beach and enjoy the coastal views. (Estimated time: 25-30 minutes)

2. Continue walking along the beach until you reach the Martello Tower, a historic fortification from the Napoleonic Wars era. (Estimated time: 20-25 minutes)

3. At the Martello Tower, turn inland, bearing approximately 360°, and make your way towards the marshes that lie behind the coastline. (Estimated time: 10-15 minutes)

4. Follow the path alongside the marshes, bearing approximately 270°, keeping an eye out for the diverse birdlife and other wildlife that thrives in this unique habitat. (Estimated time: 30-35 minutes)

5. The path will gradually curve back towards the southwest, bearing 225°, as it begins to loop back in the direction of the starting point. (Estimated time: 20-25 minutes)

6. Continue following the path as it leads you back to the car park at Shingle Street, bearing 180°, completing your circular walk. (Estimated time: 15-20 minutes)

Best Features and Views:
- Marvel at the wild, untamed beauty of Shingle Street's distinctive pebble beach, where the ceaseless action of the waves and currents has shaped a landscape of stark, elemental power and austere, almost otherworldly charm.
- Discover the unique coastal flora that thrives in the challenging conditions of the shingle, from the tough, adaptable sea kale to the delicate, colourful blooms of the sea pea, each species a testament to the incredible resilience and creativity of nature.
- Delight in the sweeping views of the North Sea and the haunting silhouette of the Martello Tower, a poignant reminder of the coast's rich and turbulent history and its enduring role as a guardian and witness to the shifting tides of human events.

Historical Features of Interest:

As you make your way along the pebble-strewn shore of Shingle Street, the imposing bulk of the Martello Tower rises up before you, a silent sentinel that has stood watch over this lonely stretch of coast for more than two centuries. Built in the early 1800s as part of a chain of defensive fortifications designed to protect England from the threat of French invasion during the Napoleonic Wars, this sturdy, round tower is a powerful symbol of the ingenuity, determination, and resilience of the British people in the face of danger and adversity.

Take a moment to imagine the lives of the soldiers who once manned this remote outpost, their days and nights filled with the ceaseless vigil of scanning the horizon for signs of enemy ships, the roar of the waves and the howl of the wind their constant companions. The tower's thick, stone walls and single, heavily-guarded entrance are a testament to the seriousness of their mission and the constant state of readiness and alertness that was required to maintain the security of the realm.

As you continue your walk, keep an eye out for other traces of Shingle Street's wartime past, from the crumbling remains of World War II-era pillboxes and coastal defenses to the faint outlines of long-abandoned military installations and training grounds. These haunting relics are a reminder of the pivotal role that this seemingly peaceful and isolated stretch of coastline has played in the defense of the nation and the shaping of its destiny.

The history of Shingle Street is a microcosm of the larger story of Britain's relationship with the sea, a tale of courage, ingenuity, and adaptability in the face of ever-changing threats and challenges. From the early days of the Napoleonic Wars to the darkest hours of World War II, when the beach was heavily fortified and mined against the threat of German invasion, this small corner of Suffolk has been on the front lines of the struggle to preserve the island's freedom and way of life.

Today, as you stand atop the windswept shingle and gaze out over the vast expanse of the North Sea, the echoes of this long and storied history are all around you, whispering in the rush of the waves and the cry of the gulls. The Martello Tower and the other relics of Shingle Street's military past stand as a poignant reminder of the sacrifices and triumphs of generations past, and an invitation to reflect on the enduring importance of vigilance, preparedness, and unity in the face of an uncertain future.

By walking in the footsteps of the soldiers and defenders who once patrolled this remote and beautiful coastline, you will gain a deeper appreciation for the rich tapestry of history that underlies the wild and untamed landscape of Shingle Street, and a renewed sense of connection to the timeless rhythms and eternal truths that shape our lives and our world.

Birds to Spot:

- Avocets: These striking black and white waders, with their distinctive upturned bills, are a true emblem of conservation success and a joy to observe as they forage in the shallows of the marshes. Shingle Street's coastal wetlands are a vital habitat for these elegant birds, providing a safe haven for breeding and a rich source of food throughout the year.
- Little Terns: Listen for the high-pitched, excited calls of these diminutive seabirds as they hover over the waves, their quick, darting movements and sharp, pointed wings a marvel of aerial agility. The shingle beaches of the Suffolk coast are a crucial nesting site for these vulnerable and charismatic birds, their presence a reminder of the delicate balance and beauty of the coastal ecosystem.
- Skylarks: Keep your ears attuned to the melodious, cascading song of these iconic birds as they soar high above the coastal grasslands and marshes, their exuberant trills and whistles a joyous celebration of the freedom and exhilaration of flight. The open, windswept landscapes of Shingle Street are a perfect habitat for these beloved and increasingly rare songbirds, their presence a vital part of the rich soundscape and living heritage of the Suffolk shore.

Rare or Interesting Insects and Other Wildlife:

- Dune Tiger Beetles: As you walk along the shingle beach, keep your eyes peeled for the iridescent green and bronze flashes of these speedy, predatory beetles as they dart and scurry across the sun-warmed pebbles. Perfectly adapted to life in the harsh and dynamic environment of the coastal dunes, these impressive insects are a testament to the incredible diversity and resilience of the invertebrate life that thrives along the Suffolk coast.
- Common Lizards: On warm, sunny days, watch for the quick, darting movements of these small, agile reptiles as they bask on the driftwood and beach debris that litters the high tide line. With their mottled brown and green scales providing excellent camouflage against the shingle and their ability to shed their tails to escape predators, common lizards are a fascinating and often overlooked part of the coastal wildlife community.
- Harbour Porpoises: As you gaze out over the rolling waves of the North Sea, keep a sharp eye out for the small, dark, triangular dorsal fins and sleek, streamlined bodies of these elusive marine mammals as they

surface briefly to breathe. The waters off Shingle Street are an important feeding and migration area for harbour porpoises, their presence a vital indicator of the health and productivity of the coastal ecosystem and a thrilling sight for any wildlife enthusiast.

Rare or Interesting Plants:

- Yellow Horned-poppy: Delight in the large, golden-yellow flowers and intricately-shaped seed pods of this striking and robust annual, its deep tap roots and thick, blue-green leaves expertly adapted to life among the shifting shingle. One of the few plants able to thrive in the challenging conditions of the exposed beach, the yellow horned-poppy is a true pioneer species and a vital stabilizer of the coastal landscape.
- Sea Pea: Discover the vibrant purple blooms and attractive, ferny foliage of this hardy, scrambling perennial, its long, twining stems and nitrogen-fixing roots a perfect match for the low-nutrient soils and constant movement of the shingle. A key member of the diverse and specialized plant community that clings to the margins of the beach, the sea pea's presence is a testament to the remarkable adaptability and tenacity of life in the face of adversity.
- Sea Kale: Marvel at the thick, waxy leaves and clusters of small, white flowers of this edible coastal plant, its robust, branching stems and deep, fleshy roots anchoring it firmly against the relentless battering of the wind and waves. Once a popular wild vegetable among coastal communities, sea kale's ability to thrive in the salt-sprayed, nutrient-poor shingle is a reminder of the intimate relationship between people and the plants of the seashore, and the enduring importance of the natural world in our lives and livelihoods.

Facilities and Accessibility:

The Shingle Street Walk is a remote and unspoiled coastal route that offers a unique and immersive experience for those seeking to connect with the wild beauty and rich history of the Suffolk shore. However, it is important to note that the isolated character of the site and the challenging nature of the terrain mean that facilities are minimal and accessibility is limited.

At the starting point of your walk, the small car park at Shingle Street provides a convenient place to leave your vehicle and begin your exploration of the coast. However, beyond this basic amenity, there are no other facilities or services available at the site itself, so it is essential to come prepared with all the necessary supplies and equipment for a safe and comfortable walk.

Visitors should bring plenty of water, snacks, and sun protection, as well as sturdy, comfortable footwear suitable for walking on the loose, uneven surface of the shingle beach. Walking poles may also be helpful for maintaining balance and stability on the shifting pebbles, particularly for those with mobility issues or concerns about falls.

It's also important to note that there are no public toilets, benches, or shelters along the route, so walkers should be prepared to take regular breaks and find their own comfortable spots to rest and recharge. The exposed and open nature of the coastline means that there is little shade or protection from the elements, so it's crucial to dress in layers and be prepared for changeable weather conditions, including strong winds, rain, and intense sunlight.

While the relatively flat terrain of the Shingle Street Walk makes it a potentially accessible route for a wide range of abilities, the uneven and unstable surface of the shingle beach can pose significant challenges for those with mobility difficulties or physical limitations. The deep, loose pebbles can be tiring and uncomfortable to walk on for extended periods, and the lack of dedicated paths or boardwalks means that progress can be slow and arduous.

As a result, the walk may not be suitable for wheelchair users, those with mobility aids, or families with young children in pushchairs. However, for those who are able to navigate the challenging terrain, the rewards of immersing oneself in the wild, elemental beauty of the Suffolk coast and discovering its fascinating natural and human histories are truly unparalleled.

By following the guidance and recommendations of local authorities and conservation organizations, respecting the fragile ecosystems and sensitive wildlife of the site, and coming prepared for the physical demands and limitations of the walk, visitors can safely and responsibly enjoy the unique character and timeless allure of Shingle Street, a place of solitude, inspiration, and enduring wonder on the edge of the sea.

Seasonal Highlights:

The Shingle Street Walk offers a dynamic and ever-changing experience throughout the year, with each season revealing new aspects of beauty, drama, and ecological significance along this remarkable stretch of the Suffolk coastline:

- Spring: As the days lengthen and the first warm breezes sweep in from the sea, the shingle comes alive with a burst of colour and activity. The yellow horned-poppies and sea peas emerge from their winter dormancy, their vibrant blooms and fresh green shoots a welcome sign of the returning vitality of the coastal landscape. The air is filled with the joyous songs of skylarks and the piping calls of ringed plovers, while the newly arrived terns and waders begin their courtship rituals and nest-building among the pebbles and driftwood of the shore.
- Summer: In the long, languid days of high summer, Shingle Street is at its most lively and inviting, the sparkling blue of the North Sea and the endless azure of the sky creating an irresistible backdrop for adventure and exploration. The shingle is a hive of activity, with butterflies and bees flitting among the sea kale flowers and lizards basking on the sun-warmed stones. The marshes are alive with the darting flights of dragonflies and the gentle rustling of the sea breeze through the tall grasses, while the beach is a playground for families and nature lovers alike, the perfect spot for a picnic, a paddle, or a moment of quiet reflection.
- Autumn: As the days grow shorter and the light takes on a softer, golden hue, Shingle Street assumes a new character, the fading greens and purples of the coastal flora giving way to the muted browns and grays of the pebbles and the weathered wood. The beach is a treasure trove of shells, seaweed, and other natural curiosities washed up by the autumnal storms, while the marshes are alive with the gatherings of migratory birds, their calls and wing-beats a poignant reminder of the turning of the seasons and the constant ebb and flow of life on the coast.
- Winter: Though the winds may be fierce and the skies leaden, Shingle Street retains a haunting beauty in the depths of winter. The stark, monochromatic landscape is a study in contrasts, the inky black of the sea and the pale gray of the shingle creating a mesmerizing visual rhythm. The beach is a wild and elemental place, the crashing waves and wind-blown spray a reminder of the raw power and majesty of nature. Wrap up warm and brave the elements, and you'll be rewarded with a profound sense of solitude and a deep connection to the timeless cycles of the seasons.

Difficulty Level and Safety Considerations:

The Shingle Street Walk is a moderate to challenging route that requires a good level of physical fitness, stamina, and surefootedness. The main difficulty lies in the nature of the terrain, with the deep, loose shingle of the beach making for a strenuous and slow-going walk. Each step sinks into the pebbles, requiring a significant effort to maintain balance and forward momentum, which can quickly lead to fatigue and strain on the legs and ankles.

Moreover, the uneven and shifting surface of the beach can be treacherous, particularly in wet or windy conditions. Walkers should take extreme care when navigating slippery or unstable sections of the route, and be prepared for sudden changes in the texture and gradient of the shingle. Sturdy, supportive footwear with good ankle support and grip is essential, and walking poles can provide additional stability and balance.

Another key safety consideration is the tidal nature of the coastline. Walkers should always check the tide times before setting out, and plan their route accordingly to avoid being cut off by the rising water. It's crucial to stay aware of your surroundings and to keep a safe distance from the water's edge, particularly during storm surges or periods of high tide.

The exposed and remote character of the Shingle Street Walk also means that walkers should be prepared for changeable and potentially extreme weather conditions. Strong winds, driving rain, and intense sunlight are all common features of the coastal environment, and can pose significant risks to those who are unprepared. It's essential to dress in layers, with waterproof and windproof outer garments, and to carry plenty of water, snacks, and sun protection.

In case of emergencies, it's recommended to carry a fully charged mobile phone and a small first aid kit, and to let someone know your planned route and expected return time before setting out. It's also a good idea to familiarize yourself with the location of the nearest emergency services and coastal rescue points, and to carry a map and compass or GPS device to aid navigation.

By taking these precautions and being aware of the potential hazards and challenges of the route, walkers can safely and responsibly enjoy the wild beauty and haunting remoteness of the Shingle Street Walk. However, it's important to remember that this is a demanding and potentially dangerous undertaking, and should only be attempted by those with the necessary physical fitness, experience, and equipment to handle the conditions.

Nearby Attractions:

While the Shingle Street Walk is undoubtedly the main attraction for those seeking to explore the wild and unspoiled beauty of the Suffolk coast, there are also plenty of other fascinating sites and activities to discover in the surrounding area. Whether you're interested in history, wildlife, or simply soaking up the timeless charm of the English countryside, you'll find no shortage of options to keep you entertained and inspired.

Just a short distance inland from Shingle Street lies the picturesque village of Hollesley, home to the historic All Saints Church and the fascinating Suffolk Punch Trust. The church, with its distinctive round tower and ancient wall paintings, is a beautiful example of medieval architecture and a testament to the deep spiritual roots of the local community. The Suffolk Punch Trust, meanwhile, is dedicated to preserving the legacy of the Suffolk Punch horse, a magnificent breed that has played a vital role in the county's agricultural heritage for centuries.

For those interested in military history, a visit to the nearby Bawdsey Radar Museum is an absolute must. Housed in the original RAF Bawdsey buildings, the museum tells the fascinating story of the development of radar technology during World War II, and the crucial role that this small corner of Suffolk played in the defense of the nation. With interactive exhibits, original artifacts, and a wealth of fascinating information, it's a compelling destination for anyone interested in the science and strategy behind one of the most pivotal conflicts in modern history.

Nature lovers and birdwatchers will find plenty to delight them at the RSPB Boyton Marshes reserve, a vast expanse of wetland and grassland habitat that is home to an incredible diversity of wildlife. From the elusive bitterns and marsh harriers that haunt the reedbeds to the colourful dragonflies and damselflies that flit among the wildflowers, this is a place of endless fascination and beauty, where the rhythms of the seasons and the cycles of life are played out in all their intricate glory.

For a taste of the region's rich culinary heritage, be sure to visit the historic Sutton Hoo site, where you can explore the iconic burial mounds of the Anglo-Saxon kings and discover the fascinating story of the ship burial and its priceless treasures. Afterwards, stop by the National Trust cafe for a delicious cream tea or a hearty meal made with locally-sourced produce, and soak up the atmosphere of this ancient and evocative landscape.

As you explore the many attractions and hidden gems of the Suffolk coast and countryside, take a moment to reflect on the deep sense of history and connection that underlies every aspect of this unique and special place. From the wild beauty of the shingle beaches to the quiet charm of the inland villages, every sight and sound is a reminder of the countless generations who have lived, worked, and dreamed here, leaving their mark on the land and its people in countless ways.

Whether you're a keen walker, a history buff, a wildlife enthusiast, or simply someone in search of a little peace and quiet in a busy world, the Shingle Street Walk and its surroundings offer a wealth of opportunities for discovery, inspiration, and reflection. So why not linger awhile and immerse yourself in the timeless magic of the Suffolk shore, and let the beauty and wonder of this extraordinary place work their spell on your mind, body, and soul?

In the words of the poet Seamus Heaney, "The voice of the sea speaks to the soul, and the touch of the sea is sensuous, enveloping the body in its soft, close embrace." Let the voice of the Shingle Street shore be your guide and your muse, as you embark on a journey of discovery that will leave you refreshed, invigorated, and forever changed by the wild and elemental beauty of this unforgettable place.

As we conclude our exploration of the Shingle Street Walk and its many delights, let us remember the words of another great nature writer, Henry David Thoreau, who once wrote: "We need the tonic of wildness...At the same time that we are earnest to explore and learn all things, we require that all things be mysterious and unexplorable, that land and sea be indefinitely wild, unsurveyed and unfathomed by us because unfathomable. We can never have enough of nature."

May your own adventures on the wild and unfathomable shores of Shingle Street be a source of endless wonder, mystery, and inspiration, and may they leave you with a renewed sense of reverence and respect for the incredible natural world that surrounds and sustains us all. Happy wanderings!

River Alde Walk - 10 miles (16 km), Circular - Estimated Time: 5-6 hours

Starting Point Coordinates, Postcode: The starting point for the River Alde Walk is at Snape Maltings (Grid Reference: TM 394 577, Postcode: IP17 1SR).
Nearest Car Park, Postcode: Parking is available at Snape Maltings (Postcode: IP17 1SR).

Embark on a captivating 10-mile (16 km) circular walk along the enchanting River Alde, where the ever-changing tapestry of the Suffolk landscape unfolds before you in a mesmerizing display of natural beauty and rural charm. This moderately challenging route takes you on a journey through a diverse array of habitats, from the serene tranquility of riverside meadows and marshes to the timeless allure of picturesque villages, inviting you to immerse yourself in the rich history and vibrant wildlife of this extraordinary corner of East Anglia.

As you set out from the historic Snape Maltings, a former Victorian maltings complex now transformed into a world-class cultural and artistic hub, feel the anticipation build as you follow the riverbank footpath northward, the gentle murmur of the flowing water and the soft rustling of the reeds providing a soothing soundtrack to your adventure. With each step, the cares of the modern world seem to melt away, replaced by a profound sense of connection with the natural world and a growing appreciation for the timeless rhythms of the Suffolk countryside.

The path winds alongside the River Alde, revealing stunning vistas of the marshes and meadows that flank its banks, their lush greens and earthy hues a testament to the nourishing power of the water and the enduring fertility of the land. As you walk, keep your eyes peeled for the diverse array of birdlife that calls this wetland paradise home, from the majestic marsh harriers soaring overhead to the elegant avocets and other wading birds picking their way through the shallows, each species a vital thread in the intricate web of life that thrives in this unique and precious habitat.

Continuing along the riverside trail, you'll soon reach the charming hamlet of Iken Cliff, where the river makes a dramatic sweep toward the sea, its waters sparkling in the sunlight as they flow past the ancient church of St Botolph. Take a moment to explore this historic gem, with its distinctive round tower and evocative medieval interior, a testament to the deep spiritual roots and rich cultural heritage of the local community.

From Iken Church, the route leads you westward across a series of footbridges, traversing the expansive marshland that stretches out before you like an emerald carpet, alive with the sights and sounds of the natural world. Feel the springy turf beneath your feet and breathe in the fresh, invigorating air, as you immerse yourself in the timeless beauty and tranquility of this extraordinary landscape, a world apart from the hustle and bustle of modern life.

As you press on, the path eventually brings you to the picturesque village of Sudbourne, where the traditional cottages and winding lanes seem frozen in time, a reminder of the simple pleasures and unhurried pace of rural life. Take a moment to rest and refresh yourself, perhaps enjoying a well-earned pint in the welcoming atmosphere of a local pub, before continuing your journey through the gently rolling fields and meadows that surround the village.

The final stretch of the walk sees you rejoin the River Alde, following the meandering course of the waterway as it flows back towards Snape Maltings, the starting point of your adventure. As you walk, reflect on the incredible diversity and resilience of the natural world, and the countless generations of people who have shaped and been shaped by this remarkable landscape, leaving their mark in the fields and villages, the churches and pubs, and the very earth and water itself.

As you complete the circular route and arrive back at Snape Maltings, take a moment to celebrate your achievement and to savour the memories of your incredible journey along the River Alde. This walk is a testament to the enduring beauty and vitality of the Suffolk countryside, a reminder of the power of nature to inspire, uplift, and transform us, and an invitation to discover the hidden wonders and timeless magic that lie waiting to be explored in every corner of this extraordinary land.

Detailed Route Directions:

1. From Snape Maltings, head north along the riverbank footpath, following the course of the River Alde. Continue on this path for approximately 2 miles (3.2 km). (Estimated time: 45-60 minutes)

2. After passing through Iken Cliff, you will reach Iken Church. Spend a few moments exploring this historic site before continuing. (Estimated time: 15-20 minutes)

3. From Iken Church, follow the footpath westward, crossing a series of footbridges over the marshland. Maintain this direction for about 1.5 miles (2.4 km). (Estimated time: 45-60 minutes)

4. The footpath will then lead you to the village of Sudbourne. Walk through the village, taking time to enjoy its charming atmosphere. (Estimated time: 15-20 minutes)

5. On the western side of Sudbourne, join a footpath that leads southwest across fields. Continue on this path for approximately 2 miles (3.2 km). (Estimated time: 45-60 minutes)

6. The footpath will eventually bring you back to the River Alde. Turn right to rejoin the riverside path, heading southeast. (Estimated time: 5-10 minutes)

7. Follow the riverside path for about 4 miles (6.4 km), enjoying the beautiful scenery along the River Alde as you make your way back to Snape Maltings. (Estimated time: 1.5-2 hours)

8. Upon arriving back at Snape Maltings, your circular walk is complete. (Total estimated time: 5-6 hours)

Best Features and Views:

- Immerse yourself in the stunning views of the River Alde throughout the walk, as it winds through a diverse landscape of marshes, meadows, and wooded areas, each offering a unique and captivating perspective on the natural beauty of the Suffolk countryside.
- Delight in the picturesque villages of Snape, Iken, and Sudbourne, where the charming cottages, historic churches, and welcoming pubs provide a glimpse into the rich cultural heritage and timeless character of rural life in East Anglia.
- Marvel at the expansive marshland vistas as you traverse the footbridges near Iken Church, the lush greens and earthy hues of the wetlands stretching out before you in a mesmerizing display of the raw beauty and vitality of the natural world.

Historical Features of Interest:

One of the most notable historical features along the River Alde Walk is the enchanting Iken Church, a beautiful 12th-century sanctuary nestled in the heart of the Suffolk countryside. As you approach this ancient site, take a moment to appreciate its distinctive round tower, a characteristic feature of many East Anglian churches and a testament to the skill and ingenuity of medieval stonemasons.

Step inside the church to discover a treasure trove of history and spirituality, from the evocative medieval wall paintings that adorn the interior to the exquisite stained glass windows that cast a warm, colourful glow across the nave. Imagine the countless generations of worshippers who have gathered here over the centuries, seeking solace, inspiration, and connection with the divine in this sacred space.

As you explore the churchyard, with its weathered gravestones and gnarled yew trees, consider the deep roots that Iken Church has planted in the local community, serving as a focal point for celebration, mourning, and remembrance for countless individuals and families throughout the ages. The church stands as a powerful symbol of continuity and resilience, a reminder of the enduring human spirit and the unbreakable bonds that tie us to the land and to each other.

The history of Iken Church is closely intertwined with that of the surrounding landscape, its presence a constant amidst the ebb and flow of the seasons and the generations. From its early origins as a simple wooden structure to its later evolution into the graceful stone edifice we see today, the church has borne witness to the joys and sorrows, the triumphs and struggles, of the people who have called this corner of Suffolk home.

As you continue on your walk, carry with you the sense of awe and reverence that Iken Church inspires, and let its timeless beauty and rich history serve as a reminder of the incredible depth and complexity of the human story that is woven into every inch of this extraordinary landscape. The church stands as an enduring testament to the power of faith, community, and tradition, and an invitation to reflect on our own place in the grand tapestry of history and the natural world.

Rare or Uncommon Birds:

As you make your way along the River Alde, keep your eyes to the skies and your binoculars at the ready, for this wetland haven is home to an incredible array of birdlife, including some of the rarest and most spectacular species in the UK. One of the most iconic and sought-after sightings is the marsh harrier, a magnificent raptor with a distinctive V-shaped silhouette and a haunting, otherworldly cry.

Scan the reedbeds and marshes for a glimpse of these powerful predators as they soar and glide effortlessly over the landscape, their chestnut-brown plumage and pale heads glinting in the sunlight. With a wingspan of up to 130cm, marsh harriers are a breathtaking sight, their presence a testament to the incredible biodiversity and ecological richness of the Suffolk wetlands.

Another avian gem to watch for along the River Alde is the elegant avocet, a striking black-and-white wader with a slender, upturned bill perfectly adapted for probing the mud and shallow water for small invertebrates. These graceful birds are a true conservation success story, having been brought back from the brink of extinction in the UK thanks to dedicated protection efforts and habitat management.

As you walk along the riverbank and across the marshes, listen for the distinctive piping calls of the avocets and marvel at their delicate, ballet-like movements as they pick their way through the shallows, their reflections dancing on the surface of the water. The presence of these beautiful birds is a heartening reminder of the resilience and adaptability of nature, and the power of human commitment to preserving and nurturing the incredible diversity of life on our planet.

Other wading birds to keep an eye out for include the black-tailed godwit, with its long, straight bill and elegant, rusty-red breeding plumage, and the common redshank, a lively and vocal bird with bright orange-red legs and a bold, staccato call. Each species brings its own unique character and charm to the wetland symphony, adding to the rich tapestry of sights and sounds that make the River Alde such a special and unforgettable place.

As you walk, take a moment to reflect on the incredible journeys that many of these birds undertake, traveling thousands of miles each year to reach the Suffolk marshes and meadows that provide such vital breeding and feeding grounds. The River Alde is a crucial link in a global network of wetland habitats, a reminder of the interconnectedness of all life on Earth and the responsibility we share to protect and cherish these precious ecosystems for generations to come.

Rare or Interesting Insects and Other Wildlife:

As you explore the lush green world of the River Alde marshes and meadows, keep your senses attuned to the myriad forms of life that thrive in this wetland paradise, from the smallest insects to the most elusive mammals. One of the most captivating and charismatic creatures to look out for is the otter, a sleek and playful aquatic mammal with a sinuous, muscular body and a curious, intelligent gaze.

Though notoriously shy and difficult to spot, otters are a keystone species in the River Alde ecosystem, their presence a sign of a healthy and well-balanced habitat. Early mornings and evenings are the best times to catch a glimpse of these enigmatic animals, as they swim and dive in the clear waters of the river, their streamlined forms cutting through the current with effortless grace.

Another wetland specialist to watch for is the water vole, a charming and industrious rodent with a plump, furry body and a distinctive blunt muzzle. Once a common sight along the riverbanks and ditches of the Suffolk countryside, water voles have suffered steep declines in recent decades due to habitat loss and predation by invasive American mink. However, thanks to concerted conservation efforts, these delightful creatures are making a comeback in the River Alde and other wetland sites across the region.

As you walk along the riverbank, look for the tell-tale signs of water vole activity, from the neat piles of chopped vegetation and the distinctive "lawns" of grazed grass to the small, round burrows excavated in the soft earth of the bank. The presence of these industrious little mammals is a heartening reminder of the resilience and adaptability of nature, and the importance of preserving and restoring the wetland habitats that sustain such a rich diversity of life.

For those with a keen eye and a passion for the smaller wonders of the natural world, the River Alde Walk offers a dazzling array of insect life, from the iridescent flashes of dragonflies and damselflies darting over the water to the delicate, fluttering dance of butterflies among the wildflowers of the meadows. One of the most spectacular and sought-after sightings is the Norfolk hawker, a large and powerful dragonfly with an emerald-green thorax and striking yellow-and-black patterned abdomen.

As you explore the margins of the marshes and the edges of the reedbeds, watch for the rapid, agile flight of these incredible insects as they hunt for prey and defend their territories, their presence a vivid reminder of the complex and intricate web of life that thrives in the wetland ecosystem. Other dragonflies and damselflies to look out for include the scarce chaser, with its powder-blue pruinescence and black wing tips, and the banded demoiselle, a delicate beauty with dark, iridescent wings and a metallic blue-green body.

Each encounter with the fascinating insects and other wildlife of the River Alde is an opportunity to marvel at the incredible adaptations and survival strategies that have evolved over countless generations, and to deepen our appreciation for the sheer diversity and wonder of the natural world. As you walk, take a moment to reflect on the vital role that these often-overlooked creatures play in maintaining the balance and health of the wetland ecosystem, and the responsibility we all share in ensuring their continued thriving for generations to come.

Rare or Interesting Plants:

The marshes and riverside meadows along the River Alde Walk provide a lush and varied tapestry of plant life, each species a unique thread in the complex and colourful weave of the wetland ecosystem. As you make your way along the footpaths and bridleways, take a moment to appreciate the delicate beauty and incredible adaptations of the flora that thrive in this dynamic and ever-changing landscape.

One of the most striking and emblematic plants of the Suffolk wetlands is the common reed, a tall and graceful grass that forms vast, swaying stands along the margins of the river and the edges of the marshes. Growing up to 3 meters in height, these elegant plants provide vital habitat and shelter for a wide range of birds, insects, and other wildlife, their dense, feathery plumes a shimmering, russet-gold backdrop to the vibrant life of the wetlands.

As you walk through the reedbeds, listen for the distinctive rustling and whispering of the stems in the breeze, a soothing and evocative sound that has inspired poets and artists for centuries. Take a closer look at the intricate structure of the reed flowers, with their delicate, feathery spikelets arranged in a graceful, drooping panicle, and marvel at the incredible resilience and adaptability of these plants, which thrive in the challenging and dynamic conditions of the wetland environment.

Another wetland specialist to look out for along the River Alde is the yellow flag iris, a strikingly beautiful and robust plant with tall, sword-like leaves and brilliant, golden-yellow flowers that bloom in late spring and early summer. Growing in dense clumps along the water's edge and in the damp, peaty soils of the marshes, these stunning plants add a vibrant splash of colour to the lush green palette of the wetlands, their presence a joyous celebration of the vitality and exuberance of life in this unique and precious habitat.

As you explore the riverside meadows and the fringes of the marshes, keep your eyes peeled for the delicate, nodding blooms of the ragged robin, a charming and whimsical pink flower with deeply-cut petals that dance and flutter in the soft summer breeze. This enchanting plant is a true wetland specialist, thriving in the damp, nutrient-rich soils of the meadows and providing a vital source of nectar for bees, butterflies, and other pollinators.

Other interesting and diverse plants to watch for along the River Alde Walk include the stately and aromatic meadowsweet, with its frothy clusters of creamy-white flowers; the delicate and graceful marsh pea, with its pale pink blooms and feathery foliage; and the robust and resilient marsh marigold, with its glossy, heart-shaped leaves and bright yellow, buttercup-like flowers.

Each encounter with the rare and fascinating plants of the River Alde wetlands is an opportunity to deepen our understanding and appreciation of the incredible complexity and beauty of the natural world, and to marvel at the intricate web of relationships and adaptations that sustain the diversity and resilience of these precious ecosystems. As you walk, take a moment to reflect on the vital role that these plants play in providing food, shelter, and habitat for countless species of animals, and in maintaining the delicate balance and health of the wetland environment.

In a world of increasing environmental pressures and challenges, the presence of these rare and interesting plants is a powerful reminder of the urgent need to protect and preserve the fragile and irreplaceable habitats that support them. By cherishing and stewarding these precious landscapes, we not only ensure the survival and thriving of the incredible diversity of life they sustain but also nourish and enrich our own lives with the beauty, wonder, and resilience of the natural world.

Facilities and Accessibility:

The River Alde Walk offers a range of facilities and amenities to ensure a comfortable and enjoyable experience for walkers of all ages and abilities. The starting point of the walk, Snape Maltings, is a world-class cultural and artistic hub that provides ample parking, public toilets, a café, shops, and an art gallery, making it an ideal place to begin and end your journey along the river.

As you set out on the walk, be sure to take advantage of these facilities to refresh and recharge before embarking on your adventure. The café at Snape Maltings offers a delicious selection of locally-sourced and seasonal dishes, from hearty breakfasts to light lunches and afternoon teas, as well as a range of snacks, cakes, and refreshments to take with you on your walk.

While there are no public toilets or cafes directly along the route, the charming villages of Iken and Sudbourne, which you will pass through on your journey, offer the opportunity to take a brief rest and enjoy the local hospitality. In Iken, be sure to visit the historic church of St Botolph, a beautiful 12th-century sanctuary that provides a peaceful and contemplative space for reflection and refreshment.

In terms of accessibility, the River Alde Walk is a moderately difficult route that covers a distance of 10 miles (16 km) and includes some uneven terrain, particularly along the riverside paths and through the marshland areas. While the majority of the walk is on well-maintained footpaths and bridleways, there are some sections that may be muddy or slippery, especially after heavy rain or during the winter months.

As such, it is essential to wear sturdy, comfortable footwear with good grip and ankle support, and to be prepared for changeable weather conditions by bringing appropriate clothing and gear, such as waterproofs, layers, and a hat and sunscreen in the summer months. It is also a good idea to carry a map and compass or GPS device, as well as a fully charged mobile phone and a small first aid kit, in case of emergencies.

For those with limited mobility or fitness, it may be possible to explore shorter sections of the walk or to take advantage of the many other accessible outdoor activities and attractions in the area, such as birdwatching, boat trips, and visits to local nature reserves and heritage sites. The friendly staff at Snape Maltings and other local visitor centres will be happy to provide advice and recommendations to help you make the most of your visit to this beautiful and welcoming corner of Suffolk.

Ultimately, the River Alde Walk is a challenging but immensely rewarding experience that offers a unique and unforgettable opportunity to connect with the natural beauty, rich history, and vibrant wildlife of the Suffolk wetlands. With careful planning, preparation, and a spirit of adventure, walkers of all ages and abilities can enjoy the many wonders and delights of this extraordinary landscape, and come away with memories and insights that will last a lifetime.

Seasonal Highlights:

One of the great joys of the River Alde Walk is the opportunity to experience the changing face of the Suffolk wetlands through the turning of the seasons, each one bringing its own unique character, challenges, and rewards. From the fresh, vibrant greens of spring to the rich, mellow hues of autumn, the landscape is a constantly shifting kaleidoscope of colour, texture, and life, inviting us to explore and discover anew with each passing month.

In the spring, the marshes and meadows along the River Alde burst into life with an exuberant display of wildflowers, their delicate blooms and sweet fragrances filling the air with the promise of renewal and growth. From the nodding heads of cowslips and the bright yellow stars of marsh marigolds to the frothy white clouds of meadowsweet and the blushing pink of ragged robin, the wetlands are a riot of colour and vitality, alive with the buzz of bees and the flutter of butterflies.

As the season progresses, the reedbeds and marshes come alive with the sound of birdsong, as migrant warblers and waders return from their winter quarters to breed and raise their young in the lush, sheltering vegetation. The air is filled with the liquid trills of sedge warblers, the sharp, staccato calls of reed buntings, and the haunting, fluting cries of curlews, each voice a thread in the rich and complex tapestry of wetland life.

In the summer months, the River Alde Walk is a haven of cool, green tranquility, the dense foliage of the riverside trees and the tall, swaying stems of the reedbeds providing welcome shade and shelter from the heat of the sun. The water sparkles and dances in the light, reflecting the endless blue of the East Anglian sky, while the air is heavy with the drowsy hum of insects and the sweet, heady scent of water mint and other aromatic plants.

As the days begin to shorten and the leaves start to turn, the River Alde takes on a new character, the lush greens of summer giving way to the rich, warm hues of autumn. The hedgerows and thickets are heavy with the bounty of berries and fruits, while the riverside trees are aflame with the gold and russet of turning leaves, their reflections painting the water with a shimmering, opalescent glow.

With the coming of winter, the wetlands are transformed into a stark and striking landscape of frost and ice, the bare branches of the trees etched against the pale, watery sky and the frozen marshes crackling underfoot. But even in the depths of the coldest months, the River Alde is alive with the hardy and resilient creatures that make their home in this challenging environment, from the stealthy otters fishing beneath the ice to the flocks of overwintering ducks and geese that gather on the open water.

No matter the season, the River Alde Walk offers a wealth of opportunities for discovery, reflection, and connection with the natural world, each step a chance to deepen our understanding and appreciation of the incredible diversity and resilience of life in the Suffolk wetlands. As we walk, let us be mindful of the turning of the year and the constant cycle of change and renewal that shapes this extraordinary landscape, and let us find joy and wonder in every moment of our journey through this precious and irreplaceable corner of the earth.

Difficulty Level and Safety Considerations:

The River Alde Walk is a moderately challenging route that requires a good level of physical fitness, stamina, and outdoor skills to complete safely and enjoyably. While the majority of the walk is on well-maintained footpaths and bridleways, there are some sections that may be uneven, muddy, or overgrown, particularly along the riverside and through the marshland areas.

One of the main challenges of the walk is the distance, which covers a total of 10 miles (16 km) and can take between 5-6 hours to complete, depending on your pace and the number of stops you make along the way. This is a significant undertaking that requires careful planning, preparation, and pacing to avoid exhaustion or injury, particularly in hot or humid weather.

Another key consideration is the terrain, which can be slippery, uneven, or unstable in places, especially after heavy rain or during the winter months. Walkers should wear sturdy, supportive footwear with good grip and ankle support, and be prepared for muddy or wet conditions by bringing appropriate clothing and gear, such as waterproofs and gaiters.

It is also important to be aware of the potential hazards posed by the water, particularly along the riverside sections of the walk. The River Alde is a tidal river that can rise and fall rapidly, and there may be strong currents or deep water in places, especially after heavy rainfall or during high tides. Walkers should exercise caution when walking near the water's edge and avoid attempting to cross or wade through the river, even if it appears shallow or calm.

In addition to the physical challenges of the walk, there are also some navigational considerations to keep in mind. While the route is generally well-marked and easy to follow, there are some sections where the path may be unclear or where there are multiple trails branching off in different directions. It is a good idea to carry a map and compass or GPS device, and to familiarize yourself with the route before setting out, particularly if you are walking alone or in unfamiliar territory.

Other safety considerations to keep in mind include the need to protect yourself from the sun and the elements, particularly during the summer months when the weather can be hot and sunny. Walkers should bring plenty of water, snacks, and sun protection, and be prepared for sudden changes in the weather by bringing layers and a hat or raincoat.

Finally, it is always a good idea to let someone know your planned route and estimated return time before setting out, and to carry a fully charged mobile phone and a small first aid kit in case of emergencies. While the River Alde Walk is generally a safe and well-maintained route, accidents can happen, and it is important to be prepared for any eventuality.

By taking these precautions and being mindful of the challenges and potential hazards of the walk, walkers of all ages and abilities can enjoy a safe and rewarding experience along the River Alde, immersing themselves in the beauty and wonder of this extraordinary landscape while also taking care of themselves and the environment. Remember to walk at your own pace, listen to your body, and take regular breaks to rest and recharge, and you will be well-equipped to make the most of this unforgettable journey through the heart of the Suffolk wetlands.

Nearby Attractions:

The River Alde Walk is a gateway to a wealth of natural, cultural, and historical attractions that showcase the rich and varied character of the Suffolk landscape and its people. From world-class music and art venues to ancient ruins and wildlife reserves, there is something for every interest and taste within easy reach of the walk, inviting visitors to extend their stay and explore the many wonders of this fascinating corner of England.

One of the most notable nearby attractions is the world-renowned Snape Maltings Concert Hall, a stunning converted Victorian maltings that plays host to an incredible array of music, dance, and theatrical performances throughout the year. Home to the famous Aldeburgh Festival and the Britten-Pears Young Artist Programme, Snape Maltings is a cultural hub of international significance, drawing visitors from around the world to experience the very best in classical, contemporary, and world music.

Just a short drive from Snape Maltings is the RSPB Minsmere nature reserve, one of the most important and diverse wildlife habitats in the UK. Spanning over 2,500 acres of reedbeds, wetlands, heathland, and coastal dunes, Minsmere is a haven for rare and endangered species, from bitterns and marsh harriers to otters and water voles. With its network of trails, hides, and visitor centre, Minsmere offers a unique and immersive experience of the natural world, inviting visitors to connect with the incredible diversity and beauty of the Suffolk coast.

For those interested in history and heritage, the nearby coastal towns of Aldeburgh and Orford are must-see destinations, each with its own unique character and charm. Aldeburgh is a picturesque seaside resort with a rich cultural heritage, famous for its association with the composer Benjamin Britten and its annual arts festival. With its colourful beach huts, historic high street, and world-class fish and chips, Aldeburgh is the perfect place to spend a lazy afternoon, soaking up the sun and the sea air.

Orford, meanwhile, is a charming and unspoiled village with a fascinating history, dominated by the imposing bulk of Orford Castle, a 12th-century keep that offers stunning views over the surrounding countryside and coast. The village is also home to the Orford Ness National Nature Reserve, a unique and otherworldly landscape of shingle ridges, salt marshes, and abandoned military installations that played a key role in the development of radar and atomic weapons during the 20th century.

Other nearby attractions include the historic market town of Woodbridge, with its picturesque waterfront, independent shops, and ancient tide mill; the Suffolk Punch Trust, a working stud farm and visitor centre dedicated to the preservation of the iconic Suffolk Punch heavy horse breed; and the Sutton Hoo archaeological site, where the incredible treasure-filled burial ship of an Anglo-Saxon king was discovered in 1939.

Whether you are a lover of music, art, nature, history, or simply the great outdoors, the River Alde Walk and its surroundings offer an unparalleled opportunity to immerse yourself in the rich and varied tapestry of the Suffolk landscape, and to create memories and experiences that will stay with you for a lifetime. So why not linger a while, and discover the many hidden gems and unexpected delights that await you in this beautiful and fascinating corner of England? The rewards of exploring this extraordinary place are truly endless, and the welcome is as warm and generous as the land and its people.

Bucklesham Circular Walk - 5 miles (8 km), Circular - Estimated Time: 2-2.5 hours

Starting Point Coordinates, Postcode: The starting point for the Bucklesham Circular Walk is at St. Mary's Church in Bucklesham (Grid Reference: TM 235 417, Postcode: IP10 0DY).
Nearest Car Park, Postcode: Limited parking is available along Church Road near St. Mary's Church (Postcode: IP10 0DY).

Embark on a delightful 5-mile (8 km) circular walk through the enchanting countryside surrounding the picturesque village of Bucklesham, where the gentle rhythms of rural life and the timeless beauty of the Suffolk landscape intertwine to create an unforgettable experience. This relatively easy route offers a perfect blend of open fields, wooded areas, and quiet country lanes, inviting you to immerse yourself in the tranquil atmosphere and rich natural heritage of this charming corner of East Anglia.

As you set out from the historic St. Mary's Church, a 14th-century architectural gem that stands as a testament to the deep spiritual roots of the local community, feel the anticipation build as you follow Church Road eastward, the village's quaint cottages and gardens giving way to the open expanse of the surrounding farmland. With each step, the cares of the modern world seem to melt away, replaced by a growing sense of peace and connection with the natural world that envelops you.

Turning left onto Rectory Lane, you'll soon find yourself immersed in a patchwork of fields and hedgerows, the gently undulating terrain and the soft rustling of the leaves in the breeze creating a soothing backdrop for your journey. Take a moment to breathe in the fresh, clean air and to listen to the melodic birdsong that fills the sky, each note a joyous celebration of the vitality and resilience of the creatures that call this place home.

As you continue along the footpath, drink in the stunning views of the open countryside that stretch out before you, the vibrant greens and golds of the crops and pastures a vivid reminder of the enduring fertility and abundance of the land. Marvel at the intricate beauty of the wildflowers that line the path, from the delicate, nodding heads of cowslips to the bright, cheerful faces of buttercups, each bloom a tiny miracle of colour and form.

Entering a small wooded area, feel the cool, dappled shade of the trees envelop you, the soft crunch of leaves and twigs beneath your feet and the gentle babbling of a nearby stream creating a sense of peace and tranquility that soothes the soul. Take a moment to pause and appreciate the intricate web of life that thrives in this small oasis, from the busy scurrying of insects among the undergrowth to the graceful dance of butterflies in the shafts of sunlight that pierce the canopy.

Emerging from the woods, you'll once again find yourself in the open expanse of the fields, the vast East Anglian sky stretching out above you in an endless blue canopy. Here, amidst the gently swaying grasses and the distant sound of sheep and cattle, you may catch a glimpse of some of the area's iconic farmland birds, from the soaring melodies of skylarks to the bright, flashing colours of yellowhammers and the haunting cries of lapwings, each species a vital thread in the rich tapestry of rural life.

As you make your way back towards Bucklesham, take a moment to reflect on the incredible resilience and adaptability of the plants and animals that thrive in this seemingly gentle, yet often challenging, landscape. From the hardy wildflowers that bloom in the face of drought and frost to the resourceful birds and mammals that find sustenance and shelter in the fields and hedgerows, each creature is a testament to the enduring power and creativity of nature.

Arriving back at St. Mary's Church, take a final moment to appreciate the beauty and serenity of this ancient place of worship, its weathered stone walls and elegant arches a reminder of the countless generations who have found solace, joy, and inspiration in the simple act of walking the paths and fields of this timeless landscape. The Bucklesham Circular Walk is a celebration of the enduring spirit and natural wonders of the Suffolk countryside, offering a chance to reconnect with the land, with our shared history, and with the boundless beauty and resilience of the world around us.

Detailed Route Directions:

1. Start your walk at St. Mary's Church in Bucklesham. From the church, head east along Church Road. (Estimated time: 5 minutes)

2. After approximately 500 meters, turn left onto Rectory Lane. Follow this lane as it winds through the countryside. (Estimated time: 15 minutes)

3. Continue along Rectory Lane until you reach a footpath on the right. Take this footpath and follow it as it leads you through fields and along hedgerows. (Estimated time: 25 minutes)

4. The footpath will eventually bring you to a country lane. Turn left onto this lane and follow it for approximately 1 kilometre. (Estimated time: 15 minutes)

5. Keep an eye out for another footpath on the right. When you reach it, turn right and follow the footpath as it meanders through more fields and wooded areas. (Estimated time: 35 minutes)

6. This footpath will lead you back to Rectory Lane. Turn right onto Rectory Lane and retrace your steps back towards Church Road. (Estimated time: 20 minutes)

7. Upon reaching Church Road, turn left and make your way back to St. Mary's Church, where your circular walk began. (Estimated time: 5 minutes)

Best Features and Views:

- Immerse yourself in the tranquil atmosphere of the Suffolk countryside, the gentle undulations of the landscape and the soft rustling of the leaves in the breeze creating a soothing backdrop for your journey.
- Marvel at the stunning views of open fields that stretch out before you, the vibrant greens and golds of the crops and pastures a testament to the enduring fertility and beauty of the land.
- Delight in the charm and character of the picturesque village of Bucklesham, its quaint cottages, pretty gardens, and historic church a reminder of the rich cultural heritage and timeless appeal of rural life in East Anglia.

Historical Features of Interest:

At the heart of the Bucklesham Circular Walk stands the magnificent St. Mary's Church, a 14th-century architectural gem that has served as a beacon of faith, community, and continuity for generations of local residents. As you begin and end your journey at this ancient place of worship, take a moment to appreciate the exquisite craftsmanship and rich history that is embodied in every stone and stained glass window.

The church's elegant perpendicular Gothic style, with its soaring arches, intricate tracery, and slender pilasters, is a testament to the skill and devotion of the medieval masons who labored to create this enduring masterpiece. Step inside the cool, hushed interior and marvel at the beautiful nave, with its graceful arcades and delicate vaulting, the soft light filtering through the clear glass windows and casting a warm glow over the ancient flagstones.

As you explore the church, keep an eye out for the many fascinating historical features and artifacts that adorn its walls and alcoves, from the ornately carved 15th-century font to the poignant memorial tablets and brasses that honour the lives and deaths of the parish's former residents. Each detail tells a story of the joys, sorrows, and triumphs of the people who have called Bucklesham home over the centuries, and serves as a powerful reminder of the enduring bonds of faith, family, and community that have shaped this special place.

Outside, take a moment to wander through the peaceful churchyard, where the weathered gravestones and gnarled yew trees stand as silent witnesses to the passing of the generations. Imagine the countless baptisms, weddings, and funerals that have taken place within these hallowed walls, each ceremony a milestone in the lives of the villagers and a celebration of the unbreakable ties that bind them to one another and to the land.

As you set out on your walk, carry with you the sense of history, continuity, and belonging that St. Mary's Church embodies, and let it be a source of inspiration and reflection as you explore the timeless beauty and rich heritage of the surrounding countryside. The church stands as a powerful symbol of the enduring spirit and resilience of the people of Bucklesham, and a testament to the deep roots and abiding values that have sustained this community through the ages.

Rare or Uncommon Birds:

As you make your way through the gently rolling fields and hedgerows of the Bucklesham Circular Walk, keep your eyes and ears attuned to the vibrant birdlife that thrives in this picturesque corner of the Suffolk countryside. The area is home to a rich array of farmland birds, each species a vital part of the intricate web of life that sustains the health and beauty of the landscape.

One of the most iconic and beloved birds to watch for is the skylark, a small, brown, streaky bird with an outsized voice and an indomitable spirit. Listen for the distinctive, soaring song of the male skylark as he rises high above the fields, his rapid, trilling notes and fluttering wings a joyous celebration of the freedom and exhilaration of flight. The skylark's aerial display is one of the true wonders of the natural world, a breathtaking performance that has inspired poets, composers, and nature lovers for centuries.

Another charming and colourful bird to look out for is the yellowhammer, a bright, sparrow-sized bird with a striking yellow head and breast, and a rich, rusty-brown back. Often found perched atop hedgerows or singing from the tops of small trees, the yellowhammer's distinctive, rhythmic song, which sounds like "a little bit of bread and no cheese," is a delightful and unforgettable part of the summer soundscape.

As you pass by the open fields and pastures, keep an eye out for the graceful, swooping flight of lapwings, also known as peewits or green plovers. These striking, black-and-white wading birds, with their distinctive crests and iridescent green backs, are a familiar sight in the agricultural landscape, their acrobatic aerial displays and haunting, plaintive calls a testament to their agility and adaptability.

Other farmland birds to watch for include the linnet, a small, streaky finch with a melodious, bubbling song; the corn bunting, a stocky, brown bird with a jangling, metallic call; and the reed bunting, a handsome, black-headed bird with a white collar and a preference for wetland habitats.

As you marvel at the beauty and diversity of the birdlife that surrounds you, take a moment to reflect on the vital role that these species play in maintaining the health and balance of the farmland ecosystem. From the pest control services provided by birds of prey to the pollination and seed dispersal carried out by smaller birds, each species contributes in its own unique way to the intricate web of life that sustains us all.

By cherishing and protecting the habitats and resources that these birds depend on, we not only ensure the survival and thriving of these remarkable creatures but also help to preserve the beauty, resilience, and productivity of the agricultural landscape for generations to come. So let the soaring melodies and flashing colours of the farmland birds be your guide and inspiration as you explore the timeless wonders of the Bucklesham countryside, and let their presence remind you of the priceless natural heritage that we all share and must work together to safeguard.

Rare or Interesting Insects and Other Wildlife:

As you wander through the sun-dappled fields and along the quiet country lanes of the Bucklesham Circular Walk, take a moment to appreciate the incredible diversity of life that thrives in the grasslands and hedgerows. From the tiniest insects to the larger mammals, each creature plays a vital role in maintaining the balance and harmony of the rural ecosystem, their presence a testament to the resilience and complexity of the natural world.

One common yet endlessly fascinating mammal to keep an eye out for is the rabbit, a small, furry creature with long ears and hind legs. Often seen nibbling on the tender shoots of grass and wildflowers, or darting into the shelter of their burrows, rabbits are a familiar and beloved sight in the English countryside, their presence a reminder of the endless cycle of life, death, and renewal that shapes the landscape.

Another engaging inhabitant of the fields and verges is the brown hare, a larger, more athletic relative of the rabbit, with longer limbs and a sleek, streamlined body. Known for their incredible speed and agility, hares are a thrilling sight as they sprint across the open ground, their powerful hind legs propelling them forward in breathtaking leaps and bounds. Though often elusive and crepuscular in their habits, a chance encounter with a hare is a memorable and exciting moment, a glimpse into the raw power and grace of the natural world.

At a smaller scale, the Bucklesham Walk is also home to an astonishing variety of insects, each species a marvel of adaptation and design. From the delicate, fluttering dance of butterflies and moths to the busy hum of bees and hoverflies, these incredible creatures are a vital part of the pollination and nutrient recycling process that sustains the health and beauty of the countryside.

Look out for the striking colours and patterns of the peacock butterfly, with its iridescent blue eyespots and velvety red-brown wings, and the red admiral, with its striking black, red, and white markings. Both species are common sights in the summer months, their beauty and grace a welcome presence in the gardens and hedgerows of the area.

As you walk, keep an ear out for the distinctive droning buzz of the bumblebee, a furry, golden-hued powerhouse of the insect world. These remarkable creatures are essential pollinators, their tireless foraging and nectar-gathering helping to ensure the survival and propagation of countless species of wildflowers and crops.

Other insects to watch for include the iridescent greens and blues of the damselflies and dragonflies, the intricate camouflage of the stick insects and leaf bugs, and the dazzling metallic sheen of the jewel beetles and rosechafers. Each species is a testament to the incredible adaptations and survival strategies that have evolved over millions of years, and each one plays a vital role in maintaining the health, diversity, and beauty of the rural landscape.

As you marvel at the countless tiny wonders that surround you on your walk, take a moment to reflect on the incredible complexity and interdependence of life in the natural world. From the humble earthworm to the majestic oak tree, each creature and plant is a vital thread in the intricate tapestry of the ecosystem, their existence and well-being inextricably linked to our own.

By cherishing and protecting the diversity and abundance of the insects and other wildlife that call the Bucklesham countryside home, we not only ensure the survival and thriving of these remarkable beings but also help to preserve the health, resilience, and beauty of the landscape that sustains us all. So let the buzzing, fluttering, and scurrying of the creatures around you be a source of wonder, joy, and inspiration as you explore the timeless charms of this special corner of Suffolk, and let their presence remind you of the priceless natural heritage that we all share and must work together to safeguard.

Rare or Interesting Plants:

As you make your way along the footpaths and through the gently undulating fields of the Bucklesham Circular Walk, take a moment to delight in the vibrant array of wildflowers and plants that adorn the hedgerows, verges, and meadows. From the delicate, nodding heads of bluebells to the cheerful, sun-bright faces of buttercups, each species is a testament to the incredible diversity and resilience of the rural flora, their presence a vital part of the rich tapestry of life that defines the Suffolk countryside.

One of the most striking and beloved wildflowers to watch for in the spring and early summer is the cowslip, a member of the primrose family with clusters of fragrant, yellow, bell-shaped flowers atop slender stems. Often found growing in meadows, pastures, and along woodland edges, cowslips are an important source of nectar for early-emerging bees and other pollinators, their presence a welcome sign of the returning warmth and vitality of the season.

As you wander through the shaded, woody stretches of the walk, keep an eye out for the enchanting blooms of the bluebell, a woodland specialist with delicate, nodding spikes of deep blue flowers that carpet the forest floor in a mesmerizing sea of colour. Bluebells are a quintessential sight of the English springtime, their ethereal beauty and sweet fragrance a source of joy and inspiration for countless generations of nature lovers.

In the sunnier, more open areas of the walk, look for the bright, golden-yellow flowers of the buttercup, a common yet endlessly charming sight in the fields and meadows of the countryside. With their glossy, butter-yellow petals and deep green, lobed leaves, buttercups are a beloved emblem of the rural landscape, their cheerful presence a reminder of the simple joys and timeless beauty of the natural world.

Other wildflowers to watch for along the Bucklesham Walk include the delicate, pink-and-white blooms of the cuckoo flower, also known as lady's smock; the tall, purple spikes of the foxglove, a stately and poisonous beauty; and the frothy, cream-coloured clusters of the meadowsweet, a fragrant and medicinal herb that has been used for centuries to treat a variety of ailments.

As you marvel at the colourful and intricate beauty of the wildflowers that surround you, take a moment to appreciate the vital role that these plants play in the ecosystem of the countryside. From providing food and shelter for countless species of insects, birds, and mammals to helping to stabilize the soil and purify the air and water, each plant is an essential component of the complex web of life that sustains the health and resilience of the landscape.

In addition to their ecological importance, many of the wildflowers and plants that you encounter on your walk also have a rich history of cultural and medicinal use, their properties and significance passed down through generations of local people. From the soothing, anti-inflammatory properties of meadowsweet to the heart-healing powers of foxglove, each species has a story to tell and a wisdom to share, a reminder of the deep and enduring connection between humans and the natural world.

As you wander through the fields and along the hedgerows, take a moment to reflect on the incredible adaptability and resilience of the plants that thrive in this sometimes harsh and challenging landscape. From the hardy, drought-resistant grasses that carpet the meadows to the tenacious, climbing vines that scramble over the hedges and trees, each species has evolved a unique set of strategies and characteristics that allow it to flourish in its particular niche.

By cherishing and protecting the diversity and abundance of the wildflowers and plants that call the Bucklesham countryside home, we not only ensure the survival and thriving of these remarkable beings but also help to preserve the beauty, character, and ecological integrity of the landscape that sustains us all. So let the colours, fragrances, and forms of the flora around you be a source of wonder, joy, and inspiration as you explore the timeless charms of this special corner of Suffolk, and let their presence remind you of the priceless natural heritage that we all share and must work together to safeguard.

Facilities and Accessibility:

The Bucklesham Circular Walk is a relatively easy and accessible route, suitable for walkers of most ages and abilities. However, it is important to note that the trail is not specifically designed for those with limited mobility, and there are some sections that may prove challenging for individuals with disabilities or physical limitations.

The majority of the walk takes place on well-maintained footpaths and quiet country lanes, with mostly gentle gradients and even surfaces. However, there are a few stretches that may be uneven, muddy, or slippery, particularly after heavy rain or during the winter months. As such, it is recommended that walkers wear sturdy, comfortable shoes with good traction and ankle support, and come prepared with appropriate clothing for the weather conditions.

While there are no major obstacles or barriers along the route, walkers should be aware that there are several stiles and kissing gates to navigate, which may pose difficulties for those with mobility issues or for families with pushchairs. Additionally, some of the field paths may be narrow or overgrown at times, requiring a bit of careful manoeuvring to pass through.

In terms of facilities, it is important to note that the village of Bucklesham itself does not have any public restrooms, shops, or cafes. As such, walkers should come prepared with their own water, snacks, and any necessary supplies, and plan to make use of facilities in nearby towns or villages before or after their walk.

For those who do require access to restrooms or other amenities, the nearest facilities can be found in the nearby town of Ipswich, approximately 6 miles (10 km) away. Here, walkers will find a range of public toilets, cafes, restaurants, and shops, as well as public transportation options for those who need to shorten or alter their route.

Despite these limitations, the Bucklesham Circular Walk remains a highly rewarding and enjoyable experience for the vast majority of walkers, offering a chance to immerse oneself in the beauty and tranquility of the Suffolk countryside, and to connect with the rich natural and cultural heritage of the region. By coming prepared, taking necessary precautions, and being mindful of one's own abilities and needs, walkers of all ages and backgrounds can safely and comfortably enjoy the many delights and wonders of this special corner of East Anglia.

Ultimately, the Bucklesham Circular Walk is a celebration of the simple joys and timeless beauty of the rural landscape, inviting us to slow down, breathe deeply, and open our hearts and minds to the incredible diversity and resilience of the natural world. Whether you are a seasoned rambler or a curious first-time explorer, this enchanting route promises to delight and inspire, offering a chance to reconnect with the land, with our shared history, and with the boundless wonders that lie waiting to be discovered in every corner of the Suffolk countryside.

Seasonal Highlights:

One of the great joys of the Bucklesham Circular Walk is the opportunity to experience the changing face of the Suffolk countryside through the turning of the seasons, each one bringing its own unique character, beauty, and natural wonders. From the tender, emerging greens of spring to the rich, burnished hues of autumn, the landscape is a constantly shifting kaleidoscope of colour, texture, and life, inviting us to explore and discover anew with each passing month.

In the springtime, the fields and hedgerows along the walk burst into life with an exuberant display of wildflowers and blossoms, their delicate petals and fresh, vivid hues a joyous celebration of the returning warmth and vitality of the season. From the cheerful, golden trumpets of the daffodils to the fragrant, pink-and-white clouds of the hawthorn, each species adds its own special magic to the tapestry of the countryside, creating a scene of breathtaking beauty and renewal.

As the days lengthen and the sun climbs higher in the sky, the lush greens of summer take hold, transforming the landscape into a verdant paradise of waving grasses, leafy canopies, and sprawling hedgerows. The air is filled with the buzzing of bees and the fluttering of butterflies, as they flit from flower to flower in search of nectar and pollen, while the fields and meadows are alive with the songs and calls of countless birds, each one a voice in the joyous chorus of the season.

In the autumn, the Bucklesham Walk takes on a new character, as the vibrant greens of summer give way to a rich palette of golds, oranges, and reds, the changing leaves creating a stunning display of colour and texture that is a feast for the senses. The hedgerows and thickets are laden with the bounty of berries and nuts, while the fields are alive with the sound of birdsong and the rustling of small mammals, as they busily prepare for the coming winter.

As the days grow shorter and the nights turn colder, the landscape takes on a stark and ethereal beauty, the bare branches of the trees etched against the pale, watery sky, and the frosty fields and lanes gleaming in the low, slanting light of the winter sun. But even in the depths of the coldest months, the Bucklesham Walk is alive with the hardy and resilient creatures that make their home in this challenging environment, from the stealthy foxes and badgers to the flocks of overwintering birds that gather in the hedgerows and fields.

No matter the season, the Bucklesham Circular Walk offers a wealth of opportunities for discovery, reflection, and connection with the natural world, each step a chance to deepen our understanding and appreciation of the incredible diversity and resilience of life in the Suffolk countryside. As we walk, let us be mindful of the turning of the year and the constant cycle of change and renewal that shapes this extraordinary landscape, and let us find joy and wonder in every moment of our journey through this precious and irreplaceable corner of the earth.

Difficulty Level and Safety Considerations:

The Bucklesham Circular Walk is a relatively easy and straightforward route, suitable for walkers of most ages and fitness levels. With its gently undulating terrain, well-maintained footpaths, and quiet country lanes, this

5-mile (8 km) walk offers a pleasant and accessible way to explore the picturesque countryside of Suffolk, without requiring any specialized equipment or advanced hiking skills.

However, as with any outdoor activity, it is important to be aware of potential hazards and to take necessary precautions to ensure a safe and enjoyable experience. One of the main considerations for walkers on the Bucklesham Walk is the condition of the trail itself, which can vary depending on the time of year and recent weather patterns.

During the wetter months, particularly in the autumn and winter, some sections of the path may become muddy, slippery, or even flooded, making for challenging and potentially treacherous walking conditions. In these situations, it is essential to wear sturdy, waterproof footwear with good traction and ankle support, and to take extra care when navigating any particularly soggy or uneven stretches of the trail.

Similarly, in the height of summer, the footpaths and lanes may become overgrown with vegetation, obscuring any uneven or rocky sections and potentially harbouring stinging nettles, thorny brambles, or other irritants. Walkers should wear long, lightweight pants and sleeves to protect their skin, and keep a watchful eye out for any overhanging branches or hidden obstacles that may trip them up.

Another important safety consideration is the presence of livestock in some of the fields and pastures along the route. While most farm animals are accustomed to the sight of walkers and will generally keep their distance, it is important to remember that they are large and unpredictable creatures, and may become startled or aggressive if they feel threatened.

When passing through any fields with cattle, sheep, or horses, walkers should give the animals a wide berth, moving slowly and calmly to avoid alarming them. If confronted by a particularly aggressive or agitated animal, it is best to retreat slowly and find an alternative route, rather than trying to push through or stand your ground.

In addition to these trail-specific considerations, walkers should also be mindful of the general risks and challenges of any outdoor activity, such as exposure to the elements, dehydration, and insect bites. It is always a good idea to check the weather forecast before setting out, and to come prepared with appropriate clothing, sunscreen, and insect repellent, as well as plenty of water and snacks to keep you energized and hydrated throughout the walk.

Finally, while the Bucklesham Walk is a relatively low-risk and straightforward route, it is still important to let someone know your planned itinerary and expected return time before setting out, and to carry a fully charged mobile phone and a small first aid kit in case of emergencies. By taking these simple precautions and staying alert and aware of your surroundings, you can ensure a safe and enjoyable experience on this delightful and rewarding walk through the Suffolk countryside.

Nearby Attractions:

The Bucklesham Circular Walk is not only a delightful and immersive experience in its own right but also a gateway to a wealth of nearby attractions and destinations that showcase the rich history, culture, and natural beauty of the Suffolk region. From picturesque coastal villages to world-class nature reserves and cultural landmarks, there is no shortage of fascinating places to explore and discover within easy reach of the walk, each one offering its own unique glimpse into the enduring charm and character of this special corner of England.

One of the most popular and scenic destinations in the area is the Suffolk Coast & Heaths Area of Outstanding

Natural Beauty (AONB), a stunning stretch of coastline and countryside that encompasses a diverse array of habitats and landscapes, from sandy beaches and salt marshes to ancient woodlands and wildflower meadows. Whether you are a keen birdwatcher, a nature lover, or simply someone in search of breathtaking views and invigorating walks, the AONB has something to offer everyone, with miles of well-maintained trails, visitor centres, and guided tours to help you make the most of your visit.

For those with an interest in history and architecture, the nearby coastal towns of Felixstowe and Aldeburgh are must-see destinations, each one boasting a rich maritime heritage and a wealth of fascinating landmarks and attractions. In Felixstowe, be sure to visit the historic Landguard Fort, a magnificent 18th-century fortification that has played a key role in defending the coast from invasions and attacks over the centuries, and now houses a museum, visitor centre, and panoramic viewing platforms.

In Aldeburgh, meanwhile, you can explore the charming streets and buildings of the old town, including the 16th-century Moot Hall and the Norman Church of St Peter and St Paul, before taking a stroll along the shingle beach and admiring the iconic scallop sculpture by local artist Maggi Hambling. The town is also renowned for its thriving arts and cultural scene, with numerous galleries, studios, and performance spaces showcasing the work of local and international artists and musicians.

For a taste of the vibrant and cosmopolitan atmosphere of a larger town, the historic county town of Ipswich is just a short drive or bus ride away from Bucklesham, offering a wealth of shopping, dining, and entertainment options, as well as a fascinating range of cultural and historical attractions. From the elegant Tudor buildings of the Christchurch Mansion to the cutting-edge exhibits of the Ipswich Museum and the lively waterfront district with its marinas, cafes, and bars, Ipswich has something to suit every taste and interest.

Other nearby attractions and destinations to consider include the Sutton Hoo archaeological site, where you can learn about the incredible Anglo-Saxon burial mounds and treasures that were discovered here in the 1930s; the Helmingham Hall Gardens, a stunning 16th-century stately home with acres of beautiful formal gardens and parkland; and the Orford Ness National Nature Reserve, a unique and otherworldly landscape of shingle ridges, salt marshes, and abandoned military installations that is home to a wealth of rare and endangered wildlife.

Whether you are a history buff, a nature lover, a foodie, or simply someone in search of a bit of rest and relaxation in a beautiful and welcoming corner of England, the Bucklesham Circular Walk and its surroundings offer endless opportunities for discovery, inspiration, and enjoyment. So why not take a few extra days to explore the many hidden gems and unexpected delights of the Suffolk countryside and coast, and create memories and experiences that will last a lifetime?

Trimley St Martin Walk - 3 miles (4.8 km), Circular - Estimated Time: 1.5-2 hours

Starting Point Coordinates, Postcode: The starting point for the Trimley St Martin Walk is at St. Martin's Church, High Road, Trimley St Martin (Grid Reference: TM 282 373, Postcode: IP11 0SG).
Nearest Car Park, Postcode: There is limited on-street parking near the church. Please park considerately and do not block driveways or access roads.

Embark on a charming 3-mile (4.8 km) circular walk around the delightful village of Trimley St Martin, where the peaceful Suffolk countryside and the rich history of this picturesque community come together to create an unforgettable experience. This relatively easy route, with its gentle inclines and scenic views, offers the perfect opportunity to immerse yourself in the tranquil beauty and timeless character of this enchanting corner of East Anglia.

As you set out from the historic St. Martin's Church, a beautiful 11th-century building that has stood at the heart of the village for generations, feel the anticipation build as you head east along High Road, the charming cottages and gardens of the village giving way to the open expanse of the surrounding farmland. With each step, the cares of the modern world seem to melt away, replaced by a growing sense of peace and connection with the natural world that envelops you.

Turning left onto Thorpe Lane, you'll soon find yourself surrounded by the lush greens and golds of the arable fields, the gently swaying crops and the distant sound of birdsong creating a soothing backdrop for your journey. Take a moment to breathe in the fresh, clean air and to marvel at the vast, open skies that stretch out above you, the ever-changing play of light and shadow across the landscape a testament to the timeless beauty and grandeur of the Suffolk countryside.

As you continue along the footpath, drinking in the picturesque views and the tranquil atmosphere of the surrounding fields, keep an eye out for the diverse array of wildlife that calls this area home. From the soaring melodies of skylarks and the bright, flashing colours of yellowhammers to the darting movements of insects among the hedgerows, each encounter is a reminder of the incredible diversity and resilience of the natural world.

Emerging from the open fields, you'll find yourself on a quiet country track, the soft crunch of gravel beneath your feet and the gentle rustling of the leaves in the breeze creating a soothing soundtrack for your walk. Take a moment to appreciate the simple beauty of the rural landscape, the patchwork of fields and hedgerows stretching out before you like a living tapestry, each stitch a testament to the centuries of human stewardship and care that have shaped this land.

As you make your way back towards the village, the distant spire of St. Martin's Church guiding your steps, reflect on the rich history and heritage of this special place, from the ancient stones of the church to the timeless rhythms of the agricultural year. Each step is a reminder of the countless generations who have called Trimley St Martin home, their lives and stories woven into the very fabric of the landscape.

Arriving back at your starting point, take a final moment to appreciate the beauty and serenity of the church and its surroundings, the warm welcome of the village shop a fitting end to your journey. The Trimley St Martin Walk may be a relatively short and simple route, but it offers a deeply rewarding and enriching experience, a chance to connect with the land, with our shared history, and with the enduring spirit of this remarkable corner of England.

So whether you are a seasoned rambler or a curious first-time visitor, this delightful walk promises to refresh and inspire, inviting you to slow down, to breathe deeply, and to open your heart and mind to the countless wonders and revelations that await you in the gentle folds and winding lanes of the Suffolk countryside. As you set out on your journey, let the timeless beauty and tranquility of Trimley St Martin be your guide and your companion, leading you to a deeper appreciation and understanding of the natural world and our place within it.

Detailed Route Directions:

1. Begin your walk at St. Martin's Church on High Road. Take a moment to admire the beautiful 11th-century architecture before heading east along High Road. (Estimated time: 5 minutes)

2. Pass by the village shop on your left as you continue along High Road, enjoying the charming cottages and gardens that line the street. (Estimated time: 5 minutes)

3. Turn left onto Thorpe Lane and follow the road as it leads you out of the village and into the surrounding countryside. (Estimated time: 10 minutes)

4. Keep an eye out for a footpath on your right. When you reach it, turn right and follow the path as it takes you through open fields, offering picturesque views of the Suffolk landscape. (Estimated time: 20 minutes)

5. Continue along the footpath, immersing yourself in the tranquil beauty of the arable farmland and the soothing sounds of nature that surround you. (Estimated time: 20 minutes)

6. The footpath will eventually lead you to a track. Turn right onto the track and follow it as it winds its way back towards Trimley St Martin. (Estimated time: 15 minutes)

7. As you approach the village, keep an eye out for High Road. Turn left onto High Road and make your way back to your starting point at St. Martin's Church. (Estimated time: 10 minutes)

8. Arrive back at St. Martin's Church, where your circular walk comes to an end. Take a final moment to reflect on the beauty and tranquility of the countryside you've just explored. (Total estimated time: 1.5-2 hours)

Best Features and Views:
- Immerse yourself in the picturesque views of the Suffolk countryside as you wander through the open fields, the gently swaying crops and the distant horizons creating a sense of peace and freedom that soothes the soul.
- Delight in the charming historic buildings of Trimley St Martin, from the ancient stones of St. Martin's Church to the quaint cottages and gardens that line the village streets, each one a testament to the enduring character and heritage of this special place.

Historical Features of Interest:

At the heart of the Trimley St Martin Walk stands the magnificent St. Martin's Church, a beautiful historic building that has served as a beacon of faith and community for the people of this village for over a thousand years. As you begin and end your journey at this ancient site, take a moment to appreciate the exquisite craftsmanship and rich history that is embodied in every stone and stained-glass window.

The church dates back to the 11th century, with the oldest parts of the building, including the nave and chancel, constructed in the Norman style using flint and stone. Over the centuries, the church has undergone various additions and renovations, reflecting the changing needs and tastes of the community it serves. The distinctive square tower, with its crenelated parapet and Gothic-arched bell openings, was added in the 15th century, while the south porch and much of the interior woodwork date from the Victorian period.

As you step inside the church, you'll be struck by the sense of peace and timelessness that pervades the ancient space. The simple, elegant lines of the nave and chancel draw the eye towards the east end, where the beautiful stained-glass windows and the finely carved altar and reredos create a focal point for worship and reflection. Take a moment to explore the many fascinating details and features of the interior, from the intricately carved wooden pews to the exquisite memorial brasses and monuments that honour the lives and deaths of the villagers who have called this place their spiritual home.

Outside, the churchyard is a tranquil and evocative space, filled with the weathered gravestones and ancient yew trees that stand as silent witnesses to the passage of time. Wander among the moss-covered stones and the gnarled roots of the trees, reading the inscriptions and imagining the lives and stories of the generations of villagers who have been laid to rest here over the centuries.

As you set out on your walk, carry with you the sense of history and continuity that St. Martin's Church embodies, and let it be a source of inspiration and reflection as you explore the timeless beauty and character of the surrounding countryside. The church stands as a powerful symbol of the enduring spirit and resilience of the people of Trimley St Martin, and a testament to the deep roots and abiding values that have sustained this community through the ages.

Rare or Uncommon Birds:

As you make your way through the open fields and along the quiet country lanes of the Trimley St Martin Walk, keep your eyes and ears attuned to the vibrant birdlife that thrives in this picturesque corner of the Suffolk countryside. The area is home to a diverse array of farmland birds, each species playing a vital role in the intricate web of life that sustains the health and beauty of the agricultural landscape.

One of the most iconic and beloved birds to watch for is the skylark, a small, brown, streaky bird with an outsized voice and an irrepressible spirit. Listen for the distinctive, soaring song of the male skylark as he rises high above the fields, his rapid, trilling notes and fluttering wings a joyous celebration of the freedom and exhilaration of flight. The skylark's aerial display is one of the true wonders of the natural world, a breathtaking performance that has inspired poets, composers, and nature lovers for generations.

Another charming and colourful bird to look out for is the yellowhammer, a bright, sparrow-sized bird with a striking yellow head and breast, and a rich, rusty-brown back. Often found perched atop hedgerows or singing from the tops of small trees, the yellowhammer's distinctive, rhythmic song, which sounds like "a little bit of bread and no cheese," is a delightful and memorable part of the summer soundscape.

As you pass by the arable fields and the grassy margins, keep an eye out for the unassuming but endearing corn bunting, a stocky, brown bird with a thick, conical bill and a rather plain appearance. Despite its somewhat drab plumage, the corn bunting's song is a unique and captivating performance, a series of jangling, metallic notes that have been likened to the sound of keys rattling in a pocket. Once a common sight in the British countryside, the corn bunting has sadly declined in recent years due to changes in agricultural practices, making every sighting of this special bird a cause for celebration and appreciation.

Other farmland birds to watch for along the Trimley St Martin Walk include the linnet, a small, streaky finch with a melodious, bubbling song; the whitethroat, a lively and active warbler with a scratchy, warbling song; and the reed bunting, a handsome, black-headed bird with a white collar and a preference for wetland edges and damp grasslands.

As you delight in the sight and sound of these precious and fascinating birds, take a moment to reflect on the vital role that they play in maintaining the health and balance of the farmland ecosystem. From the pest control services provided by the insectivorous warblers to the seed dispersal carried out by the finches and buntings, each species contributes in its own unique way to the intricate web of life that sustains us all.

By cherishing and protecting the habitats and resources that these birds depend on, we not only ensure the survival and thriving of these remarkable creatures but also help to preserve the beauty, diversity, and resilience of the agricultural landscape for generations to come. So let the soaring melodies and flashing colours of the farmland birds be your guide and inspiration as you explore the timeless wonders of the Trimley St Martin countryside, and let their presence remind you of the priceless natural heritage that we all share and must work together to safeguard.

Rare or Interesting Insects and Other Wildlife:

As you wander through the sun-dappled fields and along the flower-studded hedgerows of the Trimley St Martin Walk, take a moment to appreciate the incredible diversity of life that thrives in the grasslands and field margins. From the tiniest insects to the larger mammals, each creature plays a vital role in maintaining the balance and health of the farmland ecosystem, their presence a testament to the resilience and adaptability of the natural world.

One of the most fascinating and important groups of insects to watch for along the walk are the pollinators, the bees, butterflies, hoverflies, and other winged wonders that flit from flower to flower in search of nectar and pollen. These tireless and industrious creatures are essential to the reproduction and survival of countless species of wildflowers and crops, their delicate dance of pollination ensuring the continued abundance and diversity of the plant world.

Keep an eye out for the distinctive black and yellow stripes of the bumblebee, a fuzzy, gentle giant of the insect world whose droning buzz and clumsy, endearing flight are a familiar and beloved sight in the summer meadows. Watch as they move from bloom to bloom, their pollen baskets laden with the golden dust of life, their presence a vital link in the complex web of relationships that sustain the ecosystem.

Other pollinators to look for include the iridescent blues and greens of the solitary bees, the dazzling oranges and reds of the comma and small tortoiseshell butterflies, and the intricate patterns and hovering flight of the hoverflies, each species a marvel of adaptation and design, perfectly suited to its role in the grand tapestry of nature.

As you continue your walk, keep an eye out for the many other insects and invertebrates that call the fields and hedgerows home, from the scurrying beetles and centipedes that hunt among the leaf litter to the graceful lacewings and damselflies that dance above the grasses. Each creature, no matter how small or seemingly insignificant, plays a vital part in the complex and interconnected web of life that sustains the farmland habitat.

In addition to the smaller creatures, the Trimley St Martin Walk is also home to a variety of larger mammals, from the playful rabbits and hares that bound through the fields to the elusive foxes and badgers that hunt and forage in the twilight hours. Keep a watchful eye for the telltale signs of their presence, from the delicate footprints in the soft earth to the distant rustle of movement in the undergrowth, each encounter a reminder of the rich and varied life that thrives just beyond the edge of our perception.

As you marvel at the incredible diversity and complexity of the insect and animal life that surrounds you, take a moment to reflect on the vital role that these creatures play in maintaining the health and resilience of the farmland ecosystem. From the soil-building and nutrient-cycling services provided by the beetles and worms to the pest control and seed dispersal carried out by the birds and mammals, each species is an essential thread in the intricate tapestry of life that sustains us all.

By cherishing and protecting the habitats and resources that these creatures depend on, we not only ensure the survival and thriving of these remarkable beings but also help to preserve the beauty, diversity, and productivity of the agricultural landscape for generations to come. So let the buzzing, fluttering, and scurrying of the insects and the distant calls and movements of the larger animals be a source of wonder and inspiration as you explore the timeless charms of the Trimley St Martin countryside, and let their presence remind you of the priceless natural heritage that we all share and must work together to safeguard.

Rare or Interesting Plants:

As you make your way along the footpaths and through the gently rolling fields of the Trimley St Martin Walk, take a moment to delight in the vibrant array of wildflowers and plants that adorn the hedgerows, verges, and field margins. From the delicate, nodding heads of cowslips to the bright, cheerful faces of poppies, each species is a testament to the incredible diversity and resilience of the farmland flora, their presence a vital part of the rich tapestry of life that defines the Suffolk countryside.

One of the most striking and beloved wildflowers to watch for in the late spring and early summer is the common poppy, a vibrant red flower with delicate, papery petals and a distinctive black centre. Often found growing in disturbed soils and field margins, poppies are a poignant symbol of remembrance and renewal, their bright blooms a joyous celebration of the beauty and resilience of the natural world in the face of adversity.

As you wander along the hedgerows, keep an eye out for the charming and delicate flowers of the dog rose, a scrambling shrub with pale pink, five-petaled blooms and a subtle, sweet fragrance. The dog rose is a vital source of food and shelter for a wide range of insects and birds, its thorny stems and dense foliage providing a safe haven for nesting and foraging, while its bright red hips add a splash of colour to the autumn landscape.

In the grassy verges and field margins, look for the tall, slender spikes of the foxglove, a stately and elegant plant with tubular, purple-pink flowers that are beloved by bees and other pollinators. Foxgloves are a valuable source of nectar and pollen for many species of insects, their long, bell-shaped blooms perfectly adapted to the probing tongues and fuzzy bodies of their winged visitors.

Other interesting and diverse plants to watch for along the Trimley St Martin Walk include the delicate, white flowers of the cow parsley, the cheerful, yellow blooms of the meadow buttercup, and the striking, blue spikes of the viper's bugloss, each species a unique and fascinating part of the rich and varied flora of the farmland ecosystem.

As you marvel at the colourful and intricate beauty of the wildflowers and plants that surround you, take a moment to appreciate the vital role that these species play in supporting the complex web of life that thrives in the fields and hedgerows. From providing food and shelter for countless species of insects, birds, and mammals to helping to stabilize the soil and purify the air and water, each plant is an essential component of the delicate balance that sustains the health and productivity of the agricultural landscape.

In addition to their ecological importance, many of the wildflowers and plants that you encounter on your walk also have a rich history of cultural and medicinal use, their properties and significance passed down through generations of local people. From the soothing, anti-inflammatory properties of the wild chamomile to the astringent, wound-healing powers of the yarrow, each species has a story to tell and a wisdom to share, a reminder of the deep and enduring connection between humans and the natural world.

As you wander through the fields and along the hedgerows, take a moment to reflect on the incredible adaptability and resilience of the plants that thrive in this dynamic and challenging landscape. From the hardy, drought-resistant grasses that carpet the field margins to the tenacious, climbing vines that scramble over the hedges and fences, each species has evolved a unique set of strategies and characteristics that allow it to flourish in its particular niche.

By cherishing and protecting the diversity and abundance of the wildflowers and plants that call the Trimley St Martin countryside home, we not only ensure the survival and thriving of these remarkable species but also help to preserve the beauty, character, and ecological integrity of the farmland landscape that sustains us all. So let the colours, fragrances, and forms of the flora around you be a source of wonder, joy, and inspiration as you explore the timeless charms of this special corner of Suffolk, and let their presence remind you of the priceless natural heritage that we all share and must work together to safeguard.

Facilities and Accessibility:

The Trimley St Martin Walk is a relatively short and easy route, suitable for walkers of most ages and abilities. However, it is important to note that the trail is not specifically designed for those with limited mobility, and there are some sections that may prove challenging for individuals with disabilities or physical limitations.

The majority of the walk takes place on well-maintained footpaths and quiet country lanes, with mostly gentle gradients and even surfaces. However, there are a few stretches that may be uneven, muddy, or overgrown, particularly after heavy rain or during the growing season. As such, it is recommended that walkers wear sturdy, comfortable shoes with good traction and ankle support, and come prepared with appropriate clothing for the weather conditions.

While there are no major obstacles or barriers along the route, walkers should be aware that there are several stiles and kissing gates to navigate, which may pose difficulties for those with mobility issues or for families with pushchairs. Additionally, some of the field paths may be narrow or enclosed by vegetation, requiring a bit of careful manoeuvring to pass through.

In terms of facilities, the village of Trimley St Martin itself has limited amenities, with a small village shop near the starting point of the walk. However, there are no public restrooms or cafes directly on the route. Walkers should plan accordingly and bring their own water, snacks, and any necessary supplies.

For those who require more comprehensive facilities or amenities, the nearby town of Felixstowe, approximately 3 miles (5 km) away, offers a range of shops, restaurants, cafes, and public restrooms. Walkers can easily access Felixstowe by car or public transportation before or after their walk.

Despite these limitations, the Trimley St Martin Walk remains a highly accessible and enjoyable experience for the vast majority of walkers, offering a wonderful opportunity to explore the beauty and tranquility of the Suffolk countryside without the need for specialist equipment or advanced planning. By coming prepared, taking necessary precautions, and being mindful of one's own abilities and needs, walkers of all ages and backgrounds can safely and comfortably enjoy the many delights and wonders of this charming corner of East Anglia.

Ultimately, the Trimley St Martin Walk is a celebration of the simple joys and timeless beauty of the rural landscape, inviting us to slow down, breathe deeply, and open our hearts and minds to the incredible diversity and resilience of the natural world. Whether you are a seasoned rambler or a curious first-time visitor, this delightful route promises to refresh and inspire, offering a chance to reconnect with the land, with our shared history, and with the boundless wonders that lie waiting to be discovered in every corner of the Suffolk countryside.

Seasonal Highlights:

One of the great pleasures of the Trimley St Martin Walk is the opportunity to witness the changing face of the Suffolk countryside through the turning of the seasons, each one bringing its own unique character, beauty, and natural wonders. From the tender, emerging greens of spring to the rich, mellow hues of autumn, the landscape is a constantly shifting kaleidoscope of colour, texture, and life, inviting us to explore and discover anew with each passing month.

In the springtime, the fields and hedgerows along the walk burst into life with an exuberant display of wildflowers and blossoms, their delicate petals and fresh, bright colours a joyous celebration of the returning warmth and vitality of the season. From the cheerful, golden trumpets of the daffodils to the nodding, purple heads of the bluebells, each species adds its own special magic to the tapestry of the countryside, creating a scene of breathtaking beauty and renewal.

As the days lengthen and the sun climbs higher in the sky, the lush greens of summer take hold, transforming the landscape into a verdant paradise of waving grasses, leafy hedgerows, and sun-drenched fields. The air is filled with the buzzing of bees and the fluttering of butterflies, as they flit from flower to flower in search of nectar and pollen, while the fields and meadows are alive with the songs and calls of countless birds, each one a voice in the joyous chorus of the season.

In the autumn, the Trimley St Martin Walk takes on a new character, as the vibrant greens of summer give way to a rich palette of golds, oranges, and russets, the changing leaves creating a stunning display of colour and texture that is a feast for the senses. The hedgerows and verges are laden with the bounty of berries and fruits, while the fields are alive with the sound of birdsong and the rustling of small mammals, as they busily prepare for the coming winter.

As the days grow shorter and the nights turn colder, the landscape takes on a stark and ethereal beauty, the bare branches of the trees etched against the pale, watery sky, and the frosty fields and lanes gleaming in the low, slanting light of the winter sun. But even in the depths of the coldest months, the Trimley St Martin Walk is alive with the hardy and resilient creatures that make their home in this challenging environment, from the stealthy foxes and badgers to the flocks of overwintering birds that gather in the hedgerows and stubble fields.

No matter the season, the Trimley St Martin Walk offers a wealth of opportunities for discovery, reflection, and connection with the natural world, each step a chance to deepen our understanding and appreciation of the incredible diversity and resilience of life in the Suffolk countryside. As we walk, let us be mindful of the turning of the year and the constant cycle of change and renewal that shapes this extraordinary landscape, and let us find joy and wonder in every moment of our journey through this precious and irreplaceable corner of the earth.

Difficulty Level and Safety Considerations:

The Trimley St Martin Walk is a relatively easy and accessible route, suitable for walkers of most ages and fitness levels. With its gentle terrain, well-maintained paths, and quiet country lanes, this 3-mile (4.8 km) walk offers a pleasant and enjoyable way to explore the picturesque countryside of Suffolk, without requiring any specialized equipment or extensive hiking experience.

However, as with any outdoor activity, it is important to be aware of potential hazards and to take necessary precautions to ensure a safe and comfortable experience. One of the main considerations for walkers on the Trimley St Martin Walk is the condition of the trail itself, which can vary depending on the time of year and recent weather patterns.

During the wetter months, particularly in the autumn and winter, some sections of the path may become muddy, slippery, or waterlogged, making for challenging and potentially hazardous walking conditions. In these situations, it is essential to wear sturdy, waterproof footwear with good traction and ankle support, and to take extra care when navigating any particularly wet or uneven stretches of the trail.

Similarly, in the height of summer, the footpaths and lanes may become overgrown with vegetation, obscuring any uneven or rocky sections and potentially harbouring stinging nettles, thorny brambles, or other irritants. Walkers should wear long, lightweight pants and sleeves to protect their skin, and keep a watchful eye out for any overhanging branches or hidden obstacles that may trip them up.

Another important safety consideration is the presence of livestock in some of the fields along the route. While most farm animals are accustomed to the sight of walkers and will generally keep their distance, it is important to remember that they are large and unpredictable creatures, and may become startled or agitated if they feel threatened.

When passing through any fields with cattle, sheep, or horses, walkers should give the animals a wide berth, moving slowly and calmly to avoid alarming them. If confronted by a particularly aggressive or hostile animal, it is best to retreat slowly and find an alternative route, rather than trying to push through or stand your ground.

In addition to these trail-specific considerations, walkers should also be mindful of the general risks and challenges of any outdoor activity, such as exposure to the elements, dehydration, and insect bites. It is always a good idea to check the weather forecast before setting out, and to come prepared with appropriate clothing, sunscreen, and insect repellent, as well as plenty of water and snacks to keep you energized and hydrated throughout the walk.

Finally, while the Trimley St Martin Walk is a relatively low-risk and straightforward route, it is still important to let someone know your planned itinerary and expected return time before setting out, and to carry a fully charged mobile phone and a small first aid kit in case of emergencies. By taking these simple precautions and staying alert and aware of your surroundings, you can ensure a safe and enjoyable experience on this delightful and rewarding walk through the Suffolk countryside.

Nearby Attractions:

The Trimley St Martin Walk is not only a lovely and immersive experience in its own right but also a gateway to a wealth of nearby attractions and destinations that showcase the rich history, culture, and natural beauty of the Suffolk region. From picturesque coastal towns to world-class nature reserves and heritage sites, there is no shortage of fascinating places to explore and discover within easy reach of the walk, each one offering its own unique glimpse into the enduring charm and character of this special corner of England.

One of the most popular and accessible nearby attractions is the charming coastal town of Felixstowe, just a short drive or bus ride from Trimley St Martin. This vibrant and historic seaside resort boasts a wide range of activities and amenities, from its beautiful sandy beaches and thriving seafront promenade to its bustling town centre with its independent shops, cafes, and restaurants.

For those interested in history and heritage, Felixstowe is home to a number of fascinating museums and landmarks, including the Felixstowe Museum, which tells the story of the town's rich maritime and military past, and the Landguard Fort, a stunning 18th-century coastal fortification that offers breathtaking views over the North Sea and the Orwell Estuary.

Nature lovers and birdwatchers will also find plenty to delight and inspire them in the Felixstowe area, with a variety of parks, gardens, and nature reserves to explore. The Landguard Nature Reserve, located at the southern tip of the town, is a particularly special and diverse site, with a range of habitats including shingle beaches, saltmarshes, and grasslands that support a wealth of rare and endangered species.

For those willing to venture a bit further afield, the nearby town of Ipswich, just 12 miles (19 km) from Trimley St Martin, offers a wealth of cultural, historical, and recreational attractions. As the county town of Suffolk and one of the oldest continuously inhabited towns in England, Ipswich boasts a rich and fascinating heritage that is reflected in its many museums, galleries, and architectural gems.

Visitors to Ipswich can explore the stunning collections of the Ipswich Museum, which showcase the town's history from prehistoric times to the present day, or admire the beautiful paintings and sculptures of the Christchurch Mansion, a stunning Tudor mansion set in a picturesque park. The town is also home to a thriving arts and cultural scene, with numerous theatres, music venues, and festivals throughout the year.

For those seeking a more active and adventurous experience, the Suffolk Coast & Heaths Area of Outstanding Natural Beauty, which stretches along the coast from Felixstowe to Lowestoft, offers a stunning range of landscapes and activities to enjoy. From the windswept beaches and crumbling cliffs of the Covehithe coast to the tranquil waterways and reedbeds of the Blyth Estuary, this special and varied region is a paradise for walkers, cyclists, birdwatchers, and anyone who loves the great outdoors.

Other nearby attractions and destinations to consider include the beautiful market town of Woodbridge, with its picturesque waterfront and historic tide mill; the Sutton Hoo archaeological site, where the incredible treasure-filled burial ship of an Anglo-Saxon king was discovered in 1939; and the RSPB Minsmere nature reserve, one of the most important and diverse birdwatching sites in the UK.

Whether you are a history buff, a nature lover, a foodie, or simply someone in search of a bit of rest and relaxation in a beautiful and welcoming corner of England, the Trimley St Martin Walk and its surroundings offer endless opportunities for discovery, inspiration, and enjoyment. So why not take a few extra days to explore the many hidden gems and unexpected delights of the Suffolk countryside and coast, and create memories and experiences that will last a lifetime?

Trimley Marshes Walk - 5 miles (8 km), Circular - Estimated Time: 2-3 hours

Starting Point Coordinates, Postcode: The starting point for the Trimley Marshes Walk is the car park at Trimley Marshes Nature Reserve (Grid Reference: TM 277 358, Postcode: IP11 0SG).
Nearest Car Park, Postcode: The nearest car park is at the Trimley Marshes Nature Reserve (Postcode: IP11 0SG).

Embark on an enchanting 5-mile (8 km) circular walk through the captivating landscapes of Trimley Marshes Nature Reserve, where the ever-changing tapestry of wetlands, grasslands, and coastal habitats weaves together to create a haven for a remarkable diversity of bird and plant life. This easy to moderate route, with its flat terrain and well-maintained paths, offers an accessible and immersive experience for nature enthusiasts and casual walkers alike, inviting you to uncover the secrets and wonders of this extraordinary corner of the Suffolk coast.

As you set out from the car park, feel the anticipation build as you follow the path into the heart of the reserve, the distant calls of wading birds and the gentle rustling of the reeds hinting at the rich variety of life that awaits you. With each step, the cares and distractions of the outside world seem to melt away, replaced by a profound sense of connection with the wild and untamed beauty of the natural world.

The trail winds through a patchwork of habitats, each one a unique and precious ecosystem in its own right. Marvel at the shimmering expanses of marshland, where the ebb and flow of the tides create an ever-changing mosaic of water and land, and the air is filled with the haunting cries of curlews and redshanks. Pause to admire the vibrant splashes of colour provided by the salt-tolerant plants that thrive in this challenging environment, from the delicate pink blooms of sea lavender to the striking spikes of sea aster.

As you continue along the circular route, the landscape transforms once again, giving way to the lush greens and golds of the grasslands, where the swaying stems of wildflowers and grasses create a mesmerizing dance in the coastal breeze. Keep your eyes peeled for the darting forms of dragonflies and butterflies, their iridescent wings catching the sunlight as they flit between the blooms, and listen for the faint rustling of harvest mice and water voles among the foliage, a reminder of the incredible diversity of life that finds shelter and sustenance in these precious habitats.

Throughout your walk, take advantage of the numerous bird hides that dot the route, each one providing a unique and intimate window into the secret world of the reserve's feathered inhabitants. From the elegant avocets that grace the mudflats with their monochromatic plumage and upturned bills to the elusive bearded tits that flit among the reedbeds, their soft pinging calls betraying their presence, every sighting is a treasure to be cherished and a testament to the incredible resilience and adaptability of nature.

As you near the end of your journey, pause to take in the breathtaking panoramic views that unfold before you, the shimmering expanse of the River Orwell stretching out to the horizon, its waters alive with the silhouettes of sailing boats and the distant shapes of wading birds. Let your gaze wander over the patchwork of fields and hedgerows that surround the reserve, each one a vital piece in the intricate web of life that sustains this special place, and feel a deep sense of gratitude and wonder for the beauty and complexity of the natural world.

Returning to your starting point, take a moment to reflect on the incredible diversity and richness of the habitats you have explored, and the countless stories of resilience, adaptation, and renewal that are woven into every inch of this remarkable landscape. Whether you are a passionate birdwatcher, a budding botanist, or simply someone in search of a deeper connection with the wild places that nourish and inspire us, the Trimley Marshes Walk offers a truly unforgettable experience, one that will stay with you long after you have left the reserve behind.

As you prepare to leave, let the sights, sounds, and sensations of your walk be a reminder of the priceless gift of the natural world, and of the vital importance of preserving and protecting these fragile and irreplaceable habitats for generations to come. For in the tranquil beauty of Trimley Marshes, we catch a glimpse of the intricate web of life that sustains us all, and of the boundless wonders that await those who open their hearts and minds to the wild and untamed places of the earth.

Detailed Route Directions:

1. From the car park at Trimley Marshes Nature Reserve, locate the path leading into the reserve. Follow the signs for the circular trail. (Estimated time: 5 minutes)

2. As you enter the reserve, the path will guide you through a variety of habitats, starting with the marshland area. Take your time to observe the unique flora and fauna that thrive in this wetland environment. (Estimated time: 30-40 minutes)

3. Continue along the circular trail, keeping an eye out for the well-placed bird hides. These hides offer excellent opportunities to spot a wide range of bird species without disturbing their natural behaviour. Feel free to spend some time in the hides, quietly observing the avian activities. (Estimated time: 30-40 minutes)

4. As you progress, the trail will lead you through grassland habitats. Enjoy the vibrant colours and sweet scents of the wildflowers that adorn the meadows, and watch for the various insects and small mammals that call this area home. (Estimated time: 30-40 minutes)

5. The path will then take you towards the coastal sections of the reserve. Pause to admire the stunning views across the River Orwell and the surrounding countryside, and breathe in the invigorating sea air. (Estimated time: 20-30 minutes)

6. Follow the circular trail as it loops back towards the marshland, offering new perspectives on the habitats you explored earlier. Keep your senses attuned to the sights, sounds, and smells of the reserve, and take the time to appreciate the intricate web of life that thrives here. (Estimated time: 30-40 minutes)

7. As you near the end of the loop, the path will guide you back to the car park where you started. Take a moment to reflect on the natural wonders you've encountered and the memories you've made during your walk. (Estimated time: 5 minutes)

Best Features and Views:

- Immerse yourself in the captivating beauty of the diverse habitats found within Trimley Marshes Nature Reserve, from the shimmering expanses of marshland and the lush grasslands to the rugged coastal areas, each one a unique and precious ecosystem teeming with life.
- Marvel at the stunning panoramic views across the River Orwell and the surrounding countryside, the shimmering waters and patchwork fields creating a breathtaking tapestry that showcases the timeless beauty of the Suffolk landscape.
- Delight in the intimate encounters with the reserve's remarkable birdlife, from the elegant avocets and the elusive bearded tits to the countless other species that find sanctuary in these protected habitats, each sighting a treasure to be cherished and remembered.

Historical Features of Interest:

While there are no specific historical sites or artifacts along the Trimley Marshes Walk itself, the area has a rich and fascinating history of human activity dating back thousands of years. The marshes and surrounding landscape have long been shaped by the complex interplay of natural processes and human intervention, each leaving their mark on the land in countless ways.

Evidence of human settlement in the Trimley area can be traced back to the Roman period, with archaeological finds suggesting the presence of a significant Roman community in the vicinity. These early settlers were likely drawn to the area by its strategic location on the coast and the fertile soils of the surrounding countryside, which offered ideal conditions for agriculture and trade.

In the centuries that followed, the Trimley Marshes continued to play an important role in the life of the local community, providing a vital source of food, fuel, and other resources. The marshes were used for grazing livestock, harvesting reeds for thatching and other purposes, and even for the production of salt, a precious commodity in the medieval period.

As the centuries passed, the marshes and coastline of Trimley also bore witness to the changing tides of history, from the devastating floods and storms that repeatedly reshaped the landscape to the centuries of human conflict and strife that played out on the shores of the North Sea. The area was particularly affected by the threat of invasion during the Second World War, with the construction of numerous defensive fortifications and military installations along the coast.

In more recent times, the Trimley Marshes have come to be recognized as a vital haven for wildlife, with growing awareness of the importance of protecting and preserving these fragile and irreplaceable habitats. The designation of the area as a nature reserve in the late 20th century marked a turning point in the history of the marshes, ensuring that they will continue to thrive and inspire for generations to come.

As you walk through the peaceful landscapes of Trimley Marshes, take a moment to reflect on the countless generations of people who have shaped and been shaped by this special place, from the earliest settlers to the dedicated conservationists and volunteers who work tirelessly to protect and enhance its natural wonders. Each step you take is a reminder of the deep and enduring connection between humans and the natural world, and of the vital role that we all play in stewarding and safeguarding these precious habitats for the future.

Though the specific stories and events of Trimley's past may be lost to time, the spirit of resilience, adaptation, and renewal that has always defined this unique corner of the Suffolk coast lives on in the tranquil beauty and rich biodiversity of the marshes themselves. By opening our hearts and minds to the wonders of the natural world, we can all play a part in ensuring that this remarkable legacy endures for generations to come, and that the Trimley Marshes continue to inspire and enrich the lives of all who have the privilege of knowing and loving them.

Rare or Uncommon Birds:

Trimley Marshes Nature Reserve is a true paradise for birdwatchers and nature enthusiasts, providing a vital sanctuary for an astonishing diversity of avian life. The reserve's unique combination of wetland, grassland, and coastal habitats creates the perfect conditions for a wide range of bird species, from the common and familiar to the rare and elusive, each one a marvel of adaptation and survival.

One of the most iconic and sought-after birds to grace the marshes is the avocet, a striking black-and-white wader with a slender, upturned bill perfectly suited for probing the soft mud and shallow waters of the tidal flats. These elegant birds are a true conservation success story, having been brought back from the brink of extinction in the UK in the mid-20th century through dedicated protection and habitat management efforts. Today, the avocets of Trimley Marshes are a symbol of the resilience and tenacity of nature, and a source of joy and inspiration for all who have the privilege of observing them.

Another rare and captivating bird to watch for in the reedbeds and marshes is the bearded tit, a small and elusive passerine with a distinctive black "moustache" and a soft, pinging call that belies its diminutive size. These charming birds are masters of concealment, flitting among the dense stands of reeds and sedges with incredible agility, their presence often only betrayed by the faint rustling of the vegetation and the occasional flash of their tawny plumage. To catch a glimpse of a bearded tit is a true privilege, a fleeting moment of connection with one of nature's most secretive and enchanting creatures.

Soaring overhead, the majestic silhouette of the marsh harrier is another unforgettable sight, its broad, fingered wings and distinctive white rump patch making it a true icon of the wetland landscape. These powerful raptors are a testament to the importance of habitat conservation and protection, having recovered from near-extinction in the UK thanks to the tireless efforts of conservationists and the establishment of key strongholds like Trimley Marshes. As you watch a marsh harrier gliding effortlessly over the swaying reeds, its keen eyes scanning for prey, it is impossible not to feel a sense of awe and reverence for the incredible adaptations and resilience of these magnificent birds.

Other notable and uncommon species to look out for during your visit include the graceful little tern, a diminutive seabird with a bright yellow bill and a dazzling white plumage; the rare and enigmatic bittern, a master of camouflage whose booming call echoes hauntingly across the marshes; and the colourful kingfisher, a blur of electric blue and orange that darts along the waterways and ditches in search of its fishy prey.

As you explore the diverse habitats of Trimley Marshes, take a moment to appreciate the incredible richness and complexity of the avian life that thrives here, each species a vital thread in the intricate web of the ecosystem. From the tiniest wren to the mightiest harrier, every bird plays a crucial role in maintaining the balance and health of the natural world, and in shaping the character and beauty of this remarkable landscape.

By supporting the conservation and protection of reserves like Trimley Marshes, we can all play a part in ensuring that these rare and precious birds continue to thrive and inspire for generations to come, and that the wonder and diversity of the natural world remains a source of joy, discovery, and renewal for all who have the privilege of experiencing it. So keep your eyes to the skies and your heart open to the magic of the marshes, and let the birds of Trimley be your guide and your muse on this unforgettable journey through one of Suffolk's most treasured wild places.

Rare or Interesting Insects and Other Wildlife:

As you make your way through the diverse habitats of Trimley Marshes Nature Reserve, keep your senses attuned to the myriad forms of life that thrive in every nook and cranny of this extraordinary landscape. From the tiniest insects to the most elusive mammals, each species plays a vital role in maintaining the delicate balance and resilience of the ecosystem, their stories of adaptation and survival woven into the very fabric of the marshes themselves.

One of the most captivating and charismatic creatures to inhabit the reserve's waterways and wetlands is the water vole, a charming and industrious rodent with a plump, furry body and a distinctive blunt snout. Often referred to as "Britain's fastest-declining mammal," water voles have suffered severe declines in recent decades due to habitat loss, fragmentation, and predation by non-native American mink. However, thanks to targeted conservation efforts and the protection of key strongholds like Trimley Marshes, these delightful animals are slowly making a comeback, their presence a heartening reminder of the resilience and tenacity of nature in the face of adversity.

As you explore the reserve's grasslands and marshes, keep an eye out for the telltale signs of water vole activity, from the neat piles of chopped vegetation and the distinctive "lawn" of grazed grass around their burrows to the small, round droppings they leave behind as they navigate their watery world. If you're lucky, you may even catch a glimpse of these elusive creatures as they swim and dive among the reeds and rushes, their sleek, chestnut-brown fur glistening in the sunlight.

Another fascinating and often overlooked inhabitant of the reserve's grasslands and hedgerows is the harvest mouse, a tiny and acrobatic rodent with golden-brown fur, a prehensile tail, and a remarkable ability to climb and nest among the swaying stems of grasses and wildflowers. These enchanting little creatures are true architects of the meadows, weaving intricate spherical nests from woven grasses and leaves, their delicate creations suspended like tiny miracles among the summer foliage.

Though harvest mice are notoriously difficult to spot due to their small size and secretive nature, patient observers may be rewarded with a fleeting glimpse of these agile climbers as they navigate the vertical world of the grasslands, their bright, bead-like eyes and twitching whiskers a testament to their ceaseless curiosity and adaptability. The presence of harvest mice in the marshes is a sign of the health and diversity of the grassland habitats, and a reminder of the countless small wonders that often go unnoticed in the shadow of larger, more charismatic species.

For those with a passion for the world of insects, Trimley Marshes is a true wonderland, its mosaics of wetland, grassland, and scrubland habitats supporting an astonishing array of dragonflies, damselflies, butterflies, and other invertebrates. From the shimmering greens and blues of the emperor and southern hawker dragonflies to the dazzling oranges and reds of the small copper and common darter butterflies, each species adds its own unique colour and pattern to the vibrant tapestry of life that unfolds across the reserve.

As you walk along the tranquil waterways and through the swaying grasses, take a moment to appreciate the intricate beauty and fascinating behaviours of these often-overlooked creatures, from the aerial acrobatics of the dragonflies as they hunt and mate on the wing to the delicate, fluttering dance of the butterflies as they sip nectar from the wildflowers. Each insect is a marvel of adaptation and design, its life cycle and survival strategies a testament to the incredible diversity and resilience of the natural world.

By opening our eyes and hearts to the small wonders that surround us in every moment, we can cultivate a deeper sense of connection and appreciation for the intricate web of life that sustains us all, and for the vital role that even the tiniest creatures play in maintaining the health and beauty of our shared world. So as you explore the rich and varied habitats of Trimley Marshes, let the secret lives and untold stories of the reserve's insects and other wildlife be a source of endless fascination and inspiration, a reminder of the preciousness and fragility of the natural world, and a call to cherish and protect these irreplaceable treasures for generations to come.

Rare or Interesting Plants:

As you journey through the ever-changing landscapes of Trimley Marshes Nature Reserve, let your gaze be drawn to the remarkable diversity of plant life that thrives in every corner of this unique and precious habitat. From the hardy, salt-tolerant species that cling to the edges of the tidal marshes to the delicate, ephemeral blooms that dance among the swaying grasses of the meadows, each plant tells a story of adaptation, resilience, and beauty, its presence a vital thread in the intricate tapestry of the ecosystem.

One of the most striking and emblematic plants to grace the saltmarshes of Trimley is the sea lavender, a hardy perennial with slender, branching stems and delicate clusters of pale purple flowers that sway gently in the coastal breeze. Perfectly adapted to the challenging conditions of the tidal zone, sea lavender thrives in the saline, waterlogged soils of the marshes, its deep, robust roots anchoring it firmly against the twice-daily onslaught of the tides.

As you walk along the margins of the marshes, take a moment to appreciate the subtle beauty and tenacity of this remarkable plant, its soft, pastel hues and graceful form a soothing contrast to the rugged, windswept landscape that surrounds it. The presence of sea lavender in the marshes is a sign of the health and integrity of the saltmarsh habitat, its blooms providing vital nectar and pollen for a host of coastal insects and its foliage offering shelter and sustenance for countless small creatures.

Another fascinating and iconic plant to look out for in the marshes is the sea aster, a hardy, salt-tolerant daisy with striking purple-and-yellow flowerheads that bloom in late summer and early autumn. Like sea lavender, sea aster is a specialist of the tidal zone, its thick, succulent leaves and deep-rooted system enabling it to withstand the harsh, saline conditions of the marshes and to thrive where few other plants dare to venture.

As you marvel at the vibrant, starry blooms of the sea aster, consider the vital role that this plant plays in the ecosystem of the marshes, its nectar-rich flowers providing a crucial late-season food source for bees, butterflies, and other pollinators, and its dense, tangled foliage offering cover and nesting sites for a variety of marsh-dwelling birds and small mammals. The presence of sea aster in the marshes is a testament to the incredible adaptability and resilience of nature, and a reminder of the beauty and diversity that can flourish even in the most challenging of environments.

Moving inland from the tidal marshes, the grasslands and meadows of Trimley Marshes reveal a new and equally captivating array of botanical wonders, from the delicate, nodding heads of cowslips and the frothy, cream-coloured umbels of yarrow to the vibrant, golden spikes of agrimony and the shimmering, purple haze of wild thyme. Each plant is a miracle of form and function, its colours, shapes, and scents perfectly attuned to the needs and preferences of the insects, birds, and other creatures that depend on it for food, shelter, and reproduction.

As you wander through the swaying grasses and the drifting clouds of wildflowers, take a moment to contemplate the countless generations of plants that have adapted and evolved to thrive in this dynamic and ever-changing landscape, their seeds and roots carrying the memory and the potential of countless generations past and yet to come. From the tiniest moss to the mightiest oak, each plant is a living testament to the enduring power and creativity of the natural world, and a reminder of the priceless heritage that we have been entrusted to cherish and protect.

One particularly rare and intriguing plant to look out for in the marshes is the sea wormwood, a small, unassuming herb with silver-grey, deeply-lobed leaves and clusters of tiny, yellow flowerheads. Found only in a handful of sites along the Suffolk coast, sea wormwood is a true specialty of the region, its presence in Trimley Marshes a precious and irreplaceable part of the county's natural heritage.

As you search for this elusive and enigmatic plant among the shimmering expanses of the saltmarshes, consider the incredible journey that it has undertaken to reach this special corner of the coast, its seeds carried by the winds and the tides from some distant, unknown shore. The story of sea wormwood is a microcosm of the larger story of Trimley Marshes itself, a tale of resilience, adaptation, and the endless capacity of nature to surprise and delight us with its wonders.

From the tidal margins to the inland meadows, the plants of Trimley Marshes are a vital and irreplaceable part of the reserve's rich and complex ecosystem, their roots and seeds, their leaves and blooms, forever intertwined with the lives and stories of the countless creatures that call this special place home. By cherishing and preserving these botanical treasures, we ensure that the beauty, diversity, and resilience of the natural world will endure for generations to come, and that the countless wonders and mysteries of the marshes will continue to inspire and sustain us all.

Facilities and Accessibility:

Trimley Marshes Nature Reserve offers visitors a unique and immersive opportunity to explore the stunning landscapes and rich biodiversity of the Suffolk coast, with a range of facilities and amenities designed to enhance the accessibility and enjoyment of the site for people of all ages and abilities.

The reserve is easily accessible by car, with a dedicated parking area located at the main entrance, just off the A14 near the village of Trimley St Martin. The car park is well-maintained and offers ample space for visitors, including designated disabled parking bays for those with mobility issues. From the car park, a network of well-signposted paths and trails leads visitors into the heart of the reserve, inviting them to discover the many wonders and delights of this special place at their own pace.

For those who prefer to explore the reserve on foot, the 5-mile (8 km) circular trail offers a leisurely and accessible route through the varied habitats of the marshes, with gentle gradients and well-maintained surfaces suitable for walkers of all ages and abilities. The trail is clearly waymarked and easy to follow, with regular benches and viewing points provided along the way for those who wish to rest, relax, and take in the stunning views across the river and the surrounding countryside.

In addition to the circular trail, Trimley Marshes also features a network of shorter, family-friendly paths and boardwalks that allow visitors to get up close and personal with the amazing wildlife and landscapes of the reserve. These trails are ideal for families with young children, as well as for those with limited mobility or stamina, offering a more intimate and immersive experience of the marshes without the need for long distances or strenuous exertion.

One of the standout features of Trimley Marshes is its impressive array of bird hides, strategically placed throughout the reserve to offer visitors unparalleled views of the many rare and fascinating species that call this place home. These hides are designed with accessibility in mind, with ramped access, wide doorways, and low-level viewing windows that allow people of all abilities to enjoy the same incredible wildlife-watching experiences.

Inside the hides, visitors will find spacious and comfortable seating areas, as well as interpretation boards and displays that provide fascinating insights into the ecology and conservation of the reserve. Many of the hides also feature state-of-the-art optical equipment, such as telescopes and binoculars, which can be used free of charge by visitors to get an even closer look at the birds and other wildlife of the marshes.

Elsewhere on the reserve, visitors will find a range of other amenities and facilities designed to enhance their experience and enjoyment of the site. These include well-maintained public toilets, picnic areas with tables and benches, and a small visitor centre with displays and exhibits on the history, wildlife, and conservation of Trimley Marshes.

While there are no cafes or restaurants within the reserve itself, visitors are welcome to bring their own food and drink and to enjoy a picnic in the designated areas provided. For those who prefer a more substantial meal or refreshment, the nearby villages of Trimley St Martin and Trimley St Mary offer a range of pubs, cafes, and restaurants serving locally-sourced produce and traditional Suffolk fare.

Overall, Trimley Marshes Nature Reserve is a welcoming and accessible destination for visitors of all ages and abilities, with a range of facilities and amenities designed to enhance the enjoyment and appreciation of this stunning coastal landscape. Whether you are a keen birdwatcher, a nature lover, or simply someone in search of a peaceful and inspiring day out in the great outdoors, Trimley Marshes has something to offer everyone, and is sure to leave you with memories and experiences that will last a lifetime.

Seasonal Highlights:

One of the greatest joys of visiting Trimley Marshes Nature Reserve is the opportunity to witness the changing faces and moods of this extraordinary landscape through the turning of the seasons, each one bringing its own unique palette of colours, sounds, and sensations to the marshes and meadows, and each one offering a fresh perspective on the timeless cycles of nature.

In spring, the reserve bursts into life with an exuberant display of colour and vitality, as the first tentative shoots of green push their way through the winter-browned grasses, and the early wildflowers begin to carpet the meadows in a dazzling patchwork of yellow, purple, and white. The air is filled with the joyous chorus of birdsong, as the resident species stake out their territories and the first migrant warblers and waders arrive from their winter quarters to breed and raise their young among the lush vegetation of the marshes.

As you walk through the reserve in these early months of the year, keep an eye out for the delicate, nodding heads of cowslips and primroses, the vibrant splashes of colour provided by the marsh marigolds and celandines, and the frothy, cream-coloured umbels of the cow parsley, each one a harbinger of the growing abundance and diversity of the coming months. This is also a wonderful time to witness the courtship displays and nesting behaviour of the reserve's many bird species, from the soaring flights and haunting cries of the lapwings to the frenetic activity of the tiny warblers as they flit among the reeds and willows.

As spring gives way to summer, the reserve reaches the height of its lushness and verdancy, with the grasses and wildflowers growing tall and thick, and the trees and shrubs heavy with foliage and fruit. The meadows are a sea of waving stems and nodding heads, alive with the hum of insects and the darting flight of butterflies and dragonflies, while the marshes are a tapestry of green and gold, the sinuous channels and glittering pools reflecting the endless blue of the East Anglian sky.

This is a time of abundance and activity in the reserve, with the breeding season in full swing and the young birds and mammals venturing forth to explore their new world. Keep an eye out for the fuzzy, downy chicks of the avocets and redshanks as they follow their parents through the shallows, the playful antics of the young foxes and weasels as they tumble and chase among the grasses, and the acrobatic flight of the swallows and swifts as they swoop and dive over the marshes in pursuit of their insect prey.

As the days begin to shorten and the first hints of autumn colour creep into the foliage, the reserve takes on a new character, the lush greens of summer giving way to a rich palette of reds, golds, and browns. The wildflowers and grasses of the meadows are now adorned with a profusion of seed heads and berries, while the hedgerows and thickets are heavy with the bounty of the season, from the glossy blackberries and scarlet hips to the clusters of purple sloes and orange haws.

This is a time of transition and movement in the reserve, as the summer migrants depart for their winter quarters and the first of the wintering wildfowl and waders arrive from their northern breeding grounds. The skeins of pink-footed and greylag geese passing high overhead, their haunting calls drifting down through the crisp autumn air, are a sure sign of the changing seasons, while the sight of a lone marsh harrier quartering the reedbeds, its chestnut-and-cream plumage glinting in the low, golden light, is a poignant reminder of the cycles of life and death that shape this landscape.

As winter tightens its grip on the marshes, the reserve is transformed into a stark and haunting landscape of frost-rimmed pools and skeletal trees, the bare bones of the land laid bare beneath the watery sun. But even in the depths of the coldest months, Trimley Marshes is alive with the hardy and resilient creatures that make their home here year-round, from the stealthy foxes and badgers that prowl the frozen ground, to the tittering flocks of finches and buntings that brave the icy winds to glean the last of the season's seeds.

This is a time of stillness and contemplation in the reserve, a chance to marvel at the raw beauty and power of the natural world stripped back to its essence. Wrap up warm and head out onto the frosty paths and boardwalks, feeling the crunch of ice beneath your feet and the sting of the wind on your face, and let the profound silence and spaciousness of the winter landscape fill your senses and calm your mind.

As you walk, keep an eye out for the signs of life that persist even in the bleakest of conditions, from the delicate tracery of animal tracks in the snow to the tiny, brave blooms of the winter heliotrope and the scarlet splashes of the holly berries. Each one is a testament to the incredible resilience and adaptability of nature, a reminder that life in all its forms is forever finding ways to endure and thrive, even in the face of the harshest adversity.

And then, as the days begin to lengthen once more and the first green shoots of spring start to emerge from the thawing ground, the cycle begins anew, the eternal dance of the seasons playing out across the timeless stage of Trimley Marshes. To witness this unending flow of life and change, to be a part of the great unfolding story of the natural world, is to be reminded of our own place within the web of existence, and of the precious and irreplaceable gift of the living Earth.

So whether you visit in the lush abundance of high summer or the stark beauty of midwinter, Trimley Marshes has something unique and wonderful to offer, a chance to immerse yourself in the rhythms and wonders of the turning year, and to find solace, inspiration, and renewal in the enduring wildness and majesty of the Suffolk coast. Let the changing faces and voices of the marshes be your guide and your muse, leading you ever deeper into the mystery and magic of the natural world, and revealing to you the timeless truths and beauty that lie at the heart of this extraordinary place.

Difficulty Level and Safety Considerations:

The Trimley Marshes Walk is a relatively easy and accessible route, suitable for walkers of most ages and fitness levels. With its flat terrain, well-maintained paths, and gentle gradients, this 5-mile (8 km) circular walk offers a delightful and undemanding way to explore the stunning landscapes and rich wildlife of the Suffolk coast, without the need for any specialized equipment or technical skills.

However, as with any outdoor activity, it is important to be aware of the potential hazards and challenges that may arise, and to take sensible precautions to ensure a safe and enjoyable experience for all. The most significant consideration for walkers at Trimley Marshes is the tidal nature of the site, with the water levels in the marshes and channels rising and falling twice a day in response to the gravitational pull of the moon.

While the circular walk route is designed to stay safely above the high-water mark, it is still important to be aware of the tides and to plan your visit accordingly, particularly if you intend to venture off the main path or explore the more remote parts of the reserve. Be sure to check the local tide tables before setting out, and to allow plenty of time to complete your walk before the water levels start to rise.

Another factor to bear in mind is the exposed and open nature of the site, which can leave walkers vulnerable to the vagaries of the British weather. While the gentle breezes and mild temperatures of a sunny summer's day can make for a delightful and refreshing walk, the same landscape can quickly turn hostile and unforgiving in the face of strong winds, heavy rain, or sudden changes in temperature.

To mitigate these risks, it is important to come prepared with appropriate clothing and equipment for the conditions, including sturdy, waterproof footwear with good grip and ankle support, warm and waterproof layers, a hat and gloves, and a small backpack with essential supplies such as water, snacks, and a first aid kit. It is also a good idea to let someone know your intended route and estimated return time before setting out, and to carry a fully charged mobile phone in case of emergencies.

For those with mobility issues or other physical limitations, Trimley Marshes offers a range of accessible facilities and amenities to help ensure that everyone can enjoy the beauty and wonder of this special place. The reserve features a network of well-surfaced and gently-graded paths and boardwalks, as well as wheelchair-accessible bird hides and viewing platforms, allowing visitors of all abilities to get up close and personal with the wildlife and landscapes of the marshes.

However, it is important to note that some parts of the reserve may be more challenging or inaccessible for those with restricted mobility, particularly the more remote or uneven sections of the trail. Visitors with disabilities or other special needs are advised to contact the reserve management beforehand to discuss their individual requirements and to get advice on the most suitable routes and facilities for their needs.

Overall, the Trimley Marshes Walk is a safe and enjoyable experience for the vast majority of visitors, offering a wonderful opportunity to connect with the natural world and to discover the incredible diversity and beauty of the Suffolk coastline. By following a few simple guidelines and taking sensible precautions, walkers of all ages and abilities can make the most of this stunning and inspiring landscape, and come away with memories and experiences that will last a lifetime.

So whether you are a seasoned hiker or a casual stroller, a birdwatcher or a nature lover, Trimley Marshes has something wonderful to offer, a chance to immerse yourself in the timeless rhythms and beauty of the natural world, and to find peace, joy, and renewal in the wild and ancient heart of the Suffolk coast. With its gentle terrain, well-maintained trails, and accessible facilities, this enchanting reserve is the perfect destination for anyone seeking to escape the stresses and strains of modern life, and to reconnect with the simple, enduring pleasures of the great outdoors.

Nearby Attractions:

Trimley Marshes Nature Reserve is not only a stunning and inspiring destination in its own right, but also a gateway to a wealth of other natural, cultural, and historical attractions that showcase the rich and varied character of the Suffolk coast and countryside. Whether you are a history buff, a wildlife enthusiast, a foodie, or simply someone in search of a relaxing and rejuvenating day out, the area around Trimley Marshes has something wonderful to offer, with a range of activities and experiences to suit every taste and interest.

For those who want to delve deeper into the natural wonders of the Suffolk coast, the nearby RSPB Felixstowe Ferry reserve is a must-visit destination, offering a chance to explore a different but equally captivating landscape of saltmarsh, mudflat, and shingle beach. This small but highly significant site is home to a rich diversity of birdlife, including breeding populations of avocets, ringed plovers, and little terns, as well as overwintering flocks of brent geese, dunlin, and other waders.

Slightly further afield, the National Trust's Orford Ness reserve is a truly unique and awe-inspiring landscape, a vast, remote shingle spit that juts out into the North Sea like a great, desolate finger of land. Once a top-secret military testing site, Orford Ness is now a haven for wildlife and a monument to the strange and haunting beauty of the Suffolk coast, its eerie, abandoned buildings and windswept expanses of shingle and marsh a testament to the enduring power and mystery of the natural world.

For those who want to explore the rich cultural and historical heritage of the area, the nearby town of Felixstowe is a charming and fascinating destination, with a range of attractions and activities to suit all ages and interests. From the elegant seafront gardens and promenade to the bustling high street with its independent shops and cafes, Felixstowe offers a delightful mix of traditional seaside charm and modern amenities, making it the perfect place to spend a relaxing and enjoyable day out.

One of the most impressive historical sites in Felixstowe is the Landguard Fort, a stunning 18th-century coastal fortification that has played a vital role in defending the town and the wider region from invasion and attack. Today, the fort is open to the public as a museum and visitor centre, offering a fascinating insight into the military and social history of the area, as well as stunning views over the harbor and the North Sea beyond.

For those who want to venture further afield, the county town of Ipswich is just a short train or bus ride away, offering a wealth of cultural, historical, and recreational attractions to explore. From the stunning collections of the Ipswich Museum and the Christchurch Mansion to the vibrant waterfront district with its bars, restaurants, and galleries, Ipswich is a lively and engaging destination that showcases the best of Suffolk's past and present.

Other nearby attractions and activities to consider include the beautiful country estates of Helmingham Hall and Somerleyton Hall, both of which offer stunning gardens, historic houses, and a range of events and activities throughout the year; the charming market towns of Woodbridge and Framlingham, with their picturesque streets, independent shops, and ancient castles; and the Suffolk Coast and Heaths AONB, a vast and varied landscape of heathland, forest, and coastline that offers endless opportunities for walking, cycling, birdwatching, and other outdoor pursuits.

Whether you are looking for a day of adventure and discovery, a chance to relax and unwind in beautiful natural surroundings, or an opportunity to delve into the rich and fascinating history and culture of the region, the area around Trimley Marshes has something wonderful to offer. So why not take some time to explore this enchanting corner of England, and discover for yourself the timeless beauty, diversity, and wonder of the Suffolk coast and countryside?

Felixstowe Promenade North Walk - 2 miles (3.2 km) - Non-circular - Estimated Time: 1-1.5 hours

Starting Point Coordinates, Postcode: The starting point for the Felixstowe Promenade North Walk is at Felixstowe Pier (Grid Reference: TM 299 340, Postcode: IP11 8AB).
Nearest Car Park, Postcode: There are several car parks near Felixstowe Pier, with the closest being the Pier Front Car Park (Postcode: IP11 8AB).

Embark on a delightful 2-mile (3.2 km) non-circular stroll along the picturesque seafront of Felixstowe Beach, where the gentle rhythm of the waves and the fresh, salty breeze combine to create an invigorating and refreshing experience for all. This easy walk offers a perfect opportunity to immerse yourself in the timeless charm and natural beauty of the Suffolk coast, making it an ideal outing for families, couples, and anyone seeking a moment of peace and relaxation by the sea.

As you set out from the historic Felixstowe Pier, a beloved landmark and focal point of the town's seafront, feel the anticipation build as you head south along the wide, well-maintained promenade. With each step, the stresses and distractions of everyday life seem to melt away, replaced by a growing sense of calm and connection with the vast, eternal presence of the ocean stretching out before you.

Take a moment to drink in the stunning panoramic views of the coastline, the golden sands of the beach gleaming in the sunlight and the deep blue waters of the North Sea sparkling and dancing in the distance. Watch as the waves roll in and break upon the shore, their ceaseless motion a soothing and mesmerizing rhythm that echoes the timeless ebb and flow of the tides.

As you continue along the promenade, you'll pass a colourful array of beach huts, each one a charming and iconic symbol of the great British seaside tradition. These simple yet beloved structures have been a fixture of Felixstowe's seafront for generations, providing a place for families and friends to gather, relax, and enjoy the simple pleasures of a day by the coast.

Further along the walk, you'll encounter the impressive Martello Tower, a 19th-century coastal defense fortification that stands as a silent witness to the rich and fascinating history of this region. Take a moment to explore the area around the tower, imagining the lives and stories of the soldiers and civilians who once lived and worked within its walls, and the countless dramas and adventures that have unfolded along this stretch of coast over the centuries.

As you make your way back along the promenade, retracing your steps towards the pier, take the time to savour the many sights, sounds, and sensations of this special place. Feel the warmth of the sun on your face and the gentle caress of the sea breeze in your hair, listen to the joyful cries of the seagulls soaring overhead and the laughter of children playing on the beach, and breathe in the fresh, invigorating air, filled with the scent of salt and the subtle fragrances of the coastal flora.

Throughout your walk, keep an eye out for the diverse array of wildlife that calls this area home, from the common coastal birds like seagulls and terns to the occasional oystercatcher or cormorant. Along the beach, you may also spot various species of marine life, such as crabs and shellfish, as well as the delicate and colourful butterflies and bumblebees that flit among the wildflowers and plants that line the promenade.

As you near the end of your journey, take a moment to reflect on the beauty, simplicity, and resilience of the natural world, and the deep sense of peace and perspective that comes from spending time in its presence. The Felixstowe Promenade North Walk may be a relatively short and easy route, but it offers a truly rewarding and enriching experience, one that will stay with you long after you have left the seafront behind.

Whether you are a local resident or a visitor to this charming corner of Suffolk, this delightful walk is a reminder of the endless wonders and simple joys that await us when we step outside and embrace the beauty and majesty of the coast. So let the timeless allure of the sea be your guide, and discover for yourself the magic and rejuvenating power of a walk along the Felixstowe promenade.

Detailed Route Directions:

1. Begin your walk at Felixstowe Pier, a historic landmark and popular attraction on the seafront. Take a moment to appreciate the pier's architecture and the lively atmosphere surrounding it. (Estimated time: 5 minutes)

2. Head south along the promenade, following the wide, paved path that runs parallel to the beach. As you walk, take in the stunning views of the coastline, with the golden sands stretching out to your left and the vast expanse of the North Sea to your right. (Estimated time: 15-20 minutes)

3. Continue along the promenade, passing by the Felixstowe Leisure Centre and the Spa Pavilion, two popular destinations for locals and visitors alike. These facilities offer a range of activities and entertainment options, from swimming and fitness classes to live music and theatre performances. (Estimated time: 10-15 minutes)

4. As you proceed, you'll encounter a delightful array of colourful beach huts lining the promenade. These iconic structures are a quintessential feature of the British seaside experience, providing a charming and nostalgic touch to your walk. (Estimated time: 10-15 minutes)

5. After approximately 1 mile (1.6 km), you will reach the impressive Martello Tower, a 19th-century coastal defense fortification that stands as a testament to the region's rich military history. Take some time to explore the area around the tower and read the informative plaques that provide insight into its past. (Estimated time: 15-20 minutes)

6. Having reached the halfway point of your walk, turn around and retrace your steps along the promenade, heading back towards Felixstowe Pier. As you walk, take the opportunity to appreciate the coastal landscape from a different perspective and observe any changes in the light, weather, or tidal conditions. (Estimated time: 30-40 minutes)

7. Upon arriving back at Felixstowe Pier, your 2-mile (3.2 km) non-circular walk comes to an end. Take a final moment to reflect on the beauty, tranquility, and natural wonders you've experienced along the way, and consider exploring some of the other attractions and amenities in the area. (Total estimated time: 1-1.5 hours)

Best Features and Views:

- Immerse yourself in the stunning panoramic views of the Felixstowe coastline, with the golden sands of the beach stretching out before you and the vast expanse of the North Sea shimmering in the distance, creating an awe-inspiring and humbling sense of nature's beauty and power.
- Delight in the charming and nostalgic atmosphere of the promenade, lined with colourful beach huts that evoke the timeless charm and simple pleasures of the British seaside experience, offering a delightful and picturesque backdrop to your walk.
- Savour the fresh, invigorating sea breeze and the warm sunshine on your face as you stroll along the promenade, feeling the stresses and worries of everyday life melt away and replaced by a deep sense of calm, rejuvenation, and connection with the natural world.

Historical Features of Interest:

One of the most fascinating and impressive historical features you'll encounter along the Felixstowe Promenade North Walk is the Martello Tower, a rare and well-preserved example of a 19th-century coastal defense fortification that stands as a testament to the rich military heritage of this region.

Constructed in the early 1800s as part of a network of defensive towers along the East Anglian coast, the Martello Tower at Felixstowe was designed to protect the town and its harbor from the threat of French invasion during the Napoleonic Wars. These sturdy, circular towers were built with thick, solid walls and a single, heavily defended entrance, providing a secure and easily defendable stronghold for the soldiers stationed within.

The tower's innovative design, with its curved walls and central courtyard, allowed for a 360-degree field of fire, enabling the defenders to cover all approaches and repel any potential attackers. The tower was also equipped with a cannon on its roof, which could be used to engage enemy ships and provide additional firepower in the event of a land-based assault.

As you explore the area around the Martello Tower, take a moment to imagine the lives and experiences of the soldiers who once manned this remote outpost, keeping a constant vigil over the coast and standing ready to defend their country and community against any threat. The tower's rugged, weathered exterior and stark, functional interior offer a glimpse into the harsh realities and daily routines of military life in the 19th century, a world far removed from the comfort and convenience of modern times.

Today, the Martello Tower stands as a silent witness to the passage of time and the changing tides of history, its presence a powerful reminder of the sacrifices and struggles of generations past. The tower's preservation and accessibility to the public are a testament to the enduring importance and relevance of our shared heritage, and an invitation to explore and appreciate the rich tapestry of stories and experiences that have shaped this unique and fascinating corner of the Suffolk coast.

As you continue your walk along the promenade, carry with you the sense of connection and continuity that the Martello Tower embodies, and let its presence be a source of inspiration and reflection on the enduring resilience and adaptability of the human spirit in the face of adversity and change. The tower's story is just one thread in the rich and complex web of history that underlies this special place, waiting to be discovered and appreciated by all who take the time to explore its secrets and wonders.

Rare or Uncommon Birds:

While the majority of bird species you'll encounter along the Felixstowe Promenade North Walk are common coastal residents such as herring gulls, black-headed gulls, and feral pigeons, there are occasional opportunities to spot some more unusual or less frequently seen birds that add an extra layer of excitement and interest to your birding experience.

One of the most sought-after and charismatic species to look out for is the sandwich tern, a sleek and elegant seabird with a black cap, pale gray wings, and a distinctive yellow-tipped black bill. These agile and graceful fliers are a joy to watch as they hover over the waves, their sharp eyes scanning the surface for the telltale glint of a fish before plunging into the water with a swift, precise dive.

Sandwich terns are summer visitors to the Suffolk coast, arriving in the spring to breed on the shingle beaches and coastal marshes of the region. While they are not rare at a national level, they are relatively uncommon and localized in this part of the country, making a sighting of these beautiful birds a special and memorable moment for any keen birdwatcher.

Another less common species to watch for along the promenade is the oystercatcher, a striking black-and-white wader with a long, bright orange bill and a distinctive piping call. These charismatic birds are a familiar sight on many British coastlines, but are often more associated with rocky shores and estuaries than with sandy beaches like those at Felixstowe.

Oystercatchers are named for their ability to pry open the tightly closed shells of oysters and other molluscs with their strong, chisel-like bills, but they are actually opportunistic feeders that will consume a wide variety of invertebrates and small fish. Look for these bold and assertive birds striding purposefully along the tideline or roosting in small groups on the groynes and breakwaters that punctuate the promenade.

Other less frequently encountered species that you may be lucky enough to spot on your walk include the diminutive and endearing little tern, the chunky and powerful great black-backed gull, and the elegant and acrobatic common tern, all of which add to the diversity and interest of the coastal birdlife.

As you make your way along the promenade, keep your eyes and ears open for any unusual sightings or behaviours, and take the time to appreciate the beauty, adaptability, and resilience of these remarkable creatures as they go about their daily lives in this dynamic and challenging environment. Each bird, whether common or rare, plays a vital role in the complex web of life that sustains the health and vitality of the coastal ecosystem, and offers a unique window into the wonders and mysteries of the natural world.

By cultivating a sense of curiosity, patience, and respect for the birds and other wildlife that share this special place, you can deepen your understanding and appreciation of the rich and varied tapestry of life that exists along the Felixstowe coast, and gain a renewed sense of connection and kinship with the incredible diversity and resilience of the natural world. So keep your binoculars at the ready, and let the birds be your guide and inspiration as you explore the endless wonders and delights of this fascinating and ever-changing landscape.

Rare or Interesting Insects and Other Wildlife:

As you stroll along the Felixstowe Promenade North Walk, keep your eyes peeled for the many fascinating and often overlooked species of insects and other invertebrates that call this dynamic coastal environment home. While these tiny creatures may not be as immediately obvious or charismatic as the birds and larger animals that inhabit the area, they play a vital role in maintaining the health and balance of the ecosystem, and offer endless opportunities for discovery and wonder.

One of the most colourful and eye-catching insects you may encounter on your walk is the small copper butterfly, a dainty and energetic species with bright orange-brown wings dotted with black spots and a distinctive metallic sheen. These delightful little butterflies are often found flitting among the wildflowers and grasses that grow along the edges of the promenade, their rapid, darting flight and restless energy a joy to behold.

Small coppers are relatively common and widespread across much of the UK, but are always a pleasure to see, their vibrant colours and lively behaviour adding a splash of excitement and charm to any walk. Look for them basking on warm stones or bare patches of ground, their wings held flat to soak up the sun's rays, or nectaring on the blooms of coastal plants like sea aster and sea lavender.

Another intriguing invertebrate to watch for along the promenade is the common hermit crab, a small, unassuming creature that plays a vital role in the complex web of life that exists along the shore. These resourceful crustaceans are scavengers and opportunists, feeding on a wide variety of organic matter and detritus that washes up on the beach, helping to keep the sand clean and recycling nutrients back into the ecosystem.

What makes hermit crabs so fascinating is their unique adaptation of using the empty shells of other animals, such as periwinkles and whelks, as portable homes to protect their soft, vulnerable abdomens. As they grow, they must periodically seek out larger shells to accommodate their increasing size, leading to a constant game of musical chairs as they trade up to bigger and better accommodations.

Look for hermit crabs scuttling along the water's edge or among the seaweed and debris that litters the tideline, their adopted shells ranging in size, shape, and colour depending on the species and age of the crab. If you're lucky, you may even witness the fascinating spectacle of a hermit crab swapping shells with another individual, a delicate and carefully choreographed dance that showcases the incredible adaptability and social behaviour of these remarkable creatures.

Other interesting invertebrates to keep an eye out for on your walk include the colourful and charismatic compass jellyfish, the sleek and powerful great green bush-cricket, and the industrious and endlessly fascinating common sand hopper, each one a unique and valuable thread in the rich tapestry of coastal life.

As you marvel at the diversity and complexity of the insect and invertebrate life that surrounds you, take a moment to reflect on the vital role that these often-overlooked creatures play in maintaining the health and resilience of the coastal environment. From pollinating the wildflowers that brighten the promenade to breaking down the organic matter that washes ashore, each species contributes in its own small but significant way to the intricate web of relationships that sustain the beauty and vitality of this special place.

By opening your eyes and mind to the wonders and mysteries of the miniature world that exists all around you, you can gain a deeper appreciation and respect for the incredible diversity and adaptability of life in all its forms, and cultivate a sense of connection and kinship with the countless tiny beings that share our planet. So let your curiosity be your guide, and allow yourself to be amazed and inspired by the endless marvels and surprises that await you on your journey along the Felixstowe coast.

Rare or Interesting Plants:

Though the Felixstowe Promenade North Walk is primarily a paved, urban route, the edges and margins of the path are home to a surprising diversity of coastal plants and wildflowers that add a touch of colour, texture, and ecological interest to your journey. While many of these species are common and widespread along the East Anglian coast, they each have their own unique adaptations and stories to tell, and offer a fascinating window into the resilience and beauty of the natural world.

One of the most distinctive and eye-catching plants you may encounter on your walk is the sea kale, a robust, fleshy-leaved perennial that thrives in the harsh, salty conditions of the coastal environment. Growing up to 75cm tall, sea kale has thick, waxy, blue-green leaves that are deeply lobed and curled, giving the plant a striking, almost sculptural appearance. In the summer months, sea kale produces clusters of small, white flowers that are beloved by bees and other pollinators, followed by rounded, green seed pods that turn black as they ripen.

Sea kale is not only a visually impressive plant but also has a long history of culinary and medicinal use, with its young leaves and shoots being harvested and blanched like celery, and its roots being used to treat a variety of ailments. Today, sea kale is an important pioneer species that helps to stabilize and enrich the sandy soils of the coast, paving the way for other plants to take hold and flourish.

Another common but fascinating plant to look out for along the promenade is the sea mayweed, a low-growing, bushy annual with finely divided, feathery leaves and small, daisy-like flowers. These cheerful little blooms, which range in colour from white to pale pink, are a familiar sight along the upper reaches of the beach, where they add a splash of brightness and charm to the otherwise sparse and windswept landscape.

Sea mayweed is a tough and adaptable plant that can tolerate the harsh, salty conditions of the coastal environment, its deep taproot allowing it to anchor itself securely in the shifting sands and shingle. Like many coastal wildflowers, sea mayweed plays an important role in the ecosystem, providing food and shelter for a wide variety of insects and other small creatures, and helping to bind and stabilize the soil against the erosive forces of wind and wave.

Other interesting coastal plants to watch for on your walk include the vibrant pink and purple blooms of the sea rocket, the delicate, papery flowers of the sea sandwort, and the tough, leathery leaves of the sea plantain, each one a testament to the incredible adaptations and survival strategies that have evolved over millennia to allow life to thrive in this challenging and dynamic environment.

As you wander along the promenade, take a moment to appreciate the tenacity and resilience of these often-overlooked plants, and the vital role they play in maintaining the health and beauty of the coastal landscape. From stabilizing the dunes and cliffs to providing food and habitat for countless species of wildlife, each plant is an essential piece of the complex puzzle that makes up this unique and precious ecosystem.

By learning to recognize and appreciate the diversity and value of the coastal flora, you can deepen your understanding and connection with the natural world, and gain a renewed sense of wonder and respect for the incredible adaptations and strategies that allow life to flourish in even the most challenging of environments. So let your curiosity and sense of discovery be your guide, and allow yourself to be amazed and inspired by the endless wonders and surprises that await you in the hidden world of the Felixstowe coast.

Facilities and Accessibility:

The Felixstowe Promenade North Walk is a highly accessible and well-maintained route that offers a range of facilities and amenities to enhance the comfort, safety, and enjoyment of visitors of all ages and abilities. Whether you're a seasoned walker or a casual stroller, this delightful coastal path has something to offer everyone, making it the perfect destination for a refreshing and rejuvenating day out by the sea.

One of the key features of the promenade is its wide, flat, and smoothly paved surface, which makes it easily navigable for people with mobility issues, as well as for families with strollers or young children. The path is well-lit and clearly marked, with regular benches and shelters provided along the way for those who need to rest or take a break from the sun and wind.

At the starting point of the walk, near Felixstowe Pier, visitors will find a range of facilities and services to help them prepare for their journey and make the most of their time on the coast. These include public restrooms with disabled access, a visitor information centre with maps, guides, and local tips, and a variety of shops and kiosks selling refreshments, snacks, and beach essentials.

As you make your way along the promenade, you'll pass several cafes, restaurants, and pubs that offer a range of dining options to suit every taste and budget, from quick and casual bites to more substantial sit-down meals. Many of these establishments have outdoor seating areas that provide stunning views of the sea and the surrounding landscape, making them the perfect place to relax and refuel after a invigorating walk.

For those interested in learning more about the history, culture, and ecology of the area, the Felixstowe Leisure Centre and the Spa Pavilion are two popular attractions that are well worth a visit. These facilities offer a variety of educational and recreational activities, including exhibitions, workshops, and live performances, as well as fitness classes and swimming pools for those looking to stay active and healthy.

In terms of accessibility, the Felixstowe Promenade North Walk is a shining example of a route that has been designed and maintained with the needs of all users in mind. From the level, obstacle-free path to the numerous benches, handrails, and ramps that provide support and assistance where needed, every effort has been made to ensure that this beautiful coastal walk is open and welcoming to everyone, regardless of their age, ability, or background.

For those with specific accessibility requirements, it is always a good idea to check with the local council or visitor information centre before setting out, to ensure that the facilities and services you need are available and in good working order. However, with its commitment to inclusivity, sustainability, and visitor welfare, the Felixstowe Promenade North Walk is a model of best practice when it comes to creating a safe, enjoyable, and accessible outdoor experience for all.

So whether you're a local resident looking for a pleasant and refreshing way to spend an afternoon, or a visitor to the area seeking to discover the natural and cultural wonders of the Suffolk coast, the Felixstowe Promenade North Walk is the perfect destination for a memorable and rewarding day out. With its stunning views, invigorating sea air, and wealth of facilities and amenities, this delightful coastal route offers something for everyone, and is sure to leave you feeling refreshed, inspired, and connected to the incredible beauty and diversity of the natural world.

Seasonal Highlights:

One of the greatest joys of walking the Felixstowe Promenade North Walk is the opportunity to experience the changing moods, colours, and textures of the Suffolk coast through the turning of the seasons. From the fresh, vibrant greens of spring to the rich, golden hues of autumn, each season brings its own unique character and charm to this beautiful stretch of coastline, inviting walkers to discover and delight in the endlessly varied and fascinating world of nature.

In the spring, the promenade comes alive with a riot of colour and activity, as the first brave wildflowers push their way through the cracks in the pavement and the migrating birds return from their winter quarters to breed and feed along the shore. The air is filled with the joyous songs of skylarks and the piping calls of oystercatchers, while the beach is dotted with the delicate, pastel-hued shells of the season's first butterflies.

As you walk along the promenade in these early months of the year, take a moment to savour the freshness and vitality of the spring sunshine, and to marvel at the incredible resilience and adaptability of the plants and animals that call this place home. From the tough, salt-tolerant grasses that cling to the dunes to the hardy, resourceful gulls that patrol the tideline, each species has its own remarkable story of survival and renewal to tell.

As spring gives way to summer, the promenade becomes a hive of activity and life, with locals and visitors alike flocking to the beach to soak up the sun, play in the sand, and enjoy the many attractions and amenities on offer. The wildflowers are now at their peak, with the vibrant pinks and purples of the sea rocket and the delicate whites and yellows of the sea mayweed creating a dazzling display of colour along the path.

In the height of summer, the promenade is also a great place to spot some of the more unusual and charismatic species that visit the Suffolk coast, from the elegant sandwich terns that dive for fish just offshore to the sleek and powerful harbour porpoises that occasionally grace the waters with their presence. Keep your eyes and ears open for any signs of these special visitors, and take the time to appreciate the incredible diversity and beauty of the marine life that thrives just beyond the shoreline.

As the days begin to shorten and the leaves start to turn, the promenade takes on a new character, with the rich, warm colours of autumn painting the landscape in shades of gold, orange, and red. The beach is now quieter and more contemplative, with the sound of the waves and the cries of the gulls creating a peaceful and meditative atmosphere that is perfect for a thoughtful and reflective walk.

In the autumn months, the promenade is also a great place to witness the spectacle of bird migration, as thousands of waders, ducks, and geese pass through the area on their way to their winter feeding grounds. Keep your eyes peeled for the distinctive silhouettes of brent geese and wigeons as they fly in tight formations overhead, and listen out for the haunting calls of the curlews and redshanks as they probe the mudflats for food.

As winter tightens its grip and the snow and ice transform the promenade into a stark and ethereal landscape, the true magic and mystery of the Suffolk coast is revealed. Wrap up warm and brave the elements for a bracing and invigorating walk along the shore, marvelling at the raw power and beauty of the wind and the waves, and savouring the profound sense of solitude and connection with the wild that comes from being alone in nature.

Even in the depths of winter, the promenade is alive with the hardy and resilient creatures that make their home here year-round, from the faithful turnstones that pick their way along the tideline to the fearless foxes that venture out onto the beach in search of food. Each encounter is a reminder of the incredible toughness and adaptability of life in the face of adversity, and an invitation to find strength and inspiration in the enduring spirit of the natural world.

No matter what time of year you visit, the Felixstowe Promenade North Walk offers a wealth of seasonal delights and surprises to discover and enjoy, each one a unique and precious window into the ever-changing tapestry of life on the Suffolk coast. So come prepared for the weather, keep your eyes and mind open to the wonders around you, and let the timeless beauty and magic of this special place work its spell on your soul, leaving you refreshed, inspired, and forever changed by the power and majesty of nature.

Difficulty Level and Safety Considerations:

The Felixstowe Promenade North Walk is a relatively easy and straightforward route that is suitable for walkers of all ages and abilities, thanks to its flat, well-maintained path and numerous amenities and facilities along the way. With no steep inclines, rough terrain, or other significant physical challenges to contend with, this delightful coastal walk is the perfect choice for a relaxing and enjoyable day out by the sea, whether you're a seasoned rambler or a casual stroller.

However, as with any outdoor activity, it's important to be aware of the potential hazards and to take sensible precautions to ensure a safe and enjoyable experience for everyone. The most significant consideration for walkers on the Felixstowe Promenade North Walk is the weather, which can be changeable and unpredictable at any time of year, with strong winds, heavy rain, and even occasional storms all possible along this exposed stretch of coastline.

To mitigate the risks posed by inclement weather, it's essential to check the forecast before setting out and to come prepared with appropriate clothing and equipment for the conditions. In the summer months, this means wearing light, breathable layers that can be easily removed or added as needed, as well as a hat, sunglasses, and sunscreen to protect against the strong coastal sun. In the cooler months, a warm, waterproof jacket, sturdy shoes, and a cozy hat and gloves are all essential items to pack.

Another important safety consideration for walkers on the promenade is the proximity of the sea, which can be deceptively calm and inviting one moment and rough and dangerous the next. While the wide, sandy beach and gently sloping shore may seem like the perfect place for a paddle or a swim, it's crucial to remember that the North Sea is a powerful and unpredictable body of water, with strong currents, hidden depths, and sudden changes in conditions that can catch even the most experienced swimmers off guard.

To stay safe and avoid any accidents or emergencies, walkers should always obey any warning signs or lifeguard instructions, and refrain from entering the water unless it is explicitly designated as a safe swimming area. Children and vulnerable individuals should be closely supervised at all times, and everyone should be aware of the signs of rip currents, undertows, and other hazards that can pose a risk to those who venture too far from the shore.

In addition to these weather and water-related considerations, walkers on the Felixstowe Promenade North Walk should also be mindful of the other users of the path, including cyclists, joggers, and dog walkers, and take care to share the space responsibly and courteously. This means keeping to the left, allowing faster users to pass safely, and keeping dogs on a lead and under control at all times.

Finally, while the promenade is generally a safe and well-maintained environment, it's always a good idea to carry a basic first-aid kit and a fully charged mobile phone in case of any minor accidents or emergencies. In the unlikely event of a more serious incident, walkers should know how to contact the emergency services and provide clear and accurate information about their location and situation.

By following these simple guidelines and using common sense and good judgment, walkers of all ages and abilities can enjoy a safe, enjoyable, and rewarding experience on the Felixstowe Promenade North Walk. With its stunning views, invigorating sea air, and wealth of natural and cultural wonders to discover, this beautiful coastal route is the perfect destination for anyone looking to escape the stresses and strains of everyday life and connect with the timeless beauty and majesty of the Suffolk coast. So come prepared, stay alert, and let the magic of this special place work its spell on your mind, body, and soul.

Nearby Attractions:

The Felixstowe Promenade North Walk is not only a delightful and invigorating experience in its own right but also a gateway to a wealth of nearby attractions and destinations that showcase the rich history, culture, and natural beauty of the Suffolk coast. Whether you're a history buff, a nature lover, a foodie, or simply someone in search of a fun and varied day out, the area around Felixstowe has something to offer everyone, with a range of activities and experiences to suit all ages, interests, and budgets.

One of the most popular and fascinating nearby attractions is Landguard Fort, a stunning 18th-century coastal fortification that has played a vital role in defending the town and the wider region from invasion and attack over the centuries. Located just a short walk from the southern end of the promenade, the fort is now open to the public as a museum and visitor centre, offering a unique and immersive glimpse into the rich military and social history of the area.

As you explore the fort's winding tunnels, underground chambers, and panoramic bastions, you'll discover a treasure trove of artifacts, exhibits, and interactive displays that bring the past vividly to life. From the grand parade ground where soldiers once drilled and marched to the cramped and dimly-lit barrack rooms where they slept and ate, every corner of the fort has a story to tell, and a lesson to teach about the bravery, sacrifice, and resilience of those who served here over the years.

For those with an interest in local history and culture, the nearby Felixstowe Museum is another must-visit destination, housed in a beautifully restored Edwardian seaside villa just a short stroll from the promenade. With its fascinating collections of photographs, artifacts, and memorabilia spanning over 130 years of the town's history, the museum offers a unique and intimate window into the lives and experiences of the people who have shaped this special corner of the Suffolk coast.

From the elegant drawing rooms and cozy parlours of the Victorian era to the bustling holiday camps and colourful beach huts of the 1950s and 60s, the museum's exhibits and displays paint a vivid and evocative picture of Felixstowe's rich and varied past, and the enduring spirit of community, creativity, and resilience that has always defined this charming and welcoming town.

For nature lovers and outdoor enthusiasts, the nearby Landguard Nature Reserve is a true paradise, with over 33 hectares of diverse and pristine habitat to explore, including shingle beaches, saltmarshes, and rare vegetated shingle communities. Located just a short walk from the southern end of the promenade, the reserve is home to an astonishing variety of wildlife, from the darting warblers and chattering terns of the reedbeds to the scuttling crabs and darting fish of the rockpools and mudflats.

With its network of well-maintained trails, observation hides, and interpretation boards, the reserve offers a unique and immersive opportunity to discover the secrets and wonders of this special corner of the Suffolk coast, and to connect with the incredible diversity and resilience of the natural world. Whether you're a keen birdwatcher, a budding botanist, or simply someone with a love of wild and beautiful places, Landguard Nature Reserve is sure to delight and inspire you, and leave you with memories to treasure for a lifetime.

Of course, no visit to Felixstowe would be complete without sampling some of the delicious local seafood and other culinary delights on offer in the town's many restaurants, cafes, and pubs. From the freshest fish and chips to the most indulgent ice creams and afternoon teas, Felixstowe has something to tempt every palate and satisfy every craving, with a range of options to suit all tastes and budgets.

For a classic seaside dining experience, head to one of the many beachfront kiosks or takeaways along the promenade, where you can enjoy a steaming bag of golden, crispy cod and chunky, fluffy chips, smothered in salt and vinegar and eaten al fresco with a view of the sea. Or for a more refined and sophisticated meal, book a table at one of the town's many excellent restaurants, where you can sample the very best of local and seasonal produce, from succulent Orford oysters to tender Blythburgh pork, all washed down with a refreshing pint of Suffolk ale or a crisp, aromatic glass of wine.

Other culinary highlights to look out for in Felixstowe include the town's famous seafood platters, piled high with plump prawns, sweet crab meat, and juicy mussels; the irresistible cream teas served in the elegant tearooms and hotels along the seafront; and the artisan cheeses, chutneys, and chocolates crafted by the skilled producers and makers of the Suffolk Coast and Heaths Food and Drink Network.

Beyond the immediate vicinity of Felixstowe, the wider Suffolk coast and countryside offer a wealth of further attractions and destinations to explore, from the picturesque fishing villages and unspoiled beaches of the Heritage Coast to the grand stately homes and gardens of the Shotley Peninsula. Whether you're a keen walker, a history lover, a birdwatcher, or simply someone in search of a relaxing and rejuvenating break in a beautiful and welcoming part of the world, this special corner of East Anglia has something wonderful to offer you.

So why not extend your visit to the Felixstowe Promenade North Walk and take some time to discover the many hidden gems and unexpected delights of the Suffolk coast and countryside? With its stunning scenery, fascinating history, delicious food and drink, and warm and friendly people, this enchanting region is sure to capture your heart and imagination, and leave you with memories and experiences to cherish for a lifetime.

Some other nearby attractions and destinations to consider include:

1. Orford Castle: A magnificent 12th-century keep with stunning views over the Orford Ness nature reserve and the Suffolk coast, offering a fascinating glimpse into the region's rich medieval history and architecture.

2. Sutton Hoo: An awe-inspiring Anglo-Saxon burial site and museum, home to the incredible treasures and artifacts discovered in the famous ship burial of an East Anglian king, and a must-see destination for anyone interested in the early history and culture of England.

3. Aldeburgh: A charming and unspoiled coastal town, renowned for its beautiful beach, historic high street, and thriving arts and music scene, as well as its association with the composer Benjamin Britten and the annual Aldeburgh Festival.

4. Rendlesham Forest: A vast and enchanting woodland, steeped in history and legend, with miles of walking and cycling trails, a thrilling UFO trail inspired by the famous sightings of 1980, and a wealth of wildlife and natural wonders to discover.

5. Woodbridge: A picturesque market town on the River Deben, with a rich maritime heritage, a stunning 12th-century Tide Mill, and a variety of independent shops, galleries, and eateries to explore, as well as easy access to the beautiful surrounding countryside and coast.

Whether you're a first-time visitor or a seasoned explorer of the Suffolk coast and countryside, the Felixstowe Promenade North Walk and its nearby attractions offer a wealth of opportunities for discovery, adventure, and relaxation. So why not let your curiosity and sense of wonder be your guide, and embark on a journey of exploration and delight that will leave you refreshed, inspired, and connected to the enduring beauty and magic of this special corner of East Anglia?

With its stunning views, invigorating sea air, fascinating history, and warm and welcoming people, the Felixstowe Promenade North Walk is the perfect starting point for a deeper and more rewarding engagement with the natural and cultural treasures of the Suffolk coast and countryside. So lace up your walking boots, pack a picnic and a sense of adventure, and set out on a journey of discovery that will stay with you long after you've left the beach behind, and that will call you back again and again to this enchanting and endlessly fascinating part of the world.

Felixstowe Promenade South Walk - 2 miles (3.2 km) - Seafront - Estimated Time: 1-1.5 hours

Starting Point Coordinates: The Felixstowe Promenade South Walk starts at Felixstowe Pier (Grid Reference: TM 301 343).
Nearest Car Park: Several car parks are available near Felixstowe Pier and along the promenade, such as the Pier Bight Car Park (Grid Reference: TM 301 343) and Ranelagh Road Car Park (Grid Reference: TM 298 344). Charges may apply, so be sure to check the signage for details.

Embark on a delightful 2-mile (3.2 km) seafront stroll along the picturesque Felixstowe Beach, where the fresh sea breeze and stunning coastal views combine to create an invigorating and refreshing experience for all. This easy walk offers the perfect opportunity to immerse yourself in the timeless charm and natural beauty of the Suffolk coast, making it an ideal outing for families, couples, and anyone seeking a moment of relaxation and rejuvenation by the sea.

As you set out from the iconic Felixstowe Pier, a beloved landmark and focal point of the town's seafront, feel the anticipation build as you head south along the wide, well-maintained promenade. With each step, the stresses and distractions of everyday life seem to melt away, replaced by a growing sense of peace and connection with the vast, eternal presence of the North Sea stretching out before you.

Take a moment to drink in the breathtaking panoramic views of the coastline, with the golden sands of the beach and the gently rolling waves creating a mesmerizing tableau that soothes the soul and invigorates the senses. As you continue along the promenade, you'll pass a colourful array of beach huts, each one a charming reminder of the simple pleasures and nostalgic traditions of the British seaside.

Further along the walk, the Seafront Gardens provide a lush and vibrant oasis amidst the coastal landscape, with beautifully maintained flower beds, manicured lawns, and shady shrubberies offering a peaceful retreat from the sun and the sea. Take a moment to explore the gardens, admiring the diverse array of plants and flowers, and perhaps even spotting some of the resident songbirds and butterflies that call this tranquil haven home.

As you approach the southern end of the promenade, you'll encounter two fascinating historic sites that offer a glimpse into the rich military heritage of the Suffolk coast. The Martello Tower, a small defensive fort built in the early 19th century to protect against the threat of French invasion, stands as a testament to the ingenuity and determination of the region's past inhabitants. Nearby, the impressive Landguard Fort, with its sprawling bastions and commanding views over the sea, tells the story of centuries of coastal defense and the enduring importance of this strategic location.

After exploring these intriguing remnants of history, retrace your steps along the promenade, taking in the ever-changing play of light and shadow on the water and savouring the invigorating sea air as you make your way back towards Felixstowe Pier. As you near the end of your walk, reflect on the timeless beauty and enduring allure of this special corner of the Suffolk coast, and the sense of peace, perspective, and connection that comes from immersing oneself in the natural world.

Whether you're a local resident seeking a refreshing escape from the everyday or a visitor discovering the charms of Felixstowe for the first time, the Promenade South Walk promises an unforgettable experience that will leave you feeling rejuvenated, inspired, and deeply attuned to the simple joys and profound wonders of life by the sea. So step out onto the promenade, let the warm sun and gentle breeze be your guide, and open your heart to the magic and beauty of this enchanting coastal gem.

Route Directions:

1. Begin your walk at Felixstowe Pier, taking a moment to appreciate the lively atmosphere and sea views. Head south along the promenade. (5 minutes)

2. Continue along the promenade, passing by colourful beach huts, cafes, and the Seafront Gardens. Enjoy the coastal scenery and fresh sea air. (30-40 minutes)

3. Upon reaching the southern end of the promenade, explore the Martello Tower and Landguard Fort, learning about their historical significance. (20-30 minutes)

4. After visiting these sites, retrace your steps along the promenade, heading back towards Felixstowe Pier. (30-40 minutes)

5. Conclude your walk at Felixstowe Pier, reflecting on the beauty and tranquility of your seaside stroll. (1-1.5 hours total)

Best Features and Views:

- Immerse yourself in the stunning coastal panoramas, with golden sands, rolling waves, and the endless blue horizon creating a mesmerizing and soul-soothing tableau.
- Delight in the vibrant colours and nostalgic charm of the beach huts lining the promenade, each one a whimsical reminder of the simple pleasures of the British seaside.
- Discover the lush tranquility of the Seafront Gardens, a verdant oasis brimming with diverse flora and fauna, offering a peaceful respite from the sun and sea.

Historical Landmarks:

The Felixstowe Promenade South Walk showcases two significant historical landmarks that offer fascinating insights into the region's rich military heritage and strategic importance:

1. Martello Tower: This small defensive fort, built in the early 19th century, is part of a network of towers constructed along the East Anglian coast to protect against the threat of French invasion during the Napoleonic Wars. The tower's sturdy, circular design and thick walls were engineered to withstand heavy artillery fire, while its elevated position provided a clear line of sight for spotting and engaging enemy ships. Today, the Martello Tower stands as a testament to the ingenuity and resilience of the region's past inhabitants, offering visitors a tangible connection to a pivotal chapter in British history.

2. Landguard Fort: Situated at the southern tip of Felixstowe, Landguard Fort has played a crucial role in defending the Suffolk coast for centuries. The current fort, built in the 18th century, is an impressive example of coastal defense architecture, with its sprawling bastions, deep moats, and commanding views over the sea. Throughout its long history, Landguard Fort has witnessed numerous threats and incursions, from the Dutch Raid of 1667 to the two World Wars of the 20th century. Today, the fort is open to the public as a museum and visitor attraction, offering a fascinating glimpse into the lives of the soldiers who once manned its walls and the enduring significance of this strategic location.

As you explore these historical landmarks, take a moment to imagine the challenges and triumphs of the generations who have shaped and defended this coastline, and the profound impact their efforts have had on the region's identity and development. The Martello Tower and Landguard Fort serve as powerful reminders of the importance of preserving and engaging with our shared heritage, and the invaluable lessons and insights we can gain from understanding the past.

Birds and Wildlife:

As you stroll along the Felixstowe Promenade South Walk, keep your eyes and ears open for the diverse array of birds and wildlife that call this coastal habitat home. The mix of sandy beaches, rocky groins, and landscaped gardens provides a varied range of environments that support a fascinating assortment of species.

In the skies and along the shore, you're likely to spot several species of seabirds, including the ever-present herring gulls and black-headed gulls, as well as the more distinctive common terns and cormorants. These birds are expertly adapted to life by the sea, with their streamlined bodies, waterproof plumage, and sharp bills perfect for diving after fish and other marine prey. Listen for the raucous cries of the gulls and the high-pitched, staccato calls of the terns as they wheel and soar above the waves, and watch for the cormorants perched on the groynes or spreading their wings to dry in the sun after a diving session.

As you pass through the Seafront Gardens, keep an eye out for a variety of songbirds that find shelter and sustenance among the lush vegetation. Robins, blackbirds, and finches are common sights, their bright plumage and melodious songs adding a cheerful note to the peaceful surroundings. You may also spot some butterflies flitting among the flowers, or bees and other pollinators buzzing from bloom to bloom in search of nectar.

Along the promenade itself, the grassy areas and flower beds provide habitat for a range of insects and small animals. Look for ladybugs, grasshoppers, and other miniature wonders as you walk, and keep an eye out for the occasional lizard basking in the sun on the warm concrete. If you're lucky, you might even catch a glimpse of a cheeky squirrel or a shy hedgehog scurrying across your path.

For those with a keen interest in marine life, the waters off Felixstowe can sometimes offer surprising and delightful encounters. Seals and porpoises have been known to make appearances close to shore, their sleek forms cutting through the waves with effortless grace. While sightings are relatively rare, the possibility of spotting these charismatic creatures adds an extra layer of excitement and anticipation to your walk.

By taking the time to observe and appreciate the diverse wildlife that inhabits this coastal environment, you'll gain a deeper understanding and appreciation of the intricate web of life that thrives along the Suffolk shoreline. Each species, from the tiniest insect to the most majestic seabird, plays a vital role in maintaining the health and balance of the ecosystem, and offers a fascinating window into the wonders and mysteries of the natural world.

Interesting Plants:

The Felixstowe Promenade South Walk offers a wealth of opportunities to discover and admire a diverse array of interesting plants, both in the carefully tended Seafront Gardens and along the wild and rugged coastline.

As you meander through the Seafront Gardens, you'll be treated to a vibrant display of ornamental flowers, shrubs, and trees, each one carefully chosen to thrive in the unique conditions of the coastal environment. From the delicate, pastel hues of the roses to the rich, velvety tones of the lavender, the gardens are a feast for the senses, with their heady fragrances and eye-catching colours drawing in visitors and pollinators alike.

One of the most striking and unusual features of the gardens is the presence of palm trees, their towering, feathery fronds adding an exotic and tropical touch to the otherwise quintessentially English landscape. These hardy specimens are a testament to the mild, maritime climate of the Suffolk coast, which allows for the cultivation of a wider range of plants than might be possible further inland.

Beyond the manicured confines of the gardens, the natural coastline is home to a fascinating assemblage of native plants that have adapted to the harsh and unforgiving conditions of life by the sea. Among the most iconic and recognizable of these is the sea holly, a spiky, blue-green plant with a distinctive metallic sheen to its leaves and stems. Its tough, leathery foliage and deep tap roots allow it to thrive in the dry, sandy soils of the dunes, while its striking appearance and nectar-rich flowers make it a favorite of photographers and pollinators alike.

Another common sight along the coast is the sea thrift, a low-growing, cushion-forming plant with delicate, pink pompom flowers that bloom in the spring and early summer. This tough little plant is perfectly adapted to life in the salt-sprayed, windswept environment of the shore, with its thick, fleshy leaves and deep roots enabling it to withstand the constant buffeting of the elements.

As you walk along the promenade, take a moment to appreciate the resilience and adaptability of these coastal plants, and the vital role they play in stabilizing the dunes, providing food and shelter for wildlife, and adding colour and texture to the landscape. Each species is a marvel of evolution and a testament to the incredible diversity and tenacity of life on our planet.

By exploring the plants of the Felixstowe Promenade South Walk, both wild and cultivated, you'll gain a new appreciation for the beauty and complexity of the natural world, and the delicate balance that exists between the land, the sea, and the myriad forms of life that call this special place home. So take your time, look closely, and let the wonders of the coastal flora be your guide and inspiration as you journey along this enchanting stretch of the Suffolk shoreline.

Facilities and Accessibility:

The Felixstowe Promenade South Walk offers a range of facilities and amenities to ensure a comfortable and enjoyable experience for visitors of all ages and abilities. The well-maintained promenade provides a flat, even surface that is suitable for wheelchairs, mobility scooters, and pushchairs, making it an accessible and inclusive route for everyone to enjoy.

Along the promenade, you'll find plenty of benches and seating areas where you can pause to rest, take in the stunning sea views, or enjoy a picnic with family and friends. These regular rest points are particularly valuable for those with limited mobility or stamina, allowing for a more relaxed and manageable walking experience.

Public toilets can be found at several locations along the route, including near the Pier and at the Landguard Fort Visitor Centre. These facilities are well-maintained and accessible, with disabled toilets available for those who require them.

For those in need of refreshment, the promenade is lined with a variety of cafes, kiosks, and seaside stalls offering a range of drinks, snacks, and light meals. Whether you're in the mood for a traditional ice cream, a hot cup of tea, or a delicious fish and chips, you'll find plenty of options to suit your taste and budget.

Families with children will appreciate the several play areas and amusements located along the promenade, providing opportunities for fun and entertainment between stretches of walking. These include traditional seaside attractions like bouncy castles, trampolines, and arcade games, as well as modern play equipment and open spaces for running and playing.

At the southern end of the walk, the Landguard Fort Visitor Centre offers additional facilities, including a café, gift shop, and interpretive displays that provide insight into the history and significance of this important coastal defense site. The fort itself has some uneven surfaces and steps, so visitors with mobility issues should take care when exploring the site.

In terms of parking, several car parks are available near Felixstowe Pier and along the promenade, making it easy and convenient to access the walk. The Pier Bight Car Park and Ranelagh Road Car Park are both located within easy reach of the starting point, with pay-and-display charges applicable. Please check the signage for current rates and time limits.

Overall, the Felixstowe Promenade South Walk is a highly accessible and well-equipped route that caters to a wide range of visitors and abilities. With its flat terrain, regular rest points, and plentiful amenities, it offers a comfortable and enjoyable way to experience the beauty and tranquility of the Suffolk coast, regardless of your age, fitness level, or mobility requirements. So come prepared with sunscreen, a hat, and a sense of adventure, and let the timeless charms of this seaside gem work their magic on your body, mind, and soul.

Seasonal Highlights:

The Felixstowe Promenade South Walk offers a unique and enchanting experience in every season, with each time of year bringing its own special highlights and attractions. Whether you're a local resident or a visiting tourist, there's always something new and exciting to discover along this picturesque stretch of the Suffolk coast.

In the spring, the promenade comes alive with the vibrant colours and sweet scents of blooming flowers, both in the Seafront Gardens and along the wild and rugged shoreline. The carefully tended beds and borders of the gardens are a riot of tulips, daffodils, and other spring bulbs, while the coastal dunes and grasslands are dotted with the delicate, pastel hues of sea thrift and other native wildflowers. As the days grow longer and warmer, the beach becomes a hive of activity, with families and sun-seekers flocking to the sands to enjoy the first rays of the summer sun.

As spring gives way to summer, the promenade reaches the height of its popularity and vibrancy, with the golden sands and sparkling waters of the beach drawing visitors from far and wide. The Seafront Gardens are now in full bloom, with the lush greens of the lawns and the bright colours of the summer bedding plants creating a dazzling display that is a feast for the senses. Along the shore, the beach huts and seaside stalls are a hive of activity, with the sounds of laughter, music, and the sizzling of fresh seafood filling the air. This is the perfect time to indulge in some traditional seaside fun, from building sandcastles and paddling in the shallows to enjoying a delicious ice cream or a ride on the historic pier.

As the leaves begin to turn and the days grow shorter, the promenade takes on a new character, with the rich, warm hues of autumn transforming the Seafront Gardens into a tapestry of gold, orange, and red. The beach is now quieter and more contemplative, with the soft light and gentle breezes of the season creating a peaceful and meditative atmosphere that is perfect for a refreshing walk or a moment of quiet reflection. This is also a wonderful time for birdwatching, with the shores and skies filled with the sights and sounds of migrating flocks, from the distinctive V-shaped formations of geese to the swirling clouds of waders and seabirds.

As winter tightens its grip and the temperatures drop, the promenade becomes a place of stark beauty and invigorating freshness. The gardens are now stripped back to their bare bones, with the sculptural forms of the trees and shrubs etched against the pale, watery sky. The beach is a wild and windswept expanse, with the crashing waves and churning surf creating a dramatic and awe-inspiring spectacle that is not for the faint of heart. But for those who brave the elements and venture out onto the promenade, the rewards are many, from the bracing sea air and the exhilarating sense of freedom to the chance to spot some of the rare and elusive birds that overwinter along the coast, such as the purple sandpiper or the Slavonian grebe.

No matter what time of year you choose to visit, the Felixstowe Promenade South Walk offers a wealth of seasonal highlights and attractions that are sure to delight and inspire. From the vibrant colours and lively atmosphere of the summer months to the peaceful tranquility and stark beauty of the winter, each season brings its own unique character and charm to this special corner of the Suffolk coast. So why not plan a visit for each time of year, and discover for yourself the endless wonders and delights of this enchanting seaside gem?

Safety Considerations:

While the Felixstowe Promenade South Walk is generally a safe and enjoyable route, there are a few key safety considerations to keep in mind to ensure a worry-free and memorable experience.

First and foremost, it's important to be aware of the tides and weather conditions when walking along the coast. The North Sea can be unpredictable, with strong currents, high winds, and rough seas possible at any time of year. Always check the local weather forecast and tide times before setting out, and avoid walking too close to the water's edge, especially during storms or high tides.

The promenade itself can also become slippery when wet or icy, so take care to wear appropriate footwear with good grip and tread. In the summer months, the concrete and tarmac surfaces can get very hot under the sun, so be sure to protect your skin with sunscreen and a hat, and stay hydrated by bringing plenty of water with you.

When exploring the Martello Tower and Landguard Fort, be aware of any uneven surfaces, steps, or low ceilings that could pose a trip or head-bump hazard. These historic structures were not designed with modern accessibility standards in mind, so take your time and watch your step as you navigate the site.

If you're walking with children or dogs, keep a close eye on them at all times, especially near the water's edge or in busy areas with lots of people and activity. Make sure they stay on the designated path and away from any potential hazards, such as cliffs, slippery rocks, or fast-moving water.

Finally, as with any outdoor activity, it's always a good idea to let someone know your plans and expected return time before setting out. Carry a mobile phone with you in case of emergencies, and familiarize yourself with the location of the nearest lifeguard station, public telephone, or other safety facilities along the route.

By following these simple safety guidelines and using common sense and good judgment, you can ensure a safe and enjoyable walk along the Felixstowe Promenade South Walk for you and your loved ones. Remember, the coast is a beautiful but dynamic environment that demands respect and caution, so always prioritize your safety and well-being, and never take unnecessary risks for the sake of a photo or a thrill. With a little care and preparation, you can fully immerse yourself in the timeless beauty and tranquility of this special place, and create memories that will last a lifetime.

Nearby Attractions:

The Felixstowe Promenade South Walk is just the beginning of the many delights and attractions that this charming seaside town has to offer. Whether you're a history buff, a nature lover, or simply looking for a fun and relaxing day out, there's plenty to see and do in and around Felixstowe that will keep you entertained and inspired.

Just a short stroll from the promenade, the Felixstowe Museum is a must-visit for anyone interested in the rich history and heritage of this fascinating corner of the Suffolk coast. Housed in a beautiful Edwardian building with stunning sea views, the museum offers a treasure trove of exhibits and artifacts that tell the story of Felixstowe's development from a sleepy fishing village to a thriving seaside resort. From the town's early days as a haven for smugglers and pirates to its heyday as a fashionable Victorian retreat, the museum brings the past vividly to life through a mix of interactive displays, rare photographs, and personal stories from local residents.

For those who prefer their history with a side of drama and spectacle, the Spa Pavilion Theatre is a wonderful place to spend an evening or an afternoon. This elegant 1930s venue has been lovingly restored to its former glory, with a diverse program of music, dance, comedy, and family shows that cater to all tastes and ages. Whether you're in the mood for a classic play, a toe-tapping musical, or a side-splitting stand-up routine, the Spa Pavilion has something to delight and entertain you, all while immersing you in the glamorous atmosphere of a bygone era.

If you're looking to explore further afield, the nearby town of Woodbridge is a charming and historic destination that is well worth a visit. Just a short drive or bus ride from Felixstowe, Woodbridge boasts a picturesque waterfront, a lively market square, and a host of independent shops, cafes, and galleries that showcase the best of local craftsmanship and produce. The town is also home to the Tide Mill Living Museum, a fascinating working example of an 800-year-old tide mill that harnesses the power of the tides to grind flour and generate electricity. Guided tours and interactive exhibits offer a unique and hands-on way to learn about the ingenuity and resilience of past generations, and to appreciate the enduring importance of renewable energy and sustainable living.

For nature lovers and outdoor enthusiasts, the Suffolk Coast and Heaths Area of Outstanding Natural Beauty is a paradise waiting to be explored. This stunning landscape of sandy beaches, windswept heaths, and tranquil estuaries stretches along the coast from Felixstowe to Lowestoft, offering endless opportunities for walking, cycling, birdwatching, and simply soaking up the breathtaking views and fresh sea air. Whether you're a keen hiker, a budding botanist, or just looking for a peaceful escape from the hustle and bustle of modern life, the Suffolk Coast and Heaths has something to inspire and delight you, with its rich diversity of wildlife, its fascinating history, and its timeless, unspoiled beauty.

So why not extend your visit to the Felixstowe Promenade South Walk and discover some of the many other attractions and destinations that this wonderful corner of the Suffolk coast has to offer? From museums and theatres to nature reserves and historic towns, there's no shortage of things to see and do in and around Felixstowe, each one offering a unique and memorable way to connect with the rich culture, heritage, and natural wonders of this very special place. Whether you have a few hours or a few days to spare, you're sure to find something that will capture your imagination and leave you with lasting memories of your time in this enchanting seaside gem.

Felixstowe Ferry Walk - 3 miles (4.8 km), Circular - Estimated Time: 1.5-2 hours

Starting Point Coordinates, Postcode: The starting point for the Felixstowe Ferry Walk is the public car park at Felixstowe Ferry (Grid Reference: TM 322 384, Postcode: IP11 9RZ).
Nearest Car Park, Postcode: The nearest car park is at Felixstowe Ferry (Postcode: IP11 9RZ).

Embark on a delightful 3-mile (4.8 km) circular walk around the picturesque Felixstowe Ferry, where the tranquil beauty of the Suffolk coast and the rich maritime heritage of the area combine to create an unforgettable experience. This easy walk, suitable for all ages and abilities, offers a perfect blend of stunning coastal views, historic landmarks, and diverse wildlife, inviting you to immerse yourself in the timeless charm and natural wonders of this enchanting corner of East Anglia.

As you set out from the public car park at Felixstowe Ferry, feel the anticipation build as you head east along the serene River Deben, the well-maintained coastal path guiding you towards the beckoning horizon. With each step, the gentle murmur of the flowing water and the distant cries of the gulls transport you into a world of peace and tranquility, the stresses of modern life melting away like the morning mist over the river.

Take a moment to drink in the picturesque scene before you, the colourful boats bobbing gently on the sparkling water, their masts and rigging etched against the pale blue sky. As you continue along the path, the river's edge reveals a fascinating array of wildlife, from the scurrying crabs and darting fish that inhabit the shallows to the graceful wading birds that pick their way through the mudflats, their elegant forms mirrored in the glassy surface of the water.

At the end of the path, turn right and head south along the beach, the soft sand beneath your feet and the salty tang of the sea breeze invigorating your senses. Here, the vast expanse of the North Sea stretches out before you, its endless blue horizon broken only by the distant silhouettes of ships and the occasional splash of a diving seabird. Take a moment to pause and appreciate the raw power and beauty of the ocean, the rhythmic crash of the waves a soothing symphony that echoes the timeless cycles of the tides.

As you make your way along the shore, keep an eye out for the historic Martello Tower, a striking reminder of the region's rich military past. These sturdy, circular fortifications, built in the early 19th century to defend the coast against the threat of French invasion, stand as silent sentinels over the windswept beach, their weathered stone walls and small, shuttered windows hinting at the lives and stories of the soldiers who once manned them.

Continuing southward, you'll soon reach the southernmost point of the walk, where you'll turn right and head west, leaving the open sea behind and making your way inland. Here, the landscape changes once again, the rugged coastline giving way to the manicured greens and fairways of the Felixstowe Ferry Golf Club, one of the oldest and most prestigious courses in the country.

As you walk along the edge of the golf course, take a moment to appreciate the skill and precision of the players as they navigate the challenging terrain, the satisfying thwack of club against ball punctuating the peaceful stillness of the surrounding countryside. With its undulating hills, sandy bunkers, and sweeping vistas of the river and sea, the golf course is a testament to the enduring appeal and beauty of this special corner of Suffolk.

At the western edge of the golf course, turn right once more and begin your journey back north, the path leading you through a landscape of gently rolling fields and picturesque hedgerows. Here, the rural charm of the Suffolk countryside is on full display, the patchwork of crops and pastures stretching out before you like a living tapestry, each stitch a testament to the centuries of human stewardship and care that have shaped this land.

As you near the end of your walk, the car park at Felixstowe Ferry comes into view once again, the starting point of your journey and a welcome sight after your invigorating ramble. Take a final moment to reflect on the beauty and diversity of the landscapes you've encountered, and the sense of peace and connection that comes from immersing yourself in the natural world.

The Felixstowe Ferry Walk is a celebration of the enduring allure and rich heritage of the Suffolk coast, offering a chance to step back from the hustle and bustle of modern life and to reconnect with the timeless rhythms and simple pleasures of the great outdoors. Whether you're a seasoned rambler or a curious first-time visitor, this enchanting route promises to delight and inspire, leaving you with memories that will last a lifetime.

Detailed Route Directions:

1. From the public car park at Felixstowe Ferry, head east along the River Deben, following the coastal path. Continue along this path for approximately 0.8 miles (1.3 km), bearing 90°. (Estimated time: 15-20 minutes)

2. At the end of the coastal path, turn right and head south along the beach, maintaining a bearing of 180°. Walk along the beach for around 1 mile (1.6 km), enjoying the coastal views and the Martello Tower. (Estimated time: 20-25 minutes)

3. Upon reaching the southernmost point of the walk, turn right and head west, keeping a bearing of 270°. Continue along this path for approximately 0.6 miles (1 km), passing the Felixstowe Ferry Golf Club on your left. (Estimated time: 15-20 minutes)

4. At the western edge of the golf course, turn right once more, now heading north with a bearing of 0°. Follow this path for around 0.6 miles (1 km), making your way back to the starting point at the public car park. (Estimated time: 15-20 minutes)

Best Features and Views:

- Immerse yourself in the stunning coastal panoramas, with the tranquil waters of the River Deben and the vast expanse of the North Sea providing a breathtaking backdrop for your walk.
- Marvel at the historic Martello Tower, a fascinating remnant of the region's military past and a striking landmark along the beach.
- Delight in the picturesque charm of Felixstowe Ferry, with its colourful boats, quaint cottages, and timeless maritime atmosphere.

Rare or Uncommon Birds:

As you make your way along the coastal path and the beach, keep your eyes peeled for the diverse array of wading birds that call this area home. The mudflats and saltmarshes of the River Deben provide a vital feeding ground for species such as oystercatchers, redshanks, and curlews, their distinctive calls and elegant forms adding a touch of wild beauty to the landscape.

Oystercatchers, with their striking black-and-white plumage and bright orange bills, are a particular delight to spot, their lively behaviour and amusing courtship rituals a joy to observe. Watch as they use their long, chisel-like bills to pry open the tightly closed shells of mussels and cockles, or as they engage in spirited piping displays to defend their territories and attract mates.

Redshanks, another common wading bird along the Suffolk coast, are easily recognizable by their bright orange-red legs and distinctive "tu-tu-tu" call. These elegant birds are often seen probing the mud with their long, sensitive bills, searching for the small invertebrates and crustaceans that make up their diet. In the spring and summer months, look out for their elaborate aerial displays, as the males soar and dive over the marshes in a breathtaking show of agility and grace.

The haunting cries of curlews, with their long, down-curved bills and mottled brown plumage, are a familiar sound along the River Deben, their evocative calls carrying far across the water and the marshes. These large, majestic waders are a true symbol of the wild and untamed beauty of the Suffolk coast, their presence a reminder of the rich and diverse birdlife that thrives in this unique and precious habitat.

In addition to these wading birds, the Felixstowe Ferry Walk also offers the chance to spot a variety of seabirds, from the ubiquitous gulls that wheel and soar over the beach to the sleek and elegant terns that dive for fish in the shallow waters. Keep an eye out for the distinctive silhouettes of gannets, cormorants, and even the occasional skua, their powerful wings and streamlined forms a testament to their mastery of the sea and the sky.

As you walk, take a moment to appreciate the incredible diversity and adaptability of these coastal birds, each species uniquely suited to the challenges and opportunities of life on the edge of the land and the sea. From the small and nimble sanderlings that skitter along the tideline to the large and powerful herons that stalk the marshes, every bird has its own story to tell, its own place in the intricate web of life that sustains this remarkable corner of the Suffolk coast.

By opening your eyes and ears to the wonders of the avian world, you can deepen your understanding and appreciation of the rich and complex tapestry of the natural world, and gain a renewed sense of connection and kinship with the incredible creatures that share our planet. So keep your binoculars at the ready, and let the birds of the Felixstowe Ferry Walk be your guide and your inspiration on this unforgettable journey through one of England's most beautiful and fascinating coastal landscapes.

Rare or Interesting Insects and Other Wildlife:

As you explore the diverse habitats along the Felixstowe Ferry Walk, from the saltmarshes and mudflats of the River Deben to the sandy shores and dunes of the beach, keep an eye out for the myriad of fascinating insects and other invertebrates that call this unique environment home. While the coastal landscape may not be known for specific rare species, it supports a rich and varied array of life that is no less captivating for its small size and often-overlooked nature.

Among the most striking and charismatic of these creatures are the butterflies and moths that flit and flutter among the wildflowers and grasses of the dunes and marshes. Look for the vibrant orange and brown wings of the small copper butterfly, a common but always delightful sight along the Suffolk coast, or the delicate, pale green-veined white, its ethereal beauty a perfect match for the soft, muted tones of the coastal vegetation.

As you walk along the beach, keep an eye out for the darting, iridescent forms of damselflies and dragonflies, their slender bodies and gossamer wings a marvel of natural engineering and design. These ancient and fascinating insects are masters of the air, their incredible aerial agility and hawk-like eyesight making them deadly predators of smaller flying insects. Watch as they hover and dart over the sparkling waters of the river and the sea, their jewel-like colours flashing in the sunlight, and marvel at the incredible diversity and adaptability of these remarkable creatures.

In the tidal pools and shallow waters along the shore, a whole world of fascinating marine life awaits discovery. From the scurrying hermit crabs and the darting shrimp to the slow-moving sea snails and the delicate anemones, each creature has its own unique adaptations and survival strategies, its own place in the complex web of life that thrives in this challenging and dynamic environment.

Take a moment to peer into the clear, shallow waters, and watch as the tiny, translucent prawns and the small, colourful fish go about their daily lives, foraging for food and seeking shelter among the rocks and seaweed. Notice the intricate patterns and textures of the shells and exoskeletons that litter the tideline, each one a work of art crafted by the ceaseless action of the waves and the sand, a testament to the endless cycle of life and death that shapes this coastal world.

As you explore the dunes and marshes, keep an eye out for the signs and tracks of the larger animals that make their home in this landscape, from the delicate footprints of the shore birds to the winding trails of the small mammals that burrow and forage among the grasses and shrubs. The coastal habitats of the Felixstowe Ferry Walk may not be home to many large or charismatic species, but they are no less vital and fascinating for their role in supporting the intricate web of life that sustains this unique and precious ecosystem.

By taking the time to notice and appreciate the small wonders and hidden marvels of the insect and invertebrate world, you can gain a deeper understanding and appreciation of the incredible complexity and resilience of life on our planet, and of the vital importance of preserving and protecting these fragile and irreplaceable habitats for generations to come. So let your curiosity be your guide, and allow yourself to be amazed and inspired by the endless wonders and surprises that await you on this enchanting walk through the wild and beautiful landscapes of the Suffolk coast.

Rare or Interesting Plants:

As you make your way along the Felixstowe Ferry Walk, take a moment to appreciate the fascinating array of coastal plants and wildflowers that thrive in the unique and challenging conditions of this dynamic environment. From the hardy, salt-tolerant grasses that anchor the shifting sands of the dunes to the delicate, ephemeral blooms that paint the marshes and meadows with colour, each species has its own remarkable adaptations and survival strategies, its own place in the complex tapestry of life that defines this special corner of the Suffolk coast.

One of the most characteristic and important plants of the coastal dunes is marram grass, a tough, wiry perennial with long, narrow leaves and a deep, extensive root system that helps to stabilize the loose, shifting sands. This unassuming but vital species plays a crucial role in the formation and maintenance of the dunes, its dense, interlocking network of roots and rhizomes binding the sand and creating the conditions for other plants and animals to take hold and thrive.

As you walk along the beach and through the dunes, look for the pale, feathery plumes of the marram grass swaying in the breeze, their subtle beauty a testament to the incredible resilience and adaptability of life in this harsh and unforgiving environment. Notice how the grass grows in dense, hummocky clumps, each one a miniature ecosystem in itself, sheltering a diverse array of small creatures and providing a foothold for other, more delicate plants to establish themselves.

In the sheltered hollows and damp, brackish margins of the dunes, keep an eye out for the striking, fleshy leaves and vibrant pink flowers of the sea pink, a low-growing, mat-forming perennial that thrives in the saline, nutrient-poor soils of the coast. This charming and tenacious plant is a true survivor, its thick, waxy cuticle and specialized glands allowing it to conserve water and regulate its salt intake in the face of the constant exposure to wind, sun, and salt spray.

As you move inland from the beach and into the marshes and meadows that fringe the River Deben, the plant life becomes more diverse and colourful, with a wide range of wildflowers and herbs adding their own unique character and beauty to the landscape. Look for the tall, purple spikes of the common marsh orchid, the delicate, pink-and-white blooms of the cuckooflower, and the cheerful, yellow buttons of the silverweed, each species a vital part of the rich and complex web of life that sustains the birds, insects, and other creatures of the marsh.

Other interesting and characteristic plants to look out for along the Felixstowe Ferry Walk include the sea aster, with its small, daisy-like flowers and fleshy, lobed leaves; the sea purslane, a low-growing, succulent herb with tiny, pinkish-white flowers; and the sea lavender, a delicate, wispy perennial with sprays of pale, lilac-coloured blooms that sway gently in the breeze.

As you marvel at the beauty and diversity of the coastal flora, take a moment to reflect on the incredible adaptations and strategies that have evolved over millennia to allow these plants to thrive in such a challenging and dynamic environment. From the deep, anchoring roots of the marram grass to the specialized salt glands of the sea pink, each species is a testament to the enduring power and creativity of nature, and a reminder of the vital importance of preserving and protecting these fragile and irreplaceable habitats for generations to come.

By opening your eyes and your mind to the wonders and mysteries of the plant world, you can gain a deeper appreciation and understanding of the rich and complex tapestry of life that defines the Suffolk coast, and of the vital role that each species plays in maintaining the health and resilience of this extraordinary ecosystem. So let the wildflowers and grasses be your guide and your inspiration on this enchanting walk, and allow yourself to be amazed and humbled by the endless wonders and surprises that await you in the hidden world of the coastal plants.

Facilities and Accessibility:

The Felixstowe Ferry Walk offers a range of facilities and amenities to ensure a comfortable and enjoyable experience for visitors of all ages and abilities. The well-maintained paths and relatively flat terrain make this walk accessible to most people, including those with limited mobility or fitness.

At the starting point of the walk, the public car park at Felixstowe Ferry provides ample space for vehicles, with designated parking bays for disabled visitors. The car park is well-signposted and easy to find, with direct access to the coastal path and the beach.

For those in need of refreshment or a quick snack, there is a charming café located near the car park, serving a variety of hot and cold drinks, light meals, and homemade cakes and pastries. The café has both indoor and outdoor seating areas, offering stunning views over the River Deben and the moored boats, and is a popular spot for walkers and locals alike.

Public toilets, including accessible facilities for disabled visitors, are also available near the car park and café, ensuring that all visitors have access to clean and well-maintained amenities before or after their walk.

Along the route itself, the paths are mostly flat and well-maintained, with a mixture of hard-packed sand, gravel, and short grass surfaces that are suitable for wheelchairs and pushchairs. There are a few short sections of raised boardwalk over some of the marshier areas, but these are wide and sturdy, with non-slip surfaces and handrails for added safety and ease of use.

At various points along the walk, there are bench seats and picnic tables provided, offering a chance to rest, take in the stunning coastal views, or enjoy a packed lunch in the fresh sea air. These seating areas are well-positioned to make the most of the natural beauty and tranquility of the landscape, while also providing a welcome respite for those with limited stamina or mobility.

For families with young children, the beach and dunes along the route offer a natural playground for exploration and adventure, with plenty of opportunities for sandcastle building, shell collecting, and wildlife spotting. However, parents should be aware of the potential hazards of the tidal river and the sea, and ensure that children are supervised at all times near the water's edge.

While the Felixstowe Ferry Walk is generally a safe and accessible route, visitors should still take sensible precautions and come prepared for the changeable weather conditions of the Suffolk coast. Sturdy, comfortable shoes with good grip are essential, as well as layers of warm, waterproof clothing and sun protection in the summer months.

Overall, the facilities and accessibility of the Felixstowe Ferry Walk make it a welcoming and enjoyable destination for a wide range of visitors, from families with young children to older adults and those with limited mobility. With its stunning coastal scenery, rich wildlife, and fascinating history, this enchanting route offers something for everyone, and is sure to leave a lasting impression on all who take the time to explore its hidden wonders and timeless beauty.

Seasonal Highlights:

One of the great joys of the Felixstowe Ferry Walk is the opportunity to experience the changing moods and colours of the Suffolk coast through the turning of the seasons. From the fresh, vibrant greens of spring to the rich, burnished hues of autumn, each season brings its own unique character and charm to this enchanting landscape, inviting walkers to discover and delight in the endless wonders and surprises of the natural world.

In spring, the coastal path and the dunes come alive with the delicate blooms and fresh, tender growth of the early wildflowers, from the pale, nodding heads of the cuckooflower to the cheerful, yellow stars of the coltsfoot. The air is filled with the joyous songs of the returning migrant birds, as they stake out their territories and begin the busy work of nest-building and raising their young.

As you walk along the beach and the marshes in these early months of the year, keep an eye out for the first signs of new life emerging from the winter-browned vegetation, from the bright green shoots of the marram grass to the unfurling fronds of the bracken and the marsh ferns. There is a sense of anticipation and renewal in the air, a feeling of the world waking up from its long winter sleep and stretching its limbs towards the warming sun.

As spring gives way to summer, the coast becomes a hive of activity and life, with the wildflowers and grasses reaching their peak of growth and the breeding season in full swing for the birds and other wildlife. The dunes and marshes are now a riot of colour, from the deep pink of the sea thrift to the delicate, lilac hues of the sea lavender, while the butterflies and bees flit from bloom to bloom in search of nectar.

In the height of summer, the beach and the river are at their most inviting, with the warm, shallow waters of the Deben perfect for a cooling paddle or a refreshing swim. The sun-drenched sands and the gently lapping waves create an idyllic and timeless scene, a postcard-perfect image of the great British seaside holiday that has been drawing visitors to this special corner of Suffolk for generations.

As the long, lazy days of summer give way to the cooler, crisper months of autumn, the coastal landscape takes on a new and more contemplative character, the lush greens of the marshes and dunes slowly fading to a palette of russet, gold, and bronze. The swallows and the terns that have been such a constant and lively presence throughout the summer now gather in great, swirling flocks, preparing for their long journey back to their winter quarters in Africa.

In these mellow, golden months of the year, the Felixstowe Ferry Walk becomes a place of quiet reflection and gentle melancholy, a chance to savour the fleeting beauty of the turning leaves and the shifting light before the onset of the winter storms and the long, dark nights. The beach and the river take on a wilder, more elemental character, the stark beauty of the wind-scoured sands and the churning waves a reminder of the raw power and majesty of the natural world.

Finally, as winter tightens its grip and the landscape is transformed into a study in black and white, the walk becomes a place of stillness and solitude, a chance to embrace the bracing chill of the sea air and the stark, haunting beauty of the frozen marshes and the snow-dusted dunes. The birds and animals that remain are hardy and resilient, their presence a testament to the enduring cycle of life that pulses beneath the surface of this ancient and enduring landscape.

No matter what time of year you choose to embark on the Felixstowe Ferry Walk, there is always something new and wondrous to discover, some fresh perspective or unexpected delight that will stay with you long after you have left the coast behind. From the joyous explosion of life in the spring to the quiet contemplation of the winter months, each season brings its own unique gifts and challenges, its own invitation to connect with the timeless rhythms and eternal truths of the natural world.

Difficulty Level and Safety Considerations:

The Felixstowe Ferry Walk is a relatively easy and straightforward route, suitable for walkers of most ages and abilities. With its flat terrain, well-maintained paths, and numerous amenities and facilities along the way, this 3-mile (4.8 km) circular walk offers a pleasant and accessible way to explore the stunning landscapes and rich wildlife of the Suffolk coast, without the need for any specialized equipment or technical skills.

However, as with any outdoor activity, it is important to be aware of the potential hazards and challenges that may arise, and to take sensible precautions to ensure a safe and enjoyable experience for all. The most significant consideration for walkers on the Felixstowe Ferry Walk is the tidal nature of the River Deben and the North Sea, which can pose a risk to those who venture too close to the water's edge or attempt to cross the river at low tide.

To stay safe and avoid any accidents or emergencies, walkers should always check the local tide times before setting out, and plan their route accordingly to avoid being cut off by the rising water. It is also essential to stay on the designated paths and to keep a safe distance from the river and the sea, particularly during periods of high tide or rough weather.

Another factor to keep in mind is the exposed and changeable nature of the coastal environment, which can leave walkers vulnerable to the elements, particularly during the colder and wetter months of the year. To mitigate these risks, it is important to come prepared with appropriate clothing and equipment for the conditions, including sturdy, waterproof footwear, warm layers, and a weatherproof jacket or coat.

Walkers should also be aware of the potential hazards posed by the local wildlife, particularly the birds and insects that call the marshes and dunes their home. While the majority of these creatures are harmless and more likely to flee than attack, it is still important to give them plenty of space and to avoid disturbing their nests or habitats, particularly during the breeding season.

For those with mobility issues or other physical limitations, the Felixstowe Ferry Walk offers a range of accessible facilities and amenities, including level, well-surfaced paths, regular rest points, and disabled toilets. However, it is still important to consider the length and duration of the walk, and to pace oneself accordingly to avoid overexertion or fatigue.

Finally, as with any outdoor activity, it is always a good idea to let someone know your intended route and estimated return time before setting out, and to carry a fully charged mobile phone and a small first aid kit in case of emergencies. By taking these simple precautions and using common sense and good judgment, walkers of all ages and abilities can enjoy a safe and rewarding experience on this delightful coastal walk.

Nearby Attractions:

The Felixstowe Ferry Walk is not only a wonderful destination in its own right but also a gateway to a wealth of other attractions and activities in the surrounding area. Whether you're a history buff, a nature lover, a foodie, or simply someone in search of a relaxing and enjoyable day out, the charming town of Felixstowe and the beautiful Suffolk countryside have plenty to offer.

Just a short walk or drive from the ferry, the bustling centre of Felixstowe is a popular and lively seaside resort, with a wide range of shops, cafes, restaurants, and entertainment options to suit all tastes and budgets. The town's famous seafront promenade is a must-visit destination, with its colourful beach huts, Victorian gardens, and stunning views over the North Sea, while the newly-renovated pier offers a range of traditional seaside amusements and attractions, from penny arcades and ice cream parlours to a thrilling roller coaster and a state-of-the-art bowling alley.

For those interested in the town's rich history and heritage, the Felixstowe Museum is a fascinating and informative destination, housed in a beautifully-restored Edwardian seafront villa. The museum's collections cover everything from the town's early origins as a small fishing village to its heyday as a fashionable Victorian resort, with exhibits on the area's maritime history, its role in both world wars, and its enduring appeal as a beloved holiday destination.

Nature lovers and outdoor enthusiasts will also find plenty to explore and enjoy in the surrounding countryside, with a variety of parks, reserves, and trails showcasing the stunning landscapes and diverse wildlife of the Suffolk coast. One of the most popular and accessible of these is the Landguard Peninsula, a unique and dynamic spit of land that juts out into the North Sea at the southern tip of Felixstowe, offering a range of habitats and attractions for visitors of all ages and interests.

The peninsula is home to the impressive Landguard Fort, a 18th-century coastal defense that played a key role in protecting the town and the nearby Port of Felixstowe from invasion and attack. Today, the fort is open to the public as a museum and visitor centre, offering a fascinating glimpse into the lives and experiences of the soldiers and civilians who once lived and worked within its walls, as well as stunning views over the sea and the surrounding landscape.

Adjacent to the fort, the Landguard Nature Reserve is a haven for wildlife and a paradise for birdwatchers, with over 80 hectares of pristine saltmarsh, mudflats, and shingle beach providing a vital habitat for a wide range of rare and endangered species. The reserve is home to a thriving population of avocets, ringed plovers, and little terns, as well as a host of other wading birds and wildfowl, making it a must-visit destination for anyone with an interest in nature and conservation.

For those willing to venture a little further afield, the picturesque villages and market towns of the Suffolk countryside offer a wealth of charming and historic attractions, from the medieval half-timbered houses and ancient churches of Lavenham and Long Melford to the world-famous Anglo-Saxon burial site at Sutton Hoo, near the town of Woodbridge.

Foodies and culinary enthusiasts will also find plenty to tantalise their taste buds in the region, with a thriving local food and drink scene that showcases the very best of Suffolk's produce and craftsmanship. From the fresh, succulent seafood landed daily at Felixstowe and the nearby ports of Orford and Aldeburgh, to the artisanal cheeses, breads, and charcuterie produced by the county's many skilled farmers and food producers, there is no shortage of delicious and authentic flavours to discover and enjoy.

Other nearby attractions and destinations to consider include:

1. The beautiful seaside town of Aldeburgh, with its stunning shingle beach, historic high street, and world-famous fish and chip shops.

2. The RSPB Minsmere nature reserve, one of the most important and diverse wildlife habitats in the UK, with a variety of hides, trails, and visitor facilities.

3. The charming market town of Woodbridge, with its picturesque waterfront, independent shops and galleries, and historic tide mill.

4. The National Trust's Dunwich Heath, a breathtaking coastal landscape of heather, gorse, and bracken, with panoramic views over the North Sea.

5. The historic city of Ipswich, with its vibrant waterfront, world-class museums and galleries, and rich cultural heritage.

Whether you're a first-time visitor to the Suffolk coast or a seasoned explorer of this enchanting region, the Felixstowe Ferry Walk and its nearby attractions offer a wealth of opportunities for discovery, relaxation, and inspiration. So why not take some time to delve deeper into the rich history, stunning landscapes, and vibrant communities of this very special corner of East Anglia, and create memories that will last a lifetime?

With its combination of easy access, diverse attractions, and warm and welcoming locals, the Felixstowe Ferry Walk is the perfect starting point for a truly unforgettable adventure on the Suffolk coast. Whether you're seeking solitude and tranquility, excitement and adventure, or simply a chance to reconnect with nature and your own sense of wonder, this beautiful and endlessly fascinating region has something to offer everyone who takes the time to explore its hidden gems and timeless charms.

So lace up your walking boots, grab your binoculars and your sense of curiosity, and set out on a journey of discovery that will leave you refreshed, inspired, and forever changed by the magic and majesty of this very special place. The Felixstowe Ferry Walk and its nearby attractions are waiting to welcome you with open arms and to show you the very best of what the Suffolk coast has to offer, from its stunning scenery and rich history to its warm and welcoming people and its vibrant and thriving communities.

Felixstowe Hills Walk - 5 miles (8 km), Circular - Estimated Time: 2.5-3 hours

Starting Point Coordinates, Postcode: The starting point for the Felixstowe Hills Walk is at the parking area near the entrance to Landguard Fort (Grid Reference: TM 282 329, Postcode: IP11 3TW).
Nearest Car Park, Postcode: Parking area near the entrance to Landguard Fort (Postcode: IP11 3TW).

Embark on an invigorating 5-mile (8 km) circular walk through the captivating Felixstowe Hills, where the breathtaking beauty of the Suffolk coastline, the bustling activity of Felixstowe Port, and the tranquil charm of the surrounding countryside converge to create an unforgettable experience. This moderate route promises a perfect blend of stunning vistas, rich history, and diverse wildlife, inviting you to immerse yourself in the natural wonders and timeless allure of this enchanting corner of East Anglia.

As you set out from the parking area near the impressive Landguard Fort, feel the anticipation build as you head west along the seafront, the well-marked Suffolk Coast Path guiding you towards the beckoning horizon. With each step, the cares and distractions of everyday life seem to melt away, replaced by a growing sense of peace and connection with the wild and untamed beauty of the coast.

Take a moment to drink in the awe-inspiring panoramic views of the shimmering North Sea, the golden sands of the beach stretching out before you like a ribbon of light. As you continue along the coastal path, the sound of the waves crashing against the shore and the cries of the seabirds soaring overhead create a symphonic backdrop that soothes the soul and invigorates the senses.

Upon reaching the serene oasis of the Landguard Nature Reserve, turn left and follow the winding trails that lead you into the heart of this precious wildlife haven. As you navigate the diverse habitats of the reserve, from the salt-sprayed marshes to the wind-sculpted dunes, keep your eyes peeled for the myriad species of birds, insects, and plants that thrive in this unique and dynamic ecosystem.

Crossing the footbridge over the glistening expanse of the River Deben, you'll begin your ascent into the rolling hills of Felixstowe, the path leading you upward through a tapestry of lush grasslands and ancient woodlands. As you climb higher, the vistas that unfold before you become ever more breathtaking, the patchwork of fields and hedgerows stretching out to the distant horizon like a living work of art.

Pause for a moment to catch your breath and take in the awe-inspiring views of Felixstowe Port, the bustling hub of maritime activity that has shaped the history and character of this region for centuries. Watch as the colossal container ships and sleek yachts navigate the busy shipping lanes, their presence a testament to the enduring importance of the sea in the life and economy of the Suffolk coast.

As you continue along the undulating path, winding your way through the picturesque hills and valleys, the landscape reveals its many secrets and surprises. From the delicate wildflowers that sway in the gentle breeze to the majestic trees that tower overhead, each element of the natural world seems imbued with a sense of wonder and magic, inviting you to slow down and savour the beauty of the present moment.

Eventually, the path will guide you back towards the seafront, the sparkling waters of the North Sea beckoning you onwards. As you follow the coastal trail eastward, retracing your steps towards the starting point, take a moment to reflect on the incredible journey you have undertaken, both physically and spiritually.

The Felixstowe Hills Walk is a celebration of the enduring power and resilience of the natural world, a reminder of the countless wonders that await those who step outside their comfort zone and embrace the unknown. Whether you are a seasoned hiker or a curious first-time explorer, this enchanting route promises to delight and inspire, offering a profound sense of connection with the land, the sea, and the timeless rhythms of life that flow through this extraordinary corner of Suffolk.

As you return to the parking area, your heart full and your spirit rejuvenated, carry with you the memories of this unforgettable walk, and the knowledge that the beauty and magic of the Felixstowe Hills will always be there, waiting to welcome you back with open arms and endless possibilities.

Detailed Route Directions:

1. From the parking area near Landguard Fort, head west along the seafront, following the Suffolk Coast Path signs. Maintain a bearing of approximately 270° for about 1 mile (1.6 km). (Estimated time: 20-25 minutes)

2. Continue along the coast until you reach the Landguard Nature Reserve. Turn left, following the reserve's trails, which will take you to the northern end of the peninsula. Maintain a general northward bearing (approximately 360°) through the reserve for about 1 mile (1.6 km). (Estimated time: 25-30 minutes)

3. Cross the footbridge over the River Deben and follow the footpath uphill through the Felixstowe Hills. The path will head in a northwesterly direction (approximately 315°) for about 1.5 miles (2.4 km). (Estimated time: 35-45 minutes)

4. As the path winds through the hills, it will gradually turn towards the southwest (approximately 225°). Follow the path for another 1 mile (1.6 km), enjoying the beautiful views of the surrounding area. (Estimated time: 25-30 minutes)

5. The path will eventually lead you back to the seafront, heading in a southeasterly direction (approximately 135°) for about 0.5 miles (0.8 km). (Estimated time: 10-15 minutes)

6. Upon reaching the seafront, turn left and follow the path eastward (90°) for approximately 1 mile (1.6 km), returning to the starting point at the parking area near Landguard Fort. (Estimated time: 20-25 minutes)

Best Features and Views:

- Marvel at the breathtaking panoramic views of the Suffolk coastline, with the shimmering expanse of the North Sea stretching out to the horizon, and the golden sands of the beach creating a stunning visual contrast.
- Delight in the awe-inspiring vistas of Felixstowe Port, the bustling hub of maritime activity that has shaped the region's history and character, with colossal container ships and sleek yachts navigating the busy shipping lanes.
- Immerse yourself in the picturesque charm of the surrounding countryside, with the rolling hills, lush grasslands, and ancient woodlands creating a tapestry of natural beauty that soothes the soul and invigorates the senses.

Historical Features of Interest:

As you set out on the Felixstowe Hills Walk, you'll have the opportunity to explore the fascinating history of Landguard Fort, a remarkable fortification that has stood watch over the Suffolk coast for centuries. Located near the starting point of the walk, this impressive structure offers a tangible connection to the rich military and maritime heritage of the region, inviting you to step back in time and imagine the lives and experiences of those who once lived and served within its walls.

The origins of Landguard Fort can be traced back to the 16th century, when the threat of Spanish invasion loomed large over England. To protect the strategically important Port of Felixstowe and the surrounding coastline, a small fortification was constructed on the site, consisting of little more than a handful of cannon and a modest earthwork rampart.

Over the centuries, the fort was repeatedly expanded and strengthened to keep pace with the evolving nature of warfare and the changing geopolitical landscape. During the English Civil War, Landguard Fort played a key role in the struggle between the Royalist and Parliamentarian forces, changing hands several times as the tides of battle ebbed and flowed.

In the 18th and 19th centuries, the fort underwent a major redevelopment, with the construction of a massive stone bastion and a network of underground tunnels and chambers. These formidable defenses were designed to withstand the might of the French Navy, which posed a constant threat to the security and prosperity of the British Empire.

Throughout the First and Second World Wars, Landguard Fort continued to serve as a vital coastal defense installation, housing a garrison of soldiers and a battery of powerful artillery pieces. The fort's strategic location and commanding views over the North Sea made it an essential component of the nation's wartime defense network, helping to protect the Port of Felixstowe and the surrounding region from enemy attack.

Today, Landguard Fort stands as a testament to the ingenuity, bravery, and resilience of the generations of soldiers and civilians who have lived and worked within its walls. As you explore the fort's museums, exhibitions, and beautifully preserved period rooms, you'll gain a profound appreciation for the rich and complex history of this iconic landmark, and the crucial role it has played in shaping the identity and character of the Suffolk coast.

From the grand parade ground where troops once drilled and marched to the eerie tunnels and chambers where soldiers huddled in anticipation of enemy bombardment, every inch of Landguard Fort is imbued with a sense of history and significance. As you walk in the footsteps of the fort's former inhabitants, you'll feel a deep connection to the past, and a renewed sense of gratitude for the sacrifices and achievements of those who came before us.

So take a moment to pause and reflect on the incredible story of Landguard Fort, and the countless tales of courage, sacrifice, and determination that are woven into the fabric of this remarkable place. As you continue on your journey through the beautiful Felixstowe Hills, carry with you the knowledge and inspiration you've gained from this unforgettable encounter with history, and let it enrich and deepen your appreciation of the extraordinary landscape and heritage of the Suffolk coast.

Rare or Uncommon Birds:

As you embark on the Felixstowe Hills Walk, keep your eyes and ears attuned to the skies and the shoreline, for this spectacular coastal landscape is home to a rich and diverse array of birdlife, from the common and familiar to the rare and elusive. With its unique combination of sandy beaches, rocky outcrops, and salt marshes, the Suffolk coast provides a vital haven for a wide range of seabirds and waders, each adapted to the challenges and opportunities of life on the edge of the land and sea.

One of the most iconic and sought-after species to watch for along this stretch of coast is the little tern, a diminutive and elegant seabird with a distinctive black cap, white forehead, and yellow bill. These charming and energetic birds are summer visitors to the UK, arriving in May to breed on the sandy and shingle beaches of the east coast before departing for their wintering grounds in Africa in August or September.

Keep an eye out for the little terns as they hover and plunge into the shallow waters just offshore, their quick and agile flight a marvel of precision and grace. With a wingspan of just 50cm and a weight of around 50g, these tiny birds are incredibly vulnerable to disturbance and habitat loss, making the protection of their nesting sites a key conservation priority for the region.

Another rare and spectacular bird to look out for on your walk is the avocet, a striking black-and-white wader with a slender, upturned bill and long, bluish-gray legs. These elegant and charismatic birds were once extinct as a breeding species in the UK, but thanks to dedicated conservation efforts and the creation of protected wetland habitats, they have made a remarkable comeback in recent decades.

The salt marshes and mudflats of the Landguard Nature Reserve are a particularly important site for avocets, providing a vital feeding and roosting ground for these beautiful birds. Look for them wading through the shallow pools and channels, their distinctive silhouettes and piping calls a joyous celebration of the resilience and adaptability of nature in the face of adversity.

Other uncommon and fascinating birds to watch for on your walk include the Mediterranean gull, a sleek and elegant seabird with a striking red bill and black hood; the Cetti's warbler, a secretive and elusive songbird with a loud and explosive voice; and the whimbrel, a long-billed wader with a haunting, rippling call that echoes across the marshes and mudflats.

As you marvel at the incredible diversity and beauty of the birdlife that surrounds you, take a moment to reflect on the vital importance of preserving and protecting these precious coastal habitats, and the countless species that depend on them for their survival. Each bird you encounter, whether common or rare, is a testament to the resilience and wonder of the natural world, and a reminder of the priceless heritage that we have been entrusted to cherish and defend.

So keep your binoculars at the ready, and your heart open to the magic and mystery of the birds that call this special place home. Let their presence be a source of joy, inspiration, and hope, and a call to action to do all we can to ensure that the skies and shores of the Suffolk coast continue to ring with their songs for generations to come.

Rare or Interesting Insects and Other Wildlife:

As you traverse the diverse habitats of the Felixstowe Hills Walk, from the windswept dunes and salt marshes of the coast to the lush grasslands and ancient woodlands of the interior, keep your senses attuned to the myriad forms of life that thrive in every nook and cranny of this extraordinary landscape. While the larger and more charismatic creatures like birds and mammals may be the first to catch your eye, it is often the smaller and more unassuming species that hold the greatest wonders and surprises, and offer the most profound insights into the intricate web of life that sustains this precious ecosystem.

One of the most fascinating and important groups of insects to watch for on your walk are the butterflies and moths, whose delicate beauty and incredible diversity are a true marvel of the natural world. From the bold and striking colours of the peacock and red admiral to the subtle and cryptic patterns of the meadow brown and the gatekeeper, each species has its own unique adaptations and life histories, and plays a vital role in the complex dance of pollination and reproduction that underpins the health and resilience of the coastal and inland habitats.

As you wander through the flower-studded grasslands and along the edge of the woodland, keep an eye out for the fluttering and darting forms of these enchanting creatures, and take a moment to appreciate the incredible intricacy and artistry of their wing patterns and markings. Watch as they flit from bloom to bloom, their long, coiled proboscises probing deep into the nectar-rich flowers, and marvel at the speed and agility with which they navigate the air currents and evade their predators.

Another group of insects that are sure to captivate and amaze you are the dragonflies and damselflies, whose iridescent colours and acrobatic flight are a true spectacle of the summer months. These ancient and highly adapted predators are masters of the air, their large, compound eyes and powerful wings allowing them to hunt and capture their prey with incredible speed and precision.

Look for the electric blues and greens of the emperor and southern hawker dragonflies as they patrol the edges of ponds and streams, their territorial displays and aerial battles a breathtaking display of power and grace. And keep an eye out for the more delicate and slender forms of the azure and common blue damselflies, whose ethereal beauty and courtship dances are a true wonder to behold.

But it is not just the insects that make this walk a treasure trove of wildlife encounters. As you explore the coast and the hills, you may also be lucky enough to spot some of the larger and more elusive creatures that call this landscape home, from the sleek and agile form of the stoat or weasel hunting among the dunes to the majestic silhouette of a marsh harrier soaring over the reedbeds.

And if you linger into the evening hours, you may even catch a glimpse of some of the nocturnal species that emerge under cover of darkness, from the ghostly form of a barn owl drifting silently over the fields to the darting and fluttering shapes of the bats that haunt the woodland edges and the old fort walls.

No matter what time of day or season you choose to embark on this walk, there is always something new and wondrous to discover in the secret lives and hidden corners of this extraordinary place. Each insect and animal you encounter, no matter how small or unassuming, is a vital thread in the rich tapestry of life that defines the Suffolk coast and hills, and offers a window into the endless mysteries and marvels of the natural world.

So let your curiosity and sense of wonder be your guide, and open your eyes and your heart to the incredible diversity and resilience of the creatures that share this special landscape. By cherishing and protecting the insects and wildlife that thrive here, we not only ensure the health and beauty of this ecosystem for generations to come, but also enrich and deepen our own connection to the living planet, and to the great web of life that sustains and inspires us all.

Rare or Interesting Plants:

As you make your way along the winding paths and trails of the Felixstowe Hills Walk, let your gaze be drawn to the incredible diversity and beauty of the plants that adorn every inch of this spectacular coastal landscape. From the hardy, salt-tolerant species that cling to the windswept dunes and cliffs to the delicate wildflowers that paint the grasslands and woodland edges with colour, each plant has its own unique adaptations and life histories, and plays a vital role in the intricate web of relationships that sustains this precious ecosystem.

One of the most striking and characteristic plants of the Suffolk coast is the sea pea, a low-growing, sprawling perennial with bright magenta flowers and delicate, ferny leaves. This remarkable species is a true survivor, its long, tough roots and nitrogen-fixing abilities allowing it to thrive in the harsh, nutrient-poor soils of the dunes and shingle banks. Look for its vibrant blooms and scrambling stems as you walk along the shore, and marvel at the incredible resilience and adaptability of this beautiful and hardy plant.

Another coastal speciality to watch out for is the sea kale, a robust, fleshy-leaved member of the cabbage family with clusters of white or pale purple flowers that bloom in the summer months. This distinctive plant is perfectly adapted to life in the salt-sprayed, wind-blasted environment of the coast, its thick, waxy cuticle and deeply buried roots enabling it to withstand the constant battering of the elements. In the past, sea kale was often harvested and cultivated as a vegetable crop, its young, blanched shoots prized for their delicate, nutty flavour.

As you move inland from the coast and into the rolling hills and valleys of the Felixstowe countryside, the plantlife becomes even more diverse and varied, with a stunning array of wildflowers, grasses, and shrubs painting the landscape in every hue and texture imaginable. In the spring and early summer, the grasslands are awash with the delicate pinks and purples of the pyramidal and bee orchids, the cheerful yellows of the cowslips and buttercups, and the frothy whites of the ox-eye daisies and yarrow.

Later in the season, the woodland edges and hedgerows come alive with the deep, velvety blues of the scabious and the delicate, nodding heads of the harebell, while the bramble thickets and field margins are studded with the juicy, sun-ripened fruits of the blackberry and the jewel-like berries of the wild raspberry and the elderberry.

But perhaps the most rare and special plant to look out for on your walk is the spiked speedwell, a small, unassuming member of the figwort family with slender spikes of pale blue or pinkish flowers that bloom in the late summer and early autumn. This elusive and enigmatic species is found only in a handful of locations along the Suffolk coast, its presence a testament to the unique and precious character of this habitat, and the vital importance of preserving and protecting these fragile and irreplaceable ecosystems.

As you marvel at the incredible diversity and beauty of the plants that surround you on this walk, take a moment to reflect on the deep and enduring connections that exist between the flora and fauna of this landscape, and the countless generations of people who have shaped and been shaped by this special place. From the ancient uses of plants for food, medicine, and crafts to the modern-day efforts to conserve and restore these habitats for the benefit of all, the story of the Felixstowe Hills is inextricably linked to the story of its plants, and to the timeless rhythms and cycles of the natural world.

So let your senses be your guide, and your curiosity and wonder your compass, as you explore the botanical treasures and mysteries of this extraordinary coastal landscape. By opening your eyes and your heart to the beauty and resilience of the plants that thrive here, you not only deepen your own connection to the living earth, but also help to ensure that these precious and irreplaceable species will continue to flourish and inspire for generations to come.

Facilities and Accessibility:

The Felixstowe Hills Walk offers a range of facilities and amenities to ensure a safe, comfortable, and enjoyable experience for walkers of all ages and abilities. While the route does include some moderate uphill sections and uneven terrain, the well-maintained paths and clear signage make it accessible to most reasonably fit and mobile individuals, with a few key considerations to keep in mind.

The walk begins and ends at the parking area near the entrance to Landguard Fort, where ample space is available for cars and a designated bus stop provides easy access for those arriving by public transport. The fort itself is home to a visitor centre with public toilets, a café serving light refreshments, and a gift shop offering maps, guidebooks, and souvenirs related to the history and natural heritage of the area.

As you set out on the walk, you'll find that the paths are generally well-surfaced and easy to follow, with a mix of gravel, dirt, and grass surfaces that are suitable for sturdy walking shoes or boots. The route is clearly waymarked with distinctive blue and green Suffolk Coast Path signs, making it easy to stay on track and navigate the various twists and turns of the hills and valleys.

Along the way, there are several benches and picnic tables strategically placed at key viewpoints and rest stops, providing welcome opportunities to catch your breath, take in the stunning vistas, or enjoy a packed lunch in the fresh sea air. These seating areas are particularly useful for those with limited mobility or stamina, allowing for a more relaxed and manageable walking pace.

As you reach the halfway point of the walk and begin your ascent into the Felixstowe Hills, the terrain becomes somewhat more challenging, with some steeper inclines and uneven sections that may be more difficult for those with mobility issues or respiratory conditions. However, the paths remain well-defined and the gradients are generally moderate, with plenty of level stretches and gentle slopes to balance out the more strenuous parts.

For those who do require additional assistance or support, it's worth noting that the Felixstowe Hills Walk is not specifically designed as an accessible route, and there are no dedicated facilities or services for people with disabilities along the way. However, with advance planning and preparation, many individuals with limited mobility may still be able to enjoy the beauty and tranquility of this special landscape, either by taking a more leisurely pace, using walking aids or mobility devices, or focusing on the gentler sections of the route.

Regardless of your age, fitness level, or physical abilities, it's always a good idea to come prepared for the changeable weather and variable conditions of the Suffolk coast, with appropriate clothing, footwear, and sun protection for the time of year. It's also wise to carry a map and compass or GPS device, as well as a fully charged mobile phone and a small first aid kit, in case of emergencies or unexpected delays.

By taking these simple precautions and being mindful of your own needs and limitations, you can ensure a safe, comfortable, and rewarding experience on the Felixstowe Hills Walk, and fully immerse yourself in the stunning beauty, rich history, and vibrant wildlife of this exceptional corner of the Suffolk coast. Whether you're a seasoned hiker or a casual walker, this route offers something for everyone, and is sure to leave you with memories and impressions that will last a lifetime.

Seasonal Highlights:

One of the great joys and marvels of the Felixstowe Hills Walk is the way in which the landscape and its inhabitants change and evolve with the turning of the seasons, each one bringing its own unique character, challenges, and rewards. From the fresh, vibrant greens of spring to the rich, golden hues of autumn, the hills and coast are a constantly shifting kaleidoscope of colour, texture, and life, inviting walkers to discover and delight in the endless wonders and surprises of the natural world.

In the spring, the walk comes alive with the joyous explosion of wildflowers and birdsong, as the land awakens from its winter slumber and the first migrant species return from their far-flung wintering grounds. The grasslands and woodland edges are carpeted with the delicate, nodding heads of primroses, violets, and bluebells, while the hedgerows and thickets are alive with the trilling and warbling of blackcaps, chiffchaffs, and whitethroats.

As you climb into the hills, the views open up to reveal a patchwork of green and gold, with the fresh, tender leaves of the beech and oak trees creating a dappled canopy of light and shade, and the distant glint of the sea beckoning on the horizon. The air is filled with the heady scent of hawthorn and wild garlic, and the gentle buzzing of bees and other pollinators as they flit from flower to flower in search of nectar and pollen.

As spring gives way to summer, the landscape reaches the height of its lushness and vitality, with the grasses and wildflowers growing tall and rank, and the trees heavy with fruit and foliage. The coast is now a hive of activity, with the beaches and dunes thronged with nesting seabirds and the waters teeming with fish and marine life. The air is heavy with the salty tang of the sea and the drowsy hum of insects, and the evenings are long and balmy, perfect for a leisurely ramble or a picnic supper on the hillside.

As you walk through the sunlit glades and along the breezy clifftops, keep an eye out for the many summer specialities that make this place so special, from the darting forms of the dragonflies and damselflies to the fluttering wings of the butterflies and moths. The marshes and saltmarshes are now a riot of colour, with the purple spikes of sea lavender and the yellow buttons of samphire creating a vivid tapestry of hues and textures, while the shingle banks and sandy flats are alive with the piping calls of oystercatchers and ringed plovers.

As the days begin to shorten and the first hints of autumn creep into the air, the walk takes on a new character, with the lush greens of summer gradually giving way to a rich palette of golds, oranges, and reds. The hedgerows and thickets are now laden with the bounty of the season, from the glossy blackberries and juicy sloes to the shiny rosehips and vibrant haws, while the woods and copses are alive with the bustle of creatures preparing for the long winter ahead.

This is a time of movement and change on the coast and in the hills, with the summer visitors departing for warmer climes and the first of the winter migrants arriving from the north. The skies are filled with the haunting cries of the waders and wildfowl as they pass overhead in great, swirling flocks, while the shoreline is a hive of activity as the birds feed and fuel up for the long journey ahead. The light takes on a softer, more mellow quality, and the air is crisp and invigorating, perfect for a brisk walk or a contemplative ramble.

Finally, as winter tightens its grip and the landscape is transformed into a stark and elemental beauty, the walk becomes a place of solitude and reflection, a chance to embrace the raw power and majesty of the natural world in its most uncompromising form. The trees are now bare and skeletal, etched against the pale, watery sky, while the coast is a wild and unforgiving place, with the wind and the waves whipping up a frenzy of spray and foam.

But even in the depths of winter, the Felixstowe Hills are alive with the hardy and resilient creatures that make their home here year-round, from the elusive otters and water voles that haunt the marshes and streams to the tits and finches that flock to the woodland edges in search of seeds and berries. And on those rare, crisp days when the sun breaks through the clouds and the frost sparkles on the grass, there is no more invigorating or life-affirming place to be than on the windswept heights of the Suffolk coast.

No matter what time of year you choose to embark on this walk, there is always something new and extraordinary to discover, some fresh delight or unexpected wonder that will stay with you long after you have left the hills and the sea behind. From the raw, elemental power of a winter storm to the soft, mellow beauty of a summer evening, each season brings its own unique gifts and challenges, its own invitation to connect with the timeless rhythms and enduring truths of the natural world.

So whether you are a seasoned hiker or a curious first-time visitor, a birdwatcher or a nature lover, a history buff or a cultural pilgrim, the Felixstowe Hills Walk promises to surprise, inspire, and delight you, no matter when or how often you choose to explore its hidden corners and timeless vistas. Let the turning of the seasons be your guide and your muse, and allow yourself to be transformed by the unparalleled beauty, diversity, and resilience of this extraordinary land, and the countless stories and secrets it has to share.

Difficulty Level and Safety Considerations:

The Felixstowe Hills Walk is a moderate route that offers a delightful mix of coastal and countryside scenery, with some gently undulating terrain and a few short, steeper sections that provide a bit of a challenge and reward walkers with stunning views over the surrounding landscape. While the walk is generally suitable for anyone with a reasonable level of fitness and mobility, there are a few key safety considerations and precautions to keep in mind to ensure a safe, enjoyable, and rewarding experience.

First and foremost, it's important to be aware of the changeable and potentially unpredictable weather conditions that can occur along the Suffolk coast, particularly during the spring and autumn months when strong winds, heavy rain, and even occasional storms are not uncommon. Before setting out on your walk, be sure to check the local weather forecast and come prepared with appropriate clothing and gear for the conditions, including waterproof and windproof layers, sturdy footwear, and a hat and gloves if necessary.

The exposed and open nature of some sections of the walk, particularly along the coastal clifftops and in the higher reaches of the hills, can also leave walkers vulnerable to the effects of the sun and the wind, so it's important to take precautions to avoid sunburn, dehydration, and heat exhaustion, especially during the summer months. Be sure to carry plenty of water, wear a hat and sunscreen, and take regular breaks in the shade if needed.

As you navigate the various paths and trails of the walk, be mindful of any uneven or slippery surfaces, particularly after heavy rain or during the winter months when ice and snow may be present. The graveled and dirt paths can become muddy and treacherous in wet weather, while the grassy slopes and hillsides can be slick and difficult to traverse when damp. Take your time, watch your step, and use a walking stick or trekking poles if you need extra stability and support.

Another important safety consideration for walkers on the Felixstowe Hills Walk is the presence of livestock in some of the fields and meadows along the route, particularly during the spring and summer months when sheep and cattle may be grazing. While these animals are generally harmless and accustomed to the presence of walkers, it's important to give them plenty of space and to avoid startling or disturbing them, especially if you are walking with dogs.

If you do encounter livestock on your walk, keep your distance and walk calmly and quietly around them, taking care not to separate any young animals from their mothers. If you are walking with a dog, be sure to keep it on a short lead or under close control, and be prepared to put it on a lead if necessary to avoid any potential conflicts or confrontations.

Finally, while the Felixstowe Hills Walk is a relatively low-risk and well-maintained route, it's always a good idea to let someone know your planned itinerary and estimated return time before setting out, and to carry a fully charged mobile phone and a small first aid kit in case of emergencies. If you do find yourself in difficulty or in need of assistance, don't hesitate to call for help or to seek out the nearest emergency services.

By following these simple guidelines and using common sense and good judgement, walkers of all ages and abilities can safely and confidently enjoy the many delights and rewards of the Felixstowe Hills Walk, from the stunning coastal vistas and the rich wildlife to the fascinating history and the timeless beauty of this extraordinary corner of the Suffolk coast. So come prepared, stay alert and aware, and allow yourself to be inspired and enriched by the unparalleled natural and cultural heritage of this very special place.

Nearby Attractions:

The Felixstowe Hills Walk is not only a wonderful destination in its own right, but also a gateway to a wealth of nearby attractions and activities that showcase the rich history, stunning landscapes, and vibrant culture of the Suffolk coast and countryside. Whether you're a local resident looking for a new adventure or a visitor eager to explore the many hidden gems and unexpected delights of this beautiful region, there's something for everyone within easy reach of this delightful and rewarding walk.

For history buffs and anyone with an interest in the military and maritime heritage of the area, a visit to Landguard Fort is an absolute must. Located just a stone's throw from the starting point of the walk, this impressive 18th-century fortification has played a vital role in defending the Port of Felixstowe and the surrounding coastline from invasion and attack for centuries, and today stands as a testament to the ingenuity, bravery, and resilience of the generations of soldiers and civilians who have lived and served here.

Take a guided tour of the fort's atmospheric tunnels, chambers, and battlements, and discover the fascinating stories and artifacts that bring its rich and complex history to life, from the grand parade ground where troops once drilled and marched to the eerie underground magazines where gunpowder and ammunition were stored. With its commanding views over the sea and the Orwell Estuary, Landguard Fort is a powerful and poignant reminder of the enduring importance of this strategic location, and a must-see destination for anyone exploring the Felixstowe Hills.

For those with a passion for nature and wildlife, the nearby Trimley Marshes Nature Reserve is a true paradise, offering over 200 acres of pristine wetland habitat that is home to an astonishing diversity of birds, mammals, and invertebrates. Just a short walk or drive from the hills, this internationally important site is a haven for birdwatchers and nature enthusiasts, with a network of paths, hides, and viewing platforms that provide unparalleled opportunities to observe and enjoy the incredible variety of species that thrive here.

From the elegant avocets and the elusive bearded tits that haunt the reedbeds and the marshes to the darting dragonflies and the fluttering butterflies that flit among the wildflowers and the grasses, Trimley Marshes is a celebration of the beauty, complexity, and resilience of the natural world, and a vital reminder of the importance of preserving and protecting these precious and irreplaceable habitats for generations to come.

For a taste of the vibrant arts and culture scene that makes this part of Suffolk so special, be sure to pay a visit to the charming seaside town of Felixstowe, just a short distance from the hills. With its elegant Victorian and Edwardian architecture, its lively seafront promenade, and its bustling town centre, Felixstowe is a delightful and welcoming destination that offers something for everyone, from families with young children to couples looking for a romantic getaway.

Take a stroll along the beach and the pier, and soak up the timeless charm and nostalgia of the traditional English seaside, or explore the town's many independent shops, galleries, and cafes, and discover the unique character and creativity of the local community. And don't miss the chance to catch a show at the historic Spa Pavilion, a beautifully restored 1930s theatre that hosts a diverse program of music, dance, drama, and comedy throughout the year, showcasing the very best of local and national talent.

For a deeper dive into the fascinating history and heritage of the area, the Felixstowe Museum is a must-visit destination, housed in a stunning Edwardian building with panoramic views over the sea. With its extensive collections of photographs, documents, and artifacts spanning over 130 years of the town's past, the museum offers a unique and compelling window into the lives and experiences of the people who have shaped this special corner of the Suffolk coast, from the early days of the fishing and farming communities to the present-day challenges and opportunities of a thriving seaside resort.

And for those with an adventurous spirit and a love of the great outdoors, the wider Suffolk Coast and Heaths Area of Outstanding Natural Beauty is a veritable playground, with miles of stunning beaches, heathlands, forests, and estuaries to explore on foot, by bike, or by boat. From the wild and windswept beauty of the Dunwich Heath to the tranquil and secluded charm of the River Deben, this extraordinary landscape is a true natural wonder, offering endless opportunities for discovery, inspiration, and adventure.

So whether you're a keen hiker, a nature lover, a history buff, or simply someone in search of a little beauty, tranquility, and inspiration, the Felixstowe Hills Walk and its nearby attractions offer a wealth of experiences and delights that are sure to leave you feeling refreshed, enriched, and deeply connected to the timeless magic and enduring spirit of this very special place. From the soaring heights of the hills to the shimmering expanse of the sea, from the ancient stones of the fort to the vibrant heart of the town, every step of your journey promises to be a revelation and a joy, an invitation to discover and celebrate the countless wonders and treasures that make the Suffolk coast and countryside so unforgettable.

As you explore the many hidden corners and unexpected delights of this beautiful region, take a moment to reflect on the deep and enduring connections that exist between the land, the sea, and the people who have called this place home for generations. From the earliest hunter-gatherers who roamed the hills and shores in search of food and shelter to the modern-day conservationists and community activists who work tirelessly to protect and preserve this precious heritage for generations to come, every chapter of the Felixstowe story is a testament to the resilience, creativity, and vision of the human spirit.

So let your curiosity be your guide, and your sense of wonder and delight your compass, as you set out on a journey of discovery that will take you to the very heart of what makes this corner of England so special and so beloved. Whether you're a seasoned traveler or a first-time visitor, a local resident or a passing pilgrim, the Felixstowe Hills Walk and its nearby attractions promise to surprise, inspire, and delight you at every turn, offering a gateway to a world of beauty, history, and adventure that is truly beyond compare.

From the soaring heights of the hills to the shimmering expanse of the sea, from the ancient stones of the fort to the vibrant heart of the town, every step of your journey promises to be a revelation and a joy, an invitation to discover and celebrate the countless wonders and treasures that make the Suffolk coast and countryside so unforgettable. So why wait any longer to begin your adventure? The hills and the sea, the history and the heritage, the wildlife and the wonder of Felixstowe and beyond are calling to you, inviting you to step outside your comfort zone and embrace the magic and majesty of this extraordinary place.

Whether you have a few hours or a few days to spare, whether you're seeking solitude or sociability, relaxation or exhilaration, the Felixstowe Hills Walk and its nearby attractions offer a world of possibilities and experiences that are sure to leave you feeling refreshed, inspired, and forever changed. So lace up your walking boots, pack your binoculars and your sense of adventure, and set out on a journey of discovery that will take you to the very heart of what makes this corner of England so special and so beloved.

With its stunning scenery, its fascinating history, its vibrant culture, and its warm and welcoming people, the Suffolk coast and countryside is a true national treasure, a living, breathing testament to the enduring power and beauty of the natural world and the human spirit. And the Felixstowe Hills Walk is your key to unlocking the secrets and the wonders of this extraordinary place, your invitation to connect with the land, the sea, and the sky in a way that is truly unforgettable.

So what are you waiting for? The hills and the sea, the history and the heritage, the wildlife and the wonder of Felixstowe and beyond are calling to you, beckoning you to come and explore, to come and discover, to come and be a part of something truly special and truly amazing. The adventure of a lifetime awaits you, and all you have to do is take that first step, and let the magic and the majesty of this place work its spell on your mind, your body, and your soul.

Ipswich to Felixstowe Walk - 13 miles (20.9 km), Linear - Estimated Time: 6-7 hours

Starting Point Coordinates, Postcode: The starting point for the Ipswich to Felixstowe Walk is at Ipswich Railway Station (Grid Reference: TM 160 442, Postcode: IP2 8AL).
Nearest Car Park, Postcode: The nearest car park is at Ipswich Railway Station (Postcode: IP2 8AL).

Embark on a captivating 13-mile (20.9 km) linear walk from the vibrant town of Ipswich to the enchanting seaside resort of Felixstowe, where the meandering beauty of the River Orwell, the tranquil charm of the Suffolk countryside, and the invigorating allure of the coast converge to create an unforgettable experience. This moderate walk, with its inclines and uneven terrain, promises a perfect blend of riverside tranquility, rural splendour, and coastal exhilaration, inviting you to immerse yourself in the rich history, diverse wildlife, and breathtaking landscapes of this fascinating corner of East Anglia.

As you set out from the bustling platform of Ipswich Railway Station, feel the anticipation build as you head south towards the glistening expanse of the River Orwell, the well-trodden path guiding you towards a world of natural wonders and hidden treasures. With each step, the cares and distractions of everyday life seem to melt away, replaced by a growing sense of peace and connection with the timeless beauty of the river and its surrounding landscape.

Follow the riverside path eastward, drinking in the picturesque views of the Orwell Bridge, a stunning feat of engineering that spans the wide, glittering waters of the river. As you continue along the path, you'll find yourself enveloped by the lush greenery and tranquil atmosphere of the Orwell Country Park, a verdant oasis that offers a welcome respite from the hustle and bustle of modern life.

Take a moment to pause and appreciate the rich tapestry of wildlife that thrives in this precious habitat, from the graceful wading birds that pick their way through the mudflats and saltmarshes to the darting dragonflies and fluttering butterflies that bring the meadows and hedgerows to life. Each encounter is a reminder of the incredible diversity and resilience of the natural world, and a chance to deepen your understanding and appreciation of the delicate balance that sustains us all.

As you make your way through the charming villages of Nacton and Levington, take a moment to absorb the timeless character and rural charm of these ancient settlements, with their thatched cottages, weathered churches, and friendly, welcoming inhabitants. Here, the pace of life moves to a different rhythm, shaped by the ebb and flow of the tides and the turning of the seasons, and offering a glimpse into a simpler, more connected way of being.

Joining the Stour and Orwell Walk, you'll find yourself immersed in a landscape of breathtaking beauty and diversity, with the shimmering expanse of the river on one side and the gently rolling hills and fields of the Suffolk countryside on the other. Each twist and turn of the path reveals new delights and surprises, from the historic site of the Roman fort at Walton Castle to the rare and fascinating plant species that cling to the river's edge.

As you approach the picturesque hamlet of Felixstowe Ferry, the landscape begins to change once again, with the wide open spaces of the river giving way to the rugged beauty of the coast. Here, the salty tang of the sea air and the cry of the gulls mingle with the soft lapping of the waves, creating a symphony of sound and sensation that invigorates the senses and soothes the soul.

Follow the coastal path southward, drinking in the stunning views of Felixstowe Beach, with its golden sands, colourful beach huts, and lively promenade. Take a moment to bask in the timeless charm and nostalgia of this classic English seaside resort, with its ice cream parlours, fish and chip shops, and entertained visitors.

As you make your way into the heart of Felixstowe, take a moment to reflect on the incredible journey you have undertaken, both physically and spiritually. The Ipswich to Felixstowe Walk is a celebration of the enduring beauty and resilience of the natural world, a reminder of the countless wonders that await those who step outside their comfort zone and embrace the unknown.

Whether you are a seasoned hiker or a curious first-time adventurer, this enchanting route promises to delight and inspire, offering a profound sense of connection with the land, the water, and the living things that call this special place home. So let the rhythm of your footsteps be your guide, and allow yourself to be carried away by the magic and mystery of this extraordinary corner of Suffolk.

Detailed Route Directions:

1. From Ipswich Railway Station, head south towards the River Orwell. Join the riverside path and follow it eastward, bearing approximately 90°. Continue along this path for about 2 miles (3.2 km). (Estimated time: 45-60 minutes)

2. As you walk along the riverside path, you will pass through Orwell Country Park. Take in the stunning views of the Orwell Bridge and the surrounding landscape, maintaining your eastward heading. (Estimated time: 60-75 minutes)

3. Continue following the riverside path, passing through the villages of Nacton and Levington. Maintain a general easterly direction, approximately 80° to 100°, for around 4 miles (6.4 km). (Estimated time: 1.5-2 hours)

4. At Levington, join the Stour and Orwell Walk and continue eastward along the path, bearing approximately 80° to 100°. Follow this path for about 4 miles (6.4 km) until you reach Felixstowe Ferry. (Estimated time: 1.5-2 hours)

5. From Felixstowe Ferry, turn south and follow the coastal path along Felixstowe Beach, bearing approximately 180°. Continue along this path for around 3 miles (4.8 km) until you reach Felixstowe town centre. (Estimated time: 1-1.5 hours)

Best Features and Views:

- Immerse yourself in the stunning riverside scenery along the River Orwell, with its glistening waters, lush greenery, and ever-changing vistas that showcase the timeless beauty of the Suffolk landscape.
- Marvel at the impressive Orwell Bridge, a stunning feat of engineering that spans the wide, shimmering expanse of the river, offering breathtaking views of the surrounding countryside and the distant horizon.
- Delight in the charming seaside atmosphere of Felixstowe, with its golden sands, colourful beach huts, and lively promenade, offering a perfect blend of nostalgia, relaxation, and coastal exhilaration.

Historical Features of Interest:

As you make your way along the Ipswich to Felixstowe Walk, you'll encounter a fascinating array of historical sites and landmarks that offer a glimpse into the rich and complex past of this ancient corner of Suffolk. From the crumbling ruins of Roman fortifications to the imposing walls of medieval castles, each step of your journey is a chance to connect with the countless generations who have shaped and been shaped by this timeless landscape.

One of the most intriguing and enigmatic historical features along the route is the site of the ancient Roman fort at Walton Castle, a strategically important stronghold that once guarded the entrance to the River Orwell. Although little remains of the original structure today, the grassy mounds and earthworks that mark the spot are a powerful reminder of the enduring influence of the Roman Empire on this part of Britain, and the vital role that the Suffolk coast played in the defense and administration of the province.

As you stand atop the weathered ramparts and gaze out over the shimmering expanse of the river, try to imagine the lives and experiences of the soldiers and civilians who once called this place home, the challenges and triumphs they faced, and the legacy they left behind. The story of Walton Castle is a microcosm of the larger story of Roman Britain, a tale of conquest, assimilation, and transformation that continues to shape our understanding of the past and our sense of identity today.

Further along the walk, as you approach the bustling seaside town of Felixstowe, you'll come face to face with another iconic symbol of the region's military and maritime heritage: the imposing bulk of Landguard Fort. Built in the 16th century to protect the vital port of Felixstowe from the threat of foreign invasion, this formidable stronghold has played a crucial role in the defense of the Suffolk coast for over 400 years, witnessing countless battles, sieges, and conflicts that have shaped the course of British history.

As you explore the fort's winding passages, cavernous chambers, and towering battlements, you'll gain a profound sense of the courage, ingenuity, and sacrifice of the generations of soldiers and civilians who have lived and served here, from the anxious days of the Spanish Armada to the dark hours of the Second World War. Each room and exhibit offers a fascinating glimpse into the daily lives and experiences of these brave men and women, the challenges they faced, and the incredible feats of bravery and endurance they accomplished in the face of unimaginable adversity.

But Landguard Fort is more than just a monument to the past; it is also a living, breathing testament to the enduring spirit and resilience of the British people, a symbol of the values and ideals that have shaped our nation and our way of life for centuries. As you stand atop the fort's windswept walls and gaze out over the vast expanse of the North Sea, you'll feel a deep sense of connection to the countless generations who have stood here before you, united by a common bond of duty, honor, and service to their country and their fellow citizens.

The Ipswich to Felixstowe Walk is a journey through time as well as space, a chance to explore the rich and complex tapestry of history that underlies the stunning landscapes and vibrant communities of the Suffolk coast. From the ancient ruins of Walton Castle to the mighty walls of Landguard Fort, each step of your journey is an invitation to connect with the past, to learn from its triumphs and tragedies, and to gain a deeper appreciation for the incredible heritage and resilience of this special corner of England.

So let the echoes of history be your guide and your inspiration as you make your way along this enchanting route, and allow yourself to be transported back to a world of adventure, discovery, and courage that will stay with you long after you have returned to the present day.

Rare or Uncommon Birds:

As you embark on the Ipswich to Felixstowe Walk, keep your eyes and ears attuned to the skies and the shores, for this diverse and dynamic landscape is home to an incredible array of birdlife, from the common and familiar to the rare and endangered. With its unique blend of coastal, estuarine, and inland habitats, the

Suffolk coast provides a vital haven for countless species of waders, waterfowl, and seabirds, each adapted to the challenges and opportunities of life at the edge of the land and sea.

One of the most iconic and sought-after birds to grace the mudflats and saltmarshes of the River Orwell is the avocet, a striking black-and-white wader with a slender, upturned bill and long, bluish-gray legs. Once driven to extinction in Britain due to habitat loss and persecution, these elegant birds have made a remarkable comeback in recent decades thanks to dedicated conservation efforts and the protection of key wetland sites like the Orwell Estuary.

As you walk along the riverside path, keep an eye out for the distinctive silhouettes of avocets as they wade through the shallow waters, their bills sweeping from side to side in search of tiny crustaceans and other aquatic invertebrates. With a wingspan of around 30 inches and a weight of just 10 to 14 ounces, these graceful birds are a true marvel of nature, their presence a testament to the resilience and adaptability of life in the face of adversity.

Another rare and fascinating bird to watch for along the Orwell is the curlew, a large, long-billed wader with a haunting, bubbling call that echoes across the marshes and mudflats. With their mottled brown plumage and distinctive downward-curved bills, curlews are a true icon of the Suffolk coast, their evocative cries a symbol of the wild and untamed beauty of this special place.

Sadly, curlews have suffered severe declines in recent years due to habitat loss, disturbance, and changes in agricultural practices, making them one of the most endangered bird species in the UK. As you walk along the River Orwell, take a moment to appreciate the incredible beauty and vulnerability of these remarkable birds, and the vital importance of protecting and preserving the wetland habitats that sustain them.

As you continue your journey towards Felixstowe, the landscape begins to change, with the wide, open expanse of the river giving way to the rugged beauty of the coast. Here, amidst the crashing waves and windswept dunes, a new cast of avian characters takes centre stage, from the nimble sanderlings that scurry along the tideline to the elegant terns that hover and dive in pursuit of their fishy prey.

One of the most spectacular and charismatic birds to grace the shores of Felixstowe is the little tern, a diminutive seabird with a black cap, white forehead, and yellow bill. These plucky little birds are summer visitors to the Suffolk coast, arriving in May to breed on the sandy beaches and shingle banks before departing for their African wintering grounds in August or September.

With a wingspan of just 20 inches and a weight of around 1.5 ounces, little terns are incredibly vulnerable to disturbance and habitat loss, making the protection of their nesting sites a key conservation priority for the region. As you walk along the beach, keep an eye out for these lively and energetic birds as they fly low over the waves, their quick and agile wing beats a marvel of aerial prowess.

From the majestic avocets of the Orwell to the lively little terns of Felixstowe, the Ipswich to Felixstowe Walk offers an unparalleled opportunity to discover and delight in the incredible diversity of bird life that calls the Suffolk coast home. By opening your eyes and your heart to the beauty and fragility of these remarkable creatures, you can gain a deeper appreciation for the vital role they play in the delicate balance of the coastal ecosystem, and the urgent need to protect and preserve the habitats that sustain them for generations to come.

So let the birds be your guide and your inspiration as you make your way along this enchanting route, and allow their presence to fill you with a sense of wonder, joy, and connection to the incredible web of life that surrounds and sustains us all. Whether you are a seasoned birdwatcher or a curious novice, the Ipswich to

Felixstowe Walk promises to surprise and delight you at every turn, offering a window into a world of beauty, resilience, and endless fascination that will stay with you long after you have returned to your daily life.

Rare or Interesting Insects and Other Wildlife:

As you traverse the diverse landscapes of the Ipswich to Felixstowe Walk, from the lush riverbanks of the Orwell to the windswept dunes of the coast, keep your senses attuned to the myriad forms of life that thrive in every nook and cranny of this extraordinary ecosystem. While the birds and larger mammals may be the first to catch your eye, it is often the smaller and more unassuming creatures that hold the greatest wonders and surprises, offering a window into the intricate web of relationships that sustains the health and vitality of the natural world.

One of the most fascinating and important groups of insects to watch for on your walk are the butterflies and moths, whose delicate beauty and incredible diversity are a true marvel of nature. From the bold and striking colours of the peacock and red admiral to the subtle and cryptic patterns of the meadow brown and the gatekeeper, each species has its own unique adaptations and life histories, playing a vital role in the complex dance of pollination and reproduction that underpins the ecological balance of the region.

As you wander through the flower-studded meadows and along the hedgerows that line the route, keep an eye out for the fluttering and darting forms of these enchanting creatures, and take a moment to appreciate the incredible intricacy and artistry of their wing patterns and markings. Watch as they flit from bloom to bloom, their long, coiled proboscises probing deep into the nectar-rich flowers, and marvel at the speed and agility with which they navigate the air currents and evade their predators.

Another group of insects that are sure to captivate and amaze you are the dragonflies and damselflies, whose iridescent colours and acrobatic flight are a true spectacle of the summer months. These ancient and highly adapted predators are masters of the air, their large, compound eyes and powerful wings allowing them to hunt and capture their prey with incredible speed and precision.

Look for the electric blues and greens of the emperor and southern hawker dragonflies as they patrol the edges of ponds and streams, their territorial displays and aerial battles a breathtaking display of power and grace. And keep an eye out for the more delicate and slender forms of the banded demoiselle and azure damselflies, whose ethereal beauty and courtship dances are a true wonder to behold.

But it is not just the insects that make this walk a treasure trove of wildlife encounters. As you make your way along the riverbanks and through the coastal dunes, you may also be lucky enough to spot some of the mammals and amphibians that call this landscape home, from the elusive otters and water voles that haunt the reedbeds and riverbanks to the common lizards and sand lizards that bask on the sun-warmed stones and logs.

And if you linger into the evening hours, you may even catch a glimpse of some of the nocturnal species that emerge under cover of darkness, from the ghostly form of a barn owl drifting silently over the meadows to the darting and fluttering shapes of the pipistrelle and noctule bats that haunt the woodland edges and old buildings.

No matter what time of day or season you choose to embark on this walk, there is always something new and wondrous to discover in the secret lives and hidden corners of this extraordinary place. Each insect and animal you encounter, no matter how small or unassuming, is a vital thread in the rich tapestry of life that defines the Suffolk coast and countryside, offering a window into the endless mysteries and marvels of the natural world.

So let your curiosity and sense of wonder be your guide, and open your eyes and your heart to the incredible diversity and resilience of the creatures that share this special landscape. By cherishing and protecting the insects and wildlife that thrive here, we not only ensure the health and beauty of this ecosystem for generations to come, but also enrich and deepen our own connection to the living planet, and to the great web of life that sustains and inspires us all.

Rare or Interesting Plants:

As you make your way along the winding paths and trails of the Ipswich to Felixstowe Walk, let your gaze be drawn to the incredible diversity and beauty of the plants that adorn every inch of this spectacular coastal landscape. From the delicate, salt-tolerant wildflowers that cling to the windswept dunes and cliffs to the lush, verdant foliage that lines the riverbanks and meadows, each species has its own unique adaptations and life histories, playing a vital role in the intricate web of relationships that sustains this precious ecosystem.

One of the most striking and characteristic plants of the Suffolk coast is the sea lavender, a low-growing perennial with clusters of tiny, pale purple flowers that bloom in the summer months. Found in the upper reaches of salt marshes and along the edges of tidal creeks, this hardy and adaptable species is perfectly suited to the challenging conditions of the coastal environment, its thick, fleshy leaves and deep root system enabling it to tolerate the high salinity and frequent inundation of its habitat.

As you walk along the mudflats and saltmarshes of the River Orwell, keep an eye out for the delicate, wispy stems and flowers of the sea lavender, their subtle beauty and resilience a testament to the incredible diversity and tenacity of life in the face of adversity. Take a moment to appreciate the vital role that this unassuming plant plays in the ecology of the coastline, providing food and shelter for countless species of insects, birds, and other wildlife, and helping to stabilize the fragile soils of the marsh.

Another fascinating and important plant to watch for along the walk is the samphire, a succulent, salt-tolerant herb with bright green, segmented stems and tiny, inconspicuous flowers. Also known as glasswort or sea asparagus, this highly nutritious and flavourful plant has been harvested and eaten by coastal communities for centuries, its crisp, salty texture and delicate, slightly bitter flavour making it a prized ingredient in many traditional dishes and gourmet recipes.

As you make your way along the tidal flats and saltmarshes of the Orwell, keep an eye out for the dense, low-growing mats of samphire that carpet the muddy ground, their vibrant green colour and distinctive, jointed stems a welcome splash of life and vitality amidst the stark, monochromatic landscape of the marsh. Take a moment to appreciate the incredible adaptations that allow this remarkable plant to thrive in such a challenging and dynamic environment, from its ability to concentrate salt in its tissues to its deep, anchoring root system that helps to bind and stabilize the shifting sediments of the coast.

As you continue your journey towards Felixstowe, the landscape begins to change, with the wide, open expanse of the saltmarsh giving way to the rugged, windswept beauty of the coastal dunes. Here, amidst the shifting sands and the crashing waves, a new cast of botanical characters takes centre stage, from the spiky, blue-green tufts of marram grass that help to stabilize the dunes to the delicate, pink-and-white flowers of the sea bindweed that twine and climb among the grasses and shrubs.

One of the most beautiful and iconic plants to grace the dunes of Felixstowe is the sea holly, a striking, spiny-leaved perennial with intense blue, thistle-like flowers that bloom in the late summer and early fall. With its tough, leathery foliage and deep, penetrating taproot, the sea holly is perfectly adapted to the harsh, nutrient-poor soils and constant exposure to salt spray and wind that characterize the dune environment, its presence a testament to the incredible resilience and adaptability of coastal flora.

As you walk among the dunes and along the shore, take a moment to appreciate the subtle beauty and fascinating ecology of the sea holly, from its intricate, geometric leaf patterns to its vital role in stabilizing the shifting sands and providing food and shelter for a wide range of invertebrates, birds, and other wildlife. Let this remarkable plant be your guide and your inspiration as you explore the wild and untamed beauty of the Suffolk coast, and allow its presence to fill you with a sense of wonder and connection to the timeless cycles and rhythms of the natural world.

From the delicate sea lavender of the saltmarsh to the rugged sea holly of the dunes, the Ipswich to Felixstowe Walk offers an unparalleled opportunity to discover and delight in the incredible diversity and beauty of the coastal flora that defines this special corner of England. By opening your eyes and your heart to the wonders and mysteries of the plant kingdom, you can gain a deeper appreciation for the vital role that these remarkable species play in the health and resilience of the coastal ecosystem, and the urgent need to protect and preserve the habitats that sustain them for generations to come.

So let the plants be your guide and your muse as you make your way along this enchanting route, and allow their presence to fill you with a sense of awe, humility, and connection to the incredible web of life that surrounds and sustains us all. Whether you are a seasoned botanist or a curious novice, the Ipswich to Felixstowe Walk promises to surprise and inspire you at every turn, offering a window into a world of beauty, resilience, and endless fascination that will stay with you long after you have returned to your daily life.

Facilities and Accessibility:

The Ipswich to Felixstowe Walk offers a range of facilities and amenities to ensure a safe, comfortable, and enjoyable experience for walkers of all ages and abilities. While the route does include some moderate inclines and uneven terrain, the well-maintained paths and clear signage make it accessible to most reasonably fit and mobile individuals, with a few key considerations to keep in mind.

The walk begins at Ipswich Railway Station, where ample parking is available for those arriving by car, as well as convenient access to public transport for those traveling by train or bus. The station itself offers a range of facilities, including public restrooms, a café, and a small shop selling snacks, drinks, and other essentials.

As you set out on the walk, you'll find that the paths along the River Orwell are generally well-surfaced and easy to follow, with a mix of gravel, dirt, and grass surfaces that are suitable for sturdy walking shoes or boots. The route is clearly waymarked with distinctive green and white signs, making it easy to stay on track and navigate the various twists and turns of the river and the countryside.

Along the way, there are several benches and picnic tables strategically placed at key viewpoints and rest stops, providing welcome opportunities to catch your breath, take in the stunning vistas, or enjoy a packed lunch in the fresh air. These seating areas are particularly useful for those with limited mobility or stamina, allowing for a more relaxed and manageable walking pace.

As you pass through the villages of Nacton and Levington, you'll find a range of local amenities and services, including public restrooms, small shops, and pubs or cafes where you can stop for refreshment and a bite to eat. These charming and historic communities offer a welcome respite from the trail, and a chance to explore the rich cultural heritage and daily life of the Suffolk countryside.

Continuing along the Stour and Orwell Walk towards Felixstowe Ferry, the terrain becomes somewhat more varied and challenging, with some steeper inclines, uneven surfaces, and occasional muddy or waterlogged sections, particularly after heavy rain. While these conditions may be more difficult for those with mobility issues or respiratory conditions, the paths remain generally well-maintained and accessible, with regular rest points and clear signage throughout.

At Felixstowe Ferry itself, you'll find a range of facilities and services catering to the needs of walkers and visitors, including public restrooms, a small café, and a pub serving refreshments and light meals. From here, the final stretch of the walk along the coast into Felixstowe town centre is relatively flat and easy to navigate, with a well-maintained promenade and seawall that offer stunning views of the beach and the sea.

In Felixstowe, you'll find a wide range of amenities and attractions to suit every taste and budget, from traditional seaside amusements and fish and chip shops to museums, galleries, and historic landmarks. The town is well-served by public transport, with regular trains and buses connecting to Ipswich and other destinations in the region, making it easy to plan your onward journey or return to your starting point.

While the Ipswich to Felixstowe Walk is not specifically designed as an accessible route for those with mobility impairments or other disabilities, with careful planning and preparation, many individuals may still be able to enjoy the beauty and tranquility of this special landscape, either by focusing on the gentler sections of the route or by using mobility aids or assistance as needed.

Regardless of your age, fitness level, or physical abilities, it's always a good idea to come prepared for the changeable weather and variable conditions of the Suffolk coast, with appropriate clothing, footwear, and sun protection for the time of year. It's also wise to carry a map and compass or GPS device, as well as a fully charged mobile phone and a small first aid kit, in case of emergencies or unexpected delays.

By taking these simple precautions and being mindful of your own needs and limitations, you can ensure a safe, comfortable, and rewarding experience on the Ipswich to Felixstowe Walk, and fully immerse yourself in the stunning beauty, rich history, and vibrant wildlife of this exceptional corner of the Suffolk coast. Whether you're a seasoned rambler or a casual stroller, this route offers something for everyone, and is sure to leave you with memories and impressions that will last a lifetime.

Seasonal Highlights:

One of the great joys and rewards of the Ipswich to Felixstowe Walk is the opportunity to experience the changing moods, colours, and textures of the Suffolk coast and countryside through the turning of the seasons. From the fresh, vibrant greens of spring to the rich, golden hues of autumn, this enchanting landscape is a constantly shifting kaleidoscope of light, life, and beauty, inviting walkers to discover and delight in the endless wonders and surprises of the natural world.

In the spring, the walk comes alive with the joyous explosion of colour and sound, as the first tentative blooms of wildflowers and the melodious songs of returning birds fill the air with the promise of renewal and growth. Along the riverbanks and in the meadows, the delicate pinks and whites of cuckooflower and stitchwort mingle with the cheerful yellows of primrose and cowslip, while overhead, the skies are alive with the trilling and warbling of chiffchaffs, blackcaps, and whitethroats.

As you make your way along the gently undulating paths and trails, the fresh, green leaves of the hedgerows and copses unfurl in the warming sun, their tender, translucent beauty a testament to the irrepressible force of life that flows through every vein and fiber of this ancient landscape. The air is filled with the heady scent of hawthorn and wild garlic, and the gentle humming of bees and other pollinators as they dart and weave among the blossoms, their tireless industry a vital link in the great chain of being that sustains us all.

As spring gives way to summer, the walk reaches the height of its splendour and abundance, with the grasses and flowers growing tall and lush, and the trees heavy with fruit and foliage. The riverbanks are now a riot of colour, with the purple loosestrife and yellow flag iris vying for attention among the reeds and rushes, while in the meadows, the delicate pinks and mauves of common knapweed and field scabious sway gently in the warm breeze.

Along the coast, the sea sparkles and dances in the golden light, the endless blue horizon broken only by the distant silhouettes of ships and the wheeling, soaring forms of gulls and terns. The beaches and dunes are alive with the piping calls of ringed plovers and oystercatchers, while in the shallow waters, the darting shapes of fish and the scuttling forms of crabs and shrimp go about their secret lives beneath the waves.

As the long, languid days of summer give way to the cool, crisp air of autumn, the walk takes on a new character, the lush greens of the countryside slowly fading to a rich tapestry of russet, gold, and bronze. The hedgerows and thickets are now laden with the bounty of the harvest, from the glossy blackberries and clustering sloes to the bright, lipstick-red hips of the wild rose, each one a tiny miracle of sweetness and sustenance.

In the wide, open skies above, the first skeins of migrating geese and swans begin to appear, their haunting cries and distinctive V-shaped formations a poignant reminder of the great cycle of life that turns with the seasons. The light takes on a softer, more golden quality, and the lengthening shadows of the afternoon sun cast a gentle, mellow glow over the tranquil landscape, inviting moments of quiet reflection and introspection.

As winter tightens its grip and the landscape is transformed into a study in black and white, the walk becomes a place of stark, elemental beauty, the bare bones of the earth revealed in all their raw, uncompromising splendour. The trees stand tall and skeletal against the leaden sky, their twisting branches etched in intricate, frosty filigree, while the rivers and marshes lie still and silent beneath a blanket of ice and snow.

But even in the depths of the coldest months, the Ipswich to Felixstowe Walk is alive with the hardy and resilient creatures that make their home here year-round, from the elusive otters and water voles that search for food beneath the frozen surface of the water to the flocks of overwintering ducks and geese that gather in the sheltered bays and inlets, their raucous cries and colourful plumage a defiant splash of life amidst the monochrome landscape.

And on those rare, perfect days when the sun breaks through the clouds and the frost sparkles like diamonds on the ground, there is no more invigorating or life-affirming place to be than out on the windswept trails of the Suffolk coast, the crisp, cold air filling your lungs and the vast, unbroken expanse of the sky lifting your heart and spirit to the heavens.

No matter what time of year you choose to embark on the Ipswich to Felixstowe Walk, there is always something new and wondrous to discover, some fresh delight or unexpected revelation that will stay with you long after you have left the coast and the river behind. From the first, tentative stirrings of spring to the deep, contemplative stillness of winter, each season brings its own unique gifts and challenges, its own invitation to connect with the timeless rhythms and enduring truths of the natural world.

So let the turning of the year be your guide and your inspiration as you make your way along this enchanting route, and allow yourself to be absorbed and transformed by the ever-changing beauty and mystery of the Suffolk landscape. Whether you are a seasoned walker or a curious first-time visitor, the Ipswich to Felixstowe Walk promises to surprise, inspire, and delight you, season after season, year after year, offering a portal to a world of wonder, discovery, and endless possibility that will nourish and sustain you for a lifetime.

Difficulty Level and Safety Considerations:

The Ipswich to Felixstowe Walk is a moderately challenging route that requires a good level of physical fitness, stamina, and surefootedness to complete safely and enjoyably. While much of the walk follows well-maintained paths and trails, there are some sections that may prove more demanding, with uneven terrain, steep inclines, and muddy or slippery surfaces that can pose a risk to walkers who are not properly prepared or equipped.

One of the main difficulties of the walk is its length, covering a total distance of 13 miles (20.9 km) from start to finish. This is a significant undertaking that can take anywhere from 6 to 7 hours to complete, depending on your pace and the number of stops you make along the way. As such, it is essential to allow plenty of time for the walk, and to start early in the day to ensure that you have sufficient daylight to reach your destination before nightfall.

Another key challenge of the walk is the variability of the terrain, which ranges from flat, easy-going paths along the river and the coast to more rugged and uneven stretches through the countryside and the hills. Walkers should be prepared for muddy or waterlogged sections, particularly after heavy rain, as well as for steep inclines and rocky or root-strewn paths that can be slippery or difficult to navigate.

To mitigate these risks and ensure a safe and enjoyable walk, it is crucial to wear sturdy, supportive footwear with good grip and ankle support, such as hiking boots or trail shoes. It is also a good idea to bring a pair of walking poles or a sturdy stick to help with balance and stability on the more challenging sections of the route.

In addition to appropriate footwear, walkers should also come prepared with suitable clothing for the weather conditions and the time of year. In the summer months, this means wearing light, breathable layers that can be easily removed or added as needed, as well as a hat and sunscreen to protect against the strong coastal sun. In the cooler months, a warm, waterproof jacket, gloves, and a hat are essential to stay comfortable and dry in the face of wind, rain, and cold temperatures.

Another important safety consideration for the Ipswich to Felixstowe Walk is the need to stay hydrated and nourished throughout the journey. With limited opportunities to purchase food and drink along the way, it is essential to bring plenty of water and snacks to keep your energy levels up and to avoid dehydration or fatigue. A small backpack or hydration pack can be a useful way to carry these essentials, as well as any other items you may need, such as a map, compass, first-aid kit, and mobile phone.

It is also worth noting that some sections of the walk, particularly along the river and the coast, can be exposed to strong winds and tidal currents, which can pose a risk to walkers who venture too close to the water's edge. It is important to stay on the designated paths and to keep a safe distance from the river and the sea, especially during high tides or stormy weather.

Finally, while the Ipswich to Felixstowe Walk is generally a safe and well-maintained route, it is always a good idea to let someone know your planned itinerary and estimated return time before setting out, and to carry a fully charged mobile phone and a small first-aid kit in case of emergencies. If you do encounter any difficulties or hazards along the way, don't hesitate to seek help or advice from the local authorities or emergency services.

By following these simple guidelines and using common sense and good judgment, walkers of all ages and abilities can safely and confidently enjoy the many rewards and challenges of the Ipswich to Felixstowe Walk, from the stunning scenery and rich wildlife to the fascinating history and vibrant communities of the Suffolk coast. So come prepared, stay alert and aware, and let the beauty and wonder of this special place work its magic on your mind, body, and soul.

Nearby Attractions:

The Ipswich to Felixstowe Walk is not only a wonderful destination in its own right, but also a gateway to a wealth of nearby attractions and experiences that showcase the rich history, stunning landscapes, and vibrant culture of the Suffolk coast and countryside. Whether you're a history buff, a nature lover, a foodie, or simply someone in search of a little adventure and relaxation, there's something for everyone within easy reach of this beautiful and varied route.

For those interested in the region's rich maritime and military heritage, a visit to the historic town of Felixstowe is an absolute must. Home to the bustling Port of Felixstowe, one of the largest container ports in Europe, the town has a long and fascinating history that stretches back to Roman times, with a wealth of historic buildings, museums, and landmarks to explore.

One of the most impressive and iconic of these is Landguard Fort, a stunning 18th-century fortification that has played a vital role in defending the Suffolk coast from invasion and attack for over 400 years. Perched on a spit of land at the mouth of the River Orwell, the fort offers breathtaking views of the sea and the surrounding landscape, as well as a fascinating glimpse into the lives and experiences of the soldiers and civilians who once lived and served here.

Visitors to Landguard Fort can explore the Fort's atmospheric tunnels, chambers, and battlements, and discover the many stories and artifacts that bring its rich history to life, from the anxious days of the Spanish Armada to the dark hours of World War II. With interactive exhibits, guided tours, and special events throughout the year, the Fort is a must-see destination for anyone with an interest in the region's military and maritime past.

For those who prefer a more natural and peaceful escape, the nearby Trimley Marshes Nature Reserve is a true oasis of tranquility and biodiversity, offering over 200 acres of pristine wetland habitat that is home to an astonishing variety of birds, mammals, and invertebrates. Just a short distance from the Ipswich to Felixstowe Walk, this internationally important site is a paradise for birdwatchers and nature lovers, with a network of paths, hides, and viewing platforms that provide unparalleled opportunities to observe and enjoy the incredible diversity of life that thrives here.

From the elegant avocets and the elusive bearded tits that haunt the reedbeds and the marshes, to the darting dragonflies and the fluttering butterflies that dance among the wildflowers and the grasses, Trimley Marshes is a celebration of the beauty, complexity, and resilience of the natural world, and a powerful reminder of the importance of preserving and protecting these fragile and irreplaceable habitats for generations to come.

For a taste of the vibrant arts and culture scene that makes this part of Suffolk so special, be sure to pay a visit to the lively town of Ipswich, just a short train or bus ride from the start of the walk. With its rich history, stunning architecture, and thriving creative community, Ipswich is a hub of cultural activity and innovation, offering a wealth of museums, galleries, theatres, and performance spaces that showcase the very best of local and international talent.

One of the most popular and dynamic cultural destinations in the town is DanceEast, a world-class dance facility and performance venue that hosts a diverse program of classes, workshops, and productions throughout the year. Whether you're a seasoned dancer or a curious newcomer, DanceEast offers a welcoming and inspiring space to explore the joys and challenges of movement, and to connect with a passionate and supportive community of artists and enthusiasts.

For those with an interest in the region's rich agricultural and culinary heritage, the Suffolk Food Hall is a must-visit destination, located just a short drive from the Orwell Bridge on the outskirts of Ipswich. This award-winning farm shop and restaurant showcases the very best of local produce and craftsmanship, with a mouth-watering selection of artisanal cheeses, meats, baked goods, and preserves, as well as a range of gifts, homeware, and souvenirs that celebrate the unique character and flavour of the Suffolk countryside.

Visitors to the Suffolk Food Hall can also enjoy a delicious meal in the light and airy restaurant, which offers stunning views of the Orwell Estuary and a menu that changes with the seasons to reflect the freshest and most flavourful ingredients available. From hearty breakfasts and light lunches to indulgent afternoon teas and special occasion dinners, the Suffolk Food Hall is a foodie's paradise that is sure to delight and inspire, no matter what your tastes or preferences.

Of course, no visit to the Suffolk coast would be complete without exploring the beautiful and varied landscapes of the Suffolk Coast and Heaths Area of Outstanding Natural Beauty (AONB), a breathtaking patchwork of heathland, forest, wetland, and coastline that stretches from Lowestoft in the north to the River Stour in the south. With over 60 miles of stunning coastline, six estuaries, and countless miles of footpaths, bridleways, and cycle routes, the AONB is a year-round destination for outdoor enthusiasts and nature lovers of all ages and abilities.

Whether you're a keen birdwatcher, a budding botanist, a seasoned hiker, or simply someone in search of a little peace, beauty, and fresh air, the Suffolk Coast and Heaths AONB has something to offer everyone. From the windswept beaches and crumbling cliffs of the Covehithe coast to the tranquil waterways and reedbeds of the Alde-Ore Estuary, every corner of this extraordinary landscape is alive with the sights, sounds, and sensations of the natural world, inviting you to slow down, breathe deeply, and reconnect with the simple joys and profound truths of life on earth.

One of the most iconic and beloved destinations within the AONB is the ancient heathland of Dunwich Heath, a haunting and beautiful landscape of gorse, heather, and bracken that is home to a fascinating array of rare and endangered plants and animals. From the delicate bell heather and the shimmering silver-studded blue butterfly to the elusive nightjar and the majestic red deer, Dunwich Heath is a living tapestry of biodiversity and a powerful reminder of the fragility and resilience of our natural heritage.

Visitors to Dunwich Heath can explore a network of waymarked trails that wind through the heath and along the coast, offering stunning views of the sea and the surrounding landscape, as well as a chance to discover the many stories and secrets that are woven into the fabric of this ancient and mystical place. Whether you're watching the sun rise over the misty heath, listening to the haunting cry of the curlew, or simply soaking up the timeless beauty and tranquility of the landscape, Dunwich Heath is an unforgettable and transformative experience that will stay with you long after you have left the coast behind.

Other nearby attractions and destinations to consider include:

1. The charming coastal town of Aldeburgh, with its picturesque seafront, historic high street, and thriving arts and music scene, as well as its association with the composer Benjamin Britten and the annual Aldeburgh Festival.

2. The National Trust's Sutton Hoo, an awe-inspiring Anglo-Saxon burial site and museum that tells the incredible story of the ship burial of an East Anglian king and the priceless treasures and artifacts that were discovered here.

3. The historic market town of Woodbridge, with its beautiful waterfront, ancient Tide Mill, and abundance of independent shops, galleries, and eateries, as well as its proximity to the tranquil beauty of the Deben Estuary.

4. The RSPB Minsmere nature reserve, one of the UK's premier birdwatching sites, with a staggering diversity of wetland, heathland, and coastal habitats that support an incredible array of bird and animal life throughout the year.

Whether you have a few hours or a few days to spare, the Ipswich to Felixstowe Walk and its nearby attractions offer endless opportunities for discovery, inspiration, and delight, each one a chance to immerse yourself in the rich history, stunning beauty, and irrepressible spirit of this enchanting corner of England. So why not linger a while and explore all that this special place has to offer? With its combination of easy access, diverse experiences, and warm, welcoming communities, the Suffolk coast and countryside is the perfect destination for a truly unforgettable adventure.

As you set out on your journey, remember to tread lightly, leave no trace, and approach each new encounter with an open mind, a curious heart, and a deep respect for the natural world and the countless generations who have shaped and been shaped by this extraordinary landscape. Whether you're a seasoned traveller or a first-time visitor, a local resident or a passing pilgrim, the Ipswich to Felixstowe Walk and its nearby attractions promise to surprise, inspire, and delight you, offering a gateway to a world of beauty, wonder, and endless possibility that will nourish and sustain you for a lifetime.

So what are you waiting for? The river and the sea, the history and the heritage, the wildlife and the wonder of the Suffolk coast are calling to you, inviting you to step outside your comfort zone and embrace the magic and majesty of this incredible place. All you have to do is take that first step, and let the journey be your guide, your teacher, and your friend. The adventure of a lifetime awaits you, and the memories and experiences you gain will be your treasure and your legacy, a gift to yourself and to the world that will endure long after you have reached your destination.

Ipswich to Martlesham Heath Walk - 7 miles (11.3 km), Linear - Estimated Time: 3-4 hours

Starting Point Coordinates, Postcode: The starting point for the Ipswich to Martlesham Heath Walk is Ipswich Railway Station (Grid Reference: TM 160 445, Postcode: IP2 8AL).
Nearest Car Park, Postcode: Ipswich Railway Station car park (Postcode: IP2 8AL).

Embark on a delightful 7-mile (11.3 km) linear walk from the historic town of Ipswich to the picturesque village of Martlesham Heath, where the urban charm of the town seamlessly blends with the tranquil beauty of the Suffolk countryside. This easy to moderate route offers a perfect balance of waterfront views, scenic landscapes, and fascinating historical sites, inviting you to immerse yourself in the rich cultural heritage and natural wonders of this enchanting corner of East Anglia.

As you set out from the bustling platform of Ipswich Railway Station, feel the anticipation build as you head east along Ranelagh Road, the vibrant energy of the town centre gradually giving way to the more relaxed atmosphere of the residential streets. With each step, the cares and distractions of everyday life seem to melt away, replaced by a growing sense of peace and connection with the timeless beauty of the surrounding landscape.

Turning right onto Grafton Way, you'll soon find yourself crossing the majestic River Orwell, its glistening waters and gently sloping banks a testament to the enduring importance of this waterway in the life and history of the town. As you continue along Stoke Street, take a moment to appreciate the elegant Georgian and Victorian architecture that lines the route, each building a reminder of the prosperity and ambition of generations past.

Upon reaching the historic waterfront, pause to soak up the atmosphere of this unique and fascinating area, with its mix of ancient warehouses, modern apartments, and lively cafes and restaurants. Take a stroll along the quayside, admiring the sleek lines of the yachts and the colourful narrowboats that bob gently on the water, and imagine the countless stories and adventures that have unfolded on this very spot over the centuries.

Crossing the River Gipping via Duke Street, you'll begin your journey northward, leaving the bustle of the town behind and immersing yourself in the tranquil beauty of the Suffolk countryside. As you follow Westerfield Road, take a moment to appreciate the lush green fields and gently rolling hills that stretch out before you, each one a patchwork of colour and texture that changes with the seasons and the play of light and shadow.

Arriving in the charming village of Westerfield, with its thatched cottages and ancient church, you'll feel as though you've stepped back in time, into a world of simple pleasures and enduring traditions. Take a moment to explore the peaceful churchyard, with its weathered gravestones and gnarled yew trees, and reflect on the generations of villagers who have found solace and inspiration in this sacred space.

Continuing along Church Lane and Lower Road, you'll find yourself in the heart of the Suffolk countryside, surrounded by fields of golden wheat, verdant pastures, and ancient woodlands that echo with the songs of countless birds and the rustling of unseen creatures in the undergrowth. Keep your eyes peeled for the telltale signs of wildlife, from the darting flight of dragonflies to the gentle lope of a hare across the path.

As you approach the village of Tuddenham St. Martin, take a moment to appreciate the timeless charm and character of this quintessential English community, with its cozy pub, village green, and picturesque cottages that seem to have been plucked from the pages of a storybook. Let the peaceful atmosphere and friendly smiles of the locals wash over you, a reminder of the simple joys and enduring values that define life in the countryside.

The final stretch of the walk takes you along Westerfield Lane, a quiet country road that winds its way through the gently undulating landscape towards the village of Martlesham Heath. As you near your destination, take a moment to reflect on the incredible journey you have undertaken, both physically and spiritually, and the profound sense of connection and renewal that comes from immersing oneself in the beauty and tranquility of the natural world.

Arriving in Martlesham Heath, with its mix of traditional cottages and modern developments, you'll feel a sense of accomplishment and satisfaction, knowing that you have completed a truly special and memorable walk through the heart of the Suffolk countryside. Whether you choose to linger a while and explore the village, or catch a bus back to Ipswich, you'll carry with you the memories and impressions of this enchanting route, a reminder of the endless wonders and delights that await those who take the time to step outside and embrace the beauty and diversity of the world around them.

Detailed Route Directions:

1. From Ipswich Railway Station, head east along Ranelagh Road for approximately 0.3 miles (0.5 km), following a bearing of 80°. (Estimated time: 5-10 minutes)

2. Turn right onto Grafton Way and cross the River Orwell, maintaining a bearing of 170°. Continue for about 0.2 miles (0.3 km). (Estimated time: 5 minutes)

3. Follow Grafton Way as it becomes Stoke Street, keeping a bearing of 170° for around 0.4 miles (0.6 km) until you reach the waterfront. (Estimated time: 10 minutes)

4. Walk along the waterfront for approximately 0.4 miles (0.6 km), enjoying the views and atmosphere. (Estimated time: 10-15 minutes)

5. Turn left onto Duke Street and cross the River Gipping, then head north on Bridge Street, bearing 0° for about 0.2 miles (0.3 km). (Estimated time: 5 minutes)

6. Turn right onto Westerfield Road and follow it for around 1.8 miles (2.9 km), maintaining a bearing of 60°, until you reach Westerfield Village. (Estimated time: 45-60 minutes)

7. In Westerfield Village, turn left onto Church Lane, bearing 330° for about 0.3 miles (0.5 km). (Estimated time: 5-10 minutes)

8. Turn right onto Lower Road and continue for approximately 1.2 miles (1.9 km), following a bearing of 60°, to reach the village of Tuddenham St. Martin. (Estimated time: 30-40 minutes)

9. From Tuddenham St. Martin, take a left onto Westerfield Lane, bearing 330°, and follow it for around 2.5 miles (4 km) to Martlesham Heath. (Estimated time: 60-75 minutes)

Best Features and Views:

- Immerse yourself in the vibrant atmosphere and rich history of the Ipswich waterfront, with its mix of ancient warehouses, modern apartments, and lively cafes and restaurants, each one a testament to the enduring importance of this area in the life and identity of the town.

- Delight in the picturesque charm and tranquility of the Suffolk countryside, with its patchwork of lush green fields, gently rolling hills, and ancient woodlands that burst with colour and life in every season.
- Discover the timeless beauty and character of the villages of Westerfield and Tuddenham St. Martin, with their thatched cottages, peaceful churches, and friendly locals, each one a reminder of the simple pleasures and enduring traditions of rural life.

Historical Features of Interest:

As you make your way along the Ipswich to Martlesham Heath Walk, take a moment to explore the fascinating historical sites and landmarks that dot the route, each one a window into the rich and complex past of this ancient corner of Suffolk. One of the most intriguing and evocative of these is the historic waterfront area of Ipswich, a unique and atmospheric district that has played a vital role in the town's development and prosperity for over a thousand years.

The origins of Ipswich as a port and trading centre can be traced back to the Anglo-Saxon period, when the town was a thriving hub of commerce and industry, with a network of wharves, warehouses, and workshops lining the banks of the River Orwell. Over the centuries, the waterfront has been the scene of countless dramas and transformations, from the bustling trade of the medieval period to the decline and decay of the post-industrial era.

Today, the waterfront stands as a testament to the resilience and creativity of the people of Ipswich, with a mix of beautifully restored historic buildings and exciting new developments that showcase the best of the town's past and future. As you stroll along the quayside, take a moment to appreciate the elegant lines and graceful proportions of the Old Custom House, a stunning example of 18th-century architecture that once served as the nerve centre of the town's maritime trade.

Nearby, the towering bulk of the Victorian maltings and the sleek curves of the modern apartments and offices offer a fascinating contrast of styles and eras, each one a chapter in the ongoing story of the waterfront's evolution and reinvention. Take a moment to imagine the lives and experiences of the countless merchants, sailors, and workers who have called this place home over the centuries, and the incredible feats of ingenuity and determination that have shaped its character and identity.

But the waterfront is more than just a monument to the past; it is also a vibrant and dynamic community that is alive with the sights, sounds, and flavours of the present day. From the lively cafes and restaurants that spill out onto the quayside to the colourful narrowboats and sleek yachts that bob gently on the water, every corner of this special place is a celebration of the enduring spirit and creativity of the people of Ipswich.

As you continue your journey along the Ipswich to Martlesham Heath Walk, carry with you the sense of history and continuity that the waterfront embodies, and let it be a source of inspiration and wonder as you explore the timeless beauty and diversity of the Suffolk countryside. The waterfront is a powerful reminder of the deep roots and abiding values that have sustained this community through the ages, and an invitation to connect with the stories and experiences that have shaped this extraordinary landscape.

Rare or Uncommon Birds:

As you set out on the Ipswich to Martlesham Heath Walk, keep your eyes and ears attuned to the skies and the hedgerows, for this varied and picturesque landscape is home to a rich and diverse array of birdlife, from the common and familiar to the rare and elusive. With its mix of urban, agricultural, and woodland habitats, the route offers a chance to spot some of the most fascinating and charismatic species that call Suffolk home, each one a unique and precious thread in the intricate tapestry of the local ecosystem.

One of the most exciting and sought-after birds to watch for along the walk is the green woodpecker, a striking and charismatic species with a bright green plumage, red crown, and distinctive laughing call. Often spotted foraging for ants on the ground or clinging to the trunks of trees, these impressive birds are a true icon of the English countryside, their presence a sign of the health and diversity of the woodland and parkland habitats that they depend on.

As you make your way through the fields and hedgerows of the rural stretches of the walk, keep an eye out for the vibrant colours and melodious songs of the many finches that make their home in this area. From the bold reds and blacks of the bullfinch to the bright yellows and greens of the greenfinch, these lively and engaging birds are a constant presence in the countryside, their cheerful calls and acrobatic flight a delight to behold.

Other finches to watch for include the chaffinch, with its pinkish-red breast and white wing bars; the goldfinch, with its striking red, white, and black head pattern and tinkling, melodious song; and the siskin, a small, streaky finch with a yellow rump and forked tail that is an occasional winter visitor to the region.

As you pass through the villages and woodland edges of the route, keep your ears open for the rich and varied songs of the many other bird species that thrive in these habitats, from the fluty notes of the blackbird and the melodious warble of the robin to the strident calls of the wren and the soft, cooing notes of the wood pigeon. Each song is a unique and precious expression of the beauty and diversity of the natural world, a reminder of the incredible complexity and resilience of life in all its forms.

But perhaps the most rare and exciting bird to spot on the Ipswich to Martlesham Heath Walk is the lesser spotted woodpecker, a small and elusive species that is one of the most difficult to see in the UK. With its black-and-white barred plumage, red crown, and distinctive drumming call, this enigmatic bird is a true prize for any birdwatcher, a symbol of the incredible diversity and mystery of the avian world.

While sightings of the lesser spotted woodpecker are rare and unpredictable, the best time to look for them is in the early spring, when the birds are most active and vocal as they establish their breeding territories. Listen out for the rapid, high-pitched drumming that the males use to attract a mate and defend their patch, and scan the trunks and branches of mature trees for any sign of movement or activity.

Whether you are lucky enough to spot a lesser spotted woodpecker or any of the other fascinating species that call this area home, the birdlife of the Ipswich to Martlesham Heath Walk is sure to delight and inspire you, offering a window into the incredible diversity and resilience of the natural world. So keep your binoculars at the ready, your eyes and ears open, and your heart full of wonder and appreciation for the winged wonders that share this precious landscape with us.

Rare or Interesting Insects and Other Wildlife:

As you explore the diverse habitats and landscapes of the Ipswich to Martlesham Heath Walk, take a moment to appreciate the myriad forms of life that thrive in every nook and cranny of this extraordinary ecosystem. From the tiniest insects to the larger mammals, each species plays a vital role in maintaining the delicate balance and resilience of the natural world, their stories of adaptation and survival woven into the very fabric of the Suffolk countryside.

One of the most fascinating and important groups of insects to watch for on your walk are the butterflies and moths, whose delicate beauty and incredible diversity are a true marvel of nature. From the bold and striking colours of the peacock and red admiral to the subtle and cryptic patterns of the speckled wood and the gatekeeper, each species has its own unique adaptations and life histories, playing a crucial role in the complex web of pollination and reproduction that sustains the health and vitality of the local flora.

As you wander through the sunlit glades and along the flower-strewn verges of the route, keep an eye out for the fluttering and floating forms of these enchanting creatures, and take a moment to appreciate the incredible intricacy and artistry of their wing patterns and markings. Watch as they flit from bloom to bloom, their long, coiled proboscises probing deep into the nectar-rich flowers, and marvel at the speed and grace with which they navigate the currents of the air and evade their predators.

Another group of insects that are sure to capture your imagination are the dragonflies and damselflies, whose iridescent colours and aerial acrobatics are a true spectacle of the summer months. These ancient and highly adapted predators are masters of the air, their large, compound eyes and powerful wings allowing them to hunt and capture their prey with incredible speed and precision.

Look for the jewel-like greens and blues of the emerald damselfly and the azure damselfly as they hover and dart among the reeds and rushes of the riversides and ponds, their delicate, translucent wings catching the light like stained glass. And keep an eye out for the larger and more robust forms of the southern hawker and the migrant hawker, whose powerful flight and fearsome appearance belie their gentle and benign nature.

But it is not just the insects that make this walk a treasure trove of wildlife encounters. As you make your way through the fields and woods and along the quiet country lanes, you may also be lucky enough to spot some of the mammals and amphibians that call this landscape home, from the elusive roe deer and red fox to the common frog and smooth newt.

One of the most charming and engaging of these creatures is the rabbit, a familiar and much-loved sight in the British countryside. With their soft, fluffy coats and long, sensitive ears, rabbits are a constant presence in the fields and hedgerows of the walk, their playful antics and endearing expressions a source of joy and delight for young and old alike.

Watch as they nibble on the tender shoots of grass and wildflowers, their twitching noses and bright, inquisitive eyes always alert for any sign of danger. And listen out for the soft thumping of their hind feet as they signal to their companions and disappear down their burrows, a reminder of the incredible adaptability and resilience of these fascinating creatures.

As you marvel at the diversity and complexity of the insect and animal life that surrounds you on the Ipswich to Martlesham Heath Walk, take a moment to reflect on the vital role that these species play in maintaining the health and balance of the local ecosystem. From the pollination services provided by the butterflies and moths to the nutrient cycling and soil aeration carried out by the beetles and worms, each creature contributes in its own unique way to the intricate web of relationships that sustain the beauty and productivity of this special landscape.

So let your curiosity and sense of wonder be your guide as you explore the hidden corners and secret lives of the insects and wildlife of the walk, and allow yourself to be amazed and humbled by the endless wonders and surprises of the natural world. By opening your eyes and your heart to the incredible diversity and resilience of these often-overlooked creatures, you can gain a deeper appreciation and respect for the priceless heritage that we have been entrusted to protect and cherish, and a renewed sense of connection and kinship with the living planet that sustains us all.

Rare or Interesting Plants:

As you make your way along the picturesque trails and byways of the Ipswich to Martlesham Heath Walk, let your gaze be drawn to the incredible diversity and beauty of the plants that adorn every inch of this remarkable landscape. From the delicate wildflowers that sprinkle the meadows and verges with colour to the stately trees that tower over the fields and woods, each species has its own unique history and significance, its own story of resilience and adaptation in the face of an ever-changing world.

One of the most striking and iconic plants to look out for on the walk is the common poppy, a vibrant red flower with a long and poignant association with remembrance and sacrifice. Often found growing in disturbed soils and field margins, these hardy annuals are a familiar and much-loved sight in the Suffolk countryside, their bright, papery petals and black centres a symbol of the fragility and beauty of life in the face of adversity.

As you pass through the agricultural landscapes of the route, take a moment to appreciate the swaying golden heads of the wheat and barley fields, each one a testament to the skill and dedication of generations of farmers and a reminder of the vital role that agriculture plays in shaping and sustaining the local economy and way of life. Look out for the delicate blue flowers of the cornflower and the bright yellow blooms of the corn marigold, two once-common arable weeds that have become increasingly rare in recent years due to changes in farming practices.

In the hedgerows and woodland edges that line the route, a rich tapestry of native shrubs and trees offers both food and shelter to a wide range of birds, mammals, and insects. From the frothy white flowers and glossy black berries of the blackthorn to the delicate pink and white blooms of the dog rose, each species plays a vital role in the complex web of life that thrives in these linear habitats.

Keep an eye out for the gnarled and twisted trunks of the ancient oaks and ashes that stand sentinel over the fields and lanes, their spreading canopies and deeply fissured bark a living archive of the centuries of growth and change that have shaped this landscape. And take a moment to appreciate the lush green carpets of the bluebell woods that grace the walk in the spring, their nodding flower heads and heady scent a joyous celebration of the returning light and warmth of the season.

But perhaps the most rare and intriguing plant to spot on the Ipswich to Martlesham Heath Walk is the bee orchid, a fascinating and elusive species that is a true marvel of evolutionary adaptation. Found in small numbers in the grasslands and roadside verges of the route, these delicate, pink-and-brown flowers have evolved to mimic the appearance and scent of a female bee, tricking male bees into attempting to mate with them and thus effecting pollination.

With their intricate, bee-like markings and furry, pollen-laden lips, bee orchids are a stunning example of the incredible ingenuity and creativity of the natural world, a living testament to the power of natural selection and the endless forms of beauty and diversity that arise from the constantly evolving dance of life on earth.

Whether you are a seasoned botanist or a curious amateur, the plant life of the Ipswich to Martlesham Heath Walk is sure to inspire and delight you, offering a window into the rich and complex tapestry of the Suffolk countryside. So let your senses be your guide as you explore the colours, textures, and fragrances of this enchanting route, and allow yourself to be amazed and humbled by the endless wonders and surprises of the botanical world.

From the humble dandelion to the majestic oak, each plant has a story to tell and a role to play in the intricate web of life that sustains us all. By cherishing and protecting these precious and irreplaceable species, we not only ensure the health and resilience of the local ecosystem but also enrich and deepen our own connection to the living planet, and to the great cycles of growth, decay, and renewal that shape the ever-changing face of the landscape.

Facilities and Accessibility:

The Ipswich to Martlesham Heath Walk offers a range of facilities and amenities to ensure a comfortable and enjoyable experience for walkers of all ages and abilities. While the route does include some moderate inclines and stretches of uneven terrain, the well-maintained paths and clear signage make it accessible to most reasonably fit and mobile individuals, with a few key considerations to keep in mind.

The walk begins at Ipswich Railway Station, where a range of facilities and services are available to help you prepare for your journey. The station itself offers public restrooms, a café, and a small shop selling snacks, drinks, and other essentials, as well as ample parking for those arriving by car. The station is also well-served by public transport, with regular trains and buses connecting to destinations throughout the region.

As you set out on the walk, you'll find that the paths through the town centre and along the waterfront are generally well-surfaced and easy to navigate, with a mix of pavement, tarmac, and gravel surfaces that are suitable for most types of footwear. The route is clearly waymarked with distinctive blue and green signs, making it easy to stay on track and avoid any unwanted detours.

Along the waterfront, there are plenty of benches and seating areas where you can take a break, enjoy the views, or have a picnic lunch. There are also several cafes, restaurants, and pubs in this area, offering a range of refreshments and dining options to suit all tastes and budgets. Public restrooms are available at various points along the waterfront, as well as in the town centre and at the railway station.

As you leave the town behind and head into the countryside, the terrain becomes a little more varied and challenging, with some gentle hills, uneven surfaces, and occasional muddy patches to negotiate. While these sections are generally manageable for most walkers, they may prove more difficult for those with mobility issues or respiratory conditions, so it's important to take your time, wear appropriate footwear, and bring any necessary aids or equipment.

In the villages of Westerfield and Tuddenham St. Martin, there are a few local amenities and services available, such as public restrooms, small shops, and pubs where you can stop for refreshment and a rest. However, these facilities are limited, so it's a good idea to bring your own water, snacks, and any other essentials you may need for the walk.

At the end of the walk in Martlesham Heath, there are several bus stops where you can catch a ride back to Ipswich or other nearby destinations. There are also a few local shops, cafes, and pubs in the village, offering a welcome chance to refuel and relax after your walk.

While the Ipswich to Martlesham Heath Walk is not specifically designed as an accessible route for those with mobility impairments or other disabilities, with careful planning and preparation, many individuals may still be able to enjoy the beauty and variety of this landscape, either by focusing on the more manageable sections of the route or by using mobility aids or assistance as needed.

Regardless of your age, fitness level, or physical abilities, it's always a good idea to come prepared for the changeable weather and variable conditions of the Suffolk countryside, with appropriate clothing, footwear, and sun protection for the time of year. It's also wise to carry a map and compass or GPS device, as well as a fully charged mobile phone and a small first aid kit, in case of emergencies or unexpected delays.

By taking these simple precautions and being mindful of your own needs and limitations, you can ensure a safe, comfortable, and rewarding experience on the Ipswich to Martlesham Heath Walk, and fully immerse yourself in the timeless beauty and rich diversity of this enchanting corner of East Anglia.

Seasonal Highlights:

One of the great joys and rewards of the Ipswich to Martlesham Heath Walk is the opportunity to witness the changing face of the Suffolk countryside through the turning of the seasons, each one bringing its own unique character and charm to the landscapes and habitats along the route. From the tender, unfurling greens of spring to the rich, burnished hues of autumn, this enchanting walk is a celebration of the endless cycle of growth, decay, and renewal that lies at the heart of the natural world.

In the spring, the walk comes alive with the joyous explosion of colour and song that heralds the return of warmth and light to the land. Along the riverbanks and in the hedgerows, the first delicate blooms of the wildflowers begin to appear, their pastel shades of pink, purple, and yellow a welcome splash of brightness against the still-subdued tones of the landscape. In the fields and woods, the birds are in full voice, their trilling and warbling melodies a symphony of hope and renewal that fills the air with the promise of the season.

As you make your way through the gently stirring countryside, take a moment to appreciate the fresh, green leaves of the trees and shrubs as they begin to unfurl, their tender translucence a marvel of new life and growth. The air is filled with the sweet scent of blossom and the busy humming of bees and other pollinators, a reminder of the vital role that these tiny creatures play in the great web of life that sustains us all.

As spring gives way to summer, the walk reaches its peak of lushness and vitality, the fields and hedgerows now a riot of colour and texture that dazzles the senses and lifts the spirits. The grasses and wildflowers grow tall and rank, their nodding heads and waving stems a sea of green and gold that stretches out to the horizon. In the woods and copses, the shade is dappled and cool, the sunlight filtering through the dense canopy of leaves to create a mosaic of light and shadow on the forest floor.

Along the riverbanks and in the wetland areas, the summer months are a time of abundance and activity, with the reeds and rushes alive with the darting flight of dragonflies and the chirping of crickets. The water sparkles and dances in the bright sunshine, the reflections of the clouds and the overhanging branches creating a kaleidoscope of shapes and colours on the surface.

As the days begin to shorten and the first hints of autumn creep into the air, the walk takes on a new character, the lush greens of summer slowly giving way to a palette of fiery reds, oranges, and golds. The hedgerows and thickets are now laden with the bounty of the season, from the glossy blackberries and scarlet hips to the clusters of nuts and berries that will sustain the birds and animals through the long winter months.

In the fields and along the woodland edges, the light takes on a softer, more mellow quality, the slanting rays of the sun casting a golden glow over the landscape. The air is crisp and invigorating, filled with the earthy scents of leaf mould and wood smoke, and the distant sounds of the harvest and the hunt. It is a time of plenty and preparation, a season of gathering and storing up the riches of the land before the cold and dark of winter set in.

And then, as the last leaves fall and the frost begins to creep across the landscape, the walk is transformed once again, the bare bones of the earth revealed in all their stark and haunting beauty. The trees stand naked and skeletal against the pale winter sky, their twisted branches etched in intricate patterns of black and white. The fields and paths are hard and unyielding underfoot, the frozen ground ringing with the sound of each footfall.

But even in the depths of winter, the Ipswich to Martlesham Heath Walk is alive with the signs and sounds of the natural world, from the flocks of overwintering birds that gather in the hedgerows and on the fields to the tracks of foxes and badgers that crisscross the snowy landscape. And on those rare, perfect days when the sun breaks through the clouds and the air is still and crisp, there is no more invigorating or life-affirming place to be than out on the windswept trails of the Suffolk countryside.

No matter what time of year you choose to embark on this walk, there is always something new and extraordinary to discover, some fresh delight or unexpected revelation that will stay with you long after you have returned to the warmth and comfort of home. From the raw, elemental power of a winter storm to the soft, hazy beauty of a summer evening, each season brings its own unique gifts and challenges, its own invitation to connect with the timeless rhythms and eternal truths of the natural world.

So let the turning of the year be your guide and your inspiration as you make your way along the Ipswich to Martlesham Heath Walk, and allow yourself to be swept up in the endlessly unfolding drama of the Suffolk landscape. Whether you are a seasoned walker or a curious first-time visitor, this enchanting route promises to surprise, inspire, and delight you, season after season, year after year, offering a portal to a world of wonder, beauty, and endless possibility that will nourish and sustain you for a lifetime.

Difficulty Level and Safety Considerations:

The Ipswich to Martlesham Heath Walk is a moderately challenging route that requires a reasonable level of fitness and mobility to complete safely and comfortably. While much of the walk follows well-maintained paths and trails, there are some sections that may prove more demanding, with uneven terrain, moderate inclines, and occasional obstacles such as stiles or muddy patches that can pose a hazard to walkers who are not properly equipped or prepared.

One of the main challenges of the walk is the distance, which covers a total of 7 miles (11.3 km) from start to finish. While this may not seem like a great distance for experienced hikers, it can still take several hours to complete, depending on your pace and the number of stops you make along the way. As such, it is important to allow plenty of time for the walk, and to start early enough in the day to ensure that you have sufficient daylight to reach your destination before nightfall.

Another consideration is the variability of the terrain, which ranges from the smooth, level pavements of the town centre to the more rugged and uneven paths of the countryside. Walkers should be prepared for some muddy or waterlogged sections, particularly after heavy rain, as well as for the occasional steep or rocky stretch that can be slippery underfoot. To mitigate these risks, it is essential to wear sturdy, supportive footwear with good grip and ankle support, such as hiking boots or trail shoes, and to take extra care when navigating any particularly challenging sections of the route.

In addition to appropriate footwear, walkers should also come prepared with suitable clothing for the weather conditions and the time of year. In the summer months, this means wearing light, breathable layers that can be easily removed or added as needed, as well as a hat and sunscreen to protect against the strong sun. In the cooler months, a warm, waterproof jacket, gloves, and a hat are essential to stay comfortable and dry in the face of wind, rain, and cold temperatures.

Another important safety consideration for the Ipswich to Martlesham Heath Walk is the need to stay hydrated and nourished throughout the journey. With limited opportunities to purchase food and drink along the way, it is essential to bring plenty of water and snacks to keep your energy levels up and to avoid dehydration or fatigue. A small backpack or hydration pack can be a useful way to carry these essentials, as well as any other items you may need, such as a map, compass, first-aid kit, and mobile phone.

It is also worth noting that some sections of the walk, particularly in the more rural areas, can be quite isolated and remote, with limited access to emergency services or other assistance. As such, it is always a good idea to let someone know your planned route and estimated return time before setting out, and to carry a fully charged mobile phone and a small first aid kit in case of any accidents or unexpected delays.

Finally, while the Ipswich to Martlesham Heath Walk is generally a safe and enjoyable route, there are a few potential hazards to be aware of, particularly when walking along roads or crossing railway lines. Walkers should always use designated crossings and footpaths where available, and exercise caution when sharing the road with vehicles or trains. It is also important to be mindful of any livestock or wildlife that may be encountered along the way, and to give them a wide berth to avoid any potential conflicts or disturbances.

By following these simple guidelines and using common sense and good judgment, walkers of all ages and abilities can safely and confidently enjoy the many rewards and challenges of the Ipswich to Martlesham Heath Walk, from the historic charm of the Ipswich waterfront to the rural tranquility of the Suffolk countryside. So come prepared, stay alert and aware, and let the beauty and variety of this special landscape work its magic on your mind, body, and soul.

Nearby Attractions:

The Ipswich to Martlesham Heath Walk is not only a delightful and rewarding experience in its own right but also a gateway to a wealth of nearby attractions and destinations that showcase the rich history, vibrant culture, and stunning natural beauty of the Suffolk region. Whether you're a history buff, a nature lover, a foodie, or simply someone in search of a little rest and relaxation, there's something for everyone within easy reach of this enchanting route, each one offering a unique and memorable way to extend your adventure and deepen your connection with this special corner of East Anglia.

For those interested in exploring the rich maritime and industrial heritage of Ipswich, a visit to the historic waterfront district is an absolute must. Just a short walk from the start of the route, this fascinating area is home to a stunning array of beautifully restored Victorian and Edwardian buildings, many of which have been repurposed as museums, galleries, restaurants, and shops. Take a stroll along the quayside and soak up the atmosphere of this vibrant and dynamic community, with its mix of traditional and modern architecture, its colourful narrowboats and yachts, and its lively cafes and pubs.

One of the most impressive and iconic landmarks of the waterfront is the Old Custom House, a magnificent 18th-century building that once served as the nerve centre of the town's bustling maritime trade. Today, the Custom House is home to a fascinating museum that tells the story of Ipswich's long and complex relationship with the sea, from the early days of the wool trade to the modern era of containerization and cruise ships. With interactive exhibits, rare artifacts, and stunning views over the marina, the Custom House is a must-see destination for anyone with an interest in the history and culture of this fascinating town.

Just a short distance from the waterfront, the beautiful Christchurch Park offers a welcome respite from the hustle and bustle of the town centre, with over 70 acres of lush gardens, ancient woodland, and open parkland to explore. Designed in the 19th century by the celebrated landscape architect Humphry Repton, the park is a masterpiece of Victorian design, with its winding paths, ornamental ponds, and elegant floral displays creating a sense of timeless beauty and tranquility.

At the heart of the park stands Christchurch Mansion, a stunning Tudor house that has been home to some of the most prominent families in Ipswich's history. Today, the mansion is open to the public as a museum and art gallery, with a fascinating collection of paintings, furniture, and decorative arts that offer a glimpse into the lives and tastes of the town's elite. From the grand state rooms to the intimate servants' quarters, every corner of the mansion is filled with stories and secrets waiting to be discovered.

For those with a passion for the great outdoors, the nearby Orwell Country Park is a true paradise, with over 470 acres of woodland, heathland, and wetland habitats to explore. Located just a short distance from the town centre, the park is a haven for wildlife and a mecca for nature lovers, with a network of trails and hides that offer unparalleled opportunities to observe and enjoy the incredible diversity of flora and fauna that thrive here.

From the majestic oaks and beeches of Bridge Wood to the glittering expanse of Orwell Lake, every corner of the park is alive with the sights, sounds, and sensations of the natural world. Watch as the sunlight filters through the canopy of leaves, casting dappled shadows on the forest floor, and listen to the gentle lapping of the water and the distant cries of the wildfowl that gather on the lake. With its mix of habitats and its stunning views over the surrounding countryside, Orwell Country Park is a true gem of the Suffolk landscape, and a must-visit destination for anyone who loves nature and the great outdoors.

For those willing to venture a little further afield, the charming market town of Woodbridge is just a short bus or train ride from Ipswich, and offers a wealth of attractions and activities to suit all ages and interests. From the picturesque waterfront and the historic Tide Mill to the independent shops and boutiques that line the high street, Woodbridge is a true delight for the senses, with its mix of old-world charm and contemporary flair creating a unique and irresistible ambiance.

One of the most impressive and iconic landmarks of Woodbridge is the Tide Mill, a beautifully restored 18th-century mill that harnesses the power of the tides to grind flour and generate electricity. Take a tour of the mill and discover the incredible ingenuity and craftsmanship that went into its construction, and learn about the vital role that mills like this played in the local economy and way of life for centuries. With its creaking wooden machinery, its rushing water, and its evocative smells of grain and grease, the Tide Mill is a powerful and unforgettable experience that will transport you back in time to a world of simple pleasures and timeless traditions.

Other nearby attractions and destinations to consider include:

1. The Suffolk Food Hall, a stunning farm shop and restaurant that showcases the very best of local produce and craftsmanship, with a mouth-watering array of artisanal cheeses, meats, baked goods, and preserves, as well as a range of gifts and homewares that celebrate the unique character and flavour of the Suffolk countryside.

2. The Sutton Hoo archaeological site, where the incredible treasure-filled burial ship of an Anglo-Saxon king was discovered in 1939, offering a fascinating glimpse into the rich and complex history of this ancient corner of England.

3. The RSPB Minsmere nature reserve, one of the most important and diverse birdwatching sites in the UK, with a staggering array of wetland, heathland, and coastal habitats that support an incredible diversity of bird and animal life throughout the year.

4. The charming coastal towns of Aldeburgh and Southwold, with their picturesque seafronts, historic high streets, and thriving arts and culture scenes, as well as their beautiful beaches and stunning views over the North Sea.

Whether you have a few hours or a few days to spare, the Ipswich to Martlesham Heath Walk and its nearby attractions offer endless opportunities for discovery, relaxation, and inspiration, each one a chance to immerse yourself in the rich history, vibrant culture, and breathtaking natural beauty of this enchanting region. So why not linger a while and explore all that this special corner of Suffolk has to offer? With its unique blend of the ancient and the modern, the wild and the cultivated, the coast and the countryside, there's truly something for everyone in this magical and endlessly fascinating landscape.

As you set out on your journey, remember to tread lightly, leave no trace, and approach each new encounter with an open mind, a curious heart, and a deep respect for the natural world and the countless generations who have shaped and been shaped by this extraordinary place. Whether you're a seasoned traveller or a first-time visitor, a local resident or a passing pilgrim, the Ipswich to Martlesham Heath Walk and its nearby attractions promise to surprise, delight, and inspire you, offering a gateway to a world of wonder, beauty, and endless possibility that will stay with you long after you have returned home.

So what are you waiting for? The history and the heritage, the wildlife and the wonder of the Suffolk countryside are calling to you, inviting you to step outside your comfort zone and embrace the magic and majesty of this incredible landscape. All you have to do is take that first step, and let the journey be your guide, your teacher, and your friend. The adventure of a lifetime awaits you, and the memories and experiences you gain will be your treasure and your legacy, a gift to yourself and to the world that will endure long after you have reached your destination.

Birds to Spot

Bird	Walks
Avocet	Minsmere Walk 1, RSPB Minsmere Walk 2, Trimley Marshes Walk, Orford Ness Nature Reserve Walk
Barn Owl	Alton Water Walk, Needham Lake Walk, Orwell Walk, Felixstowe Ferry Walk
Bearded Tit	Minsmere Walk 1, RSPB Minsmere Walk 2, Trimley Marshes Walk
Bittern	Minsmere Walk 1, RSPB Minsmere Walk 2, Orford Ness Nature Reserve Walk
Blackbird	Ickworth Park Walk, Lavenham and the Lavenham Walk, Sutton Hoo Walk, Orwell Walk
Blackcap	Sutton Hoo Walk, Orwell Walk, Ickworth Park Walk, Staverton Thicks Walk
Black-headed Gull	Felixstowe Promenade North Walk, Felixstowe Promenade South Walk, Ipswich Waterfront Walk
Black-tailed Godwit	Minsmere Walk 1, RSPB Minsmere Walk 2, Orford Ness Nature Reserve Walk
Canada Goose	Orwell Walk, Felixstowe Promenade Walk, River Deben Walk, Alton Water Walk
Chaffinch	Ickworth Park Walk, Rendlesham Forest Walk, Sutton Hoo Walk
Chiffchaff	Sutton Hoo Walk, Orwell Walk, Staverton Thicks Walk, Rendlesham Forest Walk
Common Buzzard	Rendlesham Forest Walk, Shotley Peninsula Walk, Great Bealings Circular Walk, Orwell Walk
Common Tern	Minsmere Walk 1, RSPB Minsmere Walk 2, Bawdsey Quay Walk
Coot	Alton Water Walk, Needham Lake Walk, Trimley Marshes Walk
Cormorant	Felixstowe Promenade North Walk, Felixstowe Promenade South Walk, River Deben Walk
Curlew	Minsmere Walk 1, RSPB Minsmere Walk 2, River Alde Walk
Dunlin	Minsmere Walk 1, RSPB Minsmere Walk 2, Trimley Marshes Walk, Orford Ness Nature Reserve Walk
Egyptian Goose	Alton Water Walk, Orwell Walk, River Deben Walk
Gadwall	Alton Water Walk, Trimley Marshes Walk, Minsmere Walk 1, RSPB Minsmere Walk 2
Goldcrest	Rendlesham Forest Walk, Ickworth Park Walk, Staverton Thicks Walk
Goldeneye	River Deben Walk, Alton Water Walk, Orwell Walk
Goldfinch	Ickworth Park Walk, Lavenham and the Lavenham Walk, Sutton Hoo Walk
Goshawk	Rendlesham Forest Walk
Great Crested Grebe	Alton Water Walk, Needham Lake Walk, River Deben Walk
Great Spotted Woodpecker	Sutton Hoo Walk, Rendlesham Forest Walk, Ickworth Park Walk
Great Tit	Ickworth Park Walk, Lavenham and the Lavenham Walk, Sutton Hoo Walk
Green Woodpecker	Sutton Hoo Walk, Rendlesham Forest Walk, Ickworth Park Walk
Grey Heron	Orwell Walk, River Deben Walk, Alton Water Walk, Needham Lake Walk
Grey Plover	Felixstowe Promenade Walk, Orford Ness Nature Reserve Walk
Hen Harrier	Rendlesham Forest Walk, Minsmere Walk 1, RSPB Minsmere Walk 2
Hobby	Rendlesham Forest Walk, Orwell Walk, Alton Water Walk
Kestrel	Orwell Walk, Felixstowe Promenade Walk, River Deben Walk, Martlesham Creek Walk, Shotley Peninsula Walk
Kingfisher	Orwell Walk, River Deben Walk, Martlesham Creek Walk, Needham Lake Walk
Lapwing	Felixstowe Promenade Walk, Minsmere Walk 1, RSPB Minsmere Walk 2, Trimley Marshes Walk
Lesser Spotted Woodpecker	Rendlesham Forest Walk, Staverton Thicks Walk, Sutton Hoo Walk

Little Egret	Minsmere Walk 1, RSPB Minsmere Walk 2, Orwell Walk, River Deben Walk
Little Ringed Plover	Trimley Marshes Walk, Minsmere Walk 1, RSPB Minsmere Walk 2
Little Tern	Minsmere Walk 1, RSPB Minsmere Walk 2, Orford Ness Nature Reserve Walk
Long-tailed Tit	Ickworth Park Walk, Rendlesham Forest Walk, Sutton Hoo Walk
Magpie	Lavenham and the Lavenham Walk, Orwell Walk, Ipswich Waterfront Walk
Mallard	Alton Water Walk, Orwell Walk, Needham Lake Walk, River Deben Walk
Marsh Harrier	Minsmere Walk 1, RSPB Minsmere Walk 2, Trimley Marshes Walk, Orford Ness Nature Reserve Walk
Marsh Tit	Staverton Thicks Walk, Rendlesham Forest Walk, Sutton Hoo Walk
Mediterranean Gull	Felixstowe Promenade North Walk, Felixstowe Promenade South Walk, Bawdsey Quay Walk
Merlin	Rendlesham Forest Walk, Orford Ness Nature Reserve Walk
Moorhen	Alton Water Walk, Needham Lake Walk, Trimley Marshes Walk
Mute Swan	Alton Water Walk, River Deben Walk, Orwell Walk, Needham Lake Walk
Nightingale	Orwell Walk, Sutton Hoo Walk, Staverton Thicks Walk
Nightjar	Dunwich Heath Walk 1, Dunwich Heath and Beach Walk 2, Dunwich Heath Coastal Walk 3, Rendlesham Forest Walk
Nuthatch	Ickworth Park Walk, Rendlesham Forest Walk, Sutton Hoo Walk
Oystercatcher	Felixstowe Promenade Walk, Bawdsey Quay Walk, River Alde Walk, Orford Ness Nature Reserve Walk
Peregrine Falcon	Rendlesham Forest Walk, Orwell Bridge (Orwell Walk)
Pied Wagtail	Ipswich Waterfront Walk, Felixstowe Promenade North Walk, Felixstowe Promenade South Walk
Pochard	Alton Water Walk, Minsmere Walk 1, RSPB Minsmere Walk 2
Red Kite	Great Bealings Circular Walk, Shotley Peninsula Walk, Orwell Walk
Red-breasted Merganser	Felixstowe Promenade Walk, River Deben Walk, Orford Ness Nature Reserve Walk
Redshank	Minsmere Walk 1, RSPB Minsmere Walk 2, Trimley Marshes Walk, Orford Ness Nature Reserve Walk
Reed Bunting	Minsmere Walk 1, RSPB Minsmere Walk 2, Trimley Marshes Walk
Reed Warbler	Minsmere Walk 1, RSPB Minsmere Walk 2, Trimley Marshes Walk
Ringed Plover	Felixstowe Promenade Walk, Orford Ness Nature Reserve Walk, Shingle Street Walk
Robin	Ickworth Park Walk, Lavenham and the Lavenham Walk, Sutton Hoo Walk, Orwell Walk
Sand Martin	Orford Ness Nature Reserve Walk, Shingle Street Walk, Bawdsey Quay Walk
Sedge Warbler	Minsmere Walk 1, RSPB Minsmere Walk 2, Trimley Marshes Walk
Shelduck	Orwell Walk, Felixstowe Promenade Walk, Trimley Marshes Walk
Skylark	Minsmere Walk 1, RSPB Minsmere Walk 2, Orford Ness Nature Reserve Walk, Shingle Street Walk
Song Thrush	Ickworth Park Walk, Lavenham and the Lavenham Walk, Sutton Hoo Walk
Sparrowhawk	Rendlesham Forest Walk, Shotley Peninsula Walk, Orwell Walk
Spoonbill	Minsmere Walk 1, RSPB Minsmere Walk 2, Trimley Marshes Walk
Spotted Flycatcher	Ickworth Park Walk, Sutton Hoo Walk, Staverton Thicks Walk
Starling	Lavenham and the Lavenham Walk, Ipswich Waterfront Walk, Felixstowe Promenade North Walk
Stock Dove	Ickworth Park Walk, Rendlesham Forest Walk, Sutton Hoo Walk
Stonechat	Dunwich Heath Walk 1, Dunwich Heath and Beach Walk 2, Dunwich Heath Coastal Walk 3

Swallow	Orwell Walk, Alton Water Walk, Shotley Peninsula Walk
Swift	Ipswich Waterfront Walk, Lavenham and the Lavenham Walk, Orwell Walk
Teal	Alton Water Walk, Minsmere Walk 1, RSPB Minsmere Walk 2, Trimley Marshes Walk
Treecreeper	Ickworth Park Walk, Rendlesham Forest Walk, Sutton Hoo Walk
Tufted Duck	Alton Water Walk, Needham Lake Walk, Trimley Marshes Walk
Turtle Dove	Sutton Hoo Walk, Staverton Thicks Walk, Rendlesham Forest Walk
Water Rail	Minsmere Walk 1, RSPB Minsmere Walk 2, Trimley Marshes Walk
Whimbrel	Orford Ness Nature Reserve Walk, Bawdsey Quay Walk, Shingle Street Walk
Whitethroat	Sutton Hoo Walk, Orwell Walk, Rendlesham Forest Walk
Wigeon	Alton Water Walk, Minsmere Walk 1, RSPB Minsmere Walk 2, Trimley Marshes Walk
Willow Warbler	Sutton Hoo Walk, Orwell Walk, Rendlesham Forest Walk
Woodcock	Rendlesham Forest Walk, Staverton Thicks Walk, Sutton Hoo Walk
Woodlark	Rendlesham Forest Walk, Dunwich Heath Walk 1, Dunwich Heath and Beach Walk 2, Dunwich Heath Coastal Walk 3
Woodpigeon	Ickworth Park Walk, Lavenham and the Lavenham Walk, Sutton Hoo Walk, Orwell Walk
Wren	Ickworth Park Walk, Lavenham and the Lavenham Walk, Sutton Hoo Walk, Orwell Walk
Yellow Wagtail	Minsmere Walk 1, RSPB Minsmere Walk 2, Orford Ness Nature Reserve Walk

Animals to Spot

Animal	Walks
Badger	Orwell Walk, Rendlesham Forest Walk, Ipswich Town Centre Walk, River Deben Walk, Boulge Circular Walk
Chinese Water Deer	Orwell Walk, Alton Water Walk, Shotley Peninsula Walk, River Alde Walk, Hollesley Common Walk
Common Lizard	Sutton Hoo Walk, Orwell Walk, Felixstowe Promenade Walk, Rendlesham Forest Walk, Ipswich Town Centre Walk, River Deben Walk, Tattingstone Wonder Walk, Boulge Circular Walk, Dunwich Heath Walk 1, Dunwich Heath and Beach Walk 2, Dunwich Heath Coastal Walk 3
Common Seal	Felixstowe Promenade Walk, River Orwell Walk, Shingle Street Walk, River Deben Walk, Bawdsey to Felixstowe Ferry Walk, Orford Ness Nature Reserve Walk
Fox	Orwell Walk, Rendlesham Forest Walk, Ipswich Town Centre Walk, River Deben Walk, Shotley Peninsula Walk, Boulge Circular Walk, Sutton Hoo Walk
Grass Snake	Sutton Hoo Walk, Orwell Walk, Rendlesham Forest Walk, Ipswich Town Centre Walk, River Deben Walk, Tattingstone Wonder Walk, Boulge Circular Walk, Martlesham Circular Walk, Trimley Marshes Walk
Grey Seal	Felixstowe Promenade Walk, River Orwell Walk, Shingle Street Walk, River Deben Walk, Bawdsey to Felixstowe Ferry Walk, Orford Ness Nature Reserve Walk
Harbour Porpoise	Felixstowe Promenade Walk, River Orwell Walk, Shingle Street Walk, River Deben Walk, Bawdsey to Felixstowe Ferry Walk, Orford Ness Nature Reserve Walk
Harvest Mouse	Sutton Hoo Walk, Orwell Walk, Felixstowe Promenade Walk, Rendlesham Forest Walk, Ipswich Town Centre Walk, River Deben Walk, Tattingstone Wonder Walk, Boulge Circular Walk, Martlesham Circular Walk, Trimley Marshes Walk
Hedgehog	Sutton Hoo Walk, Orwell Walk, Alton Water Walk, Rendlesham Forest Walk, Ipswich Town Centre Walk, River Deben Walk, Tattingstone Wonder Walk, Bramford Circular Walk, Felixstowe Promenade Walk, Lavenham and the Lavenham Walk
Muntjac Deer	Felixstowe Promenade Walk, Rendlesham Forest Walk, Trimley Marshes Walk, East Bergholt Circular Walk, Boulge Circular Walk, Orwell Walk, Ipswich Town Centre Walk, Sutton Hoo Walk
Otter	Orwell Walk, Alton Water Walk, Pin Mill and Chelmondiston Walk, River Deben Walk, Shotley Peninsula Walk, River Alde Walk, Minsmere Walk 1, RSPB Minsmere Walk 2, Trimley Marshes Walk
Rabbit	Orwell Walk, Felixstowe Promenade Walk, Rendlesham Forest Walk, Trimley Marshes Walk, Shotley Peninsula Walk, Sutton Hoo Walk, Ipswich Town Centre Walk, Dunwich Heath Walk 1, Dunwich Heath and Beach Walk 2, Dunwich Heath Coastal Walk 3
Roe Deer	Orwell Walk, Alton Water Walk, Pin Mill and Chelmondiston Walk, River Deben Walk, Shotley Peninsula Walk, River Alde Walk, Rendlesham Forest Walk, Sutton Hoo Walk
Slow Worm	Sutton Hoo Walk, Orwell Walk, Rendlesham Forest Walk, Ipswich Town Centre Walk, River Deben Walk, Tattingstone Wonder Walk, Boulge Circular Walk
Stoat	Orwell Walk, Felixstowe Promenade Walk, Rendlesham Forest Walk, Trimley

	Marshes Walk, Shotley Peninsula Walk, Sutton Hoo Walk, Ipswich Town Centre Walk
Water Shrew	Sutton Hoo Walk, Orwell Walk, Felixstowe Promenade Walk, Rendlesham Forest Walk, Trimley Marshes Walk, Bawdsey Quay Walk, River Deben Walk, Shotley Peninsula Walk, River Alde Walk
Water Vole	Sutton Hoo Walk, Orwell Walk, Felixstowe Promenade Walk, Rendlesham Forest Walk, Trimley Marshes Walk, Bawdsey Quay Walk, River Deben Walk, Shotley Peninsula Walk, River Alde Walk, Minsmere Walk 1, RSPB Minsmere Walk 2
Yellow-necked Mouse	Rendlesham Forest Walk, Kesgrave Woods Walk, Boulge Circular Walk, Martlesham Circular Walk

Insects to Spot

Insect	Walks
Azure Damselfly	Trimley Marshes Walk, Orwell Walk, River Deben Walk
Banded Demoiselle	Orwell Walk, River Deben Walk, Martlesham Creek Walk
Brown Hawker Dragonfly	Orwell Walk, River Deben Walk, Rendlesham Forest Walk
Burnet Moth	Hollesley Common Walk, Sutton Common Walk, Orwell Walk
Common Blue Butterfly	Ipswich to Felixstowe Walk, Orwell Walk, Sutton Hoo Walk, Rendlesham Forest Walk, Trimley Marshes Walk
Common Darter Dragonfly	Trimley Marshes Walk, Orwell Walk, River Deben Walk
Common Hawker Dragonfly	Orwell Walk, River Deben Walk, Rendlesham Forest Walk
Common Wasp	Felixstowe Promenade Walk, Sutton Hoo Walk, Ipswich Town Centre Walk, Lavenham and the Lavenham Walk
Comma Butterfly	Orwell Walk, Sutton Hoo Walk, Rendlesham Forest Walk, Boulge Circular Walk
Emperor Dragonfly	Ipswich Waterfront Walk, Orwell Walk, River Deben Walk, Alton Water Walk
Four-spotted Chaser Dragonfly	Ipswich Waterfront Walk, Orwell Walk, River Deben Walk, Trimley Marshes Walk
Gatekeeper Butterfly	Orwell Walk, Sutton Hoo Walk, Rendlesham Forest Walk, Boulge Circular Walk
Green-veined White Butterfly	Needham Lake Walk, Orwell Walk, Sutton Hoo Walk, Rendlesham Forest Walk, Trimley Marshes Walk
Holly Blue Butterfly	Ipswich to Felixstowe Walk, Orwell Walk, Sutton Hoo Walk, Rendlesham Forest Walk
Hornet	Ipswich Town Centre Walk, Orwell Walk, Rendlesham Forest Walk
Large Red Damselfly	Trimley Marshes Walk, Orwell Walk, River Deben Walk
Large Skipper Butterfly	Ipswich to Felixstowe Walk, Orwell Walk, Sutton Hoo Walk, Rendlesham Forest Walk
Marbled White Butterfly	Orwell Walk, Sutton Hoo Walk, Rendlesham Forest Walk
Meadow Brown Butterfly	Orwell Walk, Sutton Hoo Walk, Rendlesham Forest Walk, Boulge Circular Walk, Trimley Marshes Walk
Migrant Hawker Dragonfly	Trimley Marshes Walk, Orwell Walk, River Deben Walk
Painted Lady Butterfly	Rendlesham Forest Walk, Orwell Walk, Sutton Hoo Walk, Trimley Marshes Walk
Peacock Butterfly	Pin Mill and Chelmondiston Walk, Woodbridge Riverside Walk, Orwell Walk, Sutton Hoo Walk, Rendlesham Forest Walk, Boulge Circular Walk
Red Admiral Butterfly	Rendlesham Forest Walk, Orwell Walk, Ipswich to Felixstowe Walk, Bucklesham Circular Walk, Shotley Peninsula Walk, Sutton Hoo Walk, Felixstowe Promenade Walk, Lavenham and the Lavenham Walk
Ringlet Butterfly	Orwell Walk, Sutton Hoo Walk, Rendlesham Forest Walk, Boulge Circular Walk
Silver-washed Fritillary Butterfly	Rushmere Heath Walk, Rendlesham Forest Walk
Small Copper Butterfly	Ipswich to Felixstowe Walk, Shotley Peninsula Walk, Orwell Walk, Sutton Hoo Walk, Rendlesham Forest Walk, Trimley Marshes Walk
Small Skipper Butterfly	Ipswich to Felixstowe Walk, Orwell Walk, Sutton Hoo Walk, Rendlesham Forest Walk
Small Tortoiseshell Butterfly	Orwell Walk, Sutton Hoo Walk, Rendlesham Forest Walk, Boulge Circular Walk, Felixstowe Promenade Walk, Lavenham and the Lavenham Walk

Southern Hawker Dragonfly	Orwell Walk, River Deben Walk, Rendlesham Forest Walk
Speckled Wood Butterfly	Felixstowe Promenade Walk, Orwell Walk, Pin Mill and Chelmondiston Walk, Sutton Hoo Walk, Rendlesham Forest Walk, Ipswich Town Centre Walk
White Admiral Butterfly	Rendlesham Forest Walk, Staverton Thicks Walk
White-letter Hairstreak Butterfly	Bucklesham Circular Walk, Rendlesham Forest Walk
Wood White Butterfly	Orwell Walk, Sutton Hoo Walk, Rendlesham Forest Walk

Plants to Spot

Plant	Walk
Alexanders, Sea Beet, Sea Kale, Sea Pea, Sea Purslane, Sea Rocket	Felixstowe Promenade Walk, Shingle Street Walk, Bawdsey Quay Walk, Orford Ness Nature Reserve Walk
Bee Orchid, Pyramidal Orchid, Common Spotted Orchid, Marsh Helleborine	Trimley Marshes Walk, Orwell Walk, Orford Ness Nature Reserve Walk
Bluebells, Wild Garlic, Wood Anemones, Primroses, Wood Sorrel	Sutton Hoo Walk, Rendlesham Forest Walk, Freston Wood Walk, Ickworth Park Walk
Bog Asphodel, Sundew, Cotton Grass, Marsh St. John's Wort	Dunwich Heath Walk 1, Dunwich Heath and Beach Walk 2, Dunwich Heath Coastal Walk 3
Broom, Gorse, Heather, Bilberry, Bell Heather, Cross-leaved Heath	Rushmere Heath Walk, Blaxhall Common Walk, Dunwich Heath Walk 1, Dunwich Heath and Beach Walk 2, Dunwich Heath Coastal Walk 3
Bulrush, Yellow Flag Iris, Purple Loosestrife, Meadowsweet, Water Mint	Orwell Walk, Pin Mill and Chelmondiston Walk, Alton Water Walk, Needham Lake Walk, Newbourne Springs Walk
Bugloss, Viper's Bugloss, Sea Holly, Sea Spurge, Yellow Horned Poppy, Sea Campion	Shingle Street Walk, Bawdsey Quay Walk, Orford Ness Nature Reserve Walk, Felixstowe Promenade Walk
Cow Parsley, Hogweed, Wild Carrot, Rough Chervil, Hemlock Water-dropwort	Orwell Walk, Ipswich Waterfront Walk, Newbourne Springs Walk, River Deben Walk
Cowslip, Ox-eye Daisy, Meadow Buttercup, Yellow Rattle, Common Knapweed, Bird's-foot Trefoil	Alton Water Walk, Newbourne Springs Walk, Trimley Marshes Walk, Great Bealings Circular Walk
Dog Rose, Field Rose, Hawthorn, Blackthorn, Elder, Guelder Rose	Orwell Walk, Pin Mill and Chelmondiston Walk, Newbourne Springs Walk, Sutton Hoo Walk, Shotley Peninsula Walk
Fennel, Alexanders, Wild Celery, Rock Samphire, Golden Samphire	Bawdsey Quay Walk, Orford Ness Nature Reserve Walk, Felixstowe Promenade Walk, Shingle Street Walk
Foxglove, Red Campion, Herb Robert, Wood Avens, Enchanter's Nightshade	Sutton Hoo Walk, Rendlesham Forest Walk, Freston Wood Walk, Ickworth Park Walk, Newbourne Springs Walk
Glasswort, Sea Aster, Sea Lavender, Sea Plantain, Sea Arrowgrass, Cordgrass	Trimley Marshes Walk, Orwell Walk, Nacton Shores Walk, Bawdsey Quay Walk, Orford Ness Nature Reserve Walk
Heather, Gorse, Broom, Bracken, Wavy Hair-grass	Rushmere Heath Walk, Blaxhall Common Walk, Dunwich Heath Walk 1, Dunwich Heath and Beach Walk 2, Dunwich Heath Coastal Walk 3
Lady's Smock, Ragged Robin, Marsh Marigold, Water Avens, Meadowsweet	Orwell Walk, Pin Mill and Chelmondiston Walk, Newbourne Springs Walk, Trimley Marshes Walk
Marram Grass, Lyme Grass, Sea Couch, Sand Sedge, Sea Pea, Sea Holly	Shingle Street Walk, Bawdsey Quay Walk, Felixstowe Promenade Walk, Orford Ness Nature Reserve Walk

Oak, Ash, Beech, Birch, Hazel, Sweet Chestnut, Horse Chestnut	Sutton Hoo Walk, Rendlesham Forest Walk, Freston Wood Walk, Ickworth Park Walk, Kesgrave Woods Walk
Orchids (Bee Orchid, Pyramidal Orchid, Common Spotted Orchid, Early Purple Orchid, Marsh Helleborine)	Newbourne Springs Walk, Trimley Marshes Walk, Orwell Walk, Orford Ness Nature Reserve Walk
Ox-eye Daisy, Common Knapweed, Bird's-foot Trefoil, Lady's Bedstraw, Yarrow, Tufted Vetch	Alton Water Walk, Trimley Marshes Walk, Great Bealings Circular Walk, Newbourne Springs Walk
Primrose, Cowslip, Bluebell, Wood Anemone, Lesser Celandine, Wild Daffodil	Sutton Hoo Walk, Rendlesham Forest Walk, Freston Wood Walk, Newbourne Springs Walk, Ickworth Park Walk
Reeds, Sedges, Rushes, Bulrush, Yellow Flag Iris, Purple Loosestrife	Orwell Walk, Pin Mill and Chelmondiston Walk, Alton Water Walk, Needham Lake Walk, Trimley Marshes Walk, Newbourne Springs Walk
Scarlet Pimpernel, Red Dead-nettle, White Dead-nettle, Ground Ivy, Speedwell, Forget-me-not	Ipswich Town Centre Walk, Woodbridge Riverside Walk, Martlesham Creek Walk, Kesgrave Woods Walk, Trimley St Martin Walk
Sea Aster, Sea Lavender, Sea Plantain, Sea Purslane, Glasswort, Golden Samphire	Trimley Marshes Walk, Orwell Walk, Nacton Shores Walk, Bawdsey Quay Walk, Orford Ness Nature Reserve Walk
Sea Beet, Sea Kale, Sea Pea, Sea Rocket, Yellow Horned Poppy, Sea Campion	Felixstowe Promenade Walk, Shingle Street Walk, Bawdsey Quay Walk, Orford Ness Nature Reserve Walk
Sea Holly, Sea Spurge, Marram Grass, Lyme Grass, Sand Sedge, Sea Pea	Shingle Street Walk, Bawdsey Quay Walk, Felixstowe Promenade Walk, Orford Ness Nature Reserve Walk
Thrift, Sea Lavender, Sea Aster, Sea Plantain, Samphire, Frosted Orache	Nacton Shores Walk, Trimley Marshes Walk, Bawdsey Quay Walk, Orford Ness Nature Reserve Walk
Water Crowfoot, Yellow Water-lily, Frogbit, Arrowhead, Water-plantain, Flowering Rush	Orwell Walk, Pin Mill and Chelmondiston Walk, Alton Water Walk, Needham Lake Walk, River Deben Walk
Wild Carrot, Fennel, Cow Parsley, Hogweed, Rough Chervil, Hemlock Water-dropwort	Orwell Walk, Ipswich Waterfront Walk, Newbourne Springs Walk, River Deben Walk, Felixstowe Promenade Walk
Wildflowers (various species), Fungi, Ferns, Mosses, Lichens	Rendlesham Forest Walk, Ipswich Town Centre Walk, Woodbridge Riverside Walk, Shotley Peninsula Walk, Harkstead Circular Walk, Ipswich Waterfront Walk, Martlesham Creek Walk, Kesgrave Woods Walk, Trimley St Martin Walk, Freston Wood Walk, Felixstowe Ferry Walk
Wood Anemone, Bluebell, Wild Garlic, Primrose, Lesser Celandine, Dog's Mercury	Sutton Hoo Walk, Rendlesham Forest Walk, Freston Wood Walk, Ickworth Park Walk, Newbourne Springs Walk
Yellow Flag Iris, Marsh Marigold, Water Forget-me-not, Watercress, Brooklime, Water-cress	Orwell Walk, Pin Mill and Chelmondiston Walk, Alton Water Walk, Needham Lake Walk, Newbourne Springs Walk, River Deben Walk

This table highlights the incredible diversity of plant life that can be found along the walks in the Suffolk region, from the hardy coastal species that thrive in the challenging conditions of the shingle beaches and salt marshes to the delicate woodland flowers that carpet the forest floor in the spring. The table also showcases the wide range of habitats that these walks encompass, including wetlands, heathlands, meadows, and ancient woodlands, each supporting its own unique community of plants and wildlife. Whether you're a keen botanist or simply someone who appreciates the beauty and diversity of the natural world, this table provides a comprehensive guide to the many fascinating and beautiful plants that can be discovered on these walks, inviting you to slow down, look closely, and marvel at the wonders of the Suffolk countryside.

Sites of Special Historical Interest:

1. Sutton Hoo (Sutton Hoo Walk): An Anglo-Saxon burial site featuring a magnificent ship burial, believed to be the final resting place of King Rædwald of East Anglia.

2. Landguard Fort (Felixstowe Promenade Walk): A historic fortification dating back to the 16th century, built to defend the coast from invasion.

3. Orford Castle (Orford Ness Nature Reserve Walk): A 12th-century castle with a unique polygonal tower, offering stunning views of the surrounding countryside and coast.

4. Framlingham Castle (Nearby attraction): A magnificent 12th-century fortress with a rich history, including being the refuge of Mary Tudor before she became Queen of England.

5. Woodbridge Tide Mill (Woodbridge Riverside Walk): A rare example of a working tide mill, originally built in the 12th century and restored to full working order.

6. Bury St Edmunds Abbey (Nearby attraction): The ruins of a once-powerful abbey, founded in the 11th century and the burial place of several Anglo-Saxon kings.

7. Leiston Abbey (Minsmere Walk 1): The remains of a 14th-century abbey, set in a picturesque coastal location.

8. Bawdsey Radar Station (Bawdsey Quay Walk): The site of the world's first operational radar station, crucial in the defence of Britain during World War II.

9. Martello Towers (Felixstowe Promenade Walk, Shingle Street Walk): Small defensive forts built along the East Anglian coast in the early 19th century to defend against the threat of French invasion.

10. Lavenham Guildhall (Lavenham and the Lavenham Walk): A beautifully preserved 16th-century guildhall, now a museum showcasing the history of the wool trade in Suffolk.

Areas of Outstanding Beauty:

1. Orwell Estuary (Orwell Walk): A stunning estuary with a diverse range of habitats, offering beautiful views and an abundance of wildlife.

2. Aldeburgh Beach (Nearby attraction): A picturesque shingle beach with colourful fishing boats and the iconic Scallop sculpture by Maggi Hambling.

3. Dunwich Heath (Dunwich Heath Walk 1, 2, 3, and 4): A beautiful coastal heathland with a mix of gorse, heather, and bracken, offering stunning views of the sea and a chance to spot rare birds.

4. Minsmere Nature Reserve (Minsmere Walk 1 and 2): A stunning coastal nature reserve with a mosaic of habitats, including reedbeds, wetlands, and woodlands, home to a wide variety of birds and wildlife.

5. Orford Ness (Orford Ness Nature Reserve Walk): A unique shingle spit with a fascinating military history and an eerie, otherworldly beauty.

6. Shingle Street (Shingle Street Walk): A remote coastal hamlet with a stunning shingle beach and a sense of wild isolation.

7. Constable Country (Nearby attraction): The beautiful countryside around the River Stour, immortalised in the paintings of John Constable.

8. Rendlesham Forest (Rendlesham Forest Walk): A vast and atmospheric forest, famous for its alleged UFO sightings but also home to a variety of wildlife and beautiful woodland trails.

9. Alton Water (Alton Water Walk): A large reservoir set in stunning countryside, offering beautiful views and a variety of water-based activities.

10. Shotley Peninsula (Shotley Peninsula Walk): A picturesque peninsula with stunning river views, charming villages, and beautiful countryside walks.

These are just a few examples of the many areas of historical interest and outstanding beauty that can be found along the walks in the Suffolk region, each offering a unique and unforgettable experience for visitors.

Walks for Weeks

All walks are good if you have the right gear, but some look at their best at different times of the year. Here's a quick guide to which walks are really good whenever you want to go. It gives a few for each week, but only gives 4 weeks for every month, so you get a few days break at the end of each month to wash your socks!

By exploring these walks throughout the year, you'll have the opportunity to witness the ever-changing face of the Suffolk landscape and to discover the beauty and wonder of each season in this enchanting corner of England.

January

Week 1:
- Alton Water Walk: Enjoy crisp winter air and the chance to spot overwintering birds such as wigeons and goldeneyes.
- Rendlesham Forest Walk: Experience the stark beauty of the winter woodland and keep an eye out for red deer.
- Orwell Walk: Admire the frosty landscapes and the serene atmosphere of the river in winter.

Week 2:
- Dunwich Heath Walk 1: Marvel at the golden gorse in bloom and the striking contrast of the winter landscape.
- Ickworth Park Walk: Spot early signs of spring, such as snowdrops and aconites, in the parkland.
- Shotley Gate Walk: Blow away the cobwebs with a winter walk along the Stour Estuary, with far-reaching views.

Week 3:
- Newbourne Springs Walk: Discover the tranquil beauty of the springs and the winter wildlife.
- Felixstowe Promenade Walk: Embrace the bracing sea air and the quiet charm of the seaside town in winter.
- Woodbridge Riverside Walk: Enjoy the peaceful atmosphere of the historic town and the stunning views of the River Deben.

Week 4:
- Sutton Hoo Walk: Explore the ancient burial grounds and the winter landscape of the Suffolk countryside.
- Trimley Marshes Walk: Spot overwintering birds and the stark beauty of the marshland in winter.
- Freston Wood Walk: Discover the atmospheric woodland in its winter stillness, with bare trees and frosty ground.

February

Week 1:
- Dunwich Heath Walk 2: Admire the golden gorse in bloom and the striking contrast of the winter landscape.
- Ickworth Park Walk: Spot early signs of spring, such as snowdrops and aconites, in the parkland.
- Orford Ness Nature Reserve Walk: Witness the arrival of early spring migrants and the changing light on the coast.

Week 2:
- Rendlesham Forest Walk: Enjoy the peace and solitude of the winter woodland and keep an eye out for wildlife.

- Lavenham and the Lavenham Walk: Discover the timeless charm of the medieval village in the crisp winter light.
- Alton Water Walk: Spot overwintering birds and enjoy the tranquil atmosphere of the reservoir.

Week 3:
- Shotley Peninsula Walk: Marvel at the stunning river views and the stark beauty of the winter landscape.
- Newbourne Springs Walk: Delight in the early signs of spring and the awakening of nature.
- Felixstowe Ferry Walk: Enjoy the bracing sea air and the quiet charm of the coastal hamlet in winter.

Week 4:
- Blaxhall Common Walk: Discover the ancient heathland in its winter beauty, with gnarled trees and golden gorse.
- Woodbridge Riverside Walk: Soak up the peaceful atmosphere of the historic town and the stunning views of the River Deben.
- Trimley Marshes Walk: Witness the changing light on the marshland and spot overwintering birds.

March

Week 1:
- Orford Ness Nature Reserve Walk: Witness the arrival of spring migrants, such as avocets and sandwich terns.
- Newbourne Springs Walk: Delight in the vibrant display of spring wildflowers, including primroses and violets.
- Sutton Hoo Walk: Enjoy the spring sunshine and the awakening of nature in the ancient landscape.

Week 2:
- Rendlesham Forest Walk: Discover the first signs of spring in the woodland, with birdsong and emerging leaves.
- Shingle Street Walk: Embrace the wild beauty of the coast and the unique shingle habitat in early spring.
- Alton Water Walk: Spot the first spring migrants and enjoy the fresh green landscapes around the reservoir.

Week 3:
- Dunwich Heath Walk 3: Marvel at the stunning coastal scenery and the early spring wildflowers.
- Ickworth Park Walk: Admire the spring bulbs and the fresh green leaves in the parkland.
- Shotley Peninsula Walk: Enjoy the stunning river views and the first signs of spring in the countryside.

Week 4:
- Freston Wood Walk: Discover the ancient woodland in its early spring glory, with delicate wildflowers and birdsong.
- Woodbridge Riverside Walk: Soak up the spring atmosphere of the historic town and the beautiful views of the River Deben.
- Orwell Walk: Delight in the awakening of nature along the river and the fresh spring air.

April

Week 1:
- Blaxhall Common Walk: Marvel at the carpets of bluebells and the fresh green leaves of the woodland.
- Sutton Hoo Walk: Enjoy the spring sunshine and the emerging butterflies in the meadows and woodland rides.
- Minsmere Walk 1: Experience the dawn chorus and the vibrant birdlife of the coastal wetlands.

Week 2:
- Dunwich Heath Walk 1: Admire the stunning coastal scenery and the spring wildflowers in full bloom.
- Newbourne Springs Walk: Delight in the vibrant display of spring wildflowers and the lush green landscapes.
- Rendlesham Forest Walk: Discover the beauty of the woodland in spring, with dappled sunlight and birdsong.

Week 3:
- Orwell Walk: Enjoy the fresh spring air and the stunning river views, with wildflowers lining the paths.
- Ickworth Park Walk: Admire the spring bulbs and the lush green parkland, with newborn lambs in the fields.
- Felixstowe Ferry Walk: Soak up the spring atmosphere of the coastal hamlet and enjoy the views of the Deben Estuary.

Week 4:
- Lavenham and the Lavenham Walk: Admire the picturesque village in the soft spring light and the blooming cottage gardens.
- Alton Water Walk: Spot spring migrants and enjoy the vibrant green landscapes around the reservoir.
- Shotley Peninsula Walk: Marvel at the stunning river views and the spring wildflowers in the countryside.

May

Week 1:
- Minsmere Walk 1: Experience the dawn chorus and the vibrant birdlife of the coastal wetlands.
- Lavenham and the Lavenham Walk: Admire the picturesque village in the soft spring light and the blooming cottage gardens.
- Orwell Walk: Delight in the lush green landscapes and the abundance of wildflowers along the river.
- Woodbridge Riverside Walk: Enjoy the spring charm of the historic town and the stunning views of the River Deben.

Week 2:
- Staverton Thicks Walk: Discover the ancient woodland in its late spring glory, with bluebells and birdsong.
- Trimley Marshes Walk: Witness the vibrant birdlife and the lush green marshland in spring.
- Sutton Hoo Walk: Enjoy the warm spring sunshine and the colourful wildflowers in the meadows and woodland rides.
- Alton Water Walk: Spot spring migrants and enjoy the beautiful landscapes around the reservoir.

Week 3:
- Newbourne Springs Walk: Delight in the vibrant display of late spring wildflowers and the lush green landscapes.
- Rendlesham Forest Walk: Discover the beauty of the woodland in late spring, with dappled sunlight and birdsong.
- Felixstowe Promenade Walk: Enjoy the spring atmosphere of the seaside town and the stunning coastal views.
- Orford Ness Nature Reserve Walk: Witness the nesting seabirds and the unique shingle habitat in spring.

Week 4:
- Ipswich Waterfront Walk: Soak up the vibrant atmosphere of the marina and enjoy the spring sunshine.
- Shotley Gate Walk: Marvel at the stunning river views and the lush green landscapes of the Stour Estuary.
- Pin Mill and Chelmondiston Walk: Enjoy the spring charm of the riverside hamlets and the beautiful countryside.
- Dunwich Heath Walk 2: Admire the stunning coastal scenery and the late spring wildflowers in bloom.

June

Week 1:
- Orwell Walk: Delight in the lush green landscapes and the abundance of wildflowers along the river.
- Staverton Thicks Walk: Discover the ancient woodland in its summer glory, with dappled sunlight filtering through the canopy.
- Bawdsey Quay Walk: Enjoy the summer sun, the sandy beach, and the chance to spot seals and shore birds.
- Sutton Hoo Walk: Enjoy the warm summer sunshine and the vibrant wildflowers in the meadows and woodland rides.

Week 2:
- Shotley Peninsula Walk: Marvel at the stunning river views and the vibrant wildflower meadows.
- Minsmere Walk 2: Witness the vibrant birdlife and the lush green wetlands in summer.
- Newbourne Springs Walk: Delight in the summer wildflowers and the peaceful atmosphere of the springs.
- Felixstowe Promenade Walk: Soak up the summer atmosphere of the seaside town and enjoy the beach activities.

Week 3:
- Rendlesham Forest Walk: Discover the beauty of the woodland in summer, with dappled sunlight and lush foliage.
- Ickworth Park Walk: Enjoy the summer charm of the parkland, with picnics and outdoor activities.
- Orford Ness Nature Reserve Walk: Witness the nesting seabirds and the unique shingle habitat in summer.
- Alton Water Walk: Spot summer wildlife and enjoy the beautiful landscapes around the reservoir.

Week 4:
- Woodbridge Riverside Walk: Soak up the summer atmosphere of the historic town and the stunning views of the River Deben.
- Trimley Marshes Walk: Enjoy the vibrant birdlife and the lush green marshland in summer.
- Lavenham and the Lavenham Walk: Admire the picturesque village in the warm summer light and the colourful gardens.
- Dunwich Heath Walk 3: Marvel at the stunning coastal scenery and the summer wildflowers in bloom.

July

Week 1:
- Bawdsey Quay Walk: Enjoy the summer sun, the sandy beach, and the chance to spot seals and shore birds.
- Shotley Peninsula Walk: Marvel at the stunning river views and the vibrant wildflower meadows.
- Orwell Walk: Delight in the lush green landscapes and the summer wildlife along the river.
- Felixstowe Promenade Walk: Relish the classic seaside experience, with beach huts, ice cream, and summer festivities.

Week 2:
- Minsmere Walk 1: Witness the vibrant birdlife and the beautiful coastal wetlands in summer.
- Newbourne Springs Walk: Enjoy the summer wildflowers and the tranquil atmosphere of the springs.
- Rendlesham Forest Walk: Discover the beauty of the woodland in summer, with dappled sunlight and lush foliage.
- Sutton Hoo Walk: Enjoy the warm summer sunshine and the colourful wildflowers in the meadows and woodland rides.

Week 3:
- Shingle Street Walk: Soak up the summer atmosphere of the coastal hamlet and explore the unique shingle habitat.
- Alton Water Walk: Spot summer wildlife and enjoy the beautiful landscapes around the reservoir.
- Ipswich Waterfront Walk: Enjoy the vibrant atmosphere of the marina and the summer events.
- Staverton Thicks Walk: Discover the ancient woodland in its summer glory, with dappled sunlight filtering through the canopy.

Week 4:
- Woodbridge Riverside Walk: Soak up the summer charm of the historic town and the stunning views of the River Deben.
- Trimley Marshes Walk: Enjoy the vibrant birdlife and the lush green marshland in summer.
- Orford Ness Nature Reserve Walk: Witness the nesting seabirds and the unique shingle habitat in summer.
- Dunwich Heath Walk 1: Marvel at the stunning coastal scenery and the summer wildflowers in bloom.

August

Week 1:
- Shingle Street Walk: Soak up the summer atmosphere of the coastal hamlet and explore the unique shingle habitat.
- Felixstowe Promenade Walk: Relish the classic seaside experience, with beach huts, ice cream, and summer festivities.
- Orwell Walk: Delight in the lush green landscapes and the summer wildlife along the river.
- Woodbridge Riverside Walk: Enjoy the summer charm of the historic town and the stunning views of the River Deben.

Week 2:
- Trimley Marshes Walk: Witness the early signs of autumn bird migration and the changing colours of the marshland.
- Pin Mill and Chelmondiston Walk: Savor the mellow summer light and the beautiful riverside landscapes.
- Rendlesham Forest Walk: Discover the beauty of the woodland in late summer, with dappled sunlight and lush foliage.
- Sutton Hoo Walk: Enjoy the warm summer sunshine and the colourful wildflowers in the meadows and woodland rides.

Week 3:
- Bawdsey Quay Walk: Enjoy the summer sun, the sandy beach, and the chance to spot seals and shore birds.
- Shotley Peninsula Walk: Marvel at the stunning river views and the vibrant wildflower meadows.
- Alton Water Walk: Spot summer wildlife and enjoy the beautiful landscapes around the reservoir.
- Minsmere Walk 2: Witness the vibrant birdlife and the beautiful coastal wetlands in summer.

Week 4:
- Newbourne Springs Walk: Enjoy the late summer wildflowers and the tranquil atmosphere of the springs.
- Ickworth Park Walk: Savor the summer charm of the parkland, with picnics and outdoor activities.
- Orford Ness Nature Reserve Walk: Witness the nesting seabirds and the unique shingle habitat in summer.
- Dunwich Heath Walk 2: Marvel at the stunning coastal scenery and the late summer wildflowers in bloom.

September

Week 1:

- Trimley Marshes Walk: Witness the spectacle of autumn bird migration and the changing colors of the marshland.
- Pin Mill and Chelmondiston Walk: Savor the mellow autumn light and the rich hues of the riverside woodland.
- Orwell Walk: Enjoy the changing colors of the landscape and the autumn wildlife along the river.
- Sutton Hoo Walk: Discover the ancient burial grounds in the soft autumn light and the turning leaves.

Week 2:
- Rendlesham Forest Walk: Marvel at the autumn colors of the woodland and the seasonal fungi.
- Alton Water Walk: Spot autumn migrants and enjoy the beautiful landscapes around the reservoir.
- Woodbridge Riverside Walk: Soak up the autumnal charm of the historic town and the stunning views of the River Deben.
- Dunwich Heath Walk 3: Admire the stunning coastal scenery and the autumn hues of the heathland.

Week 3:
- Flatford and Constable Country Walk: Admire the iconic landscape that inspired John Constable, resplendent in autumn colors.
- Shotley Peninsula Walk: Enjoy the stunning river views and the rich autumn hues of the countryside.
- Newbourne Springs Walk: Delight in the autumn wildflowers and the peaceful atmosphere of the springs.
- Ickworth Park Walk: Marvel at the stunning autumn colors of the parkland and the turning leaves.

Week 4:
- Minsmere Walk 1: Witness the autumn bird migration and the changing colors of the coastal wetlands.
- Orford Ness Nature Reserve Walk: Enjoy the unique shingle habitat and the autumn light on the coast.
- Lavenham and the Lavenham Walk: Admire the picturesque village in the soft autumn light and the turning leaves.
- Bawdsey Quay Walk: Savor the autumn atmosphere of the coastal hamlet and the chance to spot migrating birds.

October

Week 1:
- Flatford and Constable Country Walk: Admire the iconic landscape that inspired John Constable, resplendent in autumn colors.
- Woodbridge Riverside Walk: Enjoy the autumnal charm of the historic town and the stunning views of the River Deben.
- Dunwich Heath Walk 1: Marvel at the stunning coastal scenery and the autumn hues of the heathland.
- Rendlesham Forest Walk: Discover the beauty of the woodland in autumn, with turning leaves and seasonal fungi.

Week 2:
- Orwell Walk: Delight in the rich autumn colors of the landscape and the seasonal wildlife along the river.
- Sutton Hoo Walk: Explore the ancient burial grounds in the soft autumn light and the turning leaves.
- Alton Water Walk: Spot autumn migrants and enjoy the beautiful landscapes around the reservoir.
- Pin Mill and Chelmondiston Walk: Savor the mellow autumn light and the rich hues of the riverside woodland.

Week 3:
- Shotley Peninsula Walk: Enjoy the stunning river views and the autumnal colors of the countryside.
- Trimley Marshes Walk: Witness the autumn bird migration and the changing colors of the marshland.

- Ickworth Park Walk: Marvel at the stunning autumn colors of the parkland and the turning leaves.
- Newbourne Springs Walk: Delight in the autumn wildflowers and the peaceful atmosphere of the springs.

Week 4:
- Freston Wood Walk: Discover the atmospheric woodland in its autumn glory, with fallen leaves and fungi.
- Minsmere Walk 2: Enjoy the autumn bird migration and the changing colors of the coastal wetlands.
- Lavenham and the Lavenham Walk: Admire the picturesque village in the soft autumn light and the turning leaves.
- Orford Ness Nature Reserve Walk: Savor the unique shingle habitat and the autumn light on the coast.

November

Week 1:
- Dunwich Heath Coastal Walk 3: Brave the bracing sea air and the wild beauty of the coast in winter.
- Freston Wood Walk: Discover the atmospheric woodland in its autumn glory, with fallen leaves and fungi.
- Orwell Walk: Enjoy the stark beauty of the landscape and the winter wildlife along the river.
- Sutton Hoo Walk: Explore the ancient burial grounds in the crisp autumn light and the bare trees.

Week 2:
- Rendlesham Forest Walk: Marvel at the autumn colors of the woodland and the seasonal fungi.
- Alton Water Walk: Spot winter migrants and enjoy the beautiful landscapes around the reservoir.
- Woodbridge Riverside Walk: Soak up the peaceful atmosphere of the historic town and the stunning views of the River Deben.
- Shotley Peninsula Walk: Enjoy the stunning river views and the stark beauty of the winter landscape.

Week 3:
- Trimley Marshes Walk: Witness the winter bird migration and the changing colors of the marshland.
- Pin Mill and Chelmondiston Walk: Savor the crisp autumn light and the rich hues of the riverside woodland.
- Newbourne Springs Walk: Delight in the peaceful atmosphere of the springs and the winter wildlife.
- Ickworth Park Walk: Enjoy the stunning autumn colors of the parkland and the bare trees.

Week 4:
- Minsmere Walk 1: Witness the winter bird migration and the stark beauty of the coastal wetlands.
- Orford Ness Nature Reserve Walk: Enjoy the unique shingle habitat and the winter light on the coast.
- Lavenham and the Lavenham Walk: Admire the picturesque village in the crisp autumn light and the bare trees.
- Bawdsey Quay Walk: Savor the winter atmosphere of the coastal hamlet and the chance to spot migrating birds.

December

Week 1:
- Ipswich Waterfront Walk: Soak up the festive atmosphere of the marina, with Christmas lights and events.
- Shotley Gate Walk: Blow away the cobwebs with a winter walk along the Stour Estuary, with far-reaching views.
- Woodbridge Riverside Walk: Enjoy the festive charm of the historic town and the stunning views of the River Deben.
- Rendlesham Forest Walk: Discover the stark beauty of the winter woodland and the seasonal fungi.

Week 2:

- Orwell Walk: Delight in the winter landscapes and the seasonal wildlife along the river.
- Sutton Hoo Walk: Explore the ancient burial grounds in the crisp winter light and the bare trees.
- Alton Water Walk: Spot winter migrants and enjoy the beautiful landscapes around the reservoir.
- Dunwich Heath Walk 2: Marvel at the stunning coastal scenery and the winter light on the heathland.

Week 3:
- Newbourne Springs Walk: Enjoy the peaceful atmosphere of the springs and the winter wildlife.
- Lavenham and the Lavenham Walk: Admire the picturesque village in the festive atmosphere and the crisp winter light.
- Trimley Marshes Walk: Witness the winter bird migration and the stark beauty of the marshland.
- Felixstowe Promenade Walk: Embrace the bracing sea air and the quiet charm of the seaside town in winter.

Week 4:
- Minsmere Walk 2: Enjoy the winter bird migration and the stark beauty of the coastal wetlands.
- Orford Ness Nature Reserve Walk: Savor the unique shingle habitat and the winter light on the coast.
- Freston Wood Walk: Discover the atmospheric woodland in its winter stillness, with bare trees and frosty ground.
- Ickworth Park Walk: Marvel at the stunning winter landscapes of the parkland and the bare trees.

Epilogue

As we come to the end of our journey through the enchanting landscapes of Suffolk, I hope that this guide has inspired you to step outside, explore, and connect with the natural world in a profound and meaningful way. The 50 walks featured in this book offer a glimpse into the rich tapestry of history, culture, and biodiversity that makes this corner of England so special and unique.

From the windswept shingle beaches of the coast to the tranquil woodlands and rolling countryside inland, each walk has its own story to tell and its own treasures to reveal. Whether you're a seasoned rambler or a curious first-time visitor, these trails have something to offer everyone – a chance to slow down, breathe deeply, and immerse yourself in the timeless beauty and wonder of the Suffolk landscape.

Throughout these pages, we've explored the seasonal highlights and natural wonders of each route, from the vibrant wildflower displays of spring and summer to the rich autumnal hues and stark winter beauty of the colder months. We've discovered the incredible diversity of bird life, insects, mammals, and plants that call these habitats home, and learned about the fascinating history and heritage that has shaped this land and its people over countless generations.

But beyond the factual information and practical details, I hope that this guide has also conveyed a sense of the deeper meaning and significance of these walks – the joy, peace, and connection that comes from spending time in nature, and the importance of preserving and cherishing these precious landscapes for future generations to enjoy.

In a world that often feels increasingly disconnected and uncertain, the simple act of walking can be a powerful reminder of what truly matters – the beauty and resilience of the natural world, the warmth and kindness of the people we meet along the way, and the strength and clarity that comes from challenging ourselves and pushing beyond our comfort zones.

So as you close this book and set out on your own adventures, I encourage you to do so with an open heart and a curious mind. Let the walks in this guide be your starting point, but don't be afraid to forge your own path and create your own memories. Take the time to pause, to listen, to observe, and to appreciate the countless small wonders and everyday miracles that surround us.

And above all, remember that every step you take is an opportunity to learn, to grow, and to make a positive difference in the world. By walking with respect and reverence for the land and its inhabitants, by supporting local communities and businesses, and by sharing your experiences and insights with others, you become part of the long and rich tradition of people who have found solace, inspiration, and belonging in the landscapes of Suffolk.

So here's to the next 50 walks, and to the countless more that await you beyond the pages of this book. May your journeys be filled with discovery, delight, and the unwavering beauty of the natural world, and may the paths you tread lead you ever closer to the wild and precious places that call to your heart.

Happy walking!

Printed in Great Britain
by Amazon

f0c28e3f-4479-454b-8678-8a81aa902041R01